Electrical and Electronic Devices, Circuits, and Instruments

ELECTRICAL AND ELECTRONIC DEVICES, CIRCUITS, AND INSTRUMENTS

THOMAS A. DeMASSA

Professor of Electrical and Computer Engineering
Arizona State University

WEST PUBLISHING COMPANY
St. Paul · New York · Los Angeles · San Francisco

Production: *Technical Texts, Inc.*
Interior design: *Sylvia Dovner*
Illustrations: *J & R Illustrators*
Cover art: *Phillip Harrington/© The Image Bank*
Composition: *Syntax International*

Library of Congress Cataloging-in-Publication Data

DeMassa, Thomas A.
   Electrical and electronic devices, circuits, and instruments /
Thomas A. DeMassa.
     p. cm.
   ISBN 0–314–46933–8
   1. Electronics.   2. Electric engineering.   I. Title.
TK7816.D413 1989
621.3—dc19   88–28644
                                                    CIP

To my patient and loving wife, Joann;
To Karen, Tommy, Andrea, and Mary
—with a special thanks to my Mom and Dad
for their lifelong support

CONTENTS

PREFACE

Electrical and Electronic Devices, Circuits, and Instruments is intended as a textbook suitable for electrical engineering students as well as nonelectrical engineering students. Often, textbooks on these subjects become too involved in theoretical device physics and/or circuit theory, thus forcing practical-minded students to inquire, "Why are we studying this?" This text is designed to eliminate the question by leaning heavily toward practical aspects of devices and circuits. Circuit theory and device physics are included where necessary, but the discussions are kept brief.

The primary objective of the text is to furnish a basic knowledge of the operation and design of electrical and electronic devices (including transducers) and circuits. This goal is accomplished by providing easily readable descriptions of devices and circuits and by including numerous examples.

Another objective of the text is to include sufficient introductory circuit analysis in the first chapter, such that the traditional first course in circuit theory is not a prerequisite. For electrical engineering majors, the traditional circuit course would still be taken; however, it could be a corequisite. This premise provides the opportunity for students following a two-course, circuits–electronics curriculum sequence to take their first electronics course one semester earlier.

This text therefore includes introductory circuit material, but the topics are limited to those necessary for the analysis of succeeding devices and circuits. In this manner, greater emphasis on electrical and electronic devices and circuits is possible, rather than on circuit theory per se.

The topics covered in this text include a wide variety of electrical and electronic devices and circuits. The selection enables the reader to establish a broad background in this area. Additionally, basic descriptions of the operation of electrical systems that students can relate to are given throughout the text. Such examples include the electronic wristwatch, radio, television, and magnetic recorders. Further examples are given in describing the operation of basic instruments. It is believed that these examples provide an aid to the learning process by promoting reader interest.

The book can be used in several different ways. As a first course in electrical engineering and electronics, the following chapters are suggested: Chapters 1 through 9 should be covered in their entirety, with selected topics chosen from Chapters 10 to 14. For a class containing only nonelectrical engineering majors, more emphasis can be given to Chapters 11 through 14. This additional emphasis can be accomplished by only highlighting Chapters 6, 8, and 10. In instances where an introductory circuit course has already been taken, Chapter 1 can be used as a review chapter allowing more time for electronics and transducer topics.

Acknowledgments. I would like to thank my colleagues who reviewed the manuscript and individually provided helpful advice during development of this text.

Dennis L. Polla, University of Minnesota
Martin Kaliski, Calif. Poly, San Luis Obispo
Dwayne McCallister, Calif. Poly, San Luis Obispo
Frank W. Brands, Washington State University
G. W. Neudeck, Purdue University
Daniel Moore, North Carolina State
William Potter, University of Washington
Roger C. Conant, University of Illinois at Chicago
Alwyn C. Scott, University of Arizona
David Greve, Carnegie-Mellon University
Raymond Kline, Washington University

I would also like to thank the many students who suggested corrections and changes to the original notes for this text. In particular, special thanks are extended to students Edouard Baaklini, Jim Rummel, and David O'Brien for their help in providing some of the end-of-chapter problems.

1

ELECTRICAL FUNDAMENTALS WITH NETWORK EXAMPLES

In this chapter, basic electrical circuit fundamentals are introduced. Passive and active circuit elements are defined, and the methods of analyzing circuits composed of these elements are described. The topics include the network definitions and techniques that are necessary for analyzing the electronic and transducer circuits considered in succeeding chapters. Illustrative examples have been included to aid the reader in understanding the introductory material here. The reader who is familiar with electrical engineering fundamentals should use this first chapter as a review before proceeding to Chapter 2.

1.1   UNITS

In 1974, legislation was enacted in the United States to implement changeover from the American System of Units to the International System of Units (abbreviated as SI). This system has seven fundamental units, which are listed in Table 1.1 beside their corresponding physical quantities. From these fundamental units, all other units are derived. The important derived electrical quantities and their units are listed in Table 1.2. Listed in this table after each quantity are the defined units that will be used in this text and the derived units from the SI system.

In practice, many of the defined units have magnitudes that are either quite small or quite large. For convenience, numerical prefixes are used to aid in our working with these cumbersome numbers in electronics. Table 1.3 indicates important numerical prefixes and their corresponding values and symbols. Table 1.4 shows some important examples.

TABLE 1.1   SI Fundamental Units

Quantity	Unit
Length	Meter, m
Mass	Kilogram, kg
Time	Second, s
Current	Ampere, A
Temperature	Degrees Kelvin, °K
Amount of substance	Mole, mol
Luminous intensity	Candela, cd

TABLE 1.2   SI Electrical and Derived Units

Quantity	Electrical Unit	Derived Unit
Voltage	Volt, V	$kg \cdot m^2/(s^3 \cdot A)$
Power	Watt, W	$kg \cdot m^2/s^3$
Energy	Joule, J	$kg \cdot m^2/s^2$
Resistance	Ohm, Ω	$kg \cdot m^2/(s^3 \cdot A^2)$
Conductance	Siemens, S	$1/\Omega$
Capacitance	Farad, F	$A^2 \cdot s^4/(kg \cdot m^2)$
Inductance	Henry, H	$kg \cdot m^2/(s^2 \cdot A^2)$
Frequency	Hertz, Hz	$1/s$
Charge	Coulomb, C	$A \cdot s$

TABLE 1.3   Multiples and Sub-multiples of SI Units

Prefix	Factor	Symbol
Giga	10^9	G
Mega	10^6	M
Kilo	10^3	k
Centi	10^{-2}	c
Milli	10^{-3}	m
Micro	10^{-6}	μ
Nano	10^{-9}	n
Pico	10^{-12}	p

TABLE 1.4   Examples of Prefixed Units

Prefixed Unit	Magnitude without Prefix
Gigahertz, GHz	10^9 Hz
Megohm, MΩ	10^6 Ω
Milliampere, mA	10^{-3} A
Kilovolt, kV	10^3 V
Picofarad, pF	10^{-12} F
Nanohenry, nH	10^{-9} H

─────────────────────────────────── EX. 1.1

Time Unit Conversion Example

A professor in a classroom announced to the students, "My lecture this morning will last approximately one microcentury." How long did the professor mean, in minutes?

Solution: Converting the time units (accounting for an extra day each leap year), we have

$$1 \ \mu\text{century} = (10^{-6} \times 100 \ \text{yr})(365.25 \ \text{d/yr})(24 \ \text{h/d})$$
$$\times (60 \ \text{min/h}) \simeq 52.5 \ \text{min}$$

─────────────────────────────────── EX. 1.2

Electrical Unit Example

Current i, voltage v, and resistance R are related by the equation

$$v = iR$$

For a current of 1 mA and a resistance of 1 kΩ, calculate the voltage.

Solution: Substitution into the equation $v = iR$ yields

$$v = (1 \text{ mA})(1 \text{ k}\Omega) = 1 \text{ V}$$

—————————————————————————————— EX. 1.3

Another Electrical Unit Example

Current i, voltage v, and capacitance C are related by the expression

$$i = C \frac{dv}{dt}$$

For a capacitance of $1 \mu\text{F}$ and a time rate of change of voltage given by 1 kV/s, determine the current.

Solution: Substitution into the current–voltage for the capacitor yields

$$i = (10^{-6} \text{ F})(10^3 \text{ V/s}) = 1 \text{ mA}$$

1.2 CHARGE, CURRENT, AND VOLTAGE

1.2.1 Charge and Current

Electrical phenomena are described here by introducing the concept of electronic charge. In general, the electronic *charge* in a system can be positive or negative and exists as some multiple of the charge of an electron that is negative and given by $q = -1.6 \times 10^{-19}$ C. Charge is the most basic electrical quantity.

Electric current depends upon the movement of electronic charge. *Current* is defined as the time rate of change of charge, and thus

$$i = \frac{dQ}{dt} \qquad (1.2\text{--}1)$$

where

i = current in amperes (A)
Q = total charge in coulombs (C)
t = time in seconds (s)

In a system in which the flow of charge is uniform with Q coulombs passing a given point every second, $i = Q/t$.

1.2.2 Voltage

A voltage or potential difference is also associated with electronic charge. The *voltage* between two points where charges are located is defined as the energy (or work) required to move a unit charge from one point to the other. A potential difference of 1 V is produced by moving a charge of 1 C requiring an energy of 1 J. The potential difference v associated with the work or energy E necessary to move Q coulombs of charge between two points is

$$E = Qv \qquad (1.2\text{--}2)$$

Another definition of voltage is work per unit charge.

1.2.3 AC and DC Currents and Voltages

Currents and voltages existing in electronic circuits are, in general, time varying. However, for a constant flow of charge through a conductor, the current is constant and referred to as *direct current,* or DC. For the case of time-varying charge flow producing time-varying current, the simplest case is that of a current that alternates direction. This current is called *alternating current,* or AC.

Although the abbreviations AC and DC refer to alternating and direct current, they are also used for describing voltages. That is, a *DC voltage* is a constant voltage independent of time, and an *AC voltage* is an alternating or time-varying voltage.

In general, a specific current (or voltage) in an electronic circuit will consist of an AC and DC component, and the total current (or voltage) is the sum of these AC and DC parts. Additional terminology that is often used is that the AC current (or voltage) is called the *signal component*. Furthermore, if the AC component has zero average value (as in the case of sinusoids), then the DC component is the average value.

1.3   POWER AND ENERGY

Power and energy are additional important quantities in electronic circuits. In particular, the power dissipated in an electronic device must never be allowed to exceed the maximum permissible amount for that device. Similarly, there are restrictions upon the maximum current and voltage that a device can withstand. In device design, such restrictions must be taken into account by the designer.

Power is defined as the time rate of work or change of energy. Hence, if work is being done at a constant rate and if charge Q is moved through a voltage of v volts, then the work per unit time is

$$p = \frac{Qv}{t} \qquad (1.3\text{--}1)$$

where p is the power in watts (W). Substituting $i = Q/t$ yields

$$p = vi \qquad (1.3\text{--}2)$$

Equation (1.3–2) is also obtainable from basic mathematical definitions since

$$p = \frac{dE}{dt} = \frac{dE}{dQ}\frac{dQ}{dt} = vi \qquad (1.3\text{--}3)$$

Thus, the power associated with a particular device in the general case is just the voltage–current product, an extremely important result.

- EX. 1.4

Transistor Power Example

A transistor is a three-terminal device that can amplify voltage and/or current and power. This example involves a power transistor application. The power dissipation corresponding to the product of output current (i_O) and voltage (v_O) is not to exceed 100 W. Additionally, the maximum allowable current and voltage for the output of this transistor is 50 A and 30 V, respectively. Indicate the region in which the transistor can operate safely. Furthermore, indicate this region graphically using v_O as ordinate and i_O as abscissa.

Solution: Since the maximum power dissipation at the output of the transistor is 100 W, the product of output current and voltage must satisfy the following inequality:

$$i_O v_O \leq 100 \text{ W}$$

For equality, this equation becomes that of the hyperbola shown in Fig. E1.4. Thus, to satisfy the power requirement, a permissible operating point (v_O, i_O) must be on this curve or be closer to the origin. Also, to satisfy the maximum

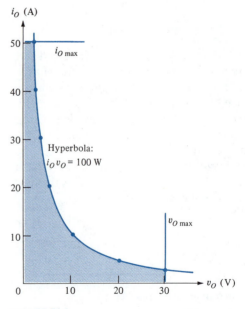

FIGURE E1.4

current and voltage requirements, this region must be terminated at 50 A and 30 V. The shaded region indicates the operating region.

1.4   BASIC CIRCUIT ELEMENTS AND IMPEDANCE

In this section, the basic circuit elements exclusive of diodes, transistors, and integrated circuits are introduced. Certain combinations of the basic circuit elements are used to represent these more complex cases, as we will see. The basic circuit elements to be described are the following:

- Current and voltage sources
- Resistors
- Capacitors
- Inductors
- Transformers

Resistors, capacitors, inductors, and transformers are referred to as *passive* elements because they do not generate electrical energy, while current sources and voltage sources are called *active* elements because they are able to generate electrical energy. Only ideal elements are focused upon in this chapter because they are adequate for our purposes.

1.4.1   Current and Voltage Sources

Current and voltage sources are active devices that are used to represent the introduction of power into a circuit. Two general categories exist for these sources in electronic circuits. They are called dependent sources and independent sources. *Dependent* sources are those that depend upon another current or voltage in the circuit, whereas *independent* sources do not. Dependent sources are sometimes also called *controlled* sources and are used in transistor equivalent circuits.

Figures 1.1a and b display the usual types of sources that arise in electronic circuits and their corresponding circuit symbols. Note that the dependent sources are denoted by using a diamond-shaped symbol, whereas the independent sources use a circular symbol. A dependent voltage source can be current or voltage controlled; the same applies for a dependent current source. The actual dependence describes the physical behavior of the source.

Quite often, independent DC voltage sources are represented by other symbols. These symbols are displayed in Figs. 1.1c and d. In Fig. 1.1c, a straight-line symbol is used to explicitly specify that the voltage between points *A* and *B* is totally DC with polarity and with magnitude V_{DC} as indicated. Another type of DC voltage symbol is

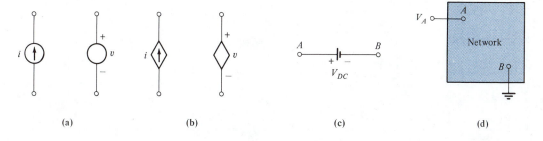

(a)   (b)   (c)   (d)

FIGURE 1.1   Independent and Dependent Current and Voltage Sources: (a) Independent current *i* and voltage *v* source symbols, (b) Dependent current *i* and voltage *v* source symbols, (c) DC voltage source symbol, (d) Another DC voltage source symbol

shown in Fig. 1.1d. The voltage next to terminal A is specified as V_A and indicates the magnitude of voltage at terminal A relative to the reference terminal B. This symbolism is also used for an AC voltage or total voltage (AC and DC) at one terminal with the ground terminal as the reference terminal. Whenever a terminal in a network has a voltage written next to it, the other terminal associated with this voltage source is the reference (or ground) terminal.

1.4.2   Resistance, Ohm's Law, and Basic Combinations of Resistors

A *resistor* is a basic circuit element that dissipates energy. It is an element that provides a resistance to charge flow (current) and converts electrical energy into thermal energy. Household appliances such as the electrical oven, dryer, and room heater utilize this energy conversion process directly.

The circuit symbol for a resistor is shown in Fig. 1.2a. The zigzag line represents the resistance to charge flow and current. Positive current i and voltage v are also defined in the figure. For the current direction and voltage polarity shown, the current–voltage expression is given by a linear relation known as *Ohm's law:*

$$v = Ri \tag{1.4-1}$$

where R is the magnitude of the *resistance*. Note that R has units of V/A, or ohms (Ω). Equation (1.4–1) is often written in the inverse form as

$$i = Gv \tag{1.4-2}$$

where G is the *conductance* and is the reciprocal of R. Note that G has units of $1/\Omega$, or siemens (S).

Equation (1.4–1) also indicates the physical mechanism by which current passing through a resistor in the direction as shown in Fig. 1.2a creates a voltage with polarity also as shown. The voltage produced always opposes the current flow. Thus, if either the defined direction of current or polar-

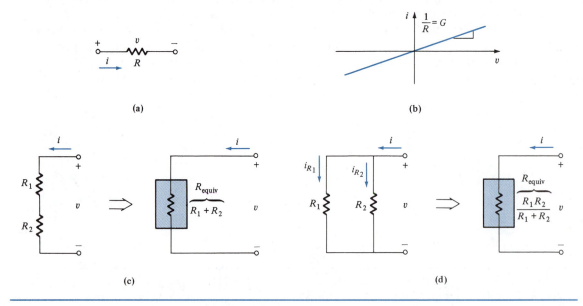

FIGURE 1.2   The Resistor and Ohm's Law: (a) Resistor symbol, (b) *i*–*v* characteristic, (c) Resistors in series, (d) Resistors in parallel

ity of voltage were reversed, a negative sign would appear in (1.4–1) and (1.4–2).

Figure 1.2b shows a graphical representation of Ohm's law. The diagram is referred to as the current–voltage, or i–v, characteristic for a resistor of magnitude $R = 1/G$. This graphical representation becomes very important in the case of nonlinear devices and forms the basis for the development of circuit models for such elements, as we will see in Chapter 2.

The power dissipated in a resistor is given by

$$p = iv = i^2 R \qquad (1.4–3)$$

which can also be written in several other forms (using Ohm's law) as follows:

$$p = \frac{i^2}{G} = \frac{v^2}{R} = v^2 G \qquad (1.4–4)$$

It should be noted that the values of resistance R and conductance G are constants, independent of the current i and voltage v. This behavior is the ideal case but is valid over wide ranges of current and voltage for actual resistors.

Quite often, certain resistor connections in circuits can be replaced with equivalent circuits that result in considerable simplification. Figures 1.2c and d indicate the two most prevalent arrangements. Figure 1.2c shows two resistors connected in *series,* an arrangement in which the same current flows through each element. Figure 1.2d shows two resistors connected in *parallel,* a situation in which the same voltage is maintained across both elements. For the case of two resistors in series, the equivalent representation is just a single resistor whose value is the sum of the two resistors. For the case of two resistors in parallel, the equivalent circuit must allow a total current to flow that is the sum of the currents through each resistor, while the identical voltage is maintained across each element. That is, $i = i_{R_1} + i_{R_2} = v/R_1 + v/R_2 = v(1/R_1 + 1/R_2) = v(R_1 + R_2)/R_1 R_2$. Therefore, the equivalent resistor is just $R_1 R_2/(R_1 + R_2)$, or $R_1 \| R_2$, where the symbol $\|$ denotes parallel combination.

———————————— EX. 1.5

Resistor Example

A 10 kΩ resistor has a voltage of 50 V across its terminals as shown in Fig. E1.5. Calculate the current through the resistor and the power dissipated in the resistor. Note that with the polarity of voltage indicated in Fig. E1.5, the positive current direction must be as shown.

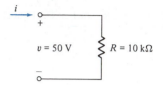

FIGURE E1.5

Solution: By application of Ohm's law, the current is given by

$$i = \frac{50 \text{ V}}{10 \text{ k}\Omega} = 5 \text{ mA}$$

Thus, the power dissipated is

$$p = iv = (5 \text{ mA})(50 \text{ V}) = 0.25 \text{ W}$$

———————————— EX. 1.6

Parallel Resistor Example

Determine the total equivalent resistance for the parallel combination of resistors shown in Fig. E1.6a.

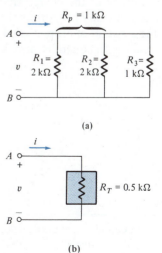

FIGURE E1.6

Solution: The solution is obtained by first combining R_1 and R_2 in parallel, where

$$R_p = R_1 \| R_2 = 1 \text{ k}\Omega$$

Then, R_p is combined with R_3 to obtain

$$R_T = R_p \| R_3 = 0.5 \text{ k}\Omega$$

The equivalent circuit is shown in Fig. E1.6b.

1.4.3   Capacitance and Inductance

Capacitors and *inductors* are passive circuit elements that store energy. An ideal capacitor consists of two parallel conducting plates separated by a dielectric (nonconducting or insulating) medium or insulator. When a voltage is applied across the capacitor plates, the insulator prohibits charge flow through it, and charge of opposite sign is developed on each plate. This separated charge has a voltage associated with it that provides the mechanism for energy storage. The circuit symbol for a capacitor is shown in Fig. 1.3a, which is very descriptive of the actual device geometry.

The magnitude of the charge that develops on each plate is directly proportional to the voltage, with the constant of proportionality defined as the *capacitance*. In equation form, we have

$$Q = Cv \tag{1.4–5}$$

where

Q = magnitude of charge on each plate in coulombs (C)
v = voltage across the plates in volts (V)
C = capacitance with derived units (A·s)/V redefined in farads (F)

Since the charge and current are related by $i = dQ/dt$, taking the time derivative of (1.4–5) and substituting for dQ/dt yield

$$i = C\frac{dv}{dt} \tag{1.4–6}$$

This equation expresses the basic i–v relation for the capacitor and can be written in integral form as follows (assuming $v = 0$, initially):

$$v = \frac{1}{C} \int i\,dt \tag{1.4–7}$$

A compact expression for the energy stored in a capacitor can be obtained from the energy–power relation as follows:

$$E = \int p\,dt = \int iv\,dt \tag{1.4–8}$$

Substituting $i = C(dv/dt)$ and integrating yield

$$E = \frac{Cv^2}{2} \tag{1.4–9}$$

Thus, the energy stored in a capacitor is directly dependent on the square of the voltage across its plates.

Note from Eq. (1.4–6) that if the voltage across the terminals of the capacitor is constant (DC), then the derivative of v is zero, and thus the current is zero. Hence for constant v, the capacitor acts as an open circuit. However, if the voltage across the capacitor varies with time (AC), a current exists. The mechanism of charge flow in this case, however, is due to the inducement of charge

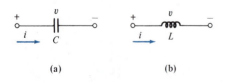

(a)   (b)

FIGURE 1.3   The Capacitor and the Inductor: (a) Capacitor symbol, (b) Inductor symbol

from one plate to the other and not to charge flow through the insulator.

An inductor consists basically of a coil of wire through which a current is passed. For an inductor, however, stored energy depends upon the current. The circuit symbol for an inductor is displayed in Fig. 1.3b, which, not unlike the capacitor, is quite descriptive of the actual geometry of an inductor. The value of *inductance* is defined by the relation

$$v = L \frac{di}{dt} \qquad (1.4-10)$$

where the unit of inductance L is $(\text{V·s})/\text{A}$ redefined as a henry (H). By integrating and solving (1.4–10) for i, in an analogous manner to obtaining (1.4–7), we have the relationship

$$i = \frac{1}{L} \int v\, dt \qquad (1.4-11)$$

The energy stored in the inductor is obtained in an analogous manner to that used in obtaining (1.4–9). The expression is

$$E = \frac{Li^2}{2} \qquad (1.4-12)$$

Thus, the energy stored in an inductor is proportional to the current squared. Note that for the case of an inductor, if the current is constant, the voltage is zero, and this current must be time varying to obtain a nonzero voltage.

1.4.4   Impedance of Passive Elements

Note that the $i–v$ relations for the capacitor given by (1.4–6) and for the inductor given by (1.4–10) involve an integral over t or a derivative with t. This fact complicates the behavior of these elements in comparison with the linear $i–v$ relation of the resistor as given by

(1.4–1). Furthermore, when inductors, capacitors, and resistors are interconnected in a circuit, differential equations arise for the currents and voltages in that circuit. In order to avoid these differential equations, it is customary to introduce the concept of impedance. Then, a sinusoidal time dependence for the voltage and current is assumed, and integrals and derivatives involving time are mathematically transformed into algebraic equations. For the development that follows, we begin by introducing basic definitions concerning sinusoids.

Sinusoidal Steady-State Conditions.   We begin by defining the expression for a *trigonometric sinusoid*. This mathematical expression for a voltage that varies with time t is given by the general relation

$$v(t) = V_m \sin(\omega t + \phi) \qquad (1.4-13)$$

where

V_m = magnitude of the voltage in volts (V)
ω = radian frequency ($\omega = 2\pi f$) in radians per second (rad/s)
ϕ = phase in radians (rad)

Figure 1.4 displays one period of this sinusoidal variation versus ωt. This function repeats itself in every other period of length 2π. The magnitude V_m, phase ϕ, and frequency ω or f completely specify the sinusoid.

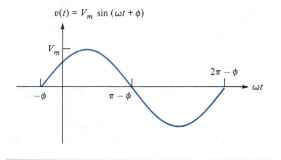

FIGURE 1.4   Sinusoidal Waveshape

We will assume that all input sources in our networks are sinusoids of a particular frequency. This condition is called the *sinusoidal steady state*. We will also assume that the networks are linear, and superposition may therefore be used. Then, all voltages and currents in these networks must also be sinusoids of the same frequency. That is, sinusoids can be integrated and/or differentiated resulting in sinusoids usually with a different amplitude and phase, but always the same frequency. Additionally, a summation of sinusoids of the same frequency results in another sinusoid of the same frequency. Therefore, the resulting sinusoids will have the same frequency as the input source. This all important result is the key to transforming the differential equations into algebraic equations.

In networks that have time-dependent sources that are not sinusoidal, such as a sawtooth waveform, the sinusoidal steady-state method involves a more detailed treatment. Since most time-dependent functions can be expanded into a unique series of sinusoids with different frequencies (a Fourier series), each such time-dependent source is expanded into this corresponding series, and each term in the series is treated as a separate sinusoid with a different frequency.

A sinusoid of unity amplitude is expressed simply as $\sin(\omega t + \phi)$. However, a more general and useful representation of a unity amplitude sinusoid is the exponential form, given by $e^{j(\omega t + \phi)}$, where ω and ϕ have the same meaning and $j = (-1)^{1/2}$. This form is called the *complex exponential representation of a sinusoid*. The trigonometric and exponential forms are related by an identity called *Euler's relation* given by

$$e^{j(\omega t + \phi)} = \cos(\omega t + \phi)$$
$$+ j \sin(\omega t + \phi) \qquad (1.4\text{--}14)$$

Note that $\sin(\omega t + \phi)$ is the imaginary part of $e^{j(\omega t + \phi)}$. This exponential form of a sinusoid is invaluable in the development of the concept of impedance.

Impedance of the Capacitor and Inductor.   We will consider a linear network under excitation by a single sinusoidal source (either voltage or current) with its time dependence given by the unity amplitude exponential $e^{j(\omega t + \phi)}$. In general, this amplitude is not unity, and so all resulting currents and voltages must be multiplied by the particular nonunity amplitude. Our goal is to analyze the network to determine all currents and voltages.

From the previous discussion, all voltages and currents in the network will have the same time dependence. The voltage and current for each element in the network can therefore be expressed generally by

$$i(t) = \mathbf{I}e^{j(\omega + \phi)} \qquad (1.4\text{--}15a)$$

and

$$v(t) = \mathbf{V}e^{j(\omega t + \phi)} \qquad (1.4\text{--}15b)$$

where $\mathbf{I}$ and $\mathbf{V}$ are independent of t, but in general are dependent upon $j\omega$. Furthermore, $\mathbf{I}$ and $\mathbf{V}$ are complex (possessing real and imaginary parts) and are called phasors. By definition, a *phasor* is a complex current or voltage that possesses both magnitude and phase and is dependent upon frequency. The determination of the phasor voltage or current for each element is a critical portion of the circuit analysis along with the conversion of these phasors back to their corresponding time functions.

If the network under consideration contains capacitors and inductors, we can substitute the exponential dependence of (1.4–15) into the i–v relations for each of these elements. Substitution of (1.4–15) into (1.4–6) and (1.4–10) yields (after canceling the time-dependent term throughout each equation)

$$\mathbf{I} = j\omega C \mathbf{V} \qquad (1.4\text{--}16)$$

and

$$\mathbf{V} = j\omega L \mathbf{I} \qquad (1.4\text{--}17)$$

Equations (1.4–16) and (1.4–17) relate the phasor voltage **V** to the phasor current **I** for each element. Additionally, each of these equations can be written as

$$\mathbf{V} = Z\mathbf{I} \qquad (1.4\text{–}18)$$

which is called the complex form of Ohm's law. Note that **I**, **V**, and therefore also Z are, in general, functions of $j\omega$.

The quantity Z is called the *impedance* and represents the complex resistance of the particular element. The impedance of the capacitor and the inductor are given, respectively, by

$$Z_C = \frac{1}{j\omega C} \qquad (1.4\text{–}19)$$

and

$$Z_L = j\omega L \qquad (1.4\text{–}20)$$

These expressions are extremely important in the analysis of circuits containing inductors and capacitors.

Quite often, it is convenient to work with the inverse of the impedance, defined as the *admittance* and given by $Y = 1/Z$. For the capacitor and inductor, the corresponding admittances are $Y_C = 1/Z_C$ and $Y_L = 1/Z_L$. These admittances are directly analogous to the conductance, which is inversely associated with resistance. In fact, the impedance of a resistor is just the resistance, and, similarly, the admittance of a resistor is the conductance.

Impedance Method.   In the impedance method of analyzing a network, we assume sinusoidal steady-state conditions. Thus, all voltages and currents in the network will have a dependence on $e^{j(\omega t + \phi)}$. Equilibrium equations are then written for the network, with each element treated as an impedance. The resulting equations are algebraic and are solved to obtain the current and voltage phasors (**I** and **V**) for each element. Equa-

tions (1.4–15a and b) then yield the corresponding time-varying current and voltage.

However, in real-life networks, the input sinusoidal sources do not physically vary as $e^{j(\omega t + \phi)}$. Instead, their actual variation is given by the simpler form of a sinusoid, expressed generally as $\sin(\omega t + \phi)$. Such variation is directly available from the output terminals of an instrument called a *sine wave generator,* or oscillator.

In order to take the actual time variation into account, note from (1.4–14) that $\sin(\omega t + \phi)$ is the imaginary part of $e^{j(\omega t + \phi)}$. Then, to obtain the resulting time-dependent current and voltage from the corresponding phasor representation, we must take the imaginary part of both sides of (1.4–15). We get, quite simply,

$$i(t) = Im[\mathbf{I}e^{j(\omega + \phi)}] \qquad (1.4\text{–}21a)$$

and

$$v(t) = Im[\mathbf{V}e^{j(\omega t + \phi)}] \qquad (1.4\text{–}21b)$$

Thus, the time-dependent current and voltage for each element is obtained for an input source possessing the time dependence $\sin(\omega t + \phi)$. Equations (1.4–21a and b) indicate the manner in which the phasors are converted back into time functions. We simply take the imaginary part of the products of **I** or **V** with the exponential $e^{j(\omega t + \phi)}$. The time-dependent current $i(t)$ and voltage $v(t)$ given by (1.4–21) will always be real quantities possessing the same time variation as the input source, but, in general, each will have a different magnitude and phase.

The time-varying current $i(t)$ and voltage $v(t)$ are said to be in the *time domain*. The corresponding phasor current $\mathbf{I}(\omega)$ and voltage $\mathbf{V}(\omega)$ are said to be in the *frequency domain*.

───────────────────────────── EX. 1.7

Phasor Conversion Example

A sinusoidal voltage $v(t) = 2 \sin \omega t$ (V) is applied to the RC circuit of Fig. E1.7a. Determine the current $i(t)$. Let $\omega = 2\pi f$, where $f = 60$ Hz, $R = 1$ kΩ, $C = 1\mu$F.

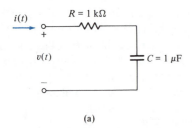

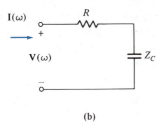

(b)

FIGURE E1.7

Solution: We begin by converting the circuit into its impedance form, as shown in Fig. E1.7b with phasor current and voltage. Using Ohm's law in complex form, we have

$$\mathbf{I} = \frac{\mathbf{V}}{R + Z_C} \tag{1}$$

and substituting $Z_C = 1/j\omega C$ yields

$$\mathbf{I} = \frac{\mathbf{V}}{R + (1/j\omega C)} = \frac{\mathbf{V}(j\omega C)}{j\omega RC + 1} = |\mathbf{I}| \angle \phi \tag{2}$$

where the latter portion of (2) represents the magnitude ($|\mathbf{I}|$) and phase angle (ϕ) of the phasor $\mathbf{I}$. Calculating the magnitude and phase, we have

$$|I| = \frac{|V||j\omega C|}{|j\omega RC + 1|} = \frac{2\omega C}{[(\omega RC)^2 + 1]^{1/2}}$$

$$= \frac{2(377)(10^{-6})}{[(0.377)^2 + 1]^{1/2}} = 0.70 \text{ mA} \tag{3}$$

and

$$\phi = \frac{\pi}{2} - \tan^{-1}\omega RC = 1.57 - 0.36 = 1.21 \text{ rad} \tag{4}$$

Thus,

$$I = 0.7 \text{ mA} \angle 1.21 \text{ rad} \tag{5}$$

and therefore

$$i(t) = Im[0.7e^{j(\omega t + 1.21)}]$$

$$= 0.7 \sin(\omega t + 1.21) \text{ mA} \tag{6}$$

Note that the phase of the current could equivalently be expressed in degrees, where 1.21 rad corresponds to (1.21)$(180/\pi) = 69.3°$.

General Impedance and Admittance Expressions. In circuits containing more than one passive element, general expressions for impedance and admittance are obtained. These general expressions have both real and imaginary parts and are a function of frequency. At a particular frequency ω, a general impedance is expressed as

$$Z(j\omega) = R(\omega) + jX(\omega) \tag{1.4–22}$$

We see from this expression that $Z(j\omega)$ is a complex number with real part $R(\omega)$ and imaginary part $X(\omega)$ defined as the *resistance* and *reactance*, respectively. In general, both R and X depend on the particular frequency ω.

This same impedance can be expressed in terms of admittance as follows:

$$Y(j\omega) = \frac{1}{Z(j\omega)} = G(\omega) + jB(\omega) \tag{1.4–23}$$

In this expression, $G(\omega)$ is defined as the *conductance*, and $B(\omega)$ is defined as the *susceptance*.

It should also be realized that impedances (like resistors) often appear in series and parallel combinations in networks. Simplification of circuits involving impedances follows resistor combinations identically. That is, impedances in series can be simplified to one impedance that is the sum of the impedances, while two impedances in parallel are equivalently equal to $Z_1 Z_2/(Z_1 + Z_2)$ and referred to as $Z_1 \| Z_2$.

Finally, the two extreme cases of impedance magnitude, zero and infinite, deserve special mention. When the impedance of an element is zero, then the voltage is zero and yields the short-circuit case. When the impedance is infinite, the current is zero and gives the open-circuit case.

_____ **EX. 1.8**

Impedance Example

Determine the equivalent resistance and reactance of the network shown in Fig. E1.8a by finding the impedance seen looking into the terminals A–B (Z_{AB}).

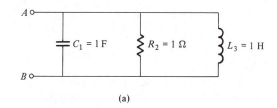

(a)

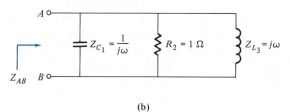

(b)

FIGURE E1.8

Solution: The first step involves visualizing the equivalent circuit in impedance form. This circuit is shown in Fig. E1.8b for a general frequency ω. The next step is to combine two of the elements in parallel. Combining the resistor and inductor in parallel yields

$$R_2 \,\|\, Z_{L_3} = \frac{j\omega}{1 + j\omega}$$

Finally, we combine Z_C in parallel with $R_2 \,\|\, Z_{L_3}$ to obtain

$$Z_{AB} = Z_{C_1} \,\|\, (R_2 \,\|\, Z_{L_3}) = \frac{\left(\dfrac{1}{j\omega}\right)\left(\dfrac{j\omega}{1 + j\omega}\right)}{\dfrac{1}{j\omega} + \dfrac{j\omega}{1 + j\omega}}$$

$$= \frac{j\omega}{1 + j\omega + (j\omega)^2}$$

After a bit of complex algebra to obtain the real and imaginary parts, we have

$$Z_{AB} = \frac{\omega^2}{(1 - \omega^2)^2 + \omega^2} + j\,\frac{\omega(1 - \omega^2)}{(1 - \omega^2)^2 + \omega^2}$$

We now recognize that

$$R(\omega) = \frac{\omega^2}{(1 - \omega^2)^2 + \omega^2}$$

and that

$$X(\omega) = \frac{\omega(1 - \omega^2)}{(1 - \omega^2)^2 + \omega^2}$$

_____ **EX. 1.9**

Admittance Example

Determine the equivalent admittance seen looking into the terminals A–B shown in Fig. E1.8b. Also, determine the construction G and the susceptance B.

Solution: Since the admittance is just the inverse of impedance, we have

$$Y_{AB} = \frac{1}{Z_{AB}} = \frac{1 + j\omega + (j\omega)^2}{j\omega}$$

or

$$Y_{AB} = j\,\frac{(\omega^2 - 1)}{\omega} + 1$$

Thus, we recognize that

$$G(\omega) = 1$$

and that

$$B(\omega) = \frac{\omega^2 - 1}{\omega}$$

Representations for Time-Varying Quantities.

Time-varying currents and voltages are represented by their instantaneous, average, and root-mean-square values. It is quite obvious that the *instantaneous value* of a time-dependent quantity is its value at a particular instant in time. Furthermore, the *average value* is that obtained by averaging this instantaneous value over a certain length of time. For periodic signals, the time average is obtained by averaging over one period.

The *root-mean-square,* or *rms, value* of a time-dependent current or voltage is defined, respectively, by the integral expressions

$$I_{\text{rms}} = \left[\frac{1}{T} \int_{t_1}^{t_1 + T} i_s^2(t) dt \right]^{1/2} \tag{1.4-24}$$

and

$$V_{\text{rms}} = \left[\frac{1}{T} \int_{t_1}^{t_1 + T} v_s^2(t) dt \right]^{1/2} \tag{1.4-25}$$

where T is the length of time over which the average is obtained and t_1 is an arbitrary time. Note from these definitions that the rms value is the square *root* of the *mean, squared* value of the current or voltage.

When we are dealing with sinusoidal currents and voltages, the rms values of these quantities are quite useful, as we will see. For example purposes, we begin by determining the rms value of a sinusoidal current (the rms voltage is obtained in an identical manner). Thus, consider a current represented by the expression

$$i_s = I_S \sin(\omega t) \tag{1.4-26}$$

Using the basic definition of (1.4–24) and by substitution and rearranging, we have (with $t_1 = 0$)

$$I_{\text{rms}} = \left[\frac{1}{T} \int_0^T I_S^2 \sin^2 \omega t \, dt \right]^{1/2}$$

$$= \left[\frac{I_S^2}{2\pi} \int_0^{2\pi} \sin^2 \omega t \, d(\omega t) \right]^{1/2} \tag{1.4-27}$$

where T is the period of the sinusoid. ($T = 1/f = 2\pi/\omega$) and a change of variables was made. Using

$\sin^2 \omega t = (1 - \cos 2\omega t)/2$ and integrating yield

$$I_{\text{rms}} = \left[\frac{I_S^2}{2\pi} \int_0^{2\pi} \frac{(1 - \cos 2\omega t)}{2} d(\omega t) \right]^{1/2}$$

$$= \left[\frac{I_S^2}{2\pi} \left(\frac{t - \frac{1}{2\omega} \sin 2\omega t}{2} \right) \Big|_0^{2\pi} \right]^{1/2} \tag{1.4-28}$$

Substituting the limits of integration and taking the square root yield

$$I_{\text{rms}} = \left[\frac{I_S^2}{2\pi} \left(\frac{2\pi - 0}{2} - 0 \right) \right]^{1/2} = \frac{I_S}{2^{1/2}}$$

$$= 0.707 I_S \tag{1.4-29}$$

Hence, the rms value of a sinusoid is the peak value reduced by the factor $2^{1/2} = 0.707$.

The primary motivation for the definition of the rms value involves the determination of power when sinusoids are involved. For example, consider a resistor R with sinusoidal current given by Eq. (1.4–26). The instantaneous power absorbed in the resistor is then obtained as follows:

$$p(t) = i^2(t) R = \frac{v^2(t)}{R}$$

$$= I_S^2 R \sin^2(\omega t) \tag{1.4-30}$$

where $v(t) = Ri(t)$. Additionally, the average power is obtained from

$$P = \frac{1}{T} \int_0^T p \, dt$$

$$= \frac{I_S^2 R}{2\pi} \int_0^{2\pi} \sin^2(\omega t) d(\omega t) \tag{1.4-31}$$

and by integration, we obtain

$$P = \frac{I_S^2 R}{2} = I_{\text{rms}}^2 R = \frac{V_{\text{rms}}^2}{R} \tag{1.4-32}$$

The last two terms in (1.4–32) are identical in form to the expressions for power when only

DC currents and voltages are associated with a resistor R. That is, for DC only, we have $P_{DC} = I_{DC}^2 R = V_{DC}^2/R$. Because of the identical form obtained when we are dealing with sinusoids, the rms values are often referred to as the *effective values*. Furthermore, for sinusoidal current and voltage associated with a resistor R, the average power is calculated directly from (1.4–32), using $I_{rms} = 0.707 I_S$ or $V_{rms} = 0.707 V_S$. For time-varying currents and voltages other than sinusoids, the general expressions of (1.4–24) and (1.4–25) must be used to obtain the rms values, which are then substituted into (1.4–32) to obtain the average power delivered to the resistor.

Power in Impedance Elements.   When general impedances are involved, the calculations of power become more involved. Consider a general impedance element defined as $Z = |Z| \angle \theta$. Additionally, let the voltage and current for this impedance be sinusoidal, given by

$$v = V_P \sin \omega t \qquad (1.4\text{–}33a)$$

and

$$i = I_P \sin(\omega t - \theta) \qquad (1.4\text{–}33b)$$

where V_P and I_P are the corresponding peak values. The instantaneous power for this element is then given by the product of $i(t)$ and $v(t)$, or

$$
\begin{aligned}
p(t) &= v(t)i(t) \\
&= V_P I_P \sin \omega t \sin(\omega t - \theta) \qquad (1.4\text{–}34)
\end{aligned}
$$

To determine the average power, we integrate the instantaneous power over one period ($T = 2\pi/\omega$), or

$$
\begin{aligned}
P &= \frac{1}{T} \int_0^T p\,dt \\
&= \frac{1}{2\pi} \int_0^{2\pi} V_P I_P \sin \omega t \sin(\omega t - \theta)\,d(\omega t)
\end{aligned}
$$
$$(1.4\text{–}35)$$

After integration, Eq. (1.4–35) becomes

$$P = \frac{V_P I_P}{2} \cos \theta = \frac{V_P}{\sqrt{2}} \frac{I_P}{\sqrt{2}} \cos \theta \qquad (1.4\text{–}36)$$

In terms of rms values (instead of peak values), (1.4–36) can be written as

$$P = V_{rms} I_{rms} \cos \theta \qquad (1.4\text{–}37)$$

Observe that $\cos \theta \le 1$; thus, $\cos \theta$ acts as a reducing factor, called the *power factor*. Observe further that for a resistor, the voltage and current are in phase ($\theta = 0$), and the average power is a maximum ($\cos \theta = 1$) given by

$$P = V_{rms} I_{rms} \qquad (1.4\text{–}38)$$

For a general impedance element, however, the average power is given by (1.4–37), which includes the power factor $\cos \theta$.

Equation (1.4–37) represents the power in watts (W) that would be measured with a wattmeter in the laboratory. This power is referred to as the *real power*.

It is useful to define other forms of power for general impedance elements. We can most conveniently define these forms by considering the general impedance triangle displayed in Fig. 1.5a, where the complex impedance Z is the sum of $R + jX$. We will modify this triangle to obtain a voltage triangle as shown in Fig. 1.5b and then a power triangle as shown in Fig. 1.5c. We begin by multiplying each of the sides of the impedance triangle by a current I (with zero phase angle) to yield a triangle with side lengths proportional to the voltage shown in Fig. 1.5b. Furthermore, by multiplying each leg of the voltage triangle again by I, we obtain another triangle with side lengths proportional to the product of current and voltage displayed in Fig. 1.5c. This triangle is known as the *power triangle,* and the magnitude of the power corresponding to each leg is defined as

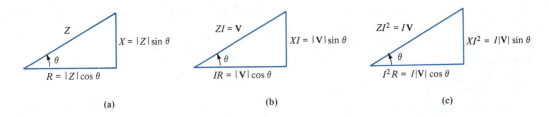

FIGURE 1.5   Power in Impedance Elements: (a) Impedance triangle, (b) Voltage triangle, (c) Power triangle

follows with corresponding units in parentheses:

$$P_T = I|V| = \text{apparent power in}$$
$$\text{volt-amperes (VA)} \qquad \text{(1.4–39)}$$

$$P = I|V| = \cos\theta = \text{real (or active) power in}$$
$$\text{watts (W)} \qquad \text{(1.4–40)}$$

$$Q = I|V| \sin\theta$$
$$= \text{imaginary (or reactive) power in volt-}$$
$$\text{amperes reactive (VAR)} \qquad \text{(1.4–41)}$$

where

$$P_T = P + jQ \qquad \text{(1.4–42)}$$

Although these newly defined units of power (VA, W, and VAR) appear to be different, they are physically equivalent while still providing a means of distinguishing one type of power from another. Note that in order for (1.4–40) to agree with (1.4–37), I and V must be the rms values and similarly in (1.4–39) and (1.4–41).

──────────────────────────── EX. 1.10

Power Calculation Example

Consider a sinusoidal voltage and current for an impedance element to be $v = 10 \sin \omega t$ V and $i = 1 \sin(\omega t - 60°)$ A. Calculate the power factor ($\cos\theta$), the real power, the reactive power, and the apparent power.

Solution: The power factor is given by

$$\cos\theta = \cos -60° = 0.5$$

with the current lagging the voltage by 60°. The various power values are obtained from

$$|IV| = \text{apparent power} = \frac{(10)(1)}{2} = 5 \text{ VA}$$

$$|IV| \cos\theta = \text{real power} = \frac{(10)(1)}{2} \cos 60° = 2.5 \text{ W}$$

$$|IV| \sin\theta = \text{reactive power} = \frac{(10)(1)}{2} \sin 60°$$

$$= 4.33 \text{ VAR}$$

1.4.5   Transformers

In this section, the lossless model for a transformer is presented. Since typical transformer losses are less than 5%, the lossless model is entirely adequate in instances that we will encounter.

A *transformer* is an electromagnetic device that may have two or more magnetically coupled windings. Figure 1.6a shows a physical representation of a transformer with two windings wrapped around a magnetic core, and Fig. 1.6b shows the corresponding circuit symbol. Each of the windings has a specific number of turns, given by N_i for the input side and N_o for the output side.

As indicated in Fig. 1.6c, an AC voltage is applied to the input terminals, which are therefore also called the *primary* terminals. Shown connected on the output side is a load impedance Z_L. The output side of a transformer is often called the *secondary* side.

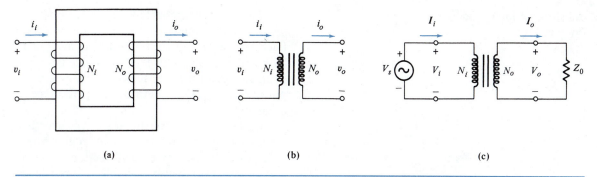

(a)   (b)   (c)

FIGURE 1.6   Iron-Core Transformer: (a) Geometry, (b) Transformer symbol, (c) Transformer with source and load

The application of an AC voltage at the primary terminals produces a time-varying magnetic field in the magnetic core. This magnetic field is confined to the magnetic core and hence passes through the secondary windings. Here, the changing magnetic field induces a voltage that produces the opposite effect as that at the primary coil. Hence, the magnetic field acts as a coupling mechanism between the two windings. Note that if the input source is constant (DC), no voltage will be induced across the output terminals. According to Faraday's law of physics, the induced voltage depends linearly upon the number of turns and the time rate of change of the magnetic field. Since the rate of change of the magnetic field is zero for DC, the induced voltage is zero. Thus, one important application of a transformer is, namely, to isolate the load element of a circuit (Z_0) from DC current and voltage.

The basic voltage transfer expression for the transformer is given by

$$\frac{\mathbf{V}_o}{\mathbf{V}_i} = \frac{N_o}{N_i} = a \qquad (1.4\text{–}43)$$

where $\mathbf{V}_o$ and $\mathbf{V}_i$ are voltage phasors and a is called the turns ratio. For a transformer, the turns ratio is an important parameter. Additionally, since we are considering an ideal transformer, without losses, the input and output power must be equal. Thus, the phasor currents and voltages

are related by

$$\mathbf{I}_i\mathbf{V}_i = \mathbf{I}_o\mathbf{V}_o$$

and substituting yields

$$\frac{\mathbf{I}_o}{\mathbf{I}_i} = \frac{1}{a} \qquad (1.4\text{–}44)$$

Equation (1.4–44) indicates that the current transfer ratio of a transformer is the inverse of the voltage transfer expression. Thus, a step-up in voltage produces a step-down in current, and vice versa. The fact that transformers can change the voltage to either step up to a larger value or step down to a smaller value is the reason that AC power systems are used exclusively for power distribution.

The physical size of a transformer varies according to the amount of power it is required to handle. High-power transformers required by electric power companies are larger than trucks, while low-power miniature units are used in electronic applications and can be hand held.

As already mentioned, another important application of a transformer is to enable an impedance transformation. From the circuit of Fig. 1.6c, the impedance seen looking into the input terminals (primary side) is

$$Z_i = \frac{\mathbf{V}_i}{\mathbf{I}_i} \qquad (1.4\text{–}45)$$

Substituting for V_i and I_i from the transfer equations yields

$$Z_i = \frac{V_o/a}{I_o a} = \frac{Z_o}{a^2} \qquad (1.4\text{--}46)$$

or

$$\frac{Z_i}{Z_o} = \frac{1}{a^2} \qquad (1.4\text{--}47)$$

Thus, as far as the input terminals are concerned, a modified impedance Z_o/a^2 is present when a transformer separates Z_o from the input. This impedance transformation property is used in some transistor amplifier circuits as an aid in circuit design.

1.5 TYPES OF PASSIVE COMPONENTS FOR ELECTRONIC CIRCUIT APPLICATIONS

Electronic circuits utilize many different types of resistors, capacitors, inductors, and transformers. These passive elements are available in a variety of packages from a variety of manufacturers. A brief description of the passive components used most often in present-day low-voltage electronic circuits will be given in this section.

1.5.1 Resistors

There are two main classes of resistors:

- Bulk resistors, which are discrete (individual) devices
- Resistor arrays, which are available in integrated circuit packages

The primary types of bulk resistors are carbon resistors, wire-wound resistors, and film resistors. Each type is named according to the method of fabrication. The various types available have differing ranges for their resistance values, maximum power consumption, and tolerance (percentage of accuracy).

Carbon resistors are the least expensive but are low-power resistors with a maximum permissible power in the 1 W range. Fraction-of-a-watt carbon resistors are also available. The resistance values vary from 1 Ω to 10 MΩ, with tolerance of $\pm 3\%$ to $\pm 20\%$.

Wire-wound resistors can be of higher power (up to 200 W) and higher accuracy (tolerances down to 1%). Values of resistance for these resistors range from a fraction of an ohm to the 100 kΩ region.

Film resistors are formed by the deposition of a film composed of a conducting layer. These resistors are available with very small resistance values (0.1 Ω) to very large resistance values (1.5 MΩ). They are also very precise, with tolerances of 0.5% or less. However, they are low-power resistors valued at 1 W or less.

Resistor arrays are also fabricated through the deposition of film layers. A large range of resistance values with small tolerances are available with low power ratings. These arrays are convenient, however, when multiple resistors are required in a circuit. The resistor arrays have the advantage of being contained in a single package.

1.5.2 Bulk Resistor Values and Tolerances

Because the techniques involved in the fabrication of a resistor do not result in a precise value of resistance (without added expense), resistor values are classified by a nominal value with a certain tolerance. Usual tolerance values used are $\pm 20\%$, $\pm 10\%$, and $\pm 5\%$, although smaller tolerances are also available. Because the resistor

TABLE 1.5   Color Code for Resistors

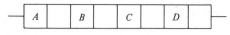

Formula: $R = (10A + B)C \pm D\%$

Band *A*		Band *B*	Band *C*		Band *D*	
Color	First Significant Figure	Second Significant Figure	Color	Multiplier	Color	Resistance Tolerance (%)
Black	0	0	Black	1	No color	± 20
Brown	1	1	Brown	10	(or black)	
Red	2	2	Red	100	Silver	± 10
Orange	3	3	Orange	1,000	Gold	± 5
Yellow	4	4	Yellow	10,000	White or	± 5
Green	5	5	Green	100,000	green (optional)	
Blue	6	6	Blue	1,000,000		
Purple	7	7				
Gray	8	8	Silver	0.01		
White	9	9	Gold	0.1		

magnitudes are not exact, only certain nominal values of resistance are available, and these depend upon the particular tolerance specified. For example, the nominal resistor values used when the tolerance is $\pm 20\%$ are 10, 15, 22, 33, 47, 68, and $100 \, \Omega$, as well as multiples of 10 of these integers. Note that 20% of each adjacent integer added together yields a value greater than or equal to the difference between the two integers. Thus, the entire range of values is covered with these nominal resistance values for a tolerance of $\pm 20\%$. For the case of a lower tolerance, such as $\pm 10\%$ or $\pm 5\%$, additional nominal resistor values are necessary.

1.5.3   Color Coding of Resistors

In order to observe visually the magnitude and tolerance for a bulk resistor, manufacturers use a standardized color code consisting of four colored bands that are placed directly on the resistor. The significance of each band and its color are displayed in Table 1.5. Also, a formula for calculating the value of resistance is indicated. Carbon resistors can be distinguished from wire-wound resistors in that band *A* is twice as large as the others for a wire-wound resistor.

───────────────────────────── **EX. 1.11**

Resistor Color Code Example

Determine the nominal value of the resistor, along with its tolerance, corresponding to the following colors of the bands: band *A*, orange; band *B*, white; band *C*, red; band *D*, silver.

Solution: Determining the numerical values corresponding to each color by using Table 1.5, we have

$$R = (39)10^2 \pm 10\%$$

or

$$R = 3.9 \, k\Omega \pm 10\%$$

1.5.4　Capacitors

There are four primary types of capacitors available:

- Ceramic
- Plastic film
- Mica
- Electrolytic

These capacitors range in capacitance from μF to pF. Each of these types is named according to the type of dielectric layer it possesses.

The magnitude of capacitance for each of these types of capacitors is based upon the parallel-plate capacitance expression, given by

$$C = \frac{\varepsilon_r \varepsilon_0 A}{d} \qquad (1.5\text{--}1)$$

where

ε_0 = dielectric permittivity of free space

ε_r = relative dielectric constant of dielectric medium

A = area of capacitor plate

d = distance between plates

From this expression, we realize that there are three variables (A, d, and ε_r) that may be adjusted to alter the magnitude of the capacitance. The first variable, A, indicates that the larger the area, the larger the value of C. However, the consequence of providing a large capacitance in this manner is increased volume of the capacitor, which is undesirable in our world of miniaturization. Furthermore, the spacing between the plates, d, can be reduced to provide a large value of C, but the limiting thickness is on the order of 1 μm. Finally, we note that the dielectric constant, ε_r, can provide wide variation in capacitance values. Thus, various materials are used for the dielectric. Although capacitors are usually made in cylindrical form, this parallel-plate capacitance expression is still approximately valid.

TABLE 1.6　Color Code for Capacitors

Formula:　$C = (10B + C)D \pm E\%$

Band A		Band B	Band C	Band D		Band E		
Color	Temp. Coeff. (ppm/°C)	First Significant Figure	Second Significant Figure	Color	Multiplier	Color	Tolerance (>10 pF)	(<10 pF)
Black	0	0	0	Black	1	Black	±20	2.0
Brown	-30	1	1	Brown	10	Brown	±1	0.1
Red	-80	2	2	Red	100	Red	±2	—
Orange	-150	3	3	Orange	1000	Green	±5	0.5
Yellow	-220	4	4	Gray	0.01	Gray	—	0.25
Green	-330	5	5	White	0.1	White	±10	1.0
Blue	-470	6	6					
Violet	-750	7	7					
Gray	$+30$	8	8					
White	$+500$	9	9					

Capacitance values in pF

Considering ceramic capacitors, the type of ceramic used can be varied widely, which results in a large range of capacitance values. Film capacitors use plastics such as polyester, polystyrene, and teflon. Often, these films are manufactured in the form of tape and are wrapped alternately with metal-film tape in cylindrical form. Mica (a combination of mineral silicates) capacitors are used in high-frequency applications; however, their capacitance values are smaller than those obtainable with the other types. Electrolytic capacitors are available with a wide range of capacitance values, but as the capacitance value is increased, the rated voltage for the capacitor is reduced. Also, some electrolytic capacitors are polar, which means that only one polarity may be used and that if the opposite polarity is used, the capacitor may destroy itself.

1.5.5   Color Coding of Capacitors

Capacitors also have a color code for distinguishing the magnitude and tolerance of the capacitance value. The significance of each band and its color are shown in Table 1.6. Note that band A indicates an additional parameter, the temperature coefficient. This parameter indicates the changes in capacitances in parts per million per degree Celsius (ppm/°C).

1.5.6   Inductors

Two common types of inductors are commercially available:

- Air-core inductors
- Iron-core inductors

Inductors are fabricated by bending a wire around a cylinder to form a number of turns, n. Then, the larger n is, the larger the inductance value. Using an iron core instead of air also increases the inductance magnitude. Additionally, in high-frequency applications where small values of L will suffice, a wire can be hand wrapped to form several turns that will provide the necessary inductance magnitude. It should be noted, however, that inductors are avoided in most electronic applications that use integrated circuits because of their bulkiness. It is currently not possible to miniaturize them to a compatible size in an integrated circuit.

1.5.7   Transformers

Transformers are used in a variety of applications in electronic circuits. Some of these applications include impedance matching in audio and video amplifiers, voltage level transforming in power supplies, and tuning in amplifier stages.

Transformers are usually constructed with an iron core by wrapping more than one winding onto the core. The iron core provides efficient magnetic coupling. The windings on the high-current, low-voltage side of the transformer are made of heavy-gauge wire to reduce the I^2R losses in this coil. The low-current, high-voltage winding can be made of finer-gauge wire, since the current is much smaller in this coil.

Transformers with one primary winding and two or more secondary windings are also used in some electronic applications. One practical example is the four-winding transformer in which the primary winding is constructed so that it is suitable for connection to a 115 V, 60 Hz power source. The three secondary voltages are typically 700 V, 6.3 V, and 5.0 V.

1.6   BASIC CIRCUIT LAWS

Basic circuit laws and circuit theory terminology are introduced in this section which will aid considerably in our future analyses of electronic circuits.

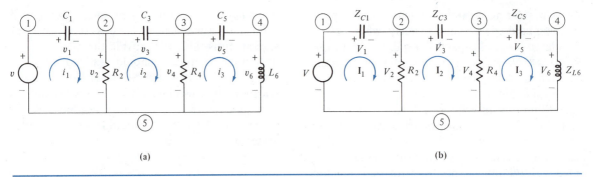

FIGURE 1.7   Simple Circuit Example: (a) Basic circuits, (b) Equivalent circuit with impedances

Figure 1.7a shows a circuit with seven elements to be used as an example. Figure 1.7b indicates the same circuit with impedances and phasor currents and voltages. These circuits contain the following elements:

- *Loops,* which are closed paths around which charge can flow resulting in current (In the example, three loops are shown with defined loop currents i_1, i_2, and i_3 or phasor loop currents $\mathbf{I}_1$, $\mathbf{I}_2$, and $\mathbf{I}_3$.)
- *Nodes,* which are junctions or points of interconnection of two or more elements (The five in the example are arbitrarily numbered with circles around the numbers.)
- *Branches,* each of which consists of an element in the circuit (In the example, there are seven branches.)

Note that in Figs. 1.7a and b, positive directions of the loop currents and branch voltages have been arbitrarily defined. Any or all of these could be oppositely directed (negative), depending upon the sources that drive the network. In Fig. 1.7a, the signal voltage source v is assumed to be sinusoidal and is given by $V\sin(\omega t + \phi)$.

1.6.1   Kirchhoff's Voltage Law

- ***Kirchhoff's Voltage Law* (KLV):**   The algebraic sum of the voltage drops and/or rises

across each element around any loop in a network is zero. Mathematically, we have

$$\sum_{j=1}^{n} v_j = 0 \qquad \text{(1.6–1)}$$

where n is the number of voltage drops and/or rises, one for each element, around the loop, and each voltage drop or rise has a specific and oppositely assigned polarity. ∎

Physically, this law indicates that if we start at one point in the circuit and add the voltage drops and/or rises algebraically around any loop, upon returning to the original point, the net voltage is zero.

———————————————————————— **EX. 1.12**

Resistor KVL Example

Determine the currents i_1 and i_2 in the resistor network of Fig. E1.12 by writing the KVL for each loop.

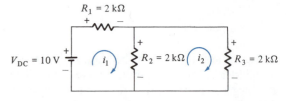

FIGURE E1.12

Solution: The polarities of the voltages across each resistor have been arbitrarily defined as indicated according to the directions chosen for i_1 and i_2, except in the case of R_2. We arbitrarily assign this polarity because the relative magnitudes of i_1 and i_2 are as yet unknown. Writing KVL for each loop, taking a rise in voltage to be positive and a drop in voltage to be negative, yields

$$V_{DC} - R_1 i_1 - R_2(i_1 - i_2) = 0$$

and

$$R_2(i_1 - i_2) - R_3 i_2 = 0$$

Solving these equations simultaneously yields

$$i_1 = \frac{(R_2 + R_3)V_{DC}}{R_1 R_2 + R_1 R_3 + R_2 R_3} = 3.33 \text{ mA}$$

and

$$i_2 = \frac{R_2 V_{DC}}{R_1 R_2 + R_1 R_3 + R_2 R_3} = 1.66 \text{ mA}$$

When capacitors and/or inductors are elements in a circuit, we use the impedance method of analysis involving phasors because the elements are not purely resistive. Nonetheless, in applying KVL, the phasor voltage polarities must also be consistently accounted for. That is, a phasor current flowing through a passive element creates a voltage that opposes the current flow. Thus, as shown in Fig. 1.8a, the phasor relation is $\mathbf{V} = \mathbf{I}_1 Z$. However, for Fig. 1.8b, where the defined direction of positive current flow has been reversed, $\mathbf{V} = -\mathbf{I}_2 Z$.

Returning to the six-element example of Fig. 1.7, note that Fig. 1.7b shows the equivalent impedance network with phasor currents and voltages. The KVL for each loop is written by inspection as follows:

Loop 1: $\mathbf{V} - \mathbf{V}_1 - \mathbf{V}_2 = 0$ (1.6–2a)

Loop 2: $\mathbf{V}_2 - \mathbf{V}_3 - \mathbf{V}_4 = 0$ (1.6–2b)

Loop 3: $\mathbf{V}_4 - \mathbf{V}_5 - \mathbf{V}_6 = 0$ (1.6–2c)

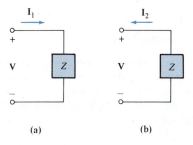

(a)   (b)

FIGURE 1.8 Element with Current Direction Defined in Opposite Manner: (a) Current inward, (b) Current outward

Substituting for the phasor voltages corresponding to the **I–V** relation of each impedance element yields

Loop 1: $V - Z_{C_1}\mathbf{I}_1 - R_2(\mathbf{I}_1 - \mathbf{I}_2) = 0$

$$(1.6\text{–}3a)$$

Loop 2: $R_2(\mathbf{I}_1 - \mathbf{I}_2) - Z_{C_3}\mathbf{I}_2$
$\qquad\qquad - R_4(\mathbf{I}_2 - \mathbf{I}_3) = 0$ (1.6–3b)

Loop 3: $R_4(\mathbf{I}_2 - \mathbf{I}_3) - Z_{C_5}\mathbf{I}_3 - Z_{L_6}\mathbf{I}_3 = 0$

$$(1.6\text{–}3c)$$

Note that only the magnitude V of the sinusoidal voltage source $[v = V\sin(\omega t + \phi)]$ appears in these phasor expressions because each of the currents and voltages in the network have the same time dependence. Furthermore, each of the impedances are directly dependent upon the original element value as indicated by (1.4–19) or (1.4–20).

Equations (1.6–3a, b, and c) are an independent set that can be solved algebraically for the phasor currents $\mathbf{I}_1$, $\mathbf{I}_2$, and $\mathbf{I}_3$. Then, to determine the voltage for any element, the specific **I–V** relation for that element is used.

Finally, to determine the corresponding time function for each phasor current and voltage, we form the product of each phasor with $e^{j(\omega t + \phi)}$ and take the imaginary part of this product. This operation is based upon the assumption that the input source in the network varies as $\sin(\omega t + \phi)$.

KVL and Impedance Method Example

Determine the phasor current and voltage as well as the time-varying current and voltage for the capacitor in Fig. E1.13. The signal voltage source is given by $v_1(t) = 10 \sin(\omega t + \phi)$, $L = 10\ \mu H$, $C = 1\ \mu F$, and $R = 10\ \Omega$. Also, let $\omega = 10^6$ rad/s and $\phi = 1.0$ rad.

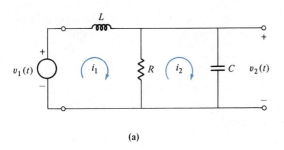

(a)

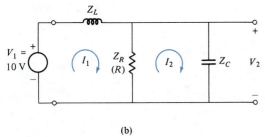

(b)

FIGURE E1.13

Solution: The first step is to transform the network into its equivalent impedance form. Replacing each element with its equivalent impedance (see Fig. E1.13b) transforms the voltages and currents into phasors.

Writing KVL for each loop of Fig. E1.13b yields

$$10 - Z_L I_1 - Z_R(I_1 - I_2) = 0 \tag{1}$$

and

$$Z_R(I_1 - I_2) - Z_C I_2 = 0 \tag{2}$$

where

$$Z_L = j\omega L$$

$$Z_C = \frac{1}{j\omega C}$$

$$Z_R = R$$

Substituting these impedance expressions and grouping terms yield

$$10 - I_1(j\omega L + R) + I_2 R = 0 \tag{3}$$

and

$$I_1 R - I_2\left(R + \frac{1}{j\omega C}\right) = 0 \tag{4}$$

Solving for I_1 in terms of I_2 using (4) gives

$$I_1 = \left[\frac{R + 1/(j\omega C)}{R}\right] I_2 \tag{5}$$

Using (5) to eliminate I_1 from (3) yields

$$10 - (j\omega L + R)\left(\frac{R + 1/(j\omega C)}{R}\right) I_2 + I_2 R = 0$$

Solving for I_2, we have

$$I_2 = \frac{10}{(j\omega L + R)[(R + 1/j\omega C)/R] - R} \tag{6}$$

This expression is the phasor current in the capacitor. Thus, the phasor voltage is

$$V_2 = Z_C I_2 = \frac{10(1/j\omega C)}{(j\omega L + R)[(R + 1/j\omega C)/R] - R} \tag{7}$$

Substituting the values of ω, R, L, and C into (6) and (7) yields the complex expressions

$$I_2 = \frac{10}{[(j10 + 10)(10 - j)/10] - 10} = \frac{10}{9j + 1} \tag{8}$$

and

$$V_2 = \frac{10/j}{[(j10 + 10)(10 - j)/10] - 10} = \frac{-10j}{9j + 1} \tag{9}$$

The equivalent magnitude and phase form for (8) and (9) are

$$I_2 = 1.1\ \angle\left(-\tan^{-1}\frac{9}{1}\right)$$

$$= 1.1\ \angle\,(-83.66°)$$

$$= 1.1\ \angle\,(-1.46\ \text{rad}) \tag{10}$$

and

$$V_2 = 1.1 \angle \left(\pm 180° + 90° - \tan^{-1} \frac{9}{1} \right)$$

$$= 1.1 \angle (186.34° \text{ or } -173.7°)$$

$$= 1.1 \angle (3.25 \text{ or } -3.03 \text{ rad}) \tag{11}$$

Finally, to obtain the corresponding current and voltage time functions, we multiply these phasors by $e^{j(\omega t + 1)}$ and take the imaginary part:

$$i_2 = Im\{[1.1 \angle -1.46]e^{j(\omega t + 1)}\} \tag{12}$$

and

$$v_2 = Im\{[1.1 \angle -3.03]e^{j(\omega t + 1)}\} \tag{13}$$

or

$$i_2 = 1.1 \sin(\omega t - 0.46) \tag{14}$$

and

$$v_2 = 1.1 \sin(\omega t - 2.03) \tag{15}$$

where the units of i_1 and v_2 are A and V, respectively. Note that the expressions for current and voltage differ in phase by $\pi/2$ rad, which was expected for the capacitor. Furthermore, in determining the phase, as in Eq. (11), either $+180°$ or $-180°$ may be selected.

1.6.2 Kirchhoff's Current Law

- **Kirchhoff's Current Law (KCL):** The algebraic sum of currents into (or out of) a node is zero. Mathematically, we have

$$\sum_{j=1}^{n} i_j = 0 \tag{1.6–4}$$

where n is the number of branch currents into the node, and each current has a consistently defined positive direction. ∎

When we apply this law, branch currents must be defined and at each node the sum of those com-

ing into (or going out of) the node must be equal to zero. This law provides an alternative method of circuit analysis to that of KVL.

——————————————————— EX. 1.14

Amplifier KCL Example

Write Kirchhoff's current law for the bipolar junction transistor (BJT) shown in the amplifier circuit of Fig. E1.14. The letters B, C, and E represent the transistor terminals called the base, collector, and emitter, respectively.

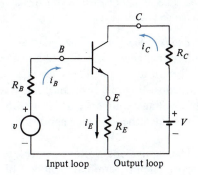

FIGURE E1.14

Solution: The KCL for the BJT is written in terms of the terminal currents as follows, with the transistor treated as a single node:

$$i_E = i_B + i_C$$

Note: This relationship is always valid and is extremely important in transistor circuit analysis.

1.6.3 Thévenin's Theorem

Thévenin's theorem is a general principle concerning a linear two-terminal network that provides a powerful technique for circuit simplification and analysis. The following is a statement of this theorem for a resistive network. It is easily modified for a network with capacitors and inductors, as we will see.

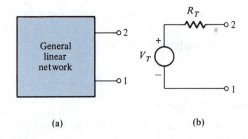

(a)   (b)

FIGURE 1.9 General Linear Network and Thévenin Equivalent Circuit: (a) Original two terminal network, (b) Thévenin equivalent circuit

- ■ **Thévenin's Theorem:** The general linear two-terminal network in Fig. 1.9a can be replaced with a unique Thévenin equivalent circuit consisting of a resistance R_T in series with a voltage source V_T as shown in Fig. 1.9b provided the following conditions exist:

  1. The equivalent voltage source is the open-circuit terminal voltage of the network.
  2. The equivalent resistance is the ratio of the open-circuit voltage of the network to the short-circuit current of the network. ■

We define the open-circuit voltage and short-circuit current as we verify Thévenin's theorem. In order for the two networks of Figs. 1.9a and b to be equivalent, we compare them at the two extreme conditions of their terminals—that is, under open-circuit and short-circuit conditions. Under open-circuit conditions, the voltage across terminals 1 and 2 in Fig. 1.9a is defined as V_{OC}, whereas in Fig. 1.9b this voltage is V_T since there is no voltage across R_T when the current is zero. Thus, for equivalence

$$V_{OC} = V_T \qquad\qquad (1.6-5)$$

Under short-circuit conditions, the short-circuit current is defined as I_{SC} and flows in the original

network, whereas V_T/R_T flows in the Thévenin equivalent circuit. Thus,

$$I_{SC} = \frac{V_T}{R_T} \qquad\qquad (1.6-6)$$

or

$$R_T = \frac{V_T}{I_{SC}} = \frac{V_{OC}}{I_{SC}} \qquad\qquad (1.6-7)$$

Hence, (1.6–5) through (1.6–7) define the basic Thévenin equivalent components.

Quite often, the equivalent Thévenin resistor R_T is obtained by determining the resistance seen looking into terminals 1 and 2 with all independent sources reduced to zero. The following example describes both methods of determining R_T.

──────────────────────────────── EX. 1.15

Thévenin Equivalent Circuit Example

Obtain the Thévenin equivalent circuit for the network of Fig. E1.15a.

Solution: We first proceed by using the two-step method in the statement of Thévenin's theorem. Thus, we begin by open-circuiting the terminals $A-B$ in Fig. E1.15a. The open-circuit voltage is then given by the product of current i and resistance R_2. Since the current under open-circuit conditions is just $i = V_1/(R_1 + R_2)$, we have

$$V_{OC} = iR_2 = \frac{R_2}{R_1 + R_2} V_1$$

The second step is to determine the short-circuit current. Placing a short across the terminals $A-B$ forces all of the current produced by V_1 to bypass R_2 and pass through the short-circuit path. The current around the loop containing the short is then obtained by writing KVL as

$$V_1 - I_{SC}R_1 = 0$$

or

$$I_{SC} = \frac{V_1}{R_1}$$

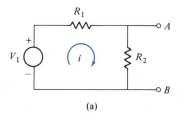

(a)

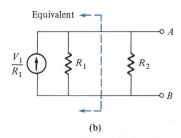

(b)

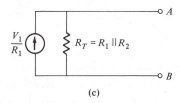

(c)

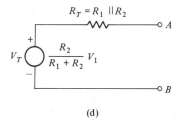

(d)

FIGURE E1.15

Thus, R_T is given by

$$R_T = \frac{V_{OC}}{I_{SC}} = \frac{R_1 R_2}{R_1 + R_2}$$

which is the parallel combination of R_1 with R_2 ($R_1 \| R_2$). The Thévenin equivalent circuit is shown in Fig. E1.15d.

Alternate Solution: The series branch containing R_1 and V_1 can be transformed into a branch with R_1 in parallel with a current source, as shown in Fig. E1.15b. Only the use of Ohm's law is necessary in this transformation pro-

cess. The resulting circuit (R_1 in parallel with a current source) is the Norton equivalent circuit (for R_1 and V_1) to be discussed in the next section. Next, in Fig. E1.15b, we combine R_1 and R_2 in parallel to obtain the circuit shown in Fig. E1.15c. Finally, we transform the current source of Fig. E1.15c back into a voltage source (using Ohm's law) to obtain the Thévenin equivalent circuit shown in Fig. E1.15d.

Note: The circuit of Fig. E1.15a is commonly used in transistor amplifiers and understanding the conversion process to obtain the Thévenin equivalent circuit is extremely important.

1.6.4   Norton's Theorem

Norton's theorem results in an alternative equivalent circuit representation of a linear network. Although Norton's equivalent circuit can be obtained in a manner analogous to obtaining Thévenin's equivalent circuit, we will instead make a circuit transformation to obtain it as was done in Ex. 1.15. The Thévenin equivalent voltage source of Fig. 1.9b is first converted into a current source by division of V_T with R_T, resulting in the equivalent circuit of Fig. 1.10a. The resistor R_T can then be replaced by its equivalent conductance G_N, resulting in Fig. 1.10b, which is the Norton equivalent circuit of the original linear network (Fig. 1.9a). Defining equations for I_N and G_N can therefore be written as follows:

$$I_N = I_{SC} \qquad\qquad (1.6\text{–}8)$$

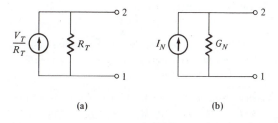

(a)   (b)

FIGURE 1.10   Development of the Norton Equivalent Circuit: (a) Circuit with V_T changed into current source, (b) Circuit with elements redefined

and

$$G_N = \frac{1}{R_T} = \frac{I_{SC}}{V_{OC}} \qquad (1.6\text{--}9)$$

The following is a statement of Norton's theorem for a resistive network.

- **Norton's Theorem:**   A general linear two-terminal network can be replaced with a unique Norton equivalent circuit consisting of a conductance G_N in parallel with a current source I_N provided the following conditions exist:

    1. The equivalent current source is the short-circuit current of the original network.
    2. The equivalent admittance is the ratio of the short-circuit current to the open-circuit voltage. ∎

Note: Under steady-state sinusoidal conditions, the Thévenin and Norton theorems can also be applied to circuits containing inductors and capacitors as well as resistors. Under these conditions, impedance replaces resistance, and admittance replaces conductance. Moreover, the sources are phasors.

———————————————————————— EX. 1.16

Norton Equivalent Circuit Example

Obtain the Norton equivalent circuit for the network of Fig. E1.15a.

Solution: The solution here is easily obtained since the work has already been done in the previous example. From Fig. E1.15c, we have $I_N = V_1/R_1$ and $G_N = 1/R_T$.

Alternate Solution: This method involves determining the short-circuit current of Fig. E1.15a by shorting the terminals $A - B$ to obtain

$$I_{SC} = \frac{V_1}{R_1} = I_N$$

Similarly, the open-circuit voltage is

$$V_{OC} = \frac{R_2}{R_1 + R_2} V_1$$

Thus, the equivalent conductance is

$$G_N = \frac{I_{SC}}{V_{OC}} = \frac{R_1 + R_2}{R_1 R_2} = \frac{1}{R_1 \| R_2}$$

———————

1.6.5   Superposition Theorem

In many circuits, more than one source of voltage and/or current is simultaneously present. If the circuit is linear and the sources are independent, the superposition theorem is applicable.

- **Superposition Theorem:**   For a linear circuit containing more than one independent voltage and/or current source, the current through or voltage across any passive element in the network is obtained by algebraically adding the contributions due to each source acting alone. ∎

Note: If controlled sources are present, they are not treated by superposition, but are included in every separate determination.

An extremely important application of this theorem comes up in AC amplifier circuits. Both DC and AC are present in these circuits, and the superposition theorem permits the separation of the DC and AC portions of the circuit from each other. Thus, the separation of the DC circuit analysis and design from the AC circuit analysis and design becomes extremely convenient.

———————————————————————— EX. 1.17

Superposition Example

Determine the current in the resistor R_2 shown in Fig. E1.17.

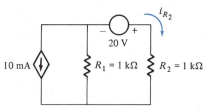

FIGURE E1.17

Solution: Using the superposition theorem, we first obtain the current component due to the current source. The voltage source is set equal to zero, and the component of current due to the 10 mA source is obtained by realizing that half of the current passes through each resistor. Thus, this component is (1/2)(10), or 5 mA in the upward direction. Then, the current due to the voltage source alone is 20/2, or 10 mA in the opposite direction. Thus, the net current is 5 mA in the indicated direction.

1.7 VOLTAGE AND CURRENT DIVIDER RULES

The voltage and current divider rules are quite useful in the analysis of a circuit. These rules might be regarded as "tricks" that considerably simplify the analysis. They provide a method for determining a voltage or current by inspection of the circuit without going through a more detailed analysis. The development of these rules comes directly from the principles of series and parallel combinations of elements along with Kirchhoff's voltage and current laws. We will develop these rules for the case of resistors, but they also apply to networks containing impedances.

1.7.1 Voltage Divider Rule

- *Voltage Divider Rule:* When a voltage exists across elements in series, the voltage across each element is a specific fraction (portion) of the total voltage. ▪

This rule is best demonstrated through an example involving resistors. Figure 1.11 shows a voltage v_A applied across three resistors in series. According to KVL, the voltage v_A must split, partly across each element R_1, R_2, and R_3. To determine the specific fraction of v_A that must exist across each resistor, we determine the current i_A by using Ohm's law. Since the three resistors are in series, their sum is the equivalent resistance, and therefore $i_A = v_A/(R_1 + R_2 + R_3)$. Each individual voltage is then obtained by multiplying by the respective resistor value. Thus,

$$v_1 = \frac{R_1}{R_1 + R_2 + R_3} v_A \qquad (1.7\text{--}1)$$

$$v_2 = \frac{R_2}{R_1 + R_2 + R_3} v_A \qquad (1.7\text{--}2)$$

$$v_3 = \frac{R_3}{R_1 + R_2 + R_3} v_A \qquad (1.7\text{--}3)$$

These expressions indicate the manner in which the voltage divider rule is applied. Namely, each individual branch voltage is a fraction of the total voltage across all of the elements in series. The fraction is given by the ratio of the resistance of the element to the sum of all the resistances in series.

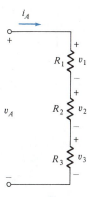

FIGURE 1.11 Series Circuit Indicating Method of Applying Voltage Divider Rule

1.7.2   Current Divider Rule

■ *Current Divider Rule:* When a current flows through several elements in parallel, the current in each element is a specific fraction (portion) of the total current. ■

To learn the method of applying the current divider rule, we again consider a resistive network using a conductance representation. For the network shown in Fig. 1.12, the total current i_A is related to each branch current (using KCL) by the relation

$$i_A = i_1 + i_2 + i_3 \qquad (1.7\text{–}4)$$

However, using Ohm's law in the form of conductances, we have by substitution

$$i_A = G_1 v_A + G_2 v_A + G_3 v_A \qquad (1.7\text{–}5)$$

or

$$v_A = \frac{i_A}{G_1 + G_2 + G_3} \qquad (1.7\text{–}6)$$

Therefore, substituting (1.7–6) into (1.7–5), we have

$$i_A = \frac{G_1}{G_1 + G_2 + G_3} i_A + \frac{G_2}{G_1 + G_2 + G_3} i_A$$
$$+ \frac{G_3}{G_1 + G_2 + G_3} i_A \qquad (1.7\text{–}7)$$

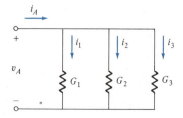

FIGURE 1.12   Parallel Circuit Indicating Method of Applying Current Divider Rule

where each term on the right side of (1.7–7) corresponds to the individual branch currents as follows:

$$i_1 = \frac{G_1}{G_1 + G_2 + G_3} i_A \qquad (1.7\text{–}8)$$

$$i_2 = \frac{G_2}{G_1 + G_2 + G_3} i_A \qquad (1.7\text{–}9)$$

$$i_3 = \frac{G_3}{G_1 + G_2 + G_3} i_A \qquad (1.7\text{–}10)$$

These expressions indicate the manner of application of the current divider rule. Each individual branch current is given by the ratio of the conductance of the branch divided by the sum of the conductances. This rule is the "dual" of the voltage divider rule, and conductance ratios are used.

—————————————————————————— EX. 1.18

Thévenin and Norton Equivalent Circuit Examples Using Voltage and Current Divider Rules

For the network shown in Fig. E1.18a, determine the Thévenin and Norton equivalent circuits.

Solution: We proceed by first finding the open-circuit voltage and the short-circuit current at the terminals A–B. The open-circuit voltage V_{ABOC} is the voltage across the 10 Ω resistor, since no current flows through the 5 Ω resistor when the terminals A–B are open circuited. This voltage is obtained by using KVL around the loop or, more simply, by the voltage divider rule as follows:

$$V_{ABOC} = \left(\frac{10}{10 + 8 + 2}\right)(10 \text{ V}) = 5 \text{ V}$$

The short-circuit current I_{ABSC} is obtained by using the current divider rule as follows:

$$I_{ABSC} = i\left(\frac{1/R_4}{1/R_4 + 1/R_3}\right) = i\left(\frac{R_3}{R_3 + R_4}\right) = i\left(\frac{2}{3}\right)$$

where

$$i = \frac{V_1}{R_1 + R_2 + R_3 \| R_4}$$

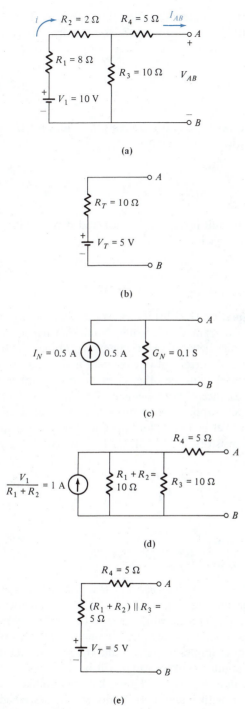

(a)

(b)

(c)

(d)

(e)

FIGURE E1.18

Therefore, we have

$$I_{ABSC} = \left[\frac{10}{10 + 5 \| 10} \right] \left(\frac{2}{3} \right) = 0.5 \text{ A}$$

Thus, the Thévenin equivalent circuit elements are

$$R_T = \frac{V_{ABOC}}{I_{ABSC}} = \frac{5 \text{ V}}{0.5 \text{ A}} = 10 \text{ }\Omega$$

and

$$V_T = V_{ABOC} = 5 \text{ V}$$

The Thévenin equivalent circuit is shown in Fig. E1.18b.

Finally, the Norton equivalent circuit elements are

$$I_N = I_{ABSC} = \frac{V_T}{R_T} = 0.5 \text{ A}$$

and

$$G_N = \frac{1}{R_T} = 0.1 \text{ S}$$

This equivalent circuit is shown in Fig. E1.18c.

Note: Perhaps an easier method to determine the Thévenin and Norton equivalent circuits is to again use source transformations and parallel–series combinations of resistors. To obtain the Thévenin equivalent circuit in this manner, we begin by combining R_1 in series with R_2 and then convert V_1 into a current source. The transformed equivalent circuit is shown in Fig. E1.18d. Next, the two 10 Ω resistors are combined in parallel, yielding 5 Ω, and the current source of 1 A is converted to a voltage source of 5 V. The resulting circuit is shown in Fig. E1.18e. Finally, by combining the two resistors in series, the Thévenin equivalent circuit is obtained as previously. The Norton equivalent circuit is then obtained by converting V_T to a current source.

EX. 1.19

Voltage Divider Example

Determine the output voltage v_O in terms of the input voltage v_I for the circuit of Fig. E1.19.

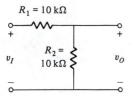

FIGURE E1.19

Solution: The output voltage is obtained using the voltage divider rule as follows:

$$v_O = \frac{R_2}{R_1 + R_2} v_I = \frac{v_I}{2}$$

――――――――――――――――――――――――― EX. 1.20

Current Divider Example

Determine the phasor current **I** (corresponding to i) in terms of the phasor source current **I**$_s$ (corresponding to i_s) for the circuit of Fig. E1.20.

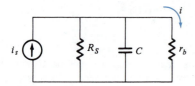

FIGURE E1.20

Solution: By inspection of the circuit and visualizing its impedance form, the phasor current **I** is obtained using the current divider rule as follows:

$$\mathbf{I} = \frac{1/r_b}{(1/R_S) + (1/r_b) + j\omega C} \mathbf{I}_s$$

Note: This example is actually the determination of the input current I of a transistor at high frequency. The expression for I indicates that as ω increases, the magnitude of I decreases. As we will see, this situation causes the current amplification of the transistor to reduce drastically at high frequency.

An alternate form of the current divider rule is sometimes desirable. To eliminate some alge-

bra, we again consider the circuit of Fig. 1.12 with $G_3 = 0$. From (1.7–8), the current i is now given by

$$i_1 = \frac{G_1}{G_1 + G_2} i_A \tag{1.7–11}$$

However, substituting resistor values for G_1 and G_2, (1.7–11) becomes

$$i_1 = \frac{1/R_1}{1/R_1 + 1/R_2} i_A \tag{1.7–12}$$

and multiplying numerator and denominator by $R_1 R_2$ yields

$$i_1 = \frac{R_2}{R_1 + R_2} i_A \tag{1.7–13}$$

Equation (1.7–13) indicates the alternate form of the current divider rule. Namely, when current divides between two elements in parallel, the fraction of current in one element is the ratio of the resistance of the other element to the sum of the resistance of both elements. This form of the rule is also applicable to more than two elements in parallel, as well as for impedance elements instead of resistors. The result for i_1 when $G_3 \neq 0$ is

$$i_1 = \frac{R_2 \| R_3}{R_1 + R_2 + R_3} i_A \tag{1.7–14}$$

1.8   FILTER NETWORKS

A *filter network* is a circuit that rejects certain signal frequencies while passing others. Simple filter networks consist of combinations of resistors, capacitors, and inductors and are excellent examples of circuits that involve impedances. The filter networks that will be described in this section are passive filters; however, active filters used with integrated circuits will be described in Chapter 7. Active filters provide the possibility of amplifying the magnitude of the signals allowed to pass through the network, whereas passive filters do not.

1.8.1   *RC* Filter Networks

Two possible arrangements for simple two-element resistance–capacitance filter networks are shown in Fig. 1.13. These *low-pass* and *high-pass* filter networks are used to filter out or reject certain ranges of high- or low-frequency signal components. In Fig. 1.13a, only a range of low-frequency signals transfers from input to output, while the circuit of Fig. 1.13b blocks low-frequency signals and allows only a range of high-frequency signals to be transmitted.

Using the impedance method, we can apply the voltage divider rule to each circuit of Fig. 1.13 to immediately obtain the output voltage in terms of the input voltage. Expressing these quantities as a ratio, we have, for the low-pass filter,

$$\frac{\mathbf{V}_o}{\mathbf{V}_i} = \frac{(1/j\omega C)}{R + (1/j\omega C)} = \frac{1}{j\omega RC + 1} \qquad (1.8\text{–}1)$$

and, for the high-pass filter,

$$\frac{V_o}{V_i} = \frac{1}{1 + (1/j\omega RC)} \qquad (1.8\text{–}2)$$

Equations (1.8–1) and (1.8–2) are referred to as *voltage transfer equations*. The magnitude of the output-to-input voltage ratio $|V_o/V_i|$ versus ω is

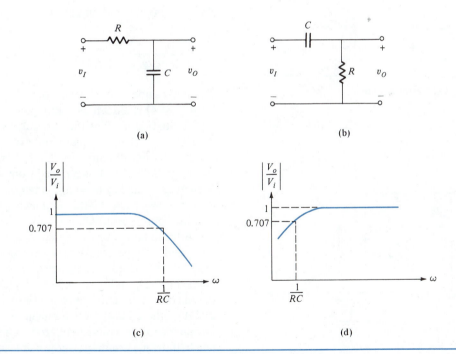

FIGURE 1.13   *RC* Filter Networks: (a) Low-pass filter network, (b) High-pass filter network, (c) Voltage transfer characteristic (low-pass), (d) Voltage transfer characteristic (high-pass)

shown in Figs. 1.13c and d. Note that (1.8–1) is unity for low frequencies ($\omega \ll 1/RC$), while (1.8–2) is unity for high frequencies ($\omega \gg 1/RC$). Thus, these two filter networks carry out their desired functions. Also note that in each case the magnitude is reduced by a factor 0.707 of the peak value at the specific frequency $\omega = 1/RC$. This frequency is referred to as the *half-power frequency* because the voltage at this value is reduced by $1/\sqrt{2} = 0.707$.

The reader should also realize that *RL* filter networks can be arranged to perform similar functions. These filter networks are formed by replacing the capacitor in Fig. 1.13 by a resistor and the resistor by an inductor. However, in modern electronic circuits, it is much more practical to use capacitors rather than inductors because of the difficulty encountered in fabricating integrated circuit inductors and because discrete inductors are larger in size.

1.8.2   *RLC* Filter Networks

RLC filters allow a middle band of frequencies to pass, but reject higher- and lower-frequency components. These filters are called *band-pass* filters.

Figure 1.14 displays an example of a circuit that performs the band-pass filter operation. Many other possibilities exist. Using the impedance method and the voltage divider rule, for the circuit of Fig. 1.14, we have

$$V_o = \frac{R}{j\omega L + (1/j\omega C) + R}\, V_i \qquad (1.8–3)$$

Rearranging and solving for V_o/V_i yield

$$\frac{V_o}{V_i} = \frac{j\omega RC}{-\omega^2 LC + j\omega RC + 1} \qquad (1.8–4)$$

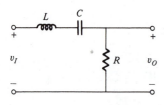

FIGURE 1.14   *RLC* Filter Network

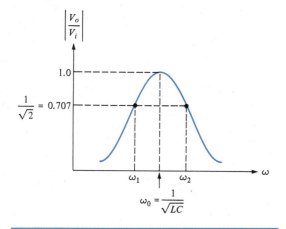

FIGURE 1.15   Voltage Transfer Characteristic for *RLC* Filter Network

The magnitude of the voltage transfer expression from (1.8–4) is shown as a function of ω in Fig. 1.15. Note that if ω has the particular value given by $1/(LC)^{1/2}$, then the magnitude of V_o/V_i has a maximum value of unity. For all other frequencies, the magnitude of V_o/V_i is reduced from unity. This particular frequency ($\omega = \omega_0$) is referred to as the *resonant frequency*. At this frequency, the circuit becomes purely resistive.

Additional concepts related to the band-pass filter and voltage transfer expressions are illustrated in Fig. 1.15. Frequencies ω_1 and ω_2 are defined corresponding to values of ω that reduce the maximum magnitude of $|V_o/V_i|$ by a factor of $2^{1/2}$. For these half-power frequencies, the power (proportional to voltage squared) is reduced by one half the maximum power possible at the resonant frequency. The difference in the half-power frequencies, $\omega_2 - \omega_1$, is a measure of the sharpness of the resonance peak. It is customary to refer to this frequency difference as the *bandwidth*.

1.9 TRANSIENT RESPONSE OF *RC* NETWORKS

We are now familiar with the impedance method of analyzing a circuit that contains resistors, capacitors, and inductors and that is excited by a sinusoidal voltage. The solution obtained using this method is the sinusoidal steady-state solution, which is the solution after all transients have reduced to zero. In some instances, however, the transient solution is the important solution. For example, in an electronic timing circuit in which a voltage is to reach a certain magnitude in a specified time, it is the transient behavior of the circuit that provides the desired result. In general, this solution is obtained by solving a differential equation.

The most important electronic network for transient applications is the series *RC* circuit. Similar behavior is obtainable with an *RL* circuit. However, inductors have the disadvantage of being larger in size and higher in cost than capacitors and are therefore seldomly used.

Figure 1.16a displays the basic *RC* circuit that we will analyze. The switch *S* is closed at *t* = 0, which initiates the transient period. The transient period ends when all voltages reach their final values (the steady-state values). The transient current *i* and voltages v_R and v_C, as indicated in Fig. 1.16a, are produced because of the current–voltage dependence of the capacitor *C*. When the switch is closed, the capacitor voltage begins to change with time, and the current around the loop is determined from $i = C(dv_C/dt)$. Writing KVL around the loop yields

$$V_{DC} - iR - v_C = 0 \qquad (1.9\text{--}1)$$

and substituting for *i*, we have

$$V_{DC} - RC\frac{dv_C}{dt} - v_C = 0 \qquad (1.9\text{--}2)$$

Rearranging yields

$$\frac{dv_C}{dt} + \frac{1}{RC}v_C = \frac{V_{DC}}{RC} \qquad (1.9\text{--}3)$$

which is the standard form of a first-order differential equation.

The solution of (1.9–3) is obtained as

$$v_C = V_{DC} - V_{DC}e^{-t/RC} \qquad (1.9\text{--}4)$$

where the voltage across v_C was taken as zero, initially. Note that the final value (as $t \to \infty$) of voltage across the capacitor from (1.9–4) is V_{DC}.

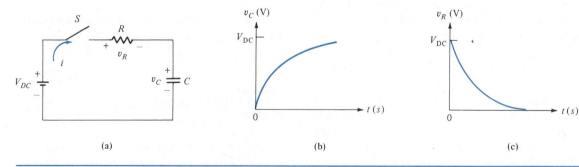

(a) (b) (c)

FIGURE 1.16 Series *RC* Transient Network (Switch Closed at *t* = 0): (a) Basic circuit, (b) Waveshape for v_C, (b) Waveshape for v_R

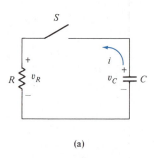

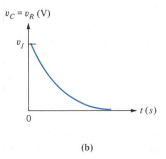

(a)   (b)

FIGURE 1.17   Parallel RC Transient Network (Switch Closed at $t = 0$): (a) Basic circuit, (b) Waveshape for $v_C = v_R$ ($v_C(0) \neq 0$)

Verification that (1.9–4) is the solution to (1.9–3) is obtained by substituting (1.9–4) into (1.9–3) and is left as a homework problem.

The voltage across the resistor in Fig. 1.16a is now easily obtained using KVL and (1.9–4). By subtracting (1.9–4) from V_{DC}, we have

$$v_R = V_{DC}e^{-t/RC} \qquad\qquad \textbf{(1.9–5)}$$

Note that the initial value ($t = 0$) of v_R is V_{DC}. The final value (as $t \to \infty$) of V_R is zero because the capacitor has charged to V_{DC} and current ceases.

The waveshapes for v_C and v_R are displayed in Figs. 1.16b and c. These waveshapes are referred to as *exponentially growing* (v_C) and *exponentially decaying* (v_R) with time. The variation of each of these curves can be expressed generally by one equation. Based upon (1.9–4) and (1.9–5), we write this expression for the voltage corresponding to a growing or decaying exponential waveshape for a passive element as follows:

$$v(t) = v_F + (v_I - v_F)e^{-t/RC} \qquad \textbf{(1.9–6)}$$

where v_I and v_F are the initial and final voltage values, respectively, for the element under consideration. Substituting the initial and final voltage values for R and C (in Fig. 1.16a) into (1.9–6) yields (1.9–4) and (1.9–5) immediately.

Equation (1.9–6) is an extremely important result that allows determination of an exponen-

tially growing or decaying voltage without solving a differential equation. This type of waveshape results when a capacitor charges or discharges through a resistor. If the RC time constant is known, along with the initial and final voltage values (v_I and v_F) for the element under consideration, the waveshape is immediately obtainable from (1.9–6).

Another important decaying waveshape case is depicted in Fig. 1.17a, where a capacitor is initially charged such that an initial voltage (v_1) appears across C, when the switch S is closed at $t = 0$. The final value of voltage across the capacitance (v_F) will be zero after sufficient current passes to completely discharge C. The waveshape for $v_C(t)$ is obtained from (1.9–6) as follows:

$$v_C(t) = v_I e^{-t/RC} \qquad\qquad \textbf{(1.9–7)}$$

Note that the voltage across R varies identically. The resulting time variation is displayed in Fig. 1.17b.

1.10   ELECTRONIC CIRCUIT NOTATION

Conventional electronic notation for currents and voltages as well as for passive elements will be used throughout the following chapters (and was also used in this chapter). This notation is specialized in a manner that not only defines the quantity uniquely but also adds meaning to the symbol. A complete understanding of this notation will provide greater insight into the circuits and analyses to come.

1.10.1   Current and Voltage Notation

It is standard notation to always denote a current by using the letters I or i, while representing voltages with V or v. Additionally, subscripts are placed on these letters in order to distinguish the various currents and voltages in a network from one another. In electronic circuits, these subscripts are usually letters from the alphabet that specifically identify the location of the current or voltage in the circuit. Furthermore, a mixture of uppercase or lowercase (capital or small) letters provides added significance, as we will see.

In defining currents, a single subscript is conventionally used. Thus, for example, we could arbitrarily define the base current of a BJT by using various combinations of letters such as I_B, i_B, or i_b. To give added significance to the notation, however, each of these has a specific meaning. In equation form, we indicate this meaning as follows:

$$i_B = I_B + i_b \qquad (1.10-1)$$

which stated in words means that the total base current (i_B) is equal to the sum of the DC base current (I_B) and the AC base current (i_b). Thus, the total current is defined using a lowercase letter (i) with an uppercase subscript (B). DC current is specified by uppercase letters, and AC current is defined using only lowercase letters.

Similarly, for the collector current of a BJT or the drain current of an FET, the general expressions for total current when both DC and AC are present are given, respectively, by

$$i_C = I_C + i_c \qquad (1.10-2)$$

and

$$i_D = I_D + i_d \qquad (1.10-3)$$

In defining voltages, a double subscript is most often used because the voltage being referred to is always that existing between two points. An important case often arises, however, when the voltage at one terminal is specified relative to ground or the reference terminal. Then, and only then, a single subscript is used, and the notation is entirely analogous to that just described for currents. For example, the total voltage associated with the base terminal of a BJT (relative to a reference terminal) can be expressed in terms of its DC and AC components as follows:

$$v_B = V_B + v_b \qquad (1.10-4)$$

where V_B is the DC base voltage and v_b is the AC base voltage, and both are relative to a reference terminal.

The double subscript case, however, is used much more extensively in electronic circuits. When a double subscript is used, the subscripts refer to the terminals between which the voltage is specified. For example, consider the equation

$$v_{BE} = V_{BE} + v_{be} \qquad (1.10-5)$$

which indicates the voltage between the base (B) and emitter (E) terminals of a BJT. In general, the total base-to-emitter voltage (v_{BE}) is equal to the sum of the DC base-to-emitter voltage (V_{BE}) and the AC base-to-emitter voltage (v_{be}). As in the case of currents, the specific combinations of uppercase and lowercase indicate greater significance.

In addition to this notation, phasor representations of currents and voltages used in analyzing circuits containing inductors and capacitors are given distinguishing notation. Phasor currents and voltages are denoted by boldface uppercase letters $\mathbf{I}$ and $\mathbf{V}$ with lowercase subscripts. For example, the phasor base current of a BJT is defined as $\mathbf{I}_b$ and the phasor base voltage (relative to ground) as $\mathbf{V}_b$.

The reason for using this notation is that we often deal with just the DC component, the AC component, or the phasor component corresponding to the total current and voltage. This notation allows us to specify exactly which component is implied.

1.10.2   Resistor Notation

In order to distinguish resistors in a network from one another, we use the conventional method of the symbol R with a subscript. Thus, the resistor symbols R_1, R_2, and R_3 denote three different resistors. However, in most instances, a letter subscript is used instead of a number. This notation is used to specifically identify the location of the resistor. For example, in a BJT circuit, resistors are connected to the base, collector, and emitter terminals of the transistor. These resistors are denoted as R_B, R_C, and R_E, respectively. Also, in an FET circuit, the resistor connected externally to the drain terminal of the FET is denoted as R_D and that connected to the gate as R_G. Furthermore, the capital letter subscripts imply that the resistor is fixed in value. This notation is the same as that used to describe DC voltages and currents.

Finally, purely AC resistors (those that conduct only AC current) arising in AC device models in electronic circuits, require specific notation. Resistors that are purely AC are denoted by a small letter r and a single small letter subscript. An example of an AC resistor is r_d, where r_d is the AC resistance associated with a PN junction diode.

CHAPTER 1 SUMMARY

- The derived units for electrical quantities are summarized in Table 1.2. Important numerical prefixes are indicated in Table 1.3.
- Power p, energy E, voltage v, and current i for any circuit element are related by

$$p = \frac{dE}{dt} = iv$$

- Passive circuit elements (no built-in source of power) are resistors, capacitors, inductors, and transformers, while active circuit elements are current and voltage sources.
- The resistance R and conductance G of an element are reciprocals given by Ohm's law:

$$v = iR = \frac{i}{G}$$

Power dissipated in a resistor is

$$p = iv = i^2R = \frac{v^2}{R} = v^2G$$

- The current–voltage i–v expression for a capacitor C is

$$i = C\frac{dv}{dt}$$

and the energy E stored in a capacitor is

$$E = \frac{Cv^2}{2}$$

- The current–voltage i–v expression for an inductor L is

$$v = L\frac{di}{dt}$$

and the energy E stored in an inductor is

$$E = \frac{Li^2}{2}$$

- For sinusoidal steady-state excitation ($e^{j\omega t}$), the impedance of a capacitor and inductor are given, respectively, by

$$Z_C = \frac{1}{j\omega C}$$

and

$$Z_L = j\omega L$$

The complex form of Ohm's law is

$$\mathbf{V} = Z\mathbf{I}$$

where $\mathbf{I}$ and $\mathbf{V}$ are phasors that, in general, possess both magnitude and phase and are dependent upon frequency.

- To obtain the time-dependent current $i(t)$ and voltage $v(t)$ from the phasor current $\mathbf{I}(\omega)$ and voltage $\mathbf{V}(\omega)$, the imaginary part is taken as follows:

$$i(t) = Im[\mathbf{I}e^{j(\omega t + \phi)}]$$

and

$$v(t) = Im[\mathbf{V}e^{j(\omega t + \phi)}]$$

where the sinusoidal excitation is assumed to be $\sin(\omega t + \phi)$ and where $\mathbf{I}$ and $\mathbf{V}$ are phasors that possess both magnitude and phase and are dependent upon frequency (ω).

- Time-varying quantities are represented by their instantaneous, average, or rms values. The rms value is the square root of the mean, squared value of the current or voltage. The rms value for a sinusoidal quantity is the peak value divided by $\sqrt{2}$.

- For a general impedance element $Z = |Z| \angle \theta$, the average real power in this element in terms of peak and rms values is

$$P = \frac{V_P I_P}{2} \cos \theta$$

$$= V_{rms} I_{rms} \cos \theta \text{ in watts (W)}$$

The apparent power is

$$P_T = V_{rms} I_{rms} \text{ in volt-amperes (VA)}$$

and the reactive power is

$$Q = V_{rms} I_{rms}$$
$$\times \sin \theta \text{ in volt-amperes reactive (VAR)}$$

- A transformer is a passive device with magnetically coupled windings. For a lossless transformer, the input and output phasor currents and voltages are related by the turns ratio a as follows:

$$\frac{\mathbf{V}_o}{\mathbf{V}_i} = \frac{\mathbf{I}_i}{\mathbf{I}_o} = a$$

- Bulk resistors are available in the form of carbon, wire-wound, and film resistors. Resistor arrays are available in IC packages. The color code for resistors is presented in Table 1.5.
- The primary types of capacitors are ceramic, plastic film, mica, and electrolytic. The color code for capacitors is presented in Table 1.6.
- Air-core and iron-core inductors are available commercially. However, inductors are avoided in modern electronic circuits because of their bulkiness.
- Kirchhoff's voltage law (KVL) states that the algebraic sum of the voltage drops and/or rises across each element around any loop in a network is zero. Kirchhoff's current law (KCL) states that the algebraic sum of currents into (or out of) a node is zero.
- Thévenin's theorem states that a two-terminal resistive linear network can be replaced with a unique equivalent circuit containing a resistance R_T in series with a voltage V_T, where V_T is the open-circuit terminal voltage of the network and R_T is the ratio of V_T to the short-circuit current of the network. Norton's theorem states that a two-terminal resistive linear network can be replaced with a unique equivalent circuit consisting of a conductance G_N in

parallel with a current source I_N, where $G_N = 1/R_T$ and $I_N = V_T/R_T$. The theorems of Thévenin and Norton are also applicable to impedance networks that are linear.

■ The superposition theorem states that, in linear networks, the total effect of several independent sources acting simultaneously is identically equal to the sum of the effects of each individual source acting alone.

■ The voltage divider rule for three resistive elements in series is for one element

$$v_1 = \frac{R_1}{R_1 + R_2 + R_3} v_A$$

The current divider rule for three resistive elements in parallel is for one element

$$i_1 = \frac{G_1}{G_1 + G_2 + G_3} i_A = \frac{R_2 \| R_3}{R_1 + R_2 + R_3} i_A$$

The voltage and current divider rules are also applicable to impedance networks.

■ Filter networks are frequency-selective circuits that reject certain signal frequencies and pass others. Simple RC circuits provide low- and high-pass action, while an RLC network allows selection of essentially one frequency.

■ The transient voltage response in an RC network due to the charging and discharging of a capacitor is given by

$$v(t) = v_F + (v_I - v_F)e^{-t/RC}$$

where v_F is the final voltage and v_I is the initial voltage.

■ The notation used for total, DC, and AC currents and voltages is

$$i_B = I_B + i_b \qquad \text{and} \qquad v_B = V_B + v_b$$

where i_B and v_B are total quantities, I_B and V_B are DC, and i_b and v_b are AC. Phasor quantities are denoted as $\mathbf{I}_b$ and $\mathbf{V}_b$.

CHAPTER 1 PROBLEMS

1.1

Current i, voltage v, and inductance L are related by the expression $v = L \, di/dt$. For an inductance $L = 10^{-3}$ H and a time rate of change in current given by 5×10^{-3} mA/μs, determine the voltage.

1.2

The current–voltage relationship in a system is given by the differential equation

$$\frac{di}{dt} = K_1 \frac{d^2v}{dt^2} + K_2 \frac{dv}{dt} + K_3 v$$

where i is the current in amperes (A), v is the voltage in volts (V), and t is time in seconds (s). Find the electrical units of the constants K_1, K_2, and K_3.

1.3

A cylindrical aluminum wire with a radius of 5 mm has a 2.2 mA current. The concentration of free electrons in the wire is 5×10^{23} electrons/cm^3. The

current is due to the motion of electrons at an average velocity v_D. Assuming a uniform distribution of free electrons throughout the wire, calculate v_D. Use the relations $l/t = v_D$, $Q = Nq$, and $N = $ (concentration)(Al), where $q = 1.6 \times 10^{-19}$ C, A is the area, l is length, and t is time.

1.4

The current–voltage relationship for a diode is given by the theoretical expression

$$I = I_s(e^{V/v_T} - 1)$$

where $I_S = 10^{-13}$ A and $v_T = 0.025$ V at room temperature. For $V = 0.7$ V, calculate the current and power. Repeat the calculation for $V = -5$ V. Is the power being absorbed or delivered by the diode?

1.5

The circuit shown in Fig. P1.5 represents two electrical networks connected together. For each differ-

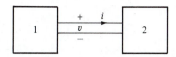

FIGURE P1.5

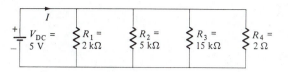

FIGURE P1.8

ent case indicated in parts (a) through (d), calculate the power and specify the direction of power flow (from 1 to 2 or vice versa).
   (a) $v = -5$ V, $i = -3$ A
   (b) $v = -2$ V, $i = 2$ A
   (c) $v = 2.5$ V, $i = -1$ A
   (d) $v = 4$ V, $i = 0.5$ A

1.6
(a) Determine the equivalent capacitance of three capacitors in parallel, as indicated in Fig. P1.6a.
(b) Determine the equivalent capacitance of three capacitors in series, as indicated in Fig. P1.6b.

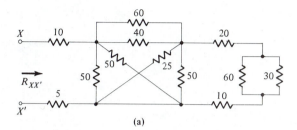

(a)

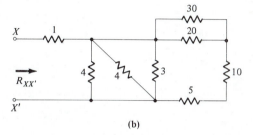

(b)

FIGURE P1.9

1.10
In the following star-to-delta resistor conversion, determine the values of R_{AB}, R_{AC}, and R_{BC} in terms of R_A, R_B, and R_C such that the two circuits are equivalent at the terminals A–B–C shown in Figs. P1.10a and b.

1.11
For the circuits in Fig. P1.10, determine R_A, R_B, and R_C in terms of R_{AB}, R_{AC}, and R_{BC} such that the two circuits are identical at the terminals A–B–C.

1.12
The voltage across the terminals of a capacitor is given by

$$v(t) = 200e^{-5 \times 10^4 t} \sin \omega t \quad \text{(for } t \geq 0)$$

and for $t \leq 0$, by $v(t) = 0$, where t is time in seconds (s). Determine the corresponding capacitor current at $t = 0$ [$i(0)$], the power delivered to the capacitor at $t = 5\pi$ μs, and the energy stored in the capacitor at 5π μs. Let $C = 1$ μF and $\omega = 10^5$ rad.

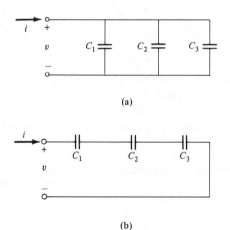

(a)

(b)

FIGURE P1.6

1.7
Repeat Problem 1.6 for the case of inductors.

1.8
Calculate the value of current I shown in Fig. P1.8. Could this result have been predicted without adding the given resistors in parallel? Explain.

1.9
Determine the equivalent resistance $R_{XX'}$ in the circuits shown in Figs. P1.9a and b. (All values are in ohms.)

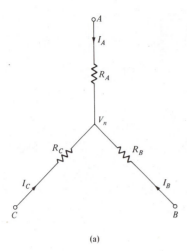

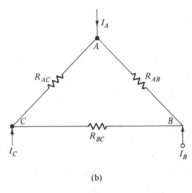

FIGURE P1.10   (a) Star, (b) Delta

1.13

The current in a 100 mH inductor is given by the time function

$$i(t) = 10 \text{ A (for } t \leq 0)$$
$$= K_1 e^{-3t} + K_2 t e^{-2t} \text{ A (for } t \geq 0)$$

If the initial voltage is -5 V, determine

(a) the initial energy stored in the inductor
(b) the coefficients K_1 and K_2
(c) the power delivered to the inductor at $t = 10$ ms

1.14

Determine the energy stored in the inductive circuit of Fig. P1.14 (all values are in henries) at $t = T/2$ s, where T is the period of the applied current given

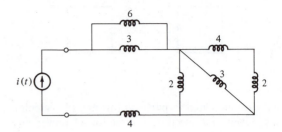

FIGURE P1.14

by

(a) $i(t) = 5 \sin 120\pi t$ A
(b) $i(t) = 5 \cos 120\pi t$ A
(c) $i(t) = 25 \sin 120\pi t \cos 120\pi t$
$\qquad = (25/2)(\sin 240\pi t)$ A

1.15

Determine the magnitude $|Z|$ and phase ϕ of the complex impedance expression given by

$$Z = \frac{-50(j10)(1 + j5)}{1 + j6}$$

1.16

Suppose $v(t) = \sin 1000t$. What is the frequency of $v(t)$ in radians and Hz.

1.17

(a) Determine the equivalent resistance and reactance seen looking into the terminals $A - B$ of the circuit in Fig. P1.17.
(b) Determine the equivalent conductance and susceptance seen looking into the terminals $A - B$ of the same circuit.

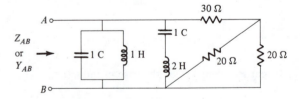

FIGURE P1.17

1.18

Determine the equivalent impedance Z_{AB} of the circuit in Fig. P1.18. If the phase angle ϕ of the impedance Z_{AB} is 45°, determine the frequency ω in terms of R, L, and C.

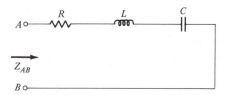

FIGURE P1.18

1.19

Determine the impedance Z_{AB} in the circuit in Fig. P1.19. A circuit is said to be in *resonance* if the impedance is real. Determine the frequency ω_0 in terms of R, L, and C at which the circuit in Fig. P1.19 is at resonance (ω_0 is the resonant frequency).

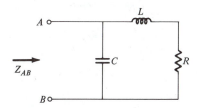

FIGURE P1.19

1.20

The circuit in Fig. P1.20 is a Wien bridge circuit and is used to measure the frequency of the signal voltage v. Resistors $R_1 = R_2$ are adjusted until the

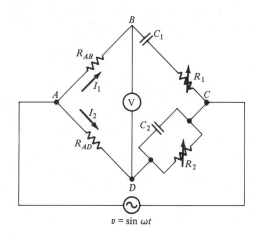

FIGURE P1.20

voltages V_b and V_d are equal (the voltmeter has a zero reading). Show that the frequency is inversely proportional to R_1.

1.21

For the circuit in Fig. P1.21, calculate the average power delivered to the load resistor $R_L = 50\ \Omega$. Assume the transformer to be ideal.

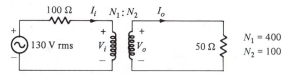

FIGURE P1.21

1.22

A general resistance expression is given by $R = \rho l / A$, where ρ is the resistivity of the material, l is the length, and A is the cross-sectional area. Determine the resistance of a copper wire with $\rho = 1.72 \times 10^{-8}$ Ω-m, $l = 10$ m, and $A = 10^{-12}$ m^2.

1.23

The resistance of a circular wire is given by $R = \rho l / \pi r^2$, where ρ is the resistivity of the material, l is the length, and r is the radius of the wire. If the relative error in l is 2% and the relative error in r is 10% (both of these are measured), determine the relative error in R in the worst case possible.

1.24

Answer the following questions:

- In the resistor color code, what does gray stand for? **(a)** 2, **(b)** 7, **(c)** 9, **(d)** 8
- What can an orange stripe on a resistor mean? **(a)** two zeros, **(b)** four zeros, **(c)** one zero, **(d)** three zeros
- Which of the following values cannot be color coded? **(a)** 220 Ω, **(b)** 2.2 kΩ, **(c)** 4.7 kΩ, **(d)** 333 kΩ
- If the third stripe is brown, what is the range of the resistance value? **(a)** 10–99 Ω, **(b)** 100–990 Ω, **(c)** 1–9.9 kΩ, **(d)** 10–99 kΩ

1.25

Consider the circuit shown in Fig. P1.25.
    (a) Write a KVL equation for the left loop.
    (b) Write a KVL equation for the right loop.
    (c) Write a KCL equation at node Z.

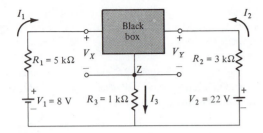

FIGURE P1.25

(d) Now suppose that $V_X = 0.7$ V and $I_3 = 5$ mA, use the equations obtained in parts (a), (b), and (c) to determine I_1, I_2, and V_Y.

1.26
Consider a new passive circuit component—a PLIR for *piecewise linear resistor*. The circuit symbol and i–v characteristics are shown in Figs. P1.26a and b. Note that if $R_1 = R_2$, the PLIR is a normal resistor. For the circuits in Figs. P1.26c and d, let $R_1 = 5$ kΩ and $R_2 = 8$ kΩ. For Fig. P1.26c, determine v if $i = 5$ mA. For Fig. P1.26d, determine v if $i = 8$ mA.

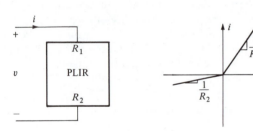

(a)   (b)

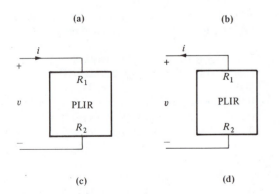

(c)   (d)

FIGURE P1.26

1.27
Use KVL and KCL to determine the currents for the circuits in Figs. P1.27a and b.

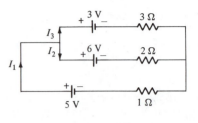

(a)

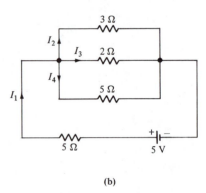

(b)

FIGURE P1.27

1.28
For the circuit in Fig. P1.28, determine the value of R that enables the maximum power to be delivered to the resistor R.

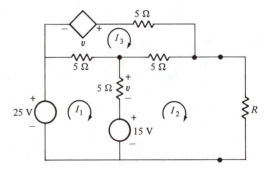

FIGURE P1.28

1.29

Use KVL and KCL to determine the voltage v for the circuit in Fig. P1.29.

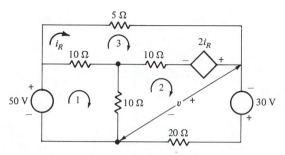

FIGURE P1.29

1.30

Repeat Problem 1.29 for the circuit in Fig. P1.30.

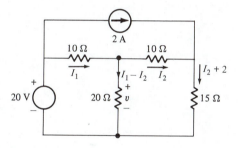

FIGURE P1.30

1.31

Looking into nodes X–X', determine the Thévenin and Norton equivalent circuits for Figs. P1.31a and b.

1.32

Repeat Problem 1.31 for the circuits in Figs. P1.32a and b.

1.33

Use the current divider rule to determine the current in the 10 Ω resistor of Fig. P1.33.

1.34

For the circuits in Fig. P1.34, determine the currents i_2 and i_3.

1.35

For the circuits in Fig. P1.35, find i_3 and v_3 (as indicated).

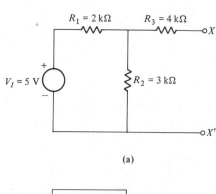

(a)

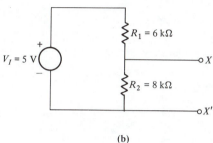

(b)

FIGURE P1.31

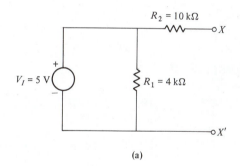

(a)

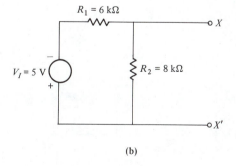

(b)

FIGURE P1.32

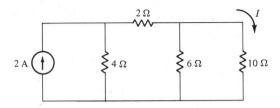

FIGURE P1.33

1.36
Determine the voltages V_1 and V_2 for the circuit in Fig. P1.36.

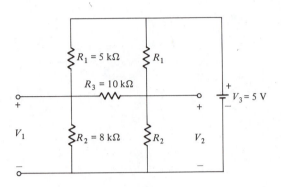

FIGURE P1.36

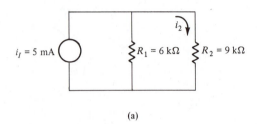

(a)

1.37
For the circuit in Fig. P1.37, use the superposition theorem to determine V_3.

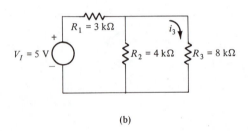

(b)

FIGURE P1.34

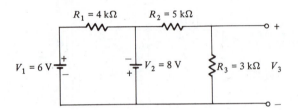

FIGURE P1.37

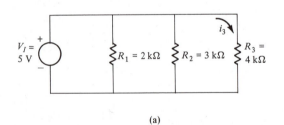

(a)

1.38
For the circuit in Fig. P1.38, determine V_3 and i_2.

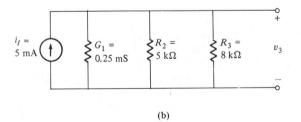

(b)

FIGURE P1.35

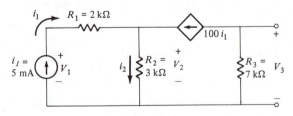

FIGURE P1.38

1.39
Continue Problem 1.38 to obtain V_2 and V_1.

1.40
Determine the ratio of phasor voltages $\mathbf{V}_2/\mathbf{V}_s$ for the circuit in Fig. P1.40.

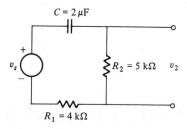

FIGURE P1.40

1.41
Verify that Eq. (1.9–4) is the solution to Eq. (1.9–3).

1.42
For the circuit in Fig. P1.42, determine the current $i_2(t)$. Let $\omega = 500$ rad/s.

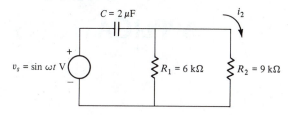

FIGURE P1.42

1.43
For the circuit in Fig. P1.43, determine the voltage $v_2(t)$. Let $\omega = 500$ rad/s.

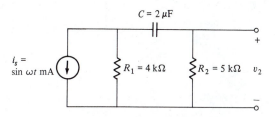

FIGURE P1.43

1.44
Parallel resonant *RLC* circuits are used to "tune in" different radio or television stations in a receiver. These circuits can select (filter out) essentially one frequency from the many that are received by the antenna. For the parallel resonant circuit in Fig. P1.44, derive an analytical expression for V_m as a function of ω. Determine the resonant frequency ω_0 and sketch V_m versus frequency. Note that V_m and I_m are real.

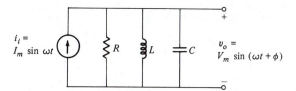

FIGURE P1.44

2

SEMICONDUCTOR DIODES AND CIRCUIT APPLICATIONS

In this chapter, a very important two-terminal semiconductor device, the PN junction diode, is introduced. This element came into being long before transistors were invented and was an exact replacement for the vacuum diode, a seldomly used element in the modern electronics world.

This chapter begins with a brief description of semiconductor materials that are commonly used in the manufacturing of semiconductor devices. Various semiconductor diodes and the circuit models used for representing their electronic behavior in the low- to medium-frequency range are then described. High-frequency effects are accounted for in a later chapter. Various electronic diode applications are then discussed, and selection guides for actual diodes are presented in an appendix to the chapter. Diode data sheets for several diodes are given in Appendix 1 at the back of the book.

2.1   SEMICONDUCTOR MATERIALS AND DIODES

2.1.1   Semiconductors

A *semiconductor material,* or simply *semiconductor,* is a solid whose conductivity lies in between that of a conductor (metal) and that of a nonconductor (insulator). Hence, the term itself denotes its intended meaning well. Semiconductors have electronic properties in the intermediate range between conductors and insulators.

We are all familiar with conductors and insulators. Examples of conductors are metals such as copper, gold, aluminum, and silver. Examples of insulators are glass, paper, and wood. Usually, however, we are not familiar with semiconductor materials because they are not present in our everyday lives.

Table 2.1 indicates typical conductivity values for conductors, insulators, and semiconductors. Values are also given for the resistivity, which is the inverse of the conductivity. Note the distinct differences in the magnitudes of conductivity (and resistivity) for the conductors, insulators, and semiconductors.

Silicon and Its Integrated Circuit Advantage. The most important semiconductor in modern-day electronics is the elemental (single-element) material *silicon* (Si). Silicon is used almost exclusively in the fabrication of integrated circuits because of its property that an oxide (silicon dioxide, or SiO_2) can be grown on its surface extremely thin and uniform in thickness. The SiO_2 provides a necessary surface-masking (shielding) layer. This layer is removed in certain regions by precise photographic methods, thus allowing a succeeding surface treatment in only the very accurately defined and desired regions. Devices are then fabricated in unison by carrying out a repetition of a series of fabrication steps involving oxide growth, selective removal, surface treatment, and total oxide removal. Each series of surface treatments must also be precisely aligned with one another,

TABLE 2.1   Conductivity and Resistivity for Various Materials (20° C)

Material	Conductivity (1/Ω-cm)	Resistivity (Ω-cm)
Conductors		
Copper	0.59×10^6	1.7×10^{-6}
Gold	0.43×10^6	2.3×10^{-6}
Aluminum	0.38×10^6	2.6×10^{-6}
Silver	0.62×10^6	1.6×10^{-6}
Tungsten	0.18×10^6	5.6×10^{-6}
Insulators		
Glasses	$10^{-16}\text{--}10^{-13}$	$10^{13}\text{--}10^{16}$
Plastics	$10^{-22}\text{--}10^{-20}$	$10^{20}\text{--}10^{22}$
Wood	$10^{-18}\text{--}10^{-16}$	$10^{16}\text{--}10^{18}$
Silicon dioxide (DC, 25° C)	$10^{-16}\text{--}10^{-14}$	$10^{14}\text{--}10^{16}$
Semiconductors		
Gallium arsenide	$10^{-7}\text{--}10^{-2}$	$10^2\text{--}10^7$
Germanium	$10^{-7}\text{--}10^{-2}$	$10^2\text{--}10^7$
Silicon	$10^{-8}\text{--}10^{-1}$	$10^1\text{--}10^8$
Intrinsic Silicon	4×10^{-12}	2.5×10^{11}

which is the reason for using photographic techniques. Since the devices are fabricated in an integrated fashion, the terminology *integrated circuit* (IC) was originated.

Gallium Arsenide and Its Optoelectronic Advantage. The binary (two-element) semiconductor *gallium arsenide* (GaAs) is also of great importance to the electronics industry. This material has particularly useful properties in the optoelectronics (light and electricity interaction) area because of its unique light-absorbing and light-emitting properties in the visible range. Usually, an additional element such as phosphorus (P) or aluminum (Al) is added to the GaAs crystal in trace amounts to improve its properties. Devices fabricated from GaAs are quite often used to convert light signals into electrical signals and vice versa. Such devices are described in Chapter 12.

High-frequency integrated circuits are also fabricated using GaAs as the semiconductor because GaAs has an advantage over Si. The crystalline structure of GaAs provides improved mobility of charges and, therefore, larger currents for the same voltages. However, GaAs does not possess a native oxide (one that can be precisely grown on its surface). Hence, this material has a built-in disadvantage insofar as the fabrication of integrated circuits is concerned.

For purely electronic applications, the semiconductor that we are most concerned with is silicon. However, we may become more concerned with gallium arsenide in the future. In the meantime, silicon is the primary semiconductor used for discrete devices as well as for integrated circuit devices, and we will concentrate on electronic devices made of silicon in this text.

Classes of Semiconductor Materials.   There are two classes of semiconductor materials:

- Intrinsic semiconductors
- Extrinsic semiconductors

The *intrinsic semiconductor* is a crystalline solid that is composed entirely of semiconductor atoms in a uniform array. The extrinsic case is essentially the same, except that impurity atoms other than semiconductor atoms are intentionally added to the crystal to alter its semiconductor properties in a desired manner. This process is called *doping* the semiconductor.

To understand the implications of the two classes of semiconductors, we will consider a portion of the periodic table of the elements as shown in Table 2.2. The Roman numerals at the top of

each column indicate the number of valence (outermost) electrons associated with each element. It is these electrons that give the solid its properties in a crystalline array.

Column IV of Table 2.2 contains the semiconductor atoms silicon (Si) and germanium (Ge). When an ordered array of only Si or Ge atoms is formed, the material is said to be *single-crystal* and the resulting conductivity is in the intrinsic semiconductor range. If the atoms are not arranged in an ordered array (are *amorphous*) or are arranged in a semiordered array (are *polycrystalline*), the conductivity is reduced to that of an insulator.

Other atoms shown in column IV of Table 2.2 are carbon (C) and tin (Sn). When an ordered array of only C or Sn atoms is formed, the resultant material has conductivity in the insulator range and the conductor range, respectively. The reason for this occurrence is due to the strength of the bonding of valence electrons to the nucleus of their parent atoms in the crystal. The attractive force due to the negative charge of the electron and the positive charge of the nucleus is quite large for a C crystal but quite small for a Sn crystal. In the case of an intrinsic semiconductor composed of only Si or Ge atoms, this force is only moderate because as we go downward from element to element in column IV, each atom has more electrons revolving about its nucleus. Moreover, for an atom with more electrons, the valence electrons are less tightly bound to the nucleus. Thus, in a crystal, some of the valence electrons are able to break their bond and move randomly in the material due to the acquisition of thermal energy supplied by room temperature becoming *mobile electrons,* or free charges.

As we proceed downward in column IV of Table 2.2 (C, Si, Ge, and then Sn), crystals for each succeeding case will have more of these mobile electrons produced at room temperature because of the weaker bonding. Furthermore, each valence electron that is able to break its bond and move randomly in the crystal leaves behind a positive charge (the absence of an electron) called

TABLE 2.2   Portion of the Periodic Table of the Elements

III	IV	V
B	C	N
Al	Si	P
Ga	Ge	As
In	Sn	Sb

a *hole*. It is precisely these mobile electrons and holes that are available for conduction of current. All of the other charges are held fixed in position and are therefore immobile.

The number per unit volume of these free charges is called the *intrinsic concentration n_i*. Values for n_i have been measured and quantified quite accurately. Table 2.3 indicates the value of n_i for intrinsic semiconductor samples of Si, Ge, and GaAs. The value of n_i for an insulator would be many orders of magnitude less; that for a conductor, many orders of magnitude larger.

The *extrinsic semiconductor* is a single-crystalline semiconductor with impurity atoms from column III or V of Table 2.2 intentionally added to the crystal. The addition of minute amounts (in comparison with the total number of semiconductor atoms) of these impurity atoms results in drastic changes in the number of mobile charges and, hence, the conductivity. When these impurity atoms are added, they replace the semiconductor atoms in the single-crystal array. The number of such impurities added must always be much less than the number of semiconductor atoms in the crystal ($\simeq 10^{22}/cm^3$) so that the resultant doped crystal will continue to be a semiconductor. Although the following discussion focuses on silicon, we should realize that other semiconductors are doped similarly.

N-Type Semiconductor. Typically, the impurity atoms from column V of Table 2.2 that are used to dope a silicon crystal are either phosphorus (P), arsenic (As), or antimony (Sb). When any of these atoms are added to the crystal, each has one additional valence electron. These additional valence electrons do not fit into the regular array of the crystalline structure and are therefore bonded very weakly to their parent impurity-atom nucleus. Hence, each of these valence electrons at room temperature can acquire the necessary small thermal energy to break their bond and become mobile. Each such action produces one mobile electron and one fixed, positively ionized, impurity atom. Only the free electron is available for

TABLE 2.3 Intrinsic Concentration at Room Temperature (Number of Charges/cm^3)

Semiconductor	n_i (300 K)
Si	1.5×10^{10}
Ge	2.4×10^{13}
GaAs	9.2×10^{6}

conduction, however, because the ionized impurity atom is fixed in position.

The atoms in column V of Table 2.2 are called *donor atoms* because each one, when replacing an Si atom in the semiconductor crystal, "donates" one mobile electron. The *concentration of donor atoms N_d* has the same units as those for n_i (number per unit volume). In order to be effective, N_d is always much greater than n_i but much less than the *total number of semiconductor atoms per unit volume N_s*. Hence, the addition of atoms from column V of Table 2.2 produces a doped semiconductor that possesses many additional free electrons that are available for conduction. This type of extrinsic semiconductor is referred to as an *N-type semiconductor* because the negative charges are primarily responsible for conduction. In an N-type semiconductor, although some free positive charges (holes) still exist, these charges are few in numbers in comparison with the many electrons. The holes in an N-type semiconductor are the *minority carriers,* whereas the mobile electrons are the *majority carriers.*

P-Type Semiconductor. A *P-type semiconductor* is one in which positive charges (holes) are responsible for its conduction properties. An intrinsic sample of Si is doped P-type through the addition of atoms from column III of Table 2.2. These atoms are called *acceptor atoms*. Typically, boron (B) is used to dope silicon.

In the single-crystal array, each acceptor atom is able to "accept" a valence electron and produce a hole somewhere else. The hole is mobile and acts like a positive charge that can conduct current. The corresponding ionized impurity atom is

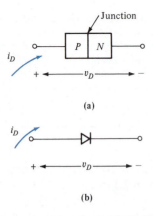

(a)

(b)

FIGURE 2.1   PN Junction Semiconductor Diode: (a) Schematic representation, (b) Circuit symbol

immobile. The *concentration of acceptor atoms* N_a has the same range as N_d, namely, $n_i \ll N_a \ll N_s$. For a P-type semiconductor, the electrons are the *minority carriers,* whereas the holes are the *majority carriers.*

2.1.2   The PN Junction Semiconductor Diode

A visual representation of the *PN junction semiconductor diode* is shown in Fig 2.1a. The P and N regions are P-type and N-type extrinsic

semiconductors that are separated by the junction between them. When terminals are attached to each region, this device becomes a two-terminal circuit element with nonlinear current–voltage behavior. The defined direction of positive diode current (i_D) and polarity of positive diode voltage (v_D) are shown in Fig. 2.1a.

Figure 2.1b displays the circuit symbol used for the PN junction semiconductor diode. The direction of positive current (i_D) and positive polarity of voltage (v_D) are the same as those in Fig. 2.1a. If the opposite current direction and voltage polarity are actually present, then the resulting current and voltage are negative and referred to as *reverse.*

The arrowhead in the circuit symbol of Fig. 2.1b always points from the P region to the N region. This direction is called the *forward* direction and is the primary direction for current flow. That is, a large current can flow in the direction of the arrowhead, whereas only a small current can normally flow in the opposite direction. When a negative voltage of critical magnitude for the particular diode is reached or exceeded, a phenomenon referred to as *breakdown* occurs. This phenomenon will be described shortly.

Typical measured and theoretical characteristic curves for i_D versus v_D for the Si PN junction diode are shown in Fig. 2.2. These curves represent a highly nonlinear relationship between i_D

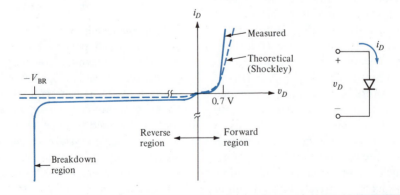

FIGURE 2.2   Measured and Theoretical i_D Versus v_D Characteristics for the Si PN Junction Semiconductor Diode

and v_D. From both of these curves, we observe that, for positive v_D, the current is positive and can be quite large. Furthermore, for negative v_D, the current is negative and quite small until v_D becomes equal to the critical voltage $(-V_{BR})$. At this value and for v_D even more negative, the current i_D is still negative but can be large.

PN Junction Diode Characteristics (Without Breakdown).   The diode i_D–v_D characteristic can be qualitatively explained by reasoning that i_D is based upon the charge flow that crosses the junction. When $v_D > 0$, the positive charges on the P side and the negative charges on the N side are attracted (and repelled) across the junction to the other side. Conventionally, these charge flows add to produce a positive diode current (in the defined positive direction of i_D). Furthermore, since there are many positive charges available on the P side and many negative charges available on the N side (the majority carriers), a large current in the positive i_D direction is produced when a positive voltage v_D is applied. An increase in v_D also results in an increased i_D. The first quadrant of Fig. 2.2, the forward region of the diode, exhibits this behavior. Positive i_D and v_D are thus also referred to as *forward current* and *forward voltage,* respectively. Additionally, this region is often referred to as a *forward bias* ("bias" implying fixed voltage) *region.*

When a negative voltage is applied, there are relatively few charges (the minority charges) available for conduction across the junction. With a negative v_D applied, positive charge flows from N to P and negative charge flows from P to N (minority charges). These charge flows combine to produce current in the negative i_D direction, and, since only a relatively few minority charges are available, the current magnitude is quite small. This reverse current is also nearly independent of the magnitude of v_D because all of the available minority charges can cross the junction even for a small reverse voltage. Thus, the reverse current reaches a saturated (constant) value. The third quadrant of Fig. 2.2, the reverse region for the diode, exhibits this behavior. Thus, in this region,

we refer to the *reverse current, reverse voltage,* and *reverse bias.*

The Breakdown Phenomenon.   As the reverse voltage is increased in magnitude, a critical voltage, called the *breakdown voltage* $-V_{BR}$, is eventually reached. This voltage is dependent upon the physical makeup of the PN diode and varies from one diode to another. When $v_D = -V_{BR}$, charge is created in the vicinity of the junction that proceeds to move in the same direction as the minority charge movement, thus resulting in the large reverse current region shown in Fig. 2.2. Either of two mechanisms is responsible for this increase in charge (and current). One mechanism is that as the magnitude of the reverse voltage is increased, the minority carriers in the vicinity of the junction attain such large velocities that impact-ionizing collisions with the valence electrons of the silicon atoms occur. These charges, in turn, can ionize additional semiconductor atoms, resulting in an "avalanching" process. The other mechanism is that as the reverse voltage is increased, the electric field associated with this voltage can become large enough (by itself) to separate electrons from their parent silicon atom. Both of these mechanisms produce a pair of charges (positive and negative) that separate from each other and move along in the direction of the already moving minority charges. Thus, the reverse current is increased, and its magnitude is controlled primarily by the magnitude of the Thévenin resistance (or impedance) and Thévenin voltage of the circuit external to the diode terminals.

Shockley's Current–Voltage Equation.   The theoretical expression for the dependence of diode current i_D upon diode voltage v_D for the PN junction diode is given exponentially by

$$i_D = I_S(e^{v_D/v_T} - 1) \qquad \text{(2.1–1)}$$

where

I_S = reverse saturation current for the diode in amperes (A)

v_T = thermal voltage in volts (V), or kT/q

k = Boltzmann constant (1.34×10^{-23} J/°K)

T = temperature in degrees Kelvin (°K)

q = electronic charge (1.6×10^{-19} C)

Equation (2.1–1) is often referred to as *Shockley's expression* after William Shockley, one of the founders of the transistor. The quantity v_T is called the *thermal voltage* because it depends upon temperature and has the units of voltage. Its magnitude at room temperature is approximately 0.025 V. The reverse saturation current I_S is given this name because for negative v_D (a reverse bias), the exponent is negligible in (2.1–1). Therefore,

$$i_D \simeq -I_S \qquad (2.1\text{–}2)$$

Note that even for $v_D = -0.05$ V, i_D from (2.1–1) is

$$i_D = I_S\,(e^{-0.05/0.025} - 1) = -I_S\,(0.865)$$

which is less than 14% different than $-I_S$. For larger magnitudes of reverse voltage, where $v_D < -0.05$ V, the approximation of (2.1–2) is even better. It is nearly exact for $v_D < -0.1$ V.

In the forward-bias region for $v_D > 0$ and $i_D > 0$, the exponential dominates in (2.1–1). For $v_D \geq 0.05$ V, we have

$$i_D \simeq I_S e^{v_D/v_T} \qquad (2.1\text{–}3)$$

For $v_D > 0.1$ V, this expression is also nearly exact.

However, Shockley's theoretical expression does not predict the excess reverse current measured in the breakdown region. Furthermore, the theoretical equation for I_S is dependent upon material and geometry quantities for the particular PN diode. In calculating I_S from this equation, I_S turns out to be quite small—much smaller than the measured reverse current. Nonetheless, as we will see, (2.1–1) is extremely important in the development of a circuit model for the PN diode.

A very useful realization for the PN junction diode is the fact that the forward voltage across the diode terminals remains approximately fixed. This fixed voltage will be denoted by V_0. For a Si PN diode, V_0 is approximately 0.7 V. Defining V_0 results in an extremely advantageous method for analyzing diode circuits (and BJT circuits as well). The following example shows that very little change occurs in the diode forward voltage, even when the current changes by a large amount.

—————————————————————————— EX. 2.1

Forward Voltage Change for a PN Diode Example

Using Shockley's expression for the current–voltage variation of a PN junction semiconductor diode, determine the change in diode forward voltage that results when the diode current changes by a factor of 100. Assume that $v_D > 0.05$ V.

Solution: Using Shockley's relation and expressing the two currents as a ratio, we have

$$\frac{i_{D_1}}{i_{D_2}} = \frac{I_S(e^{40v_{D_1}} - 1)}{I_S(e^{40v_{D_2}} - 1)} \simeq e^{40(v_{D_1} - v_{D_2})}$$

Setting this ratio equal to 100 and solving for the change in diode voltage yield

$$v_{D_1} - v_{D_2} = \frac{1}{40}\ln 100 = 0.115 \text{ V}$$

Amazingly, we observe that a diode voltage change of only 115 mV results in two orders of magnitude change in the diode current.

2.1.3   The Zener Diode

The *zener diode* is a PN junction semiconductor diode with an additional distinction. The diode is placed in an active circuit that forces the diode voltage v_D and current i_D to correspond to a point in the breakdown region. Operation in this region forces $v_D = -V_{BR}$, and the current will be a reverse current whose magnitude depends upon the Thévenin equivalent circuit external to the diode.

The special symbol used for the zener diode to denote its distinction is shown in Fig. 2.3. The symbol represents the extension of an ordinary PN diode. Additionally, the value of the breakdown voltage is ordinarily specified next to the symbol. As shown in Fig. 2.3, the voltage is V_Z, with the polarity indicated. Of course, V_Z is equal in magnitude to V_{BR} (the breakdown voltage). Furthermore, it should be understood that the direction of the current for the zener diode is always in the opposite direction from the arrowhead.

Zener diodes are available with values of V_Z ranging from one or two volts up to hundreds of volts. This range is produced by varying (in a controlled manner) the material makeup of the P and N regions during the fabrication of the device. Heavier doping on both sides of the junction reduces the magnitude of V_{BR}.

The primary application of a zener diode is as a voltage-control device. If it is desired to limit the voltage between two points in a circuit to a certain value, a zener diode with this value of V_Z can be used. Normally, the zener diode is not operated in the forward-bias region and hence does not conduct current in the forward direction. Thus, we consider the current for a zener diode to be positive in the reverse direction.

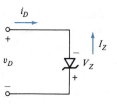

FIGURE 2.3   Zener Diode Symbol

2.1.4   The Schottky Diode

The *Schottky diode* is a modified PN junction diode. It is a two-terminal device that utilizes a specialized conducting layer to replace one of the semiconductor regions (usually the P region). It is not necessary that the conducting layer be a metal, but quite often it is. The Schottky diode is thus called a *metal-semiconductor diode,* or *MN diode.*

Figure 2.4 shows the measured and theoretical current–voltage characteristics for a Schottky Si Mn diode. The circuit symbol used is slightly altered from that of a PN diode. Shockley's theoretical expression [(2.1–1)] is still valid for this case; however, the magnitude of I_S for the Schottky diode is much larger (both theoretically and experimentally), usually by several orders of

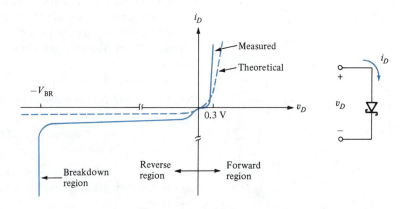

FIGURE 2.4   Measured and Theoretical i_D Versus V_D Characteristics for the Schottky Si MN Diode

magnitude. Note that the Schottky diode also exhibits a breakdown region and that the previous mechanisms described for the PN diode are also applicable here.

There are several reasons for the usefulness of this device. The first is that the device fabrication is simplified. More importantly, however, is that the fixed forward voltage V_0 across the diode is reduced. Note that the forward voltage across the Schottky Si MN diode is approximately 0.3 V as compared to 0.7 V for the Si PN diode. This difference is advantageous in certain circuit applications. An additional advantage is that the Schottky MN diode is much faster than the PN diode in switching on or off, which is the primary reason for the importance of the Schottky diode.

2.2　THE IDEAL DIODE AND ADDITIONAL CIRCUIT MODELS FOR SEMICONDUCTOR DIODES

2.2.1　The Ideal Diode

The *ideal diode* is a defined two-terminal circuit element that approximates an actual semiconductor diode. Figures 2.5a and b display the

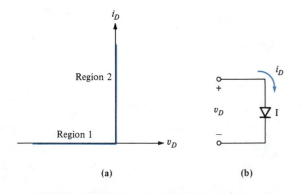

(a)　　　　　　　　**(b)**

FIGURE 2.5　Ideal Diode: (a) Current–voltage characteristic, (b) Circuit symbol

current–voltage characteristic and circuit symbol for the ideal diode. Note that in order to distinguish a semiconductor diode from an ideal diode, an "I" is included by the diode symbol to represent the ideal case. The positive diode current (i_D) direction and positive voltage (v_D) polarity are defined in the same manner as was done for the semiconductor diode cases.

The i_D–v_D characteristic is nonlinear overall; however (very importantly), it is linear in separate regions, or *piecewise linear*. The two linear regions are defined as follows:

Linear region 1:　$i_D = 0$ for $v_D \leq 0$

and

Linear region 2:　$v_D = 0$ for $i_D \geq 0$

Since region 1 corresponds to an open circuit and region 2 corresponds to a short circuit, the analysis of a circuit containing a diode can be carried out quite simply by using either a short circuit or an open circuit to replace the diode. To determine which region (or state) the diode is operating in, we consider the polarity of the voltage or current source(s). If the net effect of the source(s) is to force a current in the direction of the arrowhead (positive i_D), the ideal diode will act like a short circuit. If the net effect of the source(s) is to force a current in the opposite direction from the arrowhead (negative i_D direction, a direction in which current cannot flow), the ideal diode will act like an open circuit.

EX. 2.2

Ideal Diode and Thévenin Equivalent Circuit Example

Determine the current and voltage for the ideal diode of Fig. E2.2a by using a Thévenin equivalent network for the circuit external to the diode.

Solution:　The Thévenin equivalent resistor and voltage are obtained by using the methods of Chapter 1. These values are

$$R_T = R_1 \| R_2 = \frac{(1\ k\Omega)(4\ k\Omega)}{1\ k\Omega + 4\ k\Omega} = 0.8\ k\Omega$$

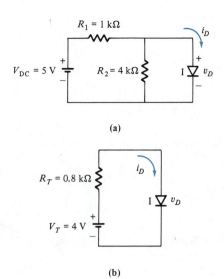

(a)

(b)

FIGURE E2.2

and

$$V_T = \frac{R_2}{R_1 + R_2} V_{DC} = \frac{4\ k\Omega}{1\ k\Omega + 4\ k\Omega} 5\ V = 4\ V$$

The equivalent circuit is shown in Fig. E2.2b. With this polarity of voltage, the diode acts like a short. Thus,

$$v_D = 0$$

and

$$i_D = \frac{V_T}{R_T} = 5\ mA$$

2.2.2 Additional Circuit Models for Semiconductor Diodes

Ideal Diode Plus Offset Voltage Model. The ideal diode is used as a replacement for a semiconductor diode if the forward voltage across the terminals of the semiconductor diode is negligible and if the semiconductor diode is not driven into the breakdown region. When the forward voltage is not negligible, we include a DC offset voltage V_0, as shown in Fig. 2.6a. Note that in Fig. 2.6b, v_D is now the voltage V_0 plus the voltage across the ideal diode, or v'_D. The translated i_D versus v_D characteristic results because for $v_D < V_0$, current is being forced in the disallowed direction and must therefore be zero since the ideal diode does not conduct current in this direction. However, if v_D is increased until $v_D = V_0$, the ideal diode voltage v'_D will be zero, and current can flow in the allowed direction of the arrowhead.

In many instances, this circuit model for a semiconductor PN junction diode is the most useful. It accounts for the forward voltage drop across the semiconductor diode. Recall that for a Si PN diode, V_0 is taken to be 0.7 V and that for a Si MN diode, this voltage is approximately 0.3 V.

We must also realize, however, that the breakdown region has not been accounted for thus far in these equivalent circuit representations. This omission is of no consequence in diode circuit

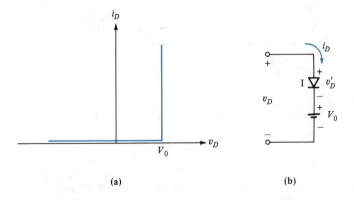

(a)

(b)

FIGURE 2.6   Ideal Diode Plus Offset Voltage: (a) Current–voltage characteristic, (b) Equivalent circuit

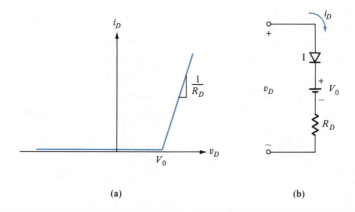

FIGURE 2.7   Ideal Diode Plus Offset Voltage and Resistor: (a) Current–voltage characteristic, (b) Equivalent circuit

applications where voltage levels remain less than $|V_{\text{BR}}|$. If the opposite is true, the breakdown region is easily modeled with an equivalent circuit by including an additional branch in parallel with the forward branch. The additional branch allows for the possibility of reverse breakdown current.

Ideal Diode Plus Offset Voltage and Resistor Model.   Figure 2.7 is a logical extension of the ideal diode plus offset voltage case. A resistor R_D is added in series, which allows the forward current i_D to increase with forward voltage v_D. In this region, with $v_D \geq V_0$, the diode is shorted so that

KVL yields

$$v_D = V_0 + i_D R_D \tag{2.2–1}$$

This is the equation of a line (slope $1/R_D$), as shown in Fig. 2.7a for $v_D \geq V_0$. For $v_D \leq V_0$, i_D must be zero because the net voltage around the loop is attempting to force current the wrong way through the diode. Hence, i_D is zero for $v_D \leq V_0$.

The circuit representation for the semiconductor diode shown in Fig. 2.7b should provide a more accurate model. However, its main drawback is that the value of R_D is not known accu-

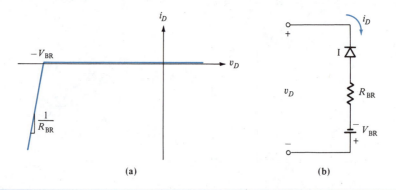

FIGURE 2.8   Ideal Diode, R_{BR}, and V_{BR} Representing the Reverse Diode Region: (a) Current-voltage characteristic, (b) Equivalent circuit

rately. In fact, R_D varies widely from one diode type to another. Thus, the simpler model of Fig. 2.6b, consisting of just the ideal diode and offset voltage V_0, is used more often.

Equivalent Circuit for the Breakdown Region.

The previous circuit models have not included a representation for diode operation in breakdown. Figures 2.8a and b show how this region may be modeled using an ideal diode, an offset voltage V_{BR}, and a resistor R_{BR}. The resulting current–voltage characteristic is another piecewise linear representation, but this time the forward current and voltage region is not accounted for.

To analyze the circuit of Fig. 2.8, we note first that if $v_D > -V_{BR}$, the diode will be open and $i_D = 0$. Secondly, for $v_D \leq -V_{BR}$ (operation in the breakdown region), current may flow in the direction of the arrowhead and i_D is negative. The expression for i_D is obtained from KVL where the ideal diode is treated like a short circuit. Thus, for $v_D \leq -V_{BR}$, we have

$$v_D = -V_{BR} + i_D R_{BR} \qquad (2.2\text{–}2)$$

or

$$i_D = \frac{1}{R_{BR}}(v_D + V_{BR}) \qquad (2.2\text{–}3)$$

For the range of $v_D \leq -V_{BR}$, (2.2–3) indicates that $i_D \leq 0$.

To include the forward voltage region with the breakdown region, we add a branch in parallel with that of Fig. 2.8b. Any of the previous models for the forward region can be used depending upon which forward region representation is desired. In Fig. 2.9a, the piecewise linear equivalent circuit for a diode is shown that includes resistors for the forward and breakdown regions R_D and R_{BR}, as well as the offset voltages V_0 and V_{BR}. Shown in Fig. 2.9b is the current–voltage characteristic, which is simply a superposition of the current–voltage characteristic for each branch.

Usually, the breakdown region is quite abrupt for a semiconductor diode, and the piecewise linear line in this region is very nearly vertical. Under these circumstances, the resistor R_{BR} is taken as zero.

―――――――――――――――――――――――――― **EX. 2.3**

Zener Diode Example

Consider a PN junction semiconductor diode operating in the breakdown region (a zener diode). Let $|V_{BR}| = 10$ V and $R_{BR} = 0$. Calculate the resulting diode current and voltage for the circuit shown in Fig. E2.3 using the values of R_T and V_T indicated.

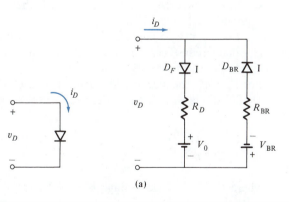

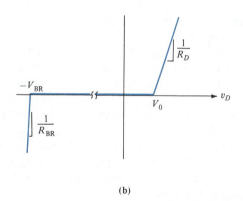

(a)   (b)

FIGURE 2.9   Complete Piecewise Linear Circuit Model for a Semiconductor Diode: (a) Circuit symbol and equivalent circuit, (b) Current–voltage characteristic

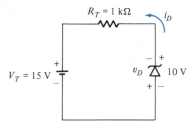

FIGURE E2.3

Solution: Since the zener diode is operating under break-down conditions, we substitute the circuit model of Fig. 2.8 to determine i_D and v_D. Since V_T is greater in magnitude than V_{BR}, the ideal diode will be shorted, and $v_D = -10\text{V}$ while $i_D = (10\text{ V} - 15\text{ V})/1\text{ k}\Omega = -5\text{ mA}$.

2.3 ANALYSIS OF DIODE–RESISTOR CIRCUITS

Since diodes have nonlinear current–voltage characteristics, special methods must be used to analyze networks containing these elements. The objective in each analysis is the same, namely, to determine the currents and voltages in the circuit.

The most powerful method of approach for the analysis of nonlinear circuits is to obtain the representative nonlinear equations and solve them using an iterative solution. This process involves guessing an initial solution, substituting this solution into the equations, and, if incorrect, changing the initial guess and substituting again. By repetition of this process, the actual solution is eventually obtained. Although this method can be successful, it is an example of overkill in almost all electronic circuit analyses. Using an iterative procedure to analyze a circuit containing a semiconductor diode is like using a sledge hammer to crack a peanut. The solutions are obtainable in a much easier manner, as we will see.

The general method of approach in analyzing diode circuits here will be to substitute linear circuit models for the diodes. This approach was

intimated in the previous section when several piecewise linear semiconductor diode models were introduced. Two distinctly different ways in which this linear model approach is used will be described. These solution methods will also be illustrated graphically to provide additional insight.

2.3.1 DC Analysis and Large-Signal Analysis

DC Analysis. When the voltage and/or current sources in a diode circuit are purely DC, the analysis of the network is greatly simplified in that all of the resulting currents and voltages will also be purely DC.

Figure 2.10a displays a diode–resistor network containing a single resistor R_T, a DC voltage source V_T, and a semiconductor diode. The resistor and voltage source are the Thévenin equivalent elements for an overall circuit external to the diode. We will consider the positive and negative regions for V_T separately.

For the range of $V_T \geq 0$, the diode current and voltage will both be positive (or zero), and thus the diode will operate in the forward-bias region. Using the linear model of Fig. 2.7 to replace the semiconductor diode in Fig. 2.10a results in the equivalent circuit shown in Fig. 2.10b. Two solutions now exist, depending upon whether V_T is greater or less than V_0.

For $V_T \leq V_0$, the ideal diode in Fig. 2.10b is an open circuit (because V_T is insufficient in magnitude to overcome V_0), and thus $i_D = I_D = 0$. For $V_T > V_0$, the ideal diode is a short circuit, and the DC diode current I_D is obtained from KVL as

$$I_D = \frac{V_T - V_0}{R_T + R_D} \tag{2.3–1}$$

where the DC diode voltage is $v_D = V_D = V_T - I_D R_T$.

Note that if the model of Fig. 2.6 ($R_D = 0$) were used, our solution would be $I_D = (V_T - V_0)/$

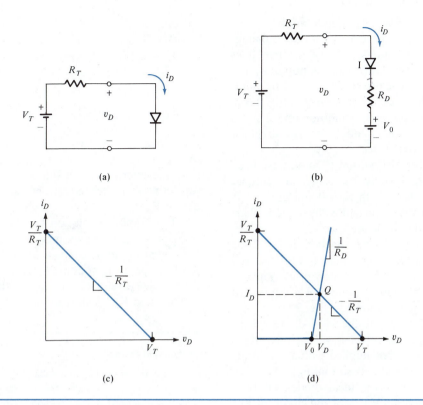

FIGURE 2.10   DC Analysis: (a) Diode–resistor network, (b) Forward-region equivalent circuit, (c) Load line for R_T and V_T, (d) Graphical determination of Q in the forward region

R_T. Similarly, using the ideal diode itself as a replacement for the semiconductor diode ($V_0 = 0$) would have resulted in $I_D = V_T/R_T$. These solutions are approximations to the solution given by (2.3–1). They are, however, valid under the conditions that $V_0 \ll V_T$ and/or $R_D \ll R_T$, which in some instances is the case.

This same solution can be obtained graphically. If we consider Fig. 2.10a, KVL for the external elements R_T and V_T may be written in terms of the total quantities v_D and i_D as

$$v_D = V_T - i_D R_T \qquad\qquad \textbf{(2.3–2)}$$

This equation expresses a linear relation between v_D and i_D and hence represents a straight line when it is plotted on the i_D–v_D axes. Customarily,

this line is called the *load line*. This line represents the terminal current–voltage relationship for the combination of a resistor and voltage source in series that is often connected as a load to the output portion of electronic circuits. The load line represents all possible combinations of v_D and i_D.

Equation (2.3–2) is displayed graphically in Fig. 2.10c. It is drawn by realizing that the voltage and current intercepts are V_T and V_T/R_T, respectively. Additionally, the slope of the line is $-1/R_T$. It is only drawn in the first quadrant since we are considering the case in which v_D and i_D are greater than zero and $V_T \geq 0$.

To obtain the DC solution graphically, it is now only necessary to sketch the current–voltage characteristic for the diode on the same axes, as is shown in Fig. 2.10d. The diode model of Fig. 2.7 is used as a replacement for the semiconductor diode. The intersection of the two lines provides the specific DC operating point (V_D, I_D). It is customary to call the DC operating point the Q point, and hence it is labeled Q in Fig. 2.10d. Q stands for the *quiescent* condition and dictates DC operation. This notation is also used for transistors, as we will see.

It should be realized that Q could have been obtained approximately by using the simpler models for the diode with $V_0 = 0$ and/or $R_D = 0$. Sometimes, measured current–voltage diode characteristics for the diode are used.

For the range of $V_T \leq 0$, the diode in Fig. 2.10a will operate under reverse-bias conditions, and the model of Fig. 2.8 is the appropriate replacement for the semiconductor diode. The equivalent circuit for analysis is shown in Fig. 2.11a.

For $-V_{BR} \leq V_T \leq 0$, the ideal diode will be an open circuit, and thus $i_D = 0$. Under these conditions, the entire applied voltage appears across the diode terminals so that $v_D = V_T$.

For $V_T \leq -V_{BR}$, the ideal diode will be a short circuit. Writing KVL for the circuit of Fig. 2.11a and solving for I_D yield

$$I_D = -\frac{V_{BR} - |V_T|}{R_D + R_T} \tag{2.3-3}$$

where

$$V_D = -|V_T| - I_D R_T \tag{2.3-4}$$

or

$$V_D = -V_{BR} + I_D R_D \tag{2.3-5}$$

Equations (2.3–4) and (2.3–5) express exactly the same value for V_D, but they are obtained from the external circuit side and the diode side, respectively. Either equation may be used to obtain V_D.

The graphical solution for the case of $V_T = -|V_T| < -V_{BR}$ is shown in Fig. 2.11b. Note that the operating point is again labeled Q, denoting the DC operating condition when V_T is the particular negative DC voltage with magnitude greater than V_{BR} as shown.

— EX. 2.4

DC Solution Example

Determine the current i_D and voltage v_D in the PN diode network of Fig. E2.4 under the conditions that $V_T = \pm 10$ V. Consider that the PN diode is made of Si with $V_{BR} = 5$ V. Also, for this diode, $R_{BR} = R_D = 0$.

Solution: The equivalent circuit model of Fig. 2.9 is applicable in this example with R_D and $R_{BR} = 0$. For the case in which $V_T = +10$ V, the breakdown branch will

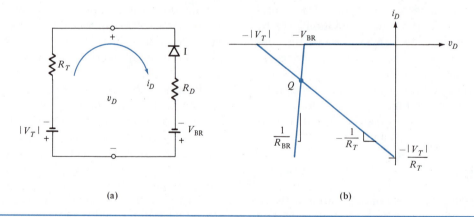

(a)   (b)

FIGURE 2.11   DC Analysis: (a) Reverse-region equivalent circuit, (b) Graphical determination of Q in the reverse region

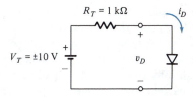

$R_T = 1\ k\Omega$

$V_T = \pm 10\ V$

i_D

v_D

FIGURE E2.4

be open and the forward branch will conduct current. Thus, replacing the diode in the forward branch with V_0 (in Fig. E2.4) and writing KVL for the loop yield

$$I_D = \frac{V_T - V_0}{R_T} = \frac{10 - 0.7}{1\ k\Omega} = 9.3\ mA$$

and

$$v_D = V_0 = 0.7\ V$$

For $V_T = -10\ V$, only the breakdown branch conducts current. Shorting the ideal diode in the breakdown branch and writing KVL for this loop yield

$$i_D = \frac{V_T + V_{BR}}{R_T} = \frac{-10 + 5}{1\ k\Omega} = -5\ mA$$

and

$$v_D = -V_{BR} = -5\ V$$

Large-Signal Analysis. We now consider a signal component of source voltage v_s in addition to the DC voltage V_T. The circuit is shown in Fig. 2.12a, and a solution is still obtainable by using a piecewise linear circuit model for the semiconductor diode. Here, however, the complete piecewise linear circuit model for the semiconductor diode (Fig. 2.9) is required since we assume that the sum of V_T and v_s has both positive and negative values as time varies. Figure 2.12b shows the resultant circuit to be analyzed; the ideal diodes have been labeled D_F (for the forward region) and D_{BR} (for the breakdown region). The branch that conducts will depend upon whether v_D is positive or negative, which, in turn, depends upon the sign of $V_T + v_s$.

For example, if $v_s + V_T$ is positive and greater than V_0, D_F will be shorted (and D_{BR} opened) so that i_D is obtained by using KVL containing the forward branch as $i_D = (v_s + V_T - V_0)/(R_D + R_T)$. Similarly, if $v_s + V_T$ is negative, D_F will be open so that i_D is found by considering KVL containing the breakdown branch. If $v_s + V_T$ is negative and greater in magnitude than V_{BR}, the resulting current is $i_D = (v_s + V_T + V_{BR})/(R_T + R_{BR})$, also negative.

This solution technique is adequate to provide what is called the *large-signal solution*, which means that the operating point shifts over a wide

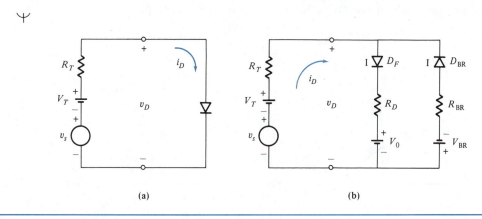

(a)   (b)

FIGURE 2.12   Large-Signal Analysis: (a) Diode circuit with DC and AC voltage sources, (b) Complete piecewise linear equivalent circuit

range of the overall diode characteristic. However, the large-signal solution can be quite approximate in small, confined regions where the actual semiconductor diode characteristic deviates from the piecewise linear approximation. In these situations, an alternate solution method is necessary.

Another reason for seeking an alternate solution method for analysis of circuits containing AC sources can be observed if we attempt to obtain our solution for currents and voltages graphically. If we assume that an AC source voltage v_s is added in series to V_T in the circuit of Fig. 2.10a, the total source voltage will then be $v_s + V_T$. The intercepts of the load line of Figs. 2.10c and d will vary with time and have the values $v_s + V_T$ and $(v_s + V_T)/R_T$. The intersection of this time-varying load line with the diode characteristic will therefore shift with time. Obtaining the solution graphically under these conditions is extremely tedious since multiple determinations of i_D and v_D as a function of time are required.

2.3.2 Small-Signal (AC) Analysis and Overall Analysis

Small-Signal Analysis. A *small-signal analysis* is a purely AC analysis that yields only the AC solution. Adding this result to the DC solution provides a complete solution comparable to that obtained by using the large-signal analysis method. However, the small-signal analysis method provides a more accurate AC solution under the conditions that the signal (AC) magnitudes of both current and voltage are much less than the DC magnitudes. This assumption is inherent in using the small-signal analysis method.

Overall Analysis. The overall analysis of a circuit, assuming that the signals are small in comparison with DC, involves a three-step procedure consisting of the following:

1. Obtain the DC solution.
2. Obtain the small-signal (AC) solution.
3. Add the results of steps 1 and 2.

To carry out this three-step procedure for the network of Fig. 2.12a, we assume, for validity, that $v_s \ll V_T$. We also assume, at least initially, that the semiconductor diode characteristic is a measured curve, as shown in Fig. 2.13a.

The DC solution is obtained graphically by setting $v_s = 0$ and sketching the load line. The intersection of the load line and the diode characteristic provides Q, as shown in Fig. 2.13a. The DC values of current and voltage (I_D and V_D) are thus obtained, and step 1 in the three-step procedure is completed.

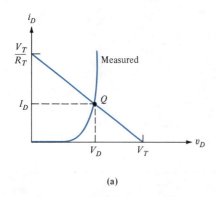

(a)

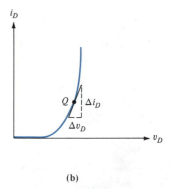

(b)

FIGURE 2.13 Overall Analysis: (a) Current–voltage characteristic, (b) Determination of r_d

Step 2, the small-signal analysis, is carried out by first developing a purely AC circuit model (valid for small signals) for the entire circuit under analysis (Fig. 2.12a). The resulting AC equivalent circuit is then analyzed to determine the AC current i_d and voltage v_d of the semiconductor diode (and all other AC quantities of interest). To obtain the AC equivalent circuit for the network of Fig. 2.12a requires substitution of an AC equivalent for each element in the circuit.

First, we consider the elements external to the semiconductor diode. The AC replacements for the resistor R_T and the signal voltage v_s are merely themselves because they have the same behavior for AC as for total voltage and current. However, the DC voltage source V_T has an AC equivalent voltage of 0 V by definition. Therefore, its equivalent AC element is a short circuit representing 0 V AC.

To determine the AC equivalent of the semiconductor diode, we consider that the diode AC current i_d and voltage v_d represent small changes in the total current i_D and voltage v_D. This consideration is the small-signal assumption and allows replacement of the actual diode current–voltage variation with a linear variation, as shown in Fig. 2.13b. The straight-line (small-signal) variation represents a resistor with its slope equal to the inverse of the resistor value. We define this AC resistor as r_d where

$$\frac{1}{r_d} = \frac{\Delta i_D}{\Delta v_D}\bigg|_Q \qquad (2.3-6)$$

It is very important to emphasize that this parameter must be evaluated at Q. For small signals, we may also write $r_d = v_d/i_d$. Furthermore, as the signal magnitudes become infinitesimal, $1/r_d$ becomes a derivative and

$$\frac{1}{r_d} = \frac{di_D}{dv_D}\bigg|_Q \qquad (2.3-7)$$

Thus, the small-signal AC equivalent of the semiconductor diode is simply a resistor (r_d) whose value depends directly upon the position of Q.

Figure 2.14 displays the AC equivalent circuit for the network under consideration (Fig. 2.12a). Writing KVL for this circuit yields, quite simply,

$$i_d = \frac{v_s}{R_T + r_d} \qquad (2.3-8)$$

and

$$v_d = i_d r_d \qquad (2.3-9)$$

Step 3 in the solution procedure now involves simply adding the DC and AC components together, which yields

$$i_D = i_d + I_D \qquad (2.3-10)$$

and

$$v_D = v_d + V_D \qquad (2.3-11)$$

Although the AC diode resistance can be determined graphically from a measured i_D–v_D characteristic in each particular case, a much more useful technique is to use an analytical expression obtained from Shockley's diode equation [(2.1-1)]. The value of r_d obtained in this manner will depend drastically upon whether Q is in the forward or reverse region.

In the forward region, i_D varies exponentially as given by (2.1–3), and taking the derivative yields

$$\frac{1}{r_d} = \left(\frac{I_S}{v_T}\right)e^{v_D/v_T}\bigg|_Q \qquad (2.3-12)$$

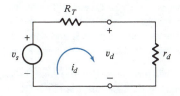

FIGURE 2.14   AC Equivalent Circuit for Figure 2.12a

However, $I_S e^{v_D/v_T}|_Q$ is the DC semiconductor diode current I_D, and thus by back substitution,

$$\frac{1}{r_d} = \frac{I_D}{v_T} \qquad (2.3\text{--}13)$$

Solving for r_d and using $v_T = 0.025$ V yield

$$r_d = \frac{0.025}{I_D} \qquad (2.3\text{--}14)$$

Note that the dependence of r_d upon the parameter I_S has been eliminated in terms of the diode DC current. Thus, accuracy not obtainable with I_S in the equation results since I_S is not known with accuracy.

The analytic relation for r_d [(2.3–14)] is extremely useful in obtaining the small-signal solution in diode networks using an AC equivalent circuit. Additionally, this AC equivalent resistor arises in transistor AC equivalent circuits and is likewise extremely useful there, as we will see.

In the reverse region (excluding breakdown), i_D is a constant $(-I_S)$ and its derivative is zero. Thus,

$$\frac{1}{r_d} = 0 \qquad (2.3\text{--}15)$$

The resistor r_d is an open circuit in this region, which implies that the AC reverse current is essentially zero.

──────────────────────────── EX. 2.5

Small-Signal Solution Example

Consider the circuit shown in Fig. E2.5, which is the same as the network of Fig. E2.4 except for the AC signal source. In Fig. E2.5, we let $V_T = +10$ V and $v_s = 1 \sin \omega t$ V. Let the PN diode be composed of Si and determine the current and voltage for the diode, assuming $R_D = 0$.

Solution: We have already determined the DC solution for this circuit in Ex. 2.4. Thus, the diode DC current and voltage are 9.3 mA and 0.7 V, respectively.

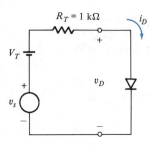

FIGURE E2.5

To carry out the AC small-signal analysis, we need to determine the AC resistance of the diode (r_d) and substitute this element into the circuit to obtain the AC diode current and voltage. Since the actual i_D–v_D characteristic curve for the diode is not given, we use the derived formula to calculate r_d as follows:

$$r_d = \frac{0.025}{I_D} = \frac{0.025 \text{ V}}{9.3 \text{ mA}} = 2.69 \ \Omega$$

Thus, with $v_s = 1 \sin \omega t$, the AC diode current and voltage are

$$i_d = \frac{v_s}{R_T + r_d} = \frac{\sin \omega t}{1002.69} = 0.997 \sin \omega t \text{ mA}$$

and

$$v_d = r_d i_d = 2.68 \sin \omega t \text{ mV}$$

Therefore, the total diode current and voltage are

$$i_D = 9.3 + 0.997 \sin \omega t \text{ mA}$$

and

$$v_D = 700 + 2.68 \sin \omega t \text{ mV}$$

──────────────────────────────

2.4   DIODE CIRCUIT APPLICATIONS

Several important semiconductor diode circuit applications are described in this section. Actual diodes used in these applications are indicated in Sections 2.5 and 2.6. This material will enable the reader to acquire additional experience in analyzing diode circuits and also to observe typical examples of semiconductor diode use. The applications considered include waveshaping examples such as rectifiers and clamper circuits. A zener diode voltage regulator and transmission gates are also described.

2.4.1   Rectifiers

A device or circuit that is used to produce a DC voltage component from an AC signal is called a *rectifier,* and the process of this conversion is known as *rectification.* Since a diode is essentially a one-way current element, it is ideal for rectifier applications, as we will see. The actual

diodes used in rectifier applications are called *rectifier diodes.* These diodes are heavy-duty devices designed with maximum current ratings of 1–50 A and even higher, with breakdown voltages of 100–1000 V. Section 2.5 contains a selection guide for rectifier diodes.

Half-Wave Rectifier.   A circuit that performs half-wave rectification is displayed in Fig. 2.15a. The AC signal voltage v_s is considered to be a sinusoid for convenience, and we let $v_s = V_S \sin \omega t$. The resistor R_S is the resistance associated with the source voltage, and R_L is the load resistor to which a DC component of voltage and current is to be delivered. The diode in the circuit is assumed to be a PN junction semiconductor diode.

The analysis of the circuit in Fig. 2.15a consists of substituting one of the piecewise linear

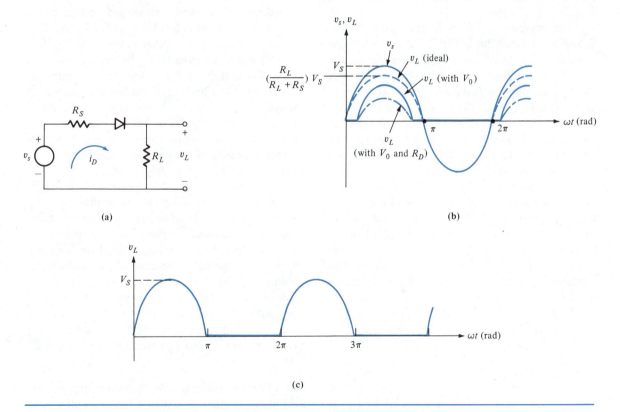

FIGURE 2.15   Half-Wave Rectifier: (a) Circuit, (b) Waveshapes of v_s and v_L, (c) Ideal half-wave sinusoid

models for the actual diode (from Fig. 2.5, 2.6, or 2.7) and determining the resulting currents and voltages in the network.

The simplest solution is obtained when we merely substitute an ideal diode (Fig. 2.5) for the semiconductor diode. The ideal diode behaves like a short circuit for $v_s \geq 0$ and like an open circuit for $v_s \leq 0$. Thus, for $v_s \geq 0$,

$$i_D = \frac{v_s}{R_S + R_L} \tag{2.4-1}$$

and

$$v_L = i_D R_L = v_s\left(\frac{R_L}{R_S + R_L}\right) \tag{2.4-2}$$

For $v_s \leq 0$, the diode is open, $i_D = 0$, and thus $v_L = 0$.

The waveshape for v_L in this ideal case is shown as a dashed line in Fig. 2.15b and is labeled v_L (ideal). Also shown is the assumed waveshape for v_s (and additional solutions to be obtained shortly). During the first half period, we see that v_L is a sine wave with a diminished amplitude from that of v_s. During the second half period, the load voltage $v_L = 0$ because the diode is an open circuit for negative v_s. This variation then repeats in each succeeding period.

If we now take into account the forward voltage V_0 for the semiconductor diode (see Fig. 2.6), two changes result in the variation of v_L. First, the amplitude of v_L is further reduced from V_S. Second, the time range for nonzero v_L is made smaller than a half period of the sinusoid. The nonzero diode current and load voltage equations for this case with $v_s \geq V_0$ are

$$i_D = \frac{v_s - V_0}{R_S + R_L} \tag{2.4-3}$$

and

$$v_L = (v_s - V_0)\left(\frac{R_L}{R_S + R_L}\right) \tag{2.4-4}$$

Equation (2.4-4) is also sketched in Fig. 2.15b, and its waveshape is labeled v_L (with V_0). Note that the diode switches from the short to the open condition and vice versa whenever $v_s = V_0$.

Finally, if we include a resistor R_D as well as V_0 to represent the forward region for the semiconductor diode, the magnitude of v_L is reduced still further. The nonzero equations for this case with $v_s \geq V_0$ are

$$i_D = \frac{v_s - V_0}{R_S + R_D + R_L} \tag{2.4-5}$$

and

$$v_L = (v_s - V_0)\left(\frac{R_L}{R_S + R_D + R_L}\right) \tag{2.4-6}$$

Equation (2.4-6) is also displayed in Fig. 2.15b, and its waveshape is labeled v_L (with V_0 and R_D).

Note that the waveshapes for v_s and v_L (for $v_s > 0$) would be essentially the same if $V_S \gg V_0$, $R_L \gg R_S$, and $R_L \gg R_D$. Often, these conditions are realized in a half-wave rectifier, and the waveshape of v_L is essentially the same as the input signal during the positive half periods of v_s. The overall ideal waveshape is shown in Fig. 2.15c and appears as half of the orginal waveshape. Thus, the term *half-wave* was originated.

To determine that this circuit performs rectification, it is only necessary to realize that the waveshape for v_L (in all cases) has a nonzero average value. Without carrying out the details, we can ascertain that since the original sinusoid has a zero average value because of its symmetrical behavior, the half wave without the negative portions of this variation must have a nonzero average value. The average value of the half wave is precisely the DC component; thus, the half-wave rectifier does indeed produce DC from a purely AC source.

Full-Wave Rectifiers.   As the name implies, the full-wave rectifier produces a full wave of rectification. Two circuits are commonly used for full-wave rectification. These circuits are shown in Figs. 2.16a and b. Each of these circuits produces the waveshape shown as a dashed line in Fig. 2.16c, under the condition that ideal diodes be used to replace the semiconductor diodes.

The analysis of the circuit of Fig. 2.16a involves first realizing that the diodes conduct current in pairs. That is, when D_1 and D_2 are shorted, D_3 and D_4 are opened and vice versa.

For the case of ideal diodes, when $v_s > 0$, D_1 and D_2 are shorted while D_3 and D_4 are opened. For $v_s > 0$, current is forced in the direction of the arrowhead for D_1 and D_2, thus causing them to short circuit and conduct current. The current that then flows through $R_L(i_L)$ produces a voltage v_L that appears directly across D_3 and D_4 as a reverse voltage. Thus, D_3 and D_4 are open circuits. When $v_s < 0$, D_3 and D_4 are shorted while D_1 and D_2 are opened. Therefore, whenever one pair of diodes (D_1D_2 or D_3D_4) is shorted, the other

pair is opened, and the signal voltage source is directly in parallel with R_L. Additionally, the current through $R_L(i_L)$ is always in the same direction as indicated.

For the case of ideal diodes, the load resistor current and voltage, therefore, are

$$i_L = \frac{|v_s|}{R_L} \qquad (2.4-7)$$

and

$$v_L = |v_s| \qquad (2.4-8)$$

thus verifying the waveshape of Fig. 2.16c. Because the same sinusoidal variation is obtained in every half period, this waveshape is called a *full-wave* sinusoid.

For cases in which the ideal diode is not an adequate representation of the semiconductor

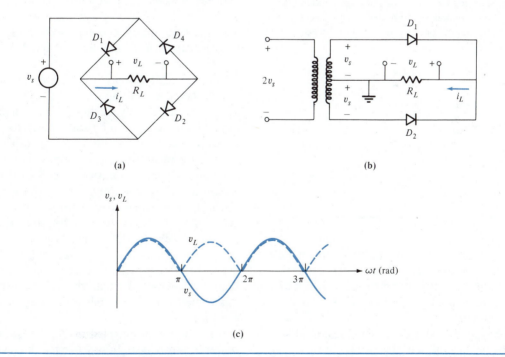

(a)

(b)

(c)

FIGURE 2.16   Full-Wave Rectifiers: (a) Four-diode rectifier, (b) Center-tab transformer rectifier, (c) Waveshapes of v_s and v_L

diode, R_D and V_0 are included in the diode equivalent circuit. The following example indicates the analysis for the general case.

EX. 2.6

Full-Wave Rectifier Example

Determine the output load voltage v_L for the full-wave rectifier of Fig. 2.16a with $v_s = 10 \sin \omega t$ V, $R_L = 1$ kΩ, $R_D = 1$ kΩ, and $V_0 = 0.7$ V.

Solution: The diode pairs $D_1 D_2$ and $D_3 D_4$ conduct current for $|v_s| > 1.4$ V. Otherwise, $i_L = 0$. When either set of diodes conduct,

$$i_L = \frac{v_s - 1.4}{2R_D + R_L} = 3.33 \sin \omega t - 0.467 \text{ mA}$$

and

$$v_L = R_L i_L = 3.33 \sin \omega t - 0.467 \text{ V}$$

The corresponding time of conduction starting during the first half-cycle is obtained from $10 \sin \omega t = 1.4$, or $\omega t \simeq 0.14$ rad. The output voltage waveshape for one period is shown in Fig. E2.6.

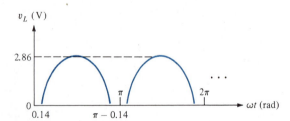

FIGURE E2.6

Figure 2.16b displays another form of full-wave rectifier that utilizes a center-tap transformer. The analysis of this circuit consists of first realizing that when D_1 is shorted, D_2 is opened and vice versa.

We again consider the case of ideal diodes to describe the operation of this rectifier. For $v_s > 0$, current is forced in the direction of the arrowhead for D_1 and in the opposite direction for D_2.

Thus, only D_1 conducts current, and D_2 is open. The direction of current flow is in the direction of i_L (as indicated in Fig. 2.16b). For $v_s < 0$, D_2 conducts while D_1 is open, and the current direction remains the same. Thus, using ideal diodes as replacements for the semiconductor diodes, when either diode shorts, v_s is directly across R_L, and (2.4–7) and (2.4–8) again provide the solution for i_L and v_L, respectively. The details of the analysis for nonideal diodes are left for the problems at the end of the chapter.

Full-wave rectifiers also produce a DC component from a purely AC signal. The average value is the DC component; however, in this case, the average value of v_L is twice that of the half-wave rectifier.

Peak Detector or Rectifier.   The *peak detector* or *rectifier* also produces a DC output voltage from an AC input voltage. Fig. 2.17a displays the basic peak rectifier circuit. The capacitor C is the load element, and a DC voltage is to be developed across its terminals. For our analysis, we begin with the assumption that the diode is ideal and that the capacitor is initially uncharged. We also assume that $v_s = V_S \sin \omega t$.

Figure 2.17b displays the time variation of v_L as a dashed curve, along with v_s for one period of v_s. Since C is initially uncharged, as v_s increases in its first quarter period, the diode shorts and v_s appears directly across the terminals of C. Since there is no resistance in the loop, C charges instantaneously to the value of v_s (see Chapter 1 for a more detailed description of the charging and discharging of capacitors). During the first half period, as v_s increases to its maximum value (V_S) and the diode is shorted, v_L increases to this value also. However, after v_s increases to V_S and then begins to decrease, v_L remains at V_S because there is no resistive path through which C can discharge this voltage. Hence, a purely DC voltage is developed for v_L.

The situation depicted in Figs. 2.17a and b, however, is too ideal. In order to use the DC voltage developed, a load resistor R_L is connected as shown in Fig. 2.17c. The value of this resistor must

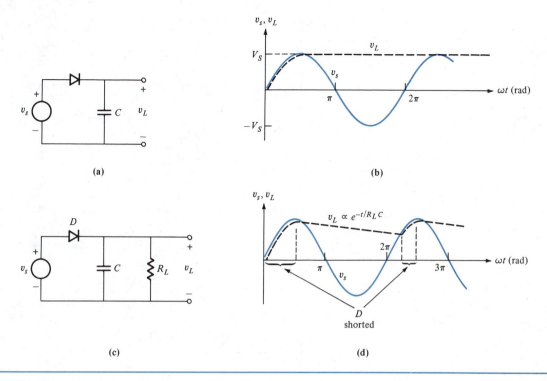

FIGURE 2.17 Peak Detector or Rectifier: (a) Basic circuit, (b) Waveshapes of v_s and v_L for basic circuit, (c) General circuit, (d) Waveshapes of v_s and v_L for general circuit

be large in order to develop DC at all, as we will see.

With R_L in the circuit, the resulting wave-shape for v_L is as shown in Fig. 2.17d (the dashed curve). During the first quarter period of v_s, the diode is shorted and the capacitor charges to the peak value of v_s at $\omega t = \pi/2$ (assuming an ideal diode). As v_s continues its cycle and begins to decrease, the diode opens and the capacitor begins to discharge through the loop containing R_L. The voltage across the capacitor then decreases exponentially, depending upon $-t/R_L C$ (as described in Chapter 1). For large R_L, the exponential decay of v_L can be approximated as a straight line, as drawn in Fig. 2.17d. In the second period of v_s, when v_s increases to v_L, the diode again shorts. The capacitor voltage once more increases with v_s as the capacitor charges until $v_L = V_S$. This process repeats itself in every succeeding period of v_s.

With D as ideal, the peak value of v_L is the peak value of v_s. If the diodes in Figs. 2.17a and c were not considered to be ideal, the dashed waveshapes of Figs. 2.17b and d would be altered. Inclusion of the forward voltage V_0 across the diode would produced a peak load voltage of $V_S - V_0$. Furthermore, the inclusion of the diode resistor R_D would slow the charging process of the capacitor, but the eventual result would be peak rectification with reduced amplitude.

2.4.2 Diode Clamper

The *diode clamper* is a waveshaping circuit that shifts the signal voltage to a desired level. Figure 2.18a displays the simplest clamper circuit. It is identical to the basic peak rectifier circuit (Fig. 2.17a), except that the capacitor and diode

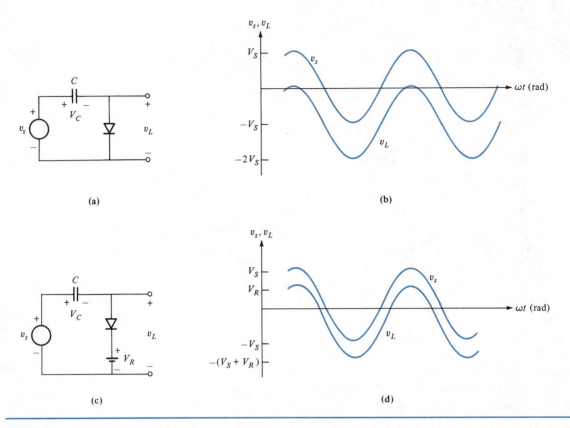

FIGURE 2.18   Diode Clamper: (a) Basic circuit, (b) Waveshapes of v_s and v_L for basic circuit, (c) General circuit, (d) Waveshapes of v_s and v_L for general circuit

are interchanged. The diode is now the load element, but the operation of the circuit remains the same as that of the peak rectifier.

From our analysis of the peak rectifier circuit, the capacitor in Fig. 2.18a will charge so that the peak value of voltage for the capacitor becomes V_S (or $V_S - V_0$, if we account for the forward voltage of the semiconductor diode). The polarity will be as indicated by V_C because current can only conduct in the direction of the arrowhead for the diode. After the capacitor has fully charged, the diode will be open circuited. Thus, the expression for the load voltage (obtained from KVL) is

$$v_L = v_s - V_C \qquad\qquad (2.4\text{--}9)$$

For the case of an ideal diode, $V_C = V_S$. The wave-

shape for v_L is shown in Fig. 2.18b along with v_s (after the capacitor has charged to V_S). The peak of this shifted voltage (v_L) is said to be "clamped at 0 V."

Figure 2.18c shows the circuit for a general clamper that includes a reference voltage V_R. In this case, after C has charged, corresponding to the peak of v_s, the diode will be open, and thereafter the capacitor voltage will be (for an ideal diode) $V_C = V_S - V_R$. From KVL, v_L is once more given by

$$v_L = v_s - V_C \qquad\qquad (2.4\text{--}10)$$

or, substituting for V_C,

$$v_L = v_s - V_S + V_R \qquad\qquad (2.4\text{--}11)$$

This waveshape is the signal voltage v_s clamped at the reference voltage V_R as shown in Fig. 2.18d. Note that V_R can be positive or negative.

If the semiconductor diode nonidealities (R_D and V_O) are considered, similar modifications to those described for the peak rectifier result. The details of the analysis are left to the problems at the end of the chapter.

It should be mentioned that the signal voltage v_s need not be sinusoidal. The capacitor in the peak rectifier and clamper circuits will then charge to the peak value of the input signal voltage, and the circuit will continue to operate in a similar manner. For the general clamper, the output voltage would consist of the nonsinusoidal signal voltage shifted to the voltage V_R.

─────────────────────────────── EX. 2.7

Diode Clamper Example

Consider the general clamper circuit of Fig. 2.18c. Let $V_R = 1$ V, and assume that C has charged to its final value at

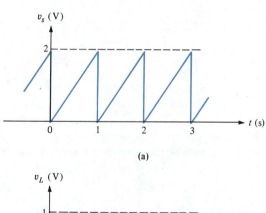

v_s (V)

(a)

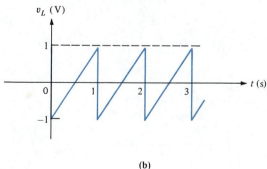

v_L (V)

(b)

FIGURE E2.7

$t = 0$. Determine and sketch the load voltage v_L for $t > 0$ for the triangular input signal voltage v_s shown in Fig. E2.7a.

Solution: The capacitor charges to $V_C = V_{peak} - V_R = 1$ V during the time $t < 0$. Thus, for $t > 0$, the load voltage is

$$v_L = v_s - V_C = v_s - 1$$

This waveshape is sketched in Fig. E2.7b.

─────────────────────────

2.4.3 Diode Limiter or Clipper

The basic diode limiter circuit is shown in Fig. 2.19a. The voltage transfer characteristic (v_O versus v_I) is displayed in Fig. 2.19b and is obtained by assuming v_I varies through all possible values and determining the corresponding values of v_O. Accounting for the forward voltage drop V_O, we observe that when D is opened ($v_I < V_R + V_O$), the current is zero and $v_I = v_O$. Additionally, when D is shorted ($v_I > V_R + V_O$), $v_O = V_R$. Since the output voltage is limited to a maximum value ($V_R + V_O$), the circuit is called a *diode limiter*.

Figure 2.19c displays voltage waveshapes for v_I and v_O corresponding to an input sinusoidal voltage with $v_I = V_S \sin \omega t$ and $V_S > V_R$. Note that the output voltage waveshape is "clipped" off for $v_I > V_R + V_O$. Hence, this circuit is also called a *diode clipper*.

Figure 2.19d shows a diode limiter that limits the negative peak value of v_O as well as the positive peak value. The voltage transfer characteristic of Fig. 2.19e is obtained by considering the state of each diode for all values of v_I. Voltage waveshapes for this case are shown in Fig. 2.19f.

2.4.4 Zener Diode Voltage Regulator

Figure 2.20 shows an example of the application of a zener diode, the *zener diode voltage regulator*. Details for various zener diodes are given in the appendix at the end of this chapter. In Fig. 2.20a, a zener diode is used to regulate the

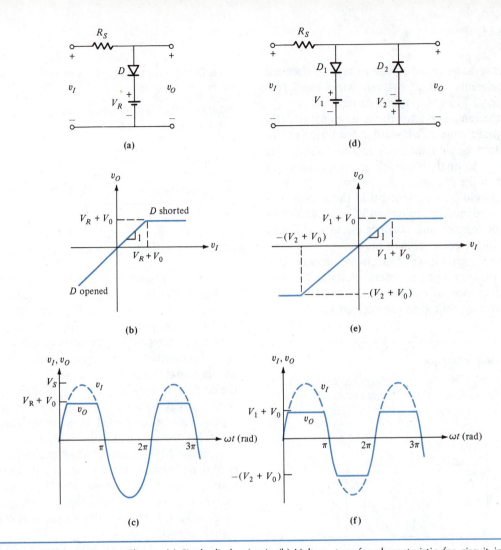

FIGURE 2.19 Diode Limiter or Clipper: (a) Single-diode circuit, (b) Voltage transfer characteristic for circuit in part a, (c) Waveshapes for circuit in part a, (d) Double-diode circuit, (e) Voltage transfer characteristic for circuit in part d, (f) Waveshapes for circuit in part d

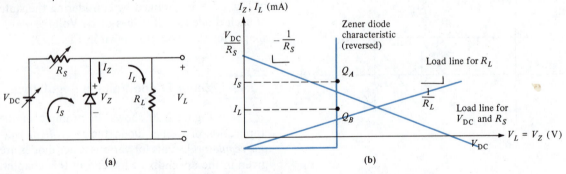

FIGURE 2.20 Zener Diode Voltage Regulator: (a) Basic circuit, (b) Current–voltage characteristics

output voltage of a DC power supply (V_{DC}) that is assumed to be variable. The voltage V_{DC} and resistor R_S are shown with arrows through their circuit symbols to indicate that these elements are variable. One restriction is that the unregulated DC supply voltage (V_{DC}) must always be greater than V_Z in order to maintain a fixed voltage across $R_L(v_L = V_Z)$. Additionally, R_S will be seen to have a restricted value.

We begin the analysis of the zener diode voltage regulator circuit by plotting the current–voltage characteristic for each of the three separate branches having $V_L = V_Z$—that is, R_S in series with V_{DC}, the zener diode, and the resistor R_L. These three characteristics are shown in Fig. 2.20b. The intersection Q_A provides the current I_S without R_L in the circuit. The intersection Q_B provides the current I_L through R_L, with V_Z maintained across the zener diode and R_L. If we consider all branches connected together, the current at $Q_B(I_L)$ must be less than or equal to that at $Q_A(I_S)$ in order for this regulator circuit to function properly. That is, I_L must be less than or equal to I_S, and the difference between these is I_Z (which cannot be negative for the zener diode and the direction chosen for I_Z).

The value of R_S necessary to achieve voltage regulation is obtained by first writing KCL as follows:

$$I_S = I_Z + I_L \tag{2.4–12}$$

Substituting $I_L = V_Z/R_L$ and $I_S = (V_{DC} - V_Z)/R_S$ yields

$$\frac{V_{DC} - V_Z}{R_S} = I_Z + \frac{V_Z}{R_L} \tag{2.4–13}$$

However, since I_Z must be greater than or equal to zero, we have

$$\frac{V_{DC} - V_Z}{R_S} \geq \frac{V_Z}{R_L} \tag{2.4–14}$$

or

$$R_S \leq R_L \frac{V_{DC} - V_Z}{V_Z} \tag{2.4–15}$$

From (2.4–15), we obtain the maximum value of R_S as

$$R_{S\,max} = \frac{R_L(V_{DC} - V_Z)}{V_Z} \tag{2.4–16}$$

For $R_S \leq R_{S\,max}$ and $V_{DC} \geq V_Z$, the load voltage will be regulated at $V_L = V_Z$.

2.4.5   Diode Transmission Gates

Diode gates are electronic switches that allow transmission of a signal voltage through a network due to the application of a control voltage or gate voltage. They are often called *transmission gates*. The gate voltage is normally digital, possessing either of two values, a high voltage or a low voltage, with precise values usually being unimportant. Transmission of the signal voltage occurs when the gate voltage has the high value; for the low value, transmission is blocked. The advantage of a transmission gate is that switching is controlled electronically. Several interesting cases that provide additional examples of diode circuit analysis are considered next.

Single-Diode Transmission Gate.   Figure 2.21a displays a single-diode transmission gate. The gate voltage V_{GG} controls the transmission of the signal voltage v_s to the load resistor R_L. We specify the high value of V_{GG} as $+V_A$ and the low value as $-V_B$ in this example (and in all other examples).

Figure 2.21b displays the equivalent circuit for analysis in this case, which is obtained by determining the Thévenin equivalent voltage corresponding to v_s and V_G and adding the results. By using the voltage divider rule for the transmitting gate voltage case ($V_G = V_A$), we have

$$v_L = \left[\left(\frac{R_G}{R_S + R_G} v_s \right) + \left(\frac{R_S}{R_S + R_G} V_A \right) - V_0 \right]$$
$$\times \left[\frac{R_L}{R_L + (R_S \parallel R_G)} \right] \tag{2.4–17}$$

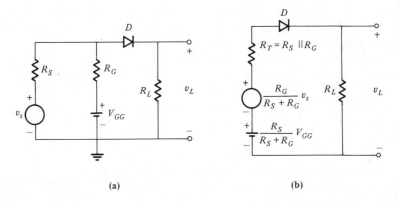

FIGURE 2.21   Single-Diode Transmission Gate: (a) Basic circuit, (b) Equivalent circuit for analysis

where the diode is assumed shorted with a voltage V_0 across its terminals. Furthermore, from Fig. 2.21b, we observe that in order for D to be shorted, the sum of the voltage sources must be greater than the diode offset voltage, or

$$\frac{R_G}{R_S + R_G} v_s + \frac{R_S}{R_S + R_G} V_A \geq V_0 \qquad (2.4\text{–}18)$$

Thus, a relationship for V_A in terms of v_s and V_0 is provided. For the low value of gate voltage ($V_{GG} = -V_B$), we desire that D be open circuited and that the following inequality be true:

$$\frac{R_G}{R_S + R_G} v_s - \frac{R_S}{R_S + R_G} V_B \leq V_0 \qquad (2.4\text{–}19)$$

Thus, a relationship for V_B in terms of v_s and V_0 is provided. The input gate voltage magnitudes (V_A and V_B) must satisfy (2.4–18) and (2.4–19) in order for transmission and blocking of v_s to occur. A restriction is also placed on the range of v_s.

A disadvantage of the single-diode gate is that a DC voltage is present in the load voltage v_L in addition to the signal voltage component. The DC voltage is the combination of the two terms involving V_A and V_0 in (2.4–17). However, this component can be eliminated by using symmetry in a circuit that utilizes two diodes, as we will see next.

Double-Diode Transmission Gate.   Figure 2.22 displays a transmission gate with two identical diodes arranged in such a manner that when both diodes conduct, v_s is transmitted to the load resistor without a DC component. Also shown are two gate voltages of identical magnitude but opposite polarity. The resistors R_D are required in this case because both diodes short (or open) at the same time.

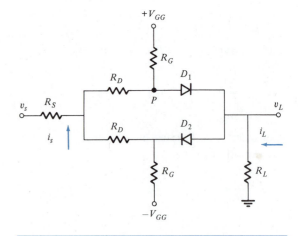

FIGURE 2.22   Double-Diode Transmission Gate

For V_{GG} equal to the high value of voltage (V_A) and considering D_1 and D_2 to be conducting (with a forward voltage V_0), we write the output voltage from Fig. 2.22 as

$$v_L = i_S(R_L \| R_G \| R_G) \qquad (2.4\text{--}20)$$

where, for a balanced circuit,

$$i_S = \frac{v_S}{R_S + (R_D/2) + R_L \| R_G \| R_G} \qquad (2.4\text{--}21)$$

Note that the current components due to $\pm V_{GG}$ and V_0 for diodes D_1 and D_2 cancel one another in the output resistance. Therefore V_0 and $\pm V_{GG}$ are effectively zero to obtain v_L in (2.4–20). Thus, v_L is dependent only upon v_s when transmission is desired.

Restrictions on the magnitudes of V_A and V_B along with the range of v_s also exist in this example. The condition on V_A relative to v_s is obtained by realizing that in order for the diodes to remain shorted, the voltage at P (see Fig. 2.22) must be greater than or equal to V_0. Similarly, the condition on V_B relative to v_s is obtained by using $V_{GG} = -V_B$ and forcing the voltage at P to be less than or equal to V_0. In each case, the voltage at P is written using voltage divider rules. Although simple, these relations are rather lengthy and are therefore omitted here.

Proper operation of the double-diode transmission gate requires matched diodes as well as matched resistors. Matched diodes are obtainable from manufacturers in pairs as well as in quads (groups of four matched diodes) and even higher numbers of pairs, as the diode selection guide indicates in the chapter appendix. These multi-pair diodes, called *diode arrays,* are fabricated in integrated-circuit form to accomplish the exact likeness of the diodes.

Diode Array Transmission Gate. Figure 2.23 shows a diode quad transmission gate. It utilizes an integrated-circuit diode array with four matched diodes. Again, two values of gate voltage

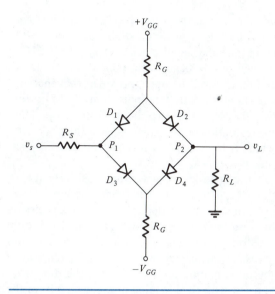

FIGURE 2.23 Diode Quad Transmission Gate

($V_{GG} = V_A$ and $-V_B$) are used to control the transmission and blocking of the signal voltage v_s to and from the load.

Under the condition in which the diodes are open circuits (for $V_{GG} = -V_B$), $v_L = 0$ and transmission of the signal voltage is blocked. For the condition in which the diodes are short circuits (for $V_{GG} = V_A$), the effects of V_A and $-V_B$ cancel and thus v_L is dependent only upon v_s. The load voltage is then easily obtained for the case in which the diodes are shorted by using the voltage divider rule for the loop in Fig. 2.23 containing v_s, R_S, D_1, D_2, and $R_L \| R_G \| R_G$. The identical voltages across D_1 and D_2 (V_0) cancel. Thus,

$$v_L = \frac{R_L \| R_G/2}{R_S + (R_L \| R_G/2)} v_s \qquad (2.4\text{--}22)$$

Note that $v_L \simeq v_s$ if $R_L \| R_G/2 \gg R_S$ or if $R_S = 0$. This result can be accomplished by using an input source arrangement with an operational amplifier, as we will see in Chapter 7.

In the diode quad transmission gate, restrictions on the magnitudes of V_A and V_B relative to v_s

prevail as in the previous examples. We first consider the case in which the diodes are supposed to be off and the signal blocked ($V_{GG} = -V_B$). If we disregard v_s to begin with, then all of the diodes of Fig. 2.23 will be open, provided that the voltage across the gate terminals from top to bottom ($2V_{GG}$) is less than two diode drops ($2V_0$). Thus, the restriction on the gate voltage is that $V_{GG} < V_0$, which is most conveniently accomplished by letting $V_{GG} = -V_B = 0$.

If we consider v_s as varying, positively at first, then as v_s increases, D_3 will eventually short. However, this condition does not create transmission because the current through D_3 and the bottom resistor R_G causes an increase in the reverse bias across D_4. Thus, when D_3 shorts, D_4 remains in the off condition (and so also are the diodes D_1 and D_2) with v_s remaining disconnected from the output. Similarly, as v_s goes negative, D_1 will eventually turn on, but D_2 will remain open along with D_3 and D_4. We note that even though D_3 or D_1 short as v_s varies, v_s is not transmitted to the load resistor. Hence, the condition that $V_{GG} = -V_B < V_0$ is sufficient to block v_s.

For the positive value of control voltage ($V_{GG} = V_A$), there is a restriction on the maximum signal voltage that can be transmitted and maintain the shorted condition of the diodes. If we consider $v_s > 0$, then as v_s increases, D_1 will eventually open. At this voltage, D_4 will also open, which is observed by realizing that the voltages at P_1 and P_2 of Fig. 2.23 are equal. Then, v_s is disconnected from the output and is no longer transmitted. To determine the relation between the required gate voltage and v_s for signal transmission, we consider the case where v_s has increased to the value that just cuts off D_1 and D_4. The maximum signal voltage that can be transmitted is defined as $v_{S\,max}$. Under these conditions, the voltage at P_1 is obtained (from KVL starting from the $-V_{GG}$ terminal with $V_{GG} = V_A$) as

$$v_{P_1} = \left[\left(\frac{v_{S\,max} - V_0 + V_A}{R_S + R_G} \right) R_G \right] - V_A + V_0 \qquad (2.4\text{-}23)$$

The voltage at P_2 is obtained from Ohm's law as

$$v_{P_2} = \left(\frac{V_A - V_0}{R_G + R_L} \right) R_L \qquad (2.4\text{-}24)$$

Since these voltages are equal, we equate them to obtain

$$\left(\frac{v_{S\,max} - V_0 + V_A}{R_S + R_G} \right) R_G - V_A + V_0$$
$$= \left(\frac{V_A - V_0}{R_G + R_L} \right) R_L \qquad (2.4\text{-}25)$$

Solving for V_A, we have

$$V_A = V_0 + \frac{R_G v_{S\,max}}{R_S + [R_L(R_S + R_G)]/(R_G + R_L)} \qquad (2.4\text{-}26)$$

This equation expresses the desired relationship between V_A and $v_{S\,max}$. If we know the value of $v_{S\,max}$ (the maximum signal voltage to be transmitted), we can then select V_A (the value of gate voltage for transmission) according to (2.4–26). A larger value is often used to ensure transmission.

Quite often the input signal voltage can be applied such that $R_S = 0$. For this special case, (2.4–26) becomes

$$V_A = V_0 + \left(\frac{R_G + R_L}{R_L} \right) v_{S\,max} \qquad (2.4\text{-}27)$$

which indicates that the value of V_A for transmission of v_s must be greater in magnitude than V_0 and also $v_{S\,max}$.

Because of the symmetry of the network, the same restrictions apply for a negative signal voltage. Equation (2.4–26) continues to hold, with $v_{S\,max}$ replaced with the maximum magnitude of the negative signal voltage.

EX. 2.8

Diode Quad Transmission Gate Example

The basic circuit of Fig. 2.23 is used with a silicon diode array. The input signal voltage is sinusoidal and given by

$v_s = 5 \sin \omega t$ V. Determine low and high values of V_{GG} for $R_S = 0$ and $R_G = R_L = 1$ kΩ.

Solution: The value for $V_{GG} = -V_B$ must be less than or equal to $V_0 = 0.7$ V. We select the convenient value of $V_B = 0$.

From (2.4–27) and the fact that the diodes are made of silicon, we have

$$V_A = 0.7 + \left[\left(\frac{1+1}{1} \right)(5) \right] = 10.7 \text{ V}$$

To ensure transmission, we chose $V_A = 15$ V. Thus, v_L for this condition, from (2.4–22), is

$$v_L = v_s = 5 \sin \omega t \text{ V}$$

2.5   SELECTION GUIDE FOR RECTIFIER DIODES, DUAL DIODES, AND DIODE ARRAYS

To exemplify the types of diodes that are commercially available, Data Sheets A2.1 and A2.2 in the appendix to this chapter display a selection guide for rectifier diodes as well as small-signal dual diodes and diode arrays. (These data sheets are used through the courtesy of Motorola, Inc.) Common diode packages are indicated in the data sheets. Note that the outer portion of diode packages consists of either glass, plastic, or metal.

The rectifier diodes in Data Sheet A2.1 are listed under four different classifications: Schottky rectifiers, ultrafast-recovery rectifiers, fast-recovery rectifiers, and general-purpose rectifiers. Parameters that require additional explanation are as follows:

- V_{RRM} is the maximum reverse recovery voltage, or reverse voltage that must not be exceeded. (Although it is similar to the zener voltage, we do not refer to the zener voltage of a diode used in rectifier applications.)
- I_{FSM} is the nonrepetitive peak surge current.

- T_C is the case temperature (°C).
- T_L is the lead temperature (°C).
- T_J is the junction temperature (°C).
- Max V_F @ $I_{FM} = I_O$ is the maximum forward voltage at maximum forward current.

Note the various packages used for the small-signal dual diodes and diode arrays, as indicated in Data Sheet A2.2. Except for the first package (TO-92 plastic), all of the other packages are standard IC packages. Note, in particular, the packages numbered case 620, 632, 646, and 648. These packages are called *dual-in-line packages* (DIP) because the pins are in line and on either side of the package.

More complete device specifications are given on data sheets that are available from manufacturers. Individual diode data sheets for several representative diodes indicating more detailed information are included in Appendix A at the back of this book.

2.6   SELECTION GUIDE FOR ZENER DIODES

To exemplify the types of zener diodes that are commercially available, Data Sheet A2.3 at the end of this chapter displays a selection guide for general-purpose regulator diodes. These devices are obtainable in various packages, as indicated in the data sheet.

For low-power devices (<5 W), glass, ceramic, or small metal can encapsulation is adequate. However, for high-power devices (>5 W), metal cases 54, 56, or 58 are required because each of these cases allows faster heat transfer away from the diode.

Note the various zener voltage values listed in Data Sheet A2.3. Nominal zener voltages are available from 1.8 V to 200 V. Complete diode data sheets for a few zener diodes are included in Appendix 1 at the back of the book.

CHAPTER 2 SUMMARY

■ Semiconductor materials have electronic properties in the intermediate range between conductors and insulators. Silicon is used in the fabrication of ICs because precise growth of SiO_2 provides a mask for subsequent surface treatments.

■ Gallium arsenide with added impurity atoms is a semiconductor with light-emitting properties and is used in optoelectronic devices and high-frequency ICs.

■ An intrinsic semiconductor consists of a uniform array of semiconductor atoms. The concentration of free electrons and holes (positive charges) is given by the intrinsic concentration n_i. For Si, $n_i = 1.5 \times 10^{10}/cm^3$ at room temperature.

■ Extrinsic materials are semiconductors with impurity atoms intentionally added. The addition of donor atoms provides additional free electrons, and the semiconductor is an N-type semiconductor. The addition of acceptor atoms provides additional holes, and the semiconductor is a P-type semiconductor.

■ A PN junction diode is a passive, two-terminal nonlinear circuit element. The theoretical current–voltage expression is given by

$$i_D = I_S(e^{v_D/v_T} - 1)$$

where i_D is defined to be positive in the direction of P to N and v_D is defined as positive on P relative to N. For v_D positive, i_D is positive and can be large, which is the forward-bias case. For v_D negative, i_D is negative and is very small, unless v_D is equal to the breakdown voltage V_{BR}. At breakdown, the reverse current is essentially independent of voltage and can be large.

■ A zener diode is a PN junction diode that operates in the breakdown region. A zener diode is used to regulate the voltage between two points in a circuit and is often called a voltage regulator.

A Schottky diode is a two-terminal metal-semiconductor diode. Si MN Schottky diodes

have a low forward voltage of approximately 0.3 V and exhibit fast switching characteristics. An ideal diode is a first-order approximation for a PN junction diode. By adding an offset

■ voltage and a resistor, a more accurate linear circuit model is obtained. For representation of the forward-bias region, an offset voltage V_0 and an ideal diode in series are usually adequate. For the reverse-bias region, an ideal diode with reversed polarity and a negative offset voltage $-V_{BR}$ are most often used.

For single-diode circuits under DC conditions, a graphical solution provides insight into ob-

■ taining the DC operating point called Q. When AC is also present, a large-signal analysis is required if the diode changes states and operates in the forward and reverse regions. For AC operation in which the diode remains in one state, a small-signal analysis is used. The diode is then represented by its AC resistance r_d, given graphically by the magnitude of the inverse slope of i_d versus v_d at Q. The diode AC resistance is also obtainable from $r_d = 0.025/I_D$. The DC diode current is obtained from the DC analysis.

Half-wave, full-wave, and peak rectifiers are diode circuits that produce DC from a purely

■ AC input. The waveshapes are distorted because of the nonidealities of the PN diode.

The diode limiter or clipper circuit restricts the output voltage to a prescribed maximum

■ and/or minimum. The diode clamper shifts the voltage waveshapes to a desired level.

Diode transmission gates are switches that allow transmission of a signal voltage through

■ a network when a control or gate voltage is applied. Single diodes, double diodes, and diode arrays are used in transmission gates.

CHAPTER 2 PROBLEMS

2.1

The formula for the resistance of a rectangular block of material is given by $R = \rho L/wt$, where ρ is the resistivity (in Ω-cm) of the material and L, w, and t are the length, width, and thickness of the block, respectively. Calculate the resistance of the block for the elemental metals listed in Table 2.1. Let $L = 1$ cm and $w = t = 0.1$ cm.

2.2

Repeat Problem 2.1 for the insulator ranges of Table 2.1.

2.3

Repeat Problem 2.1 for the semiconductor ranges of Table 2.1.

2.4

Two basic laws of physics for semiconductor materials under equilibrium conditions are (1) the law of mass action, $np = n_i^2$ and (2) the charge neutrality relation, $n + N_a = p + N_d$. Solve these relations for n and p in terms of n_i and $(N_d - N_a)$. Then, obtain approximate expressions for n and p for the following two cases: (a) $|N_d - N_a| \gg n_i$ and (b) $|N_d - N_a| \ll n_i$. Note that n and p must be positive and real.

2.5

Calculate the concentration of electrons (n) and holes (p) in an N-type Si substrate with $N_d = 10^{15}/cm^3$ and $N_a = 0$. Use the results of approximation (a) in Problem 2.4.

2.6

Repeat Problem 2.5 for a P-type Si substrate with $N_a = 7.5 \times 10^{15}/cm^3$ and $N_d = 0$. Use the results of approximation (a) in Problem 2.4.

2.7

Repeat Problems 2.5 and 2.6 for Ge substrates.

2.8

Repeat Problems 2.5 and 2.6 for GaAs substrates.

2.9

The conductivity of a semiconductor is given by the relation $\sigma = q(\mu_n n + \mu_p p)$, where q is the electronic charge (1.6×10^{-19} C) and μ_n and μ_p are the electron and hole mobilities in the semiconductor, respectively. Calculate the conductivity for an N-type Si substrate with $N_d = 10^{15}/cm^3$, and compare this result with the conductivity range in Table 2.1. [Hint: Approximate the actual (reduced) mobilities by using the intrinsic mobilities, which are given by $\mu_n = 1350$ cm^3/V·s and $\mu_p = 480$ cm^2/V·s for Si.]

2.10

Repeat Problem 2.9 for the case of a P-type Si substrate with $N_a = 7.5 \times 10^{15}/cm^3$.

2.11

Repeat Problem 2.9 for the case of an N-type GaAs substrate with $N_d = 10^{15}/cm^3$. Use the intrinsic mobility values for GaAs, which are $\mu_n = 8500$ cm^2/V·s and $\mu_p = 400$ cm^2/V·s.

2.12

Figure P2.12 displays a measured i–v characteristic along with a piecewise linear approximate characteristic. Determine the equivalent circuit for the piecewise linear representation.

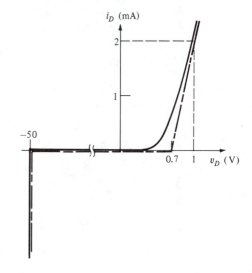

FIGURE P2.12

2.13

The diode of Problem 2.12 is used in the circuit shown in Fig. P2.13. Determine i_D and v_D using the diode piecewise linear equivalent circuit.

2.14

Repeat Problem 2.13 with $V_{DC} = -100$ V.

2.15

Repeat Problems 2.13 and 2.14 using the graphical load line method of solution.

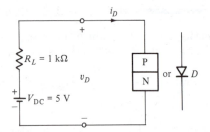

FIGURE P2.13

2.16

Determine i_D and v_D in the circuit of Fig. P2.13 using $R_L = 2\ \text{k}\Omega$ and $V_{DC} = 20$ V. Use the piecewise linear equivalent circuit representation for D under each of the following conditions:

(a) D is ideal

(b) D can be represented by an ideal diode and DC voltage $V_0 = 0.7$ V

(c) D can be represented by an ideal diode, a DC voltage (V_0), and a resistor $R_D = 150\ \Omega$

Is the diode resistance negligible?

2.17

For the circuit of Fig. P2.17, draw the voltage transfer characteristic v_L versus v_{IN}. Assume D is ideal.

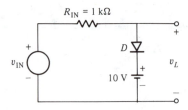

FIGURE P2.17

2.18

Repeat Problem 2.17 with D represented with an ideal diode, a DC voltage $V_0 = 0.7$ V (with proper polarity), and a resistor $R_D = 150\ \Omega$.

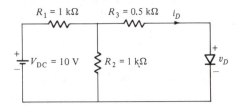

FIGURE P2.19

2.19

For the circuit of Fig. P2.19, determine R_T and V_T for the resistive network with V_{DC}. Assuming D is ideal, what is the diode current i_D?

2.20

Figure P2.20 displays a circuit in which the ideal diode and large resistor (500 kΩ) are used to represent an actual PN diode. Determine the output current i_L and output voltage v_L.

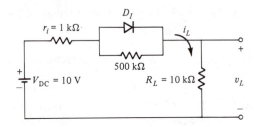

FIGURE P2.20

2.21

Determine i_L and v_L in the circuit of Fig. P2.20 for $V_{DC} = -10$ V.

2.22

For the circuit of Fig. P2.22, which uses the alternate form for representing a DC voltage, determine the output current i_L and voltage v_L, where D is ideal.

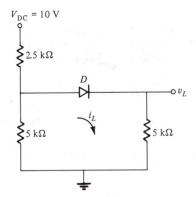

FIGURE P2.22

2.23

The circuit of Fig. P2.23 is a diode limiter circuit. Determine the transfer characteristic v_L versus v_{IN}, considering the diodes to be ideal. Repeat this anal-

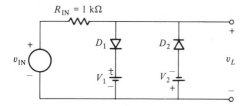

FIGURE P2.23

ysis for the case in which each diode has an offset voltage of $V_0 = 0.7$ V.

2.24

Repeat Problem 2.23 for the circuit shown in Fig. P2.24.

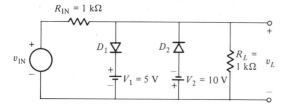

FIGURE P2.24

2.25

For the circuit of Fig. P2.23, let $V_1 = V_2 = 2$ V and $v_{IN} = 10 \sin \omega t$. Sketch the resulting load voltage waveshape versus ωt, assuming ideal diodes. Sketch the waveshape when the diode offset voltages ($V_0 = 0.7$ V) are accounted for. Repeat for $V_1 = 2$ V and $V_2 = 10$ V.

2.26

Sketch one period of the output voltage v_L for each of the circuits shown in Fig. P2.26, assuming the diodes are ideal, $V_{DC} = 5$ V, and $v_s = 10 \sin \omega t$ V.

2.27

For the rectifier circuit of Fig. P2.27, sketch v_L versus ωt, assuming that the PN diode is made of Si with $V_0 = 0.7$ V and $R_D = 100$ Ω. Use an equivalent circuit to derive equations for i_D and v_L. Let $v_s = 10 \sin \omega t$.

2.28

For the half-wave rectifier circuit of Problem 2.27, sketch v_D versus ωt.

2.29

Consider a full-wave bridge rectifier that uses four identical Si PN diodes. For $R_L = 0.5$ kΩ, determine the current and voltage for R_L, assuming ideal diodes

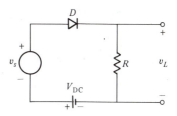

(a)

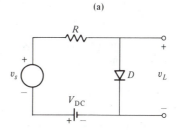

(b)

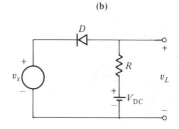

(c)

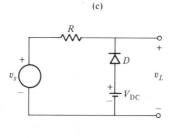

(d)

FIGURE P2.26

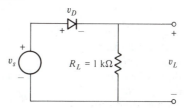

FIGURE P2.27

and $v_p = 10 \sin \omega t$ V. Sketch one period of the rectified voltage across R_L.

2.30
Repeat Problem 2.29 assuming each diode has an offset voltage $V_0 = 0.7$ V, resistance $R_D = 100\ \Omega$, and a breakdown voltage $V_{BR} = 100$ V.

2.31
Repeat Problem 2.29 for the case of a full-wave center-tap transformer rectifier circuit with two identical Si PN diodes.

2.32
Repeat Problem 2.30 for the case of a full-wave center-tap transformer rectifier.

2.33
For the peak rectifier circuit of Fig. P2.33, derive equations for i_D, v_C, and v_D, assuming the diode is ideal and C is initially uncharged. Let $v_s = 10 \sin 2\pi(60)t$ and $C = 1\ \mu\text{F}$. Sketch the waveshapes.

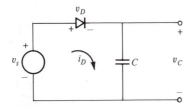

FIGURE P2.33

2.34
Repeat Problem 2.33 for the case in which the diode offset voltage is included ($V_0' = 0.7$ V).

2.35
For the circuit of Fig. P2.24, determine the output voltage v_L, for $v_{IN} = 20 + \sin \omega t$ V. Assume that the diodes D_1 and D_2 are Si diodes with an offset voltage $V_0 = 0.7$ V and an AC resistance given by $r_d = 0.025/I_D$. Obtain the DC solution by using an equivalent circuit for the diodes. From this solution, obtain the AC small-signal solution and the total result for v_L.

2.36
For the circuit of Fig. P2.36, where $v_S = 1 \sin \omega t$ V. Assume that the capacitor acts like a short circuit to AC and an open circuit to DC. Representing the diode with an offset voltage $V_0 = 0.7$ V and the AC resistance r_d, determine the following:
(a) the DC diode current

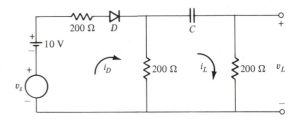

FIGURE P2.36

(b) the AC diode resistance
(c) the output load voltage

2.37
The diode in the circuit of Fig. P2.37 can be represented by Shockley's current–voltage relation. For $v_s = 0.1 \sin \omega t$ V, determine the following:
(a) the DC diode current
(b) the AC diode resistance
(c) the output voltage
Let the reverse saturation current $I_S = 1\ \mu\text{A}$ and $q/kT = 40$. Use a Thévenin equivalent circuit to obtain the DC solution as well as the AC solution.

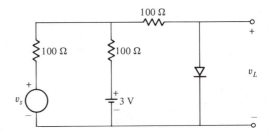

FIGURE P2.37

2.38
A zener diode is used to limit the current and voltage of the load resistor R_L in the circuit of Fig. P2.38. Determine the maximum load current i_L. Assume that the zener diode has $2\ \Omega$ resistance when it is operating under breakdown conditions.

2.39
For the single-diode transmission gate of Fig. 2.21a, let $R_S = 10$ kΩ, $R_L = 10$ kΩ, and $V_{GG} = 5$ V. Assume that the diode is shorted with an offset voltage $V_0 = 0.7$ V. Determine the DC and AC components of v_L. Assuming $v_s = V_S \sin \omega t$, what is the maximum value for V_S that will be transmitted?

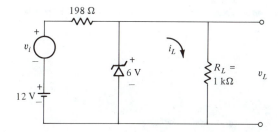

FIGURE P2.38

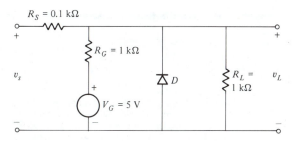

FIGURE P2.41

2.40

For the single-diode transmission gate of Fig. 2.21a, let $R_S = 100\ \Omega$, $R_L = R_G = 1\ k\Omega$, $V_G = 5\ V$, and $v_s = 5\sin \omega t$. Sketch one period of the resulting output voltage v_L using $V_0 = 0.7\ V$.

2.41

Repeat Problem 2.40 for the transmission gate displayed in Fig. P2.41. In this case, D must be cut off for transmission to occur.

2.42

For the diode gate of Fig. P2.41 with $V_G = 10\ V$, determine the minimum and maximum values of v_L. Let $v_s = 1.0\sin \omega t\ V$.

2.43

For the double-diode transmission gate of Fig. 2.22, let $R_S = R_G = R_D = R_L = 10\ k\Omega$ and $V_G = 5\ V$. Assume that the diodes are represented by the offset voltage $V_0 = 0.7\ V$ when conducting, and determine the output voltage v_L in terms of v_s.

2.44

For the diode quad transmission gate of Fig. 2.23, write the expression for v_L in terms of V_G, v_s, and the resistors. Calculate v_L for $V_{GG} = 10\ V$, $R_S = 1\ k\Omega$, $R_G = 1\ M\Omega$, $R_L = 1\ k\Omega$, and $v_s = 5\sin \omega t\ V$.

CHAPTER 2 APPENDIX

DATA SHEET A2.1   Rectifier Diodes

Rectifiers

Schottky Rectifiers

SWITCHMODE Schottky Power Rectifiers with the high speed and low forward voltage drop characteristic of Schottky's metal/silicon junctions are produced with ruggedness and temperature performance comparable to silicon-junction rectifiers. Ideal for use in low voltage, high frequency power supplies and as very fast clamping diodes, these devices feature switching times less than 10 ns, and are offered in current ranges from 0.5 to 300 amperes, and reverse voltages to 45 volts.

In some current ranges, devices are available with junction temperature specifications of 125°C, 150°C, 175°C. Devices with higher T_J ratings can have significantly lower leakage currents, but higher forward-voltage specifications. These parameter tradeoffs should be considered when selecting devices for applications that can be satisfied by more than one device type number. Detailed specifications are available on the individual data sheets.

There are many other standard features in Motorola Schottky rectifiers that give added performance and reliability.

1. GUARDRINGS are included in all Schottky die for reverse voltage stress protection from high rates of dv/dt to virtually eliminate the need for snubber networks. The guardring also operates like a zener and avalanches when subjected to voltage transients.

2. MOLYBDENUM DISCS on both sides of the die minimize fatigue from power cycling in all metal product. The plastic TO-220 devices have a special solder formulation for the same purpose.

3. QUALITY CONTROL monitors all critical fabrication operations and performs selected stress tests to assure constant processes.

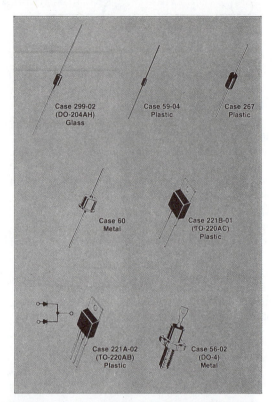

V_{RRM} (Volts)	I_O, AVERAGE RECTIFIED FORWARD CURRENT (Amperes)					
	0.5	1.0	3.0	5.0	7.5	10
	299-02 (DO-204AH) Glass	59-04 Plastic	267 Plastic	60 Metal	221B-01 (TO-220AC) Plastic	
20		1N5817	1N5820	1N5823		
30	MBR030	1N5818	1N5821	1N5824		
35					MBR735	MBR1035
40	MBR040	1N5819	1N5822	1N5825		
45					MBR745	MBR1045
I_{FSM} (Amps)	5.0	25	80	500	150	150
†T_C @ Rated I_O (°C)					105	135
†T_L @ Rated I_O (°C)	75	90	95	80		
T_J (Max) (°C)	150	125	125	125	150	150
Max V_F @ $I_{FM} = I_O$	*0.65 $T_L = 25$°C	*0.60 $T_L = 25$°C	*0.525 $T_L = 25$°C	*0.38 $T_C = 25$°C	0.57 @ 7.5A $T_C = 125$°C	0.72 @ 20 A $T_C = 125$°C

☐ TX versions available.
*   Values are for the 40-Volt units. The lower voltage parts provide lower limits.
** I_O is total device output.
†   Must be derated for reverse power dissipation. See Data Sheet.

Copyright of Motorola, Inc. Used by permission.

DATA SHEET A2.1   Continued

V_{RRM} (Volts)	I_O, AVERAGE RECTIFIED FORWARD CURRENT (Amperes)					
	15		16	20	25	
	221A-02 (TO-220AB) Plastic Dual Diode**	56-02 (DO-4) Metal	221B-01 (TO-220AC) Plastic	221A-02 (TO-220AB) Plastic Dual Diode**	56-02 (DO-4) Metal	
20		1N5826			1N5829	
30		1N5827			1N5830	1N6095
35	MBR1535CT		MBR1635	MBR2035CT		
40		1N5828			1N5831	1N6096
45	MBR1545CT		MBR1645	MBR2045CT		
I_{FSM} (Amps)	150	500	300	150	800	400
†T_C @ Rated I_O (°C)	105	85	125	135	85	70
†T_C @ Rated I_O (°C)	105	85	125	135	85	70
T_J (Max) (°C)	150	125	150	150	125	125
Max V_F @ $I_{FM} = I_O$	0.70 @ 15 A $T_C = 125°C$	*0.50 $T_C = 25°C$	0.60 @ 16 A $T_C = 125°C$	0.72 @ 20 A $T_C = 125°C$	*0.48 $T_C = 25°C$	0.86 @ 78.5 A $T_C = 70°C$

□ TX versions available.
*   Values are for the 40-Volt units. The lower voltage parts provide lower limits.
** I_O is total device output.
†   Must be derated for reverse power dissipation. See Data Sheet.

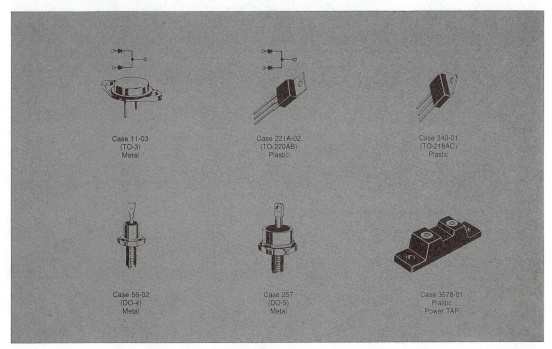

Case 11-03
(TO-3)
Metal

Case 221A-02
(TO-220AB)
Plastic

Case 340-01
(TO-218AC)
Plastic

Case 56-02
(DO-4)
Metal

Case 257
(DO-5)
Metal

Case 3578-01
Plastic
Power TAP

continues

DATA SHEET A2.1   Continued

	I_O, AVERAGE RECTIFIED FORWARD CURRENT (Amperes)					
	30			35	40	50
V_{RRM} (Volts)	11-03 (TO-3) Metal Dual Diode** (40 Mil Pins)	221A-02 (TO-220AB) Plastic Dual Diode**	340-01 (TO-218AC) Plastic Dual Diode**	56-02 (DO-4) Metal	257 (DO-5) Metal	
20					1N5832	
30					1N5833	1N6097
35	MBR3035CT	MBR2535CT	MBR3035PT	MBR3535		
40					1N5834	1N6098
45	SD241 MBR3045CT	MBR2545CT	MBR3045PT	SD41 MBR3545		
I_{FSM} (Amps)	400	300	400	600	800	800
†T_C @ Rated I_O (°C)	105	125	105	90	75	70
T_J (Max) (°C)	150	150	150	150	125	125
Max V_F @ $I_{FM} = I_O$	0.72	0.73 @ 30 A	0.72 @ 30 A	0.70 @ 78.5 A $T_C = 25$°C	0.59* $T_C = 25$°C	0.86 @ 157 A $T_C = 70$°C

	I_O, AVERAGE RECTIFIED FORWARD CURRENT (Amperes)						
	60	65	75	80	120	200	300
V_{RRM} (Volts)			257 (DO-5) Metal			357B-01 Plastic Power Tap	
35	MBR6035	MBR6535	MBR7535	MBR8035	MBR12035CT	MBR20035CT	MBR30035CT
40					1N6458	1N6460	
45	SD51 MBR6045	MBR6545	MBR7545	MBR8045	MBR12045CT 1N6457	MBR20045CT 1N6459	MBR30045CT
I_{FSM} (Amps)	800	800	1000	1000	1500	1500	2500
†T_C @ Rated I_O (°C)	90	120	90	120	140	140	140
T_J (Max) (°C)	150	175	150	175	175	175	175
Max V_F @ $I_{FM} = I_O$	0.80 @ 157 A	0.62 @ 65 A $T_C = 150$°C	0.90 @ 220 A	0.59 @ 80 A $T_C = 150$°C	0.68 @ 120 A	0.68 @ 200 A	0.165 @ 200 A

☐ TX versions available.

* Values are for the 40-Volt units. The lower voltage parts provide lower limits.

** I_O is total device output.

† Must be derated for reverse power dissipation. See Data Sheet.

DATA SHEET A2.1 Continued

Ultrafast Recovery Rectifiers

EXPANDING the SWITCHMODE Rectifier family are these ultrafast devices with reverse recovery times of 25 to 100 nanoseconds. They complement the broad Schottky offering for use in the higher voltage outputs and internal circuitry of switching power supplies as operating frequencies increase from 20 kHz to 250 kHz. Additional package styles and operating current levels are planned.

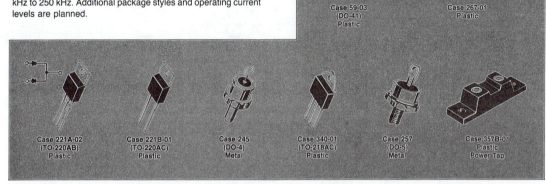

Case 59-03 (DO-41) Plastic Case 267-01 Plastic

Case 221A-02 (TO-220AB) Plastic Case 221B-01 (TO-220AC) Plastic Case 245 (DO-4) Metal Case 340-01 (TO-218AC) Plastic Case 257 (DO-5) Metal Case 357B-01 Plastic Power Tap

V_{RRM} (Volts)	I_O, AVERAGE RECTIFIED FORWARD CURRENT (Amperes)					
	1.0	4.0	6.0	8.0	15	16
	59-03 (DO-41) Plastic	267-01 Plastic	221A-02 (TO-220AB) Plastic Dual Diode**	221B-01 (TO-220AC) Plastic		221A-02 (TO-220AB) Plastic Dual Diode**
50	MUR105	MUR405	MUR605CT	MUR805	MUR1505	MUR1605CT
100	MUR110	MUR410	MUR610CT	MUR810	MUR1510	MUR1610CT
150	MUR115	MUR415	MUR615CT	MUR815	MUR1515	MUR1615CT
200	MUR120		MUR620CT	MUR820	MUR1520	MUR1620CT
400				MUR840	MUR1540	
500				MUR850	MUR1550	
600				MUR860	MUR1560	
I_{FSM} (Amps)	35	125	75	100	200	100
T_A @ Rated I_O (°C)	50	80				
T_C @ Rated I_O (°C)			130	150	150	150
T_J (Max) (°C)	175	175	175	175	175	175
t_{rr} ns	35	35	35	35/60	35/60	35

** I_O per leg is half.
Reverse Polarity (Anode-To-Case) indicated with an "R" Suffix.

continues

DATA SHEET A2.1   Continued

V_RRM (Volts)	I_O, AVERAGE RECTIFIED FORWARD CURRENT (Amperes)				
	25	30		50	100
	245 (DO-4) Metal	340-1 (TO-218AC) Plastic Dual Diode**		257 (DO-5) Metal	357B-01 Plastic Power Tap Dual Diode**
50	MUR2505	R710XPT	MUR3005PT	MUR5005	MUR10005CT
100	MUR2510	R711XPT	MUR3010PT	MUR5010	MUR10010CT
150	MUR2515		MUR3015PT	MUR5015	MUR10015CT
200	MUR2520	R712XPT	MUR3020PT	MUR5020	MUR10020CT
400		R714XPT			
I_FSM (Amps)	500	150	400	600	400
T_C @ Rate I_O (°C)	145	100	150	125	150
T_J (Max) (°C)	175	150	175	175	175
t_rr ns	50	100	35	50	50

** I_O per leg is half.
  Reverse Polarity (Anode-To-Case) indicated with an "R" Suffix.

DATA SHEET A2.1   Continued

Rectifiers – Fast Recovery

. . .available for designs requiring a power rectifier having maximum switching times ranging from 200 ns to 750 ns. These devices are offered in current ranges of 1.0 to 50 amperes and in voltages to 1000 volts.

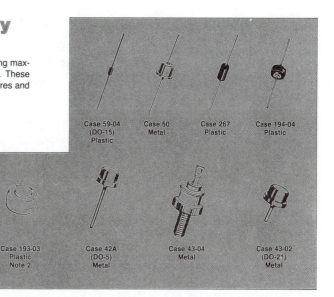

Case 59-04 (DO-15) Plastic — Case 60 Metal — Case 267 Plastic — Case 194-04 Plastic

Case 245 (DO-4) Metal — Case 339 Plastic Note 1 — Case 193-03 Plastic Note 2 — Case 42A (DO-5) Metal — Case 43-04 Metal — Case 43-02 (DO-21) Metal

	I_O, AVERAGE RECTIFIED FORWARD CURRENT (Amperes)					
	1.0		60		3.0	5.0
V_{RRM} (Volts)	59-04 Plastic		60 Metal		267-01 Plastic	194-04 Plastic
50	†1N4933	MR810	MR830	MR850	MR910	MR820
100	†1N4934	MR811	MR831	MR851	MR911	MR821
200	†1N4935	MR812	MR832	•MR852	MR912	MR822
400	†1N4936	MR814	MR834	MR854	MR914	MR824
600	†1N4937	MR816	MR836	MR856	MR916	MR826
800		MR817			MR917	
1000		MR818			MR918	
I_{FSM} (Amps)	30	30	100	100	100	300
T_A @ Rated I_O (°C)	75	75		*90	*90	*55
T_C @ Rated I_O (°C)		100	100			
T_J (Max) (°C)	150	150	150	175	175	175
t_{rr} (μs)	0.2	0.75	0.2	0.2	0.75	0.2

☐   TX versions available.
Note 1.   Meets mounting configuration of TO-220 outline.
Note 2.   Braided lead top terminal configuration available; consult your Sales Representative.
** I_O is total device output.
*   Must be derated for reverse power dissipation. See Data Sheet.
†   Package Size: 0.120″ Mx Diameter by 0.260″ Max Length.

continues

DATA SHEET A2.1   Continued

| V_RRM (Volts) | I_O. AVERAGE RECTIFIED FORWARD CURRENT (Amperes) | | | | | | |
| | 6.0 | 12 | 20 | 24 | 30 | 40 | 50 |
	245 (DO-4) Metal Note 1		42A (DO-5) Metal	339 Plastic Note 1	42A (DO-5) Metal Note 2	257 (DO-5) Metal Note 2	
50	1N3879	1N3889	1N3899	MR2400F	1N3909	MR860	MR870
100	1N3880	1N3890	1N3900	MR2401F	1N3910	MR861	MR871
200	1N3881	1N3891	1N3901	MR2402F	1N3911	MR862	MR872
400	1N3883	1N3893	1N3903	MR2404F	1N3913	MR864	MR874
600	MR1366	MR1376	MR1386	MR2406F	MR1396	MR866	MR876
I_FSM (Amps)	150	200	250	300	300	350	400
T_C @ Rated I_O (°C)	100	100	100	125	100	100	100
T_J (Max) (°C)	150	150	150	175	150	160	160
t_rr ns	0.2	0.2	0.2	0.2	0.2	0.2	0.2

☐   TX versions available.
Note 1.   Meets mounting configuration of TO-220 outline.
Note 2.   Braided lead top terminal configuration available; consult your Sales Representative.
  **   I_O is total device output.
   *   Must be derated for reverse power dissipation. See Data Sheet.
   †   Package Size: 0.120″ Max Diameter by 0.260″ Max Length.

Rectifiers – General Purpose

Motorola offers wide variety of low-cost devices, packaged to meet diverse mounting requirements. Avalanche capability is available in the axial lead 1.5,3 and 6 amp packages shown below to provide protection from transients.

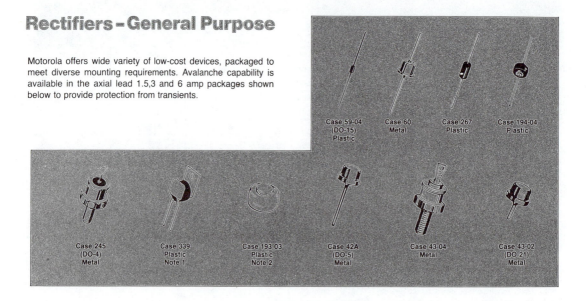

DATA SHEET A2.1   Continued

V_RRM (Volts)	I_O AVERAGE RECTIFIED FORWARD CURRENT (Amperes)					
	1.0	1.5		3.0		6.0
	59-04 (DO-15) Plastic		60 Metal	267 Plastic		194-04 Plastic
50	†1N4001	**1N5391	1N4719	**MR500	1N5400	**MR750
100	†1N4002	**1N5392	1N4720	**MR501	1N5401	**MR751
200	†1N4003	1N5393 *MR5059	1N4721	**MR502	1N5402	**MR752
400	†1N4004	1N5395 MR5060	1N4722	**MR504	1N5404	**MR754
600	†1N4005	1N5397 *MR5061	1N4723	**MR506	1N5406	**MR756
800	†1N4006	1N5398	1N4724	MR508		MR758
1000	†1N4007	1N5399	1N4725	MR510		MR760
I_FSM (Amps)	30	50	300	100	200	400
T_A @ Rated I_O (°C)	75	T_L = 70	75	95	T_L = 105	60
T_J (Max) (°C)	175	175	175	175	175	175

V_RRM (Volts)	I_O AVERAGE RECTIFIED FORWARD CURRENT (Amperes)						
	12	24	25	30		40	50
	245 (DO-4) Metal	339 Plastic Note 1	193-03 Plastic Note 2	43-02 (DO-21) Metal		42A (DO-5) Metal	43-04 Metal
50	MR1120 1N1199,A,B	MR2400	MR2500	1N3491	1N3659	1N1183A	MR5005
100	MR1121 1N1200,A,B	MR2401	MR2501	1N3492	1N3660	1N1184A	MR5010
200	MR1122 1N1202,A,B	MR2402	MR2502	1N3493	1N3661	1N1186A	MR5020
400	MR1124 1N1204,A,B	MR2404	MR2504	1N3495	1N3663	1N1188A	MR5040
600	MR1126 1N1206,A,B	MR2406	MR2506	MR328	Note 3	1N1190A	Note 3
800	MR1128 1N3988		MR2508	MR330	Note 3	Note 3	Note 3
1000	MR1130 1N3990		MR2510	MR331	Note 3	Note 3	Note 3
I_FSM (Amps)	300	400	400	300	400	800	600
T_C @ Rated I_O (°C)	150	125	150	130	100	150	150
T_J (Max) (°C)	190	175	175	175	175	190	195

† Package Size: 0.120" Max Diameter by 0.260" Max Length.
* 1N5059 series equivalent Avalanche Rectifiers.
**Avalanche versions available, consult factory.

Note 1. Meets mounting configuration of TO-220 outline.
Note 2. Request Data Sheet for Mounting Information.
Note 3. Available on special order.

continues

DATA SHEET A2.1   Continued

Rectifier-Bridges

Motorola SUPERBRIDGES offer cost effectiveness and reliability in single phase applications. Chip/leadframe techniques
are used for lower-current types, while the higher current assemblies combine pretested "button" rectifier cells for low assembly cost and high yields. Performance of four individual
diodes is achieved at the cost of only two, with reliability of the
whole assembly comparable to that of a single unit. The higher
current assemblies feature versatile slip-on/solder/wire wrap terminals.

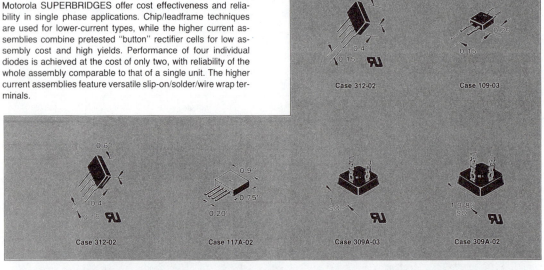

V_{RRM} (Volts)	I_O, DC OUTPUT CURRENT (Amperes)					
	1.0	1.5	2.0	4.0 8.0	25	35
	312-02	109-03	312-02	117A-02	309A-03	309A-02
50	3N246 MDA100A	MDA920A2	3N253 MDA200	MDA970A1	MDA2500	MDA3500
100	3N247 MDA101A	MDA920A3	3N254 MDA201	MDA970A2	MDA2501	MDA3501
200	3N248 MDA102A	MDA920A4	3N255 MDA202	MDA970A3	MDA2502	MDA3502
400	3N249 MDA104A	MDA920A6	3N256 MDA204	MDA970A5	MDA2504	MDA3504
600	3N250 MDA106A	MDA920A7	3N257 MDA206		MDA2506	MDA3506
800	3N251 MDA108A	MDA920A8	3N258 MDA208	CF		MDA3508
1000	3N252 MDA110A	MDA920A9	3N259 MDA210	CF		MDA3510
I_{FSM} (Amps)	30	45	60	100	400	400
T_A @ Rated I_O (°C)	75	50	55	25 @ 4 A		
T_C @ Rated I_O (°C)				55 @ 8 A	55	55
T_J (Max) (°C)	150	175	175	150	175	175

CF: Consult Factory   Dimensions given are nominal

Note 1. The MDA970A series replaces the MCDA970 in the new Case 117A-02, which has minor changes over the old Case 117.

DATA SHEET A2.2   Dual Diodes and Diode Arrays

Small-Signal – Diodes

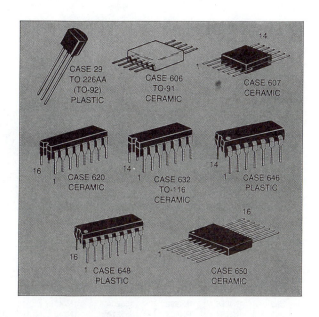

Dual Diodes

Dual diodes designed for use in low cost biasing, steering, and voltage doubler applications.

Device	Description	$V_{(BR)}$ @ Volts Min	$I_{(BR)}$ μA	I_R @ μA Max	V_R Volts	V_F @ Volts Min/Max	I_F mA	$C_{VR} = 0$ pF Max	t_{rr} ns Max	Package
MSD6100	Switching	100	100	0.1	50	0.67/0.82	10	1.5	4.0	29-02 TO-226AA
MSF6102	Common Cathode	70	100	0.1	50	0.67/1.0	10	3.0	100	
MSD6150	Common Anode	70	100	0.1	50	−/1.0	10	8.0	100	

Diode Arrays

These diode arrays are multiple diode junctions fabricated by a planar process and mounted in integrated circuit packages for use in high-current, fast-switching core-driver applications. These arrays offer many of the advantages of integrated circuits such as high-density packaging and improved reliability.

Device	Description	$V_{(BR)}$ Volts Min	I_R μA Max	V_F @ Volts Max	I_F mA	V_{FM} Volts Max	t_{fr} ns Typ	t_{rr} ns Typ	Package
MAD130	Dual 10-Diode	40	0.5	1.1/1.5	100/500	5.0	20	8.0	632, 646
MAD1103	Dual 8-Diode	50	0.5	1.1/1.5	100/500	5.0	20	8.0	606, 632, 646
MAD1107	Dual 8-Diode	50	0.5	1.1/1.5	100/500	5.0	20	8.0	607, 632, 646
MAD1108	8-Diode Array	50	0.1	1.1/1.5	100/500	5.0	20	8.0	620, 648, 650

Copyright of Motorola, Inc. Used by permission.

DATA SHEET A2.3   General-Purpose Regulators: Zener Diodes

Zener Diodes

| | Glass DO-204AH (DO-35) | Glass Case 59 (DO-41) | Metal Case 52 (DO-13) | Surmetic 30 Case 59 (DO-41) |

Nominal Zener Voltage (Note 1)	250 mW Low Level Cathode = Polarity Mark (Notes 2,3)	250 mW Low Noise Cathode = Polarity Mark (Notes 2,3,5)	400 mW Low Noise Low Leakage Cathode = Polarity Mark (Notes 2,4,5)	500 mW Cathode = Polarity Mark (Notes 2,5)	(Notes 2,6)	(Notes 1,2,13)	1 Watt Cathode = Polarity Mark (Notes 2,7)	1 Watt Cathode to Case (Notes 2,8)	1.5 Watt Cathode = Polarity Mark (Notes 2,9)
			Glass DO-204AH (DO-35)				Glass Case 59 (DO-41)	Metal Case 52 (DO-13)	Surmetic 30 Case 59 (DO-41)
1.8	1N4678	1N4614							
2.0	1N4679	1N4615							
2.2	1N4680	1N4616							
2.4	1N4681	1N4617		1N4370	1N5221	1N5985A			
2.7	1N4682	1N4618		1N4371	1N5223	1N5986A			
3.0	1N4683	1N4619		1N4372	1N5225	1N5987A			
3.3	1N4684	1N4620	1N5518A	1N746	1N5226	1N5988A	1N4728	1N3821	1N5913A
3.6	1N4685	1N4621	1N5519A	1N747	1N5227	1N5989A	1N4729	1N3822	1N5914A
3.9	1N4686	1N4622	1N5520A	1N748	1N5228	1N5990A	1N4730	1N3823	1N5915A
4.3	1N4687	1N4623	1N5521A	1N749	1N5229	1N5991A	1N4731	1N3824	1N5916A
4.7	1N4688	1N4624	1N5522A	1N750	1N5230	1N5992A	1N4732	1N3825	1N5917A
5.1	1N4689	1N4625	1N5523A	1N751	1N5231	1N5993A	1N4733	1N3826	1N5918A
5.6	1N4690	1N4626	1N5524A	1N752	1N5232	1N5594A	1N4734	1N3827	1N5919A
6.2	1N4691	1N4627	1N5525A	1N753	1N5234	1N5995A	1N4735	1N3828	1N5920A
6.8	1N4692	1N4099	1N5526A	1N754 1N957A	1N5235	1N5996A	1N4736	1N3829 1N3016A	1N5921A
7.5	1N4693	1N4100	1N5527A	1N755 1N958A	1N5236	1N5997A	1N4737	1N3830 1N3017A	1N5922A
8.2	1N4694	1N4101	1N5528A	1N756 1N959A	1N5237	1N5998A	1N4738	1N3018A	1N5923A
8.7	1N4695	1N4102			1N5238				
9.1	1N4696	1N4103	1N5529A	1N757 1N960A	1N5239	1N5999A	1N4739	1N3019A	1N5924A
10	1N4697	1N4104	1N5530A	1N758 1N961A	1N5240	1N6000A	1N4740	1N3020A	1N5925A
11	1N4698	1N4105	1N5531A	1N962A	1N5241	1N6001A	1N4741	1N3021A	1N5926A
12	1N4699	1N4106	1N5532A	1N759 1N963A	1N5242	1N6002A	1N4742	1N3022A	1N5927A
13	1N4700	1N4107	1N5533A	1N964A	1N5243	1N6003A	1N4743	1N3023A	1N5928A
14	1N4701	1N4108	1N5534A		1N5244				
15	1N4702	1N4109	1N5535A	1N965A	1N5245	1N6004A	1N4744	1N3024A	1N5929A
16	1N4703	1N4110	1N5536A	1N966A	1N5246	1N6005A	1N4745	1N3025A	1N5930A
17	1N4704	1N4111	1N5537A		1N5247				
18	1N4705	1N4112	1N5538A	1N967A	1N5248	1N6006A	1N4746	1N3026A	1N5931A
19	1N4706	1N4113	1N5539A		1N5249				
20	1N4707	1N4114	1N5540A	1N968A	1N5250	1N6007A	1N4747	1N3027A	1N5932A
22	1N4708	1N4115	1N5541A	1N969A	1N5251	1N6008A	1N4748	1N3028A	1N5933A
24	1N4709	1N4116	1N5542A	1N970A	1N5252	1N6009A	1N4749	1N3029A	1N5934A
25	1N4710	1N4117	1N5543A		1N5253				
27	1N4711	1N4118		1N971A	1N5254	1N6010A	1N4750	1N3030A	1N5935A

☐ JAN/JANTX(V) available, ±5% only.
† 1N987–1N992 supplied in DO-7 glass package.
# 1N5273–1N5281 supplied in Surmetic DO-7 plastic package.
◆ 1M110ZS10 Series supplied in Surmetic (Plastic) DO-41 package.

DATA SHEET A2.3   Continued

Nominal Zener Voltage (Note 1)	250 mW Low Level Cathode = Polarity Mark (Notes 2,3)	250 mW Low Noise Cathode = Polarity Mark (Notes 2,3,5)	400 mW Low Noise Low Leakage Cathode = Polarity Mark (Notes 2,4,5)	500 mW Cathode = Polarity Mark (Notes 2,5)	500 mW Cathode = Polarity Mark (Notes 2,6)	500 mW Cathode = Polarity Mark (Notes 1,2,13)	1 Watt Cathode = Polarity Mark (Notes 2,7) Glass Case 59 (DO-41)	1 Watt Cathode to Case (Notes 2,8) Metal Case 52 (DO-13)	1.5 Watt Cathode = Polarity Mark (Notes 2,9) Surmetic 30 Case 59 (DO-41)
	Glass DO-204AH (DO-35)								
28	1N4712	1N4119	1N5544A		1N5255				
30	1N4713	1N4120	1N5545A	1N972A	1N5256	1N6011A	1N4751	1N3031A	1N5936A
33	1N4714	1N4121	1N5546A	1N973A	1N5257	1N6012A	1N4752	1N3032A	1N5937A
36	1N4715	1N4122		1N974A	1N5258	1N6013A	1N4753	1N3033A	1N5938A
39	1N4716	1N4123		1N975A	1N5259	1N6014A	1N4754	1N3034A	1N5939A
43	1N4717	1N4124		1N976A	1N5260	1N6015A	1N4755	1N3035A	1N5940A
47		1N4125		1N977A	1N5261	1N6016A	1N4756	1N3036A	1N5941A
51		1N4126		1N978A	1N5262	1N6017A	1N4757	1N3037A	1N5942A
56		1N4127		1N979A	1N5263	1N6018A	1N4758	1N3038A	1N5943A
60		1N4128			1N5264				
62		1N4129		1N980A	1N5265	1N6019A	1N4759	1N3039A	1N5944A
68		1N4130		1N981A	1N5266	1N6020A	1N4760	1N3040A	1N5945A
75		1N4131		1N982A	1N5267	1N6021A	1N4761	1N3041A	1N5946A
82		1N4132		1N983A	1N5268	1N6022A	1N4762	1N3042A	1N5947A
87		1N4133			1N5269				
91		1N4134		1N984A	1N5270	1N6023A	1N4763	1N3043A	1N5948A
100		1N4135		1N985A	1N5271	1N6024A	1N4764	1N3044A	1N5949A
110				1N986A	1N5272	1N6025A	◆1M110ZS10	1N3045A	1N5950A
120				†1N987A	1N5273#		◆1M120ZS10	1N3046A	1N5951A
130				†1N988A	1N5274#		◆1M130ZS10	1N3047A	1N5952A
140					1N5275#				
150				†1N989A	1N5276#		◆1M150ZS10	1N3048A	1N5953A
160				†1N990A	1N5277#		◆1M160ZS10	1N3049A	1N5954A
170					1N5278#		◆1M170ZS10		
180				†1N991A	1N5279#		◆1M180ZS10	1N3050A	1N5955A
200				†1N992A	1N5281#		◆1M200ZS10	1N3051A	1N5956A

□ JAN/JANTX(V) available, ±5% only.
† 1N987–1N992 supplied in DO-7 glass package.
# 1N5273–1N5281 supplied in Surmetic DO-7 plastic package.
◆ 1M110ZS10 Series supplied in Surmetic (Plastic) DO-41 package.

Nominal Zener Voltage (Note 1)	1.5 Watt Cathode to Case (Notes 2,10) Metal Case 55	5 Watt Cathode = Polarity Mark (Notes 2,11) Surmetic 40 Case 17	10 Watt Cathode to Case = 1N3993 Series Anode to Case = 1N2970 Series (Notes 2,10,12) Metal Case 56 (DO-4)	50 Watt Anode to Case (Notes 2,10,12) Metal Case 54 (TO-3)	50 Watt Anode to Case (Notes 2,10,12) Metal Case 58 (DO-5 Type)
3.3		1N5333A			
3.6		1N5334A			
3.9		1N5335A	1N3993&R	1N4557A&RA	1N4549A&RA
4.3		1N5336A	1N3994&R	1N4558A&RA	1N4550A&RA
4.7		1N5337A	1N3995&R	1N4559A&RA	1N4551A&RA
5.1		1N5338A	1N3996&R	1N4560A&RA	1N4552A&RA
5.6		1N5339A	1N3997&R	1N4561A&RA	1N4553A&RA
6.2		1N5341A	1N3998&R	1N4562A&RA	1N4554A&RA

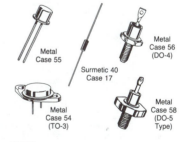

Metal Case 55

Surmetic 40 Case 17

Metal Case 56 (DO-4)

Metal Case 54 (TO-3)

Metal Case 58 (DO-5 Type)

NOTES

1. The Zener Voltage is measured at approximately 1/4 the rated power, with the following exceptions: the 1N4678–4717 is measured with I_{ZT} = 50 µAdc; the 1N4614/1N4099 is measured with I_{ZT} = 250 µAdc; the 1N4370/1N746 and the 1N5221–5242 are measured with I_{ZT} = 20 mAdc; the 1N5985A–6012A is measured with I_{ZT} = 5.0 mA; 1N6013A–6023A is measured with I_{ZT} = 2.0 mA; 1N6024–6025 is measured with I_{ZT} = 1.0 mA.

2. Contact your Motorola representative for information on intermediate voltages and tighter tolerances.

continues

DATA SHEET A2.3　Continued

Nominal Zener Voltage (Note 1)	1.5 Watt Cathode to Case (Notes 2,10) Metal Case 55	5 Watt Cathode = Polarity Mark (Notes 2,11) Surmetic 40 Case 17	10 Watt Cathode to Case = 1N3993 Series Anode to Case = 1N2970 Series (Notes 2,10,12) Metal Case 56 (DO-4)	50 Watt Anode to Case (Notes 2,10,12) Metal Case 54 (TO-3)	50 Watt (Notes 2,10,12) Metal Case 58 (DO-5 Type)
6.8	1N3785A	1N5342A	1N3999&R 1N2970A&RA	1N4563A&RA 1N2804A&RA	1N4555A&RA 1N3305A&RA
7.5	1N3786A	1N5343A	1N4000&R 1N2971A&RA	1N4564A&RA 1N2805A&RA	1N4556A&RA 1N3306A&RA
8.2	1N3787A	1N5344A	1N2972A&RA	1N2806A&RA	1N3307A&RA
8.7		1N5345A			
9.1	1N3788A	1N5346A	1N2973A&RA	1N2807A&RA	1N3308A&RA
10	1N3789A	1N5347A	1N2974A&RA	1N2808A&RA	1N3309A&RA
11	1N3790A	1N5348A	1N2975A&RA	1N2809A&RA	1N3310A&RA
12	1N3791A	1N5349A	1N2976A&RA	1N2810A&RA	1N3311A&RA
13	1N3792A	1N5350A	1N2977A&RA	1N2811A&RA	1N3312A&RA
14		1N5351A	1N2878A&RA	1N2812A&RA	1N3313A&RA
15	1N3793A	1N5352A	1N2979A&RA	1N2813A&RA	1N3314A&RA
16	1N3794A	1N5353A	1N2980A&RA	1N2814A&RA	1N3315A&RA
17		1N5354A		1N2815A&RA	1N3316A&RA
18	1N3795A	1N5355A	1N2982A&RA	1N2816A&RA	1N3317A&RA
19		1N5356A	1N2983A&RA	1N2817A&RA	1N3318A&RA
20	1N3796A	1N5357A	1N2984A&RA	1N2818A&RA	1N3319A&RA
22	1N3797A	1N5358A	1N2985A&RA	1N2819A&RA	1N3320A&RA
24	1N3798A	1N5359A	1N2986A&RA	1N2820A&RA	1N3321A&RA
25		1N5360A		1N2821A&RA	1N3222A&RA
27	1N3799A	1N5361A	1N2988A&RA	1N2822A&RA	1N3223A&RA
28		1N5362A			
30	1N3800A	1N5363A	1N2989A&RA	1N2823A&RA	1N3224A&RA
33	1N3801A	1N5364A	1N2990A&RA	1N2824A&RA	1N3225A&RA
36	1N3802A	1N5365A	1N2991A&RA	1N2825A&RA	1N3226A&RA
39	1N3803A	1N5366A	1N2992A&RA	1N2826A&RA	1N3227A&RA
43	1N3804A	1N5367A	1N2993A&RA	1N2827A&RA	1N3228A&RA
47	1N3805A	1N5368A	1N2996A&RA	1N2829A&RA	1N3330A&RA
51	1N3806A	1N5369A	1N2997A&RA	1N2831A&RA	1N3332A&RA
56	1N3807A	1N5370A	1N2999A&RA	1N2832A&RA	1N3334A&RA
60		1N5371A			
62	1N3808A	1N5372A	1N3000A&RA	1N2833A&RA	1N3335A&RA
68	1N3809A	1N5373A	1N3001A&RA	1N2834A&RA	1N3336A&RA
75	1N3810A	1N5374A	1N3002A&RA	1N2835A&RA	1N3337A&RA
82	1N3811A	1N5375A	1N3003A&RA	1N2836A&RA	1N3338A&RA
87		1N5376A			
91	1N3812A	1N5377A	1N3004A&RA	1N2837A&RA	1N3339A&RA
100	1N3813A	1N5378A	1N3005A&RA	1N2838A&RA	1N3340A&RA
110	1N3814A	1N5379A	1N3007A&RA	1N2840A&RA	1N3342A&RA
120	1N3815A	1N5380A	1N3008A&RA	1N2841A&RA	1N3343A&RA
130	1N3816A	1N5381A	1N3009A&RA	1N2842A&RA	1N3344A&RA
140					1N3345A&RA
150	1N3817A	1N5383A	1N3011A&RA	1N2843A&RA	1N3346A&RA
160	1N3818A	1N5384A	1N3012A&RA	1N2844A&RA	1N3347A&RA
170		1N5385A			
180	1N3819A	1N5386A	1N3014A&RA	1N2845A&RA	1N3349A&RA
200	1N3820A	1N5388A	1N3015A&RA	1N2846A&RA	1N3350A&RA

Tolerances

3. No suffix = ±5%

4. A suffix = ±10% — with guaranteed limits on V_Z, V_F, and I_R only

　B suffix = ±5%
　C suffix = ±2%
　D suffix = ±1%

5. 1N4370/1N746 series:　No suffix = ±10%
　　　　　　　　　　　　　A suffix = ±5%

　1N957 series:　A suffix = ±10%
　　　　　　　　B suffix = ±5%

Military parts in 1N4370/746/962 series and standard 1N987–1N992 supplied in DO-7　Military parts in 1N4370/746/962 are also available in the cost effective DO-204AH (DO-35) package as the -1 version. This version can be ordered by inserting a 1 between the part number and the JAN, JTX or JTXV suffix, ie 1N746A1JAN. MIL-STD 19500/117 and 127 state the -1 version is a direct substitute for the non -1 version. The -1 versions appear on MIL-STD 701 as the preferred parts for new designs. Military parts in 1N4614, 1N4099 and 1N5518A series supplied in DO-7.

6. No suffix = ±10% with guaranteed limits on V_Z, V_F and I_R only.
　A suffix = ±10%
　B suffix = ±5%

7. No suffix = ±10%
　A suffix = ±5%

8. 1N3821 series:　No suffix = ±10%
　　　　　　　　　A suffix = ±5%

　1N3016 series:　A suffix = ±10%
　　　　　　　　　B suffix = ±5%

9. A suffix = ±10%　C suffix = ±2%
　B suffix = ±5%　D suffix = ±1%

10. A suffix = ±10%
　B suffix = ±5%
　Exception:
　　1N3993–1N4000:　No suffix = ±10%
　　　　　　　　　　A suffix = ±5%

11. A suffix = ±10%
　B suffix = ±5%

12. RA and RB = Reverse Polarity Types Available

13. A suffix = ±10%
　B suffix = ±5%

3

BASIC TRANSISTORS AND THE GENERAL ONE-TRANSISTOR AMPLIFIER

In this chapter, transistors are introduced. These very important electronic devices are treated as discrete elements in this introduction. However, we will see that transistors are the primary devices used in integrated circuits.

The two fundamentally different types of transistors are the bipolar junction transistor (BJT) and the field effect transistor (FET). These devices have three terminals and provide the possibility of amplifying a voltage or current and, quite often, both.

The analysis and design of actual transistor amplifier circuits will be described in the next two chapters as well as in subsequent chapters. In this chapter, descriptions of the types of transistors, their structure, circuit symbols, and current–voltage relations are given. Additionally, a simple one-transistor amplifier circuit for a BJT or FET is presented and analyzed to show the basic amplification process of each. Finally, selection guides depicting actual BJTs and FETs are given. These indicate various types of discrete transistors that are available from manufacturers.

3.1   BIPOLAR JUNCTION TRANSISTORS

The *bipolar junction transistor* (BJT) was the first type of transistor to become commercially available. With the invention of this solid-state device, the demise of vacuum tubes was initiated. The word *bipolar* implies that both positive and negative charges flow in this device. The movement of these charges is controlled by voltages applied across two junctions.

3.1.1   BJT Types, Geometry, and Circuit Symbols

The two important types of BJTs are the NPN BJT and the PNP BJT. For the case of the NPN transistor (upon which most of our discussion will be focused), the ideal geometry and one possible integrated-circuit BJT geometry are shown in Fig. 3.1 along with the corresponding circuit symbol. As shown in Fig. 3.1b, there are four regions in an integrated-circuit BJT to which terminals can be attached. Voltages applied to three of these terminals, the emitter, base, and collector (labeled *E*, *B*, and *C*), establish fundamental device operation. A voltage applied to the substrate terminal (*S*) can alter this operation, but it is used primarily to reverse bias the substrate PN junction and provide electrical isolation in integrated circuits. Note that there are two junctions in the basic transistor that separate the P and N regions from one another. These junctions are called the *emitter junction* and the *collector junction* according to the one unique region next to that particular junction.

The names given to the regions have some physical meaning. The *emitter region* can be regarded as emitting carriers into the base when the emitter junction is forward biased. Furthermore, the *collector region* collects the injected emitter carriers when the collector junction is

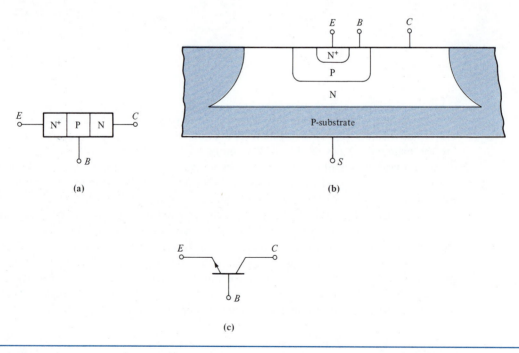

FIGURE 3.1   NPN Bipolar Junction Transistor: (a) Ideal geometry, (b) Actual geometry, (c) Circuit symbol

reverse biased or unbiased. Finally, the term *base* is used to denote the center region, which is labeled *B* in Fig. 3.1.

The circuit symbol in Fig. 3.1c distinctively describes the geometry shown in Figs. 3.1a and b. The emitter and collector parts of the symbol (labeled *E* and *C*) are drawn identically except for the arrowhead on the emitter part of the symbol. This arrowhead always points from the P material to the N material and thus indicates visually which terminal is the emitter. Omission of the arrowhead indicates the collector. The arrowhead also indicates the direction of conventional current flow when the BJT is operating as an amplifier. The base part of the symbol (labeled *B*) is different from the *E* and *C* parts and is therefore distinguishable. For the case of the PNP transistor, the P and N regions are interchanged, and the arrowhead reverses direction.

Note that the emitter N-type region in Figs. 3.1a and b is labeled N^+. This notation means that the donor impurity concentration in this region is very large (much greater than the acceptor impurity concentration in the base region and the donor concentration in the collector region). A heavier impurity concentration in the emitter region enhances the amplifying possibilities for the BJT. Also, for the transistor to be able to provide a useful magnitude of amplification, the base region must be very narrow.

In order to operate the BJT as an amplifier, the emitter junction is forward biased and the collector junction reverse biased. The combination of both of these conditions is referred to as the *amplifier region of operation* for the BJT. Such voltage conditions provide a large magnitude of current from *E* to *C* relative to a small base current, provided that the emitter impurity concentration is large and that the base width is very small. Current amplification is possible because the small base current can control a very large current from *E* to *C*. As we will see, voltage amplification can also result when resistor values at the input and output parts of the circuits are chosen properly.

3.1.2 BJT Current–Voltage Characteristics

When the BJT is operated as an amplifier, the output current depends directly upon the input current, and the dependence is very nearly linear. Typical *i–v* characteristics are shown in Figs. 3.2 and 3.3 for the NPN BJT in two important transistor configurations:

1. The BJT with the base terminal common to the input and output terminals (Fig. 3.2).
2. The BJT with the emitter terminal common to the input and output terminals (Fig. 3.3).

Positive current directions and voltage polarities are defined as shown in Figs. 3.2a and 3.3a. Note that the current directions denoting positive current are always selected in the direction of the emitter arrow and that all voltages are referenced to the common terminal.

Each of these configurations has two sets of *i–v* characteristics, one set for the input terminals and one set for the output terminals. The characteristics for the output terminals depend upon the input current, whereas the characteristics for the input terminals are essentially independent of output quantities.

For the common-base configuration, shown in Fig. 3.2, the input characteristic in Fig. 3.2b is merely that of a PN junction diode. The vertical portion of this curve is the usual region of amplfier operation for the input voltage and current of a BJT. Note that v_{EB} varies only slightly in this region, as in the case of a PN diode. Thus, if a signal voltage is applied between *E* and *B*, a large-signal current is produced even for a small-signal voltage. Hence, $v_{EB} \simeq V_0$. The output characteristics in Fig. 3.2c show an approximately linear dependence of i_C upon i_E for $v_{CB} \geq 0$. This dependence is observed by realizing that i_E is changed by an equal amount *C* in going from one characteristic to the next, which creates approximately equal changes in i_C. This region, $v_{CB} \geq 0$ and $i_C > 0$, is the amplifier region for the common-base

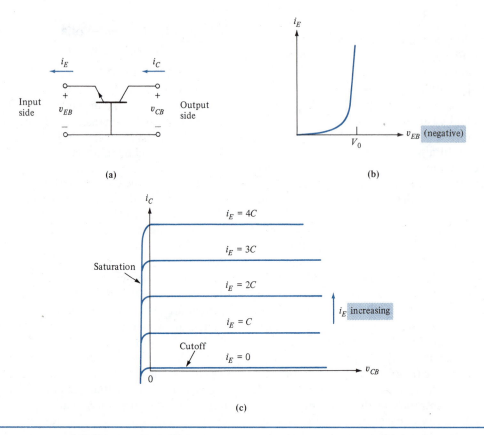

FIGURE 3.2   NPN Bipolar Junction Transistor in the Common-Base Configuration: (a) Circuit symbol, (b) Input characteristic, i_E versus v_{EB}, (c) Output characteristics, i_C versus v_{CB}, with i_E as a parameter

configuration. Note also that i_C has virtually no dependence on v_{CB} in the amplifier region.

Two other operating regions defined in Fig. 3.2c (and also Fig. 3.3c) are the saturation and cutoff regions. These regions are nonamplifying regions in which the junctions are not biased properly for amplifier operation. In the saturation region, both junctions are forward biased; in the cutoff region, both junctions are reverse biased. However, it is precisely these two regions that are used in transistor switching applications.

We see from Fig. 3.2b that v_{EB} is negative for amplifier operation of the NPN BJT. This polarity of voltage is necessary in order to forward bias

the emitter junction and use the B terminal as the voltage reference or common terminal. For the output characteristics, v_{CB} is positive for amplifier operation of NPN transistors because this polarity reverse biases the collector junction.

For the common-emitter configuration, shown in Fig. 3.3, similar i–v curves prevail. Figure 3.3b shows that the input characteristic is again that of a PN diode, except in this case the input current i_B is much smaller in magnitude than i_C or i_E. The output characteristics shown in Fig. 3.3c also indicate an approximately linear dependence of i_C upon i_B for v_{CE} slightly greater than zero. Note that here the equal incremental

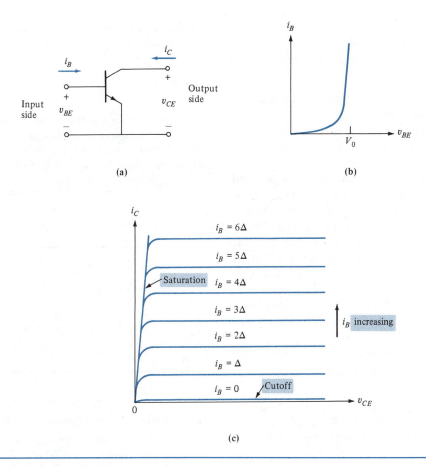

FIGURE 3.3 NPN Bipolar Junction Transistor in the Common-Emitter Configuration: (a) Circuit symbol, (b) Input characteristic, i_B versus v_{BE}, (c) Output characteristics, i_C versus v_{CE}, with i_B as a parameter

changes (Δ) in i_B in going from one characteristic to the next are small because i_B is small. The constant collector current region is the amplifier region in this case.

We see from Figs. 3.3b and c that the voltages v_{BE} and v_{CE} must both be positive for amplifier operation of the NPN BJT. Furthermore, as v_{CE} decreases toward zero, the characteristic curves of Fig. 3.3c change their form; v_{CE} becomes less than v_{BE} ($\simeq V_0$), and the collector junction becomes forward biased. In this region, $v_{BC} = v_{BE} - v_{CE}$ becomes positive. The transistor is therefore no longer operating as an amplifier. This region is the saturation region with both junctions forward biased.

3.1.3 BJT Input–Output Relation

Based upon the output current–voltage characteristics for the two BJT configurations just described, we may write a linear relationship for the output current in terms of the input current for each amplifier configuration as follows:

$$i_O = k i_I \tag{3.1-1}$$

where

> k = a different constant for each case
>
> i_O = collector current
>
> i_I = emitter current for the common-base configuration
>
> = base current for the common-emitter configuration

Furthermore, for the common-base configuration (Fig. 3.2), k is given the symbol α ($\alpha \lesssim 1$); for the common-emitter configuration (Fig. 3.3), k is given the symbol β ($\beta \gg 1$). Specific definitions of k for each case will be discussed in the next chapter. Since (3.1–1) indicates that the output current is directly proportional to the input current, the BJT is regarded as a current-controlled current amplifier. Equation (3.1–1) is often used to replace the characteristic curves of Figs. 3.2c and 3.3c.

───────────────────── **EX. 3.1a**

BJT Common-Emitter Amplification Factor Example

For the BJT output characteristics of Fig. E3.1, calculate the common-emitter current amplification factor β as a function of current for $V_{CE} = 10$ V. Use $\beta = i_C/i_B$ and carry out the calculations for $i_B = 0.02, 0.04, 0.07, 0.09,$ and 0.11 mA.

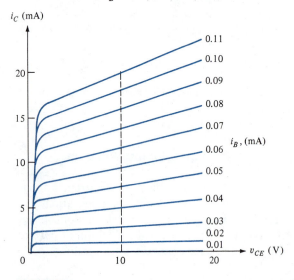

FIGURE E3.1

Solution: The common-emitter current amplification factor is defined as β and is given by

$$\beta = \frac{i_C}{i_B}$$

For $V_{CE} = 10$ V, we read the specific values of i_C from Fig. E3.1 for various values of i_B. These values are shown in Table 3.1 along with calculated ratios for β. Note that β is highly variable at low current levels but approaches a constant at higher current levels.

───────────────────────────────

TABLE 3.1   Values of i_C and i_B versus $\beta(\alpha)$ for Exs. 3.1a and b

i_B (mA)	i_C (mA)	$\beta = i_C/i_B$	$\alpha = \beta/(1+\beta)$
0.02	1.2	60	0.98
0.04	5	125	0.99
0.07	12	171	0.994
0.09	16	177	0.994
0.11	20	181	0.995

───────────────────────────────

───────────────────── **EX. 3.1b**

BJT Common-Base Amplification Factor Example

Using the data of the previous example, calculate the common-base current amplification factor α from the formula, $\alpha = \beta/(1 + \beta)$. This expression will be verified in Chapter 4.

Solution: Calculations of $\alpha = \beta/(1 + \beta)$ at the various levels of current yield the values listed in the last column of Table 3.1. Note that α changes by only a small percentage over the entire current range.

───────────────────────────────

3.2   FIELD EFFECT TRANSISTORS

Theoretically, the *field effect transistor* (FET) was the first type of transistor to be invented. However, initial material fabrication difficulties prevented its commercial availability. With time and extensive work with silicon as the

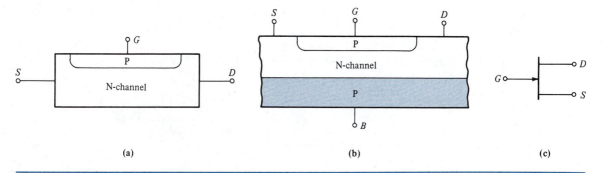

FIGURE 3.4   N-Channel Junction Field Effect Transistor: (a) Ideal geometry, (b) Actual geometry, (c) Circuit symbol

semiconductor material, these problems were overcome.

The name of this transistor is appropriate in that an electric field (with an associated applied voltage) is able to control a transverse current. Field effect transistors play a major role in today's world of integrated circuitry. They are particularly useful in digital applications because their inherent size is smaller than that of BJTs.

3.2.1   FET Types, Geometry, and Circuit Symbols

Two important types of FETs are the *junction field-effect transistor* (JFET) and the *metal-oxide-semiconductor field effect transistor*

(MOSFET or MOST). These names are quite descriptive, as we will see. Each of these types of transistors has a subset denoted by the names N-channel or P-channel transistors. Furthermore, the MOST has two variations, the enhancement-only MOST (E-O MOST) and the enhancement-depletion MOST (E-D MOST). Figures 3.4, 3.5, and 3.6 show the ideal and actual geometry for N-channel cases (which we will focus on) and the corresponding circuit symbol used for each. Note that each of the circuit symbols is different because each represents a different structure and behavior.

The electrical terminals for all FETs are the source (*S*), drain (*D*), gate (*G*), and substrate or bulk region (*B*). Voltages are applied to the terminals *S*, *D*, and *G* to provide fundamental device operation, while *B* is an additional terminal

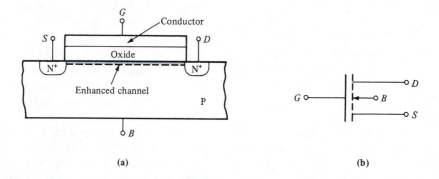

FIGURE 3.5   N-Channel Enhancement-Only MOST: (a) Actual geometry, (b) Circuit symbol

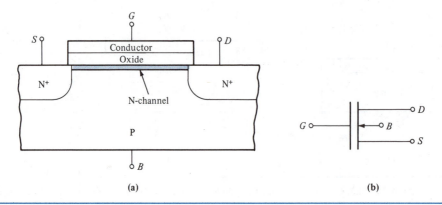

FIGURE 3.6   N-Channel Enhancement-depletion MOST: (a) Actual geometry, (b) Circuit symbol

that allows modification of this operation, but usually B and S are interconnected. In all FETs, the channel electrically connects the S and D terminals.

Note that the N-channel region for the JFET is clearly shown in Figs. 3.4a and b. However as shown in Fig. 3.5a, the channel for the E-O MOST is in the surface region directly beneath the thin oxide layer between the S and D regions and must be enhanced into the surface by applying a voltage to G. For the N-channel E-O MOST, the channel is enhanced into the surface region by applying a positive voltage on the gate relative to the source. Because the MOS layers essentially form a parallel-plate capacitor, this polarity of voltage induces (or enhances) negative charge into the surface of the semiconductor. When the magnitude of this voltage is large enough, an N-channel is enhanced into the surface and thereby connects S and D together. Only when the channel is enhanced can conduction between S and D occur; hence, the name *enhancement-only* MOST. For the P-channel case, the gate voltage polarity is reversed.

Looking at the FET cross sections of Figs. 3.4 and 3.5, we can easily understand how each obtained its particular name: For the JFET, the *junction*(s) interacts with the carrier flow from S to D; for the MOST, a set of *metal-oxide-semiconductor* (MOS) layers produces a similar inter-

action. By interaction, we mean that voltages applied to the gate (G) in each case can alter the charge in the channel region and thus the current in the channel (due to a voltage applied between S and D). Note that the MOS layers replace the junction when the JFET is compared with the MOST.

The carrier flow in the N-channel case is due to electrons that move from the source to the drain when the voltage between the drain and source (v_{DS}) is positive. Thus, the source acts like a supply of carriers, and the drain collects them; the names *source* and *drain,* then, are quite appropriate. The same meanings prevail for P-channel transistors.

One difference between the P-channel case and the N-channel case is that all semiconductor regions are interchanged. Thus, for the P-channel case, the opposite polarity for v_{DS} is required in order that the source provide a supply of positive charges that are collected at the drain. This flow also reverses the actual direction of current in the channel. Basic device operation for P- and N-channel devices, however, remains the same.

The arrowheads shown in the circuit symbols of Figs. 3.4 through 3.6 point from P to N in all cases, thus indicating the region types and, therefore, that these transistors are N-channel. For the case of P-channel transistors, the symbols are the same except that the arrowhead direction is reversed. Furthermore, the line connecting the source and drain regions in the circuit symbols of Figs. 3.4 through 3.6 represents the channel. Note that this line is shown dashed for the E-O MOST in Fig. 3.5b to indicate that the channel is not present until voltages are applied that induce (or enhance) this region. We can also realize that the gap between the vertical lines in the MOST symbol of Figs. 3.5b and 3.6b represents the oxide region and the fact that essentially an open circuit exists between the gate terminal and channel.

The other important MOST is the enhancement-depletion MOST (E-D MOST), the geometry and symbol for which are shown in Figs. 3.6a and b, respectively. Note that a channel is present without an applied voltage to G and that it is indicated by the solid line in the circuit symbol connecting S to D. The existence of this channel (without enhancement) requires additional processing steps in device fabrication. The channel in this case can be enhanced or depleted by the application of either polarity of gate voltage; hence, the name *enhancement-depletion* MOST or E-D MOST (or just depletion MOST).

3.2.2   FET Current–Voltage Characteristics

In describing the current–voltage characteristics of FETs, all of the various types will be considered simultaneously. Figure 3.7 displays a general FET with S as the reference terminal or terminal common to both the input and output sides of the transistor. Positive directions of current and polarities of voltage are defined as shown in Fig. 3.7 for the N-channel case. For the P-channel case, the drain current direction would be reversed.

Note first that the input current i_G for the FET is always very small and is therefore taken as zero. For the JFET, the junction is always reverse biased for proper operation; for the MOST, the gate current must flow across the oxide, which acts as a very high resistance. Hence, the gate current in both cases is quite small, and the

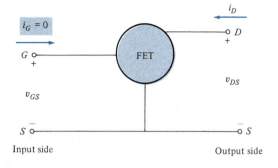

FIGURE 3.7   Voltage and Current Definitions for a General FET

input current–voltage characteristic for an FET is trivial.

Only the output current–voltage characteristics (i_D versus v_{DS}), then, are needed to describe the operation of an FET. Figure 3.8 displays a general set of output characteristic curves (i_D versus v_{DS}) with v_{GS} as the parameter. The magnitude and sign of the quantities i_D, v_{DS}, and v_{GS} are dependent upon the particular device, but the general dependence of i_D upon v_{DS} and v_{GS} is essentially the same for all FETs.

The most important difference between various FET i_D–v_{DS} characteristics is the range of

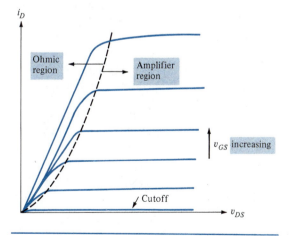

FIGURE 3.8   General Set of Output Characteristics, i_D versus v_{DS}, with v_{GS} as a Parameter for the N-Channel FET

v_{GS}. For the JFET, the junction must be reverse biased, and hence the range of v_{GS} begins at zero and continues to negative values of voltage (for the N-channel case). For the N-channel E-O MOST, the range of v_{GS} begins at some positive value of voltage, referred to as the *threshold* voltage, which is large enough to create an enhanced channel, and extends to larger positive values (more enchancement). Finally, for the N-channel E-D MOST, both positive (enhanced channel) and negative (depleted channel) values for v_{GS} are permissible.

The dashed curve in Fig. 3.8 represents a change in device operation that is obtained from a physical analysis of the FET as presented in Chapter 11. The region to the left of this curve is referred to as the *linear or ohmic region* because for small v_{DS}, i_D varies approximately linearly with v_{DS}. This region is used for FET switching applications and is the "on" condition for the FET. The region to the right of the dashed curve is referred to as the *saturation* (perhaps a misnomer) or *amplifier region*. It is in this region that the FET is operated as an amplifier. Also in this region, i_D is ideally only dependent upon v_{GS} and not upon v_{DS}. The bottom curve in Fig. 3.8 labeled *cutoff* is obtained for a particular value of v_{GS} (and all values less). The drain current is essentially zero for operation in cutoff. For switching applications, operation in cutoff is the "off" condition for the FET.

- **Square-Law Relation:**   The basic relationship for an FET operating in the amplifier region is given ideally by

$$i_D = C(v_{GS} - V_C)^2 \qquad (3.2-1)$$

where C and V_C are constants. This equation is derived by considering the physics of the device.

∎

Note: Equation (3.2–1) expresses the output current in terms of input voltage and is used as a representation for all FETs. However, the physi-

cal interpretation of C and V_C changes from one type of FET to another.

For the JFET and E-D MOST, the expression can be written as

$$i_D = I_{DSS}\left(1 - \frac{v_{GS}}{V_P}\right)^2 \qquad (3.2-2)$$

Here, V_P is the *pinch-off* voltage, and I_{DSS} is the drain-to-source saturation current.

The pinch-off voltage V_P is the required gate-to-source voltage (v_{GS}) that will entirely deplete (pinch off) the channel region of mobile charges. On transistor data sheets, V_P is called V_{GS}(OFF). For N-channel devices, the polarity required is negative on the gate relative to the source, and V_P is therefore negative for N-channel MOS transistors. For P-channel transistors, V_P is positive.

The interpretation of I_{DSS} is that it is the drain-to-source current when $v_{GS} = 0$. It is also called the saturation current because the drain-to-source current is essentially independent of v_{DS} (meaning that it has reached its maximum or saturation value). The magnitude of I_{DSS} is very geometry dependent; it is directly proportional to channel width and inversely proportional to channel length. Thus, the geometry of the FET is varied in device design to obtain a desired value for I_{DSS}. Both V_P and I_{DSS} are important JFET or E-D MOST parameters and are listed in FET data sheets.

The range of v_{GS} (and, therefore, i_D) depicts the major difference between the JFET and the E-D MOST. For the N-channel JFET, v_{GS} is in the negative voltage region with $V_P \leq v_{GS} \leq 0$ in order that the junction be reverse biased (which is a requirement for proper JFET amplifier operation). Actually, values of v_{GS} can be applied that are more negative than V_P, and the junction will still be reverse biased. Under these conditions, however, the channel is still completely depleted, and i_D must therefore remain at zero. Thus, (3.2–2) is not valid for $v_{GS} < V_P$ because it predicts non-zero values for i_D. Hence, for the N-channel JFET, the range of v_{GS} is $V_P \leq v_{GS} \leq 0$, and the corresponding range of drain current is $0 \leq i_D \leq I_{DSS}$.

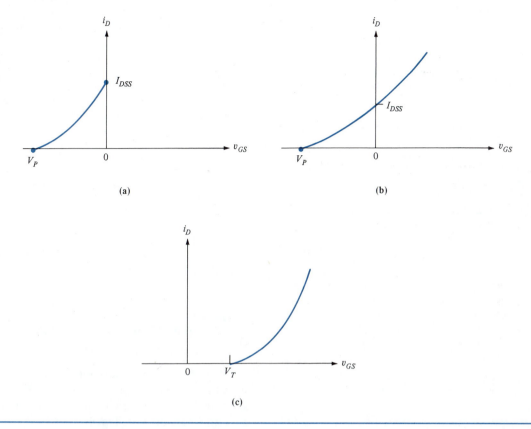

FIGURE 3.9   Current–Voltage Transfer Characteristics: (a) i_D versus v_{GS} for the N-Channel JFET, (b) i_D versus v_{GS} for the N-channel E-D MOST, (c) i_D versus v_{GS} for the N-channel E-O MOST

For the E-D MOST, this range is extended to positive values of gate voltage up to the maximum permissible v_{GS} for the particular transistor. Figures 3.9a and b show plots of (3.2–2) for each of these cases. The curves are referred to as current–voltage *transfer* characteristics.

For the E-O MOST, the i_D–v_{GS} equation is usually presented in the form

$$i_D = K(v_{GS} - V_T)^2 \qquad (3.2–3)$$

where K has units of A/V^2. This equation is valid only for $v_{GS} \geq V_T$, where V_T has the physical meaning that it is the required gate voltage (v_{GS}) for an enhancement channel to form. Note that I_{DSS} has no meaning for the E-O MOST because if $v_{GS} = 0$,

$i_D = 0$. As in the case of I_{DSS}, the quantity K is also directly proportional to channel width and inversely proportional to channel length. The corresponding transfer characteristic for the E-O MOST is shown in Fig. 3.9c.

The relationships depicted for i_D in terms of v_{GS} indicate that the FET is a voltage-controlled device (the output current depends upon the input voltage). Shortly, we will also observe that the FET is a voltage-controlled voltage amplifier.

—————————————————————— EX. 3.2

JFET Parameter Example

For the JFET characteristics shown in Fig. E3.2, determine the value of I_{DSS} and V_P.

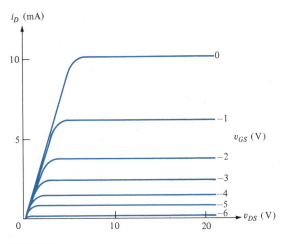

FIGURE E3.2

Solution: Reading these values directly from the curves, we have

$$I_{DSS} \simeq 10 \text{ mA}$$

and

$$V_P = -6 \text{ V}$$

EX. 3.3

E-O MOST Parameter Example

For the E-O MOST characteristics shown in Fig. E3.3, determine the parameters K and V_T. Use $v_{DS} = 10$ V and calculate K for various values of v_{GS}.

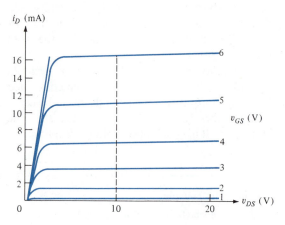

FIGURE E3.3

Solution: Reading the value of v_{GS} directly from the bottom characteristic, we have

$$V_T = 1 \text{ V}$$

The values of K for various v_{GS} (and corresponding i_D) are calculated from (3.2–3), as indicated in Table 3.2.

TABLE 3.2   Parameter Variation with v_{GS} for Ex. 3.3.

v_{GS} (V)	i_D (mA)	$K = i_D/(v_{GS} - V_T)^2$ (mA/V²)
2	1.5	1.5
4	6.5	0.72
6	16.5	0.66

3.3   THE BASIC ONE-TRANSISTOR AMPLIFIER CIRCUIT

Amplifier circuits for BJTs and FETs are, in general, similar to one another. The basic one-transistor amplifier circuit shown in Fig. 3.10 does not therefore specify a BJT or FET. In this circuit, the resistors are connected to the input and output terminals of the transistor (T) along with two DC voltage sources. This circuit arrangement allows specific DC voltage levels to be applied to each terminal of the transistor relative to each other, thus setting the Q point (DC operations point). Placement of Q (described in Chapter 4) is important since the transistor must operate in its amplifier region for signal amplification.

In general, in an amplifier circuit, an AC (signal) voltage or current is applied to the input side of the circuit (the left side of Fig. 3.10), and an amplified version of this signal is obtained in the output part of the circuit (the right side of Fig. 3.10). The primary objective of an amplifier circuit is, namely, to amplify AC. There are several different arrangements for applying the input signal voltage or current, but we will use the simplified representation of Fig. 3.10 to consider voltage amplification here.

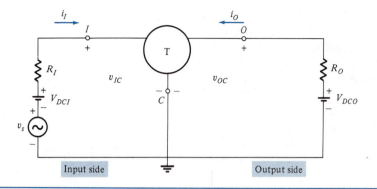

FIGURE 3.10   Amplifier Circuit with a Specific Input Signal Arrangement

The resistors in the amplifier circuit are labeled in accordance with their position in the circuit. The resistor R_I is the input resistor, and R_O is the output resistor. The input resistor is used to limit the input current, while the output resistor does likewise for the output. In this circuit, it is the output load resistor to which an amplified voltage is to be delivered.

The transistor voltages and currents are also defined in Fig. 3.10. The input current and voltage are i_I and v_{IC}, respectively; the output current and voltage are i_O and v_{OC}, respectively. The input and output voltages are thus referenced to the common terminal C. Note that it is not necessary to define the current at C, but, of course, it is a combination of i_O and i_I.

3.3.1   Amplifier Circuit Analysis Using Equations

To analyze the amplifier circuit of Fig. 3.10, we write KVL for the input and output loops, which yields

$$v_s + V_{DCI} - v_{IC} = i_I R_I \qquad (3.3\text{--}1)$$

and

$$v_{OC} = V_{DCO} - i_O R_O \qquad (3.3\text{--}2)$$

These expressions are load-line equations for the input and output sides of the amplifier circuit similar to those of the diode circuits in Chapter 2.

At this point, specific analyses for the BJT and FET differ because of the different input–output relations for the currents and voltages for these transistors. Thus, we proceed by considering the BJT and FET amplifiers individually.

BJT Analysis.   For the BJT, we utilize KVL for the input loop of Fig. 3.10, given by (3.3–1), along with the BJT current law $i_O = ki_I$ from (3.1–1). Elimination of i_I yields the output current as

$$i_O = \frac{k}{R_I}(v_s + V_{DCI} - v_{IC}) \qquad (3.3\text{--}3)$$

Using KVL for the output loop given by (3.3–2) yields the corresponding output voltage as follows:

$$v_{OC} = V_{DCO} - \frac{kR_O}{R_I}(v_s + V_{DCI} - v_{IC}) \quad (3.3\text{--}4)$$

Equations (3.3–3) and (3.3–4) show the dependence of the output current i_O and output voltage v_O upon the input signal voltage v_s, the resistors R_O and R_I, and the DC input voltage V_{DCI}. Both DC and AC components are contained in these equations.

Equation (3.3–4) shows clearly that the BJT output voltage v_{OC} has a signal component that linearly follows the input signal voltage v_s. The AC voltage amplification (between v_{OC} and v_s) is just the coefficient of v_s in (3.3–4) given by kR_O/R_I, which must be greater than 1 for useful voltage gain. From the BJT relation ($i_O = ki_I$), it can be understood that the input signal current can also be amplified provided that $k > 1$.

Note the term in (3.3–4) involving the input voltage v_{IC}. For the BJT, this voltage is directly across the forward-biased emitter junction. Thus, from the discussion of PN diodes in Chapter 2 and in Section 3.1, we know that the magnitude of this voltage is approximately equal to the forward voltage of the PN diode—that is, $v_{IC} \simeq \pm V_0$. The particular plus or minus value depends upon the type of transistor (NPN or PNP) as well as its configuration. Hence, this term is essentially a DC term, and the output current and voltage are given by

$$i_O = \frac{k}{R_I}(v_s + V_{DCI} \pm V_0) \qquad (3.3\text{–}5)$$

and

$$v_{OC} = V_{DCO} - \frac{kR_O}{R_I}(v_s + V_{DCI} \pm V_0) \qquad (3.3\text{–}6)$$

where the terms involving v_s are AC and the rest are DC components.

To complete the analysis of the BJT amplifier circuit, the input current is obtained from (3.3–1) by substituting $v_{IC} = \pm V_0$, which yields

$$i_I = \frac{v_s + V_{DCI} \pm V_0}{R_I} \qquad (3.3\text{–}7)$$

This current is also composed of AC as well as DC terms.

─────────────── EX. 3.4

BJT Amplifier Analysis Example

Determine the total currents and voltages for the NPN BJT in the common-emitter amplifier circuit shown in Fig. E3.4a. Also, calculate the AC voltage amplification (v_o/v_s). The transistor is made of silicon, and $\beta = 100$. Finally, let $v_s = 0.5 \sin \omega t$ V. Note that Fig. E3.4a uses the alternate DC

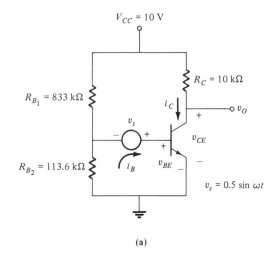

(a)

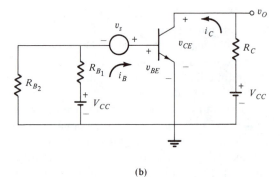

(b)

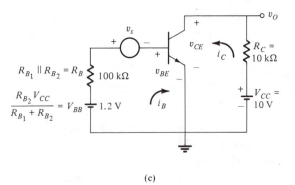

(c)

FIGURE E3.4

voltage symbol discussed in Chapter 1 and that v_O denotes the total output voltage.

Solution: Figure E3.4a is an alternate form of the one-transistor amplifier circuit of Fig. 3.10 that is very practical because only one DC voltage source is required to apply DC to the input and output sides of the circuit. An equivalent circuit with two DC voltage sources is shown in Fig. E3.4b. To obtain this equivalent (which is convenient for circuit analysis), we replace the single voltage source V_{CC} with two of equal value, one at the top of each resistor R_{B_1} and R_C. Then, the circuit is redrawn as shown in Fig. E3.4b. Next, we combine the two input branches into a Thévenin equivalent circuit (see Chapter 1) to obtain Fig. E3.4c, where $R_B = R_{B_1} \| R_{B_2}$ and $V_{BB} = [R_{B_2}/(R_{B_1} + R_{B_2})]V_{CC}$. Figure E3.4c is the equivalent circuit for analysis.

To carry out the solution, we proceed by obtaining the DC and AC solutions separately in this example. For the DC analysis, we write KVL for the input and output loops (of Fig. E3.4c) as

$$V_{BE} = V_{BB} - I_B R_B$$

and

$$V_{CE} = V_{CC} - I_C R_C$$

Moreover, since the transistor is made of silicon, $V_{BE} = V_0 = 0.7$ V. Thus, from KVL for the input loop, we have

$$I_B = \frac{V_{BB} - 0.7}{R_B} = \frac{1.2 - 0.7}{100 \text{ k}\Omega} = 5 \ \mu A$$

Knowing I_B allows the determination of I_C from the current amplification expression for this configuration ($i_C = \beta i_B$). Therefore,

$$I_C = \frac{\beta(V_{BB} - 0.7)}{R_B} = 100(5 \ \mu A) = 0.5 \text{ mA}$$

Finally, from KVL for the output loop, we have

$$V_{CE} = V_{CC} - \frac{\beta R_C(V_{BB} - 0.7)}{R_B}$$

$$= 10 - \frac{100(10 \text{ k}\Omega)(0.5)}{100 \text{ k}\Omega} = 5 \text{ V}$$

For the AC solution, we write KVL for the input and output loops of Fig. E3.4c (for AC voltages only), which yields

$$-i_b R_B + v_s = 0$$

and

$$v_{ce} + i_c R_C = 0$$

where the input-loop KVL used $v_{be} = 0$ (since $v_{BE} = V_0$). Thus,

$$i_b = \frac{v_s}{R_B} = 5 \sin \omega t \ \mu A$$

and, from the current amplification expression, we have

$$i_c = \frac{\beta v_s}{R_B} = 0.5 \sin \omega t \text{ mA}$$

Finally, from the output-loop KVL, we have

$$v_{ce} = -\left(\frac{\beta R_C}{R_B}\right)v_s = -5 \sin \omega t \text{ V}$$

Since $v_{ce} = v_o$, the voltage amplification is

$$\frac{v_O}{v_s} = \frac{-\beta R_C}{R_B} = \frac{-100(10 \text{ k}\Omega)}{100 \text{ k}\Omega} = -10$$

The total currents and voltages for the transistor are, therefore,

$$v_{BE} = 0.7 \text{ V}$$

$$i_B = 5 + 5 \sin \omega t \ \mu A$$

$$i_C = 0.5 + 0.5 \sin \omega t \text{ mA}$$

$$v_{CE} = 5 - 5 \sin \omega t \text{ V}$$

These currents and voltages are obtained graphically in Ex. 3.6.

FET Analysis. For the FET, the analysis is initially simplified because $i_I = 0$ (the gate current is essentially zero) as was discussed in Section 3.2. Thus, using KVL for the input loop of Fig. 3.10, given by (3.3–1), we have

$$v_{IC} = v_s + V_{DCI} \tag{3.3–8}$$

Substitution of this expression into the general square-law relation (3.2–1) yields

$$i_o = C(v_s + V_{DCI} - V_C)^2 \tag{3.3–9}$$

The output voltage is then given by (3.3–2) as follows:

$$v_{OC} = V_{DCO} - CR_o(v_s + V_{DCI} - V_C)^2 \tag{3.3–10}$$

Equation (3.3–10) shows that the FET output voltage v_{OC} has a signal component that follows the input signal voltage v_s in a nonlinear (square-law) manner. However, for a small input signal voltage ($v_s \ll V_{DCI} - V_C$), (3.3–9) and (3.3–10) can be approximated [using $(1 + x)^2 \simeq 1 + 2x$, for $x \ll 1$] to yield

$$i_O = C(V_{DCI} - V_C)^2 \left(1 + \frac{2v_s}{V_{DCI} - V_C}\right)$$

$$(3.3–11)$$

and

$$v_{OC} = V_{DCO} - CR_O(V_{DCI} - V_C)^2 \left(1 + \frac{2v_s}{V_{DC} - V_C}\right)$$

$$(3.3–12)$$

Thus, from (3.3–12), the AC voltage amplification (between v_s and v_{OC}) is just the coefficient of v_s given by $2CR_O(V_{DCI} - V_C)$, which must be greater than 1 for useful voltage gain. Moreover, we observe that the FET is a voltage-controlled voltage amplifier. However, we must emphasize that for the FET, the input signal must be small in order to obtain approximately linear voltage amplification unless special circuitry is used.

                 EX. 3.5

FET Amplifier Analysis Example

Consider an N-channel E-O MOST with a current–voltage transfer characteristic given by

$$i_D = 0.5(v_{GS} - 2)^2 \text{ mA}$$

which is valid only for $v_{GS} \geq V_T$ (= 2 V). This transistor is used in the amplifier circuit shown in Fig. E3.5a. Determine the currents and voltages for the MOST and the approximate AC voltage amplification (v_o/v_s). Let $v_s = V_S \sin \omega t$ V, where $V_S \ll V_{GG}$.

Solution: As in the BJT amplifier example (Ex. 3.4), we use a practical circuit with one DC voltage source. The equivalent circuit, obtained as in Ex. 3.3, is shown in Fig. E3.5b.

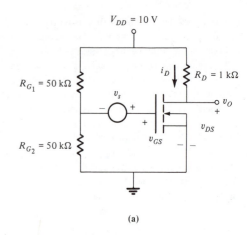

(a)

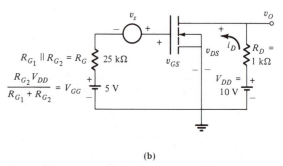

(b)

FIGURE E3.5

Using KVL for the input loop of Fig. 3.5b and with $i_G = 0$, we have (note that the value of $R_G = R_{G_1} \| R_{G_2}$ has no effect here)

$$v_{GS} = V_{GG} + v_s = 5 + V_S \sin \omega t$$

Thus, substitution into the transfer characteristic yields

$$i_D = 0.5(V_{GG} + v_s - 2)^2 = 0.5(3 + v_s)^2 \text{ mA}$$

Finally, writing KVL for the output loop of Fig. E3.5b yields

$$v_{DS} = V_{DD} - i_D R_D = 10 - i_D$$

where i_D is in mA. Substituting for i_D, we have

$$v_{DS} = V_{DD} - 0.5(V_{GG} + v_s - 2)^2 R_D$$
$$= 10 - 0.5(3 + v_s)^2$$

Expanding this equation yields

$$v_{DS} = 10 - 0.5(9 + 6v_s + v_s^2)$$

Note that the last term, dependent upon v_s^2, introduces nonlinear distortion. However, for small V_S, which was assumed, this term is negligible.

Since $v_{ds} = v_o$, the approximate AC voltage amplification for this amplifier is

$$\frac{v_o}{v_s} = -3$$

3.3.2   Graphical Solution for the BJT

In this section graphical techniques for analysis of BJT amplifier circuits are introduced. The results are also applicable to FET amplifier circuits (and are applied in Chapter 4). The graphical approach presented here is an extension of the graphical method introduced with diodes. The primary reason for considering the graphical method is to provide greater insight into the operation of amplifier circuits. In the analysis here, we will again use the one-transistor amplifier circuit of Fig. 3.10. The resulting solution will, of course, agree with that of the previous section. A general graphical solution involves graphically determining the DC solution, then the AC solution, and

adding these solutions together. To eliminate some tedium, we will use only a part of the general graphical solution. However, the entire graphical solution will be plotted for instructive purposes.

DC Solution.   The method of graphically determining the DC solution for the amplifier of Fig. 3.10 using a BJT for the transistor is shown in Fig. 3.11. First, the load-line equations, (3.3–1) and (3.3–2), are plotted on the input and output axes of Figs. 3.11a and b. Then, the DC input current I_I and voltage V_{IC} are found on the input axes by determining the intersection of the input load line with the transistor input characteristic (Fig. 3.11a). The DC input voltage V_{IC} is approximately given by V_0, and the DC input current I_I is obtained by reading its value from the vertical axis.

To obtain the position of Q on the output axes (Fig. 3.11b), we use the value of I_I obtained from the input current axis. The intersection of the particular transistor curve corresponding to I_I (the only one shown in Fig. 3.11b) with the output load line provides the position of Q on the output axes. The values of I_O and V_{OC} are then simply read from the corresponding axes.

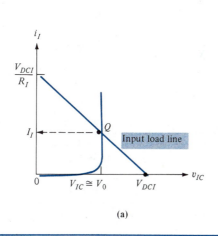

(a)

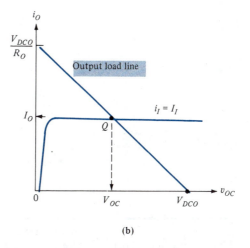

(b)

FIGURE 3.11   Graphical Determination of the DC Solution for a BJT: (a) Input DC solution, (b) Output DC solution

AC Solution.   A graphical solution is not warranted here because for the input loop, the load line is time varying. This shifting of the load line with time leads to a very tedious solution method, as we saw for the diode case. However, the input AC current is very simply obtained from the amplifier circuit of Fig. 3.10 by writing KVL for only the AC voltages around the input loop as follows:

$$i_i = \frac{v_s}{R_I} \qquad (3.3\text{–}13)$$

Note that the AC component of voltage for v_{IC} is taken as zero. This value is an approximation that will be removed in Chapter 5. Continuing with equations, the AC output current for the BJT is obtained from the current amplification equation as

$$i_o = k i_i = \left(\frac{k}{R_I}\right) v_s \qquad (3.3\text{–}14)$$

The AC output voltage is obtained by writing KVL for the AC voltages around the output loop

as

$$v_{oc} = -i_o R_O = -\left(\frac{k R_O}{R_I}\right) v_s \qquad (3.3\text{–}15)$$

Total Solution.   Before we consider the total graphical solution of the transistor characteristics, it is instructive to view the total waveshapes of the BJT currents and voltages. These waveshapes are obtained by adding the DC and AC components for the BJT amplifier under the assumption of a sinusoidal input signal voltage of $v_s = V_S \sin \omega t$. The corresponding total waveshapes of i_I and i_O, along with v_{IC} and v_{OC}, are shown in Figs. 3.12a and b, respectively, for one period of the sinusoidal variation. The numerical values used for k, R_O, and R_I have been chosen arbitrarily, but they do provide a small amount of current amplification ($i_o > i_i$). We will see in Chapter 5 that typical amplifiers provide much larger amplification. Note the defined maximum and minimum values of the currents and voltages as indicated in Fig. 3.12.

The total waveshapes of Fig. 3.12a are displayed alongside the current and voltage axes of Figs. 3.13a and b. They are shown centered about

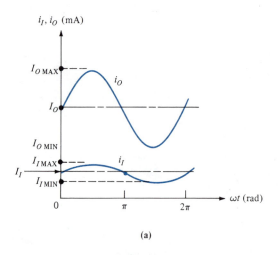

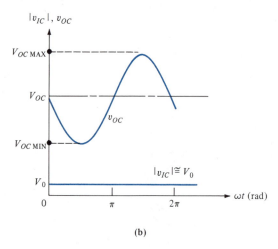

(a)   (b)

FIGURE 3.12   Waveshapes of the Currents and Voltages for the BJT Amplifier: (a) Sinusoidal input and output currents, (b) Constant input and sinusoidal output voltages

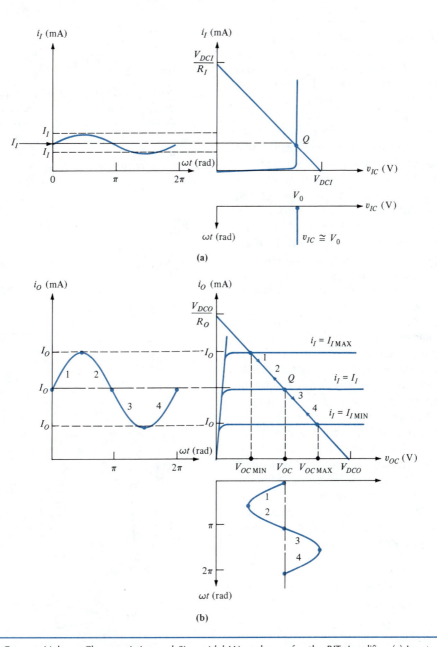

FIGURE 3.13   Current–Voltage Characteristics and Sinusoidal Waveshapes for the BJT Amplifier: (a) Input characteristics and waveshapes, (b) Output characteristics and waveshapes

their corresponding DC values. The defined maximum and minimum currents and voltages from Fig. 3.12 are also indicated in Fig. 3.13. Additionally shown are two more BJT characteristic curves, one for which the input current i_I has its maximum value and one for its minimum value.

The actual movement of the operating point on the output characteristics is described by realizing that under DC conditions the operating point is at Q. Then, as v_s increases from zero during the first quarter-cycle, the AC input current increases, and the operating point shifts upward on the load line. This upward shift continues until v_s reaches its maximum value, which is the maximum current position where $i_I = I_{I\,max}$ and $i_O = I_{O\,max}$. As time goes on, the operating point shifts back down the load line, through Q, and reaches its minimum position where $i_I = I_{I\,min}$ and $i_O = I_{O\,min}$. Finally, in the last quarter-cycle of v_s, the operating point shifts back up the load line to Q, thus completing one cycle of v_s. The arrows and numbers on the output load line indicate the shifting of the operating point between Q and the outermost positions during each successive quarter-cycle of the input sinusoid v_s.

Note that the output current waveshape in Fig. 3.13 is an amplified version of the input current waveshape. Also, the output voltage is an amplified version of the input voltage (not observable in Fig. 3.13); however, the input current and output voltage are $180°$ out of phase with each other. Thus, as the input current increases, the output voltage decreases, and vice versa.

EX. 3.6

BJT Graphical Analysis Example

Graphically determine the currents and voltages for the BJT in the common-emitter amplifier circuit of Ex. 3.4 (Fig. E3.4).

Solution: Using the equivalent circuit shown in Fig. E3.4c, the DC load-line equations (with currents in mA) are

$$V_{BE} = V_{BB} - I_B R_B = 1.2 - 100 i_B$$

and

$$V_{CE} = V_{CC} - I_C R_C = 10 - 10 i_C$$

These equations are plotted as shown in Figs. E3.6a and b. To obtain the graphical solution, we locate Q on the input axes by obtaining the intersection with the Si transistor characteristic yielding $I_B = 5\ \mu A$ and $V_{BE} = 0.7$ V. Then, the value of I_C is obtained from $i_C = \beta i_B$. Since $\beta = 100$, $I_C = 0.5$ mA. Lastly, the DC collector-to-emitter voltage is read from the voltage axis as $v_{CE} = 5$ V. The arrows in Fig.

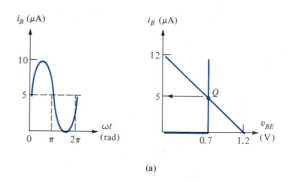

(a)

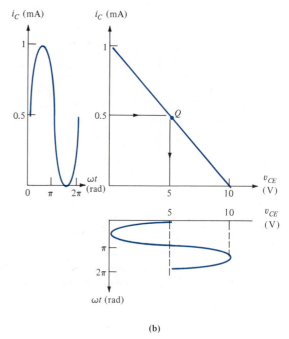

(b)

FIGURE E3.6

E3.6 indicate the manner in which the DC solution proceeds.

We obtain the AC solution by determining the AC input current (as in Ex. 3.4) from Fig. E3.4c as $i_b = v_s/R_B$. Thus,

$$i_b = 5 \sin \omega t \; \mu A$$

This variation is shown (superimposed on the DC) to the left of the input characteristic of Fig. E3.6a. Noting that the maximum and minimum values of i_B are 10 μA and 0 μA, we can immediately read off the maximum and minimum values of i_C and v_{CE} from the output characteristics. These values are $i_C = 1$ mA and 0 mA and $v_{CE} = 0$ V and 10 V, respectively. The total variation for one period is observed adjacent to each axis.

3.4   SELECTION GUIDE FOR SMALL-SIGNAL BJTS

Small-signal BJTs are used in low-power (<1 W) amplifier and switching applications. Such applications are the most popular use (by far) of discrete BJTs. Various small-signal BJTs that are commercially available are presented in Data Sheet A3.1 in the appendix to this chapter. Basic parameters for single BJTs (still called *transistors* on manufacturers' data sheets because they were the first transistors to be available) are indicated. Parameters for multiple BJT packages, consisting of two or more transistors in a single package, are also given. Included in the definition of multiple packages are Darlington-pair transistors, which are a special connection of two transistors that function as one high-gain BJT. Darlington-pair transistors will be described in Chapter 6. Parameters for multiple BJT packages consisting of dual and quad combinations of matched BJTs are given as well.

Specific uses of small-signal transistors, discussed in Chapter 5, include general-purpose, low-noise audio, and amplifiers; differential amplifiers (introduced in Chapter 4); and switching appli-

cations (discussed in Chapter 9). General-purpose transistors provide small-signal amplification from DC to low radio frequencies (<20 kHz). They are also used as oscillators (discussed in Chapter 14).

When BJTs are used as amplifiers, special designs are required for high current and/or high voltage amplification. Similarly, special designs are required for switching transistors, where transient time periods (t_{on} and t_{off}) are to be minimized. These times correspond to the turn-on and turn-off transient time periods for a BJT that are described in Chapter 9.

Other parameters listed in Data Sheet A3.1 that are not understandable by inspection are defined as follows:

- $V_{(BR)CEO}$ is the breakdown voltage associated with the collector–emitter terminals with the base terminal open (*O*). This voltage is often written simply as V_{CEO}.
- h_{FE} equals I_C/I_B, or β. However, the notation has special meaning, as we will see in Chapter 5.
- $V_{CE(sat)}$ is the collector–emitter saturation voltage (introduced in Chapter 4) and indicates the value of v_{CE} when the BJT is in saturation at the current levels stated.
- f_T is the current–gain bandwidth product, which is an indication of the magnitude of current amplification and frequency range over which amplification is possible. This parameter is described in Chapter 8.
- P_D is the maximum power dissipation for the device.

Small-signal BJTs are available in metal and plastic three-terminal packages. The plastic packages are less costly; however, the metal packages normally provide greater reliability. The multiple BJT devices are available in metal and plastic packages and with more than three terminals, except for Darlington-pair transistors. For quad transistor arrays, fourteen-pin DIP packages are commonly used, although only twelve pins are necessary (usually two pins have no connection).

More complete device specifications are given on data sheets that are available from manufacturers. More detailed BJT data sheets for several representative cases are included in Appendix B at the end of the book.

3.5 SELECTION GUIDE FOR SMALL-SIGNAL FETs

Data Sheet A3.2 in the appendix to this chapter presents parameters for a wide variety of small-signal FETs, including JFETs and E-O or E-D MOSTETs for both P- and N-channel. Applications include switching and amplifying with special designs for low frequency up to UHF.

The packages used for FETs are similar (and the same in many cases) to those used for BJTs. Both metal and less-expensive plastic packages are used.

For switching FETs, the parameters listed in Data Sheet A3.2 that require further explanation are as follows:

- $r_{DS(on)}$ is the drain-to-source resistance or the resistance of the channel in the on condition (the FET is then operating in the linear region).
- $V_{GS(off)}$ is the gate–source voltage that corresponds to cutoff operation or the off condition. This parameter is identical to V_P or V_T.
- $V_{(BR)GSS}$ is the gate–source breakdown voltage.
- C_{iss} is the input capacitance.
- C_{rss} is the reverse transfer capacitance.

The parameters t_{on} and t_{off} (as in the case of BJTs) are transient time periods for switching an FET on or off, respectively.

When FETs are used in discrete amplifiers, their primary application is as low-noise amplifiers. Different designs are used for low frequency (less than 10 MHz) and high frequency (of order 100 MHz). Parameters for various JFETs and MOSFETs used in these applications are also indicated in Data Sheet A3.2. Two of the parameters listed are different from those for the switching FETs. These parameters are defined as follows:

- $\text{Re}|Y_{fs}|$ is the real part of the forward (f) admittance (Y) parameter in the common-source (s) configuration [meaning that the source terminal is common to both the input (gate) and output (drain) sides].
- $\text{Re}|Y_{os}|$ is the real part of the output (o) admittance (Y) parameter in the common-source (s) configuration.

These AC admittance parameters are described in detail in Chapter 5. However, it is customary to define them in terms of conductances, where we use the simpler notation

$$g_m = \text{Re}|Y_{fs}|$$

and

$$g_o = \text{Re}|Y_{os}|$$

The parameter g_m is called the *transconductance* and is the gain parameter for the FET. The parameter g_o is the *output conductance* and is the conductance between the drain and the source when the FET is in the common-source configuration.

For the low-frequency/low-noise MOSFETs listed at the end of Data Sheet A3.2, $V_{GS(TH)}$ replaces $V_{GS(off)}$. This parameter is defined as follows:

- $V_{GS(TH)}$ is the gate–source threshold (TH) voltage which is the minimum applied voltage between the gate (G) and the source (S) that is necessary to form a channel.

In general, the values of $V_{GS(TH)}$ are positive for N-Channel E-O devices and negative for P-channel E-O devices.

Appendix 3 at the end of the book contains more detailed data sheets for various representative FETs.

- Two fundamentally different types of three-terminal devices called transistors are the bipolar junction transistor (BJT) and the field effect transistor (FET).

- The BJT is either NPN or PNP, depending upon the semiconductor layer arrangement. The heavily doped region is the emitter, the central region is the base, and the other region is the collector.

- The circuit symbol for the BJT always has an arrowhead on the emitter portion of this symbol. The arrowhead points in the direction of P-type semiconductor to N-type semiconductor. This arrowhead also indicates the direction of current flow for amplifier operation.

- For amplifier operation of a BJT, the emitter junction is forward biased and the collector junction is reverse biased (a forward-biased junction has positive voltage applied to P relative to N, and a reverse-biased junction has positive voltage applied to N relative to P).

- The BJT can operate in several different modes: (1) amplifier operation, (2) cutoff, and (3) saturation. When both junctions are reverse biased, the BJT is operating in cutoff. When both junctions are forward biased, the BJT is operating in saturation.

- Under forward-bias conditions, the voltage across the emitter junction is approximately V_0, where $V_0 = 0.7$ V for a silicon transistor.

- A general BJT input–output relation for amplifier operation is $i_o = k i_I$. For the common-emitter configuration, $k = \beta$ ($\beta \gg 1$); for the common-base configuration, $k = \alpha$ ($\alpha \lesssim 1$). These relations replace the current–voltage characteristics in the amplifier region.

- Two basically different types of three-terminal devices called FETs are the junction FET (JFET) and the metal-oxide-semiconductor FET (MOSFET or MOST). These types of FETs can be either N- or P-channel. The MOST has two variations, the E-O MOST and E-D MOST.

- For the JFET, voltage applied across the junction controls the channel current i_D. For the MOSFET or MOST, voltage applied across the metal-oxide-semiconductor layers controls i_D.

- The terminals of the FET are called the source, drain, and gate. The source and drain terminals are attached to either end of the channel, while the gate terminal is attached to a region separated from the channel by a junction (JFET) or by a set of metal and oxide layers (MOSFET or MOST).

- The gate of the FET conducts essentially zero current ($i_G = 0$) because of the high input impedance associated with the high resistivity of the oxide layer or that of the reverse-biased PN gate junction.

- The basic current–voltage transfer relation for an FET is given ideally by $i_D = C(v_{GS} - V_C)^2$, where V_C is the threshold voltage for the E-O MOST and C has units of A/V^2. For the JFET and E-D MOST, the relation is rearranged so that $i_D = I_{DSS}[1 - (v_{GS}/V_P)^2]$, where I_{DSS} is the drain–source saturation current for $v_{GS} = 0$, and V_P is the pinch-off voltage.

- Basic one-transistor amplifier circuits are examined and analyzed analytically and graphically. In general, these analyses involve solving four equations: KVL for the input and output portions of the amplifier circuit and two transistor relations.

- Specific circuit symbols are given throughout this chapter, along with the current–voltage characteristics and transistor parameters for both BJT and FET transistors.

3.1

Figures P3.1a and b display the symbols and defined positive current directions and voltage polarities for the NPN BJT and the PNP BJT, respectively. For each transistor, $i_C = 100i_B$. Determine the collector-to-base amplification factor (β) and the collector-to-emitter amplification factor (α).

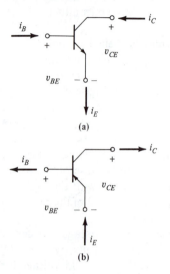

FIGURE P3.1

3.2

For the NPN BJT of Problem 3.1, sketch and dimension a family of output i–v characteristics for **(a)** the common-emitter configuration and **(b)** the common-base configuration. Indicate the required modifications for the PNP BJT. Use base current values of 0, 10, 20, 30, 40, and 50 μA.

3.3

An NPN BJT is represented by the elements inside the dashed rectangle in the circuit of Fig. P3.3. De-

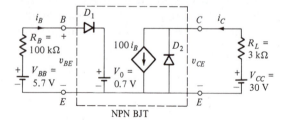

FIGURE P3.3

termine i_B and i_C. Use basic circuit analysis and assume D_1 and D_2 are ideal. Also, determine v_{CE}.

3.4

Graphically obtain the same solution for Problem 3.3. Carry this solution out in three steps. First, write KVL for the external elements of the input and output loops and plot these straight lines on the input and output i–v axes. Second, plot the input and output i–v characteristics for the BJT on the corresponding axes. Third, determine the intersections.

3.5

Repeat Problems 3.3 and 3.4 for the PNP BJT as represented by the elements inside the dashed rectangle in the circuit of Fig. P3.5.

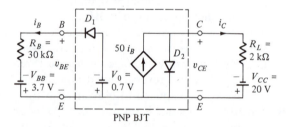

FIGURE P3.5

3.6

For the circuit of Fig. P3.3, consider a sinusoidal voltage $v_s = 10 \sin \omega t$ V to be in series with V_{BB}. Determine and sketch the waveshapes for i_B, v_{BE}, i_C, and v_{CE} for one period of v_s. What is the maximum i_B and v_s that will not distort the sinusoidal waveshapes?

3.7

For the two-port network of Fig. P3.7, the following relations were obtained: $v_A = 0.7$ V and $i_B = 100i_A$. Determine the input current, output current, and output voltage. Use basic circuit analysis and a graphical analysis for comparison.

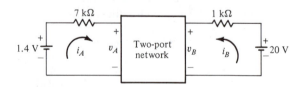

FIGURE P3.7

3.8

Analyze the circuit of Fig. P3.8 to determine the current gain (i_o/i_i) and the voltage gain (v_o/v_i). The elements inside the dashed rectangle represent a current-controlled amplifier (representative of BJTs).

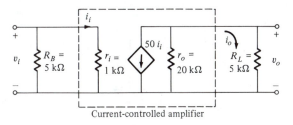

FIGURE P3.8

3.9

For the Si NPN BJT $(\beta = 50, V_0 = 0.7 \text{ V})$ in the circuit of Fig. P3.9, determine and calculate i_B, i_C, and v_{CE}.

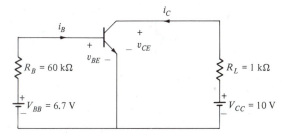

FIGURE P3.9

3.10

Argue that the circuit of Fig. P3.10b is equivalent to that of Fig. P3.10a. Using a Thévenin equivalent circuit for the input of Fig. P3.10b, determine and calculate i_B, i_C, and v_{CE}. Assume $\beta = 100$ and $V_0 = 0.7$ V for the BJT and $R_1 = 150$ kΩ, $R_2 = 75$ kΩ, $R_L = 1$ kΩ, and $V_{CC} = 5$ V.

3.11

Repeat Problem 3.10 for BJTs with $\beta = 200$ and 300. Use a graphical procedure to verify the results and note that an NPN BJT operates with $v_{CE} \geq 0$.

3.12

The NPN transistor in Fig. P3.12 is a type that is mass-produced with the values of β ranging from $50 \leq \beta \leq 300$. The DC collector current is chosen to

be 2 mA for $v_{CE} = 4$ V and $V_{CC} = 20$ V. For $\beta = 100$ and $V_0 = 0.7$ V, determine the values of R_B and R_L. Using these resistor values, determine i_C and v_{CE} for $\beta = 50$ and 300. Show the results graphically.

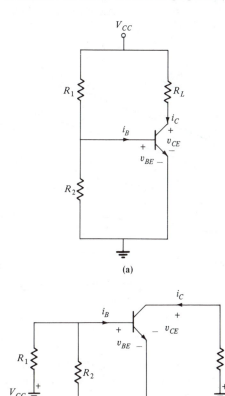

(a)

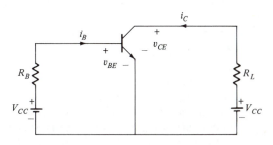

(b)

FIGURE P3.10

FIGURE P3.12

3.13

A BJT with output i_C–v_{CE} characteristics as shown in Fig. P3.13 is used in the circuit of Fig. P3.9. Assuming that the transistor is made of silicon and $V_0 = 0.7$ V, determine i_B, i_C, and v_{CE} (use a graphical procedure for i_C and v_{CE}).

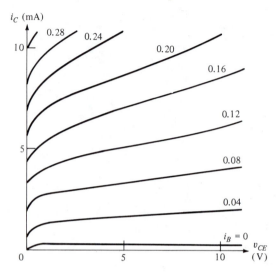

FIGURE P3.13

3.14

A BJT is used in the circuit of Fig. P3.14. Assume $\beta = 100$ and $V_0 = 0.7$ V. Determine i_E, i_C, and v_{CB}. Use the relation $i_C = \alpha i_E$, where $\alpha = \beta/(1 + \beta)$.

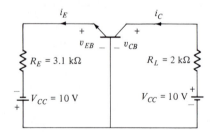

FIGURE P3.14

3.15

The circuit of Fig. P3.15 cannot be simplified any further for analysis. By writing two KVL independent equations and using the known transistor relations, determine and calculate i_B, i_C, and v_{CE}. Let $\beta = 100$ and $V_0 = 0.7$ V.

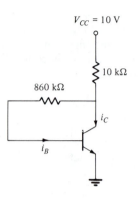

FIGURE P3.15

3.16

The N-channel JFET circuit symbol is shown in Fig. P3.16. The drain current for a particular JFET is given in mA by $i_D = 10[1 + (v_{GS}/4)]^2$. Sketch the transfer characteristic (i_D versus v_{GS}) and the output characteristics (i_D versus v_{DS}). Use $v_{GS} = 0, -1, -2, -3$, and -4 V. What are the values of I_{DSS} and V_P?

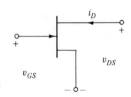

FIGURE P3.16

3.17

The JFET of Problem 3.16 is used in the circuit of Fig. P3.17. Determine v_{GS}, i_D, and v_{DS}. Use basic circuit analysis.

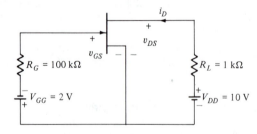

FIGURE P3.17

3.18

Graphically obtain the same solution for Problem 3.17.

3.19

The N-channel enhancement-only (E-O) MOST circuit symbol is shown in Fig. P3.19. The drain current for a particular E-O MOST is given in mA by $i_D = 0.625(v_{GS} - 1)^2$. Sketch the transfer characteristic (i_D versus v_{GS}) and the output characteristics (i_D versus v_{DS}). Use $v_{GS} = 1, 2, 3, 4,$ and 5 V. What are the values of I_{DSS} and V_T?

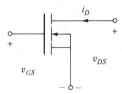

FIGURE P3.19

3.20

The E-O MOST of Problem 3.19 is used in the circuit of Fig. P3.20 (note the polarity of V_{GG}). Determine v_{GS}, i_D, and v_{DS}. Use basic circuit analysis and compare with the graphical solution.

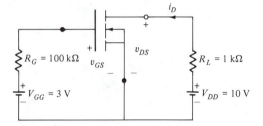

FIGURE P3.20

3.21

The drain current for a particular enhancement-depletion (E-D) MOST is given in mA by $i_D = 5[1 + (v_{GS}/2)]^2$. Sketch the transfer and output characteristics. Use values for v_{GS} of $-2, -1, 0, 1,$ and 2 V.

3.22

Draw a circuit similar in form to Figs. P3.17 and P3.20 with $V_{GG} = 0$ and using an N-channel E-D MOST. For $R_G = 100$ kΩ, $R_L = 1$ kΩ, and $V_{CC} = 20$ V, determine v_{GS}, i_D, and v_{DS}. Use basic circuit analysis and compare with the graphical solution.

3.23

Determine and sketch the voltage transfer characteristic (v_{DS} versus v_{GS}) for the circuit of Fig. P3.20 using the E-O MOST of Problem 3.19. Assume that V_{GG} is varied and use the results of Problems 3.19 and 3.20.

3.24

A P-channel E-O MOST is used in the circuit of Fig. P3.24. The drain current for this device is given in mA by $i_D = 0.8(v_{GS})^2$ for $v_{GS} \le 0$. Determine v_{GS}, i_D, and v_{DS}. Use circuit analysis and compare with the graphical solution.

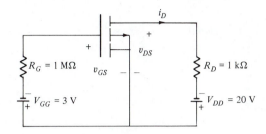

FIGURE P3.24

3.25

Analyze the circuit of Fig. P3.25 to determine the voltage gain (v_0/v_i). The elements inside the dashed rectangle represent a voltage-controlled amplifier (representative of FETs). Let $r_i = 10$ MΩ and $r_o = 20$ kΩ.

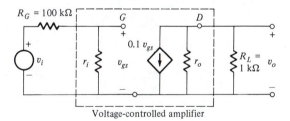

Voltage-controlled amplifier

FIGURE P3.25

3.26

The circuit of Fig. P3.26 utilizes a JFET for which $i_D = 1.65[1 + (v_{GS}/2)]^2$ in mA. If the drain current is 0.8 mA, determine v_{GS}, R_S, and v_{DS}. Hint: Write KVL for both loops.

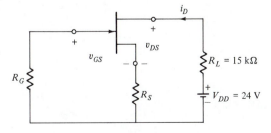

FIGURE P3.26

3.27

For the circuit of Fig. P3.27, determine v_{DS} and v_{GS} for $i_D = 10$ mA.

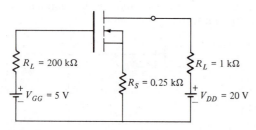

FIGURE P3.27

3.28

Show that the circuit of Fig. P3.28b is equivalent to that of P3.28a. Derive expressions for R_G and V_{GG} and calculate the values corresponding to the circuit values in Fig. P3.28a. For $i_D = 2.5$ mA, determine v_{DS} and v_{GS}.

3.29

Show that the circuit of Fig. P3.29b is equivalent to that of P3.29a. Derive expressions for R_B and V_{BB} and calculate the values corresponding to the circuit values in Fig. P3.29a. Carry out a loop analysis to determine i_B, i_C, and v_{CE}. Let $\beta = 100$ and $V_0 = 0.7$ V.

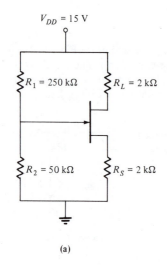

(a)

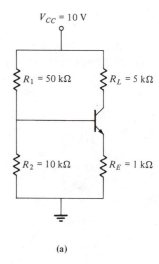

(a)

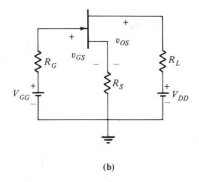

(b)

FIGURE P3.28

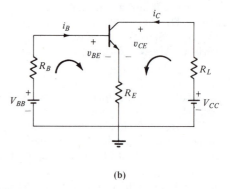

(b)

FIGURE P3.29

CHAPTER 3 APPENDIX

DATA SHEET A3.1   Metal, Plastic, and Multiple Small-Signal Transistors

Small-Signal Transistors – Metal

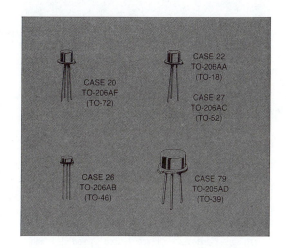

CASE 20 TO-206AF (TO-72)
CASE 22 TO-206AA (TO-18)
CASE 27 TO-206AC (TO-52)
CASE 26 TO-206AB (TO-46)
CASE 79 TO-205AD (TO-39)

Motorola Small-Signal Metal Can Transistors are designed for use as General-Purpose Amplifiers, High-Speed Switches, High-Voltage Amplifiers, Low-Level/Low-Noise Amplifiers, High-Frequency Oscillators, Choppers, and Darlingtons. These devices are manufactured in a variety of packages, i.e., TO-206AA (TO-18), TO-205AD (TO-39), TO-206AB (TO-46), TO-206AC (TO-52), and TO-206AF (TO-72).

The following selector guide tables also indicate metal can transistors which are qualified to MIL-19500 high-rel requirements. Devices are available in the JAN, JANTX, JANTXV and JANS versions as specified in the device type column.

Switching Transistors

Listed in order of decreasing turn-on time (t_{on}).

Package	Device	t_{on} ns Max	t_{off} ns Max	@ I_C mA	$V_{(BR)CEO}$ Volts Min	I_C mA Max	h_{FE} Min	@ I_C mA	$V_{CE(sat)}$ Volts Max	@ I_C mA	@ I_B mA	f_T MHz Min	@ I_C mA
NPN													
TO-206AA	2N2540	40	40	150	30		100	150	0.45	150	15	250	20
(TO-18)	2N914**	40	40	200	15	150	12	10	0.7	200	20	300	20
	2N706*	40	75		15	50	20	10	0.6	10	1.0	200	10
	2N708*	40	70		15		30	10	0.4	10	1.0	300	10
	2N4014	35	60	500	50	1000	35	500	0.52	500	50	300	50
	2N4013	35	60	500	30	1000	35	500	0.42	500	50	300	50
	2N2501	15	25	300	20		10	500	0.3	50	5.0	350	10
	2N2369	12	18	100	15	500	20	100	0.25	10	1.0	500	10
	2N2369A†	12	18	10	15	200	40	10	0.2	10	1.0	500	10
	2N3227**	12	18	100	20	50	30	100	0.25	10	1.0	500	10
TO-205AD	2N3444**	50	70	500	50		20	500	0.6	500	50	175	50
(TO-39)	2N3253**	50	70	500	40		25	500	0.6	500	50	175	50
	2N3735#	48	60	1000	50	1500	20	1000	0.5	500	50	250	50
	2N3734	48	60	1000	50	1500	30	1000	0.5	500	50	250	50
	2N3252	45	70	500	30		30	500	0.5	500	50	200	50
	2N3506#	45	90	1500	40	3000	40	1500	1.0	1500	150	60	100
	2N3507#	45	90	1500	50	3000	30	1500	1.0	1500	150	60	100
	2N3725	35	60	500	50	2000	35	500	0.52	500	50	300	50
	2N3725A	35	60	500	30	1200	35	500	0.52	500	50	300	50
	2N3724	35	60	500	30	2000	35	500	0.42	500	50	300	50
	2N3724A	35	60	500	30	1200	35	500	0.42	500	50	300	50
	MM5262	30	60	1000	50	2000	25	1000	0.8	1000	100	350(typ)	50
	2N5861	25	60	500	50	2000	25	500	0.5	500	50	200	50
	2N3303	15	25	1000	—	1000	20	10	0.7	1000	100	450	100

*JAN available   **JAN/JANTX available   #JAN/JANTX/JANTXV available   †JAN/JANTX/JANTXV/JANS available

continues

DATA SHEET A3.1   Continued

Switching Transistors (continued)

Package	Device	t_{on} ns Max	t_{off} ns Max	@ I_C mA	$V_{(BR)CEO}$ Volts Min	I_C mA Max	h_{FE} Min	@ I_C mA	$V_{CE(sat)}$ Volts Max	@ I_C mA	@ I_B mA	f_T MHz Min	@ I_C mA
NPN (continued)													
TO-206AB	2N3736	48	60	1000	30	1500	30	1000	0.5	500	50	250	50
(TO-46)	2N3737#	48	60	1000	50	1500	20	1000	0.5	500	50	250	50
	2N3647#	20	25	150	10	500	25	150	0.4	150	15	350	15
	2N3648	16	18	150	15	500	30	150	0.4	150	15	450	15
	2N3508	12	18	10	20	500	40	10	0.25	10	1.0	500	10
	2N3509	12	18	10	20	500	100	10	0.25	10	1.0	500	10
TO-206AC	MM1748	6.0	15	10	—	150	20	10	—	—	—	600	5.0
(TO-52)	MM1748A	10	15	10	—	150	20	10	—	—	—	600	5.0
	2N3014	16	25	300	20	200	30	30	0.18	30	3.0	350	30
	2N3013**	15	25	300	0.5	200	30	30	0.18	30	3.0	350	30
PNP													
TO-206AA	2N2894	60	90	30	12	200	40	30	0.2	30	3.0	400	30
(TO-18)	2N869A**	50	80	30	18	200	40	30	0.2	30	3.0	400	10
	2N3546	40	30	50	12		25	50	0.25	50	5.0	700	10
	2N4208	15	20	10	12	200	30	10	0.15	10	1.0	700	10
	MM4258	15	20	10	12	80	30	10	0.15	10	1.0	700	10
	2N4209	15	20	10	15	50	40	50	0.6	50	5.0	850	10
TO-205AD	2N3634#	400	600	50	140	1000	50	50	0.5	50	5.0	150	30
(TO-39)	2N3635#	400	600	50	140	1000	100	50	0.5	50	5.0	200	30
	2N3636#	400	600	50	175	1000	50	50	0.5	50	5.0	150	30
	2N4030	100	240(typ)	500	60	1000	15	1000	1.0	1000	100	100	50
	2N4031	100	240(typ)	500	80	1000	10	1000	0.5	500	50	100	50
	2N4032	100	240(typ)	500	60	1000	40	1000	1.0	1000	100	150	50
	2N4033#	100	240(typ)	500	80	1000	25	1000	0.5	500	50	150	50
	2N4406	75	225	1000	80	1500	20	1000	0.7	1000	100	150	50
	2N4407	75	225	1000	80	1500	30	1000	0.7	1000	100	150	50
	2N3245	55	165	500	50	1000	30	500	0.6	500	50	150	50
	2N3244	50	185	500	40	1000	50	500	0.5	500	50	175	50
	2N3467#	40	90	500	40	100	40	500	0.5	500	50	175	50
	2N3468#	40	90	500	50	1000	25	500	0.6	500	50	150	50
	2N3762#	43	115	1000	40	1500	30	1000	0.9	1000	100	180	50
	2N3763#	43	115	1000	60	1500	20	1000	0.9	1000	100	150	50
	2N4404	40	210	500	80	1000	30	500	0.5	500	50	200	50
	2N4405#	40	210	500	80	1000	50	500	0.5	500	50	200	50
	2N5022	40	90	500	—	500	25	1000	0.8	1000	100	170	50
	2N5023	40	90	500	—	500	40	1000	0.7	1000	100	200	50
TO-206AB	2N4453**	50	80	30	18	200	40	30	0.25	30	1.5	400	10
(TO-46)	2N3765#	12	115	1000	60	1500	40	150	0.9	1000	100		
	2N3764#	12	115	1000	40	1500	40	150	0.9	1000	100		

High-Gain Low-Noise Transistors

Listed in decreasing order of NF.

Package	Device	NF Wideband Typ* Max dB	$V_{(BR)CEO}$ Volts Min	I_C mA Max	h_{FE} Min	h_{FE} Max	@ I_C μA mA	f_T MHz Min	@ I_C mA
NPN									
TO-206AA	2N2484#	8.0*	60	50	100	500	10	15	0.05
(TO-18)	2N930A	3.0	45	30	100	300	10	45	0.5
	2N930**	3.0	45	30	100	300	10	30	0.5

*JAN available   **JAN/JANTX available   #JAN/JANTX/JANTXV available   †JAN/JANTX/JANTXV/JANS available

High-Gain Low-Noise Transistors (continued)

Package	Device	NF Wideband dB Typ* / Max	V(BR)CEO Volts Min	Ic mA Max	hFE Min	hFE Max	@ Ic μA/mA	fT MHz Min	@ Ic mA
PNP									
TO-206AA (TO-18)	2N3962	10	60	200	100	450	1.0	40	0.5
	2N3963	10	80	200	100	450	1.0	40	0.5
	2N3965	8.0	60	200	250	600	1.0	50	0.5
	2N3964	4.0	45	200	250	600	1.0	50	0.5
	2N3798	3.5	60	50	150	450	500	30	0.5
	2N3799	2.5	60	50	300	900	500	30	0.5
TO-206AB (TO-46)	2N2604	4.0	45	0	40	120	0.01	30	0.5
	2N2605#	4.0	45	30	100	300	0.01	30	0.5

High-Frequency Amplifiers/Oscillators

UHF, VHF

Devices are listed in decreasing order of $V_{(BR)CEO}$.

Package	Device	V(BR)CEO Volts Min	hFE Min	@ Ic mA	Gpe dB Min	NF dB Max	@ f MHz	fT MHz Min	@ Ic mA	Cobo pF Max
NPN										
TO-206AA (TO-18)	MM1941	20	25	10	7.0	—	—	600	10	2.5
TO-206AF (TO-72)	2N918†	15	20	3.0	15	6.0	60	600	4.0	1.7
PNP										
TO-206AA (TO-18)	2N3307	35	40	2.0	17	4.5	200	300	2.0	1.3
TO-206AF (TO-72)	2N4261#	15	30	10	—	—	—	1600	10	2.5
	2N4260	15	30	10	—	—	—	2000	10	2.5

High-Voltage/High-Current Amplifiers

Listed in decreasing order of $V_{(BR)CEO}$ within each package type.

Package	Device	V(BR)CEO Volts Min	Ic mA Max	hFE Min	@ Ic mA	VCE(sat) Volts Max	@ Ic mA	& IB mA	fT MHz Min	@ Ic mA
NPN										
TO-206AA (TO-18)	2N6431	300	50	50	30	0.5	20	2.0	50	10
	2N6430	200	50	50	30	0.5	20	2.0	50	10
TO-205AD (TO-39)	MM8520	500	1000	15	10	1.5	10	2.0	5.0	10
	2N3439#	350	1000	40	20	0.5	50	4.0	15	10
	MM421	325	1000	25	30	5.0	30	3.0	15	10
	2N3742	300	50	20	30	1.0	30	3.0	30	10
	2N5058	300	150	35	30	1.0	30	3.0	30	10
	MM420	250	1000	25	30	5.0	30	3.0	15	10
	2N3440#	250	1000	40	20	0.5	50	4.0	15	10
	MM3003	250	50	20	10				150	10
	2N4927	250	50	20	30	2.0	30	3.0	30	10
	2N5059	250	150	30	30	1.0	30	3.0	30	10
	MM3002	200	50	20	10				150	10
	2N4926	200	50	20	30	2.0	30	3.0	30	10
	MM3009	180	400	40	10				50	20

*JAN available   **JAN/JANTX available   #JAN/JANTX/JANTXV available   †JAN/JANTX/JANTXV/JANS available

continues

DATA SHEET A3.1   Continued

High-Voltage/High-Current Amplifiers (continued)

Package	Device	V(BR)CEO Volts Min	IC mA Max	hFE Min	@ IC mA	VCE(sat) Volts Max	@ IC mA	& IB mA	fT MHz Min	@ IC mA
NPN (continued)										
TO-205AD (TO-39)	MM3001	150	200	20	10				150	10
	2N3114	150	200	30	30	1.0	50	5.0	40	30
	2N3500#	150	300	40	150	0.4	150	15	150	20
	2N3501#	150	300	100	150	0.4	150	15	150	20
	2N3712	150	200	30	30	2.0	50	5.0	40	30
	2N5682	120	1000	40	250	0.6	250	25	30	100
	MM3008	120	400	40	10				50	20
	2N657	100		30	200	4.0	200	40		
	2N3498#	100	500	40	150	0.6	300	30	150	20
	2N3499#	100	500	100	150	0.6	300	30	150	20
	2N4924	100	200	40	150	0.4	50	5.0	100	20
	MM3007	100	2500	50	250	0.35	150	15	50	50
	2N5681	100	1000	40	250	0.6	250	25	30	100
	MM3006	80	2500	50	200	0.35	150	15	50	50
	2N4239	80	3000	30	250	0.3	500	50	2.0	100
	MM3005	60	2500	50	150	0.35	150	15	50	50
	2N656	60	—	30	200	4.0	200	40		
	2N4238	60	3000	30	250	0.3	500	50	2.0	100
	2N4237	40	3000	30	250	0.3	500	50	2.0	100
PNP										
TO-206AA (TO-18)	2N6433	300	1000	30	30	0.5	20	2.0	50	10
	2N6432	200	1000	30	30	0.5	20	2.0	50	10
	2N3497	120	100	40	10	0.35	10	1.0	150	20
	2N3496	80	100	40	10	0.3	10	1.0	200	20
TO-205AD (TO-39)	2N3743#	300	50	25	30	8.0	30	3.0	30	10
	2N5416#	300	1000	30	50	2.5	50	5.0	15	10
	MM4003	250	500	20	10	5.0	10	1.0	—	—
	2N4931#	250	500	20	20	5.0	10	1.0	20	20
	MM4002	200	500	20	10	5.0	10	1.0	—	—
	2N4930#	200	500	20	20	5.0	10	1.0	20	20
	2N5415#	200	1000	30	50	2.5	50	5.0	15	10
	2N3637#	175	1000	100	50	0.5	50	5.0	200	30
	2N3636#	175	1000	50	50	0.5	50	5.0	150	30
	2N4929	150	500	25	10	0.5	10	1.0	100	20
	MM4001	150	500	20	10	0.6	10	1.0	—	—
	2N3635#	140	1000	100	50	0.5	50	5.0	200	30
	2N3634#	140	1000	50	50	0.5	50	5.0	150	30
	2N3495	120	100	40	10	0.35	10	1.0	150	20
	2N5680	120	1000	40	250	0.6	250	25	30	100
	MM4000	100	100	20	10	0.6	10	1.0	—	—
	MM5007	100	2000	50	250	0.5	150	15	30	50
	2N4928	100	100	25	10	0.5	10	1.0	100	20
	2N5679	100	1000	40	250	0.6	250	25	30	100
	MM5006	80	2000	50	200	0.5	150	15	30	50
	2N3494	80	100	40	10	0.3	10	1.0	200	20
	2N4236	80	3000	30	250	0.6	1000	125	3.0	100
	MM5005	60	2000	50	150	0.5	150	15	30	50
	2N4235	60	3000	30	250	0.6	1000	125	3.0	100
	2N4234	40	3000	30	250	0.6	1000	125	3.0	100

*JAN available   **JAN/JANTX available   #JAN/JANTX/JANTXV available   †JAN/JANTX/JANTXV/JANS available

DATA SHEET A3.1   Continued

General-Purpose Amplifiers

Designed for dc to VHF amplifier and general-purpose switching applications, listed in decreasing order of $V_{(BR)CEO}$ within each package group.

Package	Device	$V_{(BR)CEO}$ Volts Min	f_T MHz Min	@ I_C mA	I_C mA Max	h_{FE} Min	h_{FE} Max	@ I_C mA
NPN								
TO-206AA	2N2896	90	120	50	1000	60	200	150
(TO-18)	2N3700#	80	80	1.0	1000	50		500
	2N2895	65	120	50	1000	40	120	150
	2N956	50	70	50		40	120	150
	2N2897	45	100	50	1000	50	200	150
	2N718,A	40	50	50		40	120	150
	2N2221A#	40	250	20	800	40	120	150
	2N2222A†	40	300	20	800	100	300	150
	2N3946	40	300	10	200	50	150	10
	2N3947	40	300	10	200	100	300	10
	2N2222#	30	250	20	800	100	300	150
	2N3302	30	250	50	500	100	300	150
	2N916*	25	300	10		50	200	10
TO-205AD	2N1711	80	70	50		100	300	150
(TO-39)	2N3019#	80	100	50	1000	100	300	150
	2N3020	80	80	50	1000	40	120	150
	2N1613#	50	60	50	500	40	120	150
	2N2193A	50	50	50	1000	40	120	150
	2N2270	45	100	50	1000	50	200	150
	2N697	40	50	50		40	120	150
	2N2218A#	40	250	20	800	40	120	150
	2N2219A†	40	300	20	800	100	300	150
	2N3053	40	100	50	700	50	250	150
	2N2218#	30	250	20	800	40	120	150
	2N2219#	30	250	20	800	100	300	150
	2N3300	30	250	50	500	100	300	150
TO-206AB	2N5581**	40	250	20	800	40	120	150
(TO-46)	2N5582**	40	300	20	800	100	300	150
TO-206AC	MM3903	40	250	10	200	50	150	10
(TO-52)	MM3904	40	300	10	200	100	300	10
PNP								
TO-206AA	2N4026	80	100	50	1000	15	—	100
(TO-18)	2N4027	80	100	50	1000	10	—	100
	2N4028	80	150	50	1000	40	—	100
	2N4029	80	150	50	1000	25	—	100
	2N2906A#	60	200	50	600	40	120	150
	2N2907A†	60	200	50	600	100	300	150
	2N3250A#	60	250	10	200	50	150	10
	2N3251A#	60	300	10	200	100	300	10
	2N2906#	40	200	50	600	40	120	150
	2N2907#	40	200	50	600	100	300	150
	2N3250	40	250	10	200	50	150	10
	2N3251	40	300	10	200	100	300	10
TO-205AD	MM5007	100	30	50	2000	50	250	250
(TO-39)	MM5006	80	30	50	2000	50	250	200
	2N4031	80	100	50	1000	10	—	100
	2N4033#	80	150	50	1000	25	—	100
	2N4404	80	200	50	1000	40	120	150
	2N4405**	80	200	50	1000	100	300	150

*JAN available   **JAN/JANTX available   #JAN/JANTX/JANTXV available   †JAN/JANTX/JANTXV/JANS available

continues

General-Purpose Amplifiers (continued)

Package	Device	$V_{(BR)CEO}$ Volts Min	f_T MHz Min	@ I_C mA	I_C mA Max	h_{FE} Min	Max	@ I_C mA
NPN (continued)								
TO-205AD	MM4036	65	60	50	1000	20	140	150
(TO-39)	2N4036	65	60	50	1000	40	140	150
	2N4037	65	60	50	1000	40	—	150
	MM5005	60	30	50	2000	50	250	150
	2N2904A#	60	200	50	600	40	120	150
	2N2905A†	60	200	50	600	100	300	150
	2N4030	60	100	50	1000	15	—	100
	2N4032	60	150	50	1000	40	—	100
	MM4037	40	60	50	1000	50	250	150
	2N1131A	40	50	50	600	30	90	150
	2N1132A	40	60	50	600	30	90	150
	2N2904#	40	200	50	600	40	120	150
	2N2905#	40	200	50	600	100	300	150
	2N1132*	35	60	50	600	30	90	150
TO-206AB	2N3485A**	60	200	50	600	40	120	150
(TO-46)	2N3486A**	60	200	50	600	100	300	150
	2N3673	50	200	50	600	75	225	150
	2N3486	40	200	50	600	100	300	150
TO-206AC	MM3906	40	250	10	200	100	300	10
(TO-52)	MM3905	40	200	10	200	50	150	10

*JAN available   **JAN/JANTX available   #JAN/JANTX/JANTXV available   †JAN/JANTX/JANTXV/JANS available

Choppers

Devices are listed in decreasing $V_{(BR)EBO}$.

PNP

Package	Device	$V_{(BR)EBO}$ Min	$V_{(BR)ECO}$	$h_{FE(inv)}$ Min	Offset Voltage $V_{EC(ofs)}$ Max (mV)	On-State Resistance $r_{ec(on)}$ Max (Ω)
TO-206AB	2N2946	40	35	3.0	2.0	45
(TO-46)	2N2946A	40	35	20	2.0	8.0
	2N5230	30	20	15	0.5	8.0
	2N5231	30	20	15	0.8	10
	2N2945A	25	20	30	1.0	6.0
	2N2945	25	20	4.0	1.0	35
	2N2944A	15	10	50	0.3	4.0
	2N2944	15	10	6.0	0.6	20
	2N5229	15	10	15	0.5	6.0

JAN/JANTX available

Darlington
NPN

Package	Device	$V_{(BR)CEO}$ Volts Min	I_C mA Max	h_{FE} Min	@ I_C mA	f_T MHz Min	@ I_C mA
TO-206AA (TO-18)	MM6427	40	300	5.0 k	10	125	100

DATA SHEET A3.1   Continued

Small-Signal Transistors – Plastic

Motorola's small-signal TO-226AA (TO-92) plastic transistors encompass hundreds of devices with a wide variety of characteristics for general-purpose, amplifier and switching applications. The popular high-volume TO-226AA (TO-92) package combines proven reliability, performance, economy and convenience to provide the perfect solution for industrial and consumer design problems. All Motorola TO-226AA (TO-92) devices are laser marked for ease of identification and shipped in antistatic containers, as part of Motorola's ongoing practice of maintaining the highest standards of quality and reliability.

In addition to the standard TO-226AA (TO-92) devices listed in the following tables, Motorola also offers special electrical selections of these devices. Please contact your Motorola Sales Representative regarding any special requirements you may have.

In each of the following tables, the major specifications of the TO-226AA (TO-92) transistor are given for easy comparison.

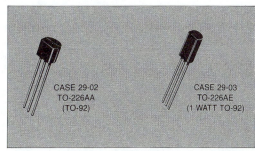

CASE 29-02
TO-226AA
(TO-92)

CASE 29-03
TO-226AE
(1 WATT TO-92)

Motorola TO-226AA (TO-92) transistors are available in the radial or axial tape and reel formats. Lead forming to fit TO-205AA (TO-5) or TO-206AA (TO-18) sockets is also available.

General-Purpose

These general-purpose transistors are designed for small-signal amplification from dc to low radio frequencies. They are also useful as oscillators and general-purpose switches. The transistors are listed in order of decreasing breakdown voltage, $V_{(BR)CEO}$.

P_D @ T_A = 25°C = 625 mW

Device and Polarity		$V_{(BR)CEO}$ Volts Min	f_T MHz Min	@ I_C mA	I_C mA Max	h_{FE}		@ I_C mA
NPN	PNP					Min	Max	
MPS8099	MPS8599	80	150	10	200	100	300	1.0
MPS-A06	MPS-A56	80	100	10	500	50	—	100
MPS8098	MPS8598	60	150	10	200	100	300	1.0
MPS-A05	MPS-A55	60	100	10	500	50	—	100
MPS651	MPS751	60	75	50	2000	40	—	2000
2N3904	2N3906	40	300	10	200	100	300	10
2N4401	2N4403	40	250	20	600	100	300	150
2N3903	2N3905	40	250	10	200	50	150	100
2N4400	2N4402	40	200	20	600	50	150	150
MPS-A20	MPS-A70	40	125	5.0	100	40	400	5.0
MPS650	MPS750	40	75	50	2000	40	—	2000
MPS6531	MPS6534	40	390†	50	600	90	270	100
MPS2222	MPS2907	30	250	20	600	100	300	150
2N4123	2N4125	30	250	10	200	50	150	2.0
MPS3704	MPS3702	30	100	50	600	100	300	50
MPS6513	MPS6517	30	330†	10	100	90	180	2.0
2N4124	2N4126	25	300	10	200	120	360	2.0
MPS6514	MPS6518	25	480†	10	100	150	300	2.0
MPS6515	MPS6519	25	480	10	100	250	500	2.0
MPS5172		25			100	100	500	10
MPS6560	MPS6562	25	60	10	500	50	200	600
MPS6601	MPS6651	25	100	50	1000	30	150	1000

1 Watt TO-92 (TO-226AE) (P_D @ T_A = 25°C = 1.0 W)

MPS6717	MPS6729	80	50	200	500	80	—	50
MPSW06	MPSW56	80	50	200	500	80	—	50
MPS6716	MPS6728	60	50	200	500	80	—	50
MPSW05	MPSW55	60	50	200	500	80	—	50
MPS6715	MPS6727	40	50	50	1000	50	—	1000
MPSW01A	MPSW51A	40	50	50	1000	50	—	1000
MPS6714	MPS6726	30	50	50	1000	50	—	1000
MPSW01	MPSW51	30	50	50	1000	50	—	1000

†Typ

continues

DATA SHEET A3.1   Continued

Low-Noise Amplifier

Listed in decreasing order of noise figure (NF).

Device	NF dB Typ	@ f*	V(BR)CEO Volts Min	hFE Min	@ IC mA	fT MHz Min	@ IC mA
NPN							
2N6428	6.0	Audio	50	250	10	100	1.0
2N4123	6.0	Audio	30	50	2.0	250	10
2N6429	5.0	Audio	45	500	10	100	1.0
2N4124	5.0	Audio	25	120	2.0	300	10
2N6428A	4.0 Max	Audio	50	250	10	100	1.0
2N6429A	3.5 Max	Audio	45	500	10	100	1.0
2N5209	3.0 Max	Audio	50	150	10	30	0.5
2N5088	3.0	Audio	30	300	10	50	0.5
MPS6520	3.0	Audio	25	200	2.0	390†	2.0
MPS6521	3.0	Audio	25	300	2.0	390†	2.0
2N5210	2.0 Max	Audio	50	250	10	30	0.5
MPS8097	2.0 Max	Audio	40	250	0.1	200	10
2N5089	2.0 Max	Audio	25	400	10	50	0.5
MPSA18	1.5 Max	Audio	45	500	10	100	1.0
MPSA09	1.4	1.0 kHz	50	100	0.1	30	0.5
PNP							
2N4125	5.0	Audio	30	50	2.0	200	10
2N4126	4.0	Audio	25	120	2.0	250	10
2N5086	3.0	Audio	50	150	10	40	0.5
MPS6522	3.0	Audio	25	200	2.0	340†	2.0
MPS6523	3.0	Audio	25	300	2.0	340†	2.0
MPS4249	3.0	1.0 kHz	60	100	10	100	1.0
2N5087	2.0	Audio	60	250	10	40	0.5
MPS4250	2.0	1.0 kHz	40	250	10	250	1.0
*MPS4250A	2.0	1.0 kHz	60	250	0.1	250	1.0

*Audio = 10 Hz to 15.7 kHz.   †Typ

High-Voltage

These high-voltage transistors are designed for driving neon bulbs and Nixie® indicator tubes, for direct line operation, and for other applications requiring high-voltage capability at relatively low collector current. Devices are listed in order of decreasing breakdown voltage, $V_{(BR)CEO}$.

Device	$V_{(BR)CEO}$ Volts Min	I_C Amp* Max	h_{FE} Min	@ I_C mA	$V_{CE(sat)}$ Volts Max	I_C mA	& I_B mA	f_T MHz Min	@ I_C mA
NPN									
MPS-A44	400	0.3	50	10	0.75	50	5.0	20	10
2N6517	350	0.5	30	30	0.30	10	1.0	40	10
MPS-A45	350	0.3	50	10	0.75	50	5.0	20	10
2N6516	300	0.5	45	30	0.30	10	1.0	40	10
MPS-A42	300	0.5	40	10	0.5	20	2.0	50	10
2N6515	250	0.5	50	30	0.30	10	1.0	40	10
MPS-A43	200	0.5	40	10	0.4	20	2.0	50	10
MPS-D01	200	0.1	20	30				40	10
2N5551	160	0.6	80	10	0.15	10	1.0	100	10
2N5550	140	0.6	60	10	0.15	10	1.0	100	10
MPS-L01	120	0.15	50	10				60	10
1 Watt TO-92 (TO-226AE)									
MPS6735	300	0.3	40	10	2.0	20	2.0	50	10
MPSW10	300	0.3	40	30	0.75	30	3.0	45	10
MPSW42	300	0.3	40	30	0.50	20	2.0	50	10
MPS6734	250	0.3	40	10	2.0	20	2.0	50	10
MPSW43	200	0.3	50	30	0.4	20	2.0	50	10
MPS6733	200	0.3	40	10	2.0	20	2.0	50	10
PNP									
2N6520	350	0.5	30	30	0.30	10	1.0	40	10
2N6519	300	0.5	45	30	0.30	10	1.0	40	10
MPS-A92	300	0.5	40	10	0.8	20	2.0	50	10
2N6518	250	0.5	50	30	0.30	10	1.0	40	10
MPS-A93	200	0.5	40	10	0.7	20	2.0	50	10
MPS-D51	200	0.1	20	30				40	10
2N5401	150	0.6	60	10	0.5	50	0.5	100	10
2N5400	120	0.6	40	10	0.5	50	0.5	100	10
MPS-L51	100	0.6	40	50	0.25	10	1.0	60	10
1 Watt TO-92 (TO-226AE)									
MPSW60	300	0.3	40	30	0.75	20	2.0	60	10
MPSW92	300	0.3	25	30	0.50	20	2.0	50	10
MPSW93	200	0.3	30	30	0.40	20	2.0	50	10

continues

DATA SHEET A3.1 Continued

Small-Signal Transistors – Multiples

The trend in electronic system design is toward the use of integrated circuits — to reduce component cost, assembly cost, and equipment cost. But ICs still aren't all things to all people, and for those circuit designs where ICs are not available, there is a noticeable swing towards the use of multiple devices.*

Motorola is reacting to this expanding market requirement by making available a large selection of Quad, Dual, and Darlington transistors for off-the-shelf delivery. The chips used in the Quad and Dual transistors are those that have emerged as the most popular ones for discrete transistor applications. But even beyond that, Motorola offers its entire vast repertoire of discrete small-signal transistors for multiple-device packaging. For special applications, special configurations can be supplied with quick turnaround time and at low premiums.

Multiple devices, as described here, encompass two or more transistor chips in a single package. Included in this definition are the Darlington transistors which consist of two interconnected devices functioning as a single-stage amplifier.

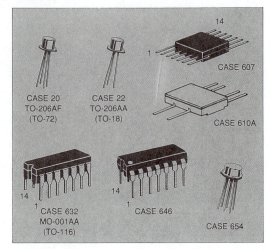

CASE 20
TO-206AF
(TO-72)

CASE 22
TO-206AA
(TO-18)

CASE 607

CASE 610A

CASE 632
MO-001AA
(TO-116)

CASE 646

CASE 654

Specification

The following short form specifications include Quad and Dual transistors listed in alphanumeric order. Some columns denote two different types of data indicated by either **bold** or *italic* typeface. See key and headings for proper identification.

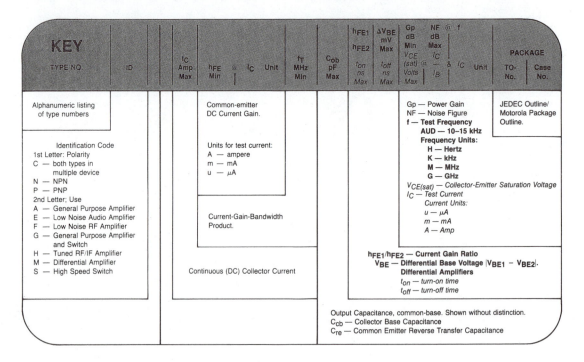

KEY TYPE NO.	ID	I_C Amp Max	h_{FE} Min	@	I_C	Unit	f_T MHz Min	C_{ob} pF Max	h_{FE1} h_{FE2} / t_{on} ns Max	ΔV_{BE} mV Max / t_{off} ns Max	Gp dB Min V_{CE} (sat) Volts Max	NF @ f dB Max I_C / I_B	&	I_C	Unit	PACKAGE TO-No.	Case No.

Alphanumeric listing of type numbers

Identification Code
1st Letter: Polarity
C — both types in multiple device
N — NPN
P — PNP
2nd Letter; Use
A — General Purpose Amplifier
E — Low Noise Audio Amplifier
F — Low Noise RF Amplifier
G — General Purpose Amplifier and Switch
H — Tuned RF/IF Amplifier
M — Differential Amplifier
S — High Speed Switch

Common-emitter DC Current Gain.

Units for test current:
A — ampere
m — mA
u — μA

Current-Gain-Bandwidth Product.

Continuous (DC) Collector Current

Gp — Power Gain
NF — Noise Figure
f — Test Frequency
AUD — 10–15 kHz
Frequency Units:
H — Hertz
K — kHz
M — MHz
G — GHz
$V_{CE(sat)}$ — Collector-Emitter Saturation Voltage
I_C — Test Current
Current Units:
u — μA
m — mA
A — Amp

h_{FE1}/h_{FE2} — Current Gain Ratio
V_{BE} — Differential Base Voltage $|V_{BE1} - V_{BE2}|$.
Differential Amplifiers
t_{on} — turn-on time
t_{off} — turn-off time

Output Capacitance, common-base. Shown without distinction.
C_{cb} — Collector Base Capacitance
C_{re} — Common Emitter Reverse Transfer Capacitance

JEDEC Outline/ Motorola Package Outline.

DATA SHEET A3.1   Continued

Quad Transistors

TYPE NO.	ID	P_D Watts T_A=25°C One Die Only	V_CEO Volts	I_C Amp Max	h_FE Min	@ I_C	f_T MHz Min Typ*	C_ob pF Max Typ*	h_FE1/h_FE2 t_on ns Max Typ*	ΔV_BE mV Max t_off ns Max Typ*	G_p dB Min V_CE(sat) Volts Max	I_C/I_B Max	NF dB Max Typ*	& I_C	PACKAGE TO-No.	Case No.
MHQ918	NF	0.65	15	0.05	20	3.0 m	600	2.0					**6.0**	**60 M**	116	632
MHQ2221	NG	0.65	40	0.5	40	150 m	200	8.0			0.4			150 m	116	632
MHQ2222†	NG	0.65	40	0.5	100	150 m	200	8.0	25*	250*	0.4	10		150 m	116	632
MHQ2369	NS	0.5	15	0.5	40	10 m	450	4.0	9.0*	15*	0.25	10		10 m	116	632
MHQ2483	NA	0.6	40	0.05	150	1.0 m	50						**3***	**AUD**	116	632
MHQ2484	NA	0.6	40	0.05	300	1.0 m	50						**2***	**AUD**	116	632
MHQ2906	PG	0.65	40	0.6	40	150 m	200	8.0	30*	100*	0.4	10		150 m	116	632
MHQ2907†	PG	0.65	40	0.6	100	150 m	200	8.0	30*	100*	0.4	10		150 m	116	632
MHQ3467†	PS	0.9	40	1.0	20	500 m	125	25	40	90	0.5	10		500 m	116	632
MHQ3546	PS	0.5	12	0.2	30	10 m	600	6.0	0.15*	25*	0.25	10		10 m	116	632
MHQ3798	PA	0.5	40	0.05	150	0.1 m	60	4.0					**3***	**AUD**	116	632
MHQ3799	PA	0.5	60	0.05	300	0.1 m	60	4.0					**2***	**AUD**	116	632
MHQ4001A	NS	0.75	40	1.5	30	500 m	200	10	40	75	0.52	10		500 m	116	632
MHQ4002A	NS	0.75	45	1.5	30	500 m	200	10	40	75	0.52	10		500 m	116	632
MHQ4013††	NS	0.75	40	1.5	35	500 m	200	10	35	60	0.52	10		500 m	116	632
MHQ4014	NS	0.75	45	1.5	35	500 m	200	10	35	60	0.52	10		500 m	116	632
MHQ6001	CA	0.65	30	0.5	40	150 m	200	8.0	30*	225*	0.4	10		150 m	116	632
MHQ6002	CA	0.65	30	0.5	100	150 m	200	8.0	30*	225*	0.4	10		150 m	116	632
MHQ6100	CA	0.5	40	0.05	75	1.0 m	175*	4.5*			0.25			1.0 m	116	632
MPQ918	NA	0.625	15	0.05	20	3.0 m	600	1.7					**6.0**	**60 M**		646
MPQ1000	NA	0.65	20	0.5	50	10 m	175	8.0			0.5	10		150 m		646
MPQ2221	NA	0.65	30	0.5	40	150 m	200	8.0	25*	250*	0.4	10		150 m		646
MPQ2221A	NA	0.65	30	0.5	40	150 m	200	8.0	25*	250*	0.4	10		150 m		646
MPQ2222	NA	0.65	30	0.5	100	150 m	200	8.0	25*	250*	0.4	10		150 m		646
MPQ2222A	NA	0.65	30	0.5	100	150 m	200	8.0	25*	250*	0.4	10		150 m		646
MPQ2369	NS	0.5	15	0.5	40	10 m	450	4.0	9.0*	15*	0.25	10		10 m		646
MPQ2483	NA	0.625	40	0.05	150	1.0 m	50						**3***	**AUD**		646
MPQ2484	NA	0.625	40	0.05	300	1.0 m	50						**2***	**AUD**		646
MPQ2906	PA	0.65	40	0.6	40	150 m	200	8.0	30*	100*	0.4	10		150 m		646
MPQ2906A	PA	0.65	60	0.6	40	150 m	200	8.0	30*	100*	0.4	10		150 m		646
MPQ2907	PA	0.65	40	0.6	100	150 m	200	8.0	30*	100*	0.4	10		150 m		646
MPQ2907A	PA	0.65	60	0.6	100	150 m	200	8.0	30*	100*	0.4	10		150 m		646
MPQ3303	NS	0.65	12	1.0	40	300 m	400	10	15	25	0.7	10		1.0 A		646
MPQ3467	PS	0.75	40	1.0	20	500 m	125	25	40	90	0.5	10		500 m		646
MPQ3546	PA	0.5	12	0.2	30	10 m	600	6.0	15*	25*	0.25	10		10 m		646
MPQ3725†	NS	1.0	40	1.0	25	500 m	250	10	35	60	0.45	10		500 m		646
MPQ3725A	NS	1.0	50	1.0	30	500 m	200	10	3.5	60	0.45			500 m		646
MPQ3762	PS	0.75	40	1.5	35	150 m	150	15	50	120	0.55	10		500 m		646
MPQ3798	PA	0.625	40	0.05	150	0.1 m	60	4.0					**3***	**AUD**		646
MPQ3799	PA	0.625	60	0.05	300	0.1 m	60	4.0					**2***	**AUD**		646
MPQ3904	NG	0.50	40	0.2	75	10 m	250	4.0	37*	136*	0.2	10		10 m		646
MPQ3906	PG	0.50	40	0.2	75	10 m	200	4.5	43*	155*	0.25	10		10 m		646

† H, HX, and HXV Suffixes also available.
†† MHQ4013 is electrically equivalent to MHQ3725.

continues

DATA SHEET A3.1   Continued

Quad Transistors (continued)

TYPE NO.	ID	P_D Watts T_A=25°C One Die Only	V_CEO Volts	I_C Amp Max	h_FE Min	@ I_C	f_T MHz Min Typ*	C_ob pF Max Typ*	h_FE1/h_FE2 (I_on ns Max Typ*)	ΔV_BE mV Max (I_off ns Max Typ*)	G_p dB Min / V_CE(sat) @ Volts Max	NF dB Max Typ* / I_C/I_B	f & I_C Unit	TO-No.	Case No.
MPQ6001	CG	0.65	30	0.5	40	150 m	200	8.0	30*	225*	0.4	10	150 m		646
MPQ6002	CG	0.65	30	0.5	100	150 m	200	8.0	30*	225*	0.4	10	150 m		646
MPQ6100	CA	0.5	40	0.05	75	1.0 m	50	4.0				4*	AUD		646
MPQ6100A	CA	0.5	45	0.05	150	1.0 m	50	4.0				4*	AUD		646
MPQ6501	CG	0.65	30	0.5	40	150 m	200	8.0	30*	225*	0.4	10	150 m		646
MPQ6502	CG	0.65	30	0.5	100	150 m	200	8.0	30*	225*	0.4	10	150 m		646
MPQ6600	CA	0.5	40	0.05	75	1.0 m	50	4.0				4*	AUD		646
MPQ6600A	CA	0.5	45	0.05	150	1.0 m	50	4.0			0.25	4.0	1.0 m		646
MPQ6700	CA	0.5	40	0.2	70	10 m	200	4.5	45	150	0.15	10	0.5 m		646
MPQ6842	CA	0.75	40	0.5	70	10 m	300	4.5	45	150	0.15	10	0.5 m		646
MPQ7041	NA	0.75	150	0.5	25	1.0 m	50	5.0			0.5	10	20 m		646
MPQ7042	NA	0.75	200	0.5	25	1.0 m	50	5.0			0.5	10	20 m		646
MPQ7043	NA	0.75	250	0.5	25	1.0 m	50	5.0			0.5	10	20 m		646
MPQ7051	CA	0.75	150	0.5	25	1.0 m	50	5.0			0.7	10	20 m		646
MPQ7052	CA	0.75	200	0.5	25	1.0 m	50	5.0			0.7	10	20 m		646
MPQ7053	CA	0.75	250	0.5	25	1.0 m	50	5.0			0.7	10	20 m		646
MPQ7091	PA	0.75	150	0.5	25	1.0 m	50	5.0			0.5	10	20 m		646
MPQ7092	PA	0.75	200	0.5	25	1.0 m	50	5.0			0.5	10	20 m		646
MPQ7093	PA	0.75	250	0.5	35	10 m	50	5.0			0.5	10	20 m		646
MQ918	NA	0.55	15	0.05	50	3.0 m	600	1.7				6.0	60 M		607
MQ930	NA	0.4	45	0.03	150	1.0 m	260*	6.0			0.5	10	150 m		607
MQ982	PA	0.4	50	0.6	40	150 m	200	8.0			0.5	10	150 m		607
MQ1120	PA	0.4	30	0.5	50	10 m	200	8.0			0.10	10	10 m		607
MQ1129	NA	0.4	30	0.5	100	10 m	200	8.0			0.15	10	10 m		607
MQ2218	NA	0.4	30	0.5	40	150 m	200	8.0			0.4	10	150 m		607
MQ2218A	NA	0.6	40	0.5	40	150 m	200	8.0			0.4	10	150 m		607
MQ2219	NA	0.6	30	0.5	100	150 m	200	8.0			0.3	10	150 m		607
MQ2219A	NA	0.4	30	0.5	100	150 m	200	8.0			0.3	10	150 m		607
MQ2369	NS	0.40	15	0.5	40	10 m	500	4.0	15	20	0.25	10	10 m		607
MQ2484	NE	0.4	60	0.03	100	10 u	260*	6.0				3.0	AUD		607
MQ2904	PG	0.4	40	0.6	40	150 m	300	8.0	42	130	0.4	10	150 m		607
MQ2905A	PG	0.4	60	0.6	100	150 m	300	8.0	42	130	0.4	10	150 m		607
MQ3251	PA	0.40	40	0.05	100	10 m	300	6.0			0.25	10	10 m		607
MQ3467	PS	0.40	40	1.0	20	500 m	150	20	40	110	0.5	10	500 m		607
MQ3725	NS	0.40	40	1.0	50	100 m	200	10	45	75	0.26	10	100 m		607
MQ3762	PS	0.40	40	1.5	20	1.0 A	150	20	40	110	1.0	10	1.0 A		607
MQ3798	PA	0.40	60	0.05	150	100 u	450*	4.0			0.2	10	1.0 m		607
MQ3799	PA	0.40	60	0.05	300	100 u	450*	4.0			0.2	10	1.0 m		607
MQ3799A	PM	0.40	60	0.05	300	100 u	450*	4.0	0.9	3.0	0.2	10	1.0 m		607
MQ6001	CG	0.40	30	0.5	40	150 m	200	8.0	60	350	0.4	10	150 m		607
MQ6002	CG	0.40	30	0.5	100	150 m	200	8.0	60	350	0.4	10	150 m		607
MQ7001	PA	0.4	30	0.6	70	1.0 m	200	8.0			0.4	10	150 m		607
MQ7003	NA	0.40	40	0.05	50	10 m	200	6.0			0.35	10	1.0 m		607
MQ7004	NA	0.40	13	0.2	30		675*	4.0			0.4	10	10 m		607
MQ7005	NA	0.4	12	0.05	30	3.0 m	400	3.0			1.0		10 m		607
MQ7007	PA	0.4	40	0.2	30	1.0 m	300	8.0			1.0	10	50 m		607
MQ7021	CG	0.40	40	0.05	50	10 m	200	6.0	28*	72*	0.35	10	10 m		607
2N5146	PA	0.4	40	1.5	20	1.0 A	150	20	40	110	1.0	10	1.0 A		607
2N6501	NS	0.6	40	1.0	50	100 m	250	10	35	60	0.3	10	100 m		607

Some columns show 2 different types of data indicated by either **bold** or *italic* typefaces. See key and headings.

DATA SHEET A3.1   Continued

Dual Transistors

TYPE NO.	ID	P_D Watts $T_A=25°C$ One Die Only	V_{CEO} Volts	I_C Amp Max	h_{FE} Min	@ I_C (Unit)	f_T MHz Min	C_{ob} pF Max	h_{FE1}/h_{FE2} | t_{on} ns Max	ΔV_{BE} mV Max | t_{off} ns Max	G_p dB Min | V_{CE}(sat)@ Volts Max	NF dB Max | I_C/I_B	f | I_C & (Unit)	PACKAGE TO-No.	Case No.
MD708	NG	0.55	15	0.2	40	10 m	300	5.0	35	75	0.20	10	10 m		654
MD708A	NM	0.55	15	0.2	40	10 m	300	5.0	0.9	5.0	0.20	10	10 m		654
MD708AF	NM	0.35	15	0.2	40	10 m	300	5.0	0.9	5.0	0.20	10	10 m		610A
MD708B	NM	0.55	15	0.2	40	10 m	300	5.0	0.8	10	0.20	10	10 m		654
MD708BF	NM	0.35	15	0.2	40	10 m	300	5.0	0.8	10	0.20	10	10 m		610A
MD708F	NG	0.35	15	0.2	40	10 m	300	5.0	35	75	0.20	10	10 m		610A
MD918	NF	0.55	15	0.05	50	3.0 m	600	1.7				6.0	60 M		654
MD918A	NM	0.55	15	0.05	50	3.0 m	600	1.7	0.9	5.0		6.0	60 M		654
MD918AF	NM	0.35	15	0.05	50	3.0 m	600	1.7	0.9	5.0		6.0	60 M		610A
MD918B	NM	0.55	15	0.05	50	3.0 m	600	1.7	0.8	10		6.0	60 M		654
MD918F,BF	NF	0.35	15	0.05	50	3.0 m	600	1.7				6.0	60 M		610A
MD982,F	PA	0.40	50	0.6	40	150 m	200	8.0			0.5	10	150 m		610A
MD984	PA	0.575	20	0.2	25	10 m	250				0.5	10	50 m		654
MD985	CA	0.575	30	0.5	40	150 m	200	8.0			0.5	10	150 m		654
MD985F	CA	0.35	30	0.5	40	150 m	200	8.0			0.5	10	150 m		610A
MD986	CA	0.55	15	0.2	25	10 m	200	4.0			0.3	10	10 m		654
MD986F	CA	0.35	15	0.2	25	10 m	200	4.0			0.3	10	10 m		610A
MD1120	NM	0.575	30	0.5	50	10 m	200	8.0	0.8	10	0.10	10	10 m		654
MD1120F	NM	0.35	30	0.5	50	10 m	200	8.0	0.8	10	0.10	10	10 m		610A
MD1121	NM	0.575	30	0.5	50	10 m	200	8.0	0.9	10	0.10	10	10 m		654
MD1121F	NM	0.35	30	0.5	50	10 m	200	8.0	0.9	10	0.10	10	10 m		654
MD1122	NM	0.575	30	0.5	50	10 m	200	8.0	0.9	5.0	0.10	10	10 m		654
MD1122F	NM	0.35	30	0.5	50	20 m	200	8.0	0.9	5.0	0.10	10	10 m		654
MD1123	PM	0.575	40	0.2	30	100 u	250	4.0	0.8	10	0.25	10	10 m		654
MD1129	NM	0.575	30	0.5	100	10 m	200	8.0	0.9	5.0	0.1	10	10 m		654
MD1129F	NM	0.35	30	0.5	100	10 m	200	8.0	0.9	5.0	0.15	10	10 m		610A
MD1130	PM	0.575	40	0.2	100	100 u	200	4.0	0.9	5.0	0.25	10	10 m		654
MD1130F	PM	0.35	40	0.2	100	100 u	200	4.0	0.9	5.0	0.25	10	10 m		610A
MD1132	NM	0.3	15	0.05	50	1.0 m	600	1.7	0.9	5.0	0.4	10	10 m		654
MD2060F	NM	0.35	60	0.5	30	0.1 m	100	15	0.9	5.0	0.10	8.0	10 m		654
MD2218	NG	0.575	30	0.5	40	150 m	200	8.0	60	350	0.4	10	150 m		654
MD2218A	NG	0.575	30	0.5	40	150 m	200	8.0	45	310	0.3	10	150 m		654
MD2218AF	NG	0.35	30	0.5	40	150 m	200	8.0	45	310	0.3	10	150 m		610A
MD2218F	NG	0.35	30	0.5	40	150 m	200	8.0	60	350	0.4	10	150 m		610A
MD2219	NG	0.575	30	0.5	100	150 m	200	8.0	60	350	0.4	10	150 m		654
MD2219A	NG	0.575	30	0.5	100	150 m	200	8.0	45	310	0.3	10	150 m		654
MD2219AF	NG	0.350	30	0.5	100	150 m	200	8.0	45	310	0.3	10	150 m		610A
MD2219F	NG	0.350	30	0.5	100	150 m	200	8.0	60	350	0.4	10	150 m		610A
MD2369	NS	0.55	15	0.5	40	10 m	500	4.0	15	20	0.25	10	10 m		654
MD2369A	NM	0.55	15	0.5	40	10 m	500	4.0	0.9	5.0	0.25	10	10 m		654
MD2369AF	NM	0.35	15	0.5	40	10 m	500	4.0	0.9	5.0	0.25	10	10 m		610A
MD2369B	NM	0.55	15	0.5	40	10 m	500	4.0	0.8	10	0.25	10	10 m		654
MD2369BF	NM	0.35	15	0.5	40	10 m	500	4.0	0.8	10	0.25	10	10 m		610A
MD2369F	NS	0.35	15	0.5	40	10 m	500	4.0	15	20	0.25	10	10 m		610A
MD2904	PG	0.575	40	0.6	40	150 m	200	8.0	45	130	0.4	10	150 m		654
MD2904A	PG	0.575	60	0.6	40	150 m	200	8.0	45	130	0.4	10	150 m		654
MD2904AF	PG	0.350	60	0.6	40	150 m	200	8.0	45	130	0.4	10	150 m		610A
MD2904F	PG	0.350	40	0.6	40	150 m	200	8.0	45	130	0.4	10	150 m		610A
MD2905	PG	0.575	40	0.6	100	150 m	200	8.0	45	130	0.4	10	150 m		654

Some columns show 2 different types of data indicated by either **bold** or *italic* typefaces. See key and headings.

continues

DATA SHEET A3.1   Continued

Dual Transistors (continued)

TYPE NO.	ID	P_D Watts $T_A=25°C$ One Die Only	V_{CEO} Volts	I_C Amp Max	h_{FE} Min	@ I_C Unit	f_T MHz Min	C_{ob} pF Max	$\dfrac{h_{FE1}}{h_{FE2}}$ / t_{on} ns Max	ΔV_{BE} mV Max / t_{off} ns Max	G_p dB Min / $V_{CE(sat)}$ @ Volts Max	NF dB Max / $\dfrac{I_C}{i_B}$	f & I_C Unit	PACKAGE TO-No.	Case No.
MD2905A	PG	0.575	60	0.6	100	150 m	200	8.0	45	130	0.4	10	150 m		654
MD2905AF	PG	0.35	60	0.6	100	150 m	200	8.0	45	130	0.4	10	150 m		610A
MD2905F	PG	0.35	40	0.6	100	150 m	200	8.0	45	130	0.4	10	150 m		610A
MD3250	PA	0.57	A40	0.20	50	1.0 m	200	6.0			0.25	10	10 m		654
MD3250A	PM	0.57	A40	0.20	50	1.0 m	200	6.0	0.9	5.0	0.25	10	10 m		654
MD3250AF	PM	0.35	40	0.20	50	1.0 m	200	6.0	0.9	5.0	0.25	10	10 m		610A
MD3250F	PA	0.35	40	0.20	50	1.0 m	200	6.0			0.25	10	10 m		610A
MD3251	PA	0.575	40	0.20	100	1.0 m	250	6.0			0.25	10	10 m		654
MD3251A	PM	0.575	40	0.20	100	1.0 m	250	6.0	0.9	5.0	0.25	10	10 m		654
MD3251AF	PM	0.35	40	0.20	100	1.0 m	250	6.0	0.9	5.0	0.25	10	10 m		610A
MD3251F	PA	0.35	40	0.20	100	1.0 m	250	6.0			0.25	10	10 m		610A
MD3409	NM	0.575	30	0.5	50	10 m	200	8.0	0.8	10	0.15	10	10 m		654
MD3410	NM	0.575	30	0.5	50	10 m	200	8.0	0.9	10	0.15	10	10 m		654
MD3467	PS	0.60	40	1.5	20	500 m	150	20	40	110	0.5	10	500 m		654
MD3467F	PS	0.35	40	1.5	20	500 m	150	20	40	110	0.5	10	500 m		610A
MD3725	NS	0.60	40	1.0	50	100 m	200	10	45	75	0.26	10	100 m		654
MD3725F	NS	0.35	40	1.0	50	100 m	200	10	45	75	0.26	10	100 m		610A
MD3762	PS	0.60	40	1.5	20	1.0 A	150	20	40	110	1.0	10	1.0 A		654
MD3762F	PS	0.35	40	1.5	20	1.0 A	150	20	40	110	1.0	10	1.0 A		610A
MD5000	PH	0.3	15	0.05	20	3.0 m	600	1.7			15		200 M		654
MD5000A	PM	0.3	15	0.05	20	3.0 m	600	1.7	0.9	5.0	15		200 M		654
MD5000B	PM	0.3	15	0.05	20	3.0 m	600	1.7	0.8	10	15		200 M		654
MD6001	CG	0.575	30	0.5	40	150 m	200	8.0	60	350	0.4	10	150 m		654
MD6001F	CG	0.35	30	0.5	40	150 m	200	8.0	60	350	0.4	10	150 m		610A
MD6002	CG	0.575	30	0.5	100	150 m	200	8.0	60	350	0.4	10	150 m		654
MD6002F	CG	0.35	30	0.5	100	150 m	200	8.0	60	350	0.4	10	150 m		610A
MD6003	CA	0.575	30	0.5	70	150 m	200	8.0			0.4	10	150 m		654
MD6003F	CA	0.35	30	0.5	70	150 m	200	8.0			0.4	10	150 m		610A
MD6100	CA	0.5	45	0.05	100	0.1 m	30	4.0			0.25	10	1.0 m		654
MD6100F	CA	0.35	45	0.05	100	0.1 m	30	4.0			0.25	10	10 m		610A
MD7000	NA	0.575	30	0.5	70	150 m	200	8.0			0.4	10	150 m		654
MD7001	PA	0.6	30	0.6	70	150 m	200	8.0			0.4	10	150 m		654
MD7001F	PA	0.350	30	0.6	70	150 m	200	8.0			0.4	10	150 m		610A
MD7002	NA	0.575	40	0.03	40	100 u	200	6.0			0.35	10	10 m		654
MD7002A	NM	0.575	40	0.03	40	100 u	200	6.0	0.75	25	0.35	10	10 m		654
MD7002B	NM	0.575	40	0.03	40	100 u	200	6.0	0.85	15	0.35	10	10 m		654
MD7003	NA	0.55	40	0.05	50	10 m	200	6.0			0.35	10	1.0 m		654
MD7003A	NM	0.55	40	0.05	50	10 m	200	6.0	0.75	25	0.35	10	1.0 m		654
MD7003AF	NM	0.35	40	0.05	50	10 m	200	6.0	0.75	25	0.35	10	1.0 m		610A
MD7003B	NM	0.55	40	0.05	50	10 m	200	6.0	0.85	15	0.35	10	1.0 m		654
MD7003F	NA	0.35	40	0.05	50	10 m	200	6.0			0.35	10	1.0 m		610A
MD7004	NA	0.55	13	0.2	30	10 m	675*	4.0			0.4	10	10 m		654
MD7004F	NA	0.35	13	0.2	30	10 m	675*	4.0			0.4	10	10 m		610A
MD7005	PA	0.55	12	0.05	30	3.0 m	650	3.0			0.4	10	10 m		654
MD7005F	PA	0.35	12	0.05	30	3.0 m	650	3.0			0.4	10	10 m		610A
MD7007	PA	0.575	40	0.2	30	1.0 m	300	8.0			1.0	10	50 m		654
MD7007A	PM	0.575	50	0.2	30	1.0 m	300	8.0	0.75	20	1.0	10	50 m		654
MD7007B	PM	0.575	60	0.2	30	1.0 m	300	8.0	0.85	10	1.0	10	50 m		654
MD7007BF	PM	0.35	40	0.2	30	1.0 m	300	8.0	0.85	10	1.0	10	50 m		610A
MD7007F	PA	0.35	40	0.2	30	1.0 m	300	8.0			1.0	10	50 m		610A

Some columns show 2 different types of data indicated by either **bold** or *italic* typefaces. See key and headings.

DATA SHEET A3.1   Continued

Dual Transistors (continued)

TYPE NO.	ID	P_D Watts $T_A=25°C$ One Die Only	V_{CEO} Volts	I_C Amp Max	h_{FE} Min	@ I_C Unit	f_T MHz Min	C_{ob} pF Max	$\frac{h_{FE1}}{h_{FE2}}$ / t_{on} ns Max	ΔV_{BE} mV Max / t_{off} ns Max	Gp dB Min / V_{CE}(sat) @ Volts Max	NF dB Max / $\frac{I_C}{I_B}$	f & I_C Unit	PACKAGE TO- No.	Case No.
MD7021	CG	0.55	40	0.05	50	10 m	200	6.0	28*	72*	0.35	10	10 m		654
MD7021F	CG	0.35	40	0.05	50	10 m	200	6.0	28*	72*	0.35	10	10 m		610A
MD8001	NM	0.575	40	0.03	100	1.0 m	260*	2.6*		15					654
MD8002	NM	0.575	40	0.03	100	1.0 m	260*	2.6*		15					654
MD8003	NM	0.575	40	0.03	100	1.0 m	260*	2.6*		15					654
2N2060	NM	0.5	60	0.5	30	100 u	60	15	0.9	5.0		8.0	1000 H	78	654
2N2060A	NM	0.5	60	0.5	30	100 u	60	15	0.9	3.0	0.6	10	50 m	78	654
2N2223	NM	0.5	60	0.5	25	100 u	50	15	0.8	15	1.2	10	50 m	78	654
2N2223A	NM	0.5	60	0.5	25	100 u	50	15	0.9	5.0	1.2	10	50 m	78	654
2N2453	NM	0.5	30	0.05	80	10 u	60	8.0	0.9	3.0		7.0	1000 H	78	654
2N2453A	NM	0.5	50	0.05	80	10 u	60	8.0	0.9	3.0		4.0	1000 H	78	654
2N2480	NM	0.3	35	0.5	30	1.0 m	50	20	0.8	10		8.0	1000 H	78	654
2N2480A	NM	0.3	40	0.5	50	1.0 m	50	18	0.8	5.0	1.3	10	50 m	78	654
2N2639	NM	0.3	45	0.03	50	10 u	80	8.0	0.9	5.0		4.0	AUD	78	654
2N2640	NM	0.3	45	0.03	50	10 u	80	8.0	0.8	10		4.0	AUD	78	654
2N2641	NE	0.3	45	0.03	50	10 u	80	8.0				4.0	AUD	78	654
2N2642	NM	0.3	45	0.03	100	10 u	80	8.0	0.9	5.0		4.0	AUD	78	654
2N2643	NM	0.3	45	0.03	100	10 u	80	8.0	0.8	10		4.0	AUD	78	654
2N2644	NE	0.3	45	0.03	100	10 u	80	8.0				4.0	AUD	78	654
2N2652	NM	0.3	60	0.5	50	1.0 m	60	15	0.85	3.0	1.2	10	50 m	78	654
2N2652A	NM	0.3	60	0.5	50	1.0 m	60	15	0.9	3.0		8.0	1000 H	78	654
2N2720	NM	0.3	60	0.04	30	0.1 m	80	6.0	0.9	5.0	1.0	10	10 m	78	654
2N2721	NM	0.3	60	0.04	30	0.1 m	80	6.0	0.8	10	1.0	10	10 m	78	654
2N2722	NM	0.3	45	0.04	50	1.0 u	100	6.0	0.9	5.0	1.0	20	10 m	78	654
2N2903	NM	0.6	30	0.05	125	1.0 m	60	8.0	0.8	10		7.0	1000 H	78	654
2N2903A	NM	0.6	30	0.05	125	1.0 m	60	8.0	0.9	5.0		7.0	1000 H	78	654
2N2913	NE	0.3	45	0.03	60	10 u	60	6.0				4.0	AUD		654
2N2914	NE	0.3	45	0.03	150	10 u	60	6.0				3.0	AUD		654
2N2915	NM	0.3	45	0.03	60	10 u	60	6.0	0.9	5.0		4.0	AUD		654
2N2916	NM	0.3	45	0.03	150	10 u	60	6.0	0.9	5.0		3.0	AUD		654
2N2917	NM	0.3	45	0.03	60	10 u	60	6.0	0.8	10		4.0	AUD		654
2N2918	NM	0.3	45	0.03	150	10 u	60	6.0	0.8	10		3.0	AUD		654
2N2919	NM	0.3	60	0.03	60	10 u	60	6.0	0.9	5.0		4.0	AUD		654
2N2920	NM	0.3	60	0.03	150	10 u	60	6.0	0.9	5.0		3.0	AUD		654
2N3043	NM	0.25	45	0.03	100	10 u	30	8.0	0.9	5.0		5.0	AUD		610A
2N3044	NM	0.25	45	0.03	100	10 u	30	8.0	0.8	10		5.0	AUD		610A
2N3045	NE	0.25	45	0.03	100	10 u	30	8.0				5.0	AUD		610A
2N3046	NM	0.25	45	0.03	50	10 u	30	8.0	0.9	5.0		5.0	AUD		610A
2N3047	NM	0.25	45	0.03	50	10 u	30	8.0	0.8	10		5.0	AUD		610A
2N3048	NE	0.25	45	0.03	50	10 u	30	8.0				5.0	AUD		610A
2N3726	PE	0.4	45	0.3	135	1.0 m	200	8.0	0.9	5.0		4.0	1000 H		654
2N3727	PE	0.4	45	0.3	135	1.0 m	200	8.0	0.9	2.5		4.0	1000 H		654
2N3806	PE	0.5	60	0.05	150	0.1 m	100	4.0				7.0	100 H		654
2N3807	PE	0.5	60	0.05	300	0.1 m	100	4.0				4.0	100 H		654
2N3808	PM	0.5	60	0.05	150	0.1 m	100	4.0	0.8	5.0		7.0	100 H		654
2N3809	PM	0.5	60	0.05	300	0.1 m	100	4.0	0.8	5.0		4.0	100 H		654
2N3810	PM	0.5	60	0.05	150	0.1 m	100	4.0	0.9	3.0		7.0	100 H		654

Some columns show 2 different types of data indicated by either **bold** or *italic* typefaces. See key and headings.

continues

DATA SHEET A3.1   Continued

Dual Transistors (continued)

TYPE NO.	ID	P_D Watts $T_A=25°C$ One Die Only	V_{CEO} Volts	I_C Amp Max	h_{FE} Min	@ I_C (Unit)	f_T MHz Min	C_{ob} pF Max	h_{FE1}/h_{FE2} · t_{on} ns Max	ΔV_{BE} mV Max · t_{off} ns Max	Gp dB Min · V_{CE}(sat) Volts Max	NF dB Max · I_C/I_B	@ f & I_C (Unit)	PACKAGE TO-No.	Case No.
2N3810A	PM	0.5	60	0.05	150	0.1 m	100	4.0	**0.95**	**1.5**		**3.0**	100 H		654
2N3811	PM	0.5	60	0.05	300	0.1 m	100	4.0	**0.9**	**3.0**		**4.0**	100 H		654
2N3811A	PM	0.5	60	0.05	300	0.1 m	100	4.0	**0.95**	**1.5**		**1.5**	100 H		654
2N3812	PM	0.5	60	0.05	150	0.1 m	100	4.0				**3.5**	AUD		610A
2N3813	PA	0.5	60	0.05	300	0.1 m	100	4.0				**2.5**	AUD		610A
2N3814	PM	0.5	60	0.05	150	0.1 m	100	4.0	**0.8**	**5.0**		**7.0**	100 H		610A
2N3815	PM	0.5	60	0.05	300	0.1 m	100	4.0	**0.8**	**5.0**		**4.0**	100 H		610A
2N3816	PM	0.5	60	0.05	150	0.1 m	100	4.0	**0.9**	**3.0**		**7.0**	100 H		610A
2N3816A	PM	0.5	60	0.05	150	0.1 m	100	4.0	**0.95**	**1.5**		**7.0**	100 H		610A
2N3817	PM	0.5	60	0.05	300	0.1 m	100	4.0	**0.9**	**3.0**		**4.0**	100 H		610A
2N3817A	PM	0.5	60	0.05	300	0.1 m	100	4.0	**0.95**	**1.5**		**4.0**	100 H		610A
2N3838	CE	0.25	40	0.6	100	150 m	200	8.0	*50*	*340*		**8.0**	1000 H		610A
2N4015	PM	0.4	60	0.3	135	1.0 m	200	8.0	**0.9**	**5.0**		**4.0**	1000 H		654
2N4016	PM	0.4	60	0.3	135	1.0 m	200	8.0	**0.9**	**2.5**		**4.0**	1000 H		654
2N4854	CE	0.3	40	0.6	100	150 m	200	8.0	*60*	*350*		**8.0**	1000 H		654
2N4855	CE	0.3	40	0.6	40	150 m	200	8.0	*60*	*350*		**8.0**	1000 H		654
2N4937	PM	0.6	40	0.05	50	1.0 m	300	5.0	**0.9**	**3.0**		**4.0**	AUD		654
2N4938	PM	0.6	40	0.05	50	1.0 m	300	5.0	**0.8**	**5.0**		**4.0**	AUD		654
2N4939	PE	0.6	40	0.05	50	1.0 m	300	5.0				**4.0**	AUD		654
2N4940	PM	0.6	40	0.05	50	1.0 m	300	5.0	**0.8**	**5.0**		**4.0**	AUD		610A
2N4941	PM	0.6	40	0.05	50	1.0 m	300	5.0	**0.9**	**3.0**		**4.0**	AUD		610A
2N4942	PE	0.6	40	0.05	50	1.0 m	300	5.0				**4.0**	AUD		610A
2N5793	NG	0.5	40	0.6	40	150 m	200	8.0	*45*	*310*	*0.3*	*10*	150 m		654
2N5794	NG	0.5	40	0.6	100	150 m	200	8.0	*45*	*310*	*0.3*	*10*	150 m		654
2N5795	NG	0.5	60	0.6	40	150 m	200	8.0	*47*	*140*	*0.4*	*10*	150 m		654
2N5796	NG	0.5	60	0.6	100	150 m	200	8.0	*47*	*140*	*0.4*	*10*	150 m		654
2N6502	NS	0.6	40	1.0	50	100 m	250	10	*35*	*60*	*0.3*	*10*	100 m		654
2N6503	NS	0.6	40	1.0	50	100 m	250	10	*35*	*60*	*0.3*	*10*	100 m		610A

Some columns show 2 different types of data indicated by either **bold** or *italic* typefaces. See key and headings.

DATA SHEET A3.2   Small-Signal FETs

Small-Signal Transistors–FETs

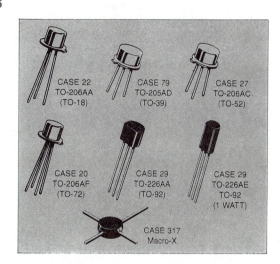

Field-Effect Transistors (FETs)

Motorola offers a line of field-effect transistors that encompasses the latest technology and covers the full range of FET applications. Included here is a wide variety of junction FETs, MOS-FETs (with P- or N-channel polarity with both single and dual gates) and TMOS FETs. These FETs include devices developed for operation across the frequency range from dc to UHF in switching and amplifying applications. Package options from low cost plastic to metal TO-206AF (TO-72) packages are available. The selector guides on the following pages are designed to emphasize those FET families and device types that, by virtue of widespread industry use, ease of manufacture and, consequently, low relative cost, merit first consideration for new equipment design.

JFETs

JFETs operate in the depletion mode. They are available in both P- and N-channel and are offered in both metal and plastic packages. Applications include general-purpose amplifiers, switches and choppers, and RF amplifiers and mixers. These devices are economical and very rugged. The drain and source are interchangeable on many typical FETs.

Switches and Choppers

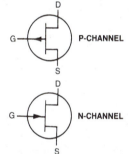

P-Channel JFETs

Package TO –	Device	$r_{ds(on)}$ Ω Max	@ I_D μA	$V_{GS(off)}$ V Min	$V_{GS(off)}$ V Max	I_{DSS} mA Min	I_{DSS} mA Max	$V_{(BR)GSS}$ V Min	C_{iss} pF Max	C_{rss} pF Max	t_{on} ns Max	t_{off} ns Max
226AA	MPF970	100	1.0	5.0	12	15	100	30	12	5.0	8.0	25
	MPF971	250	1.0	1.0	7.0	2.0	80	30	12	5.0	10	120
206AF	2N3993	150	—	4.0	9.5	10	—	25	16	4.5	—	—
	2N3994	300	—	1.0	5.5	2.0	—	25	16	4.5	—	—
	2N3994A	300	—	1.0	5.5	2.0	—	25	12	3.0	—	—

N-Channel JFETs

Package TO –	Device	$r_{ds(on)}$ Ω Max	@ I_D μA	$V_{GS(off)}$ V Min	$V_{GS(off)}$ V Max	I_{DSS} mA Min	I_{DSS} mA Max	$V_{(BR)GSS}$ V Min	C_{iss} pF Max	C_{rss} pF Max	t_{on} ns Max	t_{off} ns Max
206AA	MFE2012	10	—	3.0	10	100	—	25	50	20	16	37
	MFE2011	15	1.0	1.0	10	40	—	25	50	20	10	20
	2N4859A	25	—	2.0	6.0	50	—	30	10	4.0	8.0	20
226AA	MPF4859A	25	—	2.0	6.0	50	—	30	10	4.0	8.0	20
206AA	2N4856A	25	—	4.0	10	50	—	40	10	4.0	8.0	20
226AA	MPF4856A	25	—	4.0	10	50	—	40	10	4.0	8.0	20
206AA	2N4856	26	—	4.0	10	50	—	40	10	8.0	9.0	25
226AA	MPF4856	25	—	4.0	10	50	—	40	10	8.0	9.0	25

continues

DATA SHEET A3.2 Continued

Switches and Choppers (continued)
N-Channel JFETs (continued)

Package TO –	Device	$r_{ds(on)}$ Ω @ Max	I_D μA	$V_{GS(off)}$ V Min	$V_{GS(off)}$ V Max	I_{DSS} mA Min	I_{DSS} mA Max	$V_{(BR)GSS}$ V Min	C_{iss} pF Max	C_{rss} pF Max	t_{on} ns Max	t_{off} ns Max
206AA	2N4859	25	—	4.0	10	50	—	30	18	8.0	9.0	25
226AA	MPF4859	25	—	4.0	10	50	—	30	18	8.0	9.0	25
206AA	MFE2010	25	1.0	0.5	10	15	—	25	50	20	10	35
	2N4391	30	1.0	4.0	10	50	150	40	14	3.5	15	20
226AA	MPF4391	30	1.0	4.0	10	60	130	20	10	3.5	15	20
	2N638	30	1.0	—	(12)	50	—	30	10	4.0	9.0	15
206AA	2N4091	30	1.0	5.0	10	30	—	40	16	5.0	25	40
226AA	MPF4091	30	1.0	5.0	10	30	—	40	16	5.0	25	40
	J111	30	1.0	3.0	10	20	—	35	10^t	5.0^t	13	35
206AA	MFE2006	30	1.0	−5.0	−10	30	—	−30	16	5.0	20	40
	2N3970	30	1.0	4.0	10	50	150	40	25	6.0	20	30
226AA	MPF3970	30	1.0	4.0	10	50	150	40	25	6.0	20	30
206AA	2N4057A	40	—	2.0	6.0	20	100	40	10	3.5	10	40
226AA	MPF4857A	40	—	2.0	6.0	20	100	40	10	3.5	10	40
206AA	2N860A	40	—	2.0	6.0	20	100	30	10	3.5	10	40
226AA	MPF4860A	40	—	2.0	6.0	20	100	30	10	3.5	10	40
206AA	2N4857	40	—	2.0	6.0	20	100	40	18	8.0	10	50
226AA	MPF4857	40	—	2.0	6.0	20	100	40	18	8.0	10	50
206AA	2N4860	40	—	2.0	6.0	20	100	30	18	8.0	10	50
226AA	MPF4860	40	—	2.0	6.0	20	100	30	18	8.0	10	50
	2N5653	50	1.0	—	(12)^t	40	—	30	10	3.5	9.0	15
206AA	2N4092	50	1.0	2.0	7.0	15	—	40	16	5.0	35	60
226AA	MPF4092	50	1.0	2.0	7.0	15	—	40	16	5.0	35	60
	J112	50	1.0	1.0	5.0	5.0	—	35	10^t	5.0^t	13^t	35^t
206AA	MFE2005	50	1.0	−2.0	−8.0	15	—	−30	16	5.0	35	60
	2N4392	60	1.0	2.0	5.0	25	75	40	14	3.5	15	35
226AA	MPF4392	60	1.0	2.0	5.0	25	75	20	10	3.5	15	35
206AA	2N4858A	60	1.0	0.8	4.0	8.0	80	40	10	3.5	16	80
226AA	MPF4858A	60	1.0	0.8	4.0	8.0	80	40	10	3.5	16	80
206AA	2N4861A	60	—	0.8	4.0	8.0	80	30	10	3.5	16	80
226AA	MPF4861A	60	—	0.8	4.0	8.0	80	30	10	3.5	16	80
	2N5639	60	1.0	—	(8.0)^t	25	—	30	10	4.0	14	30
206AA	2N3971	60	1.0	2.0	5.0	25	75	40	25	6.0	30	60
226AA	MPF3971	60	1.0	2.0	5.0	25	75	40	25	6.0	30	60
206AA	2N4858	60	—	0.8	4.0	8.0	80	40	18	8.0	20	100
226AA	MPF4858	60	—	0.8	4.0	8.0	80	40	18	8.0	20	100
206AA	2N4861	60	—	0.8	4.0	8.0	80	30	18	8.0	20	100
226AA	MPF4861	60	—	0.8	4.0	8.0	80	30	18	8.0	20	100
206AA	2N4093	80	1.0	1.0	5.0	80	—	40	16	5.0	60	80
226AA	MPF4093	80	1.0	1.0	5.0	80	—	40	16	5.0	60	80
206AA	MFE2004	80	1.0	−1.0	−6.0	8.0	—	−30	16	5.0	60	80
206AF	MFE3002	100	10 V	—	3.0	—	10	15	5.0	1.5	—	—

t = typical

DATA SHEET A3.2   Continued

Switches and Choppers (continued)
N-Channel JFETs (continued)

Package TO –	Device	$r_{ds(on)}$ Ω Max	@ I_D μA	$V_{GS(off)}$ V Min	$V_{GS(off)}$ V Max	I_{DSS} mA Min	I_{DSS} mA Max	$V_{(BR)GSS}$ V Min	C_{iss} pF Max	C_{rss} pF Max	t_{on} ns Max	t_{off} ns Max
206AA	2N4393	100	1.0	0.5	3.0	5.0	30	40	14	3.5	15	50
226AA	MPF4393	100	1.0	0.5	3.0	5.0	30	20	10	3.5	15	55
	2N5654	100	1.0	—	(8.0)	15	—	25	10	3.5	14	30
	2N5640	100	1.0	—	(6.0)	5.0	—	30	10	4.0	18	45
206AA	2N3972	100	1.0	0.5	3.0	5.0	30	40	25	6.0	80	100
226AA	MPF3972	100	1.0	0.5	3.0	5.0	30	40	25	6.0	80	100
	J113	100	1.0	0.5	3.0	2.0	—	35	10^t	5.0^t	13^t	35^t
	BF246	—	—	0.5	14	10	300	25	—	—	—	—
	BF246A	35^t	1.0	1.5	4.0	30	80	25	—	—	—	—
	BF246B	50^t	1.0	3.0	7.0	60	140	25	—	—	—	—
	BF246C	65^t	1.0	5.5	12.0	110	250	25	—	—	—	—
	J107	8	—	0.5	4.5	100	—	25	—	—	—	—
	J108	8	—	3.0	10.0	80	—	25	—	—	—	—
	J109	12	—	2.0	6.0	40	—	25	—	—	—	—
	J110	18	—	0.5	4.0	10	—	25	—	—	—	—

t = typical

Low-Frequency/Low-Noise

P-Channel JFETs

Package TO –	Device	$R_e|Y_{fs}|$ mmho Min	$R_e|Y_{os}|$ μmho Max	C_{iss} pF Max	C_{rss} pF Max	$V_{(BR)GSS}$ V Min	$V_{GS(off)}$ V Min	$V_{GS(off)}$ V Max	I_{DSS} mA Min	I_{DSS} mA Max
226AA	MPF161	0.8	75	7.0	2.0	40	0.2	8.0	− 0.5	− 14
206AF	2N5265	0.9	75	7.0	2.0	60	0.3	1.5	0.5	1.0
	MFE4009	1.0	20	20	—	20	—	5.0	1.0	3.0
	MFE4012	1.0	100	20	—	20	—	8.0	5.0	15
	2N5267/8	1.0	20	20	—	20	—	6.0	1.0	6.0
	2N3909	1.0	100	32	16	20	0.3	7.9	0.3	15
206AA	MFE4007	1.0	20	25	7.0	25	0.3	1.5	0.3	1.2
	2N2608	1.0	17	—	—	30	1.0	4.0	0.9	4.5
226AA	MPF2608	1.0	—	17	—	30	1.0	4.0	0.9	4.5
	2N5460	1.0	50	7.0	2.0	40	0.75	6.0	1.0	5.0
206AF	2N5266	1.0	75	7.0	2.0	60	0.4	2.0	0.8	1.6
226AA	2N5463	1.0	75	7.0	2.0	60	0.5	4.0	1.0	5.0
206AF	2N3330	1.5	40	20	—	20	—	6.0	2.0	6.0
226AA	MPF3330	1.5	40	20	—	20	—	6.0	2.0	6.0
206AA	MFE4009	1.5	20	25	7.0	25	0.5	2.5	1.0	3.5
226AA	2N5461	1.5	50	7.0	2.0	40	1.0	7.5	2.0	9.0
206AF	2N5267	1.5	75	7.0	2.0	60	1.0	4.0	1.5	3.0
226AA	2N5464	1.5	75	7.0	2.0	60	0.8	4.5	2.0	9.0
	2N4360	2.0	100	20	5.0	20	0.4	9.0	3.0	30
	2N4342	2.0	75	20	5.0	25	—	5.5	4.0	12
	2N5462	2.0	50	7.0	2.0	40	1.8	9.0	4.0	16
206AF	2N5268	2.0	75	7.0	2.0	60	1.0	4.0	2.5	5.0

continues

Low-Frequency/Low-Noise (continued)

P-Channel JFETs (continued)

Package TO –	Device	$R_e\|Y_{fs}\|$ mmho Min	$R_e\|Y_{os}\|$ μmho Max	C_{iss} pF Max	C_{rss} pF Max	$V_{(BR)GSS}$ V Min	$V_{GS(off)}$ V Min	$V_{GS(off)}$ V Max	I_{DSS} mA Min	I_{DSS} mA Max
226AA	2N5465	2.0	75	7.0	2.0	60	1.5	6.0	4.0	16
206AF	2N3909A	2.2	100	9.0	3.0	20	0.3	7.9	1.0	15
	2N5269	2.2	75	7.0	2.0	60	2.0	6.0	4.0	8.0
206AA	2N2609	2.5	—	30	—	30	1.0	4.0	2.0	10
226AA	MPF2609	2.5	—	30	—	30	1.0	4.0	2.0	10
206AF	2N5270	2.5	75	7.0	2.0	60	2.0	6.0	7.0	14

N-Channel JFETs

Package TO –	Device	$R_e\|Y_{fs}\|$ mmho Min	@ f MHz	$R_e\|Y_{os}\|$ μmho Max	@ f MHz	C_{iss} pF Max	C_{rss} pF Max	$V_{(BR)GSS}$ V Min	$V_{GS(off)}$ V Min	$V_{GS(off)}$ V Max	I_{DSS} mA Min	I_{DSS} mA Max
206AA	2N3370	0.3	30	15	30	20	3.0	40	—	3.2	0.1	0.6
226AA	MPF111	0.5	10	200	10	—	—	20	0.5	10	0.5	20
	J201	0.5	20	1 0^t	20	5.0^t	2.0^t	40	0.3	1.5	0.2	1.0
206AA	2N3369	0.6	30	30	30	20	3.0	40	—	6.5	0.5	2.5
226AA	MPF109	0.8	15	75	15	7.0	3.0	25	0.2	8.0	0.5	24
206AA	2N4339	0.8	15	15	15	7.0	3.0	50	0.6	1.8	0.5	1.5
226AA	MPF4339	0.8	15	15	15	7.0	3.0	50	0.6	1.8	0.5	1.5
206AA	2N3460	0.8	20	5.0	30	18	6.0	50	—	1.8	0.2	1.0
	2N3438	0.8	20	5.0	30	18	6.0	50	—	2.3	0.2	1.0
206AF	2N4220	1.0	15	10	15	6.0	2.0	30	—	4.0	0.5	3.0
226AA	MPF4220	1.0	15	10	15	6.0	2.0	30	—	4.0	0.5	3.0
206AF	2N4220A	1.0	15	10	15	6.0	2.0	30	—	4.0	0.5	3.0
226AA	MPF4220A	1.0	15	10	15	6.0	2.0	30	—	4.0	0.5	3.0
206AF	2N5358	1.0	15	10	15	6.0	2.0	40	0.5	3.0	0.5	1.0
226AA	J202	1.0	20	3.5^t	20	5.0^t	2.0^t	40	0.8	4.0	0.9	4.5
206AA	2N3368	1.0	30	80	30	20	3.0	40	—	11.5	2.0	12
206AF	2N5359	1.2	15	10	15	6.0	2.0	40	0.8	4.0	0.6	1.6
206AA	2N4340	1.3	15	30	15	7.0	3.0	50	1.0	3.0	1.2	3.6
206AF	2N5360	1.4	15	20	15	6.0	2.0	40	0.8	4.0	0.5	2.5
226AA	2N5458	1.5	15	50	15	7.0	3.0	25	1.0	7.0	2.0	9.0
206AF	2N5361	1.5	15	20	15	6.0	2.0	40	1.0	6.0	2.5	5.0
226AA	J203	1.5	20	10^t	20	5.0^t	2.0^t	40	2.0	10	4.0	20
206AA	2N3459	1.5	20	20	30	18	6.0	50	—	3.4	0.8	4.0
206AF	2N3821	1.5	15	10	15	6.0	3.0	50	—	4.0	0.5	2.5
226AA	MPF3821	1.5	15	10	15	6.0	3.0	50	—	4.0	0.5	2.5
206AA	2N3437	1.5	20	20	30	18	6.0	50	—	4.8	0.8	4.0
226AA	2N5457	2.0	15	50	15	7.0	3.0	25	0.5	6.0	1.0	5.0
	2N5459	2.0	15	50	15	7.0	3.0	25	2.0	8.0	4.0	16
206AF	2N4221	2.0	15	20	15	6.0	2.0	30	—	6.0	2.0	6.0

t = typical

DATA SHEET A3.2   Continued

Low-Frequency/Low-Noise (continued)

N-Channel JFETs (continued)

Package TO –	Device	$R_e\ Y_{fs}$ @ mmho Min	$R_e\ Y_{fs}$ @ f MHz	$R_e\ Y_{os}$ @ μmho Max	$R_e\ Y_{os}$ @ f MHz	C_{iss} pF Max	C_{rss} pF Max	$V_{(BR)GSS}$ V Min	$V_{GS(off)}$ V Min	$V_{GS(off)}$ V Max	I_{DSS} mA Min	I_{DSS} mA Max
226AA	MPF4221	2.0	15	20	15	6.0	2.0	30	—	6.0	2.0	6.0
206AF	2N4221A	2.0	15	20	15	6.0	2.0	30	—	6.0	2.0	6.0
226AA	MPF4221A	2.0	15	20	15	6.0	2.0	30	—	6.0	2.0	6.0
206AF	2N5362	2.0	15	40	15	6.0	2.0	40	2.0	7.0	4.0	8.0
	2N3822	2.0	15	20	15	6.0	3.0	50	—	6.0	2.0	10
226AA	MPF3822	2.0	15	20	15	6.0	3.0	50	—	6.0	2.0	10
206AF	2N4341	2.0	15	60	15	7.0	3.0	50	2.0	6.0	3.0	9.0
206AF	2N4222	2.5	15	40	15	6.0	2.0	30	—	8.0	5.0	15
226AA	MPF4222	2.5	15	40	15	6.0	2.0	30	—	8.0	5.0	15
206AF	2N4222A	2.5	15	40	15	6.0	2.0	30	—	8.0	5.0	15
226AA	MPF4222A	2.5	15	40	15	6.0	2.0	30	—	8.0	5.0	15
206AF	2N5363	2.5	15	40	15	6.0	2.0	40	2.5	8.0	7.0	14
206AA	2N3458	2.5	20	35	30	18	6.0	50	—	7.8	3.0	15
	2N3436	2.5	20	35	30	18	6.0	50	—	9.8	3.0	15
206AF	2N5364	2.7	15	60	15	6.0	2.0	40	2.5	8.0	9.0	18
226AA	2N5670	3.0	15	75	15	7.0	3.0	25	2.0	8.0	8.0	20
206AA	2N4398	12^t	0.001	—	—	14	3.5	40	0.5	3.0	5.0	30
206AF	2N5556	6.5	0.001	20	15	6.0	3.0	30	0.2	4.0	0.5	2.5
	2N4117	20	0.001	3.0	10	3.0	1.5	40	0.6	1.8	30	90
226AA	MPF4117	20	0.001	3.0	10	3.0	1.5	40	0.6	1.8	30	90
206AF	2N4117A	70	0.001	3.0	10	3.0	1.5	40	0.6	1.8	30	90
226AA	MPF4117A	70	0.001	3.0	10	3.0	1.5	40	0.6	1.8	30	90
206AF	2N4118	80	0.001	5.0	10	3.0	1.5	40	1.0	3.0	80	240
226AA	MPF4118	80	0.001	5.0	10	3.0	1.5	40	1.0	3.0	80	240
206AF	2N4118A	80	0.001	5.0	10	3.0	1.5	40	1.0	3.0	80	240
226AA	MPF4118A	80	0.001	5.0	10	3.0	1.5	40	1.0	3.0	80	240
206AF	2N4119	100	0.001	10	10	3.0	1.5	40	2.0	6.0	200	600
226AA	MPF4119	100	0.001	10	10	3.0	1.5	40	2.0	6.0	200	600
206AF	2N4119A	100	0.001	10	10	3.0	1.5	40	2.0	6.0	200	600
226AA	MPF4119A	100	0.001	10	10	3.0	1.5	40	2.0	6.0	200	600

t = typical

continues

DATA SHEET A3.2   Continued

High-Frequency Amplifiers

N-Channel JFETs

Package TO –	Device	Re|Yfs| mmho Min	@ f MHz	Re|Yos| μmho Max	@ f MHz	Ciss pF Max	Crss pF Max	NF dB Max	@ RG=1K f MHz	V(BR)GSS V Min	VGS(off) V Min	VGS(off) V Max	IDSS mA Min	IDSS mA Max
226AA	2N5669	1.6	100	100	100	7.0	3.0	2.5	100	25	1.0	6.0	4.0	10
	MPF108	1.6	100	200	100	6.5	2.5	3.0	100	25	—	8.0	1.5	24
	MPF102	1.6	100	200	100	7.0	3.0	—	—	25	—	8.0	2.0	20
	2N3819	1.6	100	—	—	8.0	4.0	—	—	25	—	8.0	2.0	20
	2N5668	1.0	100	50	100	7.0	3.0	2.5	100	25	0.2	4.0	1.0	5.0
206AF	2N4224	1.7	200	200	200	6.0	2.0	—	—	30	0.1	8.0	20	20
226AA	MPF4224	1.7	200	200	200	6.0	2.0	—	—	30	0.1	8.0	2.0	20
	2N5484	2.5	100	75	100	5.0	1.0	3.0	100	25	0.3	3.0	1.0	5.0
206AF	MFE2000	2.5	0.001	50	0.001	5.0	1.0	4.0	400	−25	−0.5	−0.4	4.0	10
226AA	2N5670	2.5	100	150	100	7.0	3.0	2.5	100	25	2.0	8.0	8.0	20
	2N5246	2.5	400	100	400	4.5	1.0	—	—	30	0.5	4.0	1.5	7.0
206AF	2N4223	2.7	200	200	200	6.0	2.0	5.0	200	30	0.1	8.0	3.0	18
226AA	MPF4223	2.7	200	200	200	6.0	2.0	5.0	200	30	0.1	8.0	3.0	18
	2N5485	3.0	400	100	400	5.0	1.0	4.0	400	25	1.0	4.0	4.0	10
	J305	3.0t	400	80t	100	3.0t	0.8t	4.0t	400	30	0.5	3.0	1.0	8.0
206AF	2N3823	3.2	200	200	200	6.0	2.0	2.5	100	30	—	8.0	4.0	20
226AA	MPF3823	3.2	200	200	200	6.0	2.0	2.5	100	30	—	8.0	4.0	20
	2N5486	3.5	400	100	400	5.0	1.0	4.0	400	25	2.0	6.0	8.0	20
206AF	MFE2001	4.0	0.001	75	0.001	5.0	1.0	4.0	400	−25	−2.0	−6.0	8.0	20
	2N4416	4.0	400	100	400	4.0	0.8	4.0	400	30	2.0	6.0	5.0	15
226AA	MPF4416	4.0	400	100	400	4.0	0.8	4.0	400	30	2.0	6.0	5.0	15
206AF	2N4416A	4.0	400	100	400	4.0	0.8	4.0	400	30	2.0	6.0	5.0	15
226AA	MPF4416A	4.0	400	100	400	4.0	0.8	4.0	400	30	2.0	6.0	5.0	15
	2N5245	4.0	400	100	400	4.5	1.0	4.0	400	30	1.0	6.0	5.0	15
	2N5247	4.0	400	150	400	4.5	1.0	4.0	400	30	1.5	8.0	8.0	24
	J304	4.2t	400	80t	100	3.0t	0.8t	4.0t	400	30	2.0	6.0	5.0	15
206AC	U308	10	0.001	150	100	5.0	2.5	3.0t	450	25	1.0	6.0	12	60
	U309	10	0.001	150	100	5.0	2.5	3t	450	25	1.0	4.0	12	30
	U310	10	0.001	150	100	5.0	2.5	3t	450	25	2.5	6.0	24	60
226AA	J308	12t	100	250t	100	7.5	2.5	1.5t	100	25	1.0	6.5	12	60
	J309	12t	100	250t	100	7.5	2.5	1.5t	100	25	1.0	4.0	12	30
	J310	12t	100	250t	100	7.5	2.5	1.5t	100	25	2.0	6.5	24	60
206AF	MFE3004	2.0	0.001	—	—	4.5	0.4	4.5	200	20	—	−5.0	2.0	10
	3N128*	5.0	0.001	500	200	7.0	0.28	5.0	200	−50	−0.5	−8.0	5.0	25

t = typical

*N-Channel MOSFET

DATA SHEET A3.2   Continued

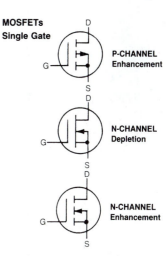

**MOSFETs
Single Gate**

P-CHANNEL
Enhancement

N-CHANNEL
Depletion

N-CHANNEL
Enhancement

MOSFETs

MOSFETs are available in either depletion/enhancement or enhancement mode (in general, depletion/enhancement devices are operated in the depletion mode and are referred to as depletion devices). They are available in both N- and P-channel, and both single gate and dual gate construction. Some MOSFETs are also offered with input diode protection which reduces the chance of damage from static charge in handling.

Low-Frequency/Low-Noise

P-Channel MOSFETs

Package TO –	Device	R_e Y_{fs} mmho Min	μmho Max	C_{iss} pF Max	C_{rss} pF Max	$V_{(BR)DSS}$ V Min	$V_{GS(TH)}$ V Min	Max	I_{DSS} mA Min	Max
206AF	3N155	1.0	60	5.0	1.3	− 35	− 1.5	− 3.2	—	− 1.0
	3N156	1.0	60	5.0	1.3	− 35	− 3.0	− 5.0	—	− 1.0
	3N157	1.0	60	5.0	1.3	− 35	− 1.5	− 3.2	—	− 1.0
	3N155A	1.0	60	5.0	1.3	− 35	− 1.5	− 3.2	—	− 0.25
	3N156A	1.0	60	5.0	1.3	− 35	− 3.0	− 5.0	—	− 0.25
	3N157A	1.0	60	5.0	1.3	− 50	− 1.5	− 3.2	—	− 0.25
	3N158	1.0	60	5.0	1.3	− 35	− 3.0	− 5.0	—	− 1.0
	3N158A	1.0	60	5.0	1.3	− 25	− 2.0	− 6.0	—	− 20
206AA	MFE823	1.0	—	6.0	1.5	− 50	− 3.0	− 5.0	—	− 0.25
206AF	MFE3003	—	—	5.0	1.0	− 15	—	− 4.0	—	10

N-Channel MOSFETs

Package TO –	Device	R_e Y_{fs} mmho Min	μmho Max	C_{iss} pF Max	C_{rss} pF Max	$V_{(BR)DSS}$ V Min	$V_{GS(TH)}$ V Min	Max	I_{DSS} mA Min	Max
206AA	2N3796	0.4	1.8	7.0	0.8	25	—	− 7.0	2.0	6.0
	MFE825	0.5	—	4.0	0.7	20	—	—	1.0	25
206AF	2N4351	1.0	—	5.0	1.3	25	1.0	5.0	—	10
	3N169	1.0	—	5.0	1.3	25	0.5	1.5	—	10
	3N170	1.0	—	5.0	1.3	25	1.0	2.0	—	10
	3N171	1.0	—	5.0	1.3	25	1.5	3.0	—	10
	MFE3002	—	—	5.0	1.0	15	—	3.0	—	10
206AA	2N3797	1.5	—	8.0	0.8	25	—	− 7.0	2.0	6.0

4

DC ANALYSIS AND DESIGN OF BASIC TRANSISTOR AMPLIFIER CIRCUITS

The objective of an amplifier circuit is to linearly amplify an incoming AC signal. The use of one or more transistors in a particular circuit arrangement can provide AC amplification. However, as we will see, amplification can be achieved only if the transistor has a suitable DC operating point, the Q point. Furthermore, amplification of the signal is possible only through conversion of a portion of the DC supplied power into AC. Hence, the DC analysis and design, often called bias circuit analysis and design, is an important part of the overall study of amplifier circuits.

This chapter deals only with the analysis and design of the bias circuit of amplifiers and proceeds, therefore, without regard to AC. The AC signal components are set equal to zero at the outset for each case, and the resulting circuits that contain DC only are then analyzed. We will examine methods for selecting a suitable position for Q for BJTs and FETs, which is the objective in the DC analysis and design of amplifier circuits.

4.1   GENERAL SELECTION OF THE QUIESCENT OPERATING POINT (Q)

4.1.1   Permissible Operating Region

As described in Chapter 3, transistors are three-terminal devices that, in general, possess both input and output current–voltage characteristics. The DC operating point Q is selected in a region of these current–voltage characteristics that allows linear amplification along with reasonable signal levels. The latter requirement necessitates a selection of Q that does not exceed the maximum current, voltage, and power capabilities of the transistor.

In general, manufacturers' data sheets provide the maximum ratings for each particular type of transistor insofar as power, current, and voltage are concerned, as we observed in the data sheets for BJT and FET transistors given in the appendix to Chapter 3. The given ratings cannot be exceeded; otherwise, destruction of the transistor may occur. If, for example, a large output power (greater than 1 W) is required of an amplifier, a special high-power transistor, or power transistor, is used. Similarly, high-current (greater than 30 mA) and high-voltage (greater than 30 V) applications require specialized devices.

The maximum transistor ratings are directly applicable to the quantities at the output terminals of the transistor since the input power, current, and voltage are usually small in comparison. The maximum values directly specify the permissible operating region in which Q will be located, as we will see.

Figure 4.1 shows the output current–voltage axes with boundaries drawn for maximum current $I_{O\,MAX}$, maximum voltage $V_{O\,MAX}$, and maximum power $P_{O\,MAX}$. In order that these maxima not be exceeded, Q must reside closer to the origin than these boundaries. In the ideal case, the *permissible operating region* is closed by the additional boundary line corresponding to zero output current and by another boundary line corresponding to zero output voltage. However, for actual BJTs and FETs, we have seen (in Chapter 3) that the output current–voltage characteristic usually displays at least a small nonzero forward voltage. This voltage causes a reduction in the permissible operating region next to the current axis as shown in Fig. 4.1.

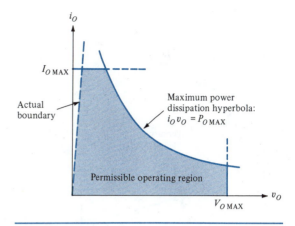

FIGURE 4.1   Permissible Amplifier Operating Region

The maximum power curve represents the equation in which the output power of the transistor, the product of i_O and v_O, is equated to the maximum permissible output power $P_{O\,MAX}$. Thus,

$$i_O v_O = P_{O\,MAX} \qquad (4.1\text{–}1)$$

Equation (4.1–1) is the equation of a hyperbola known as the *maximum power dissipation hyperbola equation*. The following example indicates the simplicity by which the operating region is obtained.

――――――――――――――――――――――――――――――― EX. 4.1

Permissible Operating Region Example

Draw the permissible operating region for a BJT with maximum output ratings given by 30 mA, 30 V, and 0.1 W. Assume that the minimum output voltage for the BJT for amplifier operation is $v_O = 0.2$ V.

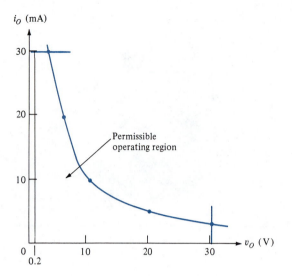

i_O (mA)

FIGURE E4.1

Solution: Figure E4.1 shows the boundaries for the permissible operating region where the equation $i_O v_O = 0.1$ W is sketched by calculating values for v_O assuming values for i_O as follows:

i_O (mA)	v_O (V)
30	3.33
20	5
10	10
5	20
3.33	30

4.1.2   Choice of Q Location

Figure 4.2 displays a circuit that represents a basic transistor amplifier. The type of transistor is not specified. The objective of the circuit is to amplify the AC input voltage (and/or current) such that a larger AC output voltage (and/or current) will be produced across the output load resistor R_O. DC voltage sources are necessary at both the input and output terminals of the amplifier in order to place Q within the permissible operating region. The specific values of R_I and R_O, along with the DC voltage source values V_{DCI} and V_{DCO}, determine the position of Q. The selection of values for these circuit elements is called the *DC bias design* and is described in Section 4.2.7 for BJTs and in Section 4.3.3 for FETs.

As we saw in Chapter 3, the analysis of an amplifier circuit is initiated by writing KVL for the input and output loops. Here, we assume DC currents and voltages only. Thus, we write the load-line equations for the amplifier of Fig. 4.2 as follows:

$$V_{IC} = V_{DCI} - I_I R_I \qquad (4.1\text{--}2)$$

and

$$V_{OC} = V_{DCO} - I_O R_O \qquad (4.1\text{--}3)$$

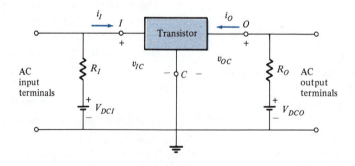

FIGURE 4.2   Basic Transistor Amplifier Circuit

The position of Q is determined by (4.1–2) and (4.1–3) along with the transistor equations (or i–v characteristics) for the specific transistor.

To determine the DC input current and voltage from transistor equations, we use the input load-line equation (4.1–2) together with the appropriate input transistor relation. For the FET, $i_I = 0$; thus, $V_{IC} = V_{DCI}$. For BJTs, $V_{IC} = \pm V_O$ (0.7 V for silicon); therefore, $I_I = (V_{DCI} \pm V_O)/R_I$.

Having obtained the DC input current and voltage in this manner, the corresponding DC output quantities are obtained by using (4.1–3) and the transistor transfer equation that relates the output current to the input current (BJT) or input voltage (FET). First, the DC output current is obtained by using the transfer equation and the known DC input quantity. Then, the DC output voltage is found from (4.1–3). Particular transistor cases will be discussed in the sections that follow.

Figure 4.3 displays the output load line representing (4.1–3). Several possibilities for the location of Q are shown on this line (Q_1, Q_2, and Q_3). The actual position of Q depends upon the particular magnitudes of the external resistors and DC voltages selected, as well as the particular transistor equations. As we will see in Chapter 5, the most desirable position for Q in amplifier applications is near the middle of the output load line, such as Q_2. One justification for this choice

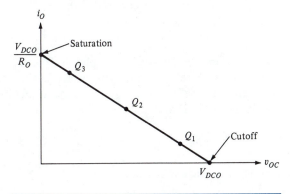

FIGURE 4.3   Output Load Line with Various Choices for Q

at present is that the ends of the load line are avoided in amplifier operation because nonlinear behavior results. Note that only the positive or negative portions of the signal can be amplified with Q at one of these positions. However, operation is exactly at either of these end-point extremes in the case of BJT digital logic circuits. Furthermore, these load-line end-points are given the names *saturation and cutoff*, as shown in Fig. 4.3. Operation in saturation requires that both junctions of the BJT be forward biased, and operation in cutoff requires that both junctions be reverse biased. These operating conditions are quite different from those for amplifier operation of a BJT where the emitter junction is forward biased and the collector junction is reverse biased (as was discussed in Chapter 3).

This section has provided a general overview of the selection of the position of Q. In sections that follow, we consider specific amplifiers and their respective DC bias design procedures.

4.2 BJT AMPLIFIERS

4.2.1 BJT Amplification Factors

Chapter 3 introduced the current–voltage characteristics for BJTs and the transfer (input–output) relation $i_O = k i_I$ for amplifier operation, where k was defined as an amplification factor. The dependence of k upon BJT design can best be understood by considering the mechanism of charge flow and current for a BJT operating under amplifier conditions.

Common-Base Amplification Factor (α).   Figure 4.4 displays a representative view of a PNP BJT operating under amplifier conditions. We use this case as an example because the emitter current is in the direction of positive charge flow. The case of an NPN BJT is essentially the same, except that the currents have the opposite directions and the DC voltage polarities are reversed. The polarities of the external DC voltages shown provide the

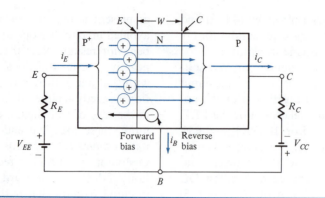

FIGURE 4.4 PNP BJT Charge Flow, Currents, and Voltages for Amplifier Operation

desired amplifier-region bias requirements to forward bias the emitter junction and reverse bias the collector junction. The resistors R_E and R_C limit the transistor current magnitudes.

When the emitter junction of the PNP BJT is forward biased, majority charges flow across this junction, holes from P^+ to N and electrons from N to P^+ (as depicted diagramatically in Fig. 4.4). Since the emitter P^+ region is heavily doped, there are many more holes crossing the emitter junction than electrons under forward-bias conditions, and i_E is therefore primarily due to this hole movement. For a narrow BJT base width (defined as W in Fig. 4.4), essentially all of the injected holes reach the collector junction as shown. The collector junction reverse bias aids these holes (which are now minority carriers in the base) in crossing the collector junction. The injected holes from the emitter that enter the collector region then flow out the collector terminal, resulting in external collector current i_C, as shown in Fig. 4.4. The external base current is due to the relatively small number of electrons crossing the emitter junction. Hence, the base current is quite small in magnitude in comparison to the emitter and collector currents. The two charge (bipolar) flows and their corresponding currents are indicated in Fig. 4.4.

By considering the flow of charges in the BJT, we can understand that the collector current is a fraction of the emitter current. We define the specific relation as follows:

$$i_C = \alpha i_E \qquad (4.2\text{–}1)$$

where α is called the *collector–emitter current amplification factor* and $\alpha < 1$. (It is customary to call α an amplification factor even though $\alpha < 1$.) In a well-designed BJT structure, with heavy emitter doping and narrow base width, α is very nearly equal to 1, a typical value being 0.99. We describe α as being less than but approximately equal to 1, or $\alpha \lesssim 1$. Thus, it should be noted, we will take $\alpha = 1$ in almost all of the amplifier analyses.

Common-Emitter Amplification Factor (β). A general relationship for every BJT is that

$$i_E = i_C + i_B \qquad (4.2\text{–}2)$$

which was first introduced in Chapter 1. This relationship is merely a statement of KCL that treats the BJT as a single node. Substituting (4.2–2) into (4.2–1) and eliminating i_E, we have

$$i_C = \alpha(i_C + i_B) \qquad (4.2\text{–}3)$$

Solving for i_C yields

$$i_C = \left(\frac{\alpha}{1 - \alpha}\right) i_B \qquad (4.2\text{–}4)$$

It is customary to define the coefficient in (4.2–4) as β; thus,

$$\beta = \frac{\alpha}{1 - \alpha} \qquad \text{(4.2–5a)}$$

where β is called the *collector–base current amplification factor*. Note that solving (4.2–5a) for α results in

$$\alpha = \frac{\beta}{1 + \beta} \qquad \text{(4.2–5b)}$$

Because of the denominator in (4.2–5a) and the fact that $\alpha \lesssim 1$, β is quite large ($\beta \gg 1$). Thus, substituting (4.2–5a) into (4.2–4) yields

$$i_C = \beta i_B \qquad \text{(4.2–6)}$$

Equations (4.2–6) and (4.2–1) are the specific current transfer relations for the BJT. These equations are valid for BJTs operating in the amplifier region. In summary, if a BJT is operating under amplifier conditions, then

$$i_O = k i_I \qquad \text{(4.2–7a)}$$

and

$$v_{IC} = \pm V_0 \qquad \text{(4.2–7b)}$$

where $k = \alpha$ or β and $V_0 = 0.7$ V for a silicon transistor.

Although we will treat the current amplification factors (α and β) as constants, in reality they vary with changes in temperature, device structure, and current level. Furthermore, the actual value of β is not known specifically for any one type of BJT, and only a range of values is provided by manufacturers. The variation of β can be problematic in the selection of the position of Q. If the design uses a particular value for β in obtaining the location of Q and the actual β-value for the transistor placed in the circuit is two or three times larger, the resultant position of Q can

be in the saturation region. This difficulty is avoided by using an additional resistor in the basic amplifier circuit, as shown in the next section.

4.2.2   Q Point Stabilization Circuit

Figure 4.5a displays a particular BJT circuit that provides stabilization of the Q point to changes in β (as well as in α). This circuit is representative of all transistor stabilizing circuits. The stabilization of Q to variations in β means that the DC output current and voltage will remain essentially unchanged even though β may vary greatly. The additional resistor in the circuit of Fig. 4.5a (as compared to Fig. 4.2) is called the *feedback resistor R_F*. This resistor forces the input current to be dependent upon the output current; hence, a portion of the output is "fed back" to the input through R_F.

To demonstrate the manner in which R_F stabilizes Q, we begin by writing KVL for the input loop of Fig. 4.5a for DC only:

$$V_{IC} = V_{DCI} - I_I R_I - (I_I + I_O)R_F \qquad \text{(4.2–8)}$$

Then, since amplifier operation of the BJT dictates that $V_{IC} = V_O$ and $I_O = k I_I$, we have

$$V_O = V_{DCI} - I_I R_I - (I_I + k I_I)R_F \qquad \text{(4.2–9)}$$

Solving algebraically for I_I yields

$$I_I = \frac{V_{DCI} - V_0}{R_I + (1 + k)R_F} \qquad \text{(4.2–10)}$$

To obtain the DC output current, we again use $I_O = k I_I$:

$$I_O = \frac{k(V_{DCI} - V_0)}{R_I + (1 + k)R_F} \qquad \text{(4.2–11)}$$

The method of stabilizing the output current I_O is implicitly contained in (4.2–11). If we choose

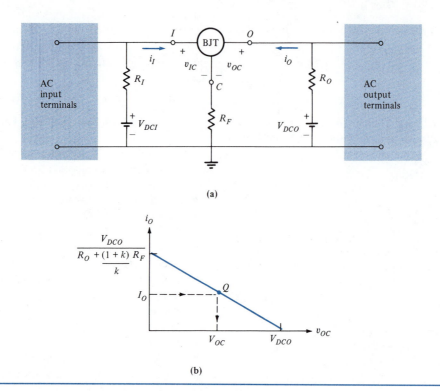

(a)

(b)

FIGURE 4.5   Q Point Stabilization: (a) General BJT amplifier circuit, (b) Load line for the output loop

values of R_I and R_F such that $(1 + k)R_F \gg R_I$, then (4.2–11) becomes

$$I_0 \simeq \frac{k(V_{DCI} - V_0)}{(1 + k)R_F} \qquad (4.2\text{--}12)$$

For the common-emitter and common-collector cases, $k = \beta \gg 1$, and the ratio $k/(k + 1)$ is essentially 1. Thus, I_0 is independent of β, which provides the desired stabilization of Q. For the common-base amplifier case in which $k = \alpha \lesssim 1$, the variation in α is very small on a percentage basis since $\alpha = \beta/(1 + \beta)$ is approximately unity. Thus, the ratio $\alpha/(1 + \alpha) \simeq 1/2$, and I_0 is again stabilized.

In the common-emitter amplifier case in which $k = \beta$, we satisfy the inequality $(1 + k)R_F \gg R_I$ by using an approximate equation given by

$$(1 + k)R_F = 10R_I \qquad (4.2\text{--}13)$$

where $R_I > 0$. (Note that choosing $R_I = 0$ would also satisfy the inequality; however, no AC current would enter the BJT.) Equation (4.2–13) is a *rule-of-thumb* design expression and is an important design relation for BJT amplifier circuits.

To determine the DC output voltage, KVL is written for the output loop as follows:

$$V_{OC} = V_{DCO} - I_0R_O - (I_0 + I_1)R_F \qquad (4.2\text{--}14)$$

Figure 4.5b displays graphically the method of determining the DC output voltage. We enter the graph with the known value of I_0 (from 4.2–12) and read the corresponding value for V_{OC} by determining the intersection of $I_0 = kI_1$, with the load line plotted using (4.2–14).

Having carried out the DC analysis for the general BJT amplifier circuit introduced in this section, the DC analysis of the specific BJT amplifier circuits that follow is considerably simplified. Most of our circuits will use NPN BJTs since this device is optimized in design.

4.2.3   DC Analysis of the Common-Emitter Amplifier

The basic DC circuit used for the *common-emitter* amplifier is shown in Fig. 4.6a. This circuit is extremely practical since only one DC supply voltage (V_{CC}) is required to bias both

junctions of the BJT. A single supply voltage is used in the common-collector and common-base circuits as well. The circuit of Fig. 4.6a stabilizes Q by proper choice of the feedback resistor R_E. As we will see in Chapter 5, amplification of both the AC input voltage and current is provided by this amplifier circuit. The circuit of Fig. 4.6a, however, is difficult to analyze; thus, we develop an equivalent circuit for analysis.

The equivalent circuit for analysis of the common-emitter amplifier is shown in Fig. 4.6b. It is obtained by replacing V_{CC} with two equivalent voltage sources of magnitude V_{CC} (in Fig. 4.6a),

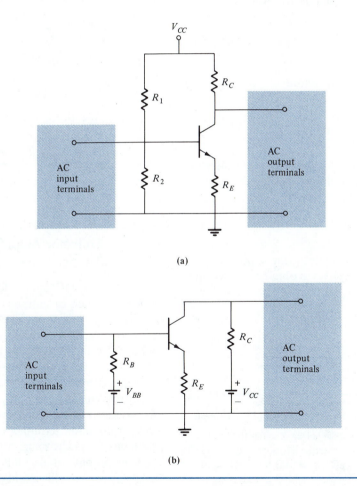

(a)

(b)

FIGURE 4.6   Common-Emitter Amplifier Circuits: (a) Basic circuit, (b) Equivalent circuit for analysis

one at the top of R_1 and the other at the top of R_C. Then, R_C and V_{CC} are redrawn equivalently on the output side of the circuit in Fig. 4.6b. The resistor R_B and DC voltage source V_{BB} (on the input side of the circuit in Fig. 4.6b) are the Thévenin equivalent elements for the combination of R_1, R_2, and V_{CC} in Fig. 4.6a with $R_B = R_1 \| R_2$ and $V_{BB} = [R_2/(R_1 + R_2)]V_{CC}$.

Note that the circuit of Fig. 4.6b is identical to that of Fig. 4.5a. The analysis is therefore identical to that of the previous section, but with the variables changed appropriately. By changing the resistor and DC voltage source symbols, the input current for the BJT is obtained as follows:

$$I_B = \frac{V_{BB} - V_0}{R_B + (1 + \beta)R_E} \tag{4.2–15a}$$

The output current is therefore

$$I_C = \beta I_B = \frac{\beta(V_{BB} - V_0)}{R_B + (1 + \beta)R_E} \tag{4.2–15b}$$

The DC output collector current I_C is stabilized by using the rule-of-thumb design expression, given specifically in this case by

$$(1 + \beta)R_E = 10R_B \tag{4.2–16}$$

To complete the analysis, the expression for the DC output voltage V_{CE} is obtained by writing KVL for the output loop of Fig. 4.6b as follows:

$$V_{CE} = V_{CC} - I_C(R_C + R_E) \tag{4.2–17}$$

where we assume that $I_E = I_C$ with $\alpha \simeq 1$.

EX. 4.2

DC Analysis for the Common-Emitter Amplifier Example

The common-emitter amplifier of Fig. 4.6a has the following values specified for the external elements: $R_1 = 20\ \text{k}\Omega$, $R_2 = 6.67\ \text{k}\Omega$, $R_C = 1.5\ \text{k}\Omega$, $R_E = 1\ \text{k}\Omega$, and $V_{CC} = 20\ \text{V}$. The transistor is made of silicon with $\beta = 100$. Determine the DC currents and voltages for the BJT.

Solution: We begin by calculating values for R_B and V_{BB} to be used in the equivalent circuit for analysis (Fig. 4.6b). Thus,

$$R_B = \frac{R_1 R_2}{R_1 + R_2} = \frac{(6.67)(20)}{6.67 + 20} = 5\ \text{k}\Omega$$

and

$$V_{BB} = \frac{R_2}{R_1 + R_2}(V_{CC}) = \frac{6.67}{20 + 6.67}(20) = 5\ \text{V}$$

Then, since $V_{BE} = 0.7\ \text{V}$, (4.2–15a) yields the value of DC input current as follows:

$$I_B = \frac{V_{BB} - V_0}{R_B + (1 + \beta)R_E} = \frac{5 - 0.7}{5 + (101)1} = 0.04\ \text{mA}$$

From (4.2–15b), the DC output current is

$$I_C = \beta I_B = 100(0.04) = 4\ \text{mA}$$

Finally, from (4.2–17), the DC output voltage is

$$V_{CE} = V_{CC} - I_C(R_C + R_E) = 20 - (4)(1.5 + 1) = 10\ \text{V}$$

4.2.4   DC Analysis of the Common-Collector Amplifier (Emitter Follower)

The DC circuit for the common-collector amplifier, or emitter follower, is displayed in Fig. 4.7. Note that the AC output terminals are across the stabilizing resistor R_E and that the resistor R_C (in Fig. 4.6) has been replaced with a short ($R_C = 0$ in Fig. 4.7). The term *common-collector* is used because the voltage at the collector terminal is purely DC. That is, there is zero AC voltage at C so that this terminal is the reference or common terminal as far as AC voltages are concerned. The name *emitter follower* implies that the output voltage "follows" the input signal ($v_i \lesssim v_o$) as we will see in Chapter 5, and this amplifier provides AC current amplification only.

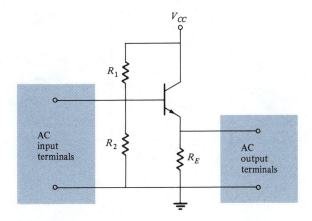

FIGURE 4.7   Common-Collector Amplifier (Emitter Follower)

In Fig. 4.7, the resistor R_E serves a dual purpose. R_E not only stabilizes the position of Q, but also it is the load resistor. We will see in Chapter 5, however, that other load resistor connections are often used. Note that the only DC biasing change in the common-collector amplifier circuit (Fig. 4.7) from that of the common-emitter amplifier circuit (Fig. 4.6) is that $R_C = 0$. Therefore, the basic equations (4.2–14) through (4.2–17) are used without change for the common-collector amplifier case.

—————————————————— EX. 4.3

DC Analysis for the Common-Collector Amplifier Example

The common-collector amplifier of Fig. 4.7 has the following values specified for the external elements: $R_1 = 8$ kΩ, $R_2 = 13.3$ kΩ, $R_E = 0.5$ kΩ, and $V_{CC} = 10$ V. The transistor is made of silicon with $\beta = 100$. Determine the DC currents and voltages for the BJT.

Solution: The values for R_B and V_{BB} are calculated in the same manner as they were in Ex. 4.2. Thus, $R_B = 5$ kΩ and $V_{BB} = 6.24$ V. Since the transistor is made of silicon, the input voltage $V_{BE} = 0.7$ V. Equation 4.2–15a then provides the DC input current:

$$I_B = \frac{6.25 - 0.7}{5 + (101)0.5} = 0.1 \text{ mA}$$

The DC output current, from $I_C = \beta I_B$, is $I_C = 10$ mA. Finally, the DC output voltage is obtained from (4.2–17) with $R_C = 0$:

$$V_{CE} = 10 - (10)0.5 = 5 \text{ V}$$

4.2.5   DC Analysis of the Common-Base Amplifier

Figure 4.8a displays the basic DC circuitry used for the *common-base* amplifier. This amplifier can amplify an AC voltage but not an AC current. Note that the resistors are denoted in exactly the same manner as those used in the common-emitter amplifier of Fig. 4.6. This notation is not a coincidence; the DC circuits are, in fact, identical.

The equivalent circuit for analysis is shown in Fig. 4.8b. It is obtained in an analogous manner to the common-emitter case. Writing KVL for the input loop and using the relations for the BJT operating under amplifier conditions, (4.2–7a) and (4.2–7b), the input current is obtained for the NPN transistor as follows:

$$I_E = \frac{V_{BB} - V_0}{R_E + [R_B/(1 + \beta)]} \qquad (4.2–18)$$

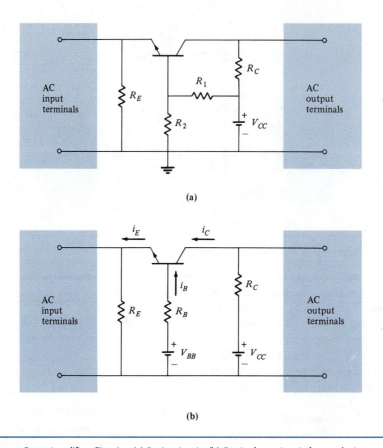

FIGURE 4.8   Common-Base Amplifier Circuits: (a) Basic circuit, (b) Equivalent circuit for analysis

The output current is thus given by

$$I_C = \alpha I_E = \frac{\alpha(V_{BB} - V_0)}{R_E + [R_B/(1 + \beta)]} \qquad (4.2\text{–}19)$$

Note that substitution for α in terms of β, where $\alpha = \beta/(\beta + 1)$, transforms (4.2–19) directly into (4.2–15b). Thus, the rule-of-thumb expression in this case is identical to the previous cases.

The expression for the DC output voltage is obtained by writing KVL for the output loop of Fig. 4.8b as follows:

$$V_{CB} = V_{CC} - V_{BB} - I_C R_C + I_B R_B \qquad (4.2\text{–}20)$$

—————————————————————————— EX. 4.4

DC Analysis for the Common-Base Amplifier Example

A silicon transistor with $\beta = 100$ is used in the common-base amplifier circuit of Fig. 4.8a. Determine the input and output currents and voltages for the transistor if the bias circuit values for the elements R_1, R_2, R_C, R_E, and V_{CC} are the same as those used in Ex. 4.2.

Solution: The simplest approach to the solution in this example is to use the results of Ex. 4.2. Since the circuit of that example is the same as the one used here, except for the configuration, the currents and voltages for the transistor remain the same. However, the DC input current and voltage for the BJT are now I_E and V_{EB}, and the DC output current and voltage for the BJT are I_C and V_{CB}.

The output quantities are then I_C, which equals 4 mA, and V_{CB}, which is determined from (4.2–20) as follows:

$$V_{CB} = 20 - 5 - 4(1.5) + \left(\frac{4}{100}\right)5 = 9.2 \text{ V}$$

The DC input current is thus

$$I_E = \frac{I_C}{\alpha} = \frac{(\beta + 1)I_C}{\beta} = 4.04 \text{ mA}$$

Note that $I_C \simeq I_E$ and differs by only 1%. Finally, the DC input voltage for the BJT is

$$V_{EB} = -V_0 = -0.7 \text{ V}$$

4.2.6 DC Analysis of Difference Amplifiers

The *difference amplifier* is a basic building block used in integrated circuits, as we will see. In many instances, difference amplifiers are used as replacements for common-emitter amplifiers.

Basic Difference Amplifier with Bias Resistor R_E. The basic circuit for the difference amplifier with a bias resistor R_E is shown in Fig. 4.9a. This circuit amplifies the difference in the input signal voltages $(v_{s_2} - v_{s_1})$ and thus derives its name. The difference amplifier has several important distinctions from all of the previous amplifiers that we have considered. First, two transistors (instead of one) are required. Also, a balanced circuit configuration is necessary, which means that along the dashed line drawn down the center of Fig. 4.9a, the circuit on either side is the mirror image of the other side. Ideally, the transistors and the resistors are identical as labeled. Additionally, the difference amplifier requires two different DC voltage sources $(V_{CC}$ and $-V_{EE})$. The voltage source $-V_{EE}$ is necessary in order to forward bias the emitter junction without including this DC voltage as a part of the input signal voltages v_{s_1} and v_{s_2}. The output signal voltage is taken from the collector terminal of either transistor. In Chapter 6, we will verify that each of these output signal voltages $(v_{c_1}$ or $v_{c_2})$ is an amplified version of the difference between the input signal voltages.

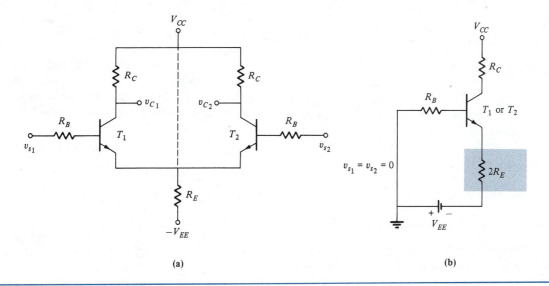

(a)

(b)

FIGURE 4.9 Difference Amplifier Circuits: (a) Basic circuit with bias resistor, (b) DC equivalent circuit for analysis

To carry out the DC analysis of the circuit in Fig. 4.9a, we first realize that under DC conditions, $v_{s_1} = v_{s_2} = 0$. Also, because of the balanced circuit, the DC currents and voltages for each transistor are identical. Because of this balance, considerable simplification is possible, as shown in the equivalent circuit for analysis, Fig. 4.9b. Here, either transistor (T_1 or T_2) is shown with the resistor $2R_E$ replacing R_E in order to account for the DC emitter current of the omitted transistor.

In analyzing the circuit, we first note that the base–emitter voltage for the BJT is V_0. Then, writing KVL for the base–emitter loop of Fig. 4.9b yields

$$V_{EE} - V_0 = I_E(2R_E) + I_B R_B \qquad (4.2-21)$$

Substituting $I_E = (1 + \beta)I_B$ and solving for I_E, we have

$$I_E = \frac{V_{EE} - V_0}{2R_E + [R_B/(1 + \beta)]} \simeq I_C \qquad (4.2-22)$$

where the usual approximation that $I_C \simeq I_E$ was used. The rule-of-thumb expression in this case is given by

$$2(1 + \beta)(R_E) = 10R_B \qquad (4.2-23)$$

which differs from the previous cases by a factor of 2.

The DC collector–emitter voltage for either transistor is then obtained by writing KVL for the output branch of Fig. 4.9b:

$$V_{CE} = V_{CC} + V_{EE} - I_C(R_C + 2R_E) \qquad (4.2-24)$$

which completes the analysis.

EX. 4.5

Difference Amplifier with Bias Resistor

Determine the DC currents and voltages for the BJTs in the difference amplifier of Fig. 4.9 for bias-element values in the circuit given by $V_{CC} = 12$ V, $V_{EE} = 8$ V, $R_C = 1$ kΩ,

$R_B = 20$ kΩ, and $R_E = 1$ kΩ. The transistor is made of silicon with $\beta = 100$.

Solution: Because of the balanced circuit, the DC currents and voltages for each transistor are equal, and Fig. 4.9b is used for analysis.

Since the transistors are made of silicon, the DC input voltage for each is $V_{BE} = 0.7$ V. The DC output current is then obtained from (4.2–22) as follows:

$$I_C \simeq \frac{8 - 0.7}{2 + (20/101)} = 3.3 \text{ mA}$$

Then, since $I_C = \beta I_B$, we have $I_B = 0.033$ mA. Finally, the DC output voltage is given by (4.2–24) as follows:

$$V_{CE} = 12 + 8 - [(3.3)(1 + 2)] \simeq 10 \text{ V}$$

Basic Difference Amplifier with DC Current Source.   Figure 4.10 displays a difference amplifier circuit with R_E of Fig. 4.9a replaced with a DC current source (I_{DC}). The actual circuitry used for the DC current source is described in Chapter 6, where we will see that its effect is to provide a large equivalent resistance without a large R_E. The value of I_{DC} is selected to provide the sum of the DC emitter current required for each transistor (twice that of one transistor).

In analyzing the circuit, the DC collector current for each transistor is specified since $I_C \simeq I_E = I_{DC}/2$. Then, the DC output voltage is obtained from the difference in voltage ($V_C - V_E$) as follows:

$$V_{CE} = V_C - V_E = (V_{CC} - I_C R_C) - (-V_0) \qquad (4.2-25)$$

or

$$V_{CE} = V_{CC} + V_0 - \frac{I_{DC} R_C}{2} \qquad (4.2-26)$$

In this case, the input voltage and input current for each transistor are V_0 and I_C/β, respectively.

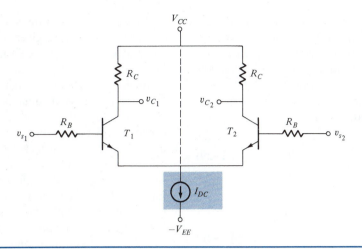

FIGURE 4.10   Difference Amplifier with DC Current Source

EX. 4.6

Difference Amplifier with DC Current Source Example

Determine the DC currents and voltages for the transistors in Fig. 4.10 with bias-element values given by $V_{CC} = 12$ V, $V_{EE} = 8$ V, $R_C = 0.8$ kΩ, and $I_{DC} = 6.6$ mA. As in Ex. 4.5, the transistors are made of silicon with $\beta = 100$.

Solution: Since I_{DC} is given as 6.6 mA, we have $I_C \simeq I_E = 3.3$ mA for each transistor. Also, for each transistor, the DC input base current is 0.033 mA and the base–emitter voltage is 0.7 V. Finally, from (4.2−26), the DC collector-emitter voltage for each transistor is

$$V_{CE} = 12 + 0.7 - [(3.3)(0.8)] \simeq 10 \text{ V}$$

Note that the DC operating point, given by $I_C = 3.3$ mA and $V_{CE} = 10$ V, is essentially the same in this example as it was in Ex. 4.5. To accomplish this result, R_C was modified here.

4.2.7   Bias Circuit Design of BJT Amplifiers

The DC, or bias circuit, design procedure generally used for electronic circuits allows flexibility and is not unique. The procedure in-volves choosing specific values for the external resistors and DC supply voltage that will provide a suitable load line and stabilized Q point. To exemplify the general method, we will consider in detail the design procedure for the common-emitter amplifier. The design procedures for the other two cases will be described briefly and follow directly from the common-emitter amplifier case.

Common-Emitter Amplifier.   Figure 4.6a shows the basic common-emitter amplifier circuit, and Fig. 4.6b displays the equivalent circuit for analysis. Recall that the Thévenin equivalent elements of Fig. 4.6b were obtained as

$$R_B = \frac{R_1 R_2}{R_1 + R_2} \qquad (4.2-27)$$

and

$$V_{BB} = \frac{R_2}{R_1 + R_2}(V_{CC}) \qquad (4.2-28)$$

The bias design is carried out by using the equivalent circuit for analysis (Fig. 4.6b) and con-sists of choosing specific values for V_{CC}, R_C, R_E,

R_B, and V_{BB} that will provide operation in the permissible operating region. However, when values for these quantities are determined, values for R_1 and R_2 of the original circuit (Fig. 4.6a) are obtained from relations developed by solving (4.2–27) and (4.2–28) simultaneously for R_1 and R_2. The resulting equations are

$$R_1 = R_B\left(\frac{V_{CC}}{V_{BB}}\right) \tag{4.2-29}$$

and

$$R_2 = \frac{R_B}{1 - (V_{BB}/V_{CC})} \tag{4.2-30}$$

Note that R_1 and R_2 are each always greater than or equal to R_B since $R_B = R_1 \| R_2$ and that $V_{BB} \leq V_{CC}$ since V_{BB} is a fraction of V_{CC} from (4.2–28).

The known relations that we have as a starting basis for design are the basic circuit and the rule-of-thumb expression (for stability of Q). Additional known quantities are the transistor ratings for maximum permissible current ($I_{C\,MAX}$), voltage ($V_{CE\,MAX}$), and power ($P_{C\,MAX}$) as well as a range of β values ($\beta_{MIN} \leq \beta \leq \beta_{MAX}$) for the particular BJT provided by the manufacturer.

The design procedure is initiated by choosing a particular load line, given analytically by (4.2–17), and DC value of collector current, as shown in Fig. 4.11. The load line selected must reside below the hyperbola of maximum power dissipation, and its intercepts must be less than the given maximum current and voltage values. Figure 4.11 shows a typical load line, which is by no means unique, with Q selected somewhere near the middle of the load line to avoid the cutoff and saturation regions. With the selection of Q, values for I_C and V_{CE} are read directly from their corresponding axes. With the selection of a specific load line, the intercepts are known and given by $I_{C\,END} = V_{CC}/(R_C + R_E)$ and V_{CC}. The rule-of-thumb expression for this circuit is given by (4.2–16), but modified here as

$$(1 + \beta_{MIN})R_E = 10R_B \tag{4.2-31}$$

where the minimum value of β (β_{MIN}) must be used. Using β_{MIN} in (4.2–31) provides the correct relation between R_E and R_B for stability of Q for the entire specified range of β ($\beta_{MIN} \leq \beta \leq \beta_{MAX}$). If the value of β for the transistor used in the circuit is, in fact, larger than β_{MIN}, then the inequality $(1 + \beta)R_E \gg 10R_B$ will still be satisfied.

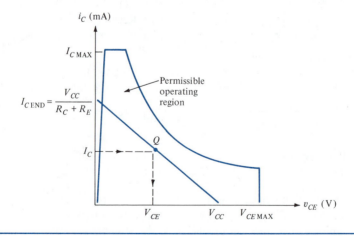

FIGURE 4.11   Selection of Load Line and Q

The equation for the DC collector current is obtained from (4.2–15b) and is approximated by

$$I_C = \frac{V_{BB} - V_0}{(R_B/\beta_{MIN}) + R_E} \tag{4.2-32}$$

where $\beta = \beta_{MIN}$ has been substituted to be consistent with (4.2–31). The approximation used is that $\beta/(\beta + 1) \simeq 1$, which is equivalent to $\alpha = 1$.

The design procedure may now proceed in either of two ways:

1. If the value of the load resistor R_C is specified, then (4.2–17) can be solved immediately for R_E, (4.2–31) yields the value of R_B, and (4.2–32) provides the value of V_{BB}, assuming that V_0 is known.

2. If the value of R_C is not known, it is customary to let the voltage across R_E be a fraction of V_{CC}. We choose this fraction to be 1/5, therefore,

$$V_{R_E} = I_E R_E = \frac{V_{CC}}{5} \tag{4.2-33}$$

Since I_C and I_E are approximately equal, we have

$$R_E \simeq \frac{V_{CC}}{5I_C} \tag{4.2-34}$$

The value of R_E is determined from (4.2–34). We then calculate the value of R_C from (4.2–17) and the value for R_B from (4.2–31). Finally, the value of V_{BB} is obtained from (4.2–32) by using the known value for V_0.

It should be noted that we did not maximize the power output in this example because it is not usually of interest, as we will see in E4.7. It is accomplished, however, by choosing a load line close and, in fact, tangent to the maximum power dissipation hyperbola. A further discussion of power amplifier circuits is given in the next chapter.

DC Bias Circuit Design for a BJT Common-Emitter Amplifier Example

A BJT is to be used in a common-emitter amplifier configuration as shown in Fig. 4.6a. The maximum ratings for the output current, voltage, and power are 50 mA, 30 V, and 1 W, respectively. Also, the range of values for β for this particular transistor is $50 \le \beta \le 150$. Determine bias elements V_{CC}, R_C, R_E, R_1, and R_2 that will provide a suitable position for Q.

Solution: Figure E4.7 shows a sketch of the permissible operating region for the specified BJT. Note that the hyperbola of maximum power dissipation plays no role in the selection of the load line. The load line and Q point shown are selected to provide operation in the permissible region, but many other choices are possible. For the load line and Q point shown, we have

$$I_{C\ END} = 40\ mA = \frac{V_{CC}}{R_E + R_C} \tag{1}$$

$$I_{CQ} = 20\ mA = \frac{V_{BB} - 0.7}{(R_B/50) + R_E} \tag{2}$$

and $V_{CC} = 20$ V. We select $R_E = V_{CC}/5I_C$ and get $R_E = 0.2$ kΩ. Then, R_C is determined by substituting this value of R_E into (1):

$$40\ mA = \frac{20}{0.2 + R_C} \tag{3}$$

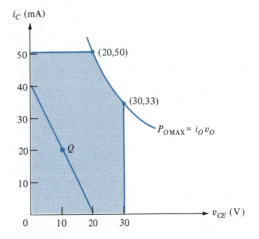

FIGURE E4.7

Thus, $R_C = 0.3$ kΩ. From the rule-of-thumb expression (4.2–31), we have

$$(1 + 50)R_E = 10R_B \qquad (4)$$

Therefore, $R_B = 10.2$ kΩ. V_{BB} is then calculated from (2), and substituting values yields

$$20 \text{ mA} = \frac{V_{BB} - 0.7}{(10.2/50) + 0.2} \qquad (5)$$

or $V_{BB} = 8.78$ V. Finally, the values of R_1 and R_2 are calculated by using (4.2–29) and (4.2–30). Substituting the values for R_B, V_{BB}, and V_{CC} yields

$$R_1 = 10.2\left(\frac{20}{8.78}\right) = 23.23 \text{ kΩ} \qquad (6)$$

and

$$R_2 = \frac{10.2}{1 - (8.78/20)} = 18.18 \text{ kΩ} \qquad (7)$$

Common-Collector Amplifier. The common-collector amplifier is identical to the common-emitter amplifier (as far as DC is concerned), except that $R_C = 0$. Thus, the DC design procedure of the common-emitter amplifier applies directly for the common-collector amplifier under the special condition that R_C is specified as zero.

Common-Base Amplifier. The DC circuit for the common-base amplifier is shown in Fig. 4.8. To carry out the bias design in this case, we must slightly modify the common-emitter design procedure.

From the equivalent circuit for analysis shown in Fig. 4.8b, we see that the DC output voltage for the transistor is V_{CB} (instead of V_{CE}). Moreover, the load-line equation for the output loop is given by (4.2–20) as

$$V_{CB} = V_{CC} - V_{BB} - I_C R_C + I_B R_B \qquad (4.2–35)$$

This equation dictates that the voltage intercept is located at $V_{CC} - V_{BB}$, and the current intercept

is located at

$$I_{C \text{ END}} = \frac{V_{CC} - V_{BB}}{R_C - (R_B/\beta)} \qquad (4.2–36)$$

The design procedure then involves selecting a suitable load line and Q point as was done for the common-emitter amplifier. This selection provides the values of I_C, V_{CB}, and the intercepts of the load line. To be consistent with the common-emitter DC design procedure (when the value of R_C is not specified), we again choose the value of R_E from $R_E = V_{CC}/5I_C$. The rule-of-thumb expression is then used to obtain R_B. With R_E and R_B known, V_{BB} is obtained from the input equation, (4.2–19) or (4.2–15b). Finally, R_C is calculated from the output load-line equation, (4.2–35).

The computations are actually more complicated when R_C is specified. Under these conditions, (4.2–19), (4.2–35), and the rule-of-thumb expression must be solved simultaneously for V_{BB}, R_B, and R_E.

Difference Amplifier with Bias Resistor R_E. The DC design procedure in the case of the difference amplifier with a bias resistor R_E is identical to the common-emitter (and common-collector) amplifier procedure, except that R_E is replaced with $2R_E$.

Difference Amplifier with DC Current Source. If we refer to the difference amplifier circuit in Fig. 4.10, we see that the DC current source shown is used to directly specify the desired value of DC collector current I_C ($I_C \simeq I_E$) for each transistor since $I_{DC} = 2I_E \simeq 2I_C$. The specific Q point is chosen inside the permissible operating region as previously described. The DC collector–emitter voltage V_{CE} is then given by (4.2–26). In this case, however, we do not have a load line with variable I_C because (4.2–26) has a fixed current. The simplest procedure to complete the DC bias design is to choose V_{CC} conveniently and to determine R_C from (4.2–26). The design of the current source is described in Chapter 6.

4.3   FET AMPLIFIERS

The bias design for FET amplifier circuits is quite similar to that for BJTs. Thus, the same general circuit was used for both BJTs and FETs at the beginning of this chapter. We now consider the DC analysis for the two basic amplifier configurations for FETs. We first determine the FET DC currents and voltages and then examine a bias design procedure for FETs.

4.3.1   DC Analysis of the Common-Source Amplifier

Figure 4.12a displays the basic *common-source* amplifier circuit. Note that the external part of this circuit is exactly the same as the BJT amplifier circuit of Fig. 4.6a. However, the resistors R_D and R_S and the DC voltage source V_{DD} are labeled according to FET terminology. The resistor R_S is the stabilizing resistor in this circuit;

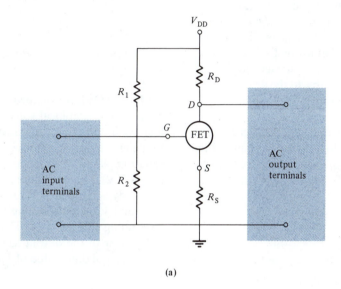

(a)

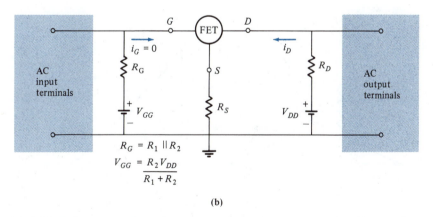

$$R_G = R_1 \parallel R_2$$
$$V_{GG} = \frac{R_2 V_{DD}}{R_1 + R_2}$$

(b)

FIGURE 4.12   Common-Source Circuits (a) Basic circuit, (b) Equivalent circuit for analysis

it provides a relatively unchanged value of I_D even though the FET parameters may change. For positive V_{DD}, the resistor R_S provides a negative component of voltage to the gate–source terminals, while R_1 and R_2 in combination with a positive V_{DD} provide a counteracting positive component. All three resistors (R_1, R_2, and R_S) are adjusted in value to obtain the desired polarity for v_{GS}, where each polarity is sometimes required depending upon the specific type of FET in the circuit. For example, the N-channel E-O MOST requires a positive gate–source voltage, whereas the N-channel JFET requires a negative gate–source voltage. The sign of V_{DD} is dictated by whether the FET is N-channel or P-channel. For each case, the sign of V_{DD} must attract the carriers in the channel from the source to the drain. Hence, for N-channel, V_{DD} is positive; for P-channel, V_{DD} is negative.

Figure 4.12b shows the equivalent circuit for analysis corresponding to the actual circuit of Fig. 4.12a, where R_G and V_{GG} are the Thévenin equivalent elements for R_1, R_2, and V_{DD}. Writing KVL (considering DC only) for the input loop and using $I_G = 0$, we have

$$V_{GS} = V_{GG} - I_D R_S \qquad (4.3–1)$$

Equation (4.3–1) is a transfer type of load-line equation expressing the relationship between output current and input voltage. Combining (4.3–1) with the general FET amplifier region transfer equation (for DC only) given by

$$I_D = C(V_{GS} - V_C)^2 \qquad (4.3–2)$$

allows determination of the DC drain current I_D and DC gate–source voltage V_{GS}. The DC drain–source voltage is then obtained by writing KVL for the output loop as follows:

$$V_{DS} = V_{DD} - I_D(R_D + R_S) \qquad (4.3–3)$$

Figure 4.13 displays the graphical determination of Q for the special case of an N-channel E-D MOST. The determination of Q for the JFET and the E-O MOST is obtained in an entirely similar manner. The graphical solution is carried out by first obtaining Q on the transfer axes (Fig. 4.13a) by determining the intersection of the load-line equation (4.3–1) with the amplifier transfer equation (4.3–2). The procedure is analogous to simultaneously solving (4.3–1) and (4.3–2). The values of I_D and V_{GS} are then simply read from the corresponding transfer axes. The value of I_D is projected over to the output axes of Fig. 4.13b, and the corresponding value of V_{DS} is obtained.

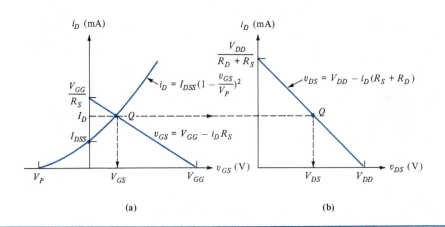

(a)

(b)

FIGURE 4.13   Graphical Determination of Q: (a) Current–voltage transfer axes, (b) Current–voltage output axes

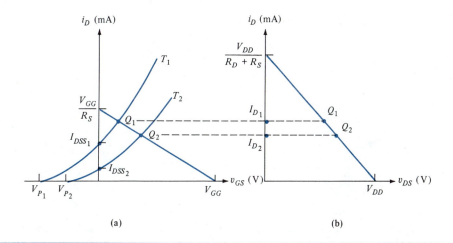

FIGURE 4.14   Stabilization of Q: (a) Shift in Q on transfer axes, (b) Shift in Q on output axes

To realize the mechanism by which stability of Q is provided in the circuits of Fig. 4.12, we consider the transfer characteristics of two different transistors, T_1 and T_2 as shown in Fig. 4.14a, and again assume E-D MOSTs. The curves are drawn assuming that the transistors have differing parameters I_{DSS} and V_P. By carefully selecting the values of V_{GG} and R_S (or R_1, R_2, and V_{DD}), the percentage change in the DC drain current I_D can be reduced to a small amount in comparison to the percentage changes in the parameters. To visualize this comparison, note that if $R_S = 0$ in the circuits of Fig. 4.12, then the transfer load line on the axes of Fig. 4.14a would be vertical, and I_D would change by the maximum amount given by the vertical separation of the two curves (at $v_{GS} = V_{GG}$). The combination of R_S with V_{GG} provides stability of I_D by tilting the transfer load line and thus reducing the percentage change in I_D. This mechanism is called *constant current biasing*.

4.3.2 DC Analysis of the Common-Drain Amplifier (Source Follower)

Figure 4.15 displays the DC circuit used for the *common-drain* amplifier, or *source follower*.

This circuit is entirely analogous to the BJT common-collector amplifier, or emitter follower. The drain terminal is the reference terminal for AC or the common terminal (as seen in Fig. 4.15), and the AC output voltage "follows" the input signal voltage almost identically for a large value of R_S.

The only difference between the circuits for the common-drain amplifier of Fig. 4.15 and the common-source amplifier of Fig. 4.12a (as far as DC is concerned) is that $R_D = 0$ for the common-drain amplifier. Therefore, (4.3–1) through (4.3–3) are applicable here with $R_D = 0$. Furthermore, the graphical method for determination of Q on the output axes of Figs. 4.13b and 4.14b can also be used by setting $R_D = 0$.

4.3.3 Bias Design of FET Amplifiers

The method of DC design for FET amplifier circuits is essentially the same as that for BJT amplifiers. The objective is to select the external resistors and DC supply voltages that will provide a suitable load line and stabilized Q point. As in the case of BJT amplifiers, the choice is not unique.

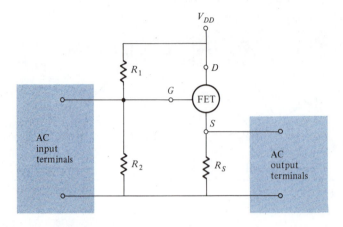

FIGURE 4.15   Common-Drain Amplifier (Source Follower)

Common-Source Amplifier.   Figure 4.12a shows the circuit configuration for the common-source amplifier. We begin by specifying a suitable output load line that satisfies the maximum current $I_{D\,MAX}$, voltage $V_{DS\,MAX}$, and power dissipation $P_{D\,MAX}$ for the particular FET. A possible choice of output load line is shown in Fig. 4.16b, where Fig. 4.16a assumes the case of an E-D MOST. The Q point is selected near the middle of the output load line to avoid the nonlinear end regions.

The equation for the output load line is that given by (4.3–3), with the current intercept at

$$I_{D\,END} = \frac{V_{DD}}{R_D + R_S} \qquad (4.3–4)$$

and the voltage intercept at V_{DD}. For the chosen load line and Q point, we now regard $I_{D\,END}$, V_{DD}, I_D, and V_{DS} as known quantities. Furthermore, the selected value of I_D (in Fig. 4.16b) is projected

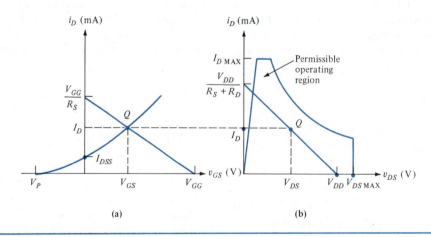

(a)

(b)

FIGURE 4.16   Selection of Load Line and Q: (a) Position of Q on transfer axes, (b) Position of Q on output axes

over to the transfer axes (Fig. 4.16a) to obtain the DC value of V_{GS}. This procedure is analogous to calculating V_{GS} from the amplifier region transfer equation given by (4.3–2).

With a specific load line and position for Q selected, the bias design procedure consists of the following: If R_D is the load resistor whose value is specified, (4.3–4) is solved to obtain the value of R_S, and (4.3–1) is used to obtain the value of V_{GG}. The last element value, R_G, is determined by selecting values for R_1 and R_2, where $R_G = R_1 \| R_2$. The resistance values for R_1 and R_2 are obtained in a somewhat arbitrary manner since only the ratio $R_2/(R_1 + R_2)$ is of importance in prescribing the value of V_{GG}. However, the magnitudes are chosen to be large (~ 100 kΩ) to correspond to the large impedances seen looking into the gate terminal of the FET ($i_G = 0$). Then, the overall input impedance of the circuit will be large, which is desirable for voltage amplifiers. Using the Thévenin equivalent voltage expression given by

$$V_{GG} = \frac{R_2}{R_1 + R_2}(V_{DD}) \tag{4.3–5}$$

and, since V_{GG} and V_{DD} are known quantities, we select the ratio $R_2/(R_1 + R_2)$ to have the desired value. We then choose $R_2 = 100$ kΩ and determine the value of R_1 from the ratio value.

If the value of R_D is not specified in the amplifier circuit, we use a design procedure similar to that of the BJT, which involves choosing the voltage across R_S to be a fraction of V_{CC}. Selecting this fraction as $1/5$, we have

$$V_{RS} = I_D R_S = \frac{V_{DD}}{5} \tag{4.3–6}$$

The value of R_S is therefore

$$R_S = \frac{V_{DD}}{5I_D} \tag{4.3–7}$$

The value of R_D is then obtained from (4.3–4), and (4.3–1) is used to obtain the value of V_{GG}. The resistor values for R_G, R_1, and R_2 are then determined as previously described.

Common-Drain Amplifier.   The bias design procedure for the common-drain amplifier (shown in Fig. 4.15) follows that of the common-source amplifier using the specific value of $R_D = 0$.

EX. 4.8

Bias Circuit Design for an FET Common-Source Amplifier Example

Figure E4.8a shows a common-source amplifier biasing circuit with an N-channel JFET. The Q point is chosen at $I_D = 10$ mA, $V_{DS} = 5$ V, and the load-line intercepts are $I_{D\,END} = 20$ mA and $V_{DD} = 10$ V. The JFET in the circuit has a transfer equation for amplifier operation given by $i_D = 20[(v_{GS}/5 + 1)^2]$ mA for $-5 \le v_{GS} \le 0$. Determine the values of the bias elements V_{DD}, R_D, R_S, R_1, and R_2.

Solution: We begin by drawing the output load line and locating Q on the output axes (i_D versus v_{DS}) as shown in Fig. E4.8c. The value of $I_D = 10$ mA is then projected over to the transfer axes (i_D versus v_{GS}), where the transfer equation for the JFET is plotted. The corresponding value of V_{GS}, calculated from the transfer equation, is -1.46 V. Note also that the load-line voltage intercept gives $V_{DD} = 10$ V.

Since the value of R_D is not given in this example, we calculate R_S from (4.3–7) as follows:

$$R_S = \frac{10}{5(10\text{ mA})} = 0.2\text{ k}\Omega$$

From (4.3–4), we have (with resistors in kΩ)

$$20\text{ mA} = \frac{10}{R_D + 0.2}$$

Calculating R_D yields $R_D = 0.3$ kΩ. Substituting values into (4.3–1) yields.

$$-1.46 = V_{GG} - 10(0.2)$$

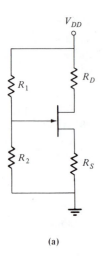

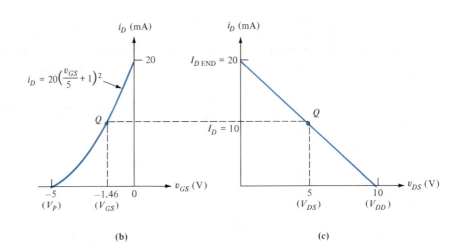

(a)

(b)

(c)

FIGURE E4.8

Solving for V_{GG}, we have $V_{GG} = 0.54$ V. Thus, $R_2/(R_1 + R_2) = V_{GG}/V_{DD} = 0.54/20 = 0.027$. Choosing $R_2 = 100$ kΩ then yields $R_1 + R_2 = 3.7$ MΩ. Therefore, $R_1 = 3.6$ MΩ.

--- **EX. 4.9**

DC Bias Circuit Design for an FET Common-Drain Amplifier Example

Repeat the bias design for Ex. 4.8 for the case in which $R_D = 0$.

Solution: Using the load line and Q point of Ex. 4.8 and

$R_D = 0$, we substitute into (4.3–4) to get

$$20 \text{ mA} = \frac{10}{R_S}$$

or $R_S = 0.5$ kΩ. Substituting values into (4.3–1) then gives

$$-1.46 = V_{GG} - 10(0.5)$$

Solving for V_{GG}, we have $V_{GG} = 3.54$ V. Thus, $R_2/(R_1 + R_2) = 3.54/20 = 0.177$. Choosing $R_2 = 100$ kΩ, we have $R_1 + R_2 = 565$ kΩ. Therefore, $R_1 = 465$ kΩ.

CHAPTER 4 SUMMARY

■ AC amplification in transistor amplifier circuits can be achieved only if the transistor has a suitable DC operating point, or Q point. This point is selected inside the permissible operating region, which is bounded by maximum current, voltage, and power. The Q point is generally selected near the middle of the (DC) load line.

■ For BJTs, $i_C = \alpha i_E$ and $i_C = \beta i_B$ (where $\alpha \lesssim 1$ and $\beta \gg 1$) are basic relations for amplifier operation only. Since $\alpha \simeq 1$, we let $i_C = i_E$. An-

other relation that is always true for BJTs is $i_E = i_C + i_B$.

■ For BJT amplifier circuits, the Q point is stabilized by using a feedback resistor such that $R_F = R_E$. The use of a feedback resistor results in the important relation $I_C = [\beta(V_{BB} - V_0)]/[R_B + (1 + \beta)R_E]$, which is true for all single-transistor BJT configurations. The rule-of-thumb design expression states that $(1 + \beta)R_E = 10R_B$, which is also valid for all single-transistor BJT configurations.

The Q point for a BJT difference amplifier with a DC current source is located at $I_C \simeq I_E = I_{DC}/2$ and $V_{CE} = V_{CC} + V_0 - (I_{DC}R_C/2)$.

The DC bias design of a BJT amplifier circuit consists of obtaining suitable values for the external resistors and DC voltages that provide Q near the middle of the DC load line. A graphical and/or analytical design procedure can be used. The design equations are given by $I_C \simeq (V_{BB} - V_0)/(R_B/\beta_{MIN} + R_E)$, $(1 + \beta)R_E = 10R_B$, and the output load-line equation. When R_C is known, the specific procedure consists of selecting Q and V_{CC} and then solving the design equations for R_E, R_B, and V_{BB}. When R_C is not known, we choose $V_{R_E} = V_{CC}/5$. The base bias resistors are then obtained from $R_1 = R_B(V_{CC}/V_{BB})$ and $R_2 = R_B/[1 - (V_{BB}/V_{CC})]$. If a range of β values is given, use $\beta = \beta_{MIN}$.

For FET amplifier circuits, the Q point is stabi-lized by the combination of a feedback resistor such that $R_F = R_S$ and a bias voltage V_{GG}.

The DC bias design for the common-source FET amplifier is similar to that for BJTs but is simplified because $I_G = 0$. The design equations are $I_D = C(V_{GS} - V_C)^2$, $V_{GS} = V_{GG} - I_D R_S$, and $V_{DS} = V_{DD} - I_D(R_D + R_S)$. When R_D is known, the specific procedure consists of selecting Q and V_{DD} and then solving the design equations for R_S, V_{GG}, and R_G. When R_D is not known, we choose $V_{R_S} = V_{DD}/5$. The value of $R_G = R_1 \| R_2$ is obtained by choosing $R_2 = 100 \text{ k}\Omega$ and solving the equation $V_{GG} = [R_2/(R_1 + R_2)]V_{DD}$ for R_1.

The bias design procedure for the common-drain, or source follower, FET amplifier follows that of the common-source amplifier using $R_D = 0$.

4.1

For the circuit of Fig. P4.1, determine v_{CE} and the state of the BJT for each of the following inputs: **(a)** $V_{IN} = 0$, **(b)** $V_{IN} = 1.7$ V, and **(c)** $V_{IN} = 5$ V. Assume $\beta = 100$ and $V_0 = 0.7$ V.

4.2

The BJT in Fig. P4.2 has $\beta = 100$ and $V_0 = 0.7$ V. Determine the DC collector current I_C and the DC collector–emitter voltage V_{CE}. Use an equivalent circuit with two DC supply voltages (the equivalent circuit for analysis).

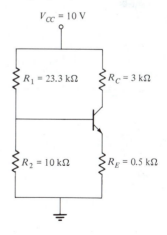

FIGURE P4.2

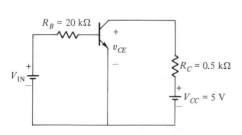

FIGURE P4.1

4.3

Determine the Q point for the Si BJT in the circuit of Fig. P4.2 with $R_1 = 3$ kΩ, $R_2 = 1$ kΩ, $R_C = 1$ kΩ, $R_E = 1$ kΩ, and $V_{CC} = 20$ V. Let $\beta = 100$ and $V_0 = 0.7$ V.

4.4

Repeat Problem 4.3 for the circuit in which $R_1 = R_2 = 20$ kΩ, $R_C = 1$ kΩ, $R_E = 2$ kΩ, and $V_{CC} = 10$ V.

4.5

For the circuit of Fig. P4.2, with $R_1 = 30$ kΩ, $R_2 = 15$ kΩ, $R_C = 5$ kΩ, $R_E = 3$ kΩ, and $V_{CC} = 20$ V, determine the Q point for $V_0 = 0.7$ V and $\beta = 50, 100,$ and 300. Sketch the output load line and show the Q points. What is the limiting value of I_C as β becomes infinite?

4.6

Repeat Problem 4.5 for the circuit of Fig. P4.6 in which $R_B = 1$ kΩ, $V_{BB} = 3.5$ V, $R_C = 0.5$ kΩ, $R_E = 0.5$ kΩ, and $V_{CC} = 10$ V.

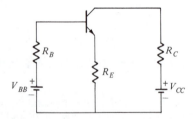

FIGURE P4.6

4.7

For the common-base circuit of Fig. P4.7, determine the following:

(a) Draw the equivalent circuit for analysis.
(b) Calculate I_C and V_{CB}.
(c) Sketch the DC load line on the output i–v characteristics and indicate Q.

Let $\beta = 100$ and $V_0 = 0.7$ V. What is the result for infinite β?

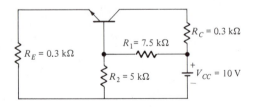

FIGURE P4.7

4.8

Repeat Problem 4.7 for the circuit in which $R_1 = 10$ kΩ, $R_2 = 5$ kΩ, $R_C = R_E = 1$ kΩ, and $V_{CC} = 30$ V.

4.9

Figure P4.9 displays a common-collector circuit. Determine I_B, I_C, and V_{CE} for $\beta = 100$ and $V_0 = 0.7$ V. What is the result for infinite β?

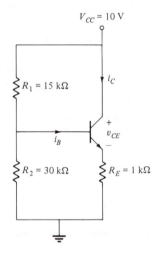

FIGURE P4.9

4.10

Repeat Problem 4.9 for the circuit in which $R_1 = 15$ kΩ, $R_2 = 20$ kΩ, $R_E = 5$ kΩ, and $V_{CC} = 20$ V.

4.11

Figure P4.11 shows the basic form of the difference amplifier circuit. For the circuit values indicated, and $T_1 = T_2$, $\beta = 100$, and $V_0 = 0.7$ V, determine I_C and V_{CE} for each transistor.

4.12

Repeat Problem 4.11 for the circuit in which $R_B = 0$, $R_C = 100$ Ω, and $R_E = 450$ Ω.

4.13

Repeat problem 4.11 for the circuit in which $R_B = 1$ kΩ, $R_C = 0.2$ kΩ, $R_E = 0.9$ kΩ, and $V_{CC} = V_{EE} = 10$ V.

4.14

Repeat Problem 4.11 for the circuit in which $R_B = 1$ kΩ, $R_C = 0.5$ kΩ, $R_E = 0.1$ kΩ, $V_{CC} = 16$ V, and $V_{EE} = 4$ V.

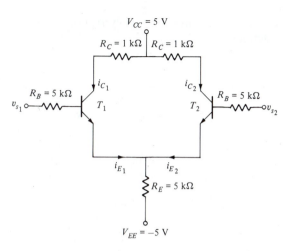

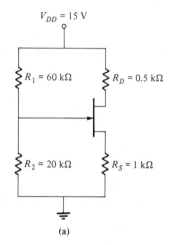

FIGURE P4.11

4.15
In the difference amplifier circuit of Fig. P4.11, the resistor R_E is replaced with a current source of magnitude 0.8 mA. Determine I_C, V_C, V_E, and V_{CE}.

4.16
For the circuit of Fig. P4.6, $R_C = 0.4$ kΩ and $V_{CC} = 10$ V. For DC values of collector current and collector–emitter voltage given by $I_C = 10$ mA and $V_{CE} = 5$ V, determine the value of R_E that places Q in the middle of the load line. Let $V_0 = 0.7$ V and $40 \le \beta \le 120$. Also, determine R_B by using the rule-of-thumb design expression and obtain R_1 and R_2 corresponding to R_B and V_{BB}.

4.17
Obtain values of R_B, V_{BB}, R_C, and R_E for the circuit of Fig. P4.6 such that $I_C = 1$ mA and $V_{CE} = 10$ V. Use the rule-of-thumb design expression and $I_E R_E = 3$ V with $V_{CC} = 20$ V, $\beta = 100$, and $V_0 = 0.7$ V.

4.18
For the common-base amplifier of Fig. 4.8, let $V_0 = 0.7$ V and $100 \le \beta \le 300$. The DC operating point is to be located at $I_C = 10$ mA and $V_{CB} = 3$ V. Using $V_{CC} = 10$ V, carry out the bias circuit design procedure for R_C unspecified and for $R_C = 0.5$ kΩ.

4.19
For the common-collector amplifier of Fig. 4.7, carry out the bias circuit design procedure based upon the following information: The BJT is made of silicon with $50 \le \beta \le 200$, $V_{CC} = 20$ V, the Q point is located at $I_C = 12$ mA, and $V_{CE} = 8$ V.

4.20
For the difference amplifier of Fig. 4.9, determine the bias circuit under the following conditions: The BJTs are made of silicon with $100 \le \beta \le 300$, $V_{CC} = 10$ V, the Q point is to be located at $I_C = 5$ mA, and $V_{CE} = 10$ V.

4.21
Consider the JFET circuit of Fig. P4.21a. Show that the circuit of Fig. P4.21b is equivalent and derive equations for R_G and V_{GG}. For $V_{GS} = -1.5$ V, determine I_D and V_{DS}. Show the solution graphically on the output i–v characteristics.

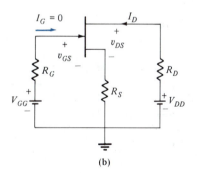

FIGURE P4.21

4.22
Another JFET is used in the circuit of Fig. P4.21a with $V_{DD} = 15$ V, $R_1 = 90$ kΩ, $R_2 = 10$ kΩ, $R_D = 1.3$ kΩ, and $R_S = 0.5$ kΩ. For $I_D = 5$ mA, determine

V_{GS} and V_{DS}. Graphically display Q on the output i–v characteristics.

4.23

Figure P4.23 displays a circuit that uses an E-D MOST. Determine the equivalent circuit for analysis with two DC supplies (corresponding to Fig. P4.21b) and calculate the values of R_G and V_{GG}. For $I_D = 5$ mA, determine V_{GS} and V_{DS}.

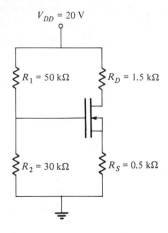

FIGURE P4.23

4.24

Consider the circuit of Fig. P4.24, which has the common-drain configuration. For $I_D = 5$ mA, determine V_{DS} and V_{GS}.

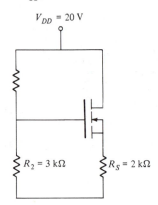

FIGURE P4.24

4.25

Figure P4.25 displays a circuit that uses an E-D N-channel MOST. Determine R_G and V_{GG} and draw

the equivalent circuit for analysis. For $V_{GS} = 2$ V, determine I_D and V_{DS}.

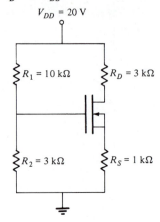

FIGURE P4.25

4.26

An N-channel E-D MOST is used in the circuit of Fig. P4.24 in the common-drain configuration. Assume that $i_D = 3[1 + (v_{GS}/2)]^2$ in mA for this device. Determine I_D, V_{GS}, and V_{DS} for $V_{DD} = 20$ V, $R_1 = 25$ kΩ, $R_2 = 50$ kΩ, and $R_S = 1$ kΩ. Use a graphical procedure involving the transfer characteristic to show the results.

4.27

An N-channel JFET has a transfer characteristic given by $i_D = 1.65[1 + (v_{GS}/2)]^2$ in mA for $-2 \le v_{GS} \le 0$ V. This transistor is used in the circuit of Fig. P4.27. For $I_D = 0.8$ mA, determine V_{GS}, R_S, and V_{DS}.

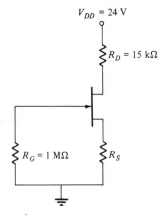

FIGURE P4.27

4.28

Consider the N-channel JFET in the common-source circuit of Fig. P4.28 that uses a stabilizing resistor R_S. Determine the values of R_S and R_D if the Q point is to be located at $V_{GS} = -2$ V, $I_D = 4$ mA, and $V_{DS} = 8$ V. Use the equivalent circuit for analysis with two DC supply voltages.

-2 V, determine R_{S_1} and R_{S_2} such that the Q point is in the middle of the output load line.

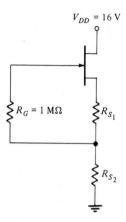

FIGURE P4.30

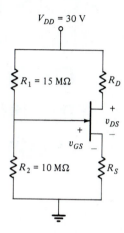

FIGURE P4.28

4.29

The N-channel E-O MOST in Fig. P4.29 is to be biased such that $I_D = 2$ mA. Determine V_{GS} and V_{DS}.

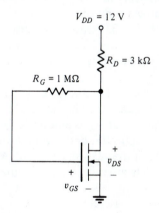

FIGURE P4.29

4.30

The JFET in Fig. P4.30 has a transfer characteristic given by $i_D = 4[1 + (v_{GS}/4)]^2$ in mA. For $V_{GS} =$

5

AC ANALYSIS AND DESIGN OF SINGLE-TRANSISTOR AMPLIFIERS

The previous chapter described various transistor amplifier circuits for the special case of DC operation. The present chapter includes the AC signals and describes overall amplifier operation. Usually, the DC and AC analyses are conducted separately, and the results are added together to obtain the complete solution.

As in the case of diodes, large-signal models for BJTs are developed from their corresponding piecewise linear transistor characteristics. The development of large-signal models is not done for FETs, however, because large signals can cause non-linear behavior for these devices.

The large-signal model for the BJT is used for analysis of a general transistor amplifier circuit without specifying the transistor configuration. The use of a general transistor amplifier circuit provides greater insight into the analysis and design of the three specific BJT amplifiers: common-emitter, common-collector, and common-base.

Following the large-signal analysis, the chapter introduces small-signal (AC) models for both the BJT and FET. The small-signal models are convenient for quantitatively determining the amount of voltage and/or current amplification possible for particular amplifier circuits. Additionally, small-signal analysis is useful in determining the dependence of the amplification upon transistor parameters and external circuit resistor values. This chapter concludes by describing power and efficiency of amplifier circuits.

5.1   LARGE-SIGNAL ANALYSIS AND DESIGN OF BJT AMPLIFIERS

Chapter 2 introduced the large-signal analysis of diode circuits as an extension of the DC analysis. Our knowledge of that material considerably simplifies the large-signal analysis of BJT amplifier circuits. We begin by considering a piecewise linear circuit model for the BJT.

5.1.1   Piecewise Linear Model for the BJT

Based upon the piecewise linear input and output current–voltage characteristics for the BJT, the large-signal circuit model is deduced. Figures 5.1a and b show the piecewise linear output current–voltage characteristics along with the

corresponding large-signal circuit model for the BJT. The output portion of the circuit model comes directly from the BJT transfer equation, indicated in Fig. 5.1c. The output characteristics indicate the corresponding ideal transfer relation and do not account for deviations such as the saturation region.

The input portion of the large-signal circuit model is drawn according to the behavior of the transistor operating under amplifier conditions. For amplifier circuits with BJTs, the input terminals in all cases are the base and emitter. Hence, since this junction is forward biased, the equivalent circuit consists of the ideal diode and offset voltage at the input of the BJT as shown in Fig. 5.1b. Note that a particular orientation for the P and N regions has been selected with the direction of the arrowhead pointing from P to N.

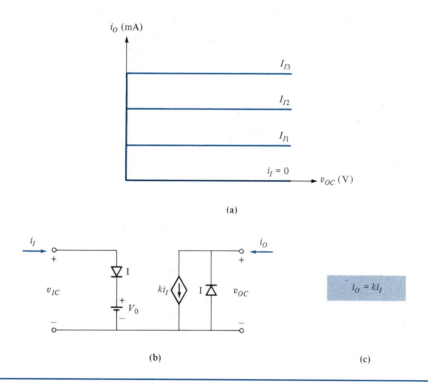

FIGURE 5.1   Piecewise Linear Model Development for the BJT: (a) Output current–voltage characteristics, (b) Large-signal circuit model, (c) Amplifier transfer equation

A large-signal model for the BJT corresponding to that of Fig. 5.1b is used in conjunction with the particular amplifier circuit to carry out the large-signal analysis and design.

5.1.2   General Amplifier Circuit (Bypass and Coupling Capacitors)

Chapter 4 introduced the general Q point stabilizing circuit (Fig. 4.5a) containing the feedback resistor R_F. This resistor stabilizes the DC output current to changes in transistor parameters and is a required element for the stabilization of Q. However, R_F also has the undesirable property that it reduces the AC current and/or voltage amplification, as we will see.

Therefore, R_F is an unwanted element as far as AC signals are concerned.

This dilemma is eliminated by placing a *bypass capacitor* C_F in parallel with R_F. If the product of capacitance and signal frequency (ωC) is large, then the capacitor will act like a short circuit (zero impedance) to the AC current. The capacitor also serves as an open circuit to the DC current since the impedance of the capacitor is $Z_{C_F} = 1/j\omega C_F$ and for DC, $\omega = 0$. Note that if the magnitude of C_F and the frequency of the AC signal (ω) are both small, then the bypass capacitor is no longer a short circuit to AC. Under these conditions, the amplification of the circuit is again reduced. In this case, the reduction is due to the parallel combination of R_F with Z_{C_F}, as we will see in Chapter 8. In this chapter, we consider

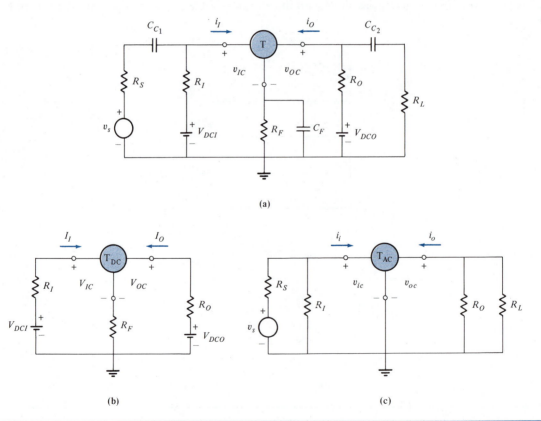

(a)

(b)

(c)

FIGURE 5.2   General Transistor Amplifier Circuit and Equivalents: (a) Basic circuit with coupling and bypass capacitors, (b) DC equivalent circuit, (c) AC equivalent circuit

all capacitors as possessing large magnitude so that $1/\omega C \to 0$.

A general amplifier circuit with bypass capacitor C_F is shown in Fig. 5.2a. In addition to C_F, *coupling capacitors* C_{C_1} and C_{C_2} are included. This circuit is a modified form of the Q point stabilizing circuit of Chapter 4 (Fig. 4.5a). For large capacitance values, the behavior of the coupling capacitors is the same as that of C_F—that is, each capacitor acts like a short circuit to AC and blocks the DC current. The reason for including C_{C_1} and C_{C_2} in the amplifier circuit is to avoid shifting the Q point because of the source resistor R_S and/or the load resistor R_L. As far as DC currents and voltages in the network are concerned, R_S and R_L are not in the circuit because C_{C_1} and C_{C_2} behave like open circuits to DC.

5.1.3   AC and DC Load Lines (ACLL and DCLL)

When capacitors are present in amplifier circuits, the DC currents and AC currents have different paths. Therefore, writing KVL for the input and output "loops" of a particular amplifier circuit for AC only and for DC only results in

entirely different load-line equations. These equations are called the *DC load-line* (DCLL) and *AC load line* (ACLL) equations.

To determine the corresponding load-line equations for the general amplifier of Fig. 5.2a, equivalent circuits are drawn for DC only and AC only as shown in Figs. 5.2b and c. The DC equivalent circuit of Fig. 5.2b is identical to Fig. 4.5a; hence, the DCLL equations are the same as those of Chapter 4. Furthermore, we also know the location of Q from the analysis of Chapter 4. The results of the DC analysis are displayed on the output current–voltage axes of Fig. 5.3a. Note that we have defined an equivalent DC output resistance as R_{DCO} to represent the inverse slope of the DCLL.

The AC equivalent circuit of Fig. 5.2c is obtained by replacing the capacitors and DC voltage sources of Fig. 5.2a with short circuits. We then write KVL for the input and output portions of the amplifier by using the AC equivalent circuit. For the input loop of Fig. 5.2c, we have (by using a Thévenin equivalent circuit)

$$v_{ic} = \left(\frac{R_I}{R_S + R_I}\right)v_s - i_i(R_S \,\|\, R_I) \qquad (5.1\text{–}1)$$

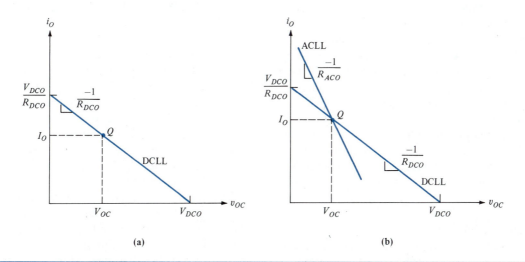

(a)   **(b)**

FIGURE 5.3   AC and DC Load Lines: (a) DCLL and Q location, (b) ACLL and DCLL intersecting at Q

For the output loop,

$$v_{oc} = -i_o R_{ACO} \qquad (5.1\text{--}2)$$

where the AC equivalent output loop resistance is R_{ACO}, or in this case $R_o \| R_L$.

The input ACLL equation given by (5.1–1) is very tedious to use graphically because of the time-varying term involving v_s. Recall this same result for the diode case in Chapter 2. However, the output ACLL equation given by (5.1–2) is exceedingly useful and is sketched on the output current–voltage axes of Fig. 5.3b. The origin for this line is at Q because here the AC current and voltage are zero ($i_o = v_{oc} = 0$) and the total current and voltage are the DC values.

The equation for the ACLL in terms of total output current and voltage is obtained by substituting for the AC components in terms of the total quantity minus the DC portion. By direct substitution into (5.1–2), we obtain the following equation for the ACLL:

$$v_{OC} - V_{OC} = -(i_O - I_O)R_{ACO} \qquad (5.1\text{--}3)$$

This equation represents a straight line on the output current–voltage axes. The slope of this line is $-1/R_{ACO}$, or in this particular case $-1/(R_o \| R_L)$. We can see immediately from the new form of the ACLL equation given by (5.1–3) that this line does indeed pass through Q because for $v_{OC} = V_{OC}$, $i_O = I_{OC}$, which satisfies (5.1–3). Figure 5.3b shows both the AC and DC load lines with different slopes assumed, which is the case when capacitors are elements in the circuit.

It should be emphasized that the ACLL is superimposed on the DCLL. Under DC conditions, the output current and voltage are given by the Q point values, and the AC current and voltage are zero. However, when a signal is applied, the operating point shifts on the AC load line. Hence, even though we call this line the AC load line, it is the operating path for total output current and voltage.

A final comment on the ACLL and the DCLL is that with no capacitors in the amplifier

circuit, these two load lines merge into a single load line.

5.1.4 Maximum Symmetrical Variation (MSV)

One of the most important considerations in large-signal amplifier design is that the magnitudes of the AC output current and voltage be maximized. This consideration requires that Q be located approximately in the middle of the ACLL. To perceive this location, we must realize that transistor amplifier operation is restricted to one quadrant of the output current–voltage characteristics. Thus, if Q is located away from the middle of the ACLL, operation is limited by the nearest intercept, and the operating point cannot go beyond the intercept. As the input signal varies, the operating point shifts to an intercept where it remains until the input signal magnitude is reduced. Note that, in actuality, the operating point can only shift close to the current intercept because the transistor characteristics exhibit a small forward voltage. For present purposes, however, we will let this small forward voltage be zero and place Q in the middle of the ACLL.

Figure 5.4 displays an ACLL with the desired position for Q. Note that under these conditions the intercepts are located at twice the DC values of current and voltage. Let us assume that the output current varies sinusoidally (due to a sinusoidal input signal) as shown beside the output current axis of Fig. 5.4. Then, the output voltage also varies sinusoidally as shown beneath the output voltage axis. Corresponding points in time are labeled A, B, and C. Note that the output current and voltage vary symmetrically about Q with maximum amplitude. The peak value for each sinusoid is just the DC (Q) value. This variation is referred to as *maximum symmetrical variation* (MSV) in output current and voltage. Under the MSV condition, the AC output current and voltage magnitudes are maximized. If Q is located at any position other than the middle of the ACLL,

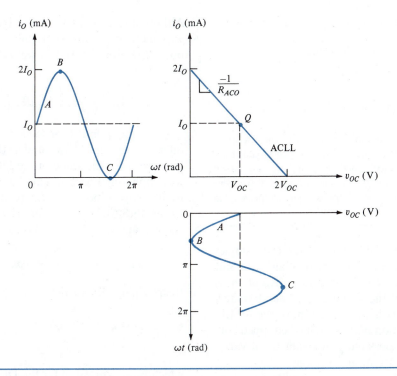

FIGURE 5.4   Maximum Symmetrical Variation (MSV) in Output Current and Voltage

the magnitude of the symmetrical variation is reduced, as the following example shows.

─────────────────── EX. 5.1

Inferior Location of Q Example

Figure E5.1 shows an ACLL for a common-emitter BJT amplifier that has a DC operating point located at $I_O = 1$ mA and $V_{OC} = 4$ V. The amplifier transfer equation for the BJT is $i_O = 100i_I$. If $i_i = 0.04 \sin \omega t$ mA is to be superimposed on the DC input current, determine and sketch the corresponding output current and voltage. What is the peak value for the output current for symmetrical variation about this position for Q?

Solution: In this example, amplifier operation is restricted by cutoff. For positive portions of the input signal, the operating point shifts upward above Q on the ACLL. For negative portions of the input signal, the operating point shifts downward below Q on the ACLL.

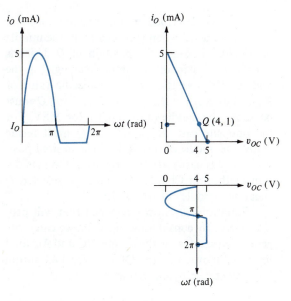

FIGURE E5.1

From basic transistor equations, we have

$$i_o = I_o + 100i_i = 1 + 4\sin\omega t \text{ mA}$$

and

$$v_{oc} = V_{oc} - i_o R_{AC} = 4 - 4\sin\omega t \text{ V}$$

However, amplifier operation is restricted to the first quadrant, and therefore i_o (and also i_i) must be greater than zero. The limited waveshapes for i_o and v_{oc} are shown next to the current and voltage axes of Fig. E5.1. For symmetrical variation about Q in this example, the peak value of output current is 1 mA.

The worst possible choices for Q in single-transistor amplifier circuits are at either end-point of the ACLL. If Q is located at either of these positions, half of the signal cannot be amplified. For example, if Q is at the bottom of the ACLL, the negative portion of the signal is "chopped" off, eliminating any possibility of symmetrical variation in output current and voltage.

5.1.5   General Design for MSV

As described in the previous section, MSV is achieved when Q is located in the middle of the ACLL. Since the position of Q depends upon the DC part of the circuit, placing Q in the middle of the ACLL requires a specific value of R_{ACO}. For $R_{ACO} < R_{DCO}$ (as in Fig. 5.3b), Q must be above the middle of the DCLL for MSV. For the opposite case, $R_{ACO} > R_{DCO}$, Q must reside below the middle of the DCLL. The desired position for Q is easily attained by using the bias design method of Chapter 4, where we selected Q "near the middle" of the DCLL.

Equations are now developed that will provide MSV. By considering Fig. 5.3b, we can write general equations for the output DC and AC load lines as follows. For the DCLL with DC output loop resistance R_{DCO}, we have

$$v_{oc} = V_{DCO} - R_{DCO}i_o \qquad (5.1-4)$$

Although (5.1–4) uses total current and voltage variables, we realize that this equation is valid only at Q. For the ACLL, we use the form given by (5.1–3), or

$$v_{oc} - V_{oc} = -(i_o - I_o)R_{ACO} \qquad (5.1-5)$$

The requirement for MSV is that the current intercept of the ACLL under these conditions must be twice the value of the DC current I_o, as was shown in Fig. 5.4. This information is used to develop important DC relations for MSV. From the ACLL equation (5.1–5), by setting $v_{oc} = 0$ and $i_o = 2I_o$, we have

$$-V_{oc} = -(2I_o - I_o)R_{ACO} \qquad (5.1-6)$$

Rearranging (5.1–6) yields

$$I_o = \frac{V_{oc}}{R_{ACO}} \qquad (5.1-7)$$

which is an expression for the DC current. Under DC conditions, by using (5.1–7) to eliminate i_o ($=I_o$) from (5.1–4) and with $v_{oc} = V_{oc}$, we obtain by direct substitution

$$V_{oc} = V_{DCO} - \left(\frac{V_{oc}}{R_{ACO}}\right)R_{DCO} \qquad (5.1-8)$$

Solving for V_{oc} yields

$$V_{oc} = \frac{R_{ACO}}{R_{ACO} + R_{DCO}} V_{DCO} \qquad (5.1-9)$$

By back substitution into (5.1–7), we have

$$I_o = \frac{V_{DCO}}{R_{ACO} + R_{DCO}} \qquad (5.1-10)$$

Equations (5.1–9) and (5.1–10) are used as design relations to place Q in the middle of the ACLL and provide MSV. Equation (5.1–10) is often equated directly to the bias design expression for I_O to obtain the additional design relationship that dictates maximum symmetrical variation in output current and voltage.

As a warning, it must be noted that (5.1–9) and (5.1–10) are not valid in all circuits and therefore cannot be used as general equations. They are valid only under MSV conditions.

5.1.6 Common-Emitter Amplifier

The common-emitter amplifier circuit shown in Fig. 5.5 includes a bypass capacitor C_E and two coupling capacitors C_{C_1} and C_{C_2}. We will see that properties of this circuit include the possibility of both current and voltage amplification.

The methodology involved in carrying out the large-signal analysis and design in all the BJT amplifier cases is to obtain the position for Q and select the load resistor that will provide MSV. In the present case, the Q point is obtained by analyzing the DC equivalent circuit corresponding to the overall circuit of Fig. 5.5. Note that under DC conditions, the capacitors are open circuits, and the remaining circuit (inside the shaded rectangle) is just the DC common-emitter amplifier circuit analyzed in Chapter 4. The DC output collector

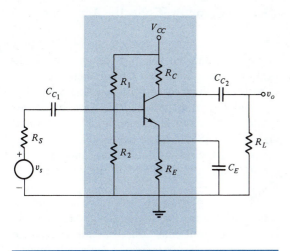

FIGURE 5.5   Common-Emitter Amplifier Circuit

current is given by

$$I_C = \frac{\beta(V_{BB} - V_O)}{R_B + (1 + \beta)R_E} \tag{5.1–11}$$

The DC collector–emitter voltage is given by (from the DCLL equation)

$$V_{CE} = V_{CC} - I_C(R_E + R_C) \tag{5.1–12}$$

where we have used $I_C \simeq I_E$ ($\beta \gg 1$). Although it was not mentioned in Chapter 4, the large-signal model for the BJT of Fig. 5.1b was used implicitly in obtaining the expressions for I_C and V_{CE}.

Upon comparison of (5.1–12) to the general expression for the DCLL, (5.1–4), the equivalent DC output resistance for the common-emitter case is $R_{DCO} = R_E + R_C$. The DCLL voltage intercept for this case is $V_{DCO} = V_{CC}$.

By writing the equation for the ACLL corresponding to the AC output loop of Fig. 5.5, we have (observing that R_C and R_L are in parallel)

$$v_{ce} = -i_c(R_C \| R_L) \tag{5.1–13}$$

By comparing (5.1–13) to (5.1–2), we obtain $R_{ACO} = R_C \| R_L$.

From the bias design for this amplifier described in Chapter 4, we may regard the values of the DC quantities (I_C, V_{CE}, R_E, R_C, R_1, R_2, and V_{CC}) as being known. Therefore, the DCLL quantities $R_{DCO} = R_E + R_C$ and $V_{DCO} = V_{CC}$ are known. The circuit element remaining to be determined is R_L, which we assume is unknown. Its value may be selected by designing the amplifier to have MSV. If R_L is specified, the amplifier circuit may not necessarily be designed for the MSV condition.

The simplest procedure for the selection of R_L is to use (5.1–7). Since $R_{ACO} = R_C \| R_L$, direct substitution into (5.1–7) yields

$$I_O = \frac{V_{OC}}{R_C \| R_L} \tag{5.1–14}$$

or

$$\frac{R_C R_L}{R_C + R_L} = \frac{V_{OC}}{I_O} \tag{5.1-15}$$

Solving for R_L results in

$$R_L = \frac{R_C}{(R_C I_O / V_{OC}) - 1} \tag{5.1-16}$$

Note that the denominator looks suspicious. However, $R_C I_O / V_{OC}$ will always be greater than unity because (5.1–14) gives $R_C \| R_L = V_{OC}/I_O$ and R_C is greater than the parallel combination of R_C with R_L.

─────────────────────────────────── EX. 5.2

MSV Common-Emitter Amplifier Example

The common-emitter amplifier circuit of Fig. 5.5 has element values as follows: $V_{CC} = 10$ V, $R_C = 0.5$ kΩ, $R_L = 0.5$ kΩ, $R_1 = 20$ kΩ, $R_2 = 5$ kΩ, $R_E = 0.1$ kΩ, and $R_S = 0$. Determine the ACLL, the DCLL, and the maximum possible symmetrical variation (peak to peak) in the output current i_C. Let $V_0 = 0.7$ V and $\beta = 100$.

Solution: The DCLL is completely specified by the value of $V_{CC} = 10$ V (which is the voltage intercept) and $R_{DCO} = R_E + R_C = 0.6$ kΩ (which is the negative inverse of the slope). The DCLL is displayed in Fig. E5.2. The DC collector current is given by (for $\beta \gg 1$)

$$I_C \simeq \frac{\beta(V_{BB} - V_0)}{R_B + \beta R_E}$$

Substituting values yields

$$I_C = \frac{100(2 - 0.7)}{4 + 100(0.1)} = 9.29 \text{ mA}$$

The DC collector–emitter voltage is then obtained by writing KVL for the output loop:

$$V_{CE} = V_{CC} - I_C(R_C + R_E)$$

Substituting values yields

$$V_{CE} = 10 - [(9.29)(0.6)] = 4.43 \text{ V}$$

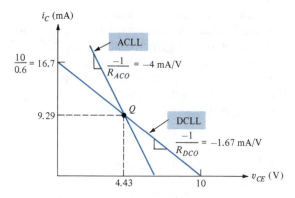

FIGURE E5.2

The ACLL is now obtained by realizing that it passes through Q with slope of $-1/R_{ACO}$ where $R_{ACO} = R_C \| R_L = 0.25$ kΩ. Hence, the magnitude of the slope is $1/0.25 = 4$ mA/V. The ACLL is also displayed in Fig. E5.2.

The maximum possible symmetrical variation for this amplifier is now observed to be dependent upon cutoff, since the load-line end-point that is the shortest distance from Q is the cutoff point. Hence, the maximum possible symmetrical variation in output current (peak to peak) is given by

$$\Delta i_C = 2(9.29) = 18.58 \text{ mA}$$

─────────────────────────────

5.1.7   Common-Collector Amplifier

The common-collector amplifier cannot amplify voltage but can provide current amplification. Figure 5.6 shows the complete common-collector amplifier circuit with coupling capacitors C_{C_1} and C_{C_2}. A bypass capacitor is not appropriate in this case. The large-signal analysis and design for MSV is again accomplished by analyzing the DC circuit to determine Q and analyzing the AC output loop to obtain the equivalent of R_{ACO}.

The DC portion of the circuit is contained inside the shaded rectangle of Fig. 5.6. The DC collector current and collector–emitter voltage are given by the expressions from Chapter 4, (4.2–15b) and (4.2–17), respectively, with $R_C = 0$.

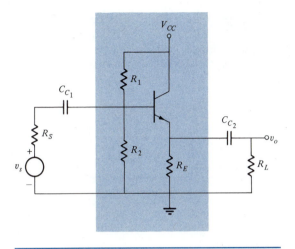

FIGURE 5.6 Common-Collector Amplifier Circuit

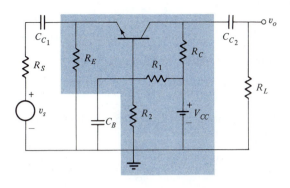

FIGURE 5.7 Common-Base Amplifier Circuit

Therefore, for this case, $R_{DCO} = R_E$ and $V_{DCO} = V_{CC}$.

The ACLL equation for the output loop of Fig. 5.6 is

$$v_{ce} = -i_e(R_E \| R_L) \simeq -i_c(R_E \| R_L) \quad (5.1\text{–}17)$$

Therefore, $R_{ACO} = R_E \| R_L$. Using this expression with (5.1–7) yields the corresponding value of R_L for MSV as follows:

$$R_L = \frac{R_E}{(R_E I_0 / V_{OC}) - 1} \quad (5.1\text{–}18)$$

5.1.8 Common-Base Amplifier

The common-base amplifier cannot amplify current but can provide voltage amplification. Figure 5.7 shows the complete circuit for the common-base amplifier. The DC portion of the circuit is contained inside the shaded region and is identical to the common-emitter amplifier, except that the input and output signals are located at different positions and Q for the BJT is located at V_{CB}, I_C. The expressions from Chapter

4, (4.2–19) and (4.2–20), are directly applicable and are given here as follows:

$$I_C = \frac{\beta(V_{BB} - V_0)}{R_B + (1 + \beta)R_E} \quad (5.1\text{–}19)$$

and

$$V_{CB} = V_{CC} - V_{BB} - I_C R_C - \left(\frac{I_C}{\beta} R_B\right) \quad (5.1\text{–}20)$$

From these expressions, $R_{DCO} = R_C - R_B/\beta$ and $V_{DCO} = V_{CC} - V_{BB}$.

The ACLL equation for the output loop of Fig. 5.7 is

$$v_{cb} = -i_c(R_C \| R_L) \quad (5.1\text{–}21)$$

Thus, $R_{ACO} = R_C \| R_L$. Using this expression with (5.1–7) yields the value of R_L for MSV as follows:

$$R_L = \frac{R_C}{(R_C I_0 / V_{OC}) - 1} \quad (5.1\text{–}22)$$

─────────── **EX. 5.3**

MSV Common-Base Amplifier Example

The common-base amplifier circuit of Fig. 5.7 has the same element values as in Ex. 5.2. For a BJT with $\beta = 100$ and $V_0 = 0.7$ V, sketch the DCLL and the ACLL and locate Q.

Also, determine the maximum possible symmetrical variation in i_C and v_{CB} for this common-base amplifier circuit.

Solution: The DCLL equation is given by

$$V_{CB} = V_{CC} - V_{BB} - i_C R_C + i_B R_B$$

Therefore, since $i_B = i_C/\beta$, the current and voltage intercepts are, respectively,

$$i_C(0) = \frac{V_{CC} - V_{BB}}{R_C - (R_B/100)} = \frac{10 - 2}{0.5 - 0.04} \simeq 17.4 \text{ mA}$$

and

$$v_{CB}(0) = V_{CC} - V_{BB} = 10 - 2 = 8 \text{ V}$$

The Q point is given by (5.1–19) and (5.1–20), and substituting values yields

$$I_C \simeq \frac{100(2 - 0.7)}{4 + 100(0.1)} = 9.29 \text{ mA}$$

and

$$V_{CB} \simeq 10 - 2 - 9.29(0.5) + \frac{9.29}{100}(4) = 3.73 \text{ V}$$

where we have included the voltage across R_B. The DCLL and Q are displayed in Fig. E5.3.

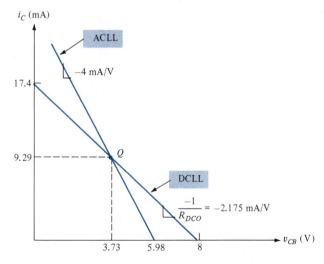

FIGURE E5.3

The ACLL passes through Q with slope of $-1/R_{ACO} = -1/(R_C \| R_L) = -1/0.25 = -4$ mA/V. The ACLL is also displayed in Fig. E5.3. The maximum possible symmetrical variation in i_C (peak to peak) is again determined by cutoff and is given by

$$\Delta i_C = 2(9.29) = 18.58 \text{ mA}$$

The maximum possible symmetrical variation in v_{CB} is

$$\Delta v_{CB} = 2(5.98 - 3.73) = 4.5 \text{ V}$$

5.2   SMALL-SIGNAL ANALYSIS OF BJT AMPLIFIERS

As we did in Chapter 2 in the case of diodes, we also develop small-signal AC models for BJTs here (and for FETs later). The transistor models of this chapter are valid only at low and medium frequencies and are extremely useful in determining the AC performance of amplifiers. At high frequencies, the small parasitic capacitances associated with the junctions of BJTs (and the gates of FETs) must be accounted for, as we will see in Chapter 8. The small-signal transistor models (like the small-signal diode model) are linearized equivalent circuits that are valid only for the conditions of small-signal operation, that is, the AC magnitudes are much less than the DC magnitudes.

Using the small-signal transistor models in AC equivalent circuits for amplifiers, analytical expressions for the AC current and voltage amplification will be developed. The amplification expressions obtained depend upon the transistor AC parameters and the external resistors in the circuit. Additional information that can be obtained from the small-signal models involves the AC input and output impedances for each amplifier circuit. These impedances provide a measure of voltage, current, or power transfer in an amplifier circuit.

5.2.1   BJT Small-Signal Hybrid Models

The small-signal models developed in this section are called *hybrid* (meaning *mixed*) *models* because the derived AC parameters associated with the models possess various mixed units. Other small-signal models are sometimes used for BJTs, but the hybrid models are much more important because each of the resulting parameters can be easily measured. Values for these parameters are listed in BJT data sheets for particular transistors.

Common-Emitter Hybrid Model. The small-signal model that we will develop for the BJT in the common-emitter configuration is also used for the BJT in the common-collector configuration, since the output currents (collector and emitter) in each configuration are very nearly equal, and the input currents (base) are identical.

Figure 5.8a shows the circuit symbol and configuration for the NPN transistor in the common-emitter configuration. A reversal of the currents converts the NPN model to the PNP case.

The total currents and voltages of Fig. 5.8a have functional dependencies upon one another that are based upon the actual behavior of the BJT under amplifier conditions as described in Chapter 3. In the common-emitter configuration, we may write these functional dependencies as follows:

$$v_{BE} = v_{BE}(i_B) \qquad (5.2-1)$$

and

$$i_C = i_C(i_B, v_{CE}) \qquad (5.2-2)$$

where it has been assumed that v_{BE} is independent of v_{CE}. Mathematically, in writing (5.2–1) and (5.2–2), we have selected i_B and v_{CE} as independent variables and i_C and v_{BE} as dependent variables. The system has four variables (v_{BE}, i_B, v_{CE}, and i_C) for which there are actually six different combinations of two independent and two dependent sets of variables. In the BJT hybrid model representation, we always choose input current and output voltage as independent quantities. Output current and input voltage are chosen as the dependent quantities. The selection of output current dependent upon input current is all important because of the fundamental current amplifier behavior of the BJT. The dependence of output current upon ouput voltage was also observed in the current–voltage characteristics of Chapter 3. Therefore, selection of output voltage as the other independent variable is justified.

To obtain the small-signal AC circuit model, we begin by mathematically expanding the functional relations of (5.2–1) and (5.2–2) in a Taylor series (from differential calculus) about the DC

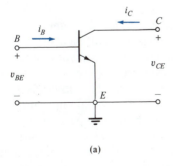

(a)

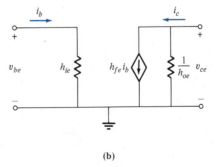

(b)

FIGURE 5.8   Common-Emitter Hybrid Model: (a) Circuit symbol and configuration, (b) Small-signal model

operating point (Q). Equations (5.2–1) and (5.2–2) then become (with i_B and v_{CE} independent)

$$v_{BE} = V_{BE} + \left.\frac{\partial v_{BE}}{\partial i_B}\right|_{\Delta v_{CE}=0} \Delta i_B \qquad (5.2\text{–}3)$$

and

$$i_C = I_C + \left.\frac{\partial i_c}{\partial i_B}\right|_{\Delta v_{CE}=0} \Delta i_B + \left.\frac{\partial i_c}{\partial v_{CE}}\right|_{\Delta i_B=0} \Delta v_{CE} \qquad (5.2\text{–}4)$$

where Δi_B and Δv_{CE} are the incremental changes in total i_B and total v_{CE}. Also, we neglect higher-order terms involving Δi_B and Δv_{CE} in the expansion. This approximation is the application of the small-signal assumption and allows linear equations to be written for the AC dependent variables in terms of the AC independent variables.

The dependent quantities, v_{BE} and i_C, can also be written in terms of the DC component and the small-signal "changing" component as follows:

$$v_{BE} = V_{BE} + \Delta v_{BE} \qquad (5.2\text{–}5)$$

and

$$i_C = I_C + \Delta i_C \qquad (5.2\text{–}6)$$

Equating the changing components of (5.2–5) with (5.2–3) yields

$$\Delta v_{BE} = \left.\frac{\partial v_{BE}}{\partial i_B}\right|_{\Delta v_{CE}=0} \Delta i_B \qquad (5.2\text{–}7)$$

Similarly, for (5.2–6) and (5.2–4), we obtain

$$\Delta i_C = \left.\frac{\partial i_c}{\partial i_B}\right|_{\Delta v_{CE}=0} \Delta i_B + \left.\frac{\partial i_c}{\partial i_{CE}}\right|_{\Delta i_B=0} \Delta v_{CE} \qquad (5.2\text{–}8)$$

We now recognize that the changes in total voltage or current given by Δv_{BE}, Δi_C, Δi_B, and Δv_{CE} must, by definition, correspond to the AC currents and voltages given by v_{be}, i_c, i_b, and v_{ce}. Furthermore, each derivative coefficient in (5.2–7) and (5.2–8) is evaluated under the condition that an AC independent variable is zero, which implies

evaluation at Q. Finally, we note that the units of the coefficients are mixed (or hybrid). The derivatives are therefore defined as the hybrid coefficients (or parameters) for the common-emitter configuration as follows:

$$h_{ie} = \left.\frac{\partial v_{BE}}{\partial i_B}\right|_Q \qquad (5.2\text{–}9)$$

$$h_{fe} = \left.\frac{\partial i_c}{\partial i_B}\right|_Q \qquad (5.2\text{–}10)$$

$$h_{oe} = \left.\frac{\partial i_c}{\partial v_{CE}}\right|_Q \qquad (5.2\text{–}11)$$

The subscripts will be defined shortly.

With these definitions for the coefficients and the fact that the incremental changes are the AC components, (5.2–7) and (5.2–8) can be written as follows:

$$v_{be} = h_{ie} i_b \qquad (5.2\text{–}12)$$

and

$$i_c = h_{fe} i_b + h_{oe} v_{ce} \qquad (5.2\text{–}13)$$

Equations (5.2–12) and (5.2–13) are the basic small-signal (AC) equations for the BJT in the common-emitter configuration. Physically, (5.2–12) and (5.2–13) are just the linearized form of the signal components for the functional relations of (5.2–1) and (5.2–2).

Equations (5.2–9) through (5.2–11) define what are called the *small-signal hybrid parameters,* or *h-parameters,* for the common-emitter configuration. Observe that each of these parameters are AC quantities that possess different units and are dependent upon the position of Q (the DC operating point). The subscripts used in these definitions have physical implications in addition to distinguishing one parameter from another. The subscript e is placed as a second subscript on each parameter to denote the common-emitter configuration.

For h_{ie} in (5.2–9), the subscript i refers to input, since this parameter is defined as the ratio of input quantities. Note also that h_{ie} has units of resistance and is therefore an AC resistor. In fact, this parameter is identical to the AC diode resistor r_d that was introduced in Chapter 2.

For h_{fe} in (5.2–10), the subscript f defines this parameter as the forward current amplification factor. This h-parameter (for linear BJT behavior) is equal to β, the collector-to-base amplification factor defined in Chapter 4. We will assume $h_{fe} = \beta$ throughout this text.

For h_{oe} in (5.2–11), the subscript o represents an output h-parameter with units of conductance. In a circuit, this conductance is represented by its resistance value, $1/h_{oe}$, to be consistent with the labeling of other resistors in the circuit.

Figure 5.8b shows the NPN BJT small-signal (AC) circuit model corresponding to the small-signal equations (5.2–12) and (5.2–13). Note that from the input side of the circuit, writing Ohm's law results in (5.2–12). From the output side of the circuit and using KCL, we observe that i_c is the sum of two components that, when added together, yield (5.2–13). We will use this circuit model to determine the properties of BJT amplifiers with common-emitter or common-collector BJT configurations.

Common-Base Hybrid Model. Figure 5.9a shows the circuit symbol and definitions of positive total currents and voltages for the NPN BJT in the common-base configuration. By analogy with the previous section, the functional relation-

ships are written as follows, based upon the current–voltage characteristics from Chapter 3 for this configuration:

$$v_{EB} = v_{EB}(i_E) \tag{5.2–14}$$

and

$$i_C = i_C(i_E, v_{CB}) \tag{5.2–15}$$

Again, the input current i_E and output voltage v_{CB} are selected as independent variables, while the output current i_C and input voltage v_{EB} are dependent variables.

From the results of the previous section, in particular (5.2–12) and (5.2–13), we immediately write the small-signal (AC) equations for the common-base configuration in a similar manner as follows:

$$v_{eb} = -h_{ib}i_e \tag{5.2–16}$$

and

$$i_c = -h_{fb}i_e + h_{ob}v_{cb} \tag{5.2–17}$$

where the minus signs in (5.2–16) and (5.2–17) result because of the current reversal of i_e at the input.

Figure 5.9b shows the small-signal circuit model corresponding to (5.2–16) and (5.2–17). On the input side of the circuit, the resistor h_{ib} relates

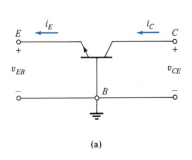

(a)

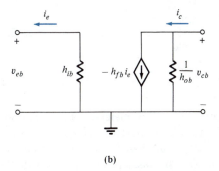

(b)

FIGURE 5.9 Common-Base Hybrid Model: (a) Circuit symbol and configuration, (b) Small-signal model

i_b and v_{eb} through Ohm's law, and thus, the negative sign in (5.2–16) results. The output portion of the circuit in Fig. 5.9b shows that the current i_c is the sum of two components, $-h_{fb}i_e$ and $h_{ob}v_{cb}$, and thus (5.2–17) results.

It is instructive to obtain the definitions of the common-base small-signal (AC) h-parameters directly from (5.2–16) and (5.2–17). Setting one independent AC variable equal to zero and solving for the remaining coefficient, we have

$$h_{ib} = -\left.\frac{v_{eb}}{i_e}\right|_{v_{cb}=0} \tag{5.2–18}$$

$$h_{fb} = -\left.\frac{i_c}{i_e}\right|_{v_{cb}=0} \tag{5.2–19}$$

$$h_{ob} = \left.\frac{i_c}{v_{cb}}\right|_{i_e=0} \tag{5.2–20}$$

where (5.2–16) was solved for h_{ib} with $v_{cb} = 0$ to be consistent. Since (5.2–18) through (5.2–20) are evaluated for an AC component equal to zero, each parameter is, in fact, evaluated at Q. Furthermore, each AC voltage and current is directly related to the incremental change in the total corresponding variable, and the ratio of each incremental change in the limit becomes a derivative. Hence, (5.2–18) through (5.2–20) can be written in the same derivative form as (5.2–9) through (5.2–11):

$$h_{ib} = -\left.\frac{\partial v_{EB}}{\partial i_E}\right|_Q \tag{5.2–21}$$

$$h_{fb} = -\left.\frac{\partial i_C}{\partial i_E}\right|_Q \tag{5.2–22}$$

$$h_{ob} = \left.\frac{\partial i_C}{\partial v_{CB}}\right|_Q \tag{5.2–23}$$

Determination of h-Parameters. The h-parameters are available in transistor data sheets. However, since looking up these values is sometimes inconvenient, we will consider quick and easy methods for determining values for these parameters.

Quite often, the output conductance for both models is nearly zero. The resistance corresponding to $1/h_{ob}$ and $1/h_{oe}$ is then infinite, and each resistor can be open circuited. Thus, each hybrid model is considerably simplified; only an AC input resistor and AC output current source are left. Expressions for the AC input resistance values are easily obtained from Shockley's equation, as in the case of diodes in Chapter 2. From the basic definitions of h_{ib} and h_{ie}, we have

$$\frac{1}{h_{ie}} = \left.\frac{\partial i_B}{\partial v_{BE}}\right|_Q \tag{5.2–24}$$

and

$$\frac{1}{h_{ib}} = -\left.\frac{\partial i_E}{\partial v_{EB}}\right|_Q = \left.\frac{\partial i_E}{\partial v_{BE}}\right|_Q \tag{5.2–25}$$

Noting that these derivative expressions involve a particular PN diode current and the forward voltage across the junction, we use the diode results of Chapter 2 ($r_d = 0.025/I_D$) to obtain

$$h_{ie} = \frac{0.025}{I_B} \tag{5.2–26}$$

and

$$h_{ib} = \frac{0.025}{I_E} \tag{5.2–27}$$

Note that these resistor values depend directly on the DC input current in each case. Furthermore, since $I_E = (1 + h_{fe})I_B$, $h_{ib} = h_{ie}/(1 + h_{fe})$.

The parameters h_{fe} and h_{fb} are given by

$$h_{fe} = \left.\frac{\partial i_C}{\partial i_B}\right|_Q \tag{5.2–28}$$

and

$$h_{fb} = -\left.\frac{\partial i_C}{\partial i_E}\right|_Q \tag{5.2–29}$$

Values for these expressions are obtained quite simply by using the current transfer relations for BJT operation under amplifier conditions. These relations are $i_C = \beta i_B$ and $i_C = \alpha i_E$. By substitution, we have

$$h_{fe} = \beta \qquad\qquad (5.2\text{--}30)$$

and

$$h_{fb} = -\alpha \qquad\qquad (5.2\text{--}31)$$

Often, as in the DC case the value of α is taken as unity, and the value of β typically used is 100.

———————————————————— EX. 5.4

Calculation of the Input h-Parameters Example

The common-emitter amplifier circuit of Fig. 5.5 has $R_1 = 28\ k\Omega$, $R_2 = 40\ k\Omega$, $R_C = 0.83\ k\Omega$, $R_E = 1.67\ k\Omega$, and $V_{CC} = 20\ V$. Calculate the input h-parameters h_{ie} and h_{ib} using $V_0 = 0.7\ V$.

Solution: Since the input h-parameters are directly dependent upon the DC input current, we begin by calculating I_B. By writing KVL for the DC input loop of Fig. 5.5 and by substituting $I_C = \beta I_B$, we have

$$I_B = \frac{V_{BB} - V_0}{R_B + (1 + \beta)R_E}$$

Substituting values yields

$$I_B = \frac{\dfrac{(40)(20)}{40 + 28} - 0.7}{\dfrac{(28)(40)}{68} + 101(1.67)} \simeq 0.06\ \text{mA}$$

Thus, the common-emitter input h-parameter is given by

$$h_{ie} = \frac{0.025}{I_B} = \frac{0.025}{0.06 \times 10^{-3}} = 0.4\ k\Omega$$

To calculate the common-base input h-parameter, recall that

$$h_{ib} = \frac{0.025}{I_E} = \frac{0.025}{(1 + h_{fe})I_B} = \frac{h_{ie}}{1 + h_{fe}}$$

Therefore, by substituting values, we have

$$h_{ib} = \frac{0.4\ k\Omega}{101} \simeq 4\ \Omega$$

Note that h_{ib} is always much less than h_{ie} and that h_{ib} is usually on the order of 1 to 10 Ω.

———————————————————————————————

Simplified small-signal BJT hybrid models with the output conductance taken as zero are displayed in Fig. 5.10. Note that we have assumed $-h_{fb} = \alpha = 1$ in the common-base hybrid model of Fig. 5.10b.

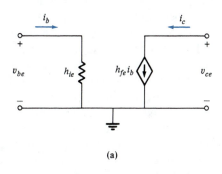
(a)

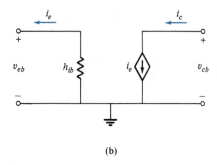
(b)

FIGURE 5.10   Simplified Small-Signal BJT Models: (a) Common-emitter hybrid model, (b) Common-base hybrid model

Graphical Determination of *h*-Parameters.   In instances in which transistor characteristics are available (either from measurement or data sheets), we can determine the parameter values graphically. Chapter 14 indicates the usual method of measuring current–voltage character-istics using a curve tracer. Here, we indicate the method of determining the parameters from avail-able characteristics.

Figure 5.11 displays a typical set of BJT out-put characteristics for the common-emitter con-figuration. The location of Q for the amplifier is also given. The following formula provides h_{fe}:

$$h_{fe} = \frac{\partial i_C}{\partial i_B}\bigg|_{v_{ce}=0} = \frac{\Delta i_C}{\Delta i_B}\bigg|_{v_{CE}=V_{CE}} = \frac{i_{C_2} - i_{C_1}}{i_{B_2} - i_{B_1}}$$

$$(5.2\text{--}32)$$

where we started with the derivative definition, substituted incremental changes, and finally eval-uated the changes at specific points P_1 and P_2.

The value of h_{oe} (as well as h_{ob}) can be ob-tained by using a similar graphical procedure. The following formula provides h_{oe}:

$$h_{oe} = \frac{\partial i_C}{\partial v_{CE}}\bigg|_{i_b=0} = \frac{\Delta i_C}{\Delta v_{CE}}\bigg|_{i_B=I_B} = \frac{i_{C_4} - i_{C_3}}{v_{CE_4} - v_{CE_3}}$$

$$(5.2\text{--}33)$$

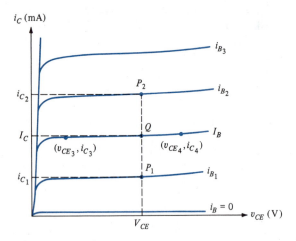

where evaluation is carried out using the points corresponding to v_{CE_4}, i_{C_4} and v_{CE_3}, i_{C_3}.

Finally, to obtain h_{ie} or h_{ib}, it is possible to use a graphical procedure based upon (5.2–24) or (5.2–25) along with the input current voltage characteristics. However, (5.2–26) and (5.2–27) are much more convenient to use.

5.2.2   Analysis of Various BJT Amplifiers

The small-signal hybrid models for the BJT are now used to determine the properties of the various BJT amplifier circuits at intermediate (or medium) frequencies. The high-frequency case, where parasitic transistor capacitors must be in-cluded, and the low-frequency case, where coup-ling and bypass capacitors are not short circuits, will be considered in Chapter 8.

In each amplifier case, we will analyze the overall AC equivalent circuit obtained by substi-tuting the AC equivalent for each element. In the medium-frequency range, we use the appropriate small-signal model for the transistor and treat by-pass and coupling capacitors as short circuits. Moreover, DC voltage sources are replaced with short circuits, and DC current sources are re-placed with open circuits (because they are the equivalent AC elements in each case). The remain-ing circuit then contains only AC currents and voltages.

Amplifier Properties in General.   The most im-portant property of an amplifier circuit is its ability to amplify the AC input signal. The input signal can be in the form of a voltage or current, and the amount of amplification of either indi-cates the superiority of the amplifier.

Additional properties that we will investigate for each case are the *AC input* and *output imped-ances*, Z_i and Z_o, respectively. These impedances are purely resistive in the medium-frequency range, and we therefore use resistor symbols R_i and R_o to denote their values.

The importance of input and output imped-ances can be understood by considering the two

FIGURE 5.11   Output Characteristics, i_C Versus V_{ce}

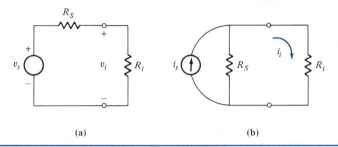

(a)   (b)

FIGURE 5.12   Source and Source Resistance Configurations: (a) Voltage source with R_S, (b) Current source with R_S

source configurations of Fig. 5.12. We assume that the resistor R_S in each case is associated with the source and is fixed in value. The resistor R_i is assumed to represent the AC input impedance of a general amplifier, which can be varied by changing the amplifier or its constituents. For Fig. 5.12a, the voltage at the input of the amplifier is given by

$$v_i = \frac{R_i}{R_i + R_S} v_s \qquad (5.2-34)$$

Then, for maximum voltage transfer from the source v_s to the input v_i, the resistor R_i would be chosen such that $R_i \gg R_S$.

In the current-source representation of the input signal shown in Fig. 5.12b, the opposite inequality for R_i and R_S ($R_i \ll R_S$) is desirable in order to maximize the current entering the amplifier. By applying the current divider rule to Fig. 5.12b, we obtain

$$i_i = \frac{R_S}{R_i + R_S} i_s \qquad (5.2-35)$$

which has a maximum value for $R_S \gg R_i$. Note that (5.2–34) and (5.2–35) indicate large and small R_i so that maximum current and voltage amplification cannot be achieved at the same time in any amplifier circuit.

For these two cases, notice that as a signal is applied to a particular amplifier, it is, in fact, reduced at the amplifier input before it even enters the amplifier. Fortunately, the amplification

achievable by the amplifier is normally larger than this reduction at its input. This discussion thus indicates that the magnitude of the input impedance of an amplifier is important.

As for the output impedance of an amplifier circuit, a similar argument involving voltage and current transfer at the output terminals is prevalent. In this case, however, the resistors involved are the load resistor R_L and the AC output impedance R_o of the amplifier. The relative magnitude of R_o with respect to R_L plays an important role in determining the amount of amplification for the amplifier.

Common-Emitter Amplifier. In Fig. 5.13a the common-emitter amplifier circuit is redrawn with the voltage-source representation of the input signal. We will determine the voltage amplification v_o/v_s for this circuit. However, this amplifier circuit can also amplify current because of the transistor configuration with its output current i_C being much greater than its input current i_B. Also shown in Fig. 5.13a is the symbolic representation of the AC input and output impedances (R_i and R_o). The arrow indicates where the corresponding impedance is located (which is seen more clearly in Figs. 5.13c and d).

To determine the AC properties of this amplifier, namely, v_o/v_s, R_i, and R_o, we begin by obtaining the AC small-signal equivalent circuit. The first version is shown in Fig. 5.13b, where we have

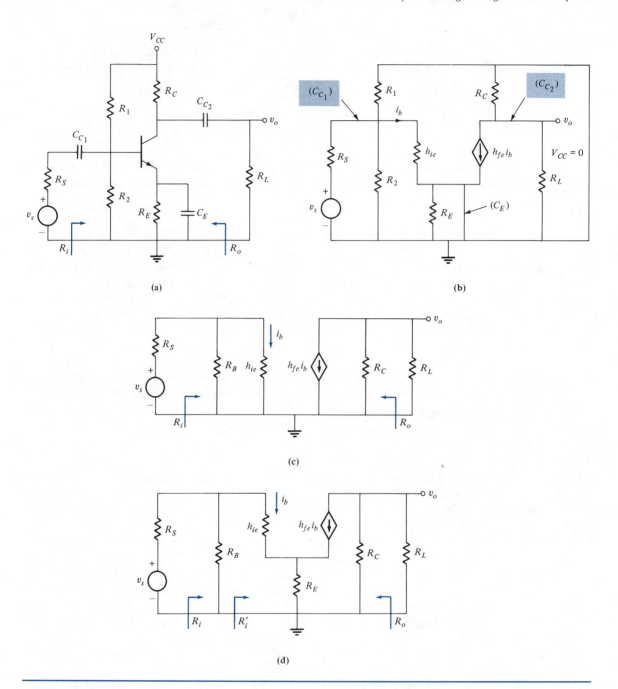

FIGURE 5.13   Common-Emitter Amplifier: (a) Amplifier circuit, (b) AC equivalent circuit, (c) Simplified AC equivalent circuit, (d) AC equivalent circuit with R_E

substituted the AC equivalent circuit for each element and used the simplified common-emitter hybrid model ($h_{oe} = 0$) for the BJT. The position of the shorted capacitors and DC voltage source are indicated in Fig. 5.13b as an aid in visualizing the formation of the AC circuit model. Further simplification of the AC equivalent circuit is carried out to obtain the circuit of Fig. 5.13c, where R_1 has been combined in parallel with R_2 ($R_B = R_1 \| R_2$) and R_E has been short circuited (with C_E). With practice, we eliminate drawing the intermediate circuit of Fig. 5.13b.

An expression for the voltage amplification v_o/v_s in terms of the external resistors and the BJT parameters is now easily obtained by analyzing the circuit of Fig. 5.13c. From the output loop using Ohm's law and combining R_L in parallel with R_C, we have

$$v_o = -(h_{fe}i_b)(R_L \| R_C) \tag{5.2-36}$$

This relation contains i_b, for which an expression is obtained in terms of v_s from the input loop of the circuit. Visualizing the combination of R_S, R_B, and v_s of Fig. 5.13c into the Thévenin equivalent circuit and then writing the loop current i_b by inspection yield

$$i_b = \frac{[R_B/(R_S + R_B)]v_s}{R_B \| R_S + h_{ie}} \tag{5.2-37}$$

Substitution of (5.2–37) into (5.2–36) provides the voltage amplification expression as follows:

$$\frac{v_o}{v_s} = \frac{-h_{fe}(R_C \| R_L)[R_B/(R_S + R_B)]}{R_B \| R_S + h_{ie}} \tag{5.2-38}$$

Note that (5.2–38) indicates that the voltage amplification depends directly upon h_{fe}. For larger h_{fe}, the amplification (voltage in this case) is increased.

The effects of the relative magnitudes of the resistors can be observed from the numerator term $R_B/(R_S + R_B)$ in (5.2–38). We realize that this ratio

has a maximum value of unity for $R_B \gg R_S$. Furthermore, choosing $R_B \gg R_S$ forces the denominator term $R_S \| R_B$ to be R_S. We must also be mindful of the rule-of-thumb bias relation from Chapter 4 where $R_B \simeq \beta R_E/10$; hence, $R_S \ll R_B \ll \beta R_E$. Thus, with $R_B \gg R_S$, (5.2–38) can be written as

$$\frac{v_o}{v_s} = -\frac{h_{fe}(R_C \| R_L)}{R_S + h_{ie}} \tag{5.2-39}$$

Next, consider the remaining resistors in the denominator of (5.2–39) and note that if one of them is much less than the other, v_o/v_s is further increased. Since we have already used $R_B \gg R_S$, if we interpret this inequality as meaning small R_S (where $R_S \ll h_{ie}$), then (5.2–39) can be written as

$$\frac{v_o}{v_s} = \frac{-h_{fe}(R_C \| R_L)}{h_{ie}} \tag{5.2-40}$$

Finally, with $R_C \| R_L$, note that this quantity is maximized for $R_C \gg R_L$ because R_L is considered fixed in value and the parallel combination of R_L with any resistor yields a smaller value of resistance. Thus, for $R_C \gg R_L$, we obtain

$$\frac{v_o}{v_s} = -h_{fe}\left(\frac{R_L}{h_{ie}}\right) \tag{5.2-41}$$

The input and output impedances for the common-emitter configuration are now easily obtained by inspection of the AC small-signal circuit of Fig. 5.13c. The input impedance is simply the parallel combination of the two input resistors, or

$$R_i = R_B \| h_{ie} \tag{5.2-42}$$

The output impedance is

$$R_o = R_C \tag{5.2-43}$$

since the current source $h_{fe}i_b$ is open for $v_S = 0$.

EX. 5.5

Use of the Common-Emitter Hybrid Model Example

Draw the small-signal equivalent circuit for the common-emitter amplifier of Fig. 5.5 and label all elements with their corresponding values. The circuit of Fig. 5.5 has the same bias resistors and DC voltage values as those in Ex. 5.4. In addition, $R_L = 20$ kΩ, $\beta = 100$, $1/h_{oe} = 20$ kΩ, and $V_O = 0.7$ V. Calculate v_o/v_s for the two cases in which $R_S = 0$ and $R_S = 10$ kΩ. Compare the results. Finally, calculate the values for R_i and R_o.

Solution: The small-signal equivalent circuit with corresponding element values is shown in Fig. E5.5. The value of $h_{ie} = 0.4$ kΩ was obtained in Ex. 5.4.

From the output portion of Fig. E5.5, the voltage v_o is given by

$$v_o = -(h_{fe}i_b)\left(R_L \| R_C \| \frac{1}{h_{oe}}\right) \tag{1}$$

where $1/h_{oe}$ is considered to be finite. The current i_b is now obtained from the input portion of Fig. E5.5, or from (5.2–37), as follows:

$$i_b = \frac{[R_B/(R_S + R_B)]v_s}{R_B \| R_s + h_{ie}} \tag{2}$$

Substituting values into (1) and (2) yields

$$v_o = -(100i_b)(20 \| 0.83 \| 20) = -76.6i_b \tag{3}$$

and

$$i_b = \frac{[16.5/(R_S + 16.5)]v_s}{(16.5)R_S/(16.5 + R_S) + h_{ie}} \tag{4}$$

For $R_S = 0$, we obtain

$$i_b = \frac{v_s}{h_{ie}} = \frac{v_s}{0.5} = 2v_s \tag{5}$$

For $R_S = 10$ kΩ, we obtain

$$i_b = \frac{[16.5/(10 + 16.5)]v_s}{(16.5)(10)/(16.5 + 10) + 0.5} = 0.093v_s \tag{6}$$

Thus, substituting (5) into (3) yields

$$\frac{v_o}{v_s} = -(76.6)(2) = -153.2$$

for the case with $R_S = 0$. Substituting (6) into (3) yields

$$\frac{v_o}{v_s} = -(76.6)(0.093) = -7.124$$

for the case with $R_S = 10$ kΩ. By comparing the two values of voltage amplification, we observe that increased R_S can severely reduce the voltage gain of the amplifier.

Finally, the input impedance is obtained directly from the circuit of Fig. E5.5, or from (5.2–42), as follows:

$$R_i = 16.5 \| 0.4 = 0.4 \text{ k}\Omega$$

The output impedance is also obtained directly from the circuit of Fig. E5.5 as follows:

$$R_o = \frac{1}{h_{oe}} \| R_C \simeq 0.8 \text{ k}\Omega$$

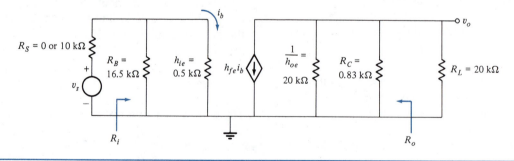

FIGURE E5.5

Common-Emitter Amplifier without Bypass Capacitor C_E.   In some instances (in particular, in direct-coupled amplifiers introduced in the next chapter), it is desirable to use the common-emitter amplifier without the capacitor C_E. The resulting small-signal AC equivalent circuit is displayed in Fig. 5.13d.

We analyze this circuit in the usual manner—that is, by first writing the expression for output voltage from the output loop, which yields

$$v_o = -(h_{fe}i_b)(R_C \| R_L) \qquad (5.2\text{--}44)$$

Next, the input loop is analyzed to determine i_b. Visualizing a Thévenin equivalent circuit for R_S, R_B, and v_s and writing KVL yield

$$\frac{v_s R_B}{R_B + R_S} = i_b R_B \| R_S + i_b h_{ie} + i_b(1 + h_{fe})R_E \qquad (5.2\text{--}45)$$

Solving for i_b yields

$$i_b = \frac{v_s R_B/(R_B + R_S)}{R_B \| R_S + h_{ie} + (1 + h_{fe})R_E} \qquad (5.2\text{--}46)$$

Substituting (5.2–46) into (5.2–44) yields the voltage amplification:

$$\frac{v_o}{v_s} = \frac{-h_{fe}R_C \| R_L[R_B/(R_B + R_S)]}{R_B \| R_S + h_{ie} + (1 + h_{fe})R_E} \qquad (5.2\text{--}47)$$

With the same resistor magnitude assumptions as in the previous section (except for R_L relative to R_C), we obtain the amplification for this amplifier as follows:

$$\frac{v_o}{v_s} = \frac{-h_{fe}(R_C \| R_L)}{(1 + h_{fe})R_E} \qquad (5.2\text{--}48)$$

Note that for large h_{fe} this expression becomes

$$\frac{v_o}{v_s} = -\frac{R_C \| R_L}{R_E} \qquad (5.2\text{--}49)$$

which indicates that the voltage amplification is independent of h_{fe} and dependent only upon the ratio of external resistors.

Note that the input impedance of this amplifier has been increased because, from Fig. 5.13d, we observe that

$$R_i = R_B \| R_i' \qquad (5.2\text{--}50)$$

where R_i' is the impedance seen looking into the base of the transistor. The expression for R_i' is obtained by forming the ratio of voltage to current as follows:

$$R_i' = \frac{v_b}{i_b} \qquad (5.2\text{--}51)$$

By substitution for v_b, we have

$$R_i' = \frac{(1 + h_{fe})i_b R_E + i_b h_{ie}}{i_b} = (1 + h_{fe})R_E + h_{ie} \qquad (5.2\text{--}52)$$

Note that the input resistance to the transistor includes an "amplified" resistor $(1 + h_{fe})R_E$, which is usually much greater than h_{ie}.

——————————————————————————— **EX. 5.6**

Effect of R_E on the Common-Emitter Amplifier Example

Calculate the voltage amplification for the circuit of Fig. 5.13d. Use the following element values: $R_S = 0$, $R_B = 5$ kΩ, $R_E = 0.5$ kΩ, $R_C = 10$ kΩ, and $R_L = \infty$. The transistor parameters are $h_{ie} = 1$ kΩ and $h_{fe} = 100$.

Solution: The voltage amplification as calculated directly from (5.2–47) is

$$\frac{v_o}{v_s} = \frac{-100(10)(1)}{0 + 1 + (101)0.5} = -19.4$$

From the approximate expression of (5.2–49), $v_o/v_s = -R_C/R_E = -10/0.5 = -20$, which is very nearly the same result.

——————————————————————————————

Common-Collector Amplifier (Emitter Follower).
Figure 5.14a displays the circuit for the common-collector amplifier, or emitter follower. Unlike the common-emitter amplifier, the voltage amplification for this case is less than unity, but usually $v_o/v_s \simeq 1$. The current amplification is much greater than unity.

The properties that make this amplifier extremely useful are high input impedance and low output impedance, as we will see. Because of these attributes, the common-collector amplifier is used as an impedance transformer and, quite often, is the output stage of a multistage amplifier system.

The small-signal AC equivalent circuit for the common-collector amplifier (obtained in the same manner as for the common-emitter case) is shown in Fig. 5.14b. By analyzing this circuit, we will determine the input impedance R_i, the output impedance R_o, and the voltage amplification v_o/v_s. Furthermore, we will determine the relative magnitudes of the resistors in the circuit that allow v_o/v_s to approach unity.

From the output portion of Fig. 5.14b, observe that using Ohm's law yields

$$v_o = i_e(R_L \| R_E) \tag{5.2-53}$$

By substituting $i_e = (1 + h_{fe})i_b$, we have

$$v_o = (1 + h_{fe})(R_L \| R_E)i_b \tag{5.2-54}$$

We next determine i_b in terms of v_s from the input portion of the circuit. Combining R_S, R_B, and v_s into a Thévenin equivalent circuit and then writing KVL for the input loop yield

$$\frac{R_B}{(R_S + R_B)} v_s = i_b(R_S \| R_B) + v_b \tag{5.2-55}$$

where v_b is the voltage at B relative to ground given by

$$v_b = i_b h_{ie} + i_e(R_L \| R_E) \tag{5.2-56}$$

Substituting for i_e yields

$$v_b = i_b h_{ie} + (1 + h_{fe})(i_b)(R_L \| R_E) \tag{5.2-57}$$

Substituting (5.2–57) into (5.2–55) and solving for i_b yield

$$i_b = \frac{R_B/[(R_S + R_B)v_s]}{R_S \| R_B + h_{ie} + (1 + h_{fe})(R_L \| R_E)} \tag{5.2-58}$$

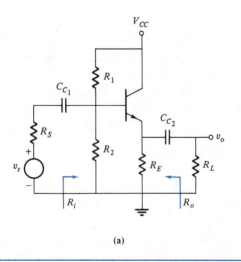

(a)

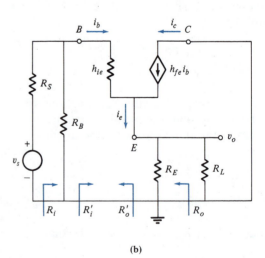

(b)

FIGURE 5.14   Common-Collector Amplifier (Emitter Follower): (a) Amplifier circuit, (b) AC equivalent circuit

Substituting (5.2–58) into (5.2–54) provides the expression for voltage amplification as follows:

$$\frac{v_o}{v_s} = \frac{[(1 + h_{fe})(R_L \| R_E)(R_B)]/(R_S + R_B)}{R_S \| R_B + h_{ie} + (1 + h_{fe})(R_L \| R_E)}$$

$$(5.2–59)$$

The last term in the denominator of (5.2–59), $(1 + h_{fe})(R_L \| R_E)$, is usually much greater than the others, and it therefore cancels directly with the identical numerator term leaving

$$\frac{v_o}{v_s} = \frac{R_B}{R_S + R_B}$$

$$(5.2–60)$$

Thus, for $R_B \gg R_S$, $v_o/v_s \simeq 1$, and the output voltage follows the input signal voltage. Hence, for v_o/v_s closer to unity, the source resistance R_S should be small.

The input impedance of the common-collector amplifier is obtained by inspection of Fig. 5.14b as follows:

$$R_i = R_B \| R_i'$$

$$(5.2–61)$$

where

$$R_i' = \frac{v_b}{i_b} = \frac{[(R_L \| R_E)(1 + h_{fe})(i_b)] + h_{ie}i_b}{i_b}$$

$$(5.2–62)$$

Thus, R_i is given by

$$R_i = R_B \| [(1 + h_{fe})(R_L \| R_E) + h_{ie}] \quad (5.2–63)$$

Note that the value of R_i will be large only if both R_B and $(1 + h_{fe})(R_L \| R_E)$ are large. Recall, however, that there is a restriction on the magnitude of R_B from bias design considerations.

The output impedance is obtained by considering Fig. 5.14b. By inspection, we have

$$R_o = R_E \| R_o'$$

$$(5.2–64)$$

The resistance R_o' is obtained by forming the ratio of voltage to current as follows:

$$R_o' = \frac{v_o}{-i_e}\bigg|_{v_s = 0} = \frac{-i_b(R_S \| R_B) - i_b h_{ie}}{-(1 + h_{fe})i_b}$$

$$(5.2–65)$$

where the numerator of (5.2–65) was obtained by setting $v_s = 0$ in the circuit and assuming a fictitious source of current at the output to provide i_b. Canceling i_b in (5.2–65) yields

$$R_o' = \frac{(R_S \| R_B) + h_{ie}}{1 + h_{fe}}$$

$$(5.2–66)$$

where R_o' is quite small since $1 + h_{fe}$ appears in the denominator. Combining R_o' in parallel with R_E then yields an even smaller value of output resistance for R_o.

EX. 5.7

Use of the Common-Emitter Model for Analysis of the Emitter Follower Example

The emitter follower circuit of Fig. 5.14a has element values of $R_L = 20$ kΩ, $R_E = 1$ kΩ, $R_1 = 7$ kΩ, $R_2 = 16$ kΩ, and $R_S = 0$. Calculate the voltage amplification and R_i and R_o for a BJT with $h_{fe} = 100$, $h_{oe} = 0$, and $h_{ie} = 0.2$ kΩ.

Solution: We begin by calculating $R_B = R_1 \| R_2$ to obtain

$$R_B = \frac{(7)(16)}{7 + 16} = 4.87 \text{ k}\Omega$$

Then, by direct substitution of the given values into (5.2–59), we have

$$\frac{v_o}{v_s} = \frac{[(101)(20 \| 1)(4.87)]/(0 + 4.87)}{0 \| 4.87 + 0.2 + (101)(20 \| 1)} = 0.998$$

or $v_o/v_s \simeq 1$. The circuit does indeed behave as an emitter follower with the output voltage following the input voltage.

By substituting the given values into (5.2–63), the input impedance is obtained as follows:

$$R_i = 4.87 \| [101(20 \| 1) + 0.2] = 4.64 \text{ k}\Omega$$

Note that this is essentially the value of $R_B = 4.87\ \text{k}\Omega$ because $R_E(1 + h_{fe}) \gg R_B$. For larger input impedance in this circuit, the value of R_B must be increased. The output impedance is obtained by substituting the given element values into (5.2–64) to obtain

$$R_o = R_E \| \frac{R_S \| R_B + h_{ie}}{1 + h_{fe}} = 1 \| \frac{0.2}{101} \simeq 2\ \Omega$$

This small value of output impedance is characteristic of the emitter follower.

Common-Base Amplifier. The common-base amplifier circuit displayed in Fig. 5.7 provides useful voltage amplification but not current amplification since the input current is less than the output current. Additional properties of the common-base amplifier are a small input impedance and a large output impedance, as we will see.

Figure 5.15 displays the small-signal AC equivalent circuit for the common-base amplifier of Fig. 5.7, where the simplified common-base hybrid model (Fig. 5.10b) for the BJT has been used. The AC small-signal analysis now includes determining the voltage amplification v_0/v_s, the input impedance R_i, and the output impedance R_o.

To determine v_0/v_s, from the output portion of the circuit of Fig. 5.15 writing Ohm's law, we have

$$v_o = -i_e(R_C \| R_L) \tag{5.2–67}$$

From the input portion of the circuit, i_e is now determined in terms of v_s. Visualizing a Thévenin equivalent circuit for R_E, R_S, and v_s and applying KVL to the input loop yield, by inspection,

$$i_e = \frac{-[R_E/(R_E + R_S)]v}{R_S \| R_E + h_{ib}} \tag{5.2–68}$$

By substitution of (5.2–68) into (5.2–67), the voltage amplification is obtained as follows:

$$\frac{v_o}{v_s} = \frac{(R_C \| R_L)[R_E/(R_E + R_S)]}{R_S \| R_E + h_{ib}} \tag{5.2–69}$$

Note that for $R_E \gg R_S$, $R_E/(R_E + R_S) \simeq 1$, and $R_S \| R_E \simeq R_S$. Under these conditions, (5.2–69) becomes

$$\frac{v_o}{v_s} = \frac{R_C \| R_L}{R_S + h_{ib}} \tag{5.2–70}$$

Note that $h_{ib} = 0.025/I_E$ and that I_E is a large current ($\simeq 1$ mA). Thus, h_{ib} will have a small value of resistance (for $I_E = 1$ mA, $h_{ib} = 25\ \Omega$). The voltage amplification thus becomes

$$\frac{v_o}{v_s} = \frac{R_C \| R_L}{R_S} \tag{5.2–71}$$

and is therefore dependent only upon external resistors. Note again that decreasing R_S results in larger voltage amplification.

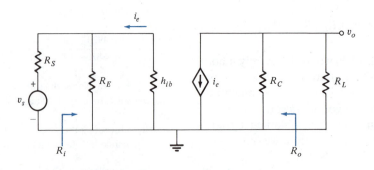

FIGURE 5.15   AC Equivalent Circuit for Common-Base Amplifier

The input impedance is obtained by inspection of Fig. 5.15 as follows:

$$R_i = R_E \| h_{ib} \qquad (5.2\text{–}72)$$

Note that R_i is essentially h_{ib} because normally $h_{ib} \ll R_E$.

The output impedance from Fig. 5.15 is simply

$$R_o = R_C \qquad (5.2\text{–}73)$$

Hence, its magnitude depends directly upon the magnitude of the resistor R_C. However, if R_C had been the load resistor (with the original load resistor R_L not in the circuit), then the output impedance from Fig. 5.15 would have been infinite. This result is strictly dependent upon the simplified BJT hybrid model that was used. If the more complete AC model for the BJT (Fig. 5.9b) were used, R_o would be $1/h_{ob}$, which is typically quite a large value of resistance ($\geq 50 \text{ k}\Omega$).

─────────────── EX. 5.8

Use of the Common-Base Hybrid Model Example

Draw the small-signal equivalent circuit for the common-base amplifier of Fig. 5.7 and label all elements with their corresponding values. Use the following element values: $R_C = 0.83 \text{ k}\Omega$, $R_L = 20 \text{ k}\Omega$, and $R_E = 1.67 \text{ k}\Omega$, with $h_{fe} = 100$, $h_{ob} = 0$, and $h_{ib} = 4 \text{ }\Omega$. Calculate v_o/v_s for the two cases in which $R_S = 0$ and $R_S = 10 \text{ k}\Omega$. Compare the results. Finally, calculate R_i and R_o.

Solution: The small-signal equivalent circuit with corresponding element values is shown in Fig. E5.8. By inspection of the output portion of this circuit, or from (5.3–67),

we have

$$v_o = -i_e(0.83 \| 20) \simeq -0.8 i_e \qquad (1)$$

From the input portion of Fig. E5.8, or from (5.2–68), we have

$$i_e = \frac{[-1.67/(R_S + 1.67)]v_s}{R \| 1.67 + 0.004} \qquad (2)$$

Substituting (2) into (1) yields

$$\frac{v_o}{v_s} = \frac{1.33/(R_S + 1.67)}{[R_S(1.67)]/(R_S + 1.67) + 0.004} \qquad (3)$$

Substituting $R_S = 0$ into (3) yields

$$\frac{v_o}{v_s} = \frac{1.33/1.67}{0.004} \simeq 199$$

and substituting $R_S = 10 \text{ k}\Omega$, we have

$$\frac{v_o}{v_s} = \frac{1.33/11.67}{[10(1.67)]/(10 + 1.67) + 0.004} = 0.079$$

Hence, the effect of increased R_S, as usual, results in decreased amplification.

The input impedance and output impedance are obtained directly from Fig. E5.8 as follows:

$$R_i = 1.67 \| 0.005 = 4.985 \text{ }\Omega$$

and

$$R_o = 0.83 \text{ k}\Omega$$

Note that the input impedance is essentially $R_i \simeq h_{ib} \simeq 4 \text{ }\Omega$ since $h_{ib} \ll R_E$.

─────────────────────────

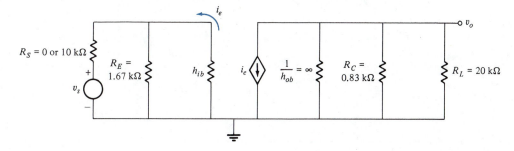

FIGURE E5.8

5.3   SMALL-SIGNAL ANALYSIS OF FET AMPLIFIERS

The development of the small-signal AC circuit model for FETs is considerably simplified now that we have examined the corresponding models for diodes and BJTs. The models of this section will be valid only at low and medium frequencies since we will omit the capacitance associated with the gate in each case.

5.3.1   FET Small-Signal Conductance Model

The general small-signal model developed in this section is valid for all FETs in the low- and medium-frequency range. The FET is assumed to have the common-source configuration as shown in Fig. 5.16a. However, the model developed is also used for an amplifier with the FET in the common-drain configuration (since the input and output currents for each configuration are identical).

The total currents and voltages of Fig. 5.16a have functional dependencies upon one another that are based on the actual behavior of the FET. From Chapter 3, we recall that for all FETs operating under amplifier conditions,

$$i_G = 0 \tag{5.3-1}$$

and

$$i_D = i_D(v_{GS}, v_{DS}) \tag{5.3-2}$$

Thus, we select v_{GS} and v_{DS} as independent variables, the current i_D as a dependent variable, and $i_G = 0$.

Expanding (5.3–2) in a Taylor series, making the small-signal assumption and retaining only the AC terms, yields

$$i_d = \frac{\partial i_D}{\partial v_{GS}}\bigg|_Q v_{gs} + \frac{\partial i_D}{\partial v_{DS}}\bigg|_Q v_{ds} \tag{5.3-3}$$

Each of the derivative coefficients in these expressions has the units of conductance because of our choice of voltages as the independent variables and currents as the dependent variables. Therefore, we define each coefficient as a conductance as follows:

$$g_m = \frac{\partial i_D}{\partial v_{GS}}\bigg|_Q \tag{5.3-4}$$

and

$$g_o = \frac{\partial i_D}{\partial v_{DS}}\bigg|_Q \tag{5.3-5}$$

The conductance g_o is the *output conductance,* and g_m is the *transconductance* because it is a "transfer" conductance (dependent upon output current and input voltage). With these definitions, (5.3–3) can

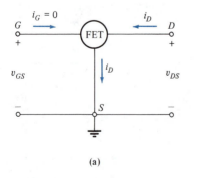

(a)

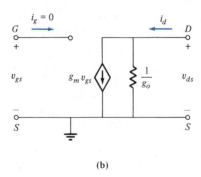

(b)

FIGURE 5.16   FET Small-Signal Conductance Model: (a) Common-source configuration, (b) Conductance model

be written as

$$i_d = g_m v_{gs} + g_o v_{ds} \qquad (5.3-6)$$

In FET data sheets, these common-source AC conductances are generalized as the admittances $Re|Y_{fs}|$ and $Re|Y_{os}|$. The specific definitions are $g_m = Re|Y_{fs}|$ and $g_o = Re|Y_{os}|$, where Re represents the real part, $|Y|$ is the magnitude of the admittance, s is for common source, f is for forward, and o is for output. Of course, the symbols g_m and g_o are much more convenient to use.

Figure 5.16b shows the small-signal AC model for the FET corresponding to (5.3–6) and $i_g = 0$. This model is called the *conductance model* because its parameters are both conductances, g_m and g_o. Also shown in the circuit of Fig. 5.16b is the output conductance labeled as $1/g_o$, which corresponds to its resistance value.

5.3.2   Determination of Conductance Model Parameters

The conductance parameters for the small-signal FET model can be determined graphically in a manner similar to that used for the BJT *h*-parameters. Output current–voltage characteristics are available from manufacturers that are sufficiently accurate to provide this determination. However, the derivative expressions of (5.3–4) and (5.3–5) can be used along with the FET current transfer equation for another estimate of the parameters.

Figure 5.17 is used to show the method of graphically determining g_m and g_o. Note that Q is located in the amplifier portion of the output characteristics. The derivative equations (5.3–4) and (5.3–5) are then evaluated in the same manner as the BJT parameters as follows:

$$g_m = \left.\frac{\partial i_D}{\partial v_{GS}}\right|_{v_{ds}=0} = \left.\frac{\Delta i_D}{\Delta v_{GS}}\right|_{v_{DS}=V_{DS}}$$

$$= \frac{I_{D_2} - I_{D_1}}{V_{GS_2} - V_{GS_1}} \qquad (5.3-7)$$

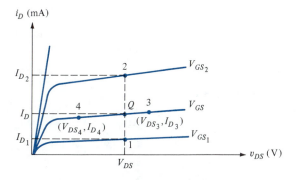

i_D (mA)

FIGURE 5.17   FET Output Characteristics, i_D Versus v_{DS}

and

$$g_o = \left.\frac{\partial i_D}{\partial v_{DS}}\right|_{v_{gs}=0} = \left.\frac{\Delta i_D}{\Delta v_{DS}}\right|_{v_{GS}=V_{GS}}$$

$$= \frac{I_{D_3} - I_{D_4}}{V_{DS_3} - V_{DS_4}} \qquad (5.3-8)$$

where points 1, 2, 3, and 4 are as indicated in Fig. 5.17. Typical values of g_m are on the order of $10^{-3}/\Omega$ (or 10^{-3} S or 1 mS), whereas typical values of $1/g_0$ are greater than 20 kΩ. Note that calculating g_o from the theoretical square-law FET equation (which shows no dependence of i_D upon v_{DS}) yields $g_o = 0$. Under these conditions, the conductance model of Fig. 5.16b consists of an open circuit at the input and a current source at the output, with $i_d = g_m v_{gs}$.

5.3.3   Common-Source Amplifier

The common-source amplifier circuit is displayed in Fig. 5.18a with its associated coupling capacitors C_{C_1} and C_{C_2} and bypass capacitor C_F. Note that the subscript F (for feedback) has been used on the stabilizing resistor R_F and on corresponding bypass capacitor C_F in order to avoid confusion with the resistance of the voltage source, denoted as R_S. Also, a specialized input circuit is used that includes an additional resistor

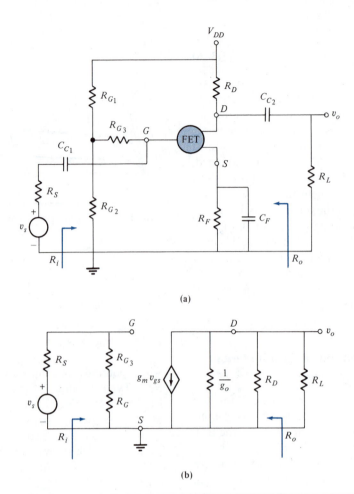

FIGURE 5.18   Common-Source Amplifier: (a) Amplifier circuit, (b) AC equivalent circuit

(R_{G_3}) that provides an increased input impedance for the amplifier. Note further that since the gate current is zero, R_{G_3} has no effect upon the bias circuit or Q.

The voltage amplification for the common-source amplifier is now easily determined from the AC equivalent circuit of Fig. 5.18b. From the output portion of the circuit, we have

$$v_o = -g_m v_{gs}\left(\frac{1}{g_o}\|R_D\|R_L\right) \qquad (5.3\text{--}9)$$

From the input portion of the circuit,

$$v_{gs} = \frac{R_{G_3} + R_G}{R_S + R_{G_3} + R_G}v_s \qquad (5.3\text{--}10)$$

By eliminating v_{gs} from (5.3–9), we obtain the voltage amplification expression as follows:

$$\frac{v_o}{v_s} = -g_m\left(\frac{1}{g_o}\|R_D\|R_L\right)\left(\frac{R_{G_3} + R_G}{R_S + R_{G_3} + R_G}\right) \qquad (5.3\text{--}11)$$

Equation (5.3–11) shows that v_o/v_s is directly dependent upon g_m; hence, g_m is referred to as the *gain parameter* of the FET.

In order to maximize the voltage amplification, we select $R_{G_3} \gg R_S$, which causes the last term in parentheses in (5.3–11) to become unity. Furthermore, for large $1/g_o$ and large R_D relative to R_L, the parallel combination term is maximized. Under these conditions, (5.3–11) becomes

$$\frac{v_o}{v_s} = -g_m R_L \qquad (5.3\text{–}12)$$

which is the voltage amplification for the common-source amplifier under ideal conditions.

The input and output impedances are obtained by inspection of Fig. 5.18b as

$$R_i = R_{G_3} + R_G \qquad (5.3\text{–}13)$$

and

$$R_o = R_D \,\Big\|\, \frac{1}{g_o} \qquad (5.3\text{–}14)$$

Note that R_i will be large provided that R_{G_3} is large and regardless of the magnitude of R_G.

———————————————————— EX. 5.9

Use of the FET Small-Signal Conductance Model Example

Determine the AC voltage amplification v_o/v_s for the common-source amplifier of Fig. 5.18a. Use the small-signal equivalent circuit of Fig. 5.18b and the following element values: $R_S = 1$ kΩ, $R_G = 10$ kΩ, $R_{G_3} = 100$ kΩ, $R_D = 10$ kΩ, and $R_L = 10$ kΩ. For the FET, let $g_m = 30$ mS and $1/g_o = 10$ kΩ. Also, calculate the input and output impedances.

Solution: By substituting directly into (5.3–9) and (5.3–10), we have

$$v_o = -(30)(10 \,\|\, 10 \,\|\, 10)(v_{gs}) = -100 v_{gs} \qquad (1)$$

and

$$v_{gs} = \frac{100 + 10}{1 + 100 + 10} v_s = 0.991 v_s \qquad (2)$$

Thus, by substitution of (2) into (1), we have

$$\frac{v_o}{v_s} = 99.1 \qquad (3)$$

Finally, the input and output impedances are calculated from (5.3–13) and (5.3–14) as

$$R_i = 100 + 10 = 110 \text{ k}\Omega$$

and

$$R_o = 10 \,\|\, 10 = 5 \text{ k}\Omega$$

———————————————————

5.3.4   Common-Drain Amplifier (Source Follower)

The common-drain amplifier, or source follower, is analogous to the BJT common-collector amplifier (or emitter follower) because it possesses similar properties—namely, the input impedance is high, the output impedance is low, and the voltage amplification is designed to be nearly equal to unity. The necessary circuit requirements to attain these properties are shown next.

Figure 5.19a displays the common-drain amplifier circuit, and Fig. 5.19b shows the associated AC equivalent circuit, which contains the complete conductance model for the FET. To determine the voltage amplification, we first write Ohm's law for the output resistor as $v_o = i_d R_L$. Then, i_d is obtained by using Fig. 5.19b, converting $g_m v_{gs}$ to a current source, and writing KVL for the output loop, which yields

$$i_d = \frac{(g_m/g_o)v_{gs}}{1/g_o + R_L} \qquad (5.3\text{–}15)$$

Thus, v_o is given by $i_d R_L$ or

$$v_o = \frac{(g_m/g_o)R_L v_{gs}}{1/g_o + R_L} \qquad (5.3\text{–}16)$$

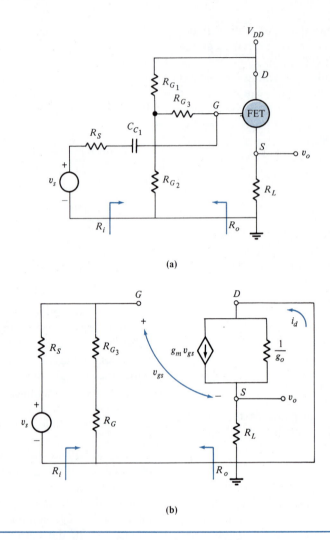

FIGURE 5.19 Common-Drain Amplifier: (a) Amplifier circuit, (b) AC equivalent circuit

Note that in (5.3–16), v_o is dependent upon v_{gs} and that from Fig. 5.19b, v_{gs} is dependent upon v_s as well as v_o. The expression for v_{gs} is obtained by combining the input portion of the circuit into a Thévenin equivalent and writing KVL for the gate–source terminals, which yields, by inspection,

$$v_{gs} = \left(\frac{R_G + R_{G_3}}{R_S + R_G + R_{G_3}} \right) v_s - v_o \qquad (5.3–17)$$

Since R_{G_3} is a large resistor (as in the previous common-source case), (5.3–17) can be written as

$$v_{gs} = v_s - v_o \qquad (5.3–18)$$

Substituting (5.3–18) into (5.3–16) yields

$$v_o = \frac{(g_m/g_o)R_L}{1/g_o + R_L}(v_s - v_o) \qquad (5.3–19)$$

Solving for the voltage amplification v_o/v_s, we have

$$\frac{v_o}{v_s} = \frac{g_m R_L}{1 + (g_m + g_o)R_L} \qquad (5.3-20)$$

Note that with $g_m R_L$ contained in the numerator and denominator, the right-hand side of (5.3–20) is a fraction; therefore, the voltage amplification is less than unity. However, for $g_m \gg g_o$ and $g_m R_L \gg 1$, we have

$$\frac{v_o}{v_s} = \frac{g_m R_L}{1 + g_m R_L} \lesssim 1 \qquad (5.3-21)$$

which gives the desired property of the output voltage following the input signal voltage with $v_o \lesssim v_s$.

The input impedance, from Fig. 5.19b, is written, by inspection, as

$$R_i = R_{G_3} + R_G \qquad (5.3-22)$$

Hence, the input impedance will be large for large R_{G_3} and is independent of R_G.

The output impedance, by definition, is given by

$$R_o = \frac{v_o}{-i_d}\bigg|_{v_s=0} \qquad (5.3-23)$$

and from the output loop of Fig. 5.19b,

$$v_o = \frac{g_m v_{gs}}{g_o} - \frac{i_d}{g_o} \qquad (5.3-24)$$

However, for v_s set equal to zero, $v_o = -v_{gs}$ so that by substituting for v_{gs} in (5.3–24), we have

$$v_o\big|_{v_s=0} = \frac{-i_d/g_o}{g_m/g_o + 1} \qquad (5.3-25)$$

Substitution of (5.3–25) into (5.3–23) thus yields cancelation of i_d, and

$$R_o = \frac{1/g_o}{g_m/g_o + 1} = \frac{1}{g_m + g_o} \qquad (5.3-26)$$

Thus, we see that the output impedance is just the output impedance of the FET $(1/g_o)$ reduced by the factor $(g_m/g_o + 1)$. Note that, quite often, $g_m \gg g_o$ so that (5.3–26) becomes

$$R_o = \frac{1}{g_m} \qquad (5.3-27)$$

We conclude that the output resistance R_o of the source follower is small because FET design maximizes g_m.

_____ EX. 5.10

Source Follower Example

Calculate v_o/v_s, R_i, and R_o for the source follower circuit of Fig. 5.19a. Use the following element values: $R_{G_1} = 20$ kΩ, $R_{G_2} = 20$ kΩ, $R_{G_3} = 100$ kΩ, $R_L = 10$ kΩ, and $R_S = 0$. FET parameters are $g_m = 30$ mS and $1/g_o = 10$ kΩ.

Solution: Substitution of the element values into (5.3–20) yields the voltage amplification as follows:

$$\frac{v_o}{v_s} = \frac{(0.03)(10^4)}{1 + [(0.03 + 10^{-4})10^4]} = 0.993$$

Thus, the output voltage follows the input voltage very closely.

The input and output impedances are calculated directly from (5.3–22) and (5.3–26), respectively, as

$$R_i = 100 + (20 \,\|\, 20) = 110 \text{ kΩ}$$

and

$$R_o = \frac{1}{0.03 + 10^{-4}} = 33.2 \text{ Ω}$$

Note that using $R_o \simeq 1/g_o = 33.3$ Ω provides a result that is less than a half percent in error.

5.3.5   MOST Amplifier

In MOS integrated circuits, resistors are replaced with MOSTs to conserve chip area, since the resistance values required are large, and the physical size of the IC resistor is directly proportional to its value of resistance. The MOST resistance value is made large by reducing the channel width. Thus, less chip area is necessary.

Figure 5.20a shows an E-O MOST amplifier that uses T_2 as a replacement for the bias resistor R_D. Note that T_2 has a special connection in which the gate and drain are connected together. This connection provides operation of T_2 in the amplifier (or FET saturation) region since $v_{DS_2} = v_{GS_2}$ and therefore $v_{DS_2} > V_{GS_2} - V_T$. Note also

that selecting the load resistor R_L to be quite large eliminates the necessity of a coupling capacitor, which is extremely advantageous because IC capacitors also require large chip area and are therefore not used unless absolutely necessary.

To determine the voltage amplification v_o/v_s for the MOST amplifier, we use the small-signal AC equivalent circuit shown in Fig. 5.20b and rearranged for analysis in Fig. 5.20c. Note that the parallel combination of the two conductances (g_{o_1} and g_{o_2}) provides a resistance value given by

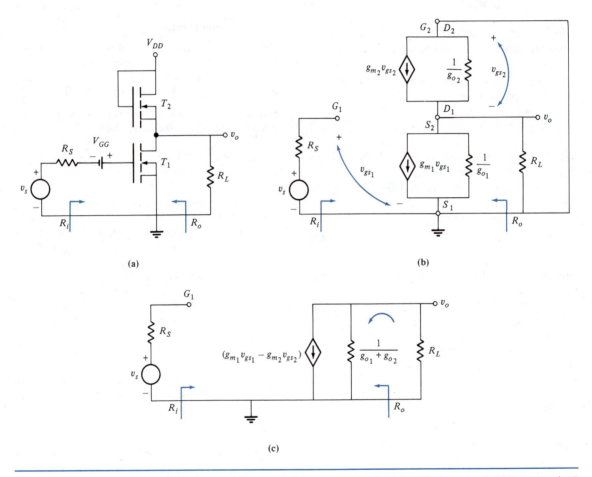

(a)   (b)

(c)

FIGURE 5.20   E-O MOST Amplifier: (a) Amplifier circuit $(V_{GG} + v_s > V_{T_1})$, (b) AC equivalent circuit, (c) Rearranged AC equivalent circuit

$1/(g_{o_1} + g_{o_2})$. The output voltage is obtained from the output loop (using Ohm's law) as

$$v_o = -(g_{m_1}v_{gs_1} - g_{m_2}v_{gs_2})\left[\frac{1}{g_{o_1} + g_{o_2} + (1/R_L)}\right] \quad (5.3\text{--}28)$$

where, from Fig. 5.20b, $v_{gs_1} = v_s$ and $v_{gs_2} = -v_o$. Substitution of these values into (5.3–28) yields

$$v_0 = -(g_{m_1}v_s + g_{m_2}v_o)\left[\frac{1}{g_{o_1} + g_{o_2} + (1/R_L)}\right] \quad (5.3\text{--}29)$$

Solving for the voltage amplification yields

$$\frac{v_o}{v_s} = \frac{-g_{m_1}}{g_{o_1} + g_{o_2} + 1/R_L + g_{m_2}} \quad (5.3\text{--}30)$$

Equation (5.3–30) is the complete expression; however, usual MOST design practice results in $g_{m_2} \gg g_{o_1} + g_{o_2}$, and since R_L is large, (5.3–30) reduces to

$$\frac{v_o}{v_s} = \frac{-g_{m_1}}{g_{m_2}} \quad (5.3\text{--}31)$$

Note that this ratio can be greater than unity because each g_m factor depends upon the physical size of the particular transistors.

The input impedance R_i of the MOST amplifier is very large and depends upon the gate–source impedance, which our model indicates as being infinite. Typical values are greater than 10 MΩ.

The output impedance R_o is obtained from the circuit of Fig. 5.20c by starting with the defining relation

$$R_o = \left.\frac{v_o}{i_o}\right|_{v_s=0} \quad (5.3\text{--}32)$$

An expression for v_o is obtained from the output loop using KVL, which yields

$$v_o = \frac{g_{m_2}v_{gs_2} - g_{m_1}v_{gs_1}}{g_{o_1} + g_{o_2}} + \frac{i_o}{g_{o_1} + g_{o_2}} \quad (5.3\text{--}33)$$

However, from Fig. 5.20b, we observe that $v_{gs_2} = -v_o$. Additionally, from Fig. 5.20c, we see that $v_{gs_1} = v_s$ and that by using the condition that $v_s = 0$, $v_{gs_1} = 0$. Thus, eliminating v_{gs_1} and v_{gs_2} from (5.3–33), solving for v_o yields

$$\left. v_o \right|_{v_s=0} = \frac{i_o}{g_{o_1} + g_{o_2} + g_{m_2}} \quad (5.3\text{--}34)$$

Finally, by comparing (5.3–34) with (5.3–32), we have

$$R_o = \frac{1}{g_{o_1} + g_{o_2} + g_{m_2}} \quad (5.3\text{--}35)$$

_____ **EX. 5.11**

MOST Amplifier Example

Calculate v_0/v_s, R_i, and R_o for the MOST amplifier circuit of Fig. 5.20a. Let $R_L = 10$ kΩ and $1/g_{o1} = 1/g_{o2} = 50$ kΩ, while $g_{m1} = 30$ mS and $g_{m2} = 1$ mS.

Solution: The voltage amplification is given by (5.3–30), and substituting values yields

$$\frac{v_o}{v_s} = \frac{-30}{2/50 + 1/10 + 1} = -26.3$$

Note that the approximate expression given by (5.3–31) yields $v_o/v_s = 30$, which is about 10% in error.

From Figs. 5.20b or c, the input impedance is seen to be ∞. The output impedance is calculated from (5.3–35), and substituting values yields

$$R_o = \frac{1}{2/50 + 1} = 0.961 \text{ kΩ}$$

_____

5.4   POWER AND EFFICIENCY FOR TRANSISTOR AMPLIFIERS

As we have seen, transistor amplifiers are used to amplify AC signals in the form of either a voltage or a current or both. As this amplification occurs, the AC power is also amplified. The additional AC power at the output is obtained through an energy conversion process in which DC power supplied by the DC source is converted into AC power.

To simplify the discussion of this conversion process, we will consider the common-emitter amplifier circuit shown in Fig. 5.21. Other types of amplifiers behave in a similar fashion. Note that we have neglected the resistance of the signal source and also used the collector bias resistor as the load resistor (comparing Fig. 5.21 to Fig. 5.13a). Thus, we have two less resistors to account for, as far as power dissipation is concerned. The effects of nonzero R_S and R_C would be the same as that of R_E, namely, to reduce the AC power delivered to R_L. We will see that the efficiency of the conversion process is less than ideal because some of the power supplied is dissipated in the remaining resistors and, in particular, the transistor.

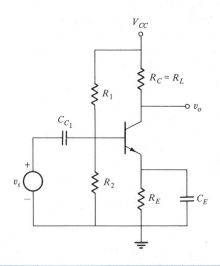

FIGURE 5.21   Simplified Common-Emitter Amplifier

5.4.1   Power Balance in Amplifier Circuits

In an amplifier circuit, we can intuitively state that the total power supplied by the sources must equal the sum of the power in all of the resistors plus the power dissipated in the transistor. Moreover, in an amplifier circuit, as we have seen, the currents and voltages are composed of DC and AC components. The DC components are just the Q point values, and we will assume that the AC components vary sinusoidally (as $\sin \omega t$). Each element in the circuit will therefore have an associated time-varying power, or *instantaneous power p*. We define the instantaneous power quantities for the circuit of Fig. 5.21 as follows:

- p_{CC} is the power supplied by the DC source.
- p_{R_E} is the power dissipated in R_E.
- p_D is the power dissipated in the transistor.
- p_{R_L} is the power delivered to R_L.

Each of these quantities varies with t, and we have intentionally not listed the AC input power or the power dissipated in R_1 and R_2. The current and voltage magnitudes on the input side of the amplifier are both much smaller than those on the output side; therefore, the power associated with these elements is neglected. Also, the power dissipated in the transistor will consist of only the output power dissipation. Thus, a power balance equation can be written as follows:

$$p_{CC} = p_{R_E} + p_D + p_{R_L} \qquad (5.4\text{–}1)$$

By substituting for each of these quantities in terms of the currents and voltages in the amplifier circuit, we have

$$i_C V_{CC} = i_E^2 R_E + i_C v_{CE} + i_C^2 R_L \qquad (5.4\text{–}2)$$

Canceling i_C throughout and using $i_C \simeq i_E$ yield

$$V_{CC} = i_E R_E + v_{CE} + i_C R_L \qquad (5.4\text{–}3)$$

which we immediately recognize as KVL for the output branch. Thus, our intuition concerning power balance is verified.

5.4.2   Efficiency of Amplifier Circuits

Basic Definitions of Efficiency and Average Power. The *efficiency* of an amplifier circuit is defined, in general, as the ratio of the useful AC output power to the total input power. For specific amplifiers, the definition is approximated as the ratio of AC output power delivered to a load resistor to the power supplied by the DC source. Both of these quantities vary with time since in each case the current varies with time. It is therefore customary to define the efficiency in terms of the ratio of the time average of these power quantities. Using η as the symbol for efficiency, the basic definition is then given by

$$\eta = \frac{\text{average AC power delivered to } R_L}{\text{average power supplied by } V_{CC}}$$

$$(5.4-4)$$

In order to develop an explicit expression for the efficiency, we consider the basic definition of average power. This definition involves averaging over time, where the *average power* P is defined in terms of the instantaneous power p as an integral over time as follows:

$$P = \frac{1}{T} \int_0^T p\, dt$$

$$(5.4-5)$$

where T is the period of time over which the average is being taken. The averaging process can be applied directly to (5.4–1), yielding

$$\frac{1}{T} \int_0^T p_{CC} dt = \frac{1}{T} \int_0^T p_{R_E} dt + \frac{1}{T} \int_0^T p_D\, dt$$

$$+ \frac{1}{T} \int_0^T p_{R_L} dt$$

$$(5.4-6)$$

or, in terms of power averages,

$$P_{CC} = P_{R_E} + P_D + P_{R_L}$$

$$(5.4-7)$$

In order to proceed further, we substitute the assumption that the currents and voltages in the amplifier vary sinusoidally. Most importantly, the collector current is written as

$$i_C = I_C + i_c = I_C + I_{C_P} \sin \omega t$$

$$(5.4-8)$$

where I_{C_P} is the peak value of the sinusoid. With this assumption, we now consider each of the terms in (5.4–7) individually, taking T to be the period of the sinusoid ($T = 2\pi/\omega$)

The average power supplied by V_{CC} is obtained by making appropriate substitutions as follows:

$$P_{CC} = \frac{1}{T} \int_0^T p_{CC} dt = \frac{1}{T} \int_0^T i_C V_{CC} dt$$

$$= \frac{1}{T} \int_0^T (I_C + i_c) V_{CC} dt$$

$$(5.4-9)$$

However, since i_c is sinusoidal, integrating (5.4–9) yields the expression for average power supplied as

$$P_{CC} = I_C V_{CC}$$

$$(5.4-10)$$

where we used the fact that the average of a sinusoid over one period is zero.

The average power dissipated in R_E is obtained by making appropriate substitutions as follows:

$$P_{R_E} = \frac{1}{T} \int_0^T p_{R_E} dt = \frac{1}{T} \int_0^T I_E^2 R_E dt \quad (5.4-11)$$

Integrating (5.4–11) then yields (using $I_C \simeq I_E$)

$$P_{R_E} \simeq I_C^2 R_E$$

$$(5.4-12)$$

Here, the power dissipated in R_E is purely DC because of the bypass capacitor C_E, and hence,

(5.4–12) indicates only DC power dissipated in R_B. However, if C_E had been omitted, a second term representing AC power dissipation in R_E results and the presence of R_E in the circuit reduces the AC power delivered to R_I.

Similarly, by appropriate substitutions, the average power dissipated in the transistor is found to be

$$P_D = \frac{1}{T}\int_0^T p_D dt = \frac{1}{T}\int_0^T v_{CE} i_C dt$$

$$= \frac{1}{T}\int_0^T (V_{CC} - i_C R_C - i_E R_E) i_C dt \quad \textbf{(5.4–13)}$$

and the average power delivered to R_L is

$$P_{R_L} = \frac{1}{T}\int_0^T p_{R_L} dt = \frac{1}{T}\int_0^T i_C^2 R_L dt \quad \textbf{(5.4–14)}$$

Note that (5.4–13) is just the power balance equation of the previous section.

Substituting (5.4–8) into (5.4–14) and carrying out the averaging process yields

$$P_{R_L} = I_C^2 R_L + \frac{1}{2}(I_{C_P}^2 R_L) \quad \textbf{(5.4–15)}$$

where we have used the additional fact that the average of a sinusoid squared is 1/2. Note that (5.4–15) consists of the sum of a DC and an AC term, where the last term is the AC component.

Finally, from the previous definition of efficiency given by (5.4–4), by substitution of (5.4–10) and the AC portion of (5.4–15), we have

$$\eta = \frac{\frac{1}{2}(I_{C_P}^2 R_L)}{I_C V_{CC}} \quad \textbf{(5.4–16)}$$

Although not immediately evident, (5.4–16) is a fraction (less than unity).

Maximum Efficiency for *RC* Amplifiers.   In order to maximize η, (5.4–16) indicates that I_{C_P} (the

peak value of i_c) should be maximized. Recall from earlier considerations in this chapter that this condition prevails for maximum symmetrical variation in collector current and voltage. For MSV, $I_C = I_{C_P}$; thus, the maximum efficiency from (5.4–16) is

$$\eta_{MAX} = \frac{1}{2}\left(\frac{I_C^2 R_L}{I_C V_{CC}}\right) = \frac{1}{2}\left(\frac{I_C R_L}{V_{CC}}\right) \quad \textbf{(5.4–17)}$$

The collector current under MSV conditions is given generally by (5.1–10), or

$$I_C = \frac{V_{CC}}{R_{ACO} + R_{DCO}} \quad \textbf{(5.4–18)}$$

Thus, if we consider the simplified common-emitter amplifier circuit of Fig. 5.21 and substitute for R_{ACO} and R_{DCO}, we obtain

$$I_C = \frac{V_{CC}}{2R_L + R_E} \quad \textbf{(5.4–19)}$$

Substitution of (5.4–19) into (5.4–17) then yields

$$\eta_{MAX} = \frac{1}{2}\left(\frac{V_{CC}}{2R_L + R_E}\right)\left(\frac{R_L}{V_{CC}}\right)$$

$$= \frac{R_L}{2(2R_L + R_E)} \quad \textbf{(5.4–20)}$$

From (5.4–20), we observe that the presence of nonzero R_E reduces the efficiency due to DC power dissipation in R_E. However, if the value of R_E is negligible with respect to that of R_L, then the maximum efficiency expression of (5.4–20) becomes

$$\eta_{MAX} = \frac{1}{4} = 25\% \quad \textbf{(5.4–21)}$$

We therefore observe that the maximum efficiency of the common-emitter *RC* amplifier is 25%.

Similar results are obtained for other *RC* amplifiers. In the case of power amplifiers, this magnitude of efficiency represents a 75% loss, which is highly undesirable. The following section describes methods of improving the efficiency.

Improvement in Efficiency Using Inductive Elements.   Figure 5.22a displays another common-emitter amplifier that uses an inductor L as a replacement for R_C and connects the load resistor R_L through a coupling capacitor C_{C_2}. Under these conditions and assuming MSV, we have

$$I_C = \frac{V_{CC}}{R_{ACO} + R_{DCO}} = \frac{V_{CC}}{R_L + R_E} \qquad (5.4\text{–}22)$$

Substitution of (5.4–22) into (5.4–17) then yields

$$\eta_{\text{MAX}} = \frac{1}{2}\left(\frac{V_{CC}}{R_L + R_E}\right)\left(\frac{R_L}{V_{CC}}\right) = \frac{R_L}{2(R_L + R_E)} \qquad (5.4\text{–}23)$$

However, if $R_E \ll R_L$, this equation becomes

$$\eta_{\text{MAX}} = \frac{1}{2} = 50\% \qquad (5.4\text{–}24)$$

Hence, the maximum efficiency is essentially doubled.

The improvement in efficiency is accomplished in the circuit of Fig. 5.22a by eliminating the DC power dissipated in R_L. It is of interest to observe this improvement graphically using the AC and DC load lines as shown in Fig. 5.22b. We observe that since R_E is small, the DCLL is nearly vertical with $V_{CE} \simeq V_{CC}$. Furthermore, since MSV exists, Q is in the middle of the ACLL, and the voltage intercept is therefore approximately located at $2V_{CC}$, with the current intercept at $2I_C$. Thus, setting up an equation for the slope of the ACLL, we have

$$\text{slope} = -\frac{1}{R_L} = \frac{2I_C - 0}{0 - 2V_{CC}} \qquad (5.4\text{–}25)$$

or, by solving for I_C, we obtain

$$I_C = \frac{V_{CC}}{R_L} \qquad (5.4\text{–}26)$$

Direct substitution of (5.4–26) into (5.4–17) again yields $\eta_{\text{MAX}} = 50\%$.

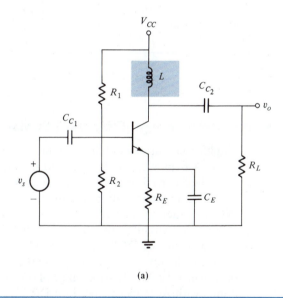

(a)

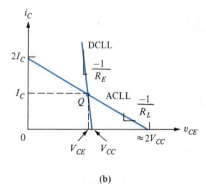

(b)

FIGURE 5.22   Common-Emitter Amplifier Using an Inductor L: (a) Amplifier circuit, (b) Graphical representation

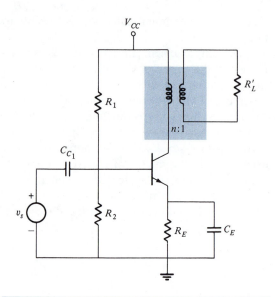

FIGURE 5.23   Transformer Coupled Common-Emitter Amplifier

A similar result is achieved using transformer coupling of the load resistor as shown in Fig. 5.23. For DC, the transformer acts like a short circuit, and $R_{DCO} = R_E$, as before. For AC, the transformer can be replaced with an equivalent AC resistance, given by $R_L = n^2 R_l'$. The circuit is then identical to the previous case. For MSV, the DC

collector current is then given by (5.4–22), and the maximum efficiency by (5.4–23), where $R_L = n^2 R_l'$ and for $R_E \ll R_L$, η_{MAX} is again 50%.

Comparison of Efficiency Example

Calculate the maximum efficiency of the circuits in Figs. 5.21 and 5.22a and compare the results. Use the following element values: $R_L = 0.8$ kΩ and $R_E = 0.2$ kΩ. Assume that the AC currents and voltages are sinusoidal and that MSV conditions prevail.

Solution: The efficiency of the RC amplifier of Fig. 5.21 is calculated directly from (5.4–20) as follows:

$$\eta_{MAX} = \frac{R_L}{2(2R_L + R_E)} = \frac{0.8}{2(1.6 + 0.2)} = 22.2\% \qquad (1)$$

Note that this value is less than the ideal amount of 25%.
The efficiency of the inductively coupled RC amplifier is calculated from (5.4–23) as follows:

$$\eta_{MAX} = \frac{R_L}{2(R_L + R_E)} = \frac{0.8}{2(0.8 + 0.2)} = 40\% \qquad (2)$$

Note that the inductively coupled amplifier offers a marked improvement in efficiency and that this value is less than the ideal amount of 50%.

CHAPTER 5 SUMMARY

- The large-signal model for a BJT consists of an ideal diode in series with the offset voltage V_0 for the base–emitter terminals and a current source in parallel with an ideal diode for the collector–emitter terminals. Large-signal models are not used for FETs because of inherent nonlinear operation.
- In discrete circuits, bypass and coupling capacitors are used. These capacitors act like short circuits to AC and open circuits to DC except at low frequency. Coupling capacitors are used to couple the AC signal into and out of the

amplifier circuit without altering the DC operation. A bypass capacitor is placed in parallel with the feedback resistor R_F to eliminate the reduction in amplification due to R_F.
- When capacitors and/or inductors are used in amplifier circuits, two different loads lines are obtained, the DCLL and the ACLL. The DCLL is the range of operating points for DC only, while the ACLL is the range of operating points with AC superimposed on DC.
- Maximum symmetrical variation (MSV) in output current and voltage is achieved when

Q is in the middle of the ACLL (neglecting the saturation region). MSV occurs when the output DC current and voltage are given, respectively, by $I_O = V_{DCO}/(R_{ACO} + R_{DCO})$ and $V_{OC} = V_{DCO}R_{ACO}/(R_{ACO} + R_{DCO})$. R_{ACO} and R_{DCO} are the output loop AC and DC resistances, respectively; V_{DCO} is the DC source voltage for the output loop. For the various amplifier circuits, R_{ACO} and R_{DCO} are obtained by inspection.

■ Small-signal AC models for transistors are used to determine the AC parameters of amplifier circuits. These parameters include voltage amplification, current amplification, input impedance, and output impedance.

■ For the BJT in the common-emitter or common-collector configuration, the hybrid model with input resistor h_{ie}, output conductance h_{oe}, and current source $h_{fe}i_b$ is used. In the common-base configuration, the corresponding hybrid model is used with h_{ib}, h_{ob}, and $h_{fb}i_b$, where $h_{fb} = -\alpha$.

■ The hybrid parameters, or h-parameters, can be determined graphically as well as experimentally. The input AC resistance is given by the expression $h_{ie} = 0.025/I_B$ for the common-emitter case and $h_{ib} = 0.025/I_E$ for the common-base case.

■ Small-signal AC analysis of an amplifier circuit involves substituting the AC equivalent for each element in the circuit. It includes substituting the small-signal model for the transistor, a short circuit for each of the capacitors, open circuits for the inductors, short circuits for DC voltages, open circuits for DC currents, and resistors unchanged. The resulting circuit is then analyzed.

■ The results for the BJT small-signal amplifier analysis are summarized in Table 5.1.

■ For FETs, the small-signal AC model consists of an open-circuited input and an output conductance g_o in parallel with a current source $g_m v_{gs}$. Values for the conductance parameters g_o and g_m are obtained graphically or experimentally.

■ The results from the small-signal analysis of the common-source amplifier show that the voltage gain is moderate with infinite input impedance and moderate output impedance. For the source follower or common-drain amplifier, similar results are obtained except that the voltage amplification is less than 1.

■ For the MOST IC amplifier, resistors are replaced with MOSTs. The AC small-signal analysis indicates that moderate voltage amplification, large input impedance, and moderate output impedance are obtained.

■ The power balance equation for amplifier circuits states that the power supplied by the DC voltage source is equal to the power dissipated in the external resistors plus the power dissipated in the transistor, where the AC and DC power on the input side of the transistor are typically quite small and are neglected.

■ The efficiency of an amplifier circuit is defined as the average AC power delivered to the load resistor R_L divided by the average power supplied by the power supply. For the common-emitter amplifier, the efficiency is obtained as $\eta = I_{C_P}^2 R_L/2I_C V_{CC}$ where AC currents and voltages are assumed to be sinusoidal and I_{C_P} is the peak value of the collector current. Under MSV conditions, $I_{C_P} = I_C$ and $\eta_{MAX} = 25\%$. By using inductive elements, the efficiency can be improved to 50%.

TABLE 5.1

Amplifier Type	Voltage Amplification	Current Amplification	R_o	R_i
Common-emitter	Good	Good	Moderate	Moderate
Common-collector	<1	Good	Small	Large
Common-base	Good	<1	Large	Small

5.1

For the circuit of Fig. P5.1, sketch the load line and locate the DC operating point Q on the i_C–v_{CE} axes for $\beta = 100$ and $V_0 = 0.7$ V. Determine the maximum possible symmetrical variation in i_C (peak to peak).

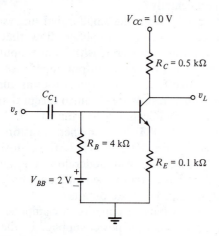

FIGURE P5.1

5.2

Repeat Problem 5.1 for a BJT with $\beta = 50$ and $\beta = 30$.

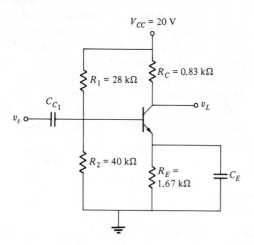

FIGURE P5.4

5.3

For the circuit of Fig. P5.1, the element values are changed to $R_B = 1$ kΩ, $V_{BB} = 1$ V, $R_C = 1$ kΩ, $V_{CC} = 6$ V, and $R_E = 0.1$ kΩ. Show that this circuit has been designed for maximum symmetrical variation in i_c for $\beta = 100$.

5.4

The circuit of Fig. P5.4 has been designed for maximum symmetrical variation in i_C. Draw the DCLL and ACLL to verify MSV. Use $\beta = 100$ and $V_0 = 0.7$ V.

5.5

For the circuit of Fig. P5.5, sketch the DCLL and locate Q. Sketch the ACLL and determine the maximum possible symmetrical variation in i_C. Let $\beta = 100$ and $V_0 = 0.7$ V.

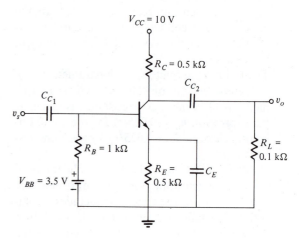

FIGURE P5.5

5.6

The common-collector amplifier circuit of Fig. P5.6 uses a BJT with $\beta = 100$ and $V_0 = 0.7$ V. Sketch the DCLL, locate the Q point, and sketch the ACLL. Determine the maximum possible symmetrical variation in i_C (peak to peak).

5.7

The circuit of Fig. P5.7 has been designed for MSV. Determine I_C, V_{CE}, and the peak-to-peak MSV for i_C.

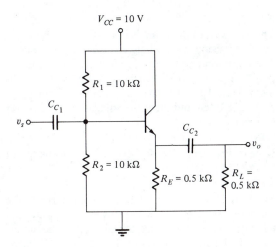

FIGURE P5.6

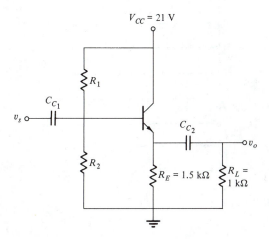

FIGURE P5.7

5.8

Determine R_1 and R_2 in the amplifier circuit of Fig. P5.7 that provide MSV. Use the rule-of-thumb expression and first obtain R_B and V_{BB}. Let $\beta = 100$, $V_0 = 0.7$ V, and use I_C from Problem 5.7.

5.9

For the amplifier circuit of Fig. P5.9, determine I_C and V_{CE} for MSV. Determine the values of R_B and V_{BB} that provide MSV. Sketch the ACLL and DCLL and locate Q. Use $\beta = 100$ and $V_0 = 0.7$ V.

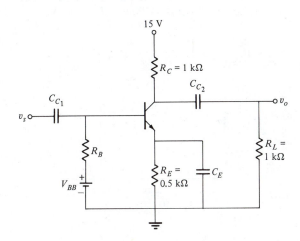

FIGURE P5.9

5.10

The common-emitter amplifier circuit of Fig. P5.10 uses a transistor with $\beta = 100$ and $V_0 = 0.7$ V. Design the overall circuit for MSV. Sketch the DCLL and ACLL and determine R_1 and R_2.

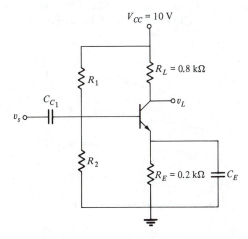

FIGURE P5.10

5.11

For the common-base amplifier of Fig. P5.11, sketch the DCLL, the ACLL, and locate Q on the i_C–v_{CB} axes. Determine the maximum possible variation in i_C (peak to peak). Let $\beta = 100$ and $V_0 = 0.7$ V.

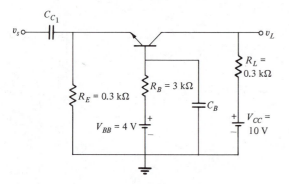

FIGURE P5.11

5.12

Repeat Problem 5.5 for the circuit of Fig. P5.5 with $R_C = 0.2$ kΩ, $R_L = 0.2$ kΩ, $R_E = 1$ kΩ, $R_B = 2$ kΩ, $V_{CC} = 20$ V, and $V_{BB} = 10$ V.

5.13

Repeat Problem 5.10 for a high-power transistor with $V_{CC} = 10$ V, $R_C = 150$ Ω, and $R_E = 100$ Ω. Use $\beta = 100$ and $V_0 = 0.7$ V.

5.14

For the circuit of Fig. P5.5, $R_C = 1$ kΩ, $R_E = 0.5$ kΩ, $R_L = 1$ kΩ, $V_{CC} = 10$ V, $\beta = 100$, and $V_0 = 0.7$ V. Determine R_B and V_{BB} for MSV. Determine Q for this condition and draw the ACLL and DCLL.

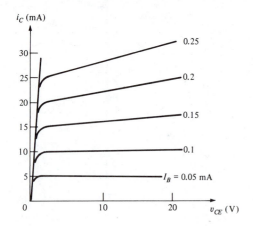

FIGURE P5.15

5.15

The AC variation of an NPN BJT is to be characterized using a small-signal h-parameter model. The common-emitter i_C–v_{CE} characteristics are displayed in Fig. P5.15. Determine h_{fe} and $1/h_{oe}$ if Q is located at (10 V, 10 mA). Calculate h_{ie} from $h_{ie} = .025/I_B$. Sketch and label the small-signal transistor equivalent circuit.

5.16

Repeat Problem 5.15 for Q located at the point (10 V, 22.5 mA).

5.17

In a manner similar to Problem 5.15, determine the common-base h-parameters from the i_C–v_{CB} characteristics of Fig. P5.17. Sketch and label the small-signal BJT equivalent circuit. The Q point is located at (5 V, 5 mA) and $h_{ib} = .025/I_E$.

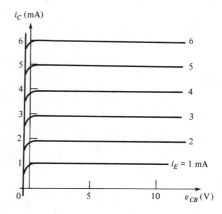

FIGURE P5.17

5.18

Convert the h-parameters for the common-emitter model of Problem 5.15 to the common-base h-parameters and draw the small-signal model.

5.19

Figure P5.19a displays the basic common-emitter voltage amplifier configuration, and Fig. P5.19b shows the equivalent circuit for analysis. Determine the voltage amplification by using a small-signal circuit model for the case $h_{fe} = 100$, $V_0 = 0.7$ V, $R_C = 5$ kΩ, $R_E = 3$ kΩ, $R_L = \infty$, $R_1 = 30$ kΩ, $R_2 = 15$ kΩ, $R_S = 0$, and $V_{CC} = 20$ V. Verify that $h_{ie} \cong 1.25$ kΩ and let $h_{oe} = 0$. Begin by drawing the small-signal circuit model.

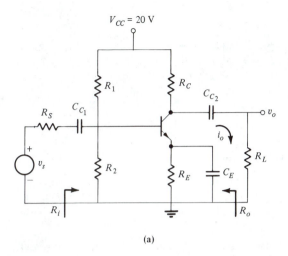

(a)

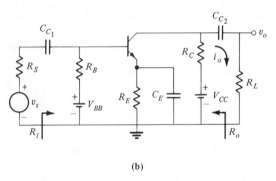

(b)

FIGURE P5.19

5.20
Repeat Problem 5.19 with $R_S = 1$ kΩ.

5.21
Repeat Problem 5.19 with $R_C = 1$ kΩ, $R_E = 2$ kΩ, $R_L = \infty$, $R_1 = 20$ kΩ, $R_2 = 20$ kΩ, $R_S = 5$ kΩ, and $V_{CC} = 10$ V. Verify that $I_B = 0.02$ mA.

5.22
For the basic voltage amplifier circuit of Fig. P5.19b, $R_C = 2.2$ kΩ, $R_E = 1.2$ kΩ, $R_L = 2.2$ kΩ, and $R_B = 200$ kΩ. For $h_{fe} = 100$, $h_{ie} = 2.35$ kΩ, and $h_{oe} = 0$, determine the voltage amplification v_o/v_s, the input impedance R_i, and the output impedance R_o. Draw the small-signal equivalent circuit and let $R_S = 1$ kΩ.

5.23
For the circuit of Fig. P5.19b, determine the current amplification i_o/i_s by replacing the series combination of v_s and R_S with the parallel combination of

i_s and R_S. Use the small-signal equivalent circuit of Problem 5.22.

5.24
For the circuit of Fig. P5.19, replace R_S and v_s with a parallel combination of R_S and i_s. Determine the input impedance R_i, the output impedance R_o, and the current amplification i_o/i_s. Use the values $R_S = 10$ kΩ, $R_1 = 50$ kΩ, $R_2 = 10$ kΩ, $R_C = 3.8$ kΩ, and $R_L = 1$ kΩ, with $h_{fe} = 100$ and $h_{ie} = 0.89$ kΩ. Begin by drawing the small-signal equivalent circuit.

5.25
Repeat Problem 5.24 with $R_S = 100$ Ω. Note the drastic reduction in current amplification.

5.26
Rework Problem 5.24 with element values of $R_S = 10$ kΩ, $R_1 = 40$ kΩ, $R_2 = 13.33$ kΩ, $R_C = 1.2$ kΩ, $R_L = 0.47$ kΩ, $h_{fe} = 100$ and $h_{ie} = 2.1$ kΩ.

5.27
Use element values of $R_C = R_S = R_1 = R_2 = R_L = 20$ kΩ, $\beta = 100$, and $h_{ie} = 1$ kΩ in the circuit of Fig. P5.19a. Determine R_i, R_o, and v_o/v_s using a small-signal equivalent circuit.

5.28
Repeat Problem 5.27 with $R_S = 100$ kΩ. Note the drastic reduction in voltage amplification.

5.29
For the amplifier circuit of Fig. P5.29, determine the voltage amplification v_o/v_s, R_i, and R_o. Use $h_{fe} = 100$ and $h_{ie} = 0.87$ kΩ and draw the small-signal equivalent circuit.

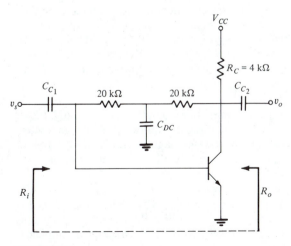

FIGURE P5.29

5.30

For the common-collector amplifier circuit of Fig. P5.30, use an AC small-signal equivalent circuit to determine R_i, R_o and i_o/i_s. Let $R_s = \infty$, $\beta = 100$, and $h_{ie} = 2$ kΩ.

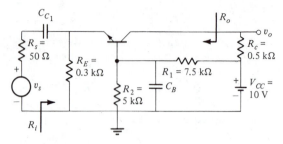

FIGURE P5.34

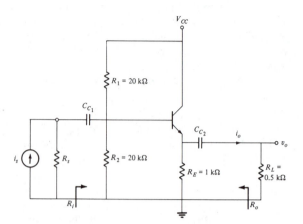

FIGURE P5.30

5.31

Repeat Problem 5.30 for $R_s = 1$ kΩ.

5.32

Determine the voltage amplification v_o/v_s for the element values of Problem 5.30 and the circuit of Fig. P5.30 by replacing i_s with v_s.

5.33

Determine the current and voltage amplification for the circuit of Fig. P5.30 using the following element values: $R_S = \infty$ or 0, $R_1 = R_2 = 200$ kΩ, $R_E = 1$ kΩ, and $R_L = 1$ kΩ, with $h_{fe} = 100$ and $V_0 = 0.7$ V. Also determine the input and output impedances, R_i and R_o.

5.34

For the common-base amplifier circuit of Fig. P5.34, draw the small-signal equivalent circuit and determine the following quantities: (a) R_i, (b) R_o, and (c) v_o/v_s. Use $h_{fe} = 100$, $V_0 = 0.7$ V and calculate h_{fb} and h_{ib}.

5.35

For the circuit of Fig. P5.34, assume that C_B is disconnected and determine the resulting voltage amplification v_o/v_s. Note: The reduction in amplifica-

tion is not as severe as the removal of C_E in the common-emitter amplifier.

5.36

For the circuit of Fig. P5.36, draw the small-signal equivalent circuit and determine the voltage amplification v_o/v_s. Use $h_{fe} \gg 1$ and $h_{ib} = 5$ Ω.

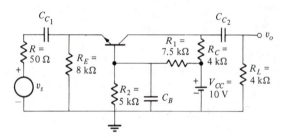

FIGURE P5.36

5.37

An E-D N-channel MOST has a current–voltage transfer characteristic given by $i_D = 5(1 + v_{GS}/2)^2$ mA. For Q located at $V_{GS} = 4$ V, determine g_m and g_o.

5.38

Figure P5.38a displays a typical set of i_{DS} versus v_{DS} characteristics for an N-channel E-O MOST. Determine g_m and g_o for this transistor for Q located at $V_{DS} = 10$ V and $V_{GS} = 3$ V. Use these parameters in the circuit of Fig. P5.38b to determine the voltage amplification v_o/v_s. Use a small-signal equivalent circuit.

5.39

A JFET has a current–voltage transfer characteristic given by $i_D = 1.65[1 + (v_{GS}/4)]^2$ mA. Determine g_m

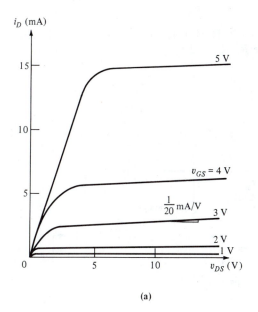

(a)

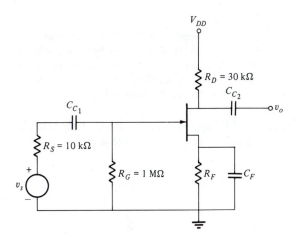

FIGURE P5.39

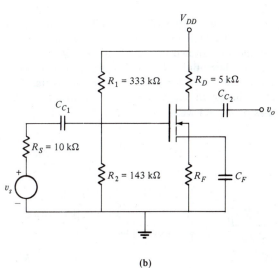

(b)

FIGURE P5.38

and g_o if Q is located at $V_{GS} = -2$ V. Use these parameters to determine the voltage amplification in the circuit of Fig. P5.39.

5.40

For the JFET amplifier of Fig. P5.40, determine R_i, R_o, and v_o/v_s. Let $g_m = 100$ mS and $1/g_o = 10$ kΩ. Sketch the small-signal equivalent circuit.

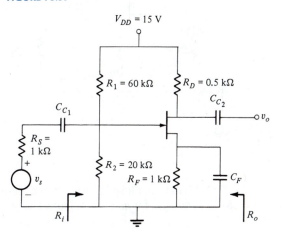

FIGURE P5.40

5.41

For the JFET amplifier of Fig. P5.41, the bias on the gate is supplied by R_F (self-bias). Using a small-signal equivalent circuit, determine R_i, R_o, and v_0/v_s. Use $g_m = 10$ mS and $1/g_0 = 10$ kΩ.

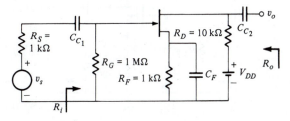

FIGURE P5.41

5.42

For the JFET amplifier of Fig. P5.41, consider that the bias resistor R_F has not been bypassed with a capacitor. Determine the reduced voltage amplification using a modified small-signal equivalent circuit. Use the parameters of Problem 5.41.

5.43

For the JFET amplifier of Fig. P5.43, determine R_i, R_o, and v_o/v_s. Use $g_m = 8$ mS and $g_0 = 0$.

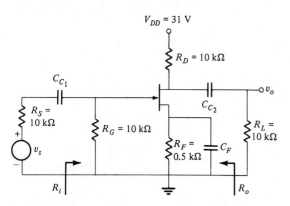

FIGURE P5.43

5.44

For the E-O MOST amplifier of Fig. P5.44, determine R_i, R_o, and v_o/v_s for $g_m = 15$ mS and $1/g_o = 100$ kΩ.

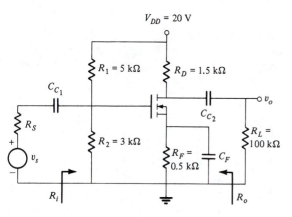

FIGURE P5.44

5.45

For the E-D MOST amplifier of Fig. P5.45, determine R_i, R_o, and v_o/v_s for $g_m = 100$ mS and $g_o = 0$.

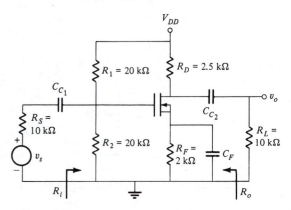

FIGURE P5.45

5.46

The input impedance of the amplifier of Fig. P5.45 can be increased (along with v_o/v_s) by the circuit modification shown in Fig. P5.46. Verify this statement by determining R_i and v_o/v_s from a small-signal equivalent circuit for that of Fig. P5.46. Note: The only change in the circuit is the inclusion of the resistor $R_3(=1$ MΩ$)$.

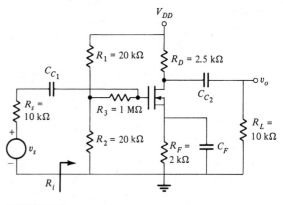

FIGURE P5.46

5.47

Sketch the small-signal equivalent circuit for the MOST amplifier of Fig. P5.47 and determine R_i, R_0,

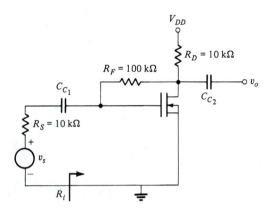

FIGURE P5.47

and v_0/v_s for $g_m = 8$ mS and $1/g_0 = 20$ kΩ. Neglect the current through R_F.

5.48
For the circuit of Fig. P5.48, determine the maximum possible collector current swing and calculate the efficiency. Note that Q is not in the middle of the ACLL and neglect $v_{CE\,SAT}$. Also, let $\beta = 100$ and $V_0 = 0.7$ V.

5.49
Consider the E-O MOST amplifier shown in Fig. P5.49. Determine the average power supplied (P_{DD}), the average AC power delivered to R_L, and the efficiency. Let $I_D = 10$ mA and $i_d = 5 \sin \omega t$ mA.

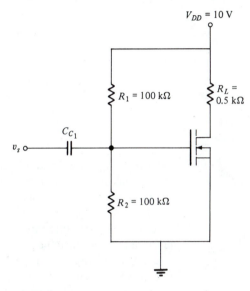

FIGURE P5.49

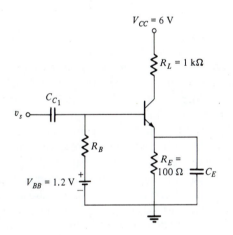

FIGURE P5.48

6

MULTIPLE-TRANSISTOR AMPLIFIER CIRCUITS

In order to increase the amplification provided by the one-transistor amplifiers described in previous chapters, two or more amplifiers are often interconnected. Various methods are used, as well as a variety of different amplifier circuits and/or connections.

In this chapter, we first consider multistage amplifiers in which the output of one amplifier is fed directly into the input of the next. This method of connecting amplifiers in cascade to increase the amplification is the technique used in discrete circuits in which coupling and bypass capacitors as well as inductors are used. These frequency-dependent elements are not used, however, in direct-coupled amplifier circuits. This chapter will describe capacitive, inductive, and direct-coupled amplifiers and explain their inadequacies.

Another amplifier that is extremely important and that is emphasized in this chapter is the difference amplifier, which was introduced in Chapter 4. This direct-coupled amplifier is particularly suitable for use in integrated circuits. In fact, it is a primary building block of linear (analog) integrated circuits.

As we will see, the difference amplifier is the first stage used in an operational amplifier, or op amp for short. The op amp is a very important integrated circuit that behaves essentially like a difference amplifier. It has many important applications, which will be described in Chapter 7. Succeeding stages in the op amp provide additional amplification and sometimes other necessary signal conditioning. Each stage is direct coupled, and capacitors and inductors are not used so that the overall circuitry can be implemented in integrated-circuit form. This chapter will describe the makeup of op amps in general and analyze a basic op amp circuit.

6.1   MULTISTAGE AMPLIFIERS

In *multistage amplifiers* using capacitive coupling, the bias circuit for each individual amplifier stage is designed separately. This design is possible since coupling capacitors are used between each stage and block any DC current. The techniques of Chapter 4 can thus be applied directly to each stage. However, at low frequencies, the coupling capacitors do not act like short circuits to AC, and direct-coupled amplifiers are sometimes used. As we will see, these amplifiers amplify AC as well as DC since each stage is not DC isolated.

6.1.1   Capacitively Coupled Amplifiers

Figure 6.1a displays a *capacitively coupled* two-stage BJT voltage amplifier that we will use as an example. Each stage consists of an identical common-emitter amplifier. As previously mentioned, the DC analysis and design of Chapter 4 apply directly to each stage.

The AC small-signal analysis is carried out using the small-signal model of Fig. 6.1b (which assumes $T_1 = T_2$). This AC equivalent circuit is obtained using the methods of Chapter 5.

We determine the AC voltage amplification v_o/v_s in a manner similar to that of Chapter 5.

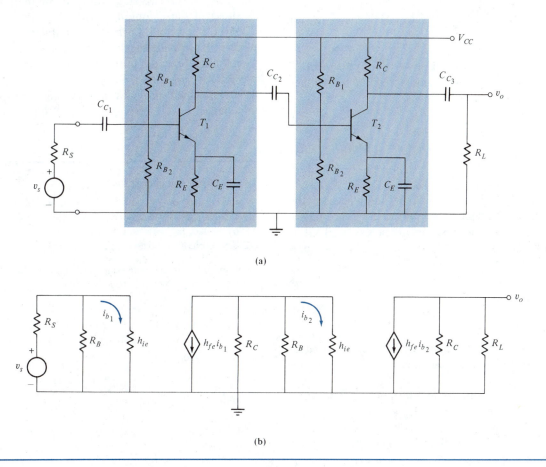

(a)

(b)

FIGURE 6.1   Capacitively Coupled Amplifier: (a) Two-stage BJT amplifier circuit, (b) Small-signal equivalent circuit

By considering the output loop and writing an expression for v_o, we have

$$v_o = -(h_{fe}i_{b_2})R_C \| R_L \qquad (6.1\text{-}1)$$

Next, i_{b_2} is obtained from the inner loop by using the current divider rule, which yields

$$i_{b_2} = -(h_{fe}i_{b_1}) \frac{R_B \| R_C}{R_B \| R_C + h_{ie}} \qquad (6.1\text{-}2)$$

From the input loop, using a Thévenin equivalent circuit for R_S, R_B, and v_s and KVL, we obtain

$$i_{b_1} = \frac{[R_B/(R_S + R_B)]v_s}{R_S \| R_B + h_{ie}} \qquad (6.1\text{-}3)$$

Finally, back substitution of (6.1-3) into (6.1-2) into (6.1-1) provides the expression for voltage amplification as follows:

$$\frac{v_o}{v_s} = h_{fe}^2(R_C \| R_L) \frac{R_B \| R_C[R_B/(R_S + R_B)]}{(R_B \| R_C + h_{ie})(R_S \| R_B + h_{ie})}$$

$$(6.1\text{-}4)$$

To obtain maximum voltage amplification, the same requirements as in Chapter 5 on the relative magnitudes of the resistors are necessary. That is, with $R_C \gg R_L$, $R_B \gg h_{ie}$, $R_C \gg R_B$, $R_B \gg R_S$, and $R_S \ll h_{ie}$, (6.1-4) becomes

$$\frac{v_o}{v_s} = \frac{h_{fe}^2 R_L}{h_{ie}} \qquad (6.1\text{-}5)$$

The important difference in this case, as compared to the single-stage amplifier, is that the voltage amplification depends upon the square of h_{fe}. Hence, much higher amplification is achievable with two stages of amplification. We should also realize that more stages can be added to increase the amplification still further. The only limitation to adding stages is that the output stage of the overall amplifier (or any stage) must not be driven into cutoff or saturation by the amplified signal.

Another observation to make is that (6.1-5) is not obtainable by multiplying together the individual voltage amplification expressions as given in (5.2-40) for each stage. This calculation would result in an amplification of $(h_{fe}R_L/h_{ie})^2$, which is too large by a factor of R_L/h_{ie}. This fact is of no serious consequence, except realizing it does avoid incorrect determination of the voltage amplification. Only in cases where the input impedance of each stage is large (in comparison to the output impedance of the preceeding stage) can we multiply the individual voltage amplification factors together to obtain the overall amplification.

Finally, note that the input and output impedances for this two-stage amplifier are the same as those of the common-emitter amplifier from Chapter 5. To obtain other values, different amplifiers that provide the desired characteristics are used for the input and output stages. For example, the output stage used most often is the emitter follower, which provides a very small output impedance.

— EX. 6.1

Two-Stage BJT Amplifier Example

Determine the voltage amplification v_o/v_s for the circuit of Fig. E6.1a. Use a small-signal equivalent circuit with $T_1 = T_2$, $h_{fe} = 100$, $h_{ie_1} = 1.3 \text{ k}\Omega$, and $h_{ie_2} = 2 \text{ k}\Omega$ (since the DC base currents are different), with $h_{oe_1} = h_{oe_2} = 0$.

Solution: The small-signal equivalent circuit is shown in Fig. E6.1b. Starting at the output, we have

$$v_o = (1 + h_{fe})i_{b_2}R_L \qquad (1)$$

From the middle portion of the circuit and using the current divider rule, we have

$$i_{b_2} = -(h_{fe}i_{b_1}) \frac{R_C \| R_B'}{R_C \| R_B' + h_{ie_2} + (1 + h_{fe})R_L} \qquad (2)$$

Finally, the use of Ohm's law on the input portion of the circuit yields

$$i_{b_1} = \frac{v_s}{h_{ie_1}} \qquad (3)$$

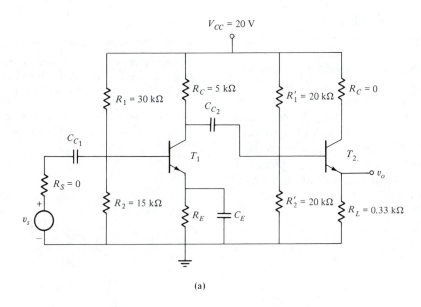

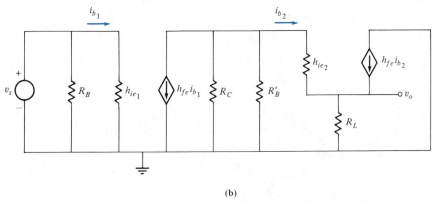

(b)

FIGURE E6.1

Substituting (3) into (2) into (1) and solving for v_o/v_s yield

$$\frac{v_o}{v_s} = \frac{-[(1 + h_{fe})R_L][h_{fe}(R_C \| R_B')]/h_{ie_1}}{R_C \| R_B' + h_{ie_2} + (1 + h_{fe})R_L} \quad (4)$$

Substituting element values, we have

$$\frac{v_o}{v_s} = \frac{-(101)(0.33)(100)(5 \| 20 \| 20)/1.3}{5 \| 20 \| 20 + 2 + (101)(0.33)}$$

$$= \frac{-8537.6}{38.66} \cong -221$$

6.1.2 Coupled Amplifiers Using Inductors

In some instances, *rf chokes* (inductors with high impedance at radio frequencies) are used to replace the collector resistors in the common-emitter amplifier configuration. The effect of the inductors is to block AC and thus prevent the loss of AC power in the collector resistors. At the same time, the inductors behave like short circuits to DC, which allows proper biasing of the transistor. Inductive coupling is used quite often in power amplifiers and in discrete amplifiers in which large physical size is unimportant.

The two-stage amplifier of Fig. 6.1a is converted to an inductively coupled amplifier by replacing each of the collector resistors R_C with inductors L. Then, the effect in the AC equivalent circuit of Fig. 6.1b is to open circuit the collector resistors. This effect ensures the desired large value of R_C, and the same result for amplification as in (6.1–5) prevails. Additionally, the output impedance of the amplifier that previously was R_C becomes infinite.

———————————————— EX. 6.2

Two-Stage Amplifier with Inductors Example

For the amplifier of Fig. E6.2a, determine the voltage amplification v_o/v_s. Use a small-signal equivalent circuit with $T_1 = T_2$, $h_{fe} = 100$, $h_{ie} = 0.4$ kΩ, and $h_{oe} = 0$.

Solution: The small-signal equivalent circuit is shown in Fig. E6.2b with the inductors replaced with open circuits. Starting at the output, we have

$$v_o = -h_{fe}i_{b_2}R_L \tag{1}$$

From the middle portion of the circuit and using the current divider rule, we have

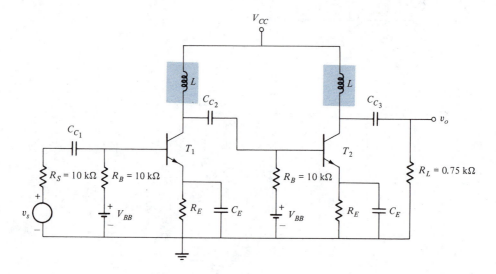

(a)

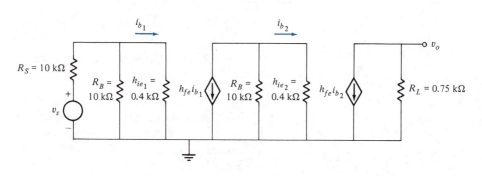

(b)

FIGURE E6.2

$$i_{b_2} = -(h_{fe}i_{b_1})\frac{R_B}{R_B + h_{ie_2}} \qquad (2)$$

Finally, from the input portion of the circuit and using a Thévenin equivalent circuit, we have

$$i_{b_1} = \frac{v_s[R_S/(R_S + R_B)]}{R_S \| R_B + h_{ie_1}} \qquad (3)$$

Substituting (3) into (2) into (1) and solving for v_o/v_s yield

$$\frac{v_o}{v_s} = \frac{h_{fe}^2 R_L[R_B/(R_B + h_{ie_2})][(R_S \| R_B)/R_S]}{R_S \| R_B + h_{ie_1}} \qquad (4)$$

Substituting element values, we have

$$\frac{v_o}{v_s} = \frac{(100)^2(0.75)(10/10.4)(5/10)}{5 + 0.4} \simeq 668$$

6.1.3   Disadvantages of Capacitive and Inductive Coupling

There are grave disadvantages in using coupling and bypass capacitors as well as chokes in amplifier circuits. First, all of these amplifiers exhibit poor performance as the frequency of the signal is reduced; that is, the capacitors do not act like short circuits to AC, and the inductors do not act like open circuits. Second, the physical size of these elements may not be desirable in many applications. Third, the physical size of capacitors and inductors prevents their use in integrated circuits. Direct-coupled amplifiers, a class of multistage amplifiers in which capacitors and inductors are not included, eliminate the problems mentioned. However, direct-coupled amplifiers have a serious associated problem, as we will see.

6.1.4   Direct-Coupled Multistage Amplifiers

To explore the operation of *direct-coupled* amplifiers, we continue to use a two-stage example, but without capacitors. The modified circuit is displayed in Fig. 6.2a, where R_{C_1} aids in

biasing both T_1 and T_2. Note that the resistor values for each stage are different since each stage must have a different Q point since the DC input voltages to each BJT are, in general, different because the DC voltage levels are also amplified by each stage.

To demonstrate the operation of a direct-coupled amplifier, we consider the specific example indicated by the element values shown in Fig. 6.2a. Note that the collector and emitter resistor values also play a role in determining the magnitude of AC voltage amplification v_o/v_s, as we will see. In the analysis for both DC and AC, we will ignore base currents because the impedance (in both cases) seen looking into the base of each transistor is quite large, $\simeq (1 + h_{fe})R_E$. In the analysis, note also that it is not necessary to specify a value for h_{fe} as long as $h_{fe} \gg 1$.

The DC analysis of Fig. 6.2a is initiated by considering the input of T_1 and determining the DC input voltage V_{B_1} produced by the resistive voltage divider. Thus,

$$V_{B_1} = \frac{R_{B_2}}{R_{B_1} + R_{B_2}} V_{CC} = \frac{1}{14.3} 30$$
$$= 2.1 \text{ V} \qquad (6.1-6)$$

where the base current of T_1 has been neglected. Next, with $V_{BE_1} = 0.7$ V, then $V_{E_1} = V_{B_1} - V_{BE_1} = 1.4$ V and

$$I_{E_1} = \frac{V_{E_1}}{R_{E_1}} = \frac{1.4 \text{ V}}{1 \text{ k}\Omega} = 1.4 \text{ mA} \qquad (6.1-7)$$

Since $I_E \simeq I_C$, we determine $V_{C_1}(= V_{B_2})$ as follows:

$$V_{C_1} = V_{CC} - I_{C_1}R_{C_1} = 30 - (1.4)(18.5)$$
$$= 4.1 \text{ V} \qquad (6.1-8)$$

where the base current of T_2 has also been neglected. Then, $V_{E_2} = V_{C_1} - V_{BE_2} = 4.1 - 0.7 = 3.4$ V and

$$I_{E_2} = \frac{V_{E_2}}{R_{E_2}} = \frac{3.4 \text{ V}}{3.6 \text{ k}\Omega} = 0.94 \text{ mA} \qquad (6.1-9)$$

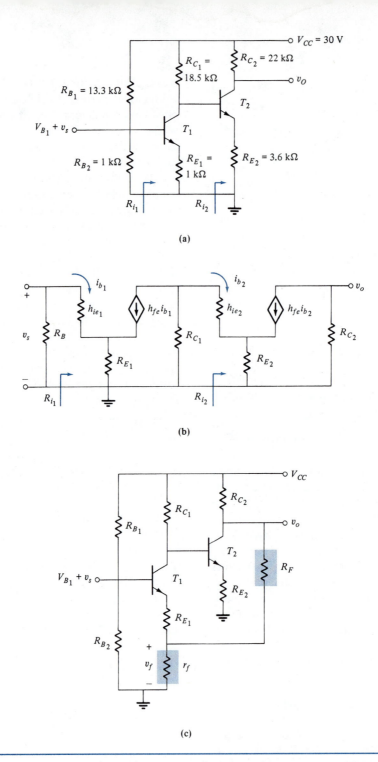

FIGURE 6.2 Direct-Coupled Amplifier: (a) Direct-coupled two-stage amplifier, (b) Small-signal equivalent circuit, (c) Direct-coupled amplifier with feedback

Again, since $I_E \simeq I_C$, the DC output voltage is obtained as follows:

$$V_{C_2} = V_{CC} - I_{C_2}R_{C_2}$$
$$= 30 - (0.94)(22) = 9.3 \text{ V} \qquad (6.1\text{--}10)$$

Finally, we can also determine the DC collector–emitter voltage for each transistor. For T_1,

$$V_{CE_1} = V_{C_1} - V_{E_1} = 4.1 - 1.4$$
$$= 2.7 \text{ V} \qquad (6.1\text{--}11)$$

For T_2,

$$V_{CE_2} = V_{C_2} - V_{E_2} = 9.3 - 3.4$$
$$= 5.9 \text{ V} \qquad (6.1\text{--}12)$$

These values of voltage (along with the specified current directions) indicate that each transistor is operating in the amplifier region. We carry out the AC analysis next.

The AC voltage amplification for this two-stage direct-coupled amplifier is determined using the small-signal equivalent circuit shown in Fig. 6.2b. Assuming again that $h_{fe} \gg 1$ for each transistor, we can use a simplified analysis. For large h_{fe}, the AC input impedance at the base of each transistor (R_{i_1} and R_{i_2} in Fig. 6.2b) is approximately $h_{fe}R_E$, which is much larger than any resistor in the circuit. For example, if $h_{fe} = 100$, $R_{i_1} = 100$ kΩ and $R_{i_2} = 360$ kΩ. Thus, each stage of amplification is isolated, and the amplification expression of (5.2–49) can be applied to each individual stage. The AC amplification for the first stage is therefore simply

$$\frac{v_{c_1}}{v_s} = \frac{R_{C_1}}{R_{E_1}} = \frac{18.5}{1} = 18.5 \qquad (6.1\text{--}13)$$

For the second stage, we have

$$\frac{v_o}{v_{c_1}} = \frac{R_{C_2}}{R_{E_2}} = \frac{22}{3.6} = 6.1 \qquad (6.1\text{--}14)$$

Thus, from (6.1–13) and (6.1–14), the overall voltage amplification is obtained as the product of these two stages, or

$$\frac{v_o}{v_s} = \left(\frac{R_{C_1}}{R_{E_1}}\right)\left(\frac{R_{C_2}}{R_{E_2}}\right) = 113 \qquad (6.1\text{--}15)$$

Hence, a significant amount of voltage amplification is obtained.

Nonetheless, direct-coupled amplifiers have a serious drawback associated with them, as previously mentioned. Since DC voltages are amplified along with AC, a small DC voltage change in an input stage can cause a large DC voltage change in a later stage. This problem is referred to as *voltage drift*. We will see that even in our two-stage amplifier, a serious drift problem exists. The small change in the DC input voltage is usually due to a slight increase in the BJT base–emitter forward voltage (V_0) with temperature. Often, this drift is sufficient to drive the last amplifier stage into saturation. Moreover, since all DC voltages are amplified, variation in the supply voltage or even a small extraneous (noise) voltage can cause voltage drift.

The following example serves to indicate the severity of this voltage drift problem in direct-coupled amplifiers.

— EX. 6.3

Direct-Coupled Amplifier Voltage Drift Example

For the direct-coupled amplifier of Fig. 6.2a, determine the shift in the DC operating point for the output stage when V_{BE_1} is increased by 100 mV.

Solution: We begin at the input stage and first determine V_{E_1}:

$$V_{E_1} = V_{B_1} - V_0 = 2.1 - 0.8 = 1.3 \text{ V}$$

which represents a reduction of only 100 mV. Next, the emitter current for T_1 is found:

$$i_{E_1} = \frac{V_{E_1}}{R_{E_1}} = 1.3 \text{ mA}$$

Hence, since $I_{E_1} \simeq I_{C_1}$, the DC voltage at B_2 is

$$V_{B_2} = V_{CC} - I_{C_1}R_{C_1} = 30 - (1.3)(18.5) = 6 \text{ V}$$

which indicates an increase of about 2 V. We then have $V_{E_2} = 6 - 0.7 = 5.3$ V, and I_{E_2} is increased to

$$I_{E_2} = \frac{V_{E_2}}{R_{E_2}} = \frac{5.3}{3.6} = 1.47 \text{ mA}$$

which is an increase of about 0.5 mA. This level of current (if possible) would create a voltage across R_{C_2} of magnitude 1.47 × 22, or 32.3 V, which would reduce the voltage at C_2 to

$$V_{C_2} = V_{CC} - I_C R_{C_2} = 30 - 32.3 = -2.3 \text{ V}$$

These values imply that the output stage has been driven into saturation. In fact, this operating point is ficticious. When T_2 is in saturation, we have neither $i_C = h_{fe}i_B$ nor $I_C \simeq I_E$; furthermore, the output stage is no longer amplifying the AC signal.

6.1.5   Multistage Feedback Amplifiers

The undesirable effect of voltage drift in direct-coupled amplifiers can be reduced through the use of a feedback resistor from the output stage to an intermediate stage, usually the input stage. In Chapter 4, we observed that the use of a negative feedback resistor R_F in a single-stage transistor amplifier provides stabilization of Q. A negative feedback resistor connected from one stage back to another has a similar effect.

Figure 6.2c displays a two-stage direct-coupled amplifier with negative feedback employed. A change in voltage in the input loop of T_1 results in the opposite change at the output terminal of T_2. For example, if v_{BE_1} is increased, then i_{B_1}, v_{E_1}, and i_{E_1} are increased. This change forces v_{C_1}, v_{E_2}, and i_{E_2} to decrease, which causes v_{C_2} to increase. Thus, feeding v_{C_2} back through R_F, v_f is increased, which causes i_{B_1} to decrease and offset the original increase. The problem of

voltage drift is thus reduced. The additional feedback, however, also has the deleterious effect of reducing the AC amplification.

To determine the AC amplification, we consider a general amplifier circuit as shown in Fig. 6.3, where negative feedback from the output stage back to the input stage is employed. The general case is considered because this type of stage-to-stage feedback is not restricted to direct-coupled amplifiers. We will see that the magnitude of the amplification is reduced from that without feedback. However, the amplification is independent of individual-stage amplification factors and can still be appreciable.

We begin the AC analysis by defining the multistage amplification factor (from Fig. 6.3) as follows:

$$A_o = \frac{v_o}{v_i'} \tag{6.1-16}$$

This factor is also called the *open-loop gain* because it is the gain without feedback. Thus, $v_o = A_o v_i'$, and since $v_i' = v_i - v_f$, where

$$v_f = \frac{r_f}{R_F + r_f} v_o \tag{6.1-17}$$

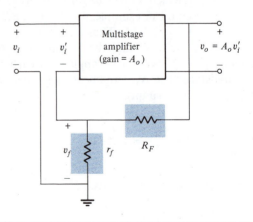

FIGURE 6.3   Multistage Feedback Amplifier

we have by substitution

$$v_o = A_o\left(v_i - \frac{r_f}{R_F + r_f}v_o\right) \qquad (6.1\text{--}18)$$

Solving for v_o/v_i yields the overall voltage amplification with feedback, known as the *closed-loop gain* and given by

$$\frac{v_o}{v_i} = \frac{A_o}{1 + A_o r_f/(r_f + R_F)} = A_C \qquad (6.1\text{--}19)$$

We observe that since the denominator of (6.1–19) is greater than 1, the closed-loop gain is less than the open-loop gain, or

$$A_C < A_O \qquad (6.1\text{--}20)$$

Thus, the addition of feedback reduces the voltage amplification.

Note that if $A_O \gg (R_F + r_f)/r_f$, then the second term in the denominator of (6.1–19) domi-

nates, and the closed-loop gain is given by

$$A_C = \frac{R_F + r_f}{r_f} \qquad (6.1\text{--}21)$$

Hence, we observe that the closed-loop gain with large A_o is dependent only upon the external resistors and not upon the amplification of individual stages. Thus, multistage amplifiers can be designed with an exact amount of voltage amplification (using precise resistors) even though h_{fe} for each discrete transistor is not known accurately. In general, negative feedback is employed in a variety of ways in linear amplifiers.

―――――――――――――――――――――――――――― EX. 6.4

Two-Stage Amplifier (with Feedback) Example

For the two identical stage RC amplifier of Fig. E6.4, derive the expression for the closed-loop voltage amplification v_o/v_s. Assume that $r_f \ll R_{oe_1}$ (the output impedance of the first amplifier stage at E_1), $R_F \gg R_C$, and $R_{B_1} \| R_{B_2} \gg R_C$.

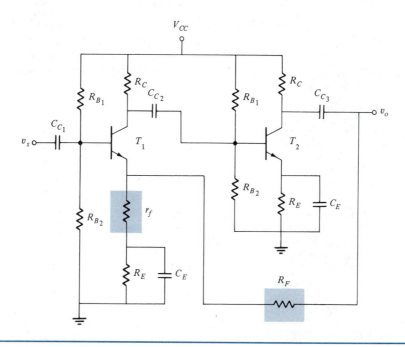

FIGURE E6.4

Solution: By visualizing a small-signal equivalent circuit for the RC amplifier, we have

$$v_{be_1} = v_s - v_f$$

where

$$v_f = \frac{r_f}{R_F + r_f} v_o$$

Thus,

$$i_{b_1} = \frac{v_{be_1}}{h_{ie}} = \frac{v_s - v_f}{h_{ie}}$$

Since $i_{c_1} = h_{fe} i_{b_1}$,

$$v_{c_1} = -i_{c_1} R_C \| h_{ie} = -\frac{h_{fe} R_C \| h_{ie}}{h_{ie}} (v_s - v_f)$$

This voltage is applied to the base of T_2 to yield

$$i_{b_2} = \frac{v_{c_1}}{h_{ie}}$$

Since $i_{c_2} = h_{fe} i_{b_2}$,

$$v_o = -i_{c_2} R_C = \frac{h_{fe} R_C}{h_{ie}} \frac{h_{fe} R_C \| h_{ie}}{h_{ie}} (v_s - v_f)$$

Substituting for $v_f = v_o r_f / (r_f + R_F)$ and rearranging yield

$$\frac{v_o}{v_s} = \frac{\left(\dfrac{h_{fe} R_C}{h_{ie}}\right)\left(\dfrac{h_{fe} R_C \| h_{ie}}{h_{ie}}\right)}{1 + \left(\dfrac{h_{fe} R_C}{h_{ie}}\right)\left(\dfrac{h_{fe} R_C \| h_{ie}}{h_{ie}}\right)\left(\dfrac{r_f}{r_f + R_F}\right)}$$

Note that if

$$\left(\frac{h_{fe} R_C}{h_{ie}}\right)\left(\frac{h_{fe} R_C \| h_{ie}}{h_{ie}}\right)\left(\frac{r_f}{r_f + R_F}\right) \gg 1$$

then the voltage amplification is given by

$$\frac{v_o}{v_s} = \frac{r_f + R_F}{r_f}$$

and is therefore dependent only upon the feedback elements.

6.2　DIFFERENCE AMPLIFIERS

Introduced in Chapter 4, *difference amplifiers* are another form of direct-coupled amplifiers that do not require the use of capacitors or inductors. One of the greatest attributes of difference amplifers is that they are particularly suitable for use in integrated circuits. These amplifiers are important because they are primary building blocks for linear (analog) integrated circuits.

A difference amplifier consists of two transistors connected in a balanced circuit arrangement that provides voltage amplification of the difference between two input signals. Direct coupling is possible in this case (without feedback between stages) because a DC drift voltage at each input cancels at the output due to the symmetry of the circuit. Various important examples of difference amplifiers are described next.

6.2.1　Basic BJT Difference Amplifier

The discussion of difference amplifiers begins here with the basic BJT circuit shown in Fig. 6.4. In this circuit, the transistors are assumed to be identical, and the collector resistors are

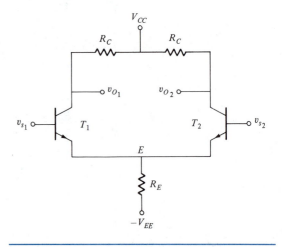

FIGURE 6.4　Basic BJT Difference Amplifier

matched and labeled as such. This balanced circuit arrangement is essential to difference amplifier operation.

Figure 6.4 is the same circuit as the one introduced in Chapter 4 where the bias analysis and design were carried out. Note that the AC input voltages v_{s_1} and v_{s_2} are applied at the input base terminal of T_1 and T_2, respectively. Also note that the AC output voltages are defined at the collector terminals of T_1 and T_2 as v_{o_1} and v_{o_2}, respectively. We will see that these output voltages are proportional to the difference in input signal voltages, where the proportionality factor is quite large and represents large amplification. Both voltage and current amplification are possible in this circuit, making it similar to the signal BJT common-emitter amplifier.

By definition, the difference amplifier amplifies the difference in the input signal voltages. Hence, it is extremely convenient to define a new voltage as the difference in input single voltages, or

$$v_d = v_{s_2} - v_{s_1} \qquad (6.2-1)$$

where v_d is the *difference-mode input signal voltage*. Furthermore, another voltage is defined as the average of the two input signals, or

$$v_c = \frac{v_{s_2} + v_{s_1}}{2} \qquad (6.2-2)$$

where v_c is *common-mode input signal voltage*. Solving (6.2-1) and (6.2-2) for v_{s_1} and v_{s_2} yields

$$v_{s_1} = v_c - \frac{v_d}{2} \qquad (6.2-3)$$

and

$$v_{s_2} = v_c + \frac{v_d}{2} \qquad (6.2-4)$$

Equations (6.2-3) and (6.2-4) indicate the relationships between the input signal voltages and the common-mode and difference-mode voltages. The design of a difference amplifier must achieve amplification of the difference-mode voltage and rejection of the common-mode voltage. Using (6.2-3) and (6.2-4) as replacements for v_{s_1} and v_{s_2} considerably simplifies the design procedure.

Figure 6.5a displays the small-signal equivalent circuit for the BJT difference amplifier of Fig. 6.4. This circuit is obtained by substituting the AC equivalent for each transistor into the original circuit of Fig. 6.4. Our initial objectives are to determine the output signal voltages, v_{o_1} and v_{o_2}, in terms of the common-mode and difference-mode voltages, v_c and v_d.

First, we note (from Fig. 6.5a) that v_{o_1} and v_{o_2} are given by Ohm's law as

$$v_{o_1} = -(h_{fe}i_{b_1})R_C \simeq -i_{e_1}R_C \qquad (6.2-5)$$

and

$$v_{o_2} = -(h_{fe}i_{b_2})R_C \simeq -i_{e_2}R_C \qquad (6.2-6)$$

where we have used $h_{fe}i_b = i_c \simeq i_e$.

Next, we substitute for i_{e_1} and i_{e_2} in terms of v_c and v_d. This task could be accomplished by writing and solving loop equations (described in Chapter 1); however, a simpler procedure is to use the equivalent circuits of Figs. 6.5c or d.

Figure 6.5b is obtained by replacing each original current source in Fig. 6.5a with two current sources that provide exactly the same currents. The collector portion of the circuit is thus separated from the emitter–base portion, as shown in Fig. 6.5b. Next, we modify this equivalent circuit as shown in Fig. 6.5c. The circuit of Fig. 6.5c is obtained by forcing i_{e_1} and i_{e_2} around the entire loop of each respective base–emitter side and changing the resistor values so that exactly the same KVL applies to the original and modified loops. Thus, since the original current through h_{ie} was i_b (for each transistor), if we increase the current magnitude by a factor of $1 + h_{fe}$ to i_e, we must decrease the resistor value by the same factor of $1 + h_{fe}$ to maintain the same voltage across the resistor. Therefore, h_{ie} is divided by $1 + h_{fe}$ in Fig. 6.5c. Note that $h_{ib} = h_{ie}/(1 + h_{fe})$ from Chapter 5.

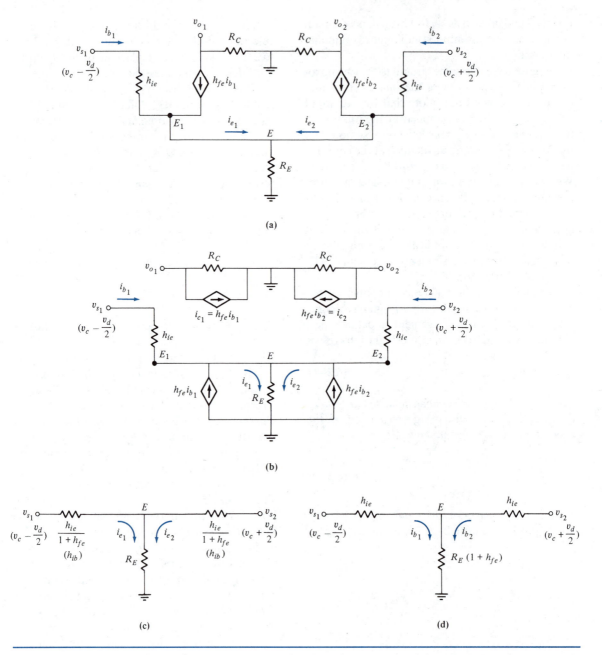

FIGURE 6.5   Difference Amplifier: (a) Small-signal equivalent circuit, (b) Separated small-signal equivalent circuit, (c) Modified equivalent circuit with emitter currents (collector portion omitted), (d) Modified equivalent circuit with base currents (collector portion omitted)

The equivalent circuit of Fig. 6.5d is obtained by forcing i_{b_1} and i_{b_2} around the entire loop corresponding to each base–emitter side. This circuit is useful in determining input impedances as seen by the sources because the correct current level prevails. Note that Fig. 6.5d involves only base currents; therefore, only the resistor R_E (with original currents i_{e_1} and i_{e_2}) must be modified. Since the current magnitudes have been reduced by $1 + h_{fe}$, the modified resistor value corresponding to R_E must be increased by this same factor (as indicated in Fig. 6.5d) to maintain the same KVL around each loop.

Expressions for the currents i_{e_1} and i_{e_2} are now obtained by using Fig. 6.5c and a three-step procedure as follows:

1. Determine i_{e_1} and i_{e_2} for $v_d = 0$ ($v_c \neq 0$).
2. Determine i_{e_1} and i_{e_2} for $v_c = 0$ ($v_d \neq 0$).
3. Add the results of steps 1 and 2.

This method is valid provided that the circuit is linear, which was assumed in substituting small-signal models for T_1 and T_2. We apply this method as follows. For the first step, where $v_d = 0$, the input signal voltages on each side of the circuit are equal ($v_{s_1} = v_{s_2} = v_c$), which, because of symmetry, implies that $i_{e_1} = i_{e_2}$. By inspection of the circuit of Fig. 6.5c for the common-mode voltage only, we write

$$i_{e_1} = \frac{v_c}{h_{ib} + 2R_E} = i_{e_2} \qquad (6.2-7)$$

which is obtained by discarding half the circuit to the right of R_E (or left of R_E) and doubling R_E to account for the discarded equal emitter current. Note that the same KVL is provided around the loop involving i_{e_1} or i_{e_2}.

For the second step, we set $v_c = 0$ and realize that the input signals are now antisymmetric. That is, $v_{s_1} = -v_d/2$ and $v_{s_2} = v_d/2$. Thus, $i_{e_2} = -i_{e_1}$, and the current through R_E is zero. Using KVL around the outer loop of Fig. 6.5c yields the emitter currents for the difference mode voltage as

follows:

$$i_{e_2} = \frac{(v_d/2) + (v_d/2)}{2h_{ib}} = -i_{e_1} \qquad (6.2-8)$$

or

$$i_{e_2} = \frac{v_d}{2h_{ib}} = -i_{e_1} \qquad (6.2-9)$$

Adding the results of (6.2–7) and (6.2–9) yields the total emitter currents as follows:

$$i_{e_1} = \frac{v_c}{h_{ib} + 2R_E} - \frac{v_d}{2h_{ib}} \qquad (6.2-10)$$

and

$$i_{e_2} = \frac{v_c}{h_{ib} + 2R_E} + \frac{v_d}{2h_{ib}} \qquad (6.2-11)$$

Substitution of (6.2–10) and (6.2–11) into (6.2–5) and (6.2–6) yields the output voltage relations as follows:

$$v_{o_1} = \frac{-R_C v_c}{h_{ib} + 2R_E} + \frac{R_C v_d}{2h_{ib}} \qquad (6.2-12)$$

and

$$v_{o_2} = \frac{-R_C v_c}{h_{ib} + 2R_E} - \frac{R_C v_d}{2h_{ib}} \qquad (6.2-13)$$

The coefficients in (6.2–12) and (6.2–13) are the amplification factors for the common-mode and difference-mode voltages. Defining these factors explicitly, we have

$$A_c = \frac{R_C}{h_{ib} + 2R_E} \qquad (6.2-14)$$

and

$$A_d = \frac{R_C}{2h_{ib}} \qquad (6.2-15)$$

Equations (6.2–12) and (6.2–13) can therefore be rewritten as follows:

$$v_{o_1} = -A_c v_c + A_d v_d \qquad (6.2\text{–}16)$$

and

$$v_{0_2} = -A_c v_c - A_d v_d \qquad (6.2\text{–}17)$$

Note that each of these output voltages is dependent upon v_d as well as v_c. Since v_d is to be amplified and v_c rejected, the difference amplifier is therefore designed so that $A_d \gg 1$ and $A_c \ll 1$ (ideally, $A_d \to \infty$ and $A_c \to 0$). It is also implied that $A_d \gg A_c$. These results suggest defining a figure of merit for the difference amplifier as a ratio of amplification factors as follows:

$$\text{CMRR} = \frac{A_d}{A_c} \qquad (6.2\text{–}18a)$$

where CMRR stands for the *common-mode rejection ratio*. Note that a CMRR $\gg 1$ is desirable. Additionally, since the CMRR is usually quite large, it is convenient to introduce a new unit, the *decibel* (dB). This unit is defined as follows:

$$\text{CMRR}|_{dB} = 20 \log \frac{A_d}{A_c} \qquad (6.2\text{–}18b)$$

A range of typical values for $\text{CMRR}|_{dB}$ is 50 to 100 dB. The corresponding range for the ratio A_d/A_c is approximately 300 to 10^5.

We now consider the relative magnitudes of the resistors in the basic BJT difference amplifier that provide $A_d \gg 1$ and $A_c \ll 1$ (and therefore also CMRR $\gg 1$). We observe from (6.2–15) that $A_d \gg 1$ is achieved if

$$R_C \gg 2h_{ib} = \frac{2h_{ie}}{1 + h_{fe}} \qquad (6.2\text{–}19)$$

This design relationship is easily satisfied for large h_{fe} and $R_C \geq h_{ie}$.

We observe from (6.2–14) that $A_c \ll 1$ is accomplished for

$$R_E \gg R_C \qquad (6.2\text{–}20)$$

with no effect on A_d [from (6.2–15)]. However, with $R_E \gg R_C \gg h_{ib}$, (6.2–14) can be written as

$$A_c \cong \frac{R_C}{2R_E} \qquad (6.2\text{–}21)$$

With the inequalities of (6.2–19) and (6.2–20) satisfied, the CMRR becomes

$$\text{CMRR} = \frac{R_C/2h_{ib}}{R_C/2R_E} = \frac{R_E}{h_{ib}} \qquad (6.2\text{–}22)$$

Thus, from (6.2–22), we observe that making $R_E \gg h_{ib}$ provides CMRR $\gg 1$. However, the magnitude of R_E is not unlimited, as we will see next.

Recall from Chapter 5 that h_{ib} is related to the DC emitter current through the relation $h_{ib} = 0.025/I_E$. Thus, substitution into (6.2–22) yields

$$\text{CMRR} = \frac{R_E I_E}{0.025} \qquad (6.2\text{–}23)$$

Hence, we observe that for maximum CMRR, $R_E I_E$ should be maximized. However, with $R_E C_E$ maximized, the DC power dissipated in $R_E(2R_E I_E^2)$ will be large, which is objectionable from the viewpoint of power dissipation. This dilemma is solved by replacing R_E with a DC current source, as described in the next section.

6.2.2   BJT Difference Amplifier with Constant Current Source

Basic Amplifier Circuit.   Figure 6.6 displays the basic circuit for a difference amplifier with a DC current source replacing R_E. Actual current source circuits used with difference amplifiers (and inte-

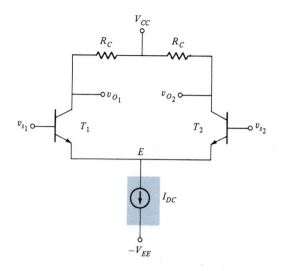

FIGURE 6.6 Difference Amplifier with Current Source

grated circuits in general) are described shortly. As far as DC is concerned, the current source provides the necessary path, whereas previously R_E performed this task. As far as AC is concerned, this replacement improves the CMRR without increased power dissipation.

First, note that the occurrence of R_E in (6.2–10) and (6.2–11) came about because R_E is the AC equivalent impedance between E and ground in Fig. 6.4. For the current source of Fig. 6.6, we define the equivalent impedance between E and ground as r_{ac}, where r_{ac} is quite large in actual circuits and infinite ideally. The equations of the previous section now apply directly with r_{ac} replacing R_E. Thus, (6.2– 14) and (6.2–15) become

$$A_c = \frac{R_C}{h_{ib} + 2r_{ac}} \simeq \frac{R_C}{2r_{ac}} \qquad (6.2\text{–}24)$$

and

$$A_d = \frac{R_C}{2h_{ib}} \qquad (6.2\text{–}25)$$

where we note that this replacement for R_E only modifies A_c. The common-mode rejection ratio is

now given by

$$\text{CMRR} = \frac{R_C/2h_{ib}}{R_C/2r_{ac}} = \frac{r_{ac}}{h_{ib}} \qquad (6.2\text{–}26)$$

Substitution for h_{ib} ($=0.025/I_E$) then yields

$$\text{CMRR} = \frac{r_{ac}I_E}{0.025} \qquad (6.2\text{–}27)$$

We observe that the common-mode rejection ratio does not depend upon the product of I_E and R_E. Hence, it can be made quite large without the previous associated large power dissipation in R_E.

—————— EX. 6.5

Difference Amplifier with DC Current Source Example

The difference amplifier of Fig. 6.6 has the following element values: $R_C = 10$ kΩ and $I_{DC} = 2$ mA. The transistor parameters are $h_{ie} = 2.5$ kΩ and $h_{fe} = 100$. For $r_{ac} = 50$ kΩ, determine the common-mode and difference-mode amplification factors and the CMRR.

Solution: Substituting directly into (6.2–24) and (6.2–25) yields

$$A_c = \frac{R_C}{2r_{ac}} = \frac{10}{100} = 0.1$$

and

$$A_d = \frac{10}{2(2.5)/101} = 202$$

Thus, the common-mode rejection ratio is

$$\text{CMRR} = \frac{202}{0.1} = 2020$$

Basic BJT Current Source. A simple constant current source used in BJT integrated circuits and its ideal representation are shown in Fig. 6.7a. The magnitude of the DC current I_{DC} is obtained by writing KVL for the base–emitter loop of the

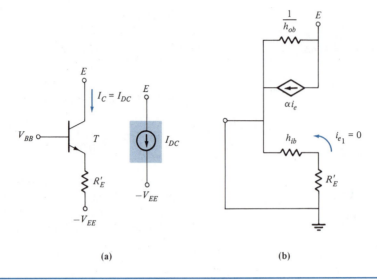

Figure 6.7   Basic BJT Current Source: (a) Circuit and symbol, (b) AC equivalent circuit

BJT, which yields

$$I_E = \frac{V_{EE} + V_{BB} - V_0}{R'_E} \qquad (6.2\text{--}28)$$

Since $I_E \simeq I_C = I_{DC}$, the constant current source magnitude is

$$I_{DC} = \frac{V_{EE} + V_{BB} - 0.7}{R'_E} \qquad (6.2\text{--}29)$$

where $V_0 = 0.7$ V has been used. The magnitude of I_{DC} is chosen according to the DC design rules for the difference amplifiers discussed in Chapter 4. The desired value is then achieved by selecting appropriate values for V_{BB} and R'_E. Since V_{EE} is usually used in other portions of the integrated circuit, its value is considered fixed in magnitude.

The AC equivalent resistance of the current source, r_{ac} between point E and ground in Fig. 6.7a, is obtained by considering the AC equivalent circuit shown in Fig. 6.7b. Note that the common-base hybrid circuit model for the BJT has been substituted. The AC equivalent resistance is most

easily obtainable using this representation. From Fig. 6.7b, we immediately observe that i_e is zero (since the resistors h_{ib} and R'_E have no voltage across them). Therefore, the current source αi_e is open, and the resistance between point E and ground is simply

$$r_{ac} = \frac{1}{h_{ob}} \qquad (6.2\text{--}30)$$

To demonstrate that r_{ac} is large, recall that the common-base hybrid parameter h_{ob} is the slope of the output characteristic (i_C versus v_{CB}) at Q. Since this configuration displays nearly horizontal characteristics, $1/h_{ob}$ is large, typically ≥ 50 kΩ.

Integrated-Circuit BJT Current Source.   When the current source of Fig. 6.7a is used in an integrated circuit with V_{CC} and V_{EE} available, V_{BB} is obtained without the necessity of an additional DC voltage source. The circuit that provides V_{BB} is shown in Fig. 6.8 and is known as the *Widlar current source* (after Robert Widlar, formerly of Fairchild and a pioneer in op amp design). The subscript R stands for "reference" in each case.

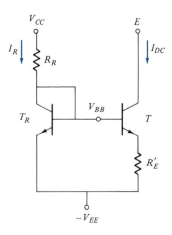

Figure 6.8   BJT Current Source Used in ICs

Note that

$$I_R = \frac{V_{CC} - V_{BB}}{R_R} \qquad (6.2\text{–}31)$$

Therefore,

$$V_{BB} = V_{CC} - I_R R_R \qquad (6.2\text{–}32)$$

The Widlar current source is used in low-current applications where I_{DC} is on the order of 0.1 mA. For higher current levels, R'_E is removed, which results in what is called a *current mirror*

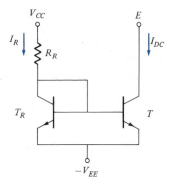

FIGURE 6.9   Higher-Current BJT Current Source Used in ICs

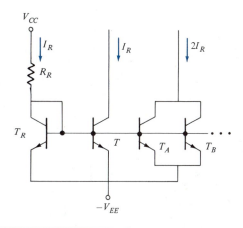

FIGURE 6.10   Current Mirrors

(the reference current is reproduced identically in the current source). Figure 6.9 displays the resulting circuit in which I_{DC} and I_R must be equal since their base–emitter voltages are equal. This circuit cannot be used at low current levels, however, because R_R becomes too large to be implemented in integrated-circuit form.

Current mirrors are extremely useful in the bias design of ICs. They eliminate the use of bias resistors that require too much chip area. The reference current is actually used to mirror currents in numerous stages, as shown in Fig. 6.10. Note that multiples of the reference current I_R may also be obtained using two transistors T_A and T_B, which produce a current source value of $2I_R$. In an actual IC, the area of the base–emitter junction of the current source transistor relative to T_R is doubled. Other ratios are also used, and the area ratios need not be an integer value.

6.2.3   Darlington Difference Amplifier

Darlington-Pair Configuration.   In order to increase the current amplification of a BJT, a special connection of two BJTs, called a *Darlington pair,* is sometimes used. Figure 6.11a displays the

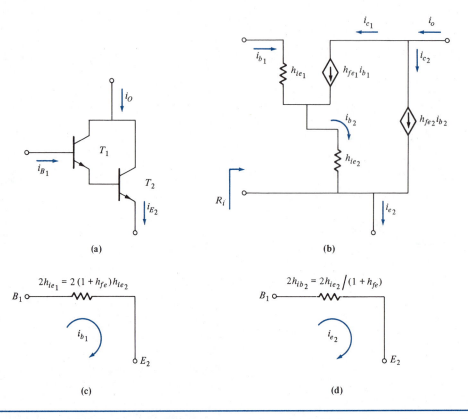

FIGURE 6.11   Darlington Pair Configuration: (a) Special two-transistor circuit, (b) Complete small-signal model, (c) Base–emitter equivalent circuit with base current, (d) Base–emitter equivalent circuit with emitter current.

special two-transistor circuit, which is easily implemented in integrated-circuit form. Note that the collector regions of T_1 and T_2 are shown connected. These regions are actually the same region in an IC.

As shown in Fig. 6.11a, the Darlington pair can be considered to be a three-terminal device, with B_1 as the input terminal and the common collector C_1 (or C_2) and E_2 as the output terminals. Not only does the Darlington pair provide much larger current amplification, but also the input impedance (between B_1 and E_2) is much larger.

The small-signal model for the Darlington pair is displayed in Fig. 6.11b. The small-signal model has been directly substituted for each tran-

sistor. This substitution assumes that each transistor is operating under amplifier conditions. We can easily determine the AC current amplification, i_o/i_{b_1} or i_{e_2}/i_{b_1}, in terms of the parameters of T_1 and T_2 by using this model.

Considering Fig. 6.11b, we begin by writing the equation for i_o as follows:

$$i_o = i_{c_1} + i_{c_2} \tag{6.2–33}$$

Substituting $i_{c_1} = h_{fe_1} i_{b_i}$ and $i_{c_2} = h_{fe_2} i_{b_2}$ yields

$$i_o = h_{fe_1} i_{b_1} + h_{fe_2} i_{b_2} \tag{6.2–34}$$

However, i_{b_2} and i_{e_1} are the same current (see Fig. 6.11a). Also, since $i_{e_1} = (1 + h_{fe_1}) i_{b_1}$, we have by substitution

$$i_o = h_{fe}i_{b_1} + h_{fe_2}(1 + h_{fe_1})i_{b_1} \qquad (6.2-35)$$

Therefore, the current amplification between the collector terminals and input terminal is

$$\frac{i_o}{i_{b_1}} = h_{fe_1} + h_{fe_2}(1 + h_{fe_1}) \qquad (6.2-36)$$

Note that the second term dominates, and the overall amplification is proportional to the product of h_{fe_1} and h_{fe_2}.

To determine the current amplification associated with the other output terminal (E_2), we write

$$i_{e_2} = (1 + h_{fe_2})i_{b_2} \qquad (6.2-37)$$

or, since $i_{b_2} = i_{e_1} = (1 + h_{fe_1})i_{b_1}$, we have

$$i_{e_2} = (1 + h_{fe_2})(1 + h_{fe_1})i_{b_1} \qquad (6.2-38)$$

The current amplification is

$$\frac{i_{e_2}}{i_{b_i}} = (1 + h_{fe_2})(1 + h_{fe_1}) \qquad (6.2-39)$$

Usually, T_1 and T_2 are designed to have $h_{fe_1} = h_{fe_2} = h_{fe}$ and also $h_{fe} \gg 1$. Under these conditions, the current amplification associated with either output terminal is approximated h_{fe}^2, or

$$\frac{i_o}{i_{b_1}} \simeq \frac{i_{e_2}}{i_{b_1}} \simeq h_{fe}^2 \qquad (6.2-40)$$

An interesting feature of the Darlington pair is observed by determining the input impedance (between B_1 and E_2). From Fig. 6.11b, this impedance is given by

$$R_i = \frac{v_{b_1}}{i_{b_1}} = \frac{h_{ie_1}i_{b_1} + h_{ie_2}i_{b_2}}{i_{b_1}} \qquad (6.2-41)$$

However, $i_{b_2} = (1 + h_{fe})i_{b_1}$. Therefore,

$$R_i = h_{ie_1} + (1 + h_{fe})h_{ie_2} \qquad (6.2-42)$$

This expression may be condensed further if we realized that $h_{ie_1} = (1 + h_{fe})h_{ie_2}$ since $h_{ie_1} = 0.025/I_{B_1}$ and $h_{ie_2} = 0.025/I_{B_2} = 0.025[(1 + h_{fe})I_{B_1}]$. Therefore, R_i may be written in either of two ways:

$$R_i = 2h_{ie_1} \qquad (6.2-43)$$

or

$$R_i = 2(1 + h_{fe})h_{ie_2} \qquad (6.2-44)$$

We now compare the magnitude of the input impedance of the Darlington pair with that of a single transistor. We must realize that if the output currents are of similar magnitude, then the base current of T_2 is of similar magnitude to that of the single transistor. This fact implies that h_{ie} for the single transistor is about equal to h_{ie_2}. Therefore, from (6.2–44), we see that the input impedance of the Darlington pair is much larger than that of the single transistor by a factor of $2(1 + h_{fe})$.

Based upon (6.2–43), Fig. 6.11c displays a very simple equivalent circuit for the Darlington pair. Note that i_{b_1} is the current around the entire loop. A similar circuit, which will be of importance shortly, is shown in Fig. 6.11d. In Fig. 6.11d, we have changed the current level around the loop to i_{e_2} $[=(1 + h_{fe})^2 i_{b_1}]$ and have therefore modified the resistor value accordingly to satisfy KVL. Furthermore, we have used the relation $h_{ie} = (1 + h_{fe})h_{ib}$ from Chapter 5.

Basic Darlington Difference Amplifier. Figure 6.12 displays the Darlington difference amplifier obtained from the basic BJT difference amplifier of Fig. 6.6 by substituting a Darlington pair for each single transistor. In order to operate as a difference amplifier, the circuit requires balance, and the transistors have been labeled accordingly. Note that the currents on the right-hand side are

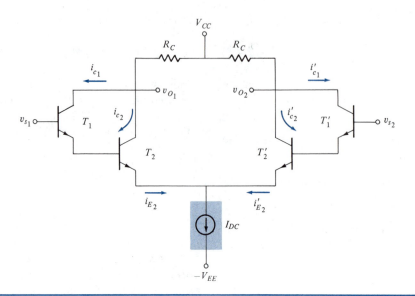

FIGURE 6.12   Darlington Difference Amplifier

different from those on the left-hand side since $v_{s_1} \neq v_{s_2}$.

As far as the DC analysis of this circuit is concerned, each Darlington pair can be treated as a single transistor with a DC voltage between B_1 and E_2 equal to $2V_0$ (1.4 V). The DC analysis of Chapter 4 then follows directly.

We now carry out the AC analysis of the Darlington difference amplifier. Figure 6.13 shows the AC equivalent circuit for the base–emitter portion of Fig. 6.12. We have substituted the AC

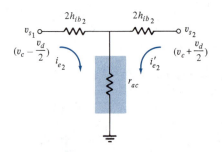

FIGURE 6.13   Equivalent Circuit for the Base–Emitter Portion Darlington Difference Amplifier

equivalent circuit of Fig. 6.11d for each Darlington pair (with emitter currents only) and r_{ac} for the AC representation of the constant current source. Note that we have also indicated the equivalent common-mode and difference-mode voltage representation for v_{s_1} and v_{s_2}.

To analyze the collector portion of the circuit, we use Fig. 6.12. By inspection, the AC output voltages are

$$v_{o_1} = -(i_{c_1} + i_{c_2})R_C \qquad (6.2-45)$$

and

$$v_{o_2} = -(i'_{c_1} + i'_{c_2})R_C \qquad (6.2-46)$$

However, for each Darlington pair, the sum of the collector currents $(i_{c_1} + i_{c_2})$ is approximately equal to the emitter current i_{e_2}. Thus,

$$v_{o_1} \simeq -i_{e_2}R_C \qquad (6.2-47)$$

and

$$v_{o_2} \simeq -i'_{e_2}R_C \qquad (6.2-48)$$

The emitter currents are now determined from Fig. 6.13 by using the three-step procedure of Section 6.2.2 in one simultaneous (giant) step. By inspection of Fig. 6.13, we have

$$i_{e_2} = \frac{v_c}{2h_{ib_2} + 2r_{ac}} - \frac{v_d}{4h_{ib_2}} \qquad (6.2\text{–}49)$$

and

$$i'_{e_2} = \frac{v_c}{2h_{ib_2} + 2r_{ac}} + \frac{v_d}{4h_{ib_2}} \qquad (6.2\text{–}50)$$

In each equation, the first term is the common-mode contribution (obtained for $v_d = 0$), and the second term is the difference-mode contribution (obtained for $v_c = 0$).

Thus, substitution into (6.2–47) and (6.2–48) yields

$$v_{o_1} = -\frac{R_C v_c}{2h_{ib_2} + 2r_{ac}} + \frac{R_C v_d}{4h_{ib_2}} \qquad (6.2\text{–}51)$$

and

$$v_{o_2} = -\frac{R_C v_c}{2h_{ib_2} + 2r_{ac}} - \frac{R_C v_d}{4h_{ib_2}} \qquad (6.2\text{–}52)$$

Defining the common-mode and difference-mode amplification factors in the same manner as was done previously yields

$$A_c = \frac{R_C}{2h_{ib_2} + 2r_{ac}} \simeq \frac{R_C}{2r_{ac}} \qquad (6.2\text{–}53)$$

and

$$A_d = \frac{R_C}{4h_{ib_2}} \qquad (6.2\text{–}54)$$

Thus, the common-mode rejection ratio is given by

$$\text{CMRR} = \frac{A_d}{A_c} = \frac{r_{ac}}{2h_{ib_2}} \qquad (6.2\text{–}55)$$

Note that the difference-mode gain along with the CMRR appear to have actually reduced by a factor of 2 as compared to the single-transistor difference amplifier. This reduction is not significant because h_{ib_2} for the Darlington-pair case is smaller than h_{ib} for the BJT case by a factor of $1 + h_{fe}$ for equivalent DC input currents. Hence, the improvement in current amplification along with input impedance warrant the use of Darlington difference amplifiers.

———————————————————————— EX. 6.6

Darlington Difference Amplifier Example

Determine A_c, A_d, and CMRR for the Darlington difference amplifier of Fig. 6.12. Use $R_C = 1\ \text{k}\Omega$, $h_{fe_1} = h_{fe_2} = 100$, $h_{ib_2} = 10\ \Omega$, and $r_{ac} = 50\ \text{k}\Omega$.

Solution: For the values stated, (6.2–53) yields

$$A_c = \frac{R_C}{2r_{ac}} = \frac{1}{2(50)} = 0.01$$

and (6.2–54) yields

$$A_d = \frac{R_C}{4h_{ib_2}} = \frac{1}{4(0.01)} = 25$$

Thus, the common-mode rejection ratio is

$$\text{CMRR} = \frac{25}{0.01} = 2500$$

6.2.4   BJT Difference Amplifier with Active Load

In order to improve the difference-mode amplification, active loads are often used as a replacement for the collector resistors. These active loads consist of BJTs and require less chip area than the collector resistors, thus providing an additional advantage.

A BJT difference amplifier with an active load is displayed in Fig. 6.14. For simplicity, the single transistors T_1 and T_2 have been used to form the differential pair. However, Darlington pairs can

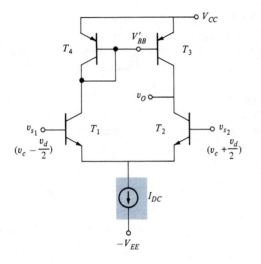

FIGURE 6.14 BJT Difference Amplifier with Active Load

also be used. Transistors T_3 and T_4 form the active load using PNP transistors in a current-mirror configuration. In this example, we have only one AC output voltage (v_o), and it is customary to refer to this output as *single-ended*. Actually, the AC voltage at the collector of T_1 is zero, as we will see.

The small-signal equivalent circuit for the difference amplifier with an active load is shown in Fig. 6.15a. All collector–emitter resistors ($1/h_{oe}$) are assumed to be the same for each BJT. Each of these resistors must be included in obtaining the AC output voltage v_o. With r_{ac} as an open circuit, Fig. 6.15b displays the corresponding small-signal equivalent circuit. Note that the AC equivalent for the active load is simply a resistor ($1/h_{oe}$) connected between the output terminal and ground. Converting each current source to a voltage source and applying the voltage divider rule to Fig. 6.15b, we obtain the output voltage, by inspection, as

$$v_o = \frac{(1/h_{oe})(h_{fe}/h_{oe})(i_{b_1} - i_{b_2})}{1/h_{oe} + (1/h_{oe}) + (1/h_{oe})} \qquad (6.2\text{–}56)$$

or

$$v_o = \frac{h_{fe}(i_{b_1} - i_{b_2})}{3h_{oe}} \qquad (6.2\text{–}57)$$

In order to substitute for $i_{b_1} - i_{b_2}$, we use the results obtained previously. From (6.2–10) and (6.2–11) with R_E replaced by r_{ac}, we have

$$i_{e_1} = \frac{v_c}{h_{ib} + 2r_{ac}} - \frac{v_d}{2h_{ib}} \qquad (6.2\text{–}58)$$

and

$$i_{e_2} = \frac{v_c}{h_{ib} + 2r_{ac}} + \frac{v_d}{2h_{ib}} \qquad (6.2\text{–}59)$$

These expressions were obtained from the base–emitter portion of the equivalent circuit, which is unchanged except for the constant current source. Thus, since

$$i_{b_1} - i_{b_2} = \frac{1}{1 + h_{fe}}(i_{e_1} - i_{e_2}) \qquad (6.2\text{–}60)$$

substitution of (6.2–58) and (6.2–59) yields, quite simply,

$$i_{b_1} - i_{b_2} = -\frac{1}{1 + h_{fe}}\frac{v_d}{h_{ib}} \qquad (6.2\text{–}61)$$

This expression is simplified further by using $h_{ie} = (1 + h_{fe})h_{ib}$. Thus,

$$i_{b_1} - i_{b_2} = -\frac{v_d}{h_{ie}} \qquad (6.2\text{–}62)$$

Substitution into (6.2–57) yields

$$v_o = \frac{h_{fe}(-v_d/h_{ie})}{3h_{oe}} \qquad (6.2\text{–}63)$$

Thus, the difference-mode voltage amplification for this case is

$$A_d = \frac{h_{fe}/h_{ie}}{3h_{oe}} \qquad (6.2\text{–}64)$$

Since $h_{ie}/h_{fe} \simeq h_{ib}$, (6.2–64) can be written as

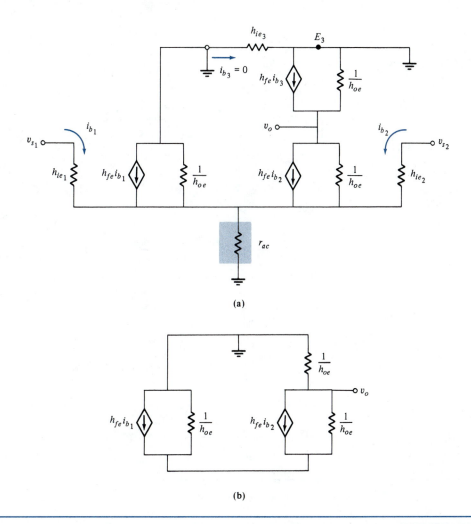

FIGURE 6.15 BJT Difference Amplifier with Active Load: (a) Small-signal equivalent circuit, (b) Equivalent circuit with r_{ac} large

$$A_d = \frac{(1/3)(1/h_{oe})}{h_{ib}} \qquad (6.2\text{--}65)$$

Comparing (6.2–65) with (6.2–25) and (6.2–54) (the expressions obtained previously using resistive loads), we see that the essential difference is that R_C is replaced with $1/h_{oe}$. Typically, h_{ib} is on the order of 1 to 10 Ω and $1/h_{oe}$ is on the order of 10 kΩ so that A_d is on the order of 10^3.

6.2.5 MOST Difference Amplifier

Basic MOST Circuit with Resistors. Figure 6.16a shows the basic circuit for the MOST difference amplifier using N-channel E-D transistors. We note that this circuit is identical in form to the basic BJT case except that the MOSTs replace the BJTs. The primary advantage of using MOSTs is that the input impedance is increased

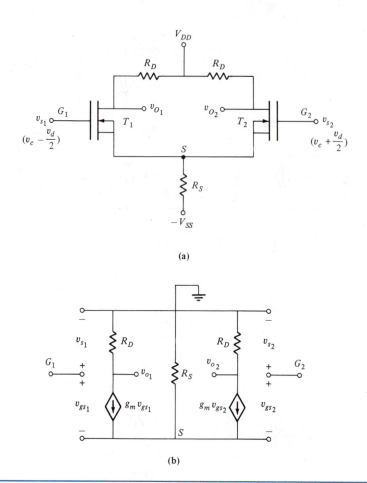

FIGURE 6.16   MOST Difference Amplifier; (a) Basic circuit with resistors, (b) Small-signal equivalent circuit

because the impedance seen looking into the input gate terminal of the MOST is enormous. However, a disadvantage of the MOST difference amplifier is that the difference-mode gain is not as large as that obtainable with BJTs.

To analyze the circuit of Fig. 6.16a, we use the small-signal equivalent circuit shown in Fig. 6.16b. This circuit has been obtained by replacing each MOST with its current source representation (with $g_o = 0$) and drawing R_S upside down.

Writing expressions for the AC output voltages v_{o_1} and v_{o_2} from Fig. 6.16b, we have

$$v_{o_1} = -g_m v_{gs_1} R_D \qquad (6.2\text{--}66)$$

and

$$v_{o_2} = -g_m v_{gs_2} R_D \qquad (6.2\text{--}67)$$

However, from Fig. 6.16b, v_{gs_1} and v_{gs_2} can be written in terms of the signal source voltages, v_{s_1} and v_{s_2}, and the voltage at the source terminal, v_s, as follows:

$$v_{gs_1} = v_{s_1} - v_s \qquad (6.2\text{--}68)$$

and

$$v_{gs_2} = v_{s_2} - v_s \qquad (6.2\text{--}69)$$

Also, from Fig. 6.16b, the voltage across R_S is

$$v_s = R_S(g_m v_{gs_1} + g_m v_{gs_2}) \qquad (6.2-70)$$

With the common-mode and difference-mode representations for v_{s_1} and v_{s_2} (shown in Fig. 6.16a), the corresponding amplification factors A_c and A_d are determined. For the common-mode case ($v_d = 0$), $v_{s_1} = v_{s_2} = v_c$, and because of the balanced circuit $v_{gs_1} = v_{gs_2}$. Therefore, from (6.2–70), we have

$$v_s = 2g_m R_S v_{gs_1} \qquad (6.2-71)$$

Substituting into (6.2–68) yields

$$v_{gs_1} = v_c - 2g_m R_S v_{gs_1} \qquad (6.2-72)$$

or, solving for v_{gs_1},

$$v_{gs_1} = \frac{v_c}{1 + 2g_m R_S} \qquad (6.2-73)$$

Hence, (6.2–66) and (6.2–67) become

$$v_{o_1} = -g_m R_D \frac{v_c}{1 + 2g_m R_S} \qquad (6.2-74)$$

and

$$v_{o_2} = -g_m R_D \frac{v_c}{1 + 2g_m R_S} \qquad (6.2-75)$$

Thus, we observe that

$$A_c = \frac{g_m R_D}{1 + 2g_m R_S} \qquad (6.2-76)$$

For the difference-mode case, $v_{s_1} = -v_d/2$ and $v_{s_2} = v_d/2$. Therefore, from (6.2–68) and (6.2–69), we have

$$v_{gs_1} = -\frac{v_d}{2} - v_s \qquad (6.2-77)$$

and

$$v_{gs_2} = \frac{v_d}{2} - v_s \qquad (6.2-78)$$

Substituting (6.2–77) and (6.2–78) into (6.2–70) yields

$$v_s = R_S\left[g_m\left(-\frac{v_d}{2} - v_s\right) + g_m\left(\frac{v_d}{2} - v_s\right)\right] \qquad (6.2-79)$$

or

$$v_s = -2g_m R_S v_s \qquad (6.2-80)$$

Hence, solving for v_s, we obtain $v_s = 0$, which means that under difference-mode conditions the current through R_S is zero. Thus, from (6.2–77) and (6.2–78), we have

$$v_{gs_1} = -\frac{v_d}{2} \qquad (6.2-81)$$

and

$$v_{gs_2} = \frac{v_d}{2} \qquad (6.2-82)$$

Substitution into (6.2–66) and (6.2–67) yields

$$v_{o_1} = \frac{g_m R_D v_d}{2} \qquad (6.2-83)$$

and

$$v_{o_2} = -\frac{g_m R_D v_d}{2} \qquad (6.2-84)$$

Thus, the difference-mode amplification is

$$A_d = \frac{g_m R_D}{2} \qquad (6.2-85)$$

Finally, the CMRR for the basic MOST difference amplifier is given by

$$\text{CMRR} = \frac{A_d}{A_c} = \frac{1 + 2g_m R_S}{2} \qquad \textbf{(6.2–86)}$$

Note that in order to maximize the CMRR, we must maximize g_m and R_S. However, g_m is characteristically low for FETs, and making R_S larger requires more chip area. This dilemma is solved by replacing R_S with a current source, as we will see.

EX. 6.7

Basic MOST Difference Amplifier Example

Consider the difference amplifier circuit of Fig. 6.16. Let the transistors be identical with $g_m = 10$ mS and $g_o = 0$. For resistor values of $R_S = R_D = 10$ kΩ, calculate values for A_c, A_d, and CMRR.

Solution: Substituting values directly into (6.2–76), (6.2–85), and (6.2–86) yields

$$A_c = \frac{g_m R_D}{1 + 2g_m R_S} = \frac{(10)(10)}{1 + 2(10)(10)} \simeq 0.5$$

$$A_d = \frac{g_m R_D}{2} = \frac{(10)(10)}{2} = 50$$

$$\text{CMRR} = \frac{50}{0.5} = 100$$

Integrated-Circuit MOST Difference Amplifier. Figure 6.17 displays a MOST difference amplifier using N-channel E-D transistors. The resistor R_S in the circuit of Fig. 6.16a has been replaced with a MOST current source, and the drain resistors have each been replaced with active loads using MOSTs. Both of these replacements conserve chip area and are essential in implementing the MOST difference amplifier in integrated-circuit form.

The MOST current source consists of T_3, T_4, and T_5 and is analogous to the BJT current source. Transistor T_5 is the active load for T_4 and sets the reference current, which is mirrored

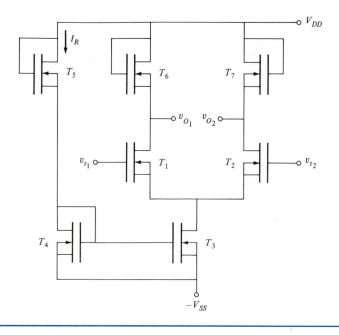

FIGURE 6.17   IC MOST Difference Amplifier

in T_3 ($I_{DC} = I_R$) since $v_{gs_3} = v_{gs_4}$. Transistors T_6 and T_7 are the active loads for T_1 and T_2, respectively.

The AC analysis of the IC MOST difference amplifier of Fig. 6.17 is now easily carried out. We use the results of the basic MOST difference amplifier of Fig. 6.16 and substitute the AC equivalent of the I_C replacement elements. First, the AC equivalent of T_3 is a resistor, $1/g_{o_3}$, corresponding to the output conductance of T_3. Second, the equivalent of T_6 and T_7 (which are identical transistors) are resistors, $1/g_{m_6}$, corresponding to the equal transconductance of T_6 and T_7, since the connection of gate to drain forces $v_{DS} = v_{GS}$. Hence, $g_m = \partial i_D / \partial v_{GS}|_Q = \partial i_D / \partial v_{DS}|_Q = g_o$.

By substitution into (6.2–76) and (6.2–85), we have

$$A_c = \frac{g_m(1/g_{m_6})}{1 + 2g_m(1/g_{o_3})} \tag{6.2–87}$$

and

$$A_d = \frac{g_m(1/g_{m_6})}{2} \tag{6.2–88}$$

where g_m remains the transconductance of T_1 and T_2. The corresponding CMRR is therefore

$$\text{CMRR} = \frac{1 + (2g_m/g_{o_3})}{2} \tag{6.2–89}$$

The primary concern in maximizing (6.2–89) is to maximize g_m/g_{o_3}. This task is conveniently accomplished in ICs by adjusting the geometries (W/L ratios) of T_1 and T_2 relative to T_3. Similarly, g_m/g_{o_3} is made large to afford $A_d \gg 1$. The geometries are also designed such that $2g_m/g_{o_3} \gg g_m/g_{m_6}$ to allow $A_c \ll 1$.

─────────────── EX. 6.8

IC MOST Difference Amplifier Example

Calculate A_c, A_d, and CMRR for the difference amplifier circuit of Fig. 6.17. Assume that T_1 and T_2 are identical

with $g_{m_1} = g_{m_2} = 10$ mS. Also, $1/g_{m_6} = 20$ kΩ and $1/g_{o_3} = 50$ kΩ.

Solution: From (6.2–87), (6.2–88), and (6.2–89) and by direct substitution, we have

$$A_c = \frac{g_m(1/g_{m_6})}{1 + 2g_m(1/g_{o_3})} = \frac{(10)(20)}{1 + 2(10)(50)} = 0.2$$

$$A_d = \frac{g_m(1/g_{m_6})}{2} = \frac{(10)(20)}{2} = 100$$

$$\text{CMRR} = \frac{100}{0.2} = 500$$

6.2.6   Basic JFET Difference Amplifier

The basic JFET difference amplifier circuit is shown in Fig. 6.18. JFETs are used instead of BJTs because of the much larger input impedance they provide, although the difference-mode amplification is reduced. Quite often, a JFET is used as the input stage of a multistage difference amplifier with succeeding stages consisting of one or more BJT difference amplifiers. The entire

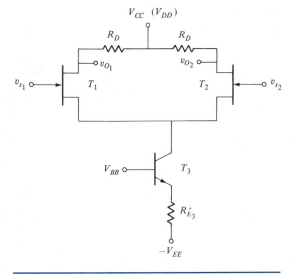

FIGURE 6.18   JFET Difference Amplifier

structure can be easily implemented in integrated-circuit form because JFET and BJT fabrication processes are compatible. Thus, higher difference-mode amplification is possible. The analysis of Fig. 6.18 is essentially the same as that for the basic MOST difference amplifier of Fig. 6.16a except that the current source has been used as a replacement for the resistor R_S. For the JFET case, we obtain by analogy

$$A_c = \frac{g_m R_D}{1 + (2g_m/h_{ob_3})} \qquad (6.2\text{--}90)$$

and

$$A_d = \frac{g_m R_D}{2} \qquad (6.2\text{--}91)$$

Once again, we observe that the difference-mode gain depends upon the product $g_m R_D$.

6.3   OPERATIONAL AMPLIFIERS

Operational amplifiers, or op amps, are integrated circuits that provide very high difference-mode gain (typically, $A_d \simeq 10^5$) through the use of multistage amplifiers that are direct coupled. Transistors are used wherever possible to replace load resistors and bias resistors in order to conserve chip area on the IC. As far as chip area is concerned, transistors occupy the least amount of space compared to resistors and capacitors. Since capacitors occupy the most space, they are used in ICs only when absolutely necessary. Inductors never appear in any IC circuit because they require an intolerable amount of chip area. Op amps also provide high input impedance and low output impedance by using input and output stages that have these properties.

The operational amplifier derives its name from the fact that in its infancy, its primary use was to perform mathematical operations (addition, subtraction, integration, and differentiation). Presently, however, op amp applications are much more widely diversified (both linear and nonlinear), as we will see in Chapter 7.

6.3.1   Basic Constituents and Circuit Symbol

Figure 6.19a shows the essential constituents of an op amp using basic IC building blocks. The first two stages are difference amplifiers, each

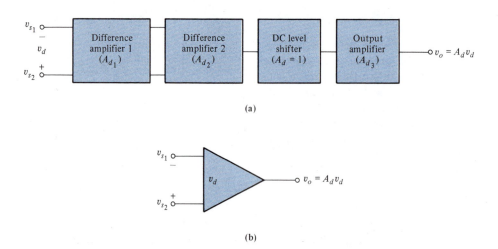

(a)

(b)

FIGURE 6.19   Operational Amplifier: (a) Block diagram, (b) Circuit symbol

of which provides difference-mode voltage amplification and common-mode voltage rejection. In special op amps, more difference amplifiers are used.

In order for the op amp to possess a large input impedance, difference amplifier 1 must have this property. A JFET or Darlington-pair difference amplifier is thus often used here. The outputs of the difference amplifiers can be either double-ended (op amp 1) or single-ended (op amp 2), as displayed in Fig. 6.19a. Since DC as well as AC is amplified by each difference amplifier, the third stage is used to remove the DC voltage. This DC level shifter, as it is called, translates the DC voltage such that the output voltage (v_o in Fig. 6.19a) is dependent only upon v_{s_1} and v_{s_2}. The level shifter is also designed such that AC voltages are unchanged from input to output, as we will see.

The output stage of the op amp usually provides additional amplification but always low output impedance. Often an emitter follower is used in the case of an IC BJT op amp or a source follower in an IC MOST op amp.

We define the overall difference-mode voltage amplification for the op amp as A_d, which is then dependent upon the individual-stage difference-mode amplifications. Similarly, we define the overall common-mode voltage amplification as A_c, which is dependent upon the individual-stage common-mode amplifications.

The output voltage is then given by

$$v_o = A_d v_d + A_c v_c \qquad (6.3\text{--}1)$$

Furthermore, if $A_c \ll A_d$ (which is the case for modern op amps), then the output voltage is expressed simply as

$$v_o = A_d v_d \qquad (6.3\text{--}2)$$

Recalling that $v_d = v_{s_2} - v_{s_1}$, we have

$$v_o = A_d(v_{s_2} - v_{s_1}) \qquad (6.3\text{--}3)$$

Considering the voltage difference in (6.3–3), we may observe an interesting result. The negative sign indicates that v_{s_1} and v_o are 180° out of phase (inverted), whereas v_o and v_{s_2} are in phase (noninverted). Actually, this result has already been indicated explicitly at the input terminals of the op amp in Fig. 6.19a by the minus and plus signs placed there to denote the polarity of v_d. The significance of this discussion is that a voltage applied to the negative input terminal of the op amp appears inverted at the output terminal, and a voltage applied to the positive input terminal appears noninverted at the output terminal.

The preceding discussion leads us to the op amp circuit symbol displayed in Fig. 6.19b. The triangular symbol is intended to imply amplification. With inputs v_{s_1} and v_{s_2} applied at the terminals indicated in Figs. 6.19a and b, the output voltage v_o is given by (6.3–2) or (6.3–3).

Note that in Fig. 6.19a each of the individual stages will require DC biasing for proper operation. However, this biasing is implicitly assumed in Fig. 6.19a as well as in Fig. 6.19b. Only the voltages applied to the negative and positive input terminals and the output voltage are kept track of using the op amp circuit symbol. This convention is extremely convenient, as we will see in Chapter 7, which describes numerous op amp applications.

6.3.2   DC Level Shifter

The third stage or building block of the basic op amp shown in Fig. 6.19a is called a *DC level shifter*. This stage performs the function of removing any undesirable DC voltage, which does not imply reducing the DC level to zero because the next stage often requires a small DC input.

A circuit that performs DC level shifting using BJTs is shown in Fig. 6.20a. This configuration is sometimes called a *cascode amplifier*. Transistor T_1, along with R'_E and V_{BB}, provide the function of a constant current source by mirroring the current in the reference transistor (not shown). Transistor T_2 is essentially connected in the emitter follower configuration.

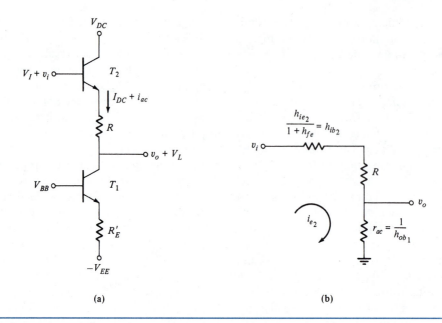

FIGURE 6.20   DC Level Shifter or Cascode Amplifier (a) Circuit configuration, (b) AC equivalent circuit

The constant current of the current source was determined previously in (6.2–29). The magnitude of I_{DC}, along with R (in Fig. 6.20a), is then used to translate the DC level of the incoming voltage V_I to the desired level V_L. From Fig. 6.20a, we observe that

$$V_L = V_I - 0.7 - I_{DC}R \qquad (6.3\text{–}4)$$

Thus, if it is desired that $V_L = 0$, the product $I_{DC}R$ is selected as $I_{DC}R = V_I - 0.7$.

As far as AC is concerned, we analyze Fig. 6.20b, which displays the AC equivalent circuit for the DC level shifter. Note that we have transformed the current through the base–emitter junction of T_2 from i_{b_2} to i_{e_2}. A corresponding reduction in input resistor value by the same amount results, maintaining KVL around the current source loop. From Fig. 6.20b, we see that the input signal voltage divides across each resistor, and by inspection

$$v_o = \frac{1/h_{ob_1}}{(1/h_{ob_1}) + h_{ib_2} + R} v_i \qquad (6.3\text{–}5)$$

Thus, if $1/h_{ob_1} \gg R + h_{ib_2}$, then $v_o \simeq v_i$. Note that another restriction is placed on R in addition to its DC level shifting function. Both restrictions must be taken into account in selecting a value for R.

DC Level Shifter Example

The bias portion of the circuit of Fig. 6.20a is to be designed such that the DC output voltage $V_L = 0$ when $V_I = 5$ V. For I_{DC} selected as 1 mA, determine the required value of resistance R. For the current source portion of Fig. 6.20a, determine R'_E for $V_{BB} = 0$ and $V_{EE} = 5$ V.

Solution: From the level shifter portion of Fig. 6.20a, we have

$$V_L = V_I - 0.7 - I_{DC}R = 0$$

Therefore,

$$R = \frac{V_I - 0.7}{I_{DC}} = \frac{5 - 0.7}{1} = 4.3 \text{ k}\Omega$$

For the current source portion of the circuit, writing KVL

yields

$$V_{EE} - 0.7 = R'_E I_E .$$

Therefore,

$$R'_E = \frac{V_{EE} - 0.7}{I_E} \simeq \frac{5 - 0.7}{1} = 4.3 \text{ k}\Omega$$

6.4 OP AMP EXAMPLE CIRCUIT

Figure 6.21 shows an example circuit for an op amp that uses two BJT difference amplifiers along with a DC level shifter and an emitter

follower for the output amplifier. The current sources, DC supplies, and collector resistors dictate the location of the Q point for each difference amplifier as well as the magnitude of the voltage amplification.

The first difference amplifier is a Darlington pair. It provides an increased input impedance from that available with a BJT difference amplifier (such as that used for difference amplifier 2). Both difference amplifiers provide approximately the same amount of voltage amplification for the element values shown in Fig. 6.21, as we will see. The current source values for T_3 and T_6 are selected as 2 mA. This choice, along with the collector resistors, provides collector–emitter bias

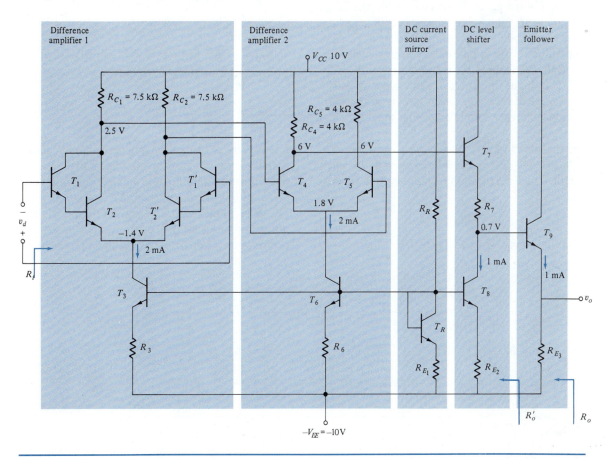

FIGURE 6.21 Basic Op Amp Circuit Using BJTs

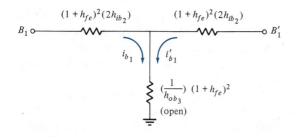

FIGURE 6.22   Equivalent Circuit for R_i

voltages that place the Q point of the BJTs in the amplifier region of operation when $v_d = 0$. This statement is easily verified using the methods of Chapter 4. We will analyze the circuit of Fig. 6.21 and determine general expressions for the input impedance R_i, the output impedance R_o, and the difference-mode voltage amplification v_o/v_d. Specific values of these parameters are calculated in Ex. 6.10. We assume h_{fe} is the same for all BJTs.

The input impedance expression is easily obtained from the equivalent circuit of the Darlington difference amplifier of Fig. 6.21, which is shown in Fig. 6.22. Since $1/h_{ob_3}$ is a large resistor (on the order of 50 kΩ), the equivalent resistor, $(1/h_{ob_3})(1 + h_{fe})^2$, is considered to be an open circuit. Thus, the input impedance (which is resistive) is

$$R_i = 2(1 + h_{fe})^2 (2h_{ib_2}) \qquad (6.4\text{–}1)$$

where $h_{ib_2} = h'_{ib_2} = 0.025/I_{E_2} \simeq 0.025/I_{C_2}$.

To obtain the expression for output impedance, the equivalent circuit for the output portion of Fig. 6.21, as shown in Fig. 6.23, is used. In this AC small-signal equivalent circuit, the resistor values are adjusted to the AC emitter current of T_9. Treating $1/h_{ob_8}$ and $1/h_{ob_5}$ as open circuits, the output impedance is, by inspection,

$$R_o = R_{E_3} \| R'_o \qquad (6.4\text{–}2)$$

where

$$R'_o = \frac{h_{ie_9}}{1 + h_{fe}} + \frac{R_7}{1 + h_{fe}} + \frac{R_{C_5} + h_{ie_7}}{(1 + h_{fe})^2} \qquad (6.4\text{–}3)$$

In practice, R_{E_3} is on the order of 1 kΩ, and R'_o is on the order of 10 Ω. Thus,

$$R_o \simeq R'_o \qquad (6.4\text{–}4)$$

Finally, to obtain the difference-mode voltage amplification v_o/v_d for the op amp of Fig. 6.21, we determine the amplification for each separate stage and multiply them together. We assume that the DC level shifter and the emitter follower have unity gain and that the common-mode amplification is zero. Therefore, the voltage amplification

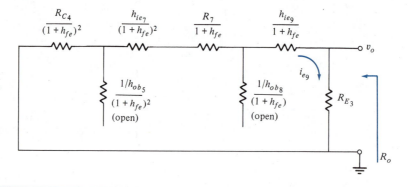

FIGURE 6.23   Equivalent Circuit for R_o

is given by

$$\frac{v_o}{v_d} = A_{d_1} A_{d_2} \tag{6.4-5}$$

where A_{d_1} and A_{d_2} are the difference-mode ampli-fication factors for the two difference amplifiers.

For the Darlington difference amplifier, the output voltage is the difference in voltage between the collector terminals of T_2 and T'_2. This double-ended voltage essentially doubles the amplifica-tion of the first difference amplifier since $v_{c_2} - v'_{c_2} = 2A_d v_d$ from (6.2–16) and (6.2–17). Also, the collector resistors R_{c_1} and R_{c_2} are in parallel with the input impedances to difference amplifier 2 ($R_{i_4} = R_{i_5}$). Employing these modifications, the amplification factor for the Darlington difference amplifier is obtained from (6.2–54) as

$$A_{d_1} = \frac{R_{C_1} \| R_{i_4}}{2h_{ib_2}} \tag{6.4-6}$$

For difference amplifier 2, the difference-mode amplification is obtained directly from (6.2–25) as

$$A_{d_2} = \frac{R_{C_4}}{2h_{ib_4}} \tag{6.4-7}$$

since the input impedance of the DC level shifter is an open circuit in comparison to R_{C_4}.

--------- EX. 6.10

Op Amp Numerical Example

For the basic op amp of Fig. 6.21, calculate the values of R_i, R_o, and v_o/v_d using the element values indicated. Let $h_{fe} = 100$ for each of the transistors. Calculate values for h_{ie} and h_{ib} based upon the DC currents and voltages indicated in Fig. 6.21.

Solution: The input impedance is calculated from (6.4–1) as follows:

$$R_i = 2(1 + h_{fe})^2[2(0.025/I_{c_2})] \tag{1}$$

Substituting values yields

$$R_i = 2(101)^2[2(0.025/1)] = 1.02 \text{ M}\Omega$$

The output impedance is obtained from (6.4–3) and (6.4–4) as follows:

$$R_o \simeq R'_o \frac{h_{ie_9}}{1 + h_{fe}} + \frac{R_7}{1 + h_{fe}} + \frac{R_{C_4} + h_{ie_7}}{(1 + h_{fe})^2} \tag{2}$$

where

$$h_{ie_7} = h_{ie_9} = \frac{0.025}{I_{B_9}} \simeq \frac{0.025 h_{fe}}{I_{C_9}}$$

$$= \frac{0.025(100)}{1} = 2.5 \text{ k}\Omega \tag{3}$$

and

$$R_7 = \frac{V_{B_7} - 1.4}{1} = \frac{6 - 1.4}{1} \simeq 4.6 \text{ k}\Omega \tag{4}$$

The voltage across R_7 reduces the DC level at the base of T_9 to 0.7 V so that T_9 operates as an emitter follower. Thus, substituting values yields

$$R_o = \frac{2.5}{101} + \frac{4.6}{101} + \frac{4 + 2.5}{101} \simeq 135 \ \Omega$$

Finally, the voltage amplification is obtained from (6.4–5), (6.4–6), and (6.4–7) as follows:

$$\frac{v_o}{v_d} = \left(\frac{R_{C_1} \| R_{i_4}}{2h_{ib_2}} \right) \left(\frac{R_{C_4}}{2h_{ib_4}} \right) \tag{5}$$

where

$$h_{ib_2} = h_{ib_4} = \frac{0.025}{I_{E_2}} \simeq \frac{0.025}{1} = 25 \ \Omega \tag{6}$$

and with $1/h_{ob_6}$ as an open circuit,

$$R_{i_4} = 2h_{ie_4} = \frac{2(0.025)}{I_B} = \frac{2(2.5)}{I_C} = 5 \text{ k}\Omega \tag{7}$$

Thus, by substitution,

$$\frac{v_o}{v_d} = \left[\frac{7.5 \| 5}{2(0.025)} \right] \left[\frac{4}{2(0.025)} \right] = (60)(80) = 4800 \tag{8}$$

Although these values seem adequate, many op amps use more elaborate circuitry and possess much improved values.

6.5   OP AMP PARAMETERS

Ideal op amps have the following characteristics:

- Infinite difference-mode voltage amplification
- Zero common-mode voltage amplification
- Infinite input impedance
- Zero output impedance
- Zero output offset voltage ($v_o = 0$ for $v_d = 0$)
- No restrictions on output voltage time rate of change

Practical op amps have parameters that depart from these ideal characteristics. As we have observed, op amps can be designed such that their values of amplification, input impedance, and output impedance approach these ideal values. Departures will always be present, and additional parameters must be defined in order to describe the performance of op amps more completely.

Typical op amp parameter values are given in Table 6.1. The values listed are conservative, and special-purpose op amps are available with improved values. Parameters listed in Table 6.1 that have not yet been defined are described in the following sections.

6.5.1   Input Bias Current and Input Current Offset

Even when the input signals applied to an op amp are zero, the input terminals of the op amp conduct DC current, or *input bias current*. This current is quite small, and if the input difference amplifier uses FETs, this bias current is negligible. However, for a BJT difference amplifier, the input bias current is on the order of 100 nA and can cause erroneous output signals. These signals can be reduced to zero by including an additional resistor in the basic amplifier circuit.

Figure 6.24 shows the basic op amp amplifier circuit with zero input signal voltage(s) and nonzero input bias currents I_{B_1} and I_{B_2}. Placing the additional resistor $R_F \| R_S$ as shown produces an input voltage at the positive terminal given by

$$v_p = -I_{B_1}(R_F \| R_S) \qquad (6.5-1)$$

Similarly, with I_{B_2} entering the op amp, at the negative terminal, the voltage there is

$$v_n = -I_{B_1}(R_S \| R_F) \qquad (6.5-2)$$

If the input bias currents I_{B_1} and I_{B_2} are equal, then the difference voltage v_d between the two in-

TABLE 6.1   Typical Op Amp Parameter Values

Parameter	Value (Conservative)
Difference-mode amplification	10^5
Common-mode rejection ratio (CMRR)	100 dB
Input impedance	50 kΩ
Output impedance	100 Ω
Input bias current	100 nA
Input current offset	10 nA
Input voltage offset	2 mV
Input voltage temperature drift	5 μV/°C
Slewing-rate limit	20 V/μs

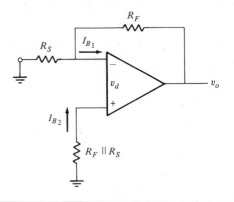

FIGURE 6.24   Basic Op Amp Circuit with No Input Signal

put terminals will be zero, and the output voltage will be zero. Hence, the additional resistor $R_F \| R_S$ cancels the erroneous output voltage due to the input bias current.

The difference between the input bias current, $I_{B_2} - I_{B_1}$, is called the *input current offset*. As indicated in Table 6.1, the input current offset is usually an order of magnitude less that the input bias current. Its effect may not be negligible, however, when the external resistor values are large.

6.5.2   Input Voltage Offset and Temperature Drift

The *input voltage offset* is defined as the amount of voltage that must be applied between the input terminals in order to reduce the output voltage offset to zero. This effect is caused by imbalances in the transistors and components. Note that if the input voltage offset is only 1 mV and the difference-mode amplification factor is 10^3, then the corresponding output voltage offset will be 1 V. If the amplification is reduced by negative feedback, the magnitude of the offset voltage is also reduced. Quite often, op amps have special terminals that are used to eliminate the input and corresponding output voltage offsets by applying a small voltage between them.

The input voltage offset also drifts with temperature changes. The amount of *temperature drift* is typically 5 μV/°C, as indicated in Table 6.1. If high precision is desired, a low-drift op amp should be used.

6.5.3   Slewing-Rate Limit

The *slewing-rate limit* (or simply, slew-rate limit) indicates how fast the output voltage can change with time. As a large-signal parameter, it defines the maximum rate at which the output can change from its most positive to its most negative values. A conservative value of 20 V/μs is indicated in Table 6.1. Thus, if the op amp has sat-

uration voltages of ± 10 V, the maximum switching time from one saturation voltage to the other will be

$$t = \frac{20}{20} = 1 \ \mu s$$

6.5.4   Frequency Compensation

Although *frequency compensation* is not listed as a parameter in Table 6.1, it deserves mention here. A more detailed description will be given in Chapter 8. Because of the internal resistors and parasitic capacitors (due to PN junctions and MOS layers) of the op amp, the voltage amplification is a function of frequency. The overall effect is that an op amp behaves in a manner similar to an *RC* filter (from Chapter 1) or actually several *RC* filters in cascade, except that the voltage gain is greater than unity. The gain falls off as the frequency is increased, and the frequency at which this begins to occur can be as low as 10 Hz. In fact, op amps are often intentionally designed with frequency compensation to prevent undesirable oscillations. Two types of frequency compensation are used. One is *internal compensation* and consists of a capacitor fabricated as part of the IC, which is allowable provided that the magnitude of C is not too large (in the pF range). The other is *external compensation* and is used with op amps that have supplementary terminals to which an external capacitor can be attached that provides the desired frequency compensation. A capacitor connected between B_4 and C_4 in Fig. 6.21 provides the desired frequency compensation, as we will see in Chapter 8.

6.6   SPECIFIC TYPES OF INTEGRATED-CIRCUIT OP AMPS

The Fairchild μA709 was the first op amp to be used extensively. It was designed by Robert Widlar and became popular in 1965. Later, while working for National Semiconductor Corporation, Widlar developed the LM301. Both of these op amps required external frequency compensation. Shortly afterward, Fairchild developed the 741 op amp with internal compensation. In most applications, the 741 could be used with only the external feedback components. This op amp thus became extremely popular with many other companies, and they developed their own version of the 741.

The μA709, LM301, and 741 all use BJTs. However, in order to improve the input impedance of op amps, a JFET difference amplifier is often used in the first stage. JFET amplifiers cannot be used throughout because of their reduced amplification in comparison with BJT amplifiers. The combination of JFET and BJT difference amplifiers on the same op amp chip creates what is called a *BiFET op amp*. Because of the high input impedance and high differential voltage amplification, BiFET op amps are used extensively.

The prefix used with an op amp numerical designation identifies the particular manufacturer. Table 6.2 lists the prefixes used for some IC manufacturers. After a specific IC is introduced by a particular manufacturer, other companies commonly begin producing an equivalent IC and thus provide multiple sourcing.

TABLE 6.2   Op Amp Prefixes

Prefix	Semiconductor Manufacturer
μA	Fairchild Semiconductor
LM, LH, NH, DM	National Semiconductor
MC, ML, M	Motorola Semiconductor
OP	Precision Monolithics
CA	RCA
RC, RM, RV	Raytheon
TL, TLC	Texas Instruments
N, NE, S, SE, SA	Signetics
SG	Silicon General

Various types of packages are used with different op amps and ICs. Standard IC packages are indicated in Fig. 6.25. Various shapes and pin configurations are displayed.

The N and J packages have 14 and 16 pins, respectively, in the dual in-line package (DIP) configuration (given this name because they have two rows of pins). The D and FH,FK packages conserve space and are referred to as SMC for surface mounted components.

Data Sheet A6.1 in the appendix to this chapter contains a selection guide for various op amps and lists the parameters for each type. The column labeled DESCRIPTION indicates the particular specialty for the particular op amp. The abbre-

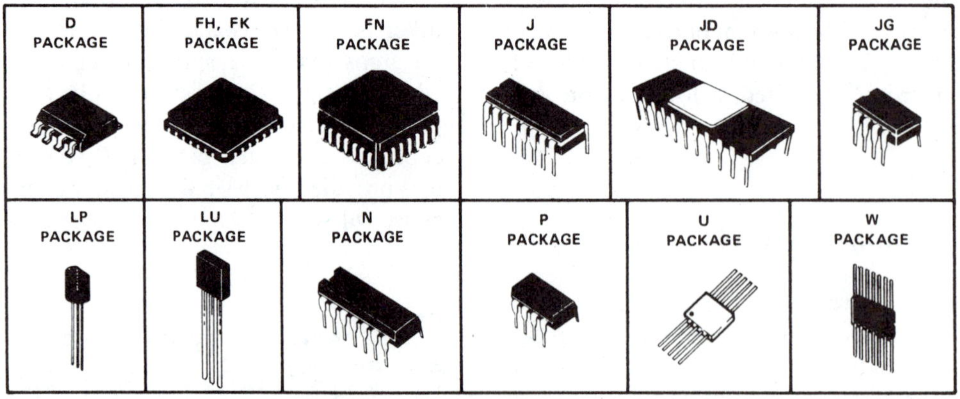

FIGURE 6.25   Standard IC Packages (Courtesy of Texas Instruments, Inc.)

viations used for columns that are not self-explanatory are defined as follows:

- B_1 is the unity gain bandwidth.
- SR is the slew rate.
- V_{IO} is the input offset voltage.

- I_{IB} is the input bias current.
- A_{VD} is the differential voltage amplification.

Appendix 4 at the end of the book provides additional op amp information and various op amp data sheets.

CHAPTER 6 SUMMARY

- Multiple-transistor amplifier circuits are used in order to improve the properties of single-transistor amplifier circuits. For voltage amplifiers, these improved properties include increased voltage amplification, increased input impedance, and decreased output impedance.
- Capacitively coupled multistage amplifiers with and without inductors (rf chokes) improve the voltage amplification considerably but have undesirable disadvantages. These disadvantages include poor performance at low frequency, physical size that is undesirably large, and physical size that prevents their use in ICs.
- Direct-coupled multistage amplifiers provide a solution to the problems associated with capacitively coupled amplifiers but introduce the intolerable problems of voltage drift and DC voltage amplification. These problems can be corrected by using negative feedback, but then the amplification is reduced.
- A difference amplifier is a unique form of direct-coupled amplifier and is a primary building block of linear (analog) integrated circuits. A signal voltage is applied at each of the two input terminals, and the voltage at the output terminal is an amplified version of the difference in input voltages. Difference amplifiers are the first and second stages used in an op amp, which behaves basically like a difference amplifier.
- The basic BJT difference amplifier is analyzed by using the small-signal equivalent circuit. The analysis is simplified by using the common-mode difference-mode three-step proce-

dure to determine i_{e_1} and i_{e_2}. This procedure involves defining $v_d = v_{s_2} - v_{s_1}$ and $v_c = (v_{s_2} + v_{s_1})/2$, which when rearranged yield $v_{s_1} = v_c - (v_d/2)$ and $v_{s_2} = v_c + (v_d/2)$. The three-step procedure involves determining i_{e_1} and i_{e_2} for $v_d = 0$, determining i_{e_1} and i_{e_2} for $v_c = 0$, and superimposing the results. The output voltages are then obtained from $v_{o_1} = -i_{c_1}R_C \simeq -i_{e_1}R_C$ and $v_{o_2} = -i_{c_2}R_C \simeq -i_{e_2}R_C$.
- The common-mode gain A_c is defined as v_o/v_c, and the difference-mode gain A_d is defined as v_o/v_d. Ideally, a difference amplifier has $A_d = \infty$ and $A_c = 0$. A figure of merit for difference amplifiers is the common-mode rejection ratio, defined as CMRR $= A_d/A_c$. For convenience, this ratio is often expressed in decibels where CMRR$|_{dB} = 20 \log A_d/A_c$.
- BJT difference amplifiers are often used with a DC current source replacing R_E. This replacement provides a smaller common-mode rejection ratio and less DC power dissipation by introducing a large AC resistance between the emitter terminal of T_1 and T_2 and ground ($r_{ac} = 1/h_{ob}$).
- DC current sources are quite prevalent in ICs and replace bias resistors that require too much chip area. The Widlar source is used in low-current (0.1 mA) applications. At higher current levels, the emitter resistor is removed, which results in a current mirror arrangement.
- The Darlington-pair configuration is a special connection of two BJTs that offers increased current amplification and larger input impedance. Two Darlington pairs are used to re-

place the BJTs of the basic difference amplifier to form a Darlington-pair difference amplifier, which provides larger A_d, R_i, and CMRR.
- MOST and JFET difference amplifiers have increased input impedance as compared to BJT difference amplifiers. However, the difference-mode voltage amplification is reduced.
- A basic op amp is composed of two difference amplifiers, a DC level shifter, and an output amplifier. The DC level shifter is necessary to remove the DC component of voltage from the output. These stages are direct coupled and provide large voltage amplification. A BiFET op amp is the combination of a FET difference amplifier as the input stage and succeeding BJT stages on the same IC chip.
- Ideal op amps are characterized by $A_d = \infty$, $A_c = 0$, $R_i = \infty$, and $R_o = 0$. Typical op amps have $A_d \simeq 10^5$, $A_c = 0.1$, $R_i = 50$ kΩ, and $R_o = 100$ Ω.
- Typical values for other op amp parameters are input bias current $= 100$ nA, input current offset $= 10$ nA, input voltage offset $= 2$ mV, input voltage temperature drift $= 5$ μV/°C, and slewing-rate limit $= 20$ V/μs.
- An op amp must have internal or external frequency compensation in order to avoid undesirable oscillations. Frequency compensation involves intentionally introducing capacitance such that the voltage amplification reduces as the frequency is increased.

CHAPTER 6 PROBLEMS

6.1
For the two-stage amplifier circuit of Fig. P6.1, sketch the small-signal equivalent circuit and determine the voltage amplification v_o/v_s. Let $h_{ie_1} = h_{ie_2} = 0.5$ kΩ and $h_{fe_1} = h_{fe_2} = 100$. Also, let $R_{B_1} = R_{C_1} = R_{B_2} = R_{C_2} = 2$ kΩ.

alent circuit. Let $h_{fe_1} = h_{fe_2} = 100$ and $h_{ie_1} = h_{ie_2} = 0.2$ kΩ. Also, let $R_1 = 20$ kΩ, $R_2 = 10$ kΩ, and $R_L = R_C = 0.5$ kΩ.

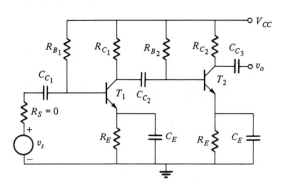

FIGURE P6.1

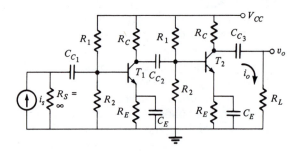

FIGURE P6.3

6.2
Repeat Problem 6.1 with $R_S = 1$ kΩ.

6.3
For the two-stage circuit of Fig. P6.3, determine the current amplification by using the small-signal equiv-

6.4
Repeat Problem 6.3 with $R_S = 1$ kΩ.

6.5
Repeat Problem 6.1 for the following circuit element values: $R_{B_1} = R_{B_2} = 10$ kΩ and $R_{C_1} = R_{C_2} = 5$ kΩ. Also, let $R_S = 5$ kΩ and $h_{ie} = 1.3$ kΩ, while $h_{fe} = 100$ for each BJT.

6.6

The inductively coupled amplifier of Fig. P6.6 uses BJTs with $h_{fe} = 100$ and $h_{ie} = 1$ kΩ. Using the small-signal equivalent circuit, determine the voltage amplification v_o/v_s for $R_s = 0$ and 10 kΩ.

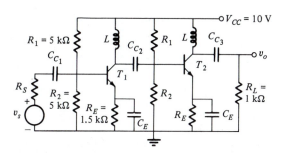

FIGURE P6.6

6.7

Repeat Problem 6.6 for the following circuit element values: $R_1 = 13.7$ kΩ, $R_2 = 37$ kΩ, $R_E = 1$ kΩ, and $R_L = 1$ kΩ. Use $h_{fe} = 100$ and $h_{ie} = 0.5$ kΩ.

6.8

Use a small-signal equivalent circuit to determine the voltage amplification v_o/v_s for the two-stage JFET amplifier of Fig. P6.8. Use $g_m = 50$ mS and $g_o = 0$ with $R_G = R_D = 1$ kΩ. Calculate v_o/v_s for $R_s = 0$ and 10 kΩ.

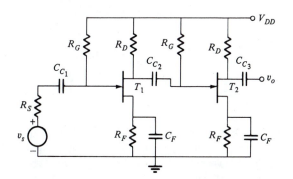

FIGURE P6.8

6.9

Repeat Problem 6.8 with $R_G = 100$ kΩ.

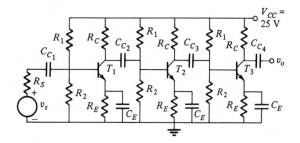

FIGURE P6.10

6.10

Determine the voltage amplification v_o/v_s for the three-stage BJT amplifier of Fig. P6.10. Use element values given by $R_1 = 16$ kΩ, $R_2 = 6$ kΩ, $R_C = 1.6$ kΩ, $R_S = 0$, and $R_E = 1$ kΩ with $h_{fe} = 100$ and $h_{ie} = 0.43$ kΩ.

6.11

Repeat Problem 6.10 with $R_S = 10$ kΩ.

6.12

A BJT amplifier is shown in Fig. P6.12. Using the element values indicated, calculate the voltage amplification for $R_S = 0$. Let $h_{fe_1} = h_{fe_2} = 100$ and $h_{ie_1} = h_{ie_2} = 0.26$ kΩ (this may be verified from $h_{ie} = 0.025/I_B$ with v_s removed).

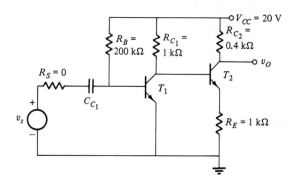

FIGURE P6.12

6.13

Repeat Problem 6.12 with $R_S = 10$ kΩ.

6.14

For the direct-coupled JFET amplifier of Fig. P6.14, determine the voltage amplification v_o/v_s. Use a small-signal equivalent circuit with $g_m = 10$ mS and $g_o = 0$ for each JFET. Note that since the DC is amplified, R_{F_2} must be increased (in order to provide the correct polarity of DC bias to G_2) and the amplification is disasterously reduced.

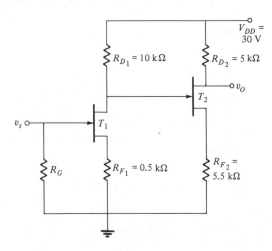

FIGURE P6.14

6.15

Repeat Problem 6.12, but refer to Fig. P6.12 and add a negative feedback loop as shown in Fig. P6.15. Note that $r_f = 200 \, \Omega$ and $R_F = 5$ kΩ. This results in the same Q point for T_1 and T_2 in both problems.

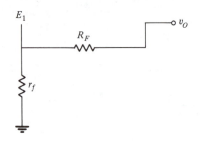

FIGURE P6.15

6.16

Determine the voltage amplification for the two-stage amplifier of Fig. P6.16. Let $h_{fe_1} = h_{fe_2} = 100$. Calculate h_{ie_1} and h_{ie_2} using $h_{ie} = 0.025/I_B$.

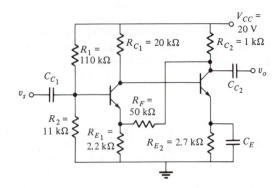

FIGURE P6.16

6.17

For the BJT difference amplifier of Fig. P6.17, determine the common-mode and difference-mode voltage gain and the CMRR. Let $T_i = T_2$, $h_{fe} = 100$, and $h_{ie} = 528 \, \Omega$ (calculated from $h_{ie} = 0.025/I_B$). Use a small-signal equivalent circuit to carry out the three-step procedure using v_c and v_d.

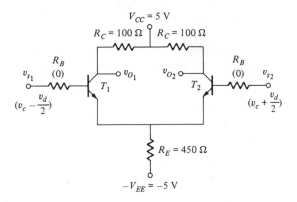

FIGURE P6.17

6.18

Replace R_E in the circuit of Fig. P6.17 with a constant current source that has an equivalent AC resistance between E and ground of value 100 kΩ (r_{ac}). Calculate the resulting values of A_c and A_d by using a modified small-signal equivalent circuit.

6.19

A difference amplifier with a constant current source is shown in Fig. P6.19. Determine A_c, A_d, and the CMRR by using a small-signal equivalent circuit and $T_1 = T_2$, $h_{fe} = 100$, $h_{ie} = 2.5$ kΩ, and $r_{ac} = 100$ kΩ (for T_3).

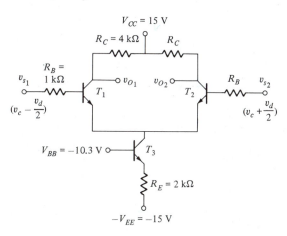

FIGURE P6.19

6.20
Repeat Problem 6.17 with different resistors of value $R_B = R_E = 5$ kΩ and $R_C = 10$ kΩ. Let $T_1 = T_2$ with $h_{fe} = 100$ and $h_{ie} = 5.8$ kΩ.

6.21
Analyze the difference amplifier of Fig. P6.21 to determine A_c, A_d, and the CMRR. Let $T_1 = T_2 = T_A$, $h_{fe} = 100$, and $h_{ie} = 0.5$ kΩ. Hint: As an intermediate step, obtain the common-mode and difference-mode gain between v_{b_1} and v_{o_1}. Then, multiply this gain by the voltage amplification provided by transistors T_A.

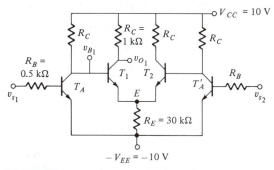

FIGURE P6.21

6.22
Repeat Problem 6.21 for the circuit in which R_E is replaced with a current source that maintains the

same Q points for T_1 and T_2 but that provides an equivalent AC resistance between E and ground of value 50 MΩ. Also, compare the CMRR for each case and comment on the improvement.

6.23
Using a small-signal equivalent circuit for the Darlington amplifier of Fig. P6.23, determine R_i, R_o, and v_o/v_s. Let $h_{ie_1} = 1$ kΩ and $h_{fe_1} = h_{fe_2} = 100$. Also determine i_o/i_i [use the relation $h_{ie_1} = (1 + h_{fe})h_{ie_2}$].

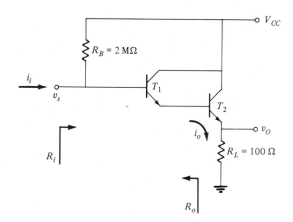

FIGURE P6.23

6.24
For the voltage amplifier circuit of Fig. P6.24, determine R_i, R_o, and v_o/v_s by using a small-signal equivalent circuit. Let $g_m = 100$ mS, $g_o = 0$, $h_{fe} = 100$, and $h_{ie} = 1$ kΩ. Also, let $R_G = 1$ MΩ and calculate v_o/v_s for $R_S = 0$ and 10 kΩ.

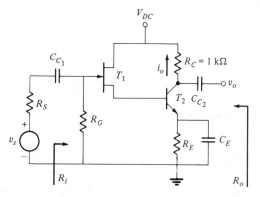

FIGURE P6.24

6.25

For the Darlington amplifier of Fig. P6.25, determine R_i, R_o, v_o/v_s, and i_o/i_i by using a small-signal equivalent circuit. Let $h_{fe_1} = h_{fe_2} = 99$ and $h_{ie_1} = 1$ kΩ.

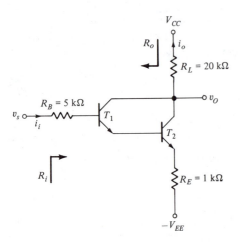

FIGURE P6.25

6.26

Repeat Problem 6.25 for the circuit of Fig. P6.26.

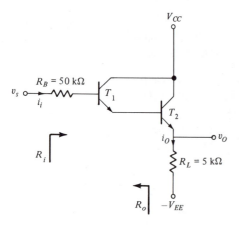

FIGURE P6.26

6.27

For the amplifier of Fig. P6.27, determine R_i, R_o, and v_o/v_s by using a small-signal equivalent circuit. Let $h_{fe_1} = h_{fe_2} = 100$, $h_{ie_2} = 50$ kΩ, and $R_S = 0$ and 10 kΩ.

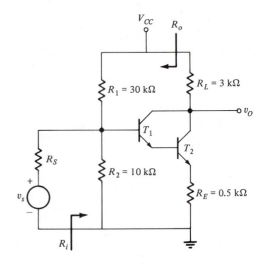

FIGURE P6.27

6.28

For the Darlington difference amplifier of Fig. P6.28, determine v_{o_1} in terms of v_c and v_d, A_c, A_d, and the CMRR by using a small-signal equivalent circuit. Let $h_{fe_1} = h_{fe_2} = 100$ and $h_{ie_1} = 5$ kΩ.

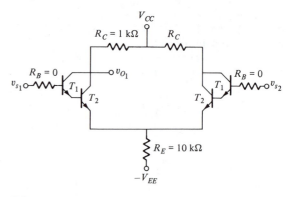

FIGURE P6.28

6.29

Repeat Problem 6.28 with element values given by $R_B = 0$, $R_C = 4$ kΩ, and $R_E = 2$ kΩ. Let $h_{fe_1} = h_{fe_2} = 100$ and $h_{ie_2} = 1$ kΩ.

6.30

Figure P6.30 shows a Darlington difference amplifier with a current source (replacing R_E from pre-

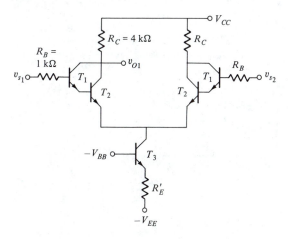

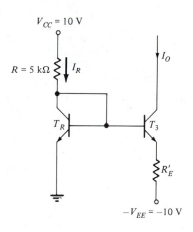

FIGURE P6.30

FIGURE P6.32

vious examples). For $h_{fe_1} = h_{fe_2} = 100$ and $h_{ie_2} = 1\ \text{k}\Omega$, along with $r_{ac} = 50\ \text{k}\Omega$, determine v_{O_1}, A_c, and A_d. Use a small-signal equivalent circuit.

6.31
For the current source shown in Fig. P6.31, determine the current $I_{E_3}\ (\simeq I_{C_3})$ for $V_{BB} = 3\ \text{V}$, $V_{EE} = 6.7\ \text{V}$, and three different values of $R'_E = 0.3\ \text{k}\Omega$, $3\ \text{k}\Omega$, and $30\ \text{k}\Omega$. Let $h_{fe} = 100$ and $V_0 = 0.7\ \text{V}$.

6.33
For the difference amplifier circuit of Fig. P6.19, determine the current source value.

6.34
Figure P6.34 shows a DC voltage level shifter. For $h_{fe} = 100$ and $V_o = 0.7\ \text{V}$, determine the following:
  (a) I_{E_3}
  (b) R_C (in kΩ) that forces $V_L = 0$
  (c) the corresponding V_{CE_1}

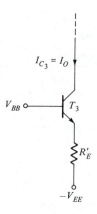

FIGURE P6.31

6.32
Analyze the circuit of Fig. P6.32 to determine I_R and R'_E. Let $V_0 = 0.7\ \text{V}$ and $I_O = 1\ \text{mA}$. Suppose R'_E is doubled. What is the resulting current I_O?

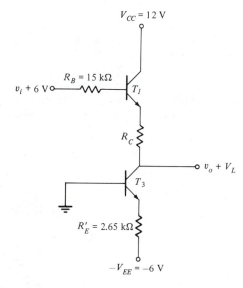

FIGURE P6.34

6.35

Analyze the BJT difference amplifier with an active load as shown in Fig. P6.35 by using a small-signal equivalent circuit. Let $h_{fe} = 100$, $h_{ie} = 0.5\ k\Omega$, and $1/h_{oe} = 50\ k\Omega$. All transistors are identical. The AC resistance of the current source is ∞. Determine the AC output voltage v_o, A_c, A_d, and the CMRR

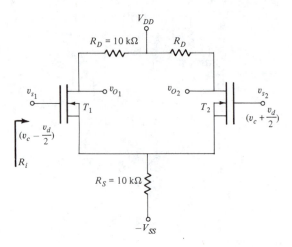

FIGURE P6.37

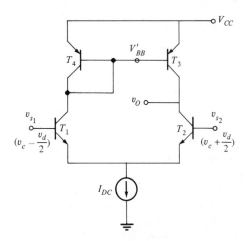

FIGURE P6.35

6.36

Repeat Problem 6.35 with $h_{ie} = 1\ k\Omega$ and $1/h_{oe} = 100\ k\Omega$.

6.37

Consider the MOST difference amplifier of Fig. P6.37. Determine A_c, A_d, and the CMRR by using a small-signal equivalent circuit. Let $g_m = 100\ mS$. What is R_i?

6.38

Repeat Problem 6.37 for $R_D = 20\ k\Omega$ and $R_S = 100\ k\Omega$.

6.39

For the IC MOST difference amplifier of Fig. 6.17, determine A_r, A_d, and the CMRR. Use $g_m = 100\ mS$ (the transconductance of $T_1 = T_2$), $1/g_{o3} = 50\ k\Omega$, and $1/g_{m6} = 50\ k\Omega$. What is R_i?

6.40

Repeat Problem 6.39 for $1/g_{o3} = 200\ k\Omega$ and $1/g_{m6} = 200\ k\Omega$. Maintain g_m for T_1 and T_2 at $g_m = 100\ mS$.

6.41

For the JFET difference amplifier of Fig. 6.18, determine A_c, A_d, and the CMRR. Use $R_D = 10\ k\Omega$, $g_m = 100\ mS$, and $1/h_{ob_3} = 50\ k\Omega$. What is R_i?

6.42

Repeat Problem 6.41 for $R_D = 50\ k\Omega$ and $1/h_{ob_3} = 200\ k\Omega$.

DATA SHEET A6.1 Selection Guide for Operational Amplifiers

noncompensated, single

military temperature range

(values specified for $T_A = 25°C$)

DEVICE NUMBER	DESCRIPTION	B_1 (MHz) TYP	SR (V/µs) TYP	V_{IO} (mV) MAX	I_{IB} (nA) MAX	A_{VD} (V/mV) MIN	SUPPLY VOLTAGE (V) MIN	SUPPLY VOLTAGE (V) MAX	PACKAGES	PAGE
TL080M	BIFET, Low Power	3	13	6	0.2	25	± 3.5	± 18	JG	3-135
uA709AM	General Purpose	1	0.3	2	200	45 Typ	± 5	± 18	J,JG,U,W	3-215
uA709M	General Purpose	1	0.3	5	500	45 Typ		± 18	J,JG,U,W	3-215
LM108	High Performance	1	0.3	2	2	50	± 5	± 18	JG	3-25
LM101A	High Performance	1	0.5	2	75	50	± 5	± 22	FH,FK,JG,U,W	3-17
uA748M	General Purpose	1	0.5	5	500	± 10	± 2	± 22	JG,U	3-229
TL060M	BIFET, Low Power	1	3.5	6	0.2	4	± 1.5	± 18	JG	3-103

industrial temperature range

(values specified for $T_A = 25°C$)

DEVICE NUMBER	DESCRIPTION	B_1 (MHz) TYP	SR (V/µs) TYP	V_{IO} (mV) MAX	I_{IB} (nA) MAX	A_{VD} (V/mV) MIN	SUPPLY VOLTAGE (V) MIN	SUPPLY VOLTAGE (V) MAX	PACKAGES	PAGE
TL070I	BIFET, Low Noise	13	13	6	200	50	± 3.5	± 18	D,JG,P	3-125
TL080I	BIFET, Low Power	3	13	6	200	50	± 3.5	± 18	JG,P	3-135
LM208	High Performance	1	0.3	2	2	50	± 5	± 18	D,JG,P	3-25
LM201A	High Performance	1	0.5	2	75	50	± 5	± 22	D,JG,P,W	3-17
TL060I	BIFET, Low Power	1	3.5	6	0.2	4	± 1.5	± 18	D,JG,P	3-103

commercial temperature range

(values specified for $T_A = 25°C$)

DEVICE NUMBER	DESCRIPTION	B_1 (MHz) TYP	SR V/µs TYP	V_{IO} (mV) MAX	I_{IB} (nA) MAX	A_{VD} (V/mV) MIN	SUPPLY VOLTAGE (V) MIN	SUPPLY VOLTAGE (V) MAX	PACKAGES	PAGE
TL070AC	BIFET, Low Noise	3	13	6	0.2	50	± 3.5	± 18	D,JG,P	3-125
TL070C	BIFET, Low Noise	3	13	10	0.2	25	± 3.5	± 18	D,JG,P	3-125
TL080AC	BIFET, Low Power	3	13	6	0.2	50	± 3.5	± 18	JG,P	3-135
TL080C	BIFET, Low Power	3	13	15	0.4	25	± 3.5	± 18	JG,P	3-135
uA709C	General Purpose	1	0.3	7.5	1500	15		± 18	JG,P	3-215
LM308	High Performance	1	0.3	7.5	7	25	± 5	± 18	D,JG,P	3-25
uA748C	General Purpose	1	0.5	6	500	20	± 2	± 18	D,JG,P	3-229
TL060AC	BIFET, Low Power	1	3.5	6	0.2	4	± 1.5	± 18	D,JG,P	3-103
TL060BC	BIFET, Low Power	1	3.5	3	0.2	4	± 1.5	± 18	D,JG,P	3-103
TL060C	BIFET, Low Power	1	3.5	15	0.2	3	± 1.5	± 18	D,JG,P	3-103
LM301A	High Performance	1	7.5	50	75	25	± 5	± 18	D,JG,P,W	3-17

continues

DATA SHEET A6.1   Continued

internally compensated, single

military temperature range   (values specified for $T_A = 25°C$)

DEVICE NUMBER	DESCRIPTION	B₁ (MHz) TYP	SR (V/μs) TYP	V_IO (mV) MAX	I_IB (nA) MAX	A_VD (V/mV) MIN	SUPPLY VOLTAGE (V) MIN	SUPPLY VOLTAGE (V) MAX	PACKAGES	PAGE
TL291M	High Speed	20	50				± 4	± 18	JG	3-155
SE5534	High Performance	10	13	2	800	50	± 3	± 20	FH,FK,JG,U	3-91
SE5534A	High Performance	10	13	2	800	50	± 3	± 20	FH,FK,JG,U	3-91
LM110	Unity-Gain Voltage Follower	10	30	4	3	1	± 5	± 18	JG	3-27
TL081M	BIFET, General Purpose	3	13	6	0.2	25	± 4	± 18	FH,FK,JG	3-135
TL088M	BIFET, Low V_IO	3	13	3	0.4	50	± 3.5	± 18	JG,U	3-145
TL071M	BIFET, Low Noise	3	13	6	0.2	35	± 5	± 18	FH,FK,JG	3-125
TLC271AM	LinCMOS, Programmable	2.3	4.5	5	0.001 Typ	10	4	16	FH,FK,JG	3-165
TLC271BM	LinCMOS, Programmable	2.3	4.5	2	0.001 Typ	10	4	16	FH,FK,JG	3-165
TLC271M	LinCMOS, Programmable	2.3	4.5	10	0.001 Typ	10	4	16	FH,FK,JG	3-165
TLC277M	LinCMOS, High Bias	2.3	4.5	0.5	0.001 Typ	10	4	16	FH,FK,JG	3-201
uA741M	General Purpose	1	0.5	5	500	50	± 2	± 22	FH,FK,J,JG,U	3-223
LM107	High Performance	1	0.5	2	75	50	± 2	± 22	J,JG,U,W	3-21
TL061M	BIFET, Low Power	1	3.5	6	0.2	4	± 1.5	± 18	FH,FK,JG,U	3-103
TL066M	BIFET, Adjustable Low-Power	1	3.5	6	0.2	4	± 1.2	± 18	FH,FK,JG	3-113
OP-012A	Precision Low-Input Current	0.8	0.12	0.15	2	80	± 5	± 20	JG	3-71
OP-012B	Precision Low-Input Current	0.8	0.12	0.3	2	80	± 5	± 20	JG	3-71
OP-012C	Precision Low-Input Current	0.8	0.12	1	5	40	± 5	± 20	JG	3-71
TLC27M7M	LinCMOS, Programmable	0.7	0.6	0.5	0.001 Typ	20	4	16	FH,FK,JG	3-201
TL321M	High Performance	0.6	0.3	5	− 150	50	3	30	FH,FK,JG	3-157
uA702M	General Purpose	0.5	11	2	500	6000	6 / −3	14 / − 7	JG,U	3-201
TLC27L7M	LinCMOS, Programmable	0.1	0.04	0.5	0.001 Typ	30	4	16	FH,FK,JG	3-201

automotive temperature range   (values specified for $T_A = 25°C$)

DEVICE NUMBER	DESCRIPTION	B₁ (MHz) TYP	SR (V/μs) TYP	V_IO (mV) MAX	I_IB (nA) MAX	A_VD (V/mV) MIN	SUPPLY VOLTAGE (V) MIN	SUPPLY VOLTAGE (V) MAX	PACKAGES	PAGE
LM218	High Performance	15	70	4	250	50	± 5	± 20	D,JG,P	3-43
TLC271AI	LinCMOS, Programmable	2.3	4.5	5	0.001 Typ	7	3	16	D,JG,P	3-165
TLC271BI	LinCMOS, Programmable	2.3	4.5	2	0.001 TYP	7	3	16	D,JG,P	3-165
TLC271I	LinCMOS, Programmable	2.3	4.5	10	0.001 Typ	7	3	16	D,JG,P	3-165
TLC277I	LinCMOS, Programmable	2.3	4.5	0.5	0.001 Typ	10	3	16	D,JG,P	3-201
LM207	High Performance	1	0.5	2	75	50	± 2	± 22	J,N,W	3-21
TLC27M7I	LinCMOS, Medium Bias	0.7	0.6	0.5	0.001 Typ	20	3	16	D,JG,P	3-201
TLC27L7I	LinCMOS, Low Bias	0.1	0.04	0.5	0.001 Typ	30	3	16	D,JG,P	3-201
TLC261AI	LinCMOS, Programmable		12	5	0.001 Typ		2	16	D,JG,P	3-199
TLC261BI	LinCMOS, Programmable		12	2	0.001 Typ		2	16	D,JG,P	3-199
TLC261I	LinCMOS, Programmable		12	10	0.001 Typ		2	16	D,JG,P	3-199

DATA SHEET A6.1   Continued

internally compensated, single

industrial temperature range   (values specified for $T_A = 25°C$)

DEVICE NUMBER	DESCRIPTION	B$_I$ (MHz) TYP	SR (V/µs) TYP	V$_{IO}$ (mV) MAX	I$_{IB}$ (nA) MAX	A$_{VD}$ (V/mV) MIN	SUPPLY VOLTAGE (V) MIN	SUPPLY VOLTAGE (V) MAX	PACKAGES	PAGE
TL087I	BIFET, Low Offset	25	13	0.5	0.2	50	± 4	± 18	D,JG,P	3-145
LM210	Unity-gain Voltage Follower	20	30	4	3		± 5	± 18	JG,P	3-27
TL071I	BIFET, Low Noise	3	13	6	0.2	50	± 3.5	± 18	D,JG,P	3-125
TL081I	BIFET, General Purpose	3	13	6	0.2	50	± 3.5	± 18	JG,P	3-135
TL088I	BIFET, Low V$_{IO}$	3	13	1	0.2	50	± 4	± 18	D,JG,P	3-145
TL066I	BIFET, Adjustable Low-Power	1	3.5	6	0.2	4	± 1.2	± 18	D,JG,P	3-113
TL061I	BIFET, Low Power	1	3.5	6	0.2	4	± 1.5	± 18	D,JG,P	3-103
TL321I	High Performance	0.6	0.3	5	- 150	50	3	30	JG,P	3-157

commercial temperature range   (values specified for $T_A = 25°C$)

DEVICE NUMBER	DESCRIPTION	B$_I$ (MHz) TYP	SR (V/µs) TYP	V$_{IO}$ (mV) MAX	I$_{IB}$ (nA) MAX	A$_{VD}$ (V/mV) MIN	SUPPLY VOLTAGE (V) MIN	SUPPLY VOLTAGE (V) MAX	PACKAGES	PAGE
LM310	Unity-Gain Voltage Follower	20	30	7.5	7		± 5	± 18	JG,P	3-27
TL291C	High Speed	20	50				± 4	± 18	JG,P	3-155
LM318	High Performance	15	70	10	250	25	± 5	± 20	D,JG,P	3-43
NE5534	High Performance	10	6	4	1500	25	± 3	± 20	JG,P	3-91
NE5534A	High Performance	10	6	4	1500	25	± 3	± 20	JG,P	3-91
OP-227E	Low Noise	8	2.8	0.08	± 40	800			J,N	3-77
OP-227F	Low Noise	8	2.8	0.12	± 55	800			J,N	3-77
OP-227G	Low Noise	8	2.8	0.18	± 80	700			J,N	3-77
TL087C	BIFET, General Purpose	3	13	0.5	0.2	50	± 5	± 18	D,JG,P	3-145
TL071AC	BIFET, Low Noise	3	13	6	0.2	50	± 3.5	± 18	D,JG,P	3-125
TL071BC	BIFET, Low Noise	3	13	3	0.2	50	+ 3.5	± 18	D,JG,P	3-125
TL071C	BIFET, Low Noise	3	13	10	0.2	25	± 3.5	± 18	D,JG,P	3-125
TL081AC	BIFET, General Purpose	3	13	6	0.2	50	± 3.5	± 18	JG,P	3-135
TL081BC	BIFET, General Purpose	3	13	3	0.2	50	± 3.5	± 18	JG,P	3-135
TL081C	BIFET, General Purpose	3	13	15	0.4	25	± 3.5	± 18	JG,P	3-135
TL088C	BIFET, Low V$_{IO}$	3	13	1	0.2	50	± 4	± 18	D,JG,P	3-145
TLC277C	LinCMOS, High Bias	2.3	4.5	0.5	0.001 Typ	10 Typ	3	16	D,JG,P	3-201
uA741C	General Purpose	1	0.5	6	500	20	± 2	± 18	D,JG,P	3-223
LM307	High Performance	1	0.5	7.5	250	25	± 2	± 18	D,J,JG,N,P,W	3-21
TL061AC	BIFET, Low Power	1	3.5	6	0.2	4	± 1.5	± 18	D,JG,P	3-103
TL061BC	BIFET, Low Power	1	3.5	3	0.2	4	± 1.5	± 18	D,JG,P	3-103
TL061C	BIFET, Low Power	1	3.5	15	0.2	3	± 1.5	± 18	D,JG,P	3-103
TL066AC	BIFET, Adjustable Low-Power	1	3.5	6	0.2	4	± 1.2	± 18	D,JG,P	3-113
TL066BC	BIFET, Adjustable Low-Power	1	3.5	3	0.2	4	± 1.2	± 18	D,JG,P	3-113
TL066C	BIFET, Adjustable Low-Power	1	3.5	15	0.4	3	± 1.2	± 18	D,JG,P	3-113
TL068C	BIFET, Buffer	1	7	15	0.4		± 1.5	18	LP	3-123

continues

DATA SHEET A6.1   Continued

internally compensated, single

commercial temperature range (continued)   (values specified for T_A = 25°C)

DEVICE NUMBER	DESCRIPTION	B_I (MHz) TYP	SR (V/μs) TYP	V_{IO} (mV) MAX	I_{IB} (nA) MAX	A_{VD} (V/mV) MIN	SUPPLY VOLTAGE (V)		PACKAGES	PAGE
							MIN	MAX		
TLC251AC	LinCMOS, Programmable	0.7	0.6	5	0.001 Typ	10 Typ	1	16	D,JG,P	3-165
TLC251BC	LinCMOS, Programmable	0.7	0.04	2	0.001 Typ	10 Typ	1	16	D,JG,P	3-165
TLC251C	LinCMOS, Programmable	0.7	0.04	10	0.001 Typ	10 Typ	1	16	D,JG,P	3-165
TLC271AC	LinCMOS, Programmable	0.7	0.04	5	0.001 Typ	10 Typ	3	16	D,JG,P	3-165
TLC271BC	LinCMOS, Programmable	0.7	0.04	3	0.001 Typ	10 Typ	3	16	D,JG,P	3-165
TLC271C	LinCMOS, Programmable	0.7	0.04	10	0.001 Typ	10 Typ	3	16	D,JG,P	3-165
TLC27M7C	LinCMOS, Medium Bias	0.7	0.6	0.5	0.001 Typ	20	3	16	D,JG,P	3-201
uA714C	Ultra-Low Offset Voltage	0.6	0.17	0.15	±7	100	±3	±18	JG,P	3-219
uA714E	Ultra-Low Offset Voltage	0.6	0.17	0.075	±4	200	±3	±18	JG,P	3-219
uA714L	Ultra-Low Offset Voltage	0.6	0.17	0.25	±30	50	±3	±18	JG,P	3-219
TL321C	High Performance	0.6	0.3	7	−250	25	3	30	JG,P	3-157
OP-07C	Ultra-Low Offset	0.6	0.3	0.15	±7	100	±3	±18	JG,P	3-67
OP-07D	Ultra-Low Offset	0.6	0.3	0.15	±12	400 Typ	±3	±18	JG,P	3-67
OP-07E	Ultra-Low Offset	0.6	0.3	0.75	+4	150	±3	±18	JG,P	3-67
OP-12E	Precision Low-Input Current	0.2	0.12	0.15	2	80	±5	±20	D,JG,P	3-71
OP-12F	Precision Low-Input Current	0.2	0.12	0.3	2	80	±5	±20	D,JG,P	3-71
OP-12G	Precision Low-Input Current	0.2	0.12	1	5	40	±5	±20	D,JG,P	3-71
TLC27L7C	LinCMOS, Low Bias	0.1	0.04	0.5	0.001 Typ	30	3	16	D,JG,P	3-201

internally compensated, dual

military temperature range   (values specified for T_A = 25°C)

DEVICE NUMBER	DESCRIPTION	B_I (MHz) TYP	SR (V/μs) TYP	V_{IO} (mV) MAX	I_{IB} (nA) MAX	A_{VD} (V/mV) MIN	SUPPLY VOLTAGE (V)		PACKAGES	PAGE
							MIN	MAX		
TL292M	High Frequency	20	50						JG	3-155
RM4558	High Performance	3	1.7	5	500	50		±22	JG	3-87
TL072M	BIFET, Low Noise	3	13	6	0.2	35	±3.5	±18	FH,FK,JG	3-125
TL082M	BIFET, General Purpose	3	13	16	0.2	25	±3.5	±18	FH,FK,JG	3-135
TL083M	BIFET, General Purpose	3	13	6	0.2	25	±3.5	±18	FH,FK,J	3-135
TL088M	BIFET, General Purpose	3	13	3	0.4	50	±3.5	±18	JG,U	3-145
TL288M	BIFET, General Purpose	3	13	3	0.4	50	±3.5	±18	JG,U	3-145
TLC272AM	LinCMOS, High Bias	2.3	4.5	5	0.001 Typ	10	4	16	FH,FK,JG	3-175
TLC272BM	LinCMOS, High Bias	2.3	4.5	2	0.001 Typ	10	4	16	FH,FK,JG	3-175
TLC272M	LinCMOS, High Bias	2.3	4.5	10	0.001 Typ	10	4	16	FH,FK,JG	3-175
TLC27L2AM	LinCMOS, Low Bias	2.3	4.5	5	0.001 Typ	30	4	16	FH,FK,JG	3-175
TLC27L2BM	LinCMOS, Low Bias	2.3	4.5	2	0.001 Typ	30	4	16	FH,FK,JG	3-175
TLC27L2M	LinCMOS, Low Bias	2.3	4.5	10	0.001 Typ	30	4	16	FH,FK,JG	3-175
TLC27M2AM	LinCMOS, Medium Bias	2.3	4.5	5	0.001 Typ	20	4	16	FH,FK,JG	3-175
TLC27M2BM	LinCMOS, Medium Bias	2.3	4.5	2	0.001 Typ	20	4	16	FH,FK,JG	3-175
TLC27M2M	LinCMOS, Medium Bias	2.3	4.5	10	0.001 Typ	20	4	16	FH,FK,JG	3-175
MC1558	General Purpose	1	0.5	5	500	50	±2	±22	FH,FK,JG,U	3-53
TL322M	Low Power	1	0.6	8	−500	200 Typ	±1.5	±18	JG	3-161
TL062M	BIFET, Low Power	1	3.5	6	0.2	4	±1.5	±18	FH,FK,JG,U	3-103
LM158	High Gain	0.6		5	−150	50	3	30	FH,FK,JG,U	3-36
TL022M	Low Power	0.5	0.5	5	100	72	±2	±22	U	3-95

DATA SHEET A6.1   Continued

internally compensated, dual

automotive temperature range

(values specified for T_A = 25°C)

DEVICE NUMBER	DESCRIPTION	B_1 (MHz) TYP	SR (V/μs) TYP	V_{IO} (mV) MAX	I_{IB} (nA) MAX	A_{VD} (V/mV) MIN	SUPPLY VOLTAGE (V) MIN	MAX	PACKAGES	PAGE
LM2904	High Gain	5	1	7	-250	100 Typ	± 3	± 26	D,JG,P,U	3-36
RV4558	High Performance	3	1.7	6	500	50	± 5	± 18	D,JG,P	3-87
TLC272AI	LinCMOS, High Bias	2.3	4.5	5	0.002 Typ	10	3	16	D,JG,P	3-175
TLC272BI	LinCMOS, High Bias	2.3	4.5	2	0.002 Typ	10	3	16	D,JG,P	3-175
TLC272I	LinCMOS, High Bias	2.3	4.5	10	0.002 Typ	10	3	16	D,JG,P	3-175
TLC27L2AI	LinCMOS, Low Bias	2.3	4.5	5	0.002 Typ	30	3	16	D,JG,P	3-175
TLC27L2BI	LinCMOS, Low Bias	2.3	4.5	2	0.002 Typ	30	3	16	D,JG,P	3-175
TLC27L2I	LinCMOS, Low Bias	2.3	4.5	10	0.002 Typ	30	3	16	D,JG,P	3-175
TLC27M2AI	LinCMOS, Medium Bias	2.3	4.5	5	0.002 Typ	20	3	16	D,JG,P	3-175
TLC27M2BI	LinCMOS, Medium Bias	2.3	4.5	2	0.002 Typ	20	3	16	D,JG,P	3-175
TLC27M2I	LinCMOS, Medium Bias	2.3	4.5	10	0.002 Typ	20	3	16	D,JG,P	3-175
TL322I	Low Power	1	0.6	8	0.5	20	± 1.5	± 18	D,JG,P	3-157

industrial temperature range

(values specified for T_A = 25°C)

DEVICE NUMBER	DESCRIPTION	B_1 (MHz) TYP	SR (V/μs) TYP	V_{IO} (mV) MAX	I_{IB} (nA) MAX	A_{VD} (V/mV) MIN	SUPPLY VOLTAGE (V) MIN	MAX	PACKAGES	PAGE
TL072I	BIFET, Low Noise	3	13	6	0.2	50	± 3.5	± 18	D,JG,P	3-125
TL082I	BIFET, General Purpose	3	13	6	0.2	50	± 3.5	± 18	JG,P	3-135
TL083I	BIFET, General Purpose	3	13	6	0.2	50	± 3.5	± 18	J,N	3-135
TL287I	BIFET, General Purpose	3	13	0.5	0.2	50	± 3.5	± 18	D,JG,P	3-145
TL288I	BIFET, General Purpose	3	13	0.5	0.2	50	± 3.5	± 18	D,JG,P	3-145
TL062I	BIFET, Low Power	1	3.5	6	0.2	4	± 1.5	± 18	D,JG,P	3-103
LM258	High Gain	0.6		5	-150	50	3	30	D,JG,P,U	3-36
LM258A	High Gain	0.6		3	-80	50	3	30	D,JG,P,U	3-36
TLC262AI	LinCMOS, Programmable		12	5	0.001 Typ		2	16	D,JG,P	3-199
TLC262BI	LinCMOS, Programmable		12	2	0.001 Typ		2	16	D,JG,P	3-199
TLC262I	LinCMOS, Programmable		12	10	0.001 Typ		2	16	D,JG,P	3-199

continues

DATA SHEET A6.1   Continued

internally compensated, dual

commercial temperature range   (values specified for $T_A = 25°C$)

DEVICE NUMBER	DESCRIPTION	B_1 (MHz) TYP	SR (V/μs) TYP	V_{IO} (mV) MAX	I_{IB} (nA) MAX	A_{VD} (V/mV) MIN	SUPPLY VOLTAGE (V) MIN	MAX	PACKAGES	PAGE
TL292C	High Frequency	20	50				±4	±18	JG,P	3-155
NE5532	Low Noise	10	9	4	800	15	±3	±20	JG,P	3-63
NE5532A	Low Noise	10	9	4	800	15	±3	±20	JG,P	3-63
RC4559	High Performance	4	2	6	250	20		±18	D,P	3-101
RC4558	High Performance	3.5	1.7	5	500	50		±18	D,JG, P	3-103
TL072AC	BIFET, Low Noise	3	13	6	0.2	50	±3.5	±18	D,JG,P	3-145
TL072BC	BIFET, Low Noise	3	13	3	0.2	50	±3.5	±18	D,JG,P	3-145
TL072C	BIFET, Low Noise	3	13	10	0.2	25	±3.5	±18	D,JG,P	3-145
TL082AC	BIFET, General Purpose	3	13	6	0.2	50	±3.5	±18	JG,P	3-155
TL082BC	BIFET, General Purpose	3	13	3	0.2	50	±3.5	±18	JG,P	3-155
TL082C	BIFET, General Purpose	3	13	15	0.4	25	±3.5	±18	JG,P	3-155
TL083AC	BIFET, General Purpose	3	13	6	0.2	50	±3.5	±18	J,N	3-155
TL083C	BIFET, General Purpose	3	13	15	0.4	25	±3.5	±18	J,N	3-155
TL287C	BIFET, General Purpose	3	13	0.5	0.2	50	±3.5	±18	D,JG,P	3-155
TL288C	BIFET, General Purpose	3	13	0.5	0.2	50	±3.5	±18	D,JG,P	3-155
TLC252AC	LinCMOS, High Bias	2.3	4.5	5	0.001 Typ	10	1	16	D,JG,P	3-175
TLC252BC	LinCMOS, High Bias	2.3	4.5	2	0.001 Typ	10	1	16	D,JG,P	3-175
TLC252C	LinCMOS, High Bias	2.3	4.5	10	0.001 Typ	10	1	16	D,JG,P	3-175
TLC25L2AC	LinCMOS, Low Bias	2.3	4.5	5	0.001 Typ	30	1	16	D,JG,P	3-175
TLC25L2BC	LinCMOS, Low Bias	2.3	4.5	2	0.001 Typ	30	1	16	D,JG,P	3-175
TLC25L2C	LinCMOS, Low Bias	2.3	4.5	10	0.001 Typ	30	1	16	D,JG,P	3-175
TLC25M2AC	LinCMOS, Medium Bias	2.3	4.5	5	0.001 Typ	20	1	16	D,JG,P	3-175
TLC25M2BC	LinCMOS, Medium Bias	2.3	4.5	2	0.001 Typ	20	1	16	D,JG,P	3-175
TLC25M2C	LinCMOS, Medium Bias	2.3	4.5	10	0.001 Typ	20	1	16	D,JG,P	3-175
TLC272AC	LinCMOS, High Bias	2.3	4.5	5	0.001 Typ	10	3	16	D,JG,P	3-175
TLC272BC	LinCMOS, High Bias	2.3	4.5	2	0.001 Typ	10	3	16	D,JG,P	3-175
TLC272C	LinCMOS, High Bias	2.3	4.5	10	0.001 Typ	10	3	16	D,JG,P	3-175
TLC27L2AC	LinCMOS, Low Bias	2.3	4.5	5	0.001 Typ	20	3	16	D,JG,P	3-175
TLC27L2BC	LinCMOS, Low Bias	2.3	4.5	2	0.001 Typ	20	3	16	D,JG,P	3-175
TLC27L2C	LinCMOS, Low Bias	2.3	4.5	10	0.001 Typ	20	3	16	D,JG,P	3-175
TLC27M2AC	LinCMOS, Medium Bias	2.3	4.5	5	0.001 Typ	20	3	16	D,JG,P	3-175
TLC27M2BC	LinCMOS, Medium Bias	2.3	4.5	2	0.001 Typ	20	3	16	D,JG,P	3-175
TLC27M2C	LinCMOS, Medium Bias	2.3	4.5	10	0.001 Typ	20	3	16	D,JG,P	3-175
MC1458	General Purpose	1	0.5	6	500	20	±1.5	±18	D,JG,P,U	3-53
TL322C	Low Power	1	0.6	10	−500	20	±1.5	±18	D,JG,P	3-161
TL062AC	BIFET, Low Power	1	3.5	6	0.2	4	±1.2	±18	D,JG,P	3-103
TL062BC	BIFET, Low Power	1	3.5	3	0.2	4	±1.2	±18	D,JG,P	3-103
TL062C	BIFET, Low Power	1	3.5	15	0.2	3	±1.2	±18	D,JG,P	3-103
LM358	High Gain	0.6		7	−250	25	3	30	D,JG,P,U	3-36
LM358A	High Gain	0.6		3	−100	25	3	30	D,JG,P,U	3-36
TL022C	Low Power	0.5	0.5	5	250	60	±2	±18	JG,P	3-95

DATA SHEET A6.1   Continued

internally compensated, quad

military temperature range   (values specified for T_A = 25°C)

DEVICE NUMBER	DESCRIPTION	B₁ (MHz) TYP	SR (V/μs) TYP	V_IO (mV) MAX	I_IB (nA) MAX	A_VD (V/mV) MIN	SUPPLY VOLTAGE (V) MIN	SUPPLY VOLTAGE (V) MAX	PACKAGES	PAGE
TL294M	High Frequency	20	50				± 4	± 18	J	3-155
RM4136	High Performance	3.5	1.7	4	400	50	± 4	± 22	FH,FK,J,W	3-83
TL074M	BIFET, Low Noise	3	13	9	0.2	35	± 3.5	± 18	FH,FK,J,W	3-125
TL084M	BIFET, General Purpose	3	13	9	0.2	25	± 3.5	± 18	FH,FK,J,W	3-135
TLC274AM	LinCMOS, High Bias	2.3	4.5	5	0.001 Typ	10	1	16	FH,FK,J	3-187
TLC274BM	LinCMOS, High Bias	2.3	4.5	2	0.001 Typ	10	1	16	FH,FK,J	3-187
TLC274M	LinCMOS, High Bias	2.3	4.5	10	0.001 Typ	10	1	16	FH,FK,J	3-187
LM148	General Purpose	1	0.5	5	100	50		± 22	FH,FK,J	3-33
MC3503	General Purpose	1	0.6	5	−500	50	± 1.5	± 18	J	3-57
TL064M	BIFET, Low Power	1	3.5	9	0.2	4	± 1.5	± 18	FH,FK,J,W	3-103
TLC27M4AM	LinCMOS, Medium Bias	0.7	0.6	5	0.001 Typ	20	1	16	FH,FK,J	3-187
TLC27M4BM	LinCMOS, Medium Bias	0.7	0.6	2	0.001 Typ	20	1	16	FH,FK,J	3-187
TLC27M4M	LinCMOS, Medium Bias	0.7	0.6	10	0.001 Typ	20	1	16	FH,FK,J	3-187
LM124	General Purpose	0.6	0.5	5	−150	50	3	30	FH,FK,J,W	3-29
TL044M	Low Power	0.5	0.5	5	100	72	± 2	± 22	FH,FK,J,W	3-99
TLC27L4AM	LinCMOS, Low Bias	0.1	0.04	5	0.001 Typ	30	1	16	FH,FK,J	3-187
TLC27L4BM	LinCMOS, Low Bias	0.1	0.04	2	0.001 Typ	30	1	16	FH,FK,J	3-187
TLC27L4M	LinCMOS, Low Bias	0.1	0.04	10	0.001 Typ	30	1	16	FH,FK,J	3-187

automotive temperature range   (values specified for T_A = T_A 25°C)

DEVICE NUMBER	DESCRIPTION	B₁ (MHz) TYP	SR (V/μs) TYP	V_IO (mV) MAX	I_IB (nA) MAX	A_VD (V/mV) MIN	SUPPLY VOLTAGE (V) MIN	SUPPLY VOLTAGE (V) MAX	PACKAGES	PAGE
RV4136	High Performance	3	1.7	6	500	20	± 4.5	± 32	D,J,N,W	3-83
LM2900	General Purpose	2.5	0.5		200	1.2	± 4.5	± 32	J,N	3-47
TLC274AI	LinCMOS, High Bias	2.3	4.5	5	0.001 Typ	10	1	16	D,J,N	3-187
TLC274BI	LinCMOS, High Bias	2.3	4.5	2	0.001 Typ	10	1	16	D,J,N	3-187
TLC274I	LinCMOS, High Bias	2.3	4.5	10	0.001 Typ	10	1	16	D,J,N	3-187
MC3303	General Purpose	1	0.6	8	−500	20	3	36	D,J,N	3-187
TLC27M4AI	LinCMOS, Medium Bias	0.7	0.6	5	0.001 Typ	20	1	16	D,J,N	3-187
TLC27M4BI	LinCMOS, Medium Bias	0.7	0.6	2	0.001 Typ	20	1	16	D,J,N	3-187
TLC27M4I	LinCMOS, Medium Bias	0.7	0.6	10	0.001 Typ	20	1	16	D,J,N	3-187
LM2902	General Purpose	0.6		7	−250	100 Typ	3	26	D,J,N,W	3-29
TLC27L4AI	LinCMOS, Low Bias	0.1	0.04	5	0.001 Typ	30	1	16	D,J,N	3-187
TLC27L4BI	LinCMOS, Low Bias	0.1	0.04	2	0.001 Typ	30	1	16	D,J,N	3-187
TLC27L4I	LinCMOS, Low Bias	0.1	0.04	10	0.001 Typ	30	1	16	D,J,N	3-187
TLC264AI	LinCMOS, Programmable		12	5	0.001 Typ		2	16	D,J,N	3-199
TLC264BI	LinCMOS, Programmable		12	2	0.001 Typ		2	16	D,J,N	3-199
TLC264I	LinCMOS, Programmable		12	10	0.001 Typ		2	16	D,J,N	3-199

continues

DATA SHEET A6.1   Continued

internally compensated, quad

industrial temperature range   (values specified for T_A = T_A 25°C)

DEVICE NUMBER	DESCRIPTION	B$_1$ (MHz) TYP	SR (V/μs) TYP	V$_{IO}$ (mV) MAX	I$_{IB}$ (nA) MAX	A$_{VD}$ (V/mV) MIN	SUPPLY VOLTAGE (V) MIN	SUPPLY VOLTAGE (V) MAX	PACKAGES	PAGE
TLO74I	BIFET, Low Noise	3	13	6	0.2	50	± 3.5	± 18	D,J,N	3-125
TLO84I	BIFET, General Purpose	3	13	6	0.2	50	± 3.5	± 18	J,N	3-125
LM248	General Purpose	1	0.5	6	200	25		± 18	D,J,N	3-33
TLO64I	BIFET, Low Power	1	3.5	6	0.2	4	± 1.5	± 18	D,J,N	3-103
LM224	General Purpose	0.6		5	− 150	50	3	30	D,J,N,W	3-135

internally compensated, quad

commercial temperature range   (values specified for T_A = 25°C)

DEVICE NUMBER	DESCRIPTION	B$_1$ (MHz) TYP	SR (V/μs) TYP	V$_{IO}$ (mV) MAX	I$_{IB}$ (nA) MAX	A$_{VD}$ (V/mV) MIN	SUPPLY VOLTAGE (V) MIN	SUPPLY VOLTAGE (V) MAX	PACKAGES	PAGE
TL294C	High Frequency	20	50				± 4	± 18	J,N	3-155
RC4136	High Performance	3	1.7	6	500	20	± 4	± 18	D,J,N,W	3-83
TL074AC	BIFET, Low Noise	3	13	6	0.2	50	± 3.5	± 18	D,J,N	3-125
TL074BC	BIFET, Low Noise	3	13	3	0.2	50	± 3.5	± 18	D,J,N	3-125
TL074C	BIFET, Low Noise	3	13	10	0.2	25	± 3.5	± 18	D,J,N	3-125
TL075C	BIFET, Low Noise	3	13	13	0.2	25	± 3.5	± 18	N	3-125
TL084AC	BIFET, General Purpose	3	13	6	0.2	50	± 3.5	± 18	J,N	3-135
TL084BC	BIFET, General Purpose	3	13	3	0.2	50	± 3.5	± 18	J,N	3-135
TL084C	BIFET, General Purpose	3	13	15	0.4	25	± 3.5	± 18	J,N	3-135
TL085C	BIFET, General Purpose	3	13	15	0.4	25	± 3.5	± 18	N	3-135
TL136C	High Performance	3	2	6	500	3 Typ	± 4	± 18	D,J,N	3-151
LM3900	General Purpose	2.5	0.5			1.2	± 4.5	± 18	J,N	3-47
TLC254AC	LinCMOS, High Bias	2.3	4.5	5	0.001 Typ	10	1	16	D,J,N	3-187
TLC254BC	LinCMOS, High Bias	2.3	4.5	2	0.001 Typ	10	1	16	D,J,N	3-187
TLC254C	LinCMOS, High Bias	2.3	4.5	10	0.001 Typ	10	1	16	D,J,N	3-187
TLC274AC	LinCMOS, High Bias	2.3	4.5	5	0.001 Typ	10	3	16	D,J,N	3-187
TLC274BC	LinCMOS, High Bias	2.3	4.5	2	0.001 Typ	10	3	16	D,J,N	3-187
TLC274C	LinCMOS, High Bias	2.3	4.5	10	0.001 Typ	10	3	16	D,J,N	3-187
LM348	General Purpose	1	0.5	6	200	25		± 18	D,J,N	3-187
MC3403	General Purpose	1	0.6	10	− 500	20	± 1.5	± 18	D,J,N	3-57
TL064AC	BIFET, Low Power	1	3.5	6	0.2	4	± 1.5	± 18	D,J,N	3-103
TL064BC	BIFET, Low Power	1	3.5	3	0.2	4	± 1.5	± 18	D,J,N	3-103
TL064C	BIFET, Low Power	1	3.5	15	0.2	3	± 1.5	± 18	D,J,N	3-103
TLC25M4AC	LinCMOS, Medium Bias	0.7	0.6	5	0.001 Typ	20	1	16	D,J,N	3-187
TLC25M4BC	LinCMOS, Medium Bias	0.7	0.6	2	0.001 Typ	20	1	16	D,J,N	3-187
TLC25M4C	LinCMOS, Medium Bias	0.7	0.6	10	0.001 Typ	20	1	16	D,J,N	3-187
TLC27M4AC	LinCMOS, Medium Bias	0.7	0.6	5	0.001 Typ	20	3	16	D,J,N	3-187
TLC27M4BC	LinCMOS, Medium Bias	0.7	0.6	2	0.001 Typ	20	3	16	D,J,N	3-187
TLC27M4C	LinCMOS, Medium Bias	0.7	0.6	10	0.001 Typ	20	3	16	D,J,N	3-187
LM324	General Purpose	0.6		7	− 250	25	3	30	D,J,N,W	3-29
LM324A	General Purpose	0.6		7	− 100	25	3	30	D,J,N,W	3-29
TL044C	General Purpose	0.5	0.5	5	250	60	± 2	± 18	J,N,W	3-99
TLC25L4AC	LinCMOS, Low Bias	0.1	0.04	5	0.001 Typ	30	1	16	D,J,N	3-187
TLC25L4BC	LinCMOS, Low Bias	0.1	0.04	2	0.001 Typ	30	1	16	D,J,N	3-187
TLC25L4C	LinCMOS, Low Bias	0.1	0.04	10	0.001 Typ	30	1	16	D,J,N	3-187
TLC27L4AC	LinCMOS, Low Bias	0.1	0.04	5	0.001 Typ	30	3	16	D,J,N	3-187
TLC27L4BC	LinCMOS, Low Bias	0.1	0.04	2	0.001 Typ	30	3	16	D,J,N	3-187
TLC27L4C	LinCMOS, Low Bias	0.1	0.04	10	0.001 Typ	30	3	16	D,J,N	3-187

7

OPERATIONAL AMPLIFIER AND COMPARATOR APPLICATIONS

Similar to the replacement of vacuum tubes with transistors, operational amplifiers (op amps) are now replacing transistors in many electronic applications. Op amps are inexpensive integrated circuits that are used in a wide variety of applications. The use of op amp circuits was not previously feasible in discrete-component form because too many components as well as highly matched resistors and transistors in the difference amplifier stages were required. We will consider both linear and nonlinear applications of op amps in this chapter. In general, any of the op amps from the selection guide in Chapter 6 can be used in these applications. However, most op amps are designed specifically for one application and are most suitable for that particular use.

In addition to op amp applications, we will consider various applications of comparators. A more complete name for this type of IC is voltage comparator, a name that uniquely describes the type of nonlinear application in which this IC is used. Comparators are essentially op amps that are designed with increased A_d, which is achieved through the use of positive feedback. A selection guide for various commercially available comparators is presented at the end of this chapter.

7.1 OP AMP AMPLIFIER CIRCUITS

In this section, we consider the two special circuit configurations used with op amps that provide linear voltage amplification. We begin by examining the amplification properties of the op amp without additional circuitry.

7.1.1 The Op Amp

Reason for Special Op Amp Amplifier Circuits. Chapter 6 described specific internal circuitry for operational amplifiers. Additionally, it mentioned that the primary property of an op amp is to provide large difference voltage amplification. However, we did not indicate that this property is valid only for a finite (and rather small) difference voltage v_d.

To show this restriction, we consider Fig. 7.1, which displays the circuit symbol for the op amp (introduced in Chapter 6) and its corresponding voltage transfer characteristic (where the voltage scales used for v_o and v_d are different). The transfer characteristic indicates that the *amplifier region* for the op amp in which $v_o = A_d v_d$ is restricted to an output voltage range of $V_{o\,min} \leq v_o \leq V_{o\,max}$. In this region, each individual stage of the op amp is restricted to amplifier operation. If the signal level drives the output stage (or any stage) into saturation or cutoff, the output voltage of the op amp (v_o) becomes fixed, as indicated in Fig. 7.1b, outside the amplifier region. Furthermore, *positive saturation* of the op amp occurs when $v_o = V_{o\,max}$, and *negative saturation* occurs when $v_o = V_{o\,min}$.

The restriction on output voltage of the op amp indicates that the input difference voltage of the op amp (v_d) has similar restrictions. In fact, the range of v_d that provides amplifier operation is easily determined if we realize that the slope of the transfer characteristic in this region is A_d. Thus, the range of v_d for amplifier operation is given by

$$\frac{V_{o\,min}}{A_d} \leq v_d \leq \frac{V_{o\,max}}{A_d} \tag{7.1-1}$$

Although not immediately evident, the result of (7.1–1) is the primary reason for using special circuit configurations for op amp amplifier applications. Note that in Chapter 6, we emphasized that A_d is quite large. Furthermore, the values of $V_{o\,max}$ and $V_{o\,min}$ are on the order of the power supply voltage (as we recall from the load-line discussions of Chapter 4). Thus, if $A_d = 10^5$ and $V_{o\,max} = 10\ V$ (which are typical values), we obtain an astound-

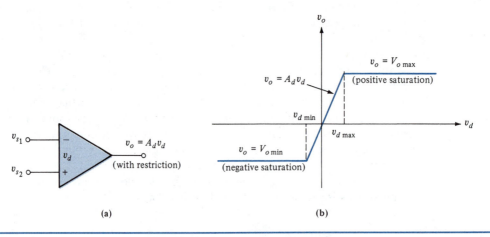

(a)   (b)

FIGURE 7.1   The Op Amp: (a) Circuit symbol, (b) Voltage transfer characteristic

ingly low value for the maximum difference voltage that can be amplified: $v_{d\,max} = V_{o\,max}/A_d = 10^{-5}$ V, or 0.1 mV. This limit on the maximum input signal voltage is definitely over-restricting in amplifier applications.

We must also realize that the value of A_d for any particular op amp is not known precisely. The specialized amplifier circuits used to overcome the restriction on signal voltage magnitude also provide fixed voltage difference amplification that is independent of the op amp A_d. These op amp amplifier circuits utilize external negative feedback that provides greater stability and hence an additional advantage.

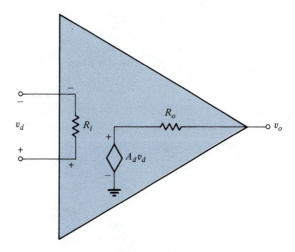

FIGURE 7.2   Equivalent Circuit for the Op Amp

in Fig. 7.2. The input impedance R_i is actually the resistance between the inverting (negative) and noninverting (positive) terminals. The output impedance R_o is the resistance seen looking into the output terminal relative to ground. We will use this equivalent circuit (when necessary) in the analyses that follow.

── EX. 7.1

Op Amp Saturation Example

Consider an op amp with $A_d = 10^5$, $R_i = \infty$, $R_o = 0$, and $V_{o\,max} = -V_{o\,min} = 10$ V. If a sinusoidal voltage with 1 V amplitude is applied directly across the input terminals of the op amp, determine the corresponding output voltage v_o. Also determine the length of time during which the op amp is operating in the amplifier region if the period of the sinusoid is 1 s.

Solution: The op amp will operate in positive and negative saturation over nearly the entire period (T) of the sinusoid. Only during the range $V_{o\,min} \leq v_o \leq V_{o\,max}$ will operation be maintained in the amplifier region, which corresponds to v_d between the limits of $\pm V_{o\,max}/A_d = \pm 10/10^5 = \pm 0.1$ mV. The resulting output voltage is therefore essentially a square wave with peaks at ± 10 V and a sinusoidally varying dependence between peaks. For a sinusoidal period T of 1 s, the amplifier region duration of each period is four times the initial duration or

$$t = \frac{2}{\pi}\sin^{-1}\frac{V_{o\,max}}{A_d} = \frac{2}{\pi}\sin^{-1}\frac{10}{10^5} = 0.0037 \text{ s}$$

Equivalent Circuit for the Op Amp.   In Chapter 6, we also emphasized that op amps possess large input impedance and small output impedance along with large difference voltage amplification (and small common-mode rejection, hence we neglect A_c). An equivalent circuit model that represents these characteristics of the op amp is shown

7.1.2   Inverting Amplifier Circuit

The basic op amp inverting amplifier circuit in Fig. 7.3a consists of only two resistors, the op amp, and the input voltage source v_s, which in general may contain AC or DC or both. However, we will continue to use v_s to represent the source voltage. The resistor R_F is the feedback resistor that together with R_S provides negative feedback to the inverting or negative terminal n. Very importantly, the voltage amplification v_o/v_s for this amplifier configuration is

$$\frac{v_o}{v_s} = -\frac{R_F}{R_S} \qquad (7.1-2)$$

with certain tolerable restrictions, as we will see. This amplifier configuration is said to be *inverting* because of the negative sign in (7.1–2).

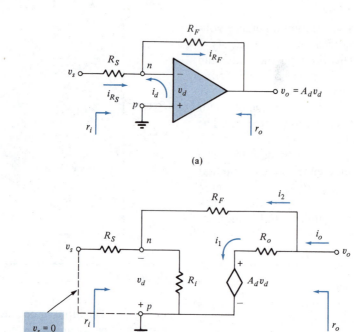

(a)

(b)

FIGURE 7.3   Inverting Amplifier: (a) Basic configuration, (b) Equivalent circuit

To derive the expression for the voltage amplification v_o/v_s, we begin by writing KCL at node n in terms of the defined currents in Fig. 7.3a:

$$i_{R_S} + i_d = i_{R_F} \tag{7.1-3}$$

However, based upon the assumption that the input impedance of the op amp is large ($i_d \cong 0$) and substituting for the remaining currents in terms of the voltages across each resistor, we have

$$\frac{v_s - v_n}{R_S} = \frac{v_n - v_o}{R_F} \tag{7.1-4}$$

Finally, since $v_o = A_d v_d$ (provided that $R_o \ll R_L$), substitution for $v_n = -v_d = -v_o/A_d$ yields

$$\frac{v_s + (v_o/A_d)}{R_S} = \frac{(-v_o/A_d) - v_o}{R_F} \tag{7.1-5}$$

We may note that (7.1–5) immediately yields (7.1–2) if A_d approaches infinity. However, A_d need not be that large in order for (7.1–2) to still be valid, as we observe by solving (7.1–5) for v_o/v_s. Rearranging (7.1–5) by collecting all of the v_o terms yields

$$\frac{v_s}{R_S} = -v_o \left(\frac{1}{A_d R_S} + \frac{1}{A_d R_F} + \frac{1}{R_F} \right) \tag{7.1-6}$$

By factoring, we have

$$\frac{v_s}{R_S} = -\frac{v_o}{R_F} \left[\frac{(R_F/R_S) + 1 + A_d}{A_d} \right] \tag{7.1-7}$$

Note that from (7.1–7) the voltage amplification v_o/v_s is, in general, a function of A_d. However, for large A_d, specifically $A_d \gg (R_F/R_S) + 1$, the term in brackets in (7.1–7) is approximately unity, and (7.1–2) is verified.

To reiterate, three assumptions are necessary for (7.1–2) to be valid. These assumptions are as follows:

- Large input impedance, or $i_d = 0$
- Small output impedance, or $R_o \ll R_L$
- Large difference voltage amplification, or $A_d \gg (R_F/R_S) + 1$

By considering Fig. 7.3a, the input impedance r_i of the inverting amplifier circuit is easily obtained. This impedance, as seen by the source, is given by the ratio of voltage to current at the input terminal, or

$$r_i = \frac{-v_d + i_{R_S} R_S}{i_{R_S}} \tag{7.1–8}$$

Under the assumption of large A_d, or $A_d \gg (R_F/R_S) + 1$, $v_d \cong 0$, and the input impedance is simply

$$r_i = R_S \tag{7.1–9}$$

Hence, an important effect of the op amp is observed: Since $v_d \cong 0$, the negative terminal is essentially at ground and is called a *virtual ground* since the negative terminal is not physically connected to ground. Note also that $v_d \cong 0$ is consistent with the discussion in the previous section where it was emphasized that v_d is small. Thus, we observe that negative feedback provides the required small value for v_d.

To determine the output impedance r_o of this amplifier circuit, it is convenient to obtain its inverse, which is defined by

$$\frac{1}{r_o} = \frac{i_o}{v_o}\bigg|_{v_s = 0} \tag{7.1–10}$$

From Fig. 7.3b and by setting $v_s = 0$ and assuming ficticiously that v_o is applied, the resulting current i_o is found:

$$i_o = i_1 + i_2 = \frac{v_o - A_d v_d}{R_o} + \frac{v_o}{R_F + (R_S \| R_i)} \tag{7.1–11}$$

However, from Fig. 7.3b and using the voltage divider rule,

$$v_n = -v_d = \frac{R_S \| R_i}{R_F + (R_S \| R_i)} v_o \tag{7.1–12}$$

By eliminating v_d, (7.1–11) becomes

$$i_o = \frac{v_o + \dfrac{A_d R_S \| R_i}{R_F + (R_S \| R_i)} v_o}{R_o} + \frac{v_o}{R_F + (R_S \| R_i)} \tag{7.1–13}$$

By dividing through by v_o, we have

$$\frac{1}{r_o} = \frac{i_o}{v_o} = \frac{1 + \dfrac{R_S \| R_i A_d}{R_F + (R_S \| R_i)}}{R_o} + \frac{1}{R_F + (R_S \| R_i)} \tag{7.1–14}$$

Since this expression represents conductance, the two terms added together are conductances in parallel. Therefore, the corresponding resistance value is the inverse of this conductance. In usual cases, the conductance of the first term is much larger than the second because R_o is small. Thus,

$$r_o \cong \frac{R_o}{1 + \dfrac{R_S \| R_i A_d}{R_F + (R_S \| R_i)}} \tag{7.1–15}$$

Thus, we observe that the output impedance of the inverting amplifier circuit is even smaller than that of the op amp itself since the denominator is greater than unity. Equations (7.1–14) and (7.1–15) are also valid for the noninverting amplifier configuration, as we will see.

—————————————————————— EX. 7.2

Inverting Amplifier Example

Determine the input impedance, the output impedance, and the voltage amplification for an inverting amplifier with $R_S = 1\ M\Omega$ and $R_F = 10\ M\Omega$. The op amp parameters are given by $R_i = 100\ k\Omega$, $R_o = 100\ \Omega$, and $A_d = 10^5$.

Solution: Since $(R_F/R_S) + 1 \ll A_d$, $R_o \ll R_L$ ($R_L = \infty$ in this example), and R_i is large, we use (7.1–2), (7.1–9), and (7.1–14) directly to obtain

$$\frac{v_o}{v_s} = -10$$

$$r_i = R_S = 1\ M\Omega$$

$$\frac{1}{r_o} = \frac{1 + \dfrac{(10^3 \| 10)(10^5)}{10^4 + 10^3 \| 10^2}}{0.1} + \frac{1}{10^4 + (10^3 \| 10^2)}$$

$$\cong \frac{1 + \dfrac{(90.9)(10^5)}{10^4}}{0.1} + \frac{1}{10^4} = 9100 + 10^{-4}$$

Note that the first term is much larger than the second, and r_o is thus obtained as follows:

$$r_o = \frac{1}{9100} = 0.11\ m\Omega$$

—————————————————————

7.1.3   Noninverting Amplifier Circuit

The basic circuit for the noninverting amplifier configuration is shown in Fig. 7.4. Note that this circuit is identical to the inverting amplifier configuration. However, the input signal terminal and ground terminal are interchanged.

We determine the voltage amplification in a simpler manner here by first noting that the voltage at n can be written as

$$v_n = \frac{R_S}{R_S + R_F} v_o \qquad (7.1-16)$$

which assumes that the input impedance of the op amp (R_i) is large.

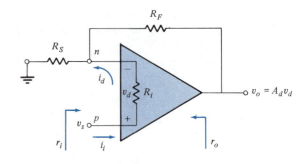

FIGURE 7.4   Noninverting Amplifier Configuration

Next, since $v_d \cong 0$, which assumes large A_d, or $A_d \gg (R_F/R_S) + 1$, we have

$$v_s = v_n = \frac{R_S}{R_S + R_F} v_o \qquad (7.1-17)$$

By rearranging (7.1–17), we obtain the voltage amplification for the noninverting amplifier as follows:

$$\frac{v_o}{v_s} = \frac{R_F}{R_S} + 1 \qquad (7.1-18)$$

This expression is equivalent to (7.1–2) and is valid for the same three assumptions. Since v_o/v_s is positive, the amplifier is said to be *noninverting*.

The input impedance (r_i) for this case is obtained by using Fig. 7.4 and writing the ratio of input voltage to current as

$$r_i = \frac{v_s}{i_i} \qquad (7.1-19)$$

However, $i_i = i_d = v_d/R_i$, where $v_d = v_o/A_d$. Therefore, by substitution,

$$r_i = \frac{v_s}{v_o/A_d R_i} \qquad (7.1-20)$$

Substituting the noninverting amplification expression (7.1–18) and rearranging, we have

$$r_i = \frac{A_d R_i}{(R_F/R_S) + 1} \qquad \text{(7.1–21)}$$

Note that since $A_d \gg (R_F/R_S) + 1$, the input impedance of the inverting amplifier circuit (r_i) is much greater than the input impedance of the op amp (R_i).

The output impedance (r_o) for this case is given identically by the inverting amplifier expressions (7.1–14) or (7.1–15). Note that r_o is obtained with $v_s = 0$ and that, under these conditions, the two op amp amplifier circuits are exactly the same.

7.1.4   Op Amp Voltage Follower

A specific noninverting amplifer called the *voltage follower* is displayed in Fig. 7.5. This amplifier circuit performs the same operation as the BJT emitter follower or FET source follower in that it acts like an isolating element and provides unity voltage amplification.

Note that by comparing Fig. 7.5 to Fig. 7.4, we see that the resistor values for the voltage follower are $R_F = 0$ and $R_S = \infty$. By substituting these resistor values into (7.1–18), (7.1–21), and (7.1–14), we obtain

$$\frac{v_o}{v_s} = 1 \qquad \text{(7.1–22)}$$

$$r_i = A_d R_i \qquad \text{(7.1–23)}$$

$$r_o = \frac{R_o}{1 + A_d} \cong \frac{R_o}{A_d} \qquad \text{(7.1–24)}$$

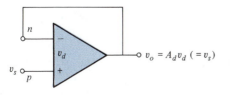

FIGURE 7.5   Voltage Follower

Note also that in the voltage follower circuit of Fig. 7.5, even if the signal voltage has resistance associated with it, we obtain the same results because the input current to the op amp (i_d) is negligible and therefore the voltage across the source resistor is negligible. The op amp voltage follower is used quite often in isolating a source of voltage with its associated resistance from a load resistor.

───────────────────────────────── EX. 7.3

Noninverting Amplifier Example

Determine the input impedance, the output impedance, and the voltage amplification for a noninverting amplifier with $R_S = 100$ kΩ and $R_F = 10$ MΩ. The op amp parameters are given by $R_i = 100$ kΩ, $R_o = 100$ Ω, and $A_d = 10^5$.

Solution: Based upon the three op amp assumptions, we use (7.1–18), (7.1–21), and (7.1–15) to obtain

$$\frac{v_o}{v_s} = \frac{R_F}{R_s} + 1 = 100 + 1 = 101$$

$$r_i = \frac{A_d R_i}{(R_F/R_s) + 1} = \frac{(10^5)(100)}{100 + 1} \cong 10^5 \text{ k}\Omega$$

$$r_o = \frac{R_o}{1 + \dfrac{(R_S \| R_i) A_d}{R_F + (R_S \| R_i)}} = \frac{0.1}{1 + \dfrac{(50)(10^5)}{10^4 + 50}} \cong \frac{0.1}{1 + 500}$$

$$\cong 0.2 \text{ m}\Omega$$

───────────────────────────────────────

7.2   OTHER LINEAR APPLICATIONS

7.2.1   Mathematical Operations

As mentioned previously, the operational amplifier received its name because of the mathematical operations it is capable of performing. Circuits used to perform the operations of subtraction, addition, integration, and differentiation are considered next. All of these operations are possible based upon the three op amp assumptions.

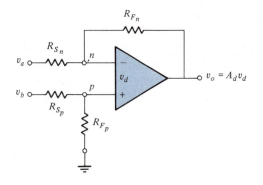

FIGURE 7.6   Difference Amplifier, or Subtractor

Difference Amplifier, or Subtractor.   Figure 7.6 shows a unique circuit that performs the subtraction of one signal from another with amplification of the difference (if desired). To show that the output voltage is proportional to the difference in input voltages, we will use superposition since the amplifier behaves linearly.

The output voltage due to v_a alone ($v_b = 0$) is simply that of the inverting amplifier case given by (7.1–2), or

$$v_o = -\left(\frac{R_{F_n}}{R_{S_n}}\right)v_a \qquad (7.2–1)$$

The contribution to v_o from v_b (with $v_a = 0$) is obtained by first realizing that the voltage at the positive terminal is (for large R_i)

$$v_p = \left(\frac{R_{F_p}}{R_{F_p} + R_{S_p}}\right)v_b \qquad (7.2–2)$$

Thus, using (7.1–18) written as $v_0 = [(R_{F_n} + R_{S_n})/R_{S_n}]v_p$ and substituting v_p from (7.2–2), we obtain

$$v_o = \left(\frac{R_{F_n} + R_{S_n}}{R_{S_n}}\right)\left(\frac{R_{F_p}}{R_{F_p} + R_{S_p}}\right)v_b \qquad (7.2–3)$$

However, if $R_{F_n} = R_{F_p} = R_F$ and $R_{S_n} = R_{S_p} = R_S$,

then (7.2–3) becomes

$$v_o = \left(\frac{R_F}{R_S}\right)v_b \qquad (7.2–4)$$

Hence, using superposition by adding (7.2–1) and (7.2–4) yields

$$v_o = \left(\frac{R_F}{R_S}\right)(v_b - v_a) \qquad (7.2–5)$$

Note that if $R_F = R_S$, the circuit will perform subtraction without amplification.

———————————————————— EX. 7.4

Op Amp Difference Amplifier Example

Consider the difference amplifier of Fig. E7.4. Determine the output voltage v_o assuming that the op amp is ideal and is operating in the amplifier region.

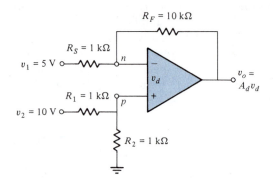

FIGURE E7.4

Solution: By superposition, the output voltage is given by

$$v_o = v_1\left(\frac{-R_F}{R_S}\right) + v_2\left(\frac{R_2}{R_1 + R_2}\right)\left(\frac{R_F}{R_S} + 1\right)$$

Substituting the resistor and voltage values yields

$$v_o = 5\left(\frac{-10}{1}\right) + 10\left(\frac{1}{2}\right)\left(\frac{10}{1} + 1\right) = 5 \text{ V}$$

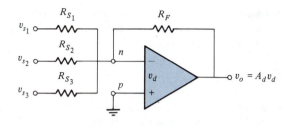

FIGURE 7.7   Summing Amplifier, or Adder

Summing Amplifier, or Adder.   The circuit that performs the addition of signals with amplification (if desired) is shown in Fig. 7.7. We again use superposition to obtain the output voltage v_o.

First, note that the contribution from each separate input signal (with the other signals zero) will have the same effect since the n terminal is a virtual ground. Therefore, by applying (7.1–2) directly to each input signal and adding the results together, we obtain

$$v_o = -\left(\frac{R_F}{R_{S_1}} v_{s_1} + \frac{R_F}{R_{S_2}} v_{s_2} + \frac{R_F}{R_{S_3}} v_{s_3}\right) \tag{7.2–6}$$

where inversion occurs and cannot be avoided. Note that the input signals can be added together with different weighting factors, if desired. However, if all of the resistors are chosen to be equal, the circuit adds the input voltages together at the output.

Integrating the Differentiating Amplifiers.   A circuit in which the output voltage is proportional to

the integral of the input voltage is shown in Fig. 7.8. An alternate circuit that performs the same mathematical operation is obtained by interchanging the capacitor and resistor and then replacing the capacitor with an inductor.

To show that Fig. 7.8 can be used as an integrator, we use KCL and equate the currents at node n (neglecting i_d):

$$i_{R_S} = i_{C_F} \tag{7.2–7}$$

In terms of the voltages across each element,

$$\frac{v_s - v_n}{R_S} = C_F \frac{d}{dt}(v_n - v_o) \tag{7.2–8}$$

Once again, however, $v_n = -v_d = -v_o/A_d = 0$ for large A_d. Thus,

$$\frac{v_s}{R_S} = -C_F \frac{dv_o}{dt} \tag{7.2–9}$$

Integrating (7.2–8) and solving for v_o yield

$$v_o = -\frac{1}{R_S C_F} \int v_s dt \tag{7.2–10}$$

Convenient magnitudes of resistance and capacitance typically used are in the MΩ and μFd range.

To obtain a differentiating amplifier circuit, the capacitor and resistor are interchanged. We thus denote the capacitor as C_S and the resistor as R_F. Writing KCL for the n terminal (neglecting

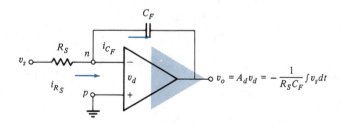

FIGURE 7.8   Integrating Amplifier

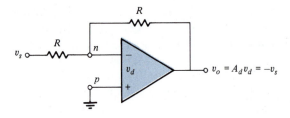

FIGURE 7.9   Analog Inverter

i_d) and substituting voltages, we have

$$C_S \frac{d}{dt}(v_S - v_n) = \frac{v_n - v_o}{R_F} \qquad (7.2\text{--}11)$$

Finally, neglecting v_n and rearranging (7.2–11), we obtain

$$v_o = -R_F C_S \frac{dv_S}{dt} \qquad (7.2\text{--}12)$$

Thus, (7.2–10) and (7.2–12) show the desired mathematical operations. The negative sign in each of these equations indicates that the output is inverted. To eliminate the negative sign, another op amp can be used at the input or output and is connected as shown in Fig. 7.9. The values of resistance are equal in this circuit and thus provide (sign) inversion between input and output. The circuit of Fig. 7.9 is called an *analog inverter* as opposed to a digital inverter. Digital inverters will be described in Chapter 9.

Note that all of the circuits for linear operation utilize the three basic op amp assumptions of large A_d, large R_i, and small R_o, which is precisely the reason that op amps are designed with these properties. The use of integrated circuits is an absolute necessity to obtain these circuit qualities.

──────────────────────────────── EX. 7.5

Op Amp Integrator Example

Combine the integrating amplifier and analog inverter to produce an output voltage that is equal to (and in phase with) the integral of the input voltage.

Solution: The solution for this example is by no means unique. One solution is to choose $R_s = 1\,\text{M}\Omega$ and $C_F = 1\,\mu\text{F}$ and to substitute into (7.2–10) to obtain

$$v_o = -\int v_s\, dt$$

To eliminate the negative sign, an analog inverter (as shown in Fig. 7.9) is connected at either the input or the output portion of the circuit.

7.2.2   Active Filters

If general impedances are used in the inverting amplifier of Fig. 7.3a, as shown in Fig. 7.10, the basic inverting amplifier relationship of (7.1–2) becomes (in terms of the phasors discussed in Chapter 1)

$$\frac{\mathbf{V}_o}{\mathbf{V}_s} = -\frac{Z_F}{Z_S} \qquad (7.2\text{--}13)$$

This relationship can be exploited to obtain desired filter action. Since the voltage can also be amplified, these circuits are called *active filters*.

Figures 7.11a and c show examples of low-pass and high-pass filters that utilize the relationship given by (7.2–13). The output voltage expression for the low-pass filter is

$$\mathbf{V}_o = \frac{-R_{FH}\,\|\,(1/j\omega C_{FH})}{R_{SH}}\,\mathbf{V}_s = \frac{-R_{FH}/R_{SH}}{j\omega C_{FH} R_{FH} + 1}\,\mathbf{V}_s$$
$$(7.2\text{--}14)$$

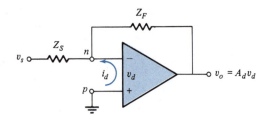

FIGURE 7.10   General Inverting Amplifier

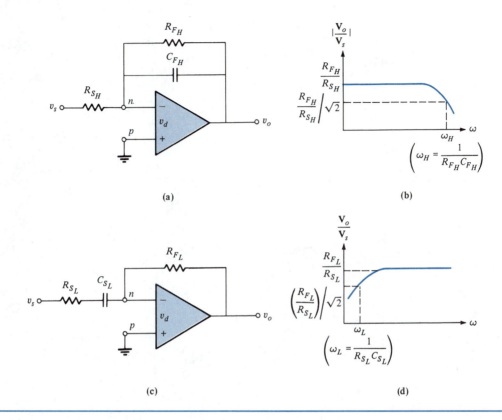

FIGURE 7.11   Active Filters: (a) Low-pass filter, (b) Low-pass filter response, (c) High-pass filter, (d) High-pass filter response

where the subscript H refers to the characteristic high frequency $\omega_H = 1/R_{F_H}C_{F_H}$. For the high-pass filter of Fig. 7.11c,

$$\mathbf{V}_o = \frac{-R_{F_L}}{R_{S_L} + 1/j\omega C_{S_L}}\mathbf{V}_s = -\frac{j\omega C_{S_L}R_{F_L}}{j\omega C_{S_L}R_{S_L} + 1}\mathbf{V}_s$$

$$(7.2\text{--}15)$$

where the subscript L refers to the characteristic low frequency $\omega_L = 1/R_{S_L}C_{S_L}$. The corresponding voltage amplification variation with frequency is displayed in Figs. 7.11b and d. The half-power frequencies are located at $\omega_H = 1/R_{F_H}C_{F_H}$ and $\omega_L = 1/R_{S_L}C_{S_L}$. Simple methods of determining

these half-power frequencies are described in Chapter 8.

Additional practical configurations used to provide low- and high-pass filtering action are shown in Figs. 7.12a and c. These circuit configurations explicitly carry out the noninverting amplifier operation in which amplification is provided by R_F and R_S. In addition, however, the input signal is filtered by the RC network at the positive input terminal.

Figure 7.12a provides low-pass filter action and voltage amplification as follows. The voltage at the positive terminal is

$$\mathbf{V}_p = \frac{1}{1 + j\omega R_H C_H}\mathbf{V}_s \qquad (7.2\text{--}16)$$

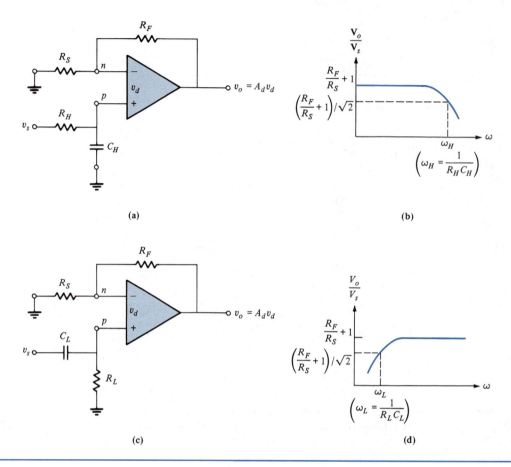

Figure 7.12   Additional Active Filters: (a) Low-pass filter, (b) Low-pass filter response, (c) High-pass filter, (d) High-pass filter response

which is amplified in passing through the noninverting configuration to obtain

$$V_o = \left(\frac{R_F}{R_S} + 1\right)V_p = \frac{(R_F/R_S) + 1}{1 + j\omega R_H C_H} V_s$$

$$(7.2-17)$$

Figure 7.12b displays the variation in the magnitude of the voltage amplification with frequency. The amplification provided in the low-frequency region is $(R_F/R_S) + 1$, and the half-power frequency in the high-frequency range is given by $\omega_H = 1/R_H C_H$.

Figure 7.12c displays the corresponding high-pass active filter case. The output voltage here is

$$V_o = \left(\frac{j\omega R_L C_L}{1 + j\omega R_L C_L}\right)\left(\frac{R_F}{R_S} + 1\right)V_s \quad (7.2-18)$$

The corresponding voltage amplification response is displayed in Fig. 7.12d. In this case, the half-power frequency is in the low-frequency range and is given by $\omega_L = 1/R_L C_L$.

By combining the actions of the low- and high-pass filters, a band-pass active filter is achieved as shown in Fig. 7.13a. The output volt-

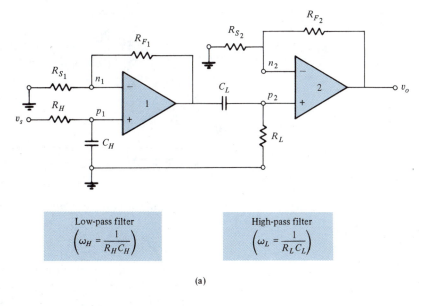

Low-pass filter
$$\left(\omega_H = \frac{1}{R_H C_H}\right)$$

High-pass filter
$$\left(\omega_L = \frac{1}{R_L C_L}\right)$$

(a)

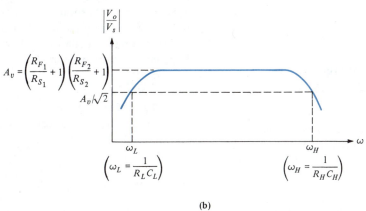

$$A_v = \left(\frac{R_{F_1}}{R_{S_1}} + 1\right)\left(\frac{R_{F_2}}{R_{S_2}} + 1\right)$$

$$A_v/\sqrt{2}$$

$$\left(\omega_L = \frac{1}{R_L C_L}\right)$$

$$\left(\omega_H = \frac{1}{R_H C_H}\right)$$

(b)

FIGURE 7.13   Band-Pass Filter: (a) Low- and high-pass filters combined, (b) Band-pass filter response

age is given by the product of the results of (7.2–17) and (7.2–18), and the magnitude as a function of frequency is shown in Fig. 7.13b.

EX. 7.6

Band-Pass Filter Example

Calculate the upper and lower half-power frequencies and the voltage amplification between these frequencies for the band-pass filter of Fig. 7.13a in which $R_{S_1} = R_{S_2} = 10\ k\Omega$, $R_{F_1} = R_{F_2} = 100\ k\Omega$, $R_L = R_H = 10\ k\Omega$, $C_L = 1\ \mu F$, and $C_H = 1\ pF$.

Solution: The expression for voltage amplification is obtained by multiplying the coefficients of (7.2–17) and (7.2–18) to obtain

$$\frac{v_o}{v_s} = \frac{[(R_F/R_S) + 1][(R_F/R_S) + 1]j\omega C_L R_L}{(1 + j\omega C_L R_L)(1 + j\omega C_H R_H)} \tag{1}$$

Therefore, at mid-band (between ω_L and ω_H),

$$\frac{v_o}{v_s} = \left(\frac{R_{F_1}}{R_{S_1}} + 1\right)\left(\frac{R_{F_2}}{R_{S_2}} + 1\right) \qquad (2)$$

and

$$\omega_1 = \frac{1}{R_L C_L} \qquad (3)$$

$$\omega_2 = \frac{1}{R_H C_H} \qquad (4)$$

Substituting values yields

$$\frac{v_o}{v_s} = (10 + 1)(10 + 1) = 121$$

$$\omega_1 = \frac{1}{(10^4)(10^{-6})} = 10^2 \text{ rad/s}$$

$$\omega_2 = \frac{1}{(10^4)(10^{-12})} = 10^8 \text{ rad/s}$$

7.2.3   Current-to-Voltage and Voltage-to-Current Converters

A *current-to-voltage converter* is a circuit that provides an output voltage that is directly proportional to the input current. Figure 7.14 shows this type of converter with an op amp connected in the inverting amplifier configuration and $R_s = 0$.

Since the op amp input current i_d is essentially zero, the input current flows directly through R_F. Also, since the n terminal is a virtual ground,

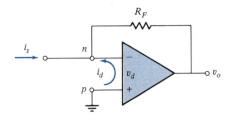

FIGURE 7.14   Current-to-Voltage Converter

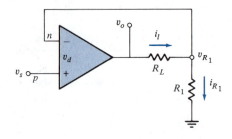

FIGURE 7.15   Voltage-to-Current Converter

the output voltage is given by

$$v_o = -i_s R_F \qquad (7.2\text{--}19)$$

Thus, the desired result is obtained.

Note that the input impedance of the current-to-voltage converter of Fig. 7.14 is approximately zero since the n terminal is a virtual ground. A more careful examination of this circuit would yield an input impedance of R_F/A_d. The output impedance is also quite small and is given by (7.1–14) with $R_S = 0$.

A *voltage-to-current converter* is displayed in Fig. 7.15. Since v_d is quite small, v_s is directly across R_1, and $i_{R_1} = v_s/R_1$. However, this current must be supplied by the output of the op amp. Therefore,

$$i_l = i_{R_1} = \frac{v_s}{R_1} \qquad (7.2\text{--}20)$$

Note that (7.2–20) indicates that the current through R_L is independent of R_L. Hence, this circuit can be used as a constant current source that is voltage controlled. Note also that the input impedance is quite large (essentially given by R_i of the op amp). A disadvantage of this converter, however, is that neither end of R_L may be grounded.

This restriction is removed when the circuit arrangement shown in Fig. 7.16 is used. Here, we will show that i_L is independent of R_L and is given by $i_L = v_s/R_S$. Note that the quantity K is a constant. We begin by equating the currents at

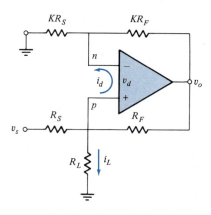

FIGURE 7.16   Constant Current Source (constant if $v_s = V_{SS}$)

the n and p terminals, respectively, to obtain

$$\frac{0 - v_n}{KR_S} = \frac{v_n - v_o}{KR_F} \tag{7.2-21}$$

and

$$\frac{v_s - v_p}{R_S} = \frac{v_p - v_o}{R_F} + i_L \tag{7.2-22}$$

where the input current to the op amp was neglected ($i_d = 0$). Canceling K in (7.2–21) and multiplying through by the resistors in the denominators of (7.2–21) and (7.2–22) yield, respectively,

$$-R_F v_n = R_S(v_n - v_o) \tag{7.2-23}$$

and

$$R_F(v_s - v_p) = R_S(v_p - v_o) + R_S R_F i_L \tag{7.2-24}$$

Subtracting (7.2–23) from (7.2–24) to cancel the v_o terms yields

$$R_F v_s + R_F(v_n - v_p) = R_S(v_p - v_n) + R_S R_F i_L \tag{7.2-25}$$

Finally, solving for i_L and realizing that $v_n \cong v_p$

(since v_d is small) yield the desired result

$$i_l = \frac{v_s}{R_S} \tag{7.2-26}$$

Recalling that the input signal to an op amp circuit may contain DC, we note that for $v_s = V_{SS}$ (a DC value), the voltage-to-current converter circuit of Fig. 7.16 becomes a constant current source.

7.2.4   Amplifiers with Automatic Gain Control (AGC)

Figure 7.17 displays an application of an op amp in which a control voltage V_C can be used to vary the amplification of the input voltage v_s. The arrangement of the circuit is identical to that of the noninverting amplifier for which $v_o = [(R_F/R_S) + 1]v_s$, with R_S replaced by the JFET source-to-drain channel resistance R_{CH}. Thus, the output voltage of Fig. 7.17 is given by

$$v_o = \left(\frac{R_F}{R_{CH}} + 1\right)v_s \tag{7.2-27}$$

Note that since $v_n \cong v_p$, the input voltage v_s is also the drain-to-source voltage of the JFET, which places a restriction on the magnitude of v_s relative to V_C. Additionally, the drain-to-source resistance R_{CH} is controlled in a nonlinear fashion by V_C.

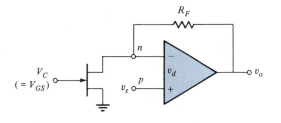

FIGURE 7.17   Single JFET AGC Amplifier

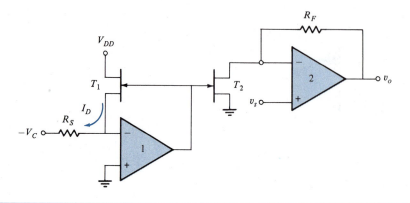

These restrictions are removed using the AGC amplifier of Fig. 7.18. Matched JFETs are used in a FET current mirror arrangement (introduced and discussed in Chapter 6) along with two op amps. Since the n terminal of op amp 1 is a virtual ground, $v_{GS_1} = v_{GS_2}$ and $R_{CH_1} = R_{CH_2}$ (which we define as R_{CH}). The values of these resistances are then given by

$$R_{CH} = \frac{V_{DD}}{I_D} \qquad (7.2\text{–}28)$$

where I_D is provided by the control voltage V_C and the resistor R_S since

$$I_D = \frac{V_C}{R_S} \qquad (7.2\text{–}29)$$

Substituting (7.2–29) into (7.2–28) yields

$$R_{CH} = \frac{V_{DD}R_S}{V_C} \qquad (7.2\text{–}30)$$

Finally, since the output voltage of the AGC amplifier of Fig. 7.18 is also given by

$$v_o = \left(\frac{R_F}{R_{CH}} + 1\right)v_s \qquad (7.2\text{–}31)$$

substitution of (7.2–30) into (7.2–31) yields

$$v_o = \left(\frac{R_F V_C}{R_S V_{DD}} + 1\right)v_s \qquad (7.2\text{–}32)$$

Note that the variation of V_C thus produces linear gain control for the AGC amplifier of Fig. 7.18. The amplification factor multiplying v_s in (7.2–32) is often called the *gain control factor*.

——————————————————————— EX. 7.7

JFET AGC Amplifier Example

Matched JFETs are used in the AGC amplifier circuit of Fig. 7.18 along with the following elements: $R_F = 100$ kΩ, $R_S = 1$ kΩ, $V_C = 5$ V, and $V_{DD} = 10$ V. Calculate the gain control factor for this amplifier.

Solution: Substituting directly into (7.2–32), we obtain

$$\frac{v_o}{v_s} = \frac{(100)(5)}{(1)(10)} = 50$$

——————————————————————————

7.2.5   AGC in Receiver Systems

The receivers in communication systems have a potential problem because the incoming signals vary widely in magnitude. For example, in audio (radio) systems, the signal power at the receiving antenna is inversely proportional to the square of the distance to the radio station. Thus, in changing the station, each different station can have a vastly different signal level that changes the volume level considerably. Furthermore, a high enough signal level can damage the speaker, in addition to being annoying. This problem is eliminated by using AGC and negative feedback from the output to provide the control voltage. The output signal level can then be maintained essentially constant even though the received signal at the antenna varies.

Note that AGC amplifiers are available in IC form from various manufacturers. One popular audio AGC amplifier is the LM170. Further description and example circuits that use negative feedback to provide the AGC control voltage will be given in Chapter 14.

7.3   NONLINEAR APPLICATIONS

Various nonlinear applications of op amps are presented in this section. The first several applications, such as op amp rectifiers and the op amp clamper, provide more ideal behavior than the simple PN diode circuits of Chapter 2, as we will see. Other important nonlinear applications that are described include the comparator, Schmitt trigger, logarithmic amplifier, and exponential amplifier.

7.3.1   Op Amp Diode Rectifiers

Active Half-Wave Rectifiers.   In Chapter 2, we considered diode rectifier circuits. It was noted there that the resulting output voltage of the diode half-wave rectifier exhibits amplitude reduction and shortening of pulse width such that an exact half wave corresponding to the input is not reproduced.

Figures 7.19 and 7.20 show two possibilities for active half-wave rectifiers that produce very accurate half-wave rectification without distortion. Also shown are the corresponding voltage transfer characteristics. The operation of these circuits is understood by considering the conditions necessary to open or short the diode. The rectifier of Fig. 7.19 provides voltage amplification, whereas that of Fig. 7.20 does not.

In order to demonstrate the transfer characteristics shown, we begin by considering Fig. 7.19 and $v_s < 0$. Under this condition, the output of the op amp ($A_d v_d$) will be negative, and therefore, the diode will be reverse biased and open. Thus, the output voltage v_o must be zero, and this portion ($v_s < 0$) of the transfer characteristic in Fig. 7.19b is verified.

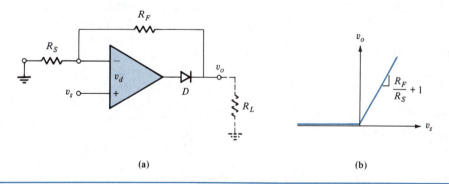

(a)   (b)

FIGURE 7.19   Active Half-Wave Noninverting Rectifier: (a) Circuit, (b) Transfer characteristic

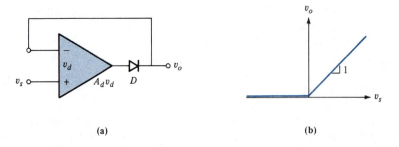

(a)   (b)

FIGURE 7.20   Active Half-Wave Voltage Follower Rectifier: (a) Circuit, (b) Transfer characteristic

For $v_s > 0$, the output of the op amp ($A_d v_d$) is positive, and the diode is forward biased. Thus, the forward voltage is 0.7 V across the diode terminals. Under this condition, we write the output voltage v_o from Fig. 7.19 as

$$v_o = A_d v_d - 0.7 \qquad (7.3\text{–}1)$$

However, v_d is the difference in voltage from the positive to the negative terminal, or $v_d = v_p - v_n$ where $v_p = v_s$ and

$$v_n = \frac{R_S}{R_F + R_S} v_o \qquad (7.3\text{–}2)$$

Substitution into (7.3–1) yields

$$v_o = A_d \left(v_s - \frac{R_S}{R_F + R_S} v_o \right) - 0.7 \qquad (7.3\text{–}3)$$

Rearranging this expression and solving for v_o yield

$$v_o = \frac{A_d v_s}{A_d \left(\dfrac{R_S}{R_F + R_S} \right) + 1} - \frac{0.7}{A_d \left(\dfrac{R_S}{R_F + R_S} \right) + 1} \qquad (7.3\text{–}4)$$

Hence, for $A_d \gg (R_F + R_S)/R_S$, we neglect the 1

in the denominator of each term to obtain

$$v_o = \left(\frac{R_F + R_S}{R_S} \right) v_s - \left(\frac{R_F + R_S}{R_S A_d} \right) 0.7 \qquad (7.3\text{–}5)$$

Notice that the last term in (7.3–5) is negligible because A_d is large. Thus,

$$v_o = \left(\frac{R_F}{R_S} + 1 \right) v_s \qquad (7.3\text{–}6)$$

This expression verifies the region of Fig. 7.19 in which $v_s > 0$. Furthermore, we have accounted for the diode forward voltage as $V_0 = 0.7$ V; however, it has absolutely no effect on the output voltage v_o because of the large voltage amplification of the op amp. Figure 7.20 is thus also verified since it is the special case in which $R_F = 0$ and $R_S = \infty$.

Figure 7.21 shows the inverting amplifier version of the active half-wave rectifier. Note that an extra diode (D_F) is included and is necessary for half-wave rectification. This diode has several useful purposes, as we will see.

For the circuit of Fig. 7.21, with $v_s > 0$, v_n is positive and the output of the op amp is negative. Thus, D_F shorts, providing a feedback path around the op amp. Moreover, v_n is a virtual ground and because of D_F, the output of the op amp is fixed at -0.7 V, which keeps D_1 off and $v_o = 0$. Note that without D_F in the circuit (Fig. 7.21a) and for $v_s > 0$, a resistive voltage divider exists involving R_S, R_F, and the output load resistor (not shown).

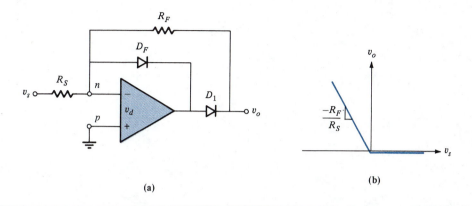

(a)

(b)

FIGURE 7.21   Active Half-Wave Inverting Rectifier: (a) Circuit, (b) Transfer characteristic

This divider network would result in a nonzero output voltage (v_o). Hence, D_F has the effect of bypassing R_F and maintaining $v_o = 0$. This feedback diode also keeps the op amp out of negative saturation, which requires a certain time delay in switching back to amplifier operation. For $v_s < 0$, v_n is negative and the output of the op amp is positive. Thus, D_F is open and D_1 is shorted, providing $v_o = -(R_F/R_S)v_s$.

Thus, the voltage transfer characteristic of Fig. 7.21b is verified. This transfer characteristic corresponds to rectification with sign inversion. Furthermore, R_F and/or R_S may be varied to change the slope.

EX. 7.8

Active Half-Wave Rectifier Example

Sketch the output voltage for the circuit of Fig. 7.19a with $v_s = 5 \sin \omega t$ V, $R_F = 100$ kΩ, and $R_s = 1$ kΩ.

v_s, v_o (V)

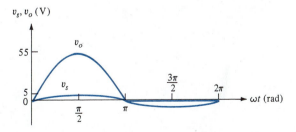

FIGURE E7.8

Solution: From the transfer characteristic of Fig. 7.19b, we realize immediately that the output voltage is an exact rectified and amplified sine wave, as displayed in Fig. E7.8. Note that the amplitude of this sine wave is amplified by $(R_f/R_s) + 1 = 11$.

Active Full-Wave Rectifiers.   By using half-wave rectifiers in tandem, full-wave rectification can be achieved. Several alternatives exist, and we will consider the one shown in Fig. 7.22a.

To describe the operation of this circuit, we use the results of the active half-wave rectifiers just obtained. Note that the upper and lower op amp circuits of Fig. 7.22a are set up so that they will each provide unity gain. Also, the top portion of the circuit (containing op amp 2) is on when the bottom portion (containing op amp 1) is off, and vice versa.

For positive v_s, op amp 2 produces an output voltage $v_o = v_s$. Meanwhile, in the bottom portion of the circuit, diode D_1 is open and zero current flows through the equal resistors (R_F) since the voltage at either end is v_s. Note that if the feedback diode (D_F) of Fig. 7.21 were included in Fig. 7.22a, a current path would be provided through the feedback resistor, D_F, and the output of op amp 1. Hence, D_F must be left out in this case.

For negative v_s, the bottom portion of the circuit of Fig. 7.22a provides an output voltage

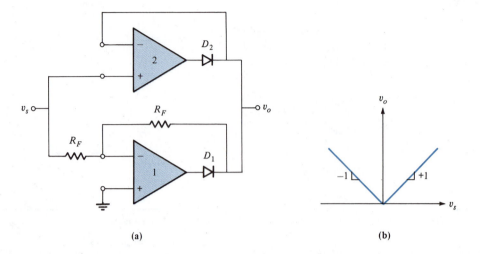

(a)   **(b)**

FIGURE 7.22   Active Full-Wave Rectifier: (a) Circuit, (b) Transfer characteristic

$v_o = -v_s = +|v_s|$. D_2 is forced open, and op amp 2 is in negative saturation $(v_{d_2} = -2|v_s|)$. Also, with negative v_s applied to the positive terminal of op amp 2, the output voltage of op amp 2 is large and negative, which maintains D_2 open. Thus, the full-wave voltage transfer characteristic of Fig. 7.22b is verified.

Active Peak Rectifier.   An active peak rectifier circuit is shown in Fig. 7.23a. Note that in comparison to the diode peak rectifier circuit of Chapter 2 (redrawn in Fig. 7.23b), the op amp and diode replace the single diode. The effect of adding the op amp allows the capacitor to charge to the peak values of v_s without a reduction due to the forward voltage across the diode.

As described in Chapter 2, the diode peak rectifier has an additional disadvantage. When a load resistor R_L is connected at its output, the capacitor will discharge through R_L causing v_o to decay with time. This effect is eliminated, as shown in Fig. 7.23c, by connecting the load resistor through an additional op amp connected in the voltage follower configuration. Since the voltage follower has an enormous input impedance $(A_d R_i)$, the capacitor is unable to discharge and R_L has no effect upon the capacitor voltage. Thus, v_o is maintained at the peak value of v_s.

─────────────────────────────────── EX. 7.9

Active Peak Rectifier Example

Sketch and compare the output voltage waveshapes for the diode peak rectifiers of Figs. 7.23b and c for $v_s = 2 \sin 2\pi f t$ ($f = 100$ Hz) and with a load resistor $R_L = 100$ kΩ connected to each output. Let $C = 1$ μF in each case and treat the op amps as ideal. Replace each diode with an ideal diode and an offset voltage of $V_0 = 0.7$ V. Also, let the capacitors be initially uncharged.

Solution: For the diode peak rectifier of Fig. 7.23b, C charges to the peak value of v_s minus the offset voltage of the diode, or 1.3 V. The charging of C continues during the range $V_0 \le v_s \le V_s = 2$ V, after which C discharges through R_L. During this portion of time, v_o decays exponentially as $v_o = 1.3 \exp(-t/R_L C)$. Since the exponential decay continues for approximately 3/4 of each period of length, $1/f = 2\pi/\omega$, the time of decay is $t = (3/4)(2\pi/\omega)$. The output voltage therefore decays to

$$v_o = 1.3 e^{-t/R_L C} = 1.3 e^{-(3/4)[2\pi/(\omega R_L C)]}$$

or

$$v_o = 1.3 e^{-(3/4)/[(100)(10^5)(10^{-6})]} = 1.2 \text{ V}$$

However, the active peak rectifier behaves in an ideal manner with C charging to the peak of v_s and unable to discharge such that the output voltage $v_o = 2$ V.

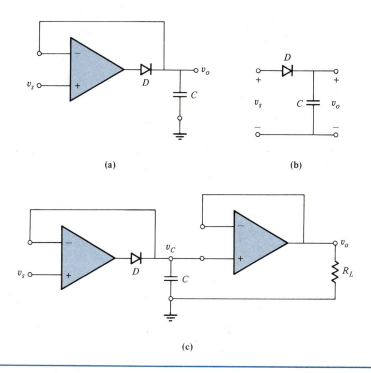

(a)   (b)

(c)

FIGURE 7.23 Peak Rectifiers: (a) Active peak rectifier with op amp and diode, (b) Diode peak rectifier, (c) Active peak rectifier with active load

7.3.2 Comparators

Comparators are similar to op amps except that A_d is made larger for these integrated circuits by including positive feedback in the internal circuitry. The same circuit symbol as that of the op amp is used to represent the comparator, and op amps are often used for comparator applications with external positive feedback.

The basic comparator circuit is shown in Fig. 7.24a; its corresponding transfer characteristic is shown in Fig. 7.24b. The extremely high voltage

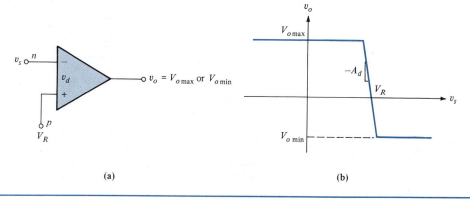

(a)   (b)

FIGURE 7.24 The Comparator: (a) Basic circuit, (b) Transfer characteristic

amplification (A_d) leads directly to a transfer characteristic that displays an abrupt change in v_o when $v_s = V_R$. The output voltage essentially provides digital operation in that only two possible values exist, $V_{o\,min}$ and $V_{o\,max}$, since A_d is very large. These values are provided when $v_s < V_R$ and $v_s > V_R$, respectively. Note that one possibility for the reference voltage is $V_R = 0$.

In this basic comparator circuit, we see that there is no external feedback and that the reference voltage V_R is fixed. We consider next the case in which positive feedback is used externally to provide a reference voltage that depends upon v_o. Depending upon the amount of feedback, an interesting result occurs.

7.3.3   Schmitt Trigger

Figure 7.25a displays a comparator (or op amp) circuit with positive feedback. The voltage at the positive terminal is given by

$$v_p = \frac{R_{S_p}}{R_{S_p} + R_{F_p}} v_o \qquad (7.3-7)$$

The voltage at the negative terminal is $v_n = v_s$. Assuming amplifier operation ($v_o = A_d v_d$) for the time being, we have

$$v_o = A_d v_d = A_d(v_p - v_n)$$
$$= A_d\left[\left(\frac{R_{S_p}}{R_{S_p} + R_{F_p}} v_o\right) - v_s\right] \qquad (7.3-8)$$

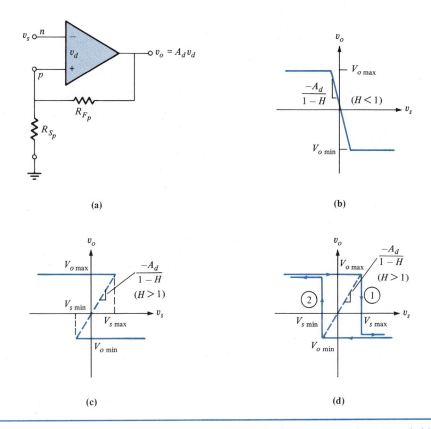

(a)

(b)

(c)

(d)

FIGURE 7.25   Comparator with External Positive Feedback: (a) Circuit, (b) Comparator ($0 \le H < 1$), (c) Schmitt trigger ($H > 1$), (d) Operating path for the Schmitt trigger

Thus, solving for v_o yields

$$v_o = \frac{-A_d v_s}{1 - \dfrac{A_d R_{S_p}}{R_{S_p} + R_{F_p}}} \qquad (7.3\text{-}9)$$

Since amplifier operation has been assumed, (7.3–9) represents the variation in v_o versus v_s in the transition region between positive and negative saturation. The output voltage is observed to be very dependent upon the amount of feedback through the ratio $A_d R_{S_p}/(R_{S_p} + R_{F_p})$, which we defined as H. The range of this ratio is $0 \leq H \leq A_d$ as R_{S_p} is varied from zero to a large value much greater than R_{F_p} (where H may also be varied by changing R_{F_p}). Rewriting (7.3–9) in terms of H yields

$$v_o = \frac{-A_d v_s}{1 - H} \qquad (7.3\text{-}10)$$

We note that the denominator, $1 - H$, can be either positive or negative.

If the denominator of (7.3–10) is positive, or $0 \leq H < 1$, we essentially have our simplest comparator example, and the voltage transfer characteristic of Fig. 7.24b is valid, with negative slope, as shown in Fig. 7.25b. Note that for $0 \leq H < 1$, only a small amount of positive feedback is present or no feedback is present. Also, for $H = 0$, ($R_S = 0$ or $R_F = \infty$), Fig. 7.25b is the transfer characteristic with $V_R = 0$. This range of $H < 1$ is the comparator case with the corresponding transfer characteristic as shown in Fig. 7.25b.

However, for $H > 1$, the denominator of (7.3–10) changes sign, and v_o becomes positive. The circuit is then referred to as a *Schmitt trigger*. Rearranging (7.3–4) for $H > 1$, we have

$$v_o = \frac{A_d}{H - 1} v_s \qquad (7.3\text{-}11)$$

The voltage transfer characteristic for this case

is displayed in Fig. 7.25c, where we observe that the amplifier region has positive slope. It is also drawn as a dashed line because this region is unstable, as we will see.

To determine the input signal voltage v_s at the end-points of the amplifier region in Fig. 7.25c, we use (7.3–11) and solve for the values of v_s that correspond to the output voltage values $V_{o\,max}$ and $V_{o\,min}$. Thus,

$$V_{s\,max} = \frac{H - 1}{A_d} V_{o\,max} \qquad (7.3\text{-}12)$$

and

$$V_{s\,min} = \frac{H - 1}{A_d} V_{o\,min} \qquad (7.3\text{-}13)$$

To observe that the dashed amplifier region of Fig. 7.25c is not an actual operating region, we will vary v_s and determine the resulting v_o. If we look at Fig. 7.25c, we note that as v_s is increased from a large negative value ($v_s < V_{s\,min}$), the output voltage is $V_{o\,max}$ until v_s becomes positive and greater than $V_{s\,max}$. Then, the output voltage switches to $V_{o\,min}$. But, starting with $v_s > V_{s\,max}$, $v_o = V_{o\,min}$. Then, decreasing v_s until v_s is negative and $v_s < V_{o\,min}$, v_o switches to $V_{o\,max}$. These two different operating paths are shown in Fig. 7.25d and are separated slightly when they are coincident. We note that transitions from high-to-low and from low-to-high output voltage occur at different values of v_s. This condition is referred to as *hysteresis*, and the circuit exhibits memory. The switching occurs at an input voltage dependent upon the past state of the output.

Note also that a reference voltage V_R can be used to shift the hysteresis loop to another desired range of v_s. Figure 7.26a shows the modified Schmitt trigger, and Fig. 7.26b shows the resulting transfer characteristic. The new values of $V'_{s\,max}$ and $V'_{s\,min}$ are then obtained from (7.3–12) and (7.3–13) by adding V_R to each.

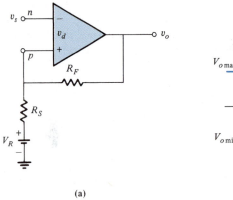

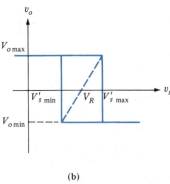

(a)   (b)

FIGURE 7.26   Schmitt Trigger with Hysteresis Loop around V_R: (a) Circuit, (b) Transfer characteristic

---

EX. 7.10

Comparator Example

An op amp is connected as shown in Fig. 7.25a with $A_d = 10^4$, $V_{om} = V_{o\,max} = -V_{o\,min} = 10$ V, $R_{S_p} = 1$ kΩ, and $R_{F_p} = 100$ MΩ. Calculate the parameter H and determine the output voltage waveshape for $V_R = 0$ V and $v_s = \sin \omega t$ V.

Solution: The parameter H is calculated from

$$H = \frac{A_d R_{S_p}}{R_{S_p} + R_{F_p}}$$

which yields

$$H = \frac{(10^4)(1)}{1 + 10^5} = 0.1$$

Thus, the circuit behaves like a comparator. The output voltage is almost a true square wave because as v_s changes from positive to negative (and vice versa), the output switches from negative saturation to positive saturation almost instantly. Switching into saturation occurs when the magnitude of the input difference voltage, given by $|v_d| = |v_s - V_R| = |v_s|$, causes saturation and the output saturates when $|A_d v_d|$ becomes equal to V_{om}, or $|v_d| = |v_s| = V_{om}/A_d = 10^{-3}$ V $= 1$mV.

EX. 7.11

Schmitt Trigger Example

The op amp in the circuit of Fig. 7.25a has $V_{o\,max} = -V_{o\,min} = 10$ V. Design the Schmitt trigger circuit so that $V_{s\,max} = -V_{s\,min} = 2$ V. To obtain a unique result, let $R_{S_p} = 1$ kΩ. Assume that $A_d \gg (R_{S_p} + R_{F_p})/R_{S_p}$.

Solution: Substitution of $H = A_d R_{S_p}/(R_{S_p} + R_{F_p}) \gg 1$ into (7.3–12) yields

$$V_{s\,max} = \frac{A_d R_{S_p}/(R_{S_p} + R_{F_p})}{A_d} V_{o\,max} \qquad (1)$$

Thus, canceling values of A_d and solving for R_{F_p} yield

$$R_{F_p} = \left(\frac{V_{o\,max}}{V_{s\,max}}\right) R_{S_p} - R_{S_p} \qquad (2)$$

Substituting values yields

$$R_{F_p} = \left(\frac{10}{2}\right)(1) - 1 = 4 \text{ k}\Omega$$

Comparators are circuit elements that are used basically to detect zero crossings of an input signal waveform. Using a comparator in the form of a Schmitt trigger can eliminate the erroneous detection of multiple crossings associated with a primary signal and a noise signal superimposed. A more specific application of a comparator in the detection of zero crossings of an input waveform is a square wave generator. The next section indicates one example of a square wave (as well as a triangular wave) generator.

7.3.4 Astable Multivibrator

The *astable multivibrator* is essentially a square wave generator. This circuit is obtained by modifying the basic Schmitt trigger of Fig. 7.25a to include feedback at the n terminal that varies with time. Figure 7.27a shows the modified circuit in which $H = A_d R_{S_p}/(R_{S_p} + R_{F_p}) > 1$. We will as-

sume that symmetry exists for the hysteresis loop such that $V_{o\,max} = -V_{o\,min} = V_{om}$. The elements that have been added to the Schmitt trigger are R_{F_n} and C_{S_n}. As we will see, these elements provide astable operation—that is, two operating states that are each stable for a finite period of time. To analyze the circuit of Fig. 7.27a, we must first realize that v_n varies if current passes through the negative feedback resistor R_{F_n} and the capacitor C_{S_n}. The capacitor voltage varies with current, and the direction of the current through C_{S_n} determines whether v_n is increasing or decreasing. If we consider v_n increasing because $v_o = +V_{om}$, then as v_n increases and becomes equal to $v_p = [R_{S_p}/(R_{S_p} + R_{F_p})]V_{om}$ and minutely larger, v_d becomes negative and v_o switches to $-V_{om}$. The current through C_{S_n} is forced to reverse direction,

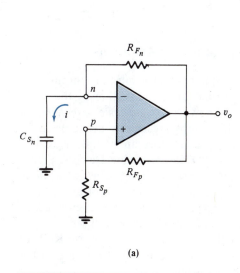

(a)

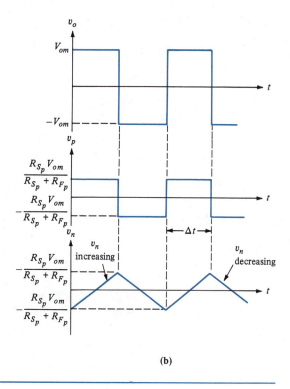

(b)

FIGURE 7.27 Astable Multivibrator: (a) Circuit, (b) Voltage waveshapes

and the voltage v_n begins to decrease. As v_n decreases, it approaches the negative voltage at p—that is, $v_p = -[R_{S_p}/(R_{S_p} + R_{F_p})]V_{om}$. When v_n becomes minutely less than this negative value, v_d becomes positive and v_o switches back to $+V_{om}$. This process then repeats and repeats as the multivibrator switches back and forth from one semi-stable (or quasi-stable) state to the other. The corresponding voltage waveshapes are displayed in Fig. 7.27b.

We quantitatively determine the period of the waveshapes in Fig. 7.27b by considering the current–voltage relation for the capacitor, which is given by

$$i_C = C_{S_n} \frac{dv_n}{dt} \qquad (7.3\text{--}14)$$

For the half-cycles in which $v_o = V_{om}$, the capacitor current is given by

$$i_C = \frac{V_{om} - v_n}{R_{F_n}} \cong \frac{V_{om}}{R_{F_n}} \qquad (7.3\text{--}15)$$

since v_n is small in comparison to V_{om}. Thus, equating (7.3–14) with (7.3–15) and rearranging, we have

$$\frac{dv_n}{dt} = \frac{V_{om}}{R_{F_n}C_{S_n}} \qquad (7.3\text{--}16)$$

Integrating yields

$$v_n(t) = \frac{V_{om}}{R_{F_n}C_{S_n}} t + C \qquad (7.3\text{--}17)$$

where C is a constant. The length of time for which (7.3–17) is valid must correspond to the range of v_n given by

$$-\frac{R_{S_p}}{R_{S_p} + R_{F_p}} V_{om} \leq v_n \leq \frac{R_{S_p}}{R_{S_p} + R_{F_p}} V_{om} \qquad (7.3\text{--}18)$$

By substituting the extremes for v_n in (7.3–18) into

(7.3–17), we obtain equations for the corresponding times as follows:

$$-\frac{R_{S_p}}{R_{S_p} + R_{F_p}} V_{om} = \frac{V_{om}}{R_{F_n}C_{S_n}} t_1 + C \qquad (7.3\text{--}19)$$

and

$$\frac{R_{S_p}}{R_{S_p} + R_{F_p}} V_{om} = \frac{V_{om}}{R_{F_n}C_{S_n}} t_2 + C \qquad (7.3\text{--}20)$$

Subtracting (7.3–19) from (7.3–20) and solving for $\Delta t = t_2 - t_1$ yield

$$\Delta t = \frac{2R_{S_p}}{R_{S_p} + R_{F_p}} R_{F_n}C_{S_n} \qquad (7.3\text{--}21)$$

Finally, because of the symmetry of the circuit, the period of the waveshapes shown in Fig. 7.27b is $T = 2\Delta t$, or

$$T = 4 \frac{R_{S_p}}{R_{S_p} + R_{F_p}} R_{F_n}C_{S_n} \qquad (7.3\text{--}22)$$

Note that the frequency corresponding to this period is $f = 1/T$.

7.3.5 Logarithmic Amplifiers

Amplifiers that exhibit an output voltage proportional to the logarithm of the input voltage are known as *logarithmic amplifiers*, or *log amps*. Log amps are readily available in IC form. Two popular types are the Burr–Brown 4127 and the AD755 by Analog Devices. Figure 7.28 displays a basic log amp in which a diode is placed in the feedback branch of an op amp circuit. An alternative basic log amp circuit uses a BJT instead of a diode, as we will see.

To obtain v_o in terms of v_s, we use KCL at the negative terminal in Fig. 7.28 and neglect the op amp input current. Thus,

$$i_{Rs} = i_D \qquad (7.3\text{--}23)$$

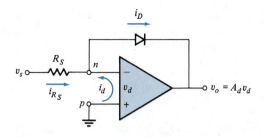

FIGURE 7.28   Basic Log Amp

However, $i_{R_S} = v_s/R_S$ since the negative terminal is a virtual ground. The diode current is given by Shockley's relation of Chapter 2, (2.1–2). Thus, by substitution on either side of (7.3–23), we obtain

$$\frac{v_s}{R_S} = I_S(e^{(v_n - v_o)/v_T} - 1) \qquad (7.3–24)$$

where I_S is the reverse current parameter and v_T is the thermal voltage. We now neglect -1 relative to the exponential, which is equivalent to assuming $-v_o$ is positive or that $v_s > 0$. Hence, only positive input voltages will provide logarithmic operation, which is appropriate since the logarithm of a negative quantity is undefined and v_s must therefore be positive. Additionally, since

the n terminal is a virtual ground, $v_n = 0$. Thus, taking the natural log (ln) of each side of (7.3–24) and solving for v_o, we have

$$v_o = -v_T \ln \frac{v_s}{R_S I_S} \qquad (7.3–25)$$

Hence, the output voltage of the circuit in Fig. 7.28 is proportional to the natural logarithm of the input voltage v_s for positive v_s. A difficulty for this case, however, is that v_o, as in (7.3–25), depends upon I_S, whose value is not known accurately (as first mentioned in Chapter 2).

This disadvantage can be eliminated by using the circuit of Fig. 7.29, which utilizes two matched BJTs (T_1 and T_2) and two op amps. Note that the grounded base transistor T_1 has replaced the diode in the original circuit of Fig. 7.28. A wider range of voltage over which the current varies exponentially with voltage is thus provided. The matched transistors produce cancelation of the reverse current parameter, as we will see.

Assuming $i_B \ll i_E$ and large R_i, the emitter currents i_{E_1} and i_{E_2} in Fig. 7.29 are given by

$$i_{E_1} = \frac{v_s}{R_{S_1}} \qquad (7.3–26)$$

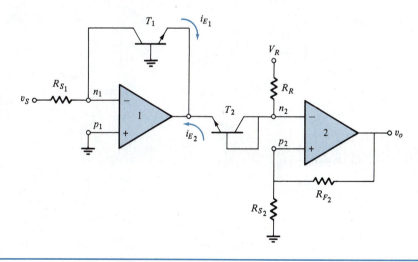

FIGURE 7.29   Log Amp Using Matched Transistors

and

$$i_{E_2} = \frac{V_R}{R_R} \qquad (7.3\text{-}27)$$

If we start from ground at the base of T_1 and use KVL to reach the n_2 terminal, we have

$$v_{n_2} = -v_{BE_1} + v_{BE_2} \qquad (7.3\text{-}28)$$

However, the base–emitter voltage and emitter current for each transistor are related through Shockley's relation as $i_E = I_S(e^{v_{BE}/v_T})$, or $v_{BE} = v_T \ln(i_E/I_S)$. Thus, (7.3–28) can be written as

$$v_{p_2} \cong v_{n_2} = -v_T \ln \frac{i_{E_1}}{I_S} + v_T \ln \frac{i_{E_2}}{I_S} \qquad (7.3\text{-}29)$$

Combining the terms in (7.3–29) using the properties of logarithms yields

$$v_{p_2} = -v_T \ln \frac{i_{E_1}}{i_{E_2}} \qquad (7.3\text{-}30)$$

where cancelation of I_S has resulted. Furthermore, considering the resistive divider at the output, we have

$$v_{p_2} = \frac{R_{S_2}}{R_{S_2} + R_{F_2}} v_o \qquad (7.3\text{-}31)$$

Thus, equating (7.3–30) and (7.3–31) and solving for v_o yield

$$v_o = -\frac{R_{S_2} + R_{F_2}}{R_{S_2}} v_T \ln \frac{i_{E_1}}{i_{E_2}} \qquad (7.3\text{-}32)$$

Finally, substituting for i_{E_1} and i_{E_2} from (7.3–26) and (7.3–27) yields the desired logarithmic relation:

$$v_o = -\frac{R_{S_2} + R_{F_2}}{R_{S_2}} v_T \ln \frac{v_s R_R}{V_R R_{S_1}} \qquad (7.3\text{-}33)$$

Note that v_o is dependent upon temperature through the thermal voltage v_T $(= kT/q)$. This dependence can be removed by using a temperature-

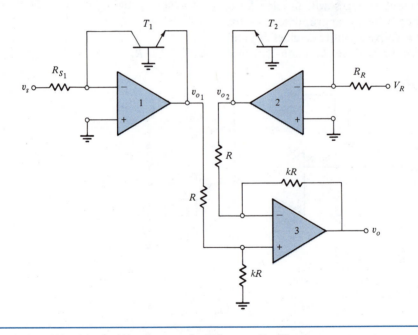

FIGURE 7.30   Log Amp Using Matched Transistors and a Difference Amplifier

compensated resistor for R_{S_2} such that R_{S_2} varies linearly with T. Additionally, R_{F_2} is chosen to be much larger than R_{S_2} and thus also provides amplification of the output voltage.

An alternative logarithmic amplifier is displayed in Fig. 7.30. Op amps 1 and 2 operate in the same manner as that of the basic log amp of Fig. 7.28. Since transistors T_1 and T_2 are matched, taking the difference signal using op amp 3 again provides cancelation of the reverse current parameter. Op amp 3 also furnishes amplification.

The output voltage of Fig. 7.30 is given by superposition of the inputs as

$$v_o = -kv_{o_2} + (k+1)\left(\frac{k}{k+1}\right)(v_{o_1})$$

$$= k(v_{o_1} - v_{o_2}) \qquad (7.3\text{-}34)$$

where the resistor values have been selected to provide the same amplification for each input and k is a constant. Note that the outputs of op amps 1 and 2 are given by expressions similar to (7.3–25):

$$v_{o_1} = -v_T \ln \frac{v_s}{R_{S_1} I_S} \qquad (7.3\text{-}35)$$

and

$$v_{o_2} = -v_T \ln \frac{V_R}{R_R I_S} \qquad (7.3\text{-}36)$$

Substitution into (7.3–34) yields the log amp expression for the circuit of Fig. 7.30 as

$$v_o = -kv_T \ln \frac{v_s R_R}{V_R R_{S_1}} \qquad (7.3\text{-}37)$$

--- EX. 7.12

Logarithmic Amplifier Example

Consider the log amp of Fig. 7.30 and assume that the transistors are matched. For $k = 100$, $R_R = 100$ kΩ, $R_{S_1} = 1$ kΩ, $R = 10$ kΩ, and $V_R = 5$ V, obtain the expression for v_o versus v_s.

Solution: Direct substitution into (7.3–37) yields

$$v_o = -100(0.025) \ln\left(\frac{v_s}{5}\right)\left(\frac{100}{1}\right) = -2.5 \ln 20v_s$$

7.3.6   Exponential Amplifiers

The inverse of the logarithmic function, the exponential, can also be achieved with special op amp circuits. Figure 7.31 displays a basic *exponential amplifier* using a PNP transistor, an op amp, and a feedback resistor. A similar dependence is obtained if the BJT is replaced with a diode as in Problem 4.45 at the end of the chapter.

Analysis of the circuit of Fig. 7.31 simply involves realizing that the output voltage is given by

$$v_o = -i_c R_F \qquad (7.3\text{-}38)$$

since the n terminal is a virtual ground. Thus, since the emitter current of a BJT varies exponentially with the base–emitter voltage and $i_c = \alpha i_E$, we have

$$i_c = \alpha I_S e^{v_s/v_T} \qquad (7.3\text{-}39)$$

by substitution into (7.3–38), the desired exponential relation is obtained as

$$v_o = -\alpha I_S R_F e^{v_s/v_T} \qquad (7.3\text{-}40)$$

As in the case of the basic log amp, we observe that transistor parameter I_S also arises in the

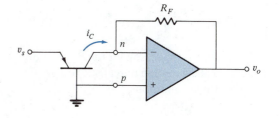

FIGURE 7.31   Basic Exponential Amplifier

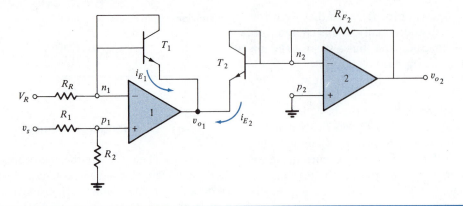

FIGURE 7.32   Exponential Amplifier Using Matched Transistors

basic exponential amplifier expression of (7.3–40). By using the exponential amplifier of Fig. 7.32 that utilizes matched transistors, cancellation of this transistor parameter is achieved.

To analyze the circuit of Fig. 7.32, we first note that i_{E_1} and i_{E_2} must be positive. However, no such restriction is placed upon v_s; both positive and negative values are allowable.

From the circuit of Fig. 7.32, we observe that

$$i_{E_1} = \frac{v_R - v_{n_1}}{R_R} \qquad (7.3\text{–}41)$$

where

$$v_{n_1} \cong v_{p_1} = \frac{v_s R_2}{R_1 + R_2} \qquad (7.3\text{–}42)$$

To ensure that i_{E_1} is greater than zero the following inequality is satisfied:

$$V_R \gg v_s\left(\frac{R_2}{R_1 + R_2}\right) \qquad (7.3\text{–}43)$$

This inequality is the reason for the R_1–R_2 voltage divider and alleviates the unrealistic restriction on v_s. Under these conditions,

$$i_{E_1} = \frac{V_R}{R_R} \qquad (7.3\text{–}44)$$

and the output voltage of op amp 1 (v_{o_1}) is negative (V_{ref} is positive). Since v_{o_1} is negative, the output voltage of the circuit (v_{o_2}) is positive, and thus

$$i_{E_2} = \frac{v_{o_2} - v_{n_2}}{R_{F_2}} \cong \frac{v_{o_2}}{R_{F_2}} \qquad (7.3\text{–}45)$$

since n_2 is a virtual ground.

However, i_{E_1} and i_{E_2} are also given by Shockley's current–voltage expression. Under forward-bias conditions (which we have established), i_{E_1} and i_{E_2} can be written as

$$i_{E_1} = \frac{V_R}{R_R} = I_S e^{v_{BE_1}/v_T} \qquad (7.3\text{–}46)$$

and

$$i_{E_2} = \frac{v_{o_2}}{R_{F_2}} = I_S e^{v_{BE_2}/v_T} \qquad (7.3\text{–}47)$$

From the circuit of Fig. 7.32, the base–emitter voltages are given by

$$\begin{aligned} v_{BE_1} &= v_{n_1} - v_{o_1} \\ &\cong v_s\left(\frac{R_2}{R_1 + R_2}\right) - v_{o_1} \end{aligned} \qquad (7.3\text{–}48)$$

which uses (7.3–42), and by

$$v_{BE_2} = v_{n_2} - v_{o_1} \cong -v_{o_1} \qquad (7.3\text{–}49)$$

since the n_2 terminal is a virtual ground. By dividing (7.3–46) into (7.3–47) and substituting for the base–emitter voltages given by (7.3–48) and (7.3–49), we obtain (after algebra)

$$\left(\frac{v_{o_2}}{R_{F_2}}\right)\left(\frac{R_R}{V_R}\right) = e^{-v_s[R_2/(R_1 + R_2)]/v_T} \quad \textbf{(7.3–50)}$$

where $I_S e^{-v_{o_1}/v_T}$ cancels. Solving for v_{o_2} yields the final result:

$$v_{o_2} = V_R\left(\frac{R_{F_2}}{R_R}\right)e^{-v_s[R_2/(R_1 + R_2)]v_T} \quad \textbf{(7.3–51)}$$

This expression is valid for either polarity of v_s.

7.4 SELECTION GUIDE FOR VOLTAGE COMPARATORS

As described, comparators are special ICs with internal positive feedback. Unlike op amps, comparators are not frequency compensated and therefore will oscillate if used as amplifiers. However, uncompensated op amps are sometimes used in comparator applications.

Comparators are designed to have very fast response times, typically on the order of 100 ns. Additionally, comparators usually require both positive and negative supply voltages.

Data Sheet A7.1 at the end of the chapter provides a selection guide for several voltage comparators. The parameters are varied from one type to another by special IC designs.

The abbreviations used for columns that are not entirely self-explanatory are defined as follows:

- $V_{IO(MAX)}$ is the maximum input offset voltage.
- $I_{IB(MAX)}$ is the maximum input bias current.
- $I_{OL(MIN)}$ is the minimum low-level output current.

Note that the packages used are standard and are similar to those used for op amps. Refer back to Fig. 6.25 for the definitions of the standard packages. Also, refer to Appendix 5 at the end of the book for more detailed comparator information and various comparator data sheets.

CHAPTER 7 SUMMARY

- The voltage transfer characteristic for an op amp has three operating regions: the amplifier region, $v_o = A_d v_d$; positive saturation, $v_o = V_{o\ max}$; and negative saturation, $v_o = V_{o\ min}$.
- The equivalent circuit of an op amp contains three elements: the input impedance R_i, the output impedance R_o, and a voltage source $A_d v_d$.
- For amplifier operation, $v_d = v_p - v_n$ is restricted to small values, and special amplifier circuits with negative feedback are therefore used. Two of these circuits, called the inverting and the noninverting amplifier circuits, provide a fixed amount of voltage amplification under the assumptions of large R_i, small R_o, and large A_d.
- Negative feedback in the inverting and the noninverting amplifier circuits is provided by feedback resistor R_F and series resistor R_S. Large A_d implies that $A_d \gg (R_S + R_F)/R_S$.
- For the inverting amplifier, $v_o/v_s = -R_F/R_S$ and $r_i = R_S$. For the noninverting amplifier, $v_o/v_s = (R_F + R_S)/R_S$ and $r_i = A_d R_i/(R_F/R_S + 1)$. For both amplifiers, $r_o \cong R_o/[1 + (R_S\|R_i A_d)/(R_F + R_S\|R_i)]$.
- The properties of the op amp voltage follower are $v_o/v_s = 1$, $r_i = A_d R_i$, and $r_o = R_o/A_d$.
- Op amps can be used to perform the mathematical operations of subtraction, addition, integration, and differentiation.
- When impedances are used for the feedback and series elements in the inverting and noninverting amplifier connections, the voltage amplification expressions are $\mathbf{V}_o/\mathbf{V}_s = -Z_F/Z_S$ and $\mathbf{V}_o/\mathbf{V}_s = Z_F/Z_S + 1$, respectively. By using capacitors and resistors for Z_F and Z_S,

low-pass, high-pass and band-pass filters can be obtained.

■ The basic current-to-voltage converter uses a single resistor R_F.

■ The basic voltage-to-current converter uses a resistor R_1 to control the current magnitude through R_L in a simple circuit with negative feedback.

■ The basic AGC amplifier uses the noninverting configuration with feedback resistor R_F and a series resistor that is the channel resistance (R_{CH}) of a JFET. The magnitude of R_{CH} is controlled by V_C. When negative feedback is used to provide V_C, the voltage amplification varies inversely with the magnitude of input signal voltage, and an approximately constant output voltage magnitude is maintained.

■ Active half-wave, full-wave, and peak rectifiers are obtained through specific connections of diodes with op amps. Very accurate rectification is achieved without distortion. Voltage amplification is also provided.

■ Voltage comparators are ICs that are identical to op amps except that A_d has been increased

through the use of positive feedback. An extremely narrow amplifier region is thus created, and only two outputs are possible, $V_{o\,max}$ and $V_{o\,min}$, depending upon whether v_d is positive or negative. Comparators are used in nonlinear applications.

■ When external positive feedback is used with a comparator, the sign of the voltage amplification depends upon the amount of feedback. With the ratio $A_d R_{S_p}/(R_{S_p} + R_{F_p})$ defined as H, the amplification is negative for $H < 1$ and positive for $H > 1$, where $H < 1$ corresponds to comparator operation and $H > 1$ corresponds to Schmitt trigger operation. The Schmitt trigger is a comparator that exhibits hysteresis.

■ Comparators can be used to generate square and triangular waveshapes. The period of these waveshapes is adjustable by varying the magnitude of external elements.

■ The basic logarithmic amplifier uses a diode as the feedback element and a resistor R_S as the series element. Interchanging these two elements yields the basic exponential amplifier.

CHAPTER 7 PROBLEMS

7.1

For the op amp amplifier circuit of Fig. P7.1, derive expressions for the input impedance r_i, the output impedance r_o, and the voltage amplification v_o/v_{s_2} for $v_{s_1} = 0$. Assume that v_{s_2} does not drive the op amp into saturation and calculate these quantities for $R_i = 1$ MΩ, $R_o = 1$ kΩ, and $A_d = 10^4$.

7.2

Repeat Problem 7.1 for $v_{s_1} \neq 0$ and $v_{s_2} = 0$.

7.3

Use superposition to determine the output voltage of the amplifier of Fig. P7.1 with v_{s_1} and $v_{s_2} \neq 0$. Calculate v_o for $v_{s_1} = 2$ V and $v_{s_2} = 0.4$ V.

7.4

Calculate r_i, r_o, and v_o/v_s for Problems 7.1 and 7.2 with $R_S = 100$ kΩ and $R_F = 0.9$ MΩ.

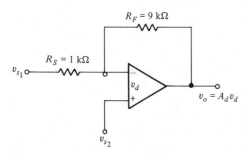

FIGURE P7.1

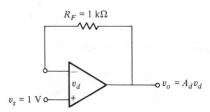

FIGURE P7.5

7.5
For the op amp voltage follower circuit of Fig. P7.5, calculate r_i, r_o, and v_o. Use $A_d = 10^4$, $R_i = 1$ MΩ, and $R_o = 1$ kΩ.

7.6
For the op amp circuit of Fig. P7.6, calculate the output voltage v_o.

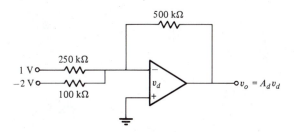

FIGURE P7.6

7.7
Repeat Problem 7.6 for the circuit of Fig. P7.7.

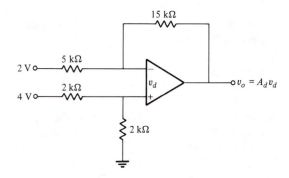

FIGURE P7.7

7.8
For the op amp circuit of Fig. P7.8, calculate v_o.

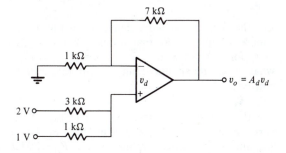

FIGURE P7.8

7.9
For the op amp circuit of Fig. P7.9, calculate v_o.

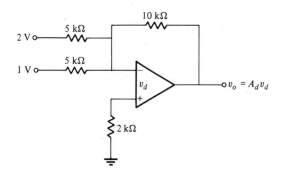

FIGURE P7.9

7.10
Two op amps are used in the circuit of Fig. P7.10. Determine the overall voltage amplification v_o/v_s and calculate v_o for $v_s = 0.05$ V.

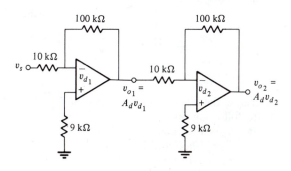

FIGURE P7.10

7.11
Design an op amp circuit that provides an output voltage $v_o = -5v_{s_1} + 6v_{s_2}$ for inputs v_{s_1} and v_{s_2}. Use resistors in the 1 to 10 kΩ range and only one op amp.

7.12

Determine the output voltage in terms of the input voltages for the circuit of Fig. P7.12. What mathematical operation is performed?

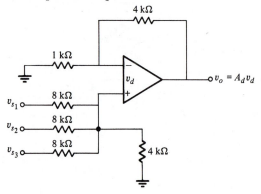

FIGURE P7.12

7.13

For the circuit of Fig. P7.13, using KCL at the negative terminal and the i–v relations for L and R, determine v_o in terms of v_s for the circuit of Fig. P7.13. Assume A_d and R_i are large and R_o is small.

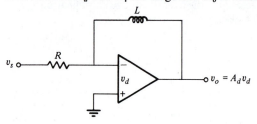

FIGURE P7.13

7.14

Determine the function of the circuit of Fig. P7.14 for the following cases:

(a) $Z_S = 1$ kΩ and $Z_F = 100$ kΩ
(b) $Z_S = 1/j\omega C$ and $Z_F = R_F$
(c) $Z_S = j\omega L$ and $Z_F = R_F$.

Make the usual assumptions.

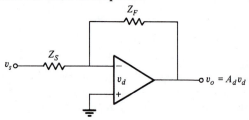

FIGURE P7.14

7.15

For the circuit of Fig. P7.15, determine the expression for V_o/V_s as a function of frequency (ω). Sketch the magnitude of V_o/V_s versus ω for $R_1 = 1$ kΩ, $R_2 = 10$ kΩ, $C_1 = 1$ μF, and $C_2 = 10$ PF. Indicate the half-power frequencies.

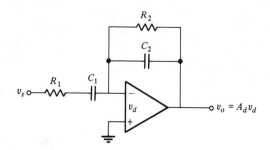

FIGURE P7.15

7.16

For the circuit of Fig. P7.15, replace C_1 with a short and repeat Problem 7.15.

7.17

For the circuit of Fig. P7.15, remove C_2 from the circuit and repeat Problem 7.15.

7.18

For the circuit of Fig. 7.12a, determine the expression for V_o/V_s as a function of ω. Sketch the magnitude of V_o/V_s versus ω for $R_S = 1$ kΩ, $R_F = 100$ kΩ, $R_H = 10$ kΩ, and $C_H = 0.1$ μF.

7.19

Analyze the circuit of Fig. 7.13a to determine V_o/V_s as a function of ω. Sketch the magnitude of V_o/V_s for $R_{S_1} = R_{S_2} = 1$ kΩ, $R_{F_1} = R_{F_2} = 100$ kΩ, $R_L = 1$ kΩ, $R_H = 100$ kΩ, $C_L = 0.1$ μF, and $C_H = 10$ PF.

7.20

For the current-to-voltage converter of Fig. P7.20, choose a value for R_F that provides a 1 V change in v_o for a 1 mA change in i_s. For $V_{o\,max} = 15$ V, what is the maximum input current that can be converted?

7.21

For the current-to-voltage converter of Fig. P7.20, derive expressions for r_i and r_o in terms of A_d, R_i, R_o, and R_F. Calculate r_i and r_o for $R_i = 1$ MΩ, $R_o = 1$ kΩ, $A_d = 10^5$, and $R_F = 1$ kΩ (use the equivalent circuit for the op amp).

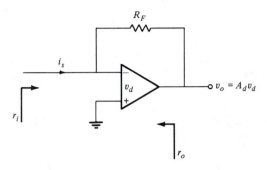

FIGURE P7.20

7.22

Analyze the constant current source circuit of Fig. 7.16 to determine i_L. For $v_s = 10$ V, $R_S = 1$ kΩ, $K = 10$, and $R_F = 1$ kΩ, calculate i_L, Make the usual op amp assumptions.

7.23

Determine the values of i_L and i_{R_1} in the voltage-to-current converter shown in Fig. P7.23.

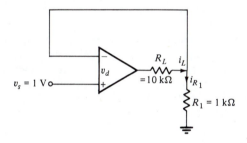

FIGURE P7.23

7.24

The AGC amplifier of Fig. 7.17 is used with a JFET for which $i_D = 10(v_{GS} + 4)v_{DS}$ mA for $v_{DS} \le 1$ V and -4 V $\le v_{GS} \le 0$ V. Determine the voltage amplification v_o/v_s for $V_C = -4, -3, -2,$ and -1 V with $R_F = 1$ kΩ. What restrictions must be placed on v_s?

7.25

For the matched JFET AGC amplifier of Fig. 7.18, derive the expression for voltage amplification v_o/v_s making the usual op amp assumptions. For $R_F = 100$ kΩ, $R_S = 1$ kΩ, and $V_{OD} = 2$ V, determine v_o/v_s for $V_C = 0, 1, 2, 3,$ and 4 V.

7.26

Repeat the calculations of Problem 7.25 for $R_F = 1$ MΩ.

7.27

A sinusoidal voltage of unit amplitude is applied to the rectifier circuits of Figs. 7.19 through 7.22 ($v_s = \sin \omega t$). Sketch v_o versus ωt for one period for each case with $R_F = 10$ kΩ and $R_S = 1$ kΩ.

7.28

Determine the voltage transfer characteristic for the active half-wave rectifier of Fig. 7.20 with the diode reversed. Sketch the transfer characteristic and consider that R_L is attached at the output, where $R_L \ll R_i$.

7.29

Determine the voltage transfer characteristic for the circuit of Fig. P7.29. Let the diodes be 5 V zeners with $V_0 = 0.7$ V and sketch v_o versus v_s.

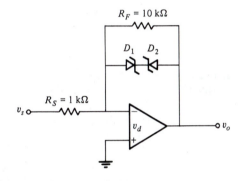

FIGURE P7.29

7.30

Repeat Problem 7.29 with D, replaced with a different zener diode having $V_z = 10$ V.

7.31

Derive equations for i_D, v_C, and v_D for the peak rectifier of Fig. P7.31. Let $v_s = 10 \sin \omega t$ V, $\omega = 2\pi(60)$ rad/s, and $C = 1$ μF. Assume that C is initially uncharged and sketch the waveshapes versus ωt.

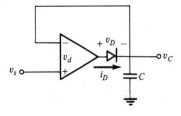

FIGURE P7.31

7.32

For the op amp circuit of Fig. P7.32, determine the transfer characteristic v_o versus v_s for $A_d = 10^4$, $R_{S_p} = 50\ \Omega$, and $R_{F_p} = 1\ M\Omega$, which results in comparator operation, $A_d R_{S_p}/(R_{S_p} + R_{F_p}) < 1$. The op amp saturation voltages are $V_{o\,max} = 5\ V$ and $V_{o\,min} = -5\ V$.

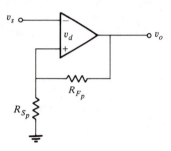

FIGURE P7.32

7.33

Repeat Problem 7.32 for a Schmitt trigger application in which $R_{S_p} = R_{F_p} = 10\ k\Omega$.

7.34

For $v_s = 10 \sin \omega t$, sketch the output voltages (v_o) versus ωt for the circuits of Problems 7.32 and 7.33. Compare the rise and fall times.

7.35

A Schmitt trigger application of the circuit of Fig. P7.32 uses $R_{S_p} = R_{F_p}$. For $A_d = 10^5$, $V_{o\,max} = 20\ V$, and $V_{o\,min} = -20\ V$, sketch the transfer characteristic. What is v_o for $v_s = 15, 0,$ and $-15\ V$?

7.36

Repeat Problem 7.35 for R_{S_p} disconnected, $V_{o\,max} = 10\ V$, and $V_{o\,min} = 0$.

7.37

The astable multivibrator of Fig. 7.27 is used as a square wave generator. Determine and sketch the output voltage v_o for $C_{S_n} = 1\ \mu F$, $R_{F_n} = 1\ M\Omega$, $R_{F_p} = R_{S_p}$, and $V_{o\,max} = -V_{o\,min} = 5\ V$.

7.38

A logarithmic amplifier of the type shown in Fig. 7.28 uses a PN diode with an i–v characteristic approximated by $i_D = 10(e^{40v_D} - 1)\ \mu A$. Derive the expression for the output voltage v_o in terms of the input voltage. For $R_S = 10\ k\Omega$, calculate v_o for $v_s = 1, 10, 100,$ and $200\ V$. Sketch v_o versus v_s using semilog axes.

7.39

Derive the expression for v_o in terms of v_s for the log amp of Fig. 7.29. For $R_{F_2} = 10\ k\Omega$, $R_{S_2} = 1\ k\Omega$, $R_R = 100\ k\Omega$, $R_{S_1} = 1\ k\Omega$, and $V_R = 5\ V$, calculate v_o for $v_s = 1, 10, 100,$ and $200\ V$. Sketch v_o versus v_s using semilog axes.

7.40

Repeat Problem 7.38 for a diode whose i–v characteristic is approximated by $i_D = 100(e^{40v_D} - 1)\ \mu A$.

7.41

Derive the expression for v_o in terms of v_s for the log amp of Fig. 7.30. For $k = 10$, $R_R = 100\ k\Omega$, $R_{S_1} = 1\ k\Omega$, $R = 1\ k\Omega$, and $V_R = 10\ V$, calculate v_o for $v_s = 1, 10, 100,$ and $200\ V$. Sketch v_o versus v_s using semilog axes.

7.42

Derive the expression for v_o versus v_s for the exponential amplifier of Fig. 7.31. Assume that the transistor collector current is given by $i_C = 1.0e^{40v_s}\ mA$, $R_F = 1\ k\Omega$, and $V_{o\,max} = -V_{o\,min} = 10\ V$. Calculate and sketch v_o versus v_s for v_s in the mV range. What is the minimum value of v_s that forces the op amp into saturation?

7.43

Derive the expression for v_{o_2} versus v_s for the exponential amplifier of Fig. 7.32. Calculate and sketch v_o versus v_s for v_s in the mV range. Let each transistor have $i_C = 10e^{40v_s}\ mA$ and circuit elements as follows: $V_R = 10\ V$, $R_R = 40\ k\Omega$, $R_{F_2} = 1\ k\Omega$, $R_1 = 3\ k\Omega$, and $R_2 = 1\ k\Omega$.

7.44

Design an exponential amplifier of the type shown in Fig. 7.32 that will provide values of output voltage in the range 0 to 10 V for v_s in the range -0.1 to $0.1\ V$. Note that the solution is by no means unique. Thus, assume that $V_R = 10\ V$ and available resistors have values of $1\ k\Omega, 9\ k\Omega, 39\ k\Omega, 90\ k\Omega,$ and $100\ k\Omega$.

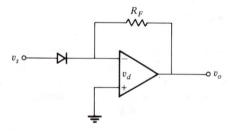

FIGURE P7.45

7.45

For the circuit of Fig. P7.45, determine the output voltage v_o in terms of the input voltage v_s.

7.46

Show that the circuit of Fig. P7.46 can also be used as a half-wave rectifier, where v_{out} is the rectified voltage.

7.47

Show that the circuit of Fig. P7.47 performs the operation of a clipper circuit (from Chapter 2) by sketching the voltage transfer characteristic and the waveshape for v_{out} with $v_i = V_I \sin \omega t$ and $V_I > V_R$.

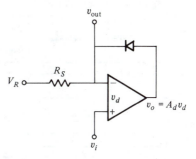

FIGURE P7.47

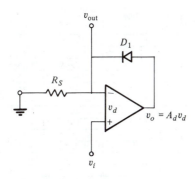

FIGURE P7.46

DATA SHEET A7.1   Selection Guide for Voltage Comparators

VOLTAGE COMPARATORS

military temperature range   (values specified at T_A = 25 °C)

DEVICE NUMBER	TYPE	REMARKS	V_{IO} MAX (mV)	I_{IB} MAX (µA)	I_{OL} MIN (mA)	RESPONSE TIME TYP (ns)	POWER SUPPLIES		PACKAGES	PAGE
							$V_{CC}+$ NOM (V)	$V_{CC}-$ NOM (V)		
uA710M			2	20	2	40	12	−6	J,JG,U	4-87
LM106		Strobe	2	45	100	28	12	−6	J,JG,W	4-9
LM111		Strobe	3	0.1	8	115	15	−15	FH,FK,J,JG,U	4-15
TL510M	Single	Strobe	2	15	2	30			FH,FK,JG,U	4-45
TL810M		Improved TL710M	2	15	2	30			JG,U	4-67
TL710M			5	75	1.6	40	15	−6	FH,FK,JG,U	4-59
TL331M		V_{CC}: 12 V to 36 V	5	−0.1	6	300	5	0	JG	4-37
TL506M		Strobes	2	20	100	28	12	−6	J,W	4-39
TL820M		Dual TL810M	2	15	2	30	12	−6	J	4-79
TL514M	Dual	Dual TL510M	2	15	2	30	12	−6	FH,FK,J,W	4-51
uA711M		Strobes	3.5	75	0.5	40	12	−6	J,U	4-91
TL811M		Strobes	3.5	20	0.5	33	12	−6	J,U	4-73
LM193		V_{CC}: 2 V to 36 V	5	0.1	6	300	5	0	FH,FK,JG	4-29
TLC372M		LinCMOS	10			200	5	0	JG	4-83
LM139A	Quad	V_{CC}: 2 V to 36 V	2	−0.1	6	300	5	0	D,J,N	4-25
LM139		V_{CC}: 2 V to 36 V	5	−0.1	6	300	5	0	FH,FK,J	4-25
TLC374M		LinCMOS	10			200	5	0	J	4-85

automotive temperature range   (values specified at T_A = 25 °C)

DEVICE NUMBER	TYPE	REMARKS	V_{IO} MAX (mV)	I_{IB} MAX (µA)	I_{OL} MIN (mA)	RESPONSE TIME TYP (ns)	POWER SUPPLIES		PACKAGES	PAGE
							$V_{CC}+$ NOM (V)	$V_{CC}-$ NOM (V)		
LM2903	Dual	V_{CC}: 2 V to 36 V	7	0.25	6	300	5	0	D,JG,P	4-29
LM3302	Quad	V_{CC}: 2 V to 36 V	20	−0.5	6	300	5	0	D,J,N	4-35
LM2901		V_{CC}: 2 V to 36 V	15	−0.4	6	300	5	0	D,J,N	4-25

industrial temperature range   (values specified at T_A = 25 °C)

DEVICE NUMBER	TYPE	REMARKS	V_{IO} MAX (mV)	I_{IB} MAX (µA)	I_{OL} MIN (mA)	RESPONSE TIME TYP (ns)	POWER SUPPLIES		PACKAGES	PAGE
							$V_{CC}+$ NOM (V)	$V_{CC}-$ NOM (V)		
LM206		Strobe	2	45	100	28	12	−6	D,J,JG,N,P	4-9
LM211	Single	Strobe	3	0.1	8	115	15	−15	D,JG,P	4-15
TL331I			5	−0.1	6	300	5	0	D,JG,P	4-37
LM293A			2	0.25	6	300	5	0	D,JG,P	4-29
LM219	Dual		4	0.5	3.2	80	5	0	J,N	4-33
LM293			5	0.25	6	300	5	0	D,JG,P	4-29
LM239A	Quad		4	−0.4	6	300	5	0	D,J,N	4-25
LM239			9	−0.4	6	300	5	0	D,J,N	4-25

DATA SHEET A7.1 (Continued)

commercial temperature range

(values specified at $T_A = 25\,°C$)

DEVICE NUMBER	TYPE	REMARKS	V_{IO} MAX (mV)	I_{IB} MAX (μA)	I_{OL} MIN (mA)	RESPONSE TIME TYP (ns)	$V_{CC}+$ NOM (V)	$V_{CC}-$ NOM (V)	PACKAGES	PAGE
TL510C		Strobe	3.5	20	1.6	30			JG,P	4-45
TL810C		Improved TL710C	3.5	20	1.6	30			JG,P	4-67
LM306		Strobe	5	40	100	28	12	-6	D,J,JG,N,P	4-9
TL331C	Single	V_{CC}: 2 V to 36 V	5	-0.25	6	300	5	0	D,JG,P	4-37
LM311		Strobe	7.5	0.25	8	115	15	-15	D,JG,P	4-15
TL710C			7.5	100		40	12	-6	J,JG,N,P,U	4-59
TL721			±100			12 Max	0	-5.2	JG,P	4-65
TL712		Output enable				25			JG,P	4-63
LM393A		V_{CC}: 2 V to 36 V	2	0.25	6	300	5	0	D,JG,P	4-29
TL820C	Dual	Dual TL810C	3.5	20	1.6	30	12	-6	J,N	4-79
TL514C		Dual TL510C	3.5	20	1.6	30	12	-6	J,N	4-51
uA711C		Strobes	5	100	0.5	40	12	-6	J,N	4-91
TL811C	Dual	Improved uA711C	5	30	0.5	33	12	-6	J,N	4-73
TL506C		Strobes	5	25	100	28	12	-6	J,N	4-39
LM393		V_{CC}: 2 V to 36 V	5	0.25	6	300	5	0	D,JG,P	4-29
LM319			8	1	3.2	80	5	0	J,N	4-33
TLC372C		LinCMOS	10			200	5	0	JG,P	4-83
LM339	Quad		5	-0.15	6	300	5	0	D,J,N	4-25
LM339A			2	-0.15	6	300	5	0	D,J,N	4-25
TLC374C		LinCMOS	10			200	5	0	D,J,N	4-85

8

FREQUENCY LIMITATIONS OF ELECTRONIC AMPLIFIERS

Previous chapters did not take into account the frequency dependence of electronic amplifier circuits. All amplifier circuits were considered as operating in the medium-frequency range, where low- and high-frequency effects are negligible.

In this chapter, we will observe the detrimental low-frequency effects of coupling and bypass capacitors as well as inductors. We will also observe the detrimental high-frequency effects of transistor parasitic capacitors. The primary effect of the coupling and bypass capacitors, the inductors, and the parasitic capacitors of the transistors is to reduce the amplification in certain ranges of frequency. We will analyze small-signal circuits containing capacitors and inductors to determine these detrimental frequency effects.

This chapter begins with a general description of the frequency dependence of amplifier circuits and introduces a simulated amplifier model to show this dependence. Next, techniques for simplifying the analysis of frequency-dependent circuits are introduced. These techniques reduce the amount of algebra required in the analysis and allow greater insight. We will apply these techniques to various BJT and FET amplifier circuits.

8.1 GENERAL DESCRIPTION

The behavior of electronic circuits is affected by the frequency of the input signal source. The effects are detrimental at both low and high frequencies. The most prominent effect is the reduction in voltage and current amplification at these extremes.

At low frequencies, a reduction in the magnitude of amplification occurs because the equivalent impedances of coupling and bypass capacitors used in discrete circuits are no longer short circuits to AC. Similarly, inductors and transformers are not open circuits. Furthermore, corresponding occurrences are associated with these elements when used with integrated circuits.

At high frequencies, reduction in amplification is due to the small parasitic capacitances associated with PN junctions and MOS layers. These equivalent elements are in parallel with the junction and the MOS layer and are intentionally made quite small by miniaturizing the device size. Regardless, as the frequency is increased to a large enough value, these capacitors will act like short circuits. Since the amplification is controlled by the junction of a BJT (or JFET) or the MOS layer of the MOST, placing a short circuit across these regions is indeed detrimental.

8.1.1 General Dependence of Amplification on Frequency

Our previous analyses of electronic circuits were carried out in what is called the *medium-frequency,* or *mid-frequency, range,* where high- and low-frequency effects are not present. This medium-frequency range, in which the amplification is essentially constant, is defined as the *bandwidth* (*BW*) of the amplifier and depends upon the magnitudes of the capacitors and inductors as well as the types of transistors used in the circuit. Figure 8.1 displays the typical variation in the magnitude of voltage amplification ($|A_v|$) with

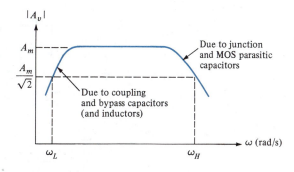

FIGURE 8.1 Typical Voltage Amplification Response, $|A_v|$ versus ω

frequency (ω) for a voltage amplifier. Current amplification ($|A_i|$) for a current amplifier is entirely similar. The amplification in the medium-frequency range is defined as A_m and is seen to be relatively constant between the low and high half-power frequencies ω_L and ω_H, respectively (where the amplification is $A_m/\sqrt{2}$). The difference $\omega_H - \omega_L$ is the *BW* in this case.

In addition to the reduction in amplification at the low- and high-frequency extremes, the amplification is complex in these ranges (due to the impedances), and the phase angle between the input and output current or voltage is also frequency dependent. For example, in any of the voltage follower circuits, in the medium-frequency range, the output and input voltages are in phase. However, at low and high frequencies, the phase of the output voltage is altered from that of the input voltage, as we will see. The phase of the output relative to the input is another important factor that we will consider.

8.1.2 Simulated Amplifier Model

The concept of the variation in amplifier amplification with frequency can be grasped by considering the simulated amplifier circuit of Fig. 8.2a. The noninverting op amp configuration provides the amplification, ($R_F/R_S + 1$), which we will

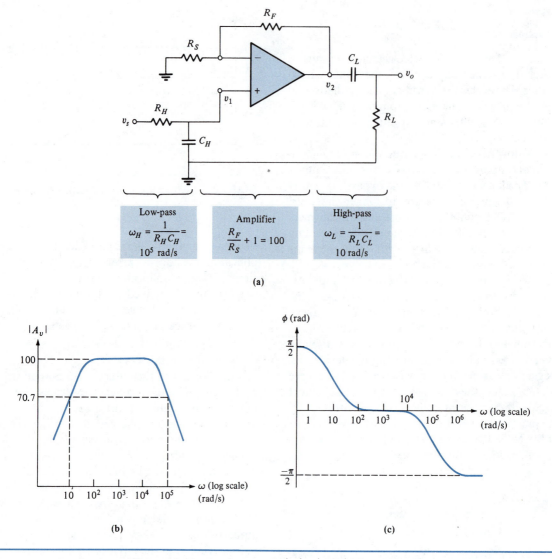

Low-pass
$$\omega_H = \frac{1}{R_H C_H} = 10^5 \text{ rad/s}$$

Amplifier
$$\frac{R_F}{R_S} + 1 = 100$$

High-pass
$$\omega_L = \frac{1}{R_L C_L} = 10 \text{ rad/s}$$

(a)

(b)

(c)

FIGURE 8.2   Simulated Amplifier Model: (a) Overall circuit, (b) $|A_v| = |\mathbf{V}_o/\mathbf{V}_s|$ versus ω, (c) Phase of A_v versus ω

consider here to be independent of frequency. This restriction is plausible because the op amp is designed to operate at low frequencies (even DC), and we assume that high-frequency effects occur at much higher frequencies than those under consideration. A low-pass filter (R_H, C_H) is connected at the input, a high-pass filter (R_L, C_L), at the

output. However, the same results would be obtained if these filters were interchanged.

To analyze the circuit of Fig. 8.2a and determine the complex voltage amplification $\mathbf{V}_o/\mathbf{V}_s$, we start at the low-pass filter section. The combination of R_H and C_H provides a voltage divider to the input signal voltage. Thus, the voltage at the positive terminal of the op amp (in terms of phasors) is given by

$$\mathbf{V}_1 = \mathbf{V}_s \left[\frac{1/j\omega C_H}{R_H + (1/j\omega C_H)} \right] \qquad (8.1-1)$$

or

$$\mathbf{V}_1 = \frac{\mathbf{V}_s}{1 + j\omega R_H C_H} \qquad (8.1-2)$$

where the input impedance of the op amp is assumed to be much greater than the impedance of the capacitor C_H. The voltage $\mathbf{V}_1$ in (8.1–2) is then amplified by the noninverting amplifier to yield

$$\mathbf{V}_2 = \left(\frac{R_F}{R_S} + 1 \right) \mathbf{V}_1 \qquad (8.1-3)$$

By substituting for $\mathbf{V}_1$ from (8.1–2) and defining $A_m = (R_F/R_S + 1)$, we have

$$\mathbf{V}_2 = A_m \left(\frac{\mathbf{V}_s}{1 + j\omega R_H C_H} \right) \qquad (8.1-4)$$

The combination of R_L and C_L in Fig. 8.2a provides a high-pass filter such that

$$\mathbf{V}_o = \left(\frac{j\omega R_L C_L}{1 + j\omega R_L C_L} \right) \mathbf{V}_2 \qquad (8.1-5)$$

Thus, by substituting for $\mathbf{V}_2$ and dividing by $\mathbf{V}_s$, we obtain

$$\frac{\mathbf{V}_o}{\mathbf{V}_s} = A_m \left(\frac{1}{1 + j\omega R_H C_H} \right) \left(\frac{j\omega R_L C_L}{1 + j\omega R_L C_L} \right) \qquad (8.1-6)$$

This expression for amplification is a typical (although simplified) response of a single-transistor amplifier circuit. Note that each term in (8.1–6) is implicitly dimensionless. We define the characteristic frequencies as $\omega_H = 1/R_H C_H$ and $\omega_L = 1/R_L C_L$, where these frequencies turn out to be

the high (ω_H) and low (ω_L) half-power frequencies for the magnitude of the voltage amplification. Thus, substitution of these definitions into (8.1–6) yields

$$\frac{\mathbf{V}_o}{\mathbf{V}_s} = A_m \left(\frac{1}{1 + j\omega/\omega_H} \right) \left(\frac{j\omega/\omega_L}{1 + j\omega/\omega_L} \right) \qquad (8.1-7)$$

We refer to this type of equation as the *standard form* of the amplification; each term is explicitly dimensionless.

Note that the amplification is complex and may be expressed in terms of magnitude and phase angle as follows:

$$\frac{\mathbf{V}_o}{\mathbf{V}_s} = \left| \frac{\mathbf{V}_o}{\mathbf{V}_s} \right| \measuredangle \phi \qquad (8.1-8)$$

where ϕ is the phase angle between $\mathbf{V}_o$ and $\mathbf{V}_s$. Both the magnitude and phase vary with frequency in the low- and high-frequency ranges.

The magnitude of $\mathbf{V}_o/\mathbf{V}_s$ as a function of frequency can be explicitly written from (8.1–7) by individually substituting the magnitude for each numerator and denominator term as follows:

$$\left| \frac{\mathbf{V}_o}{\mathbf{V}_s} \right| = A_m \left[\frac{1}{[1 + (\omega/\omega_H)^2]^{1/2}} \right]$$
$$\times \left[\frac{\omega/\omega_L}{[1 + (\omega/\omega_L)^2]^{1/2}} \right] \qquad (8.1-9)$$

The equation for the phase of a complex ratio is obtained, in general, by adding the contributions of the numerator terms and subtracting those of the denominator. Each individual component contributes a phase angle corresponding to $\tan^{-1}$ (imaginary part/real part). The phase of (8.1–7) is therefore given by

$$\phi = \frac{\pi}{2} - \tan^{-1} \frac{\omega}{\omega_H} - \tan^{-1} \frac{\omega}{\omega_L} \qquad (8.1-10)$$

where the $\pi/2$ contribution is obtained from j in the numerator ($j = e^{j\pi/2}$).

Equations (8.1–9) and (8.1–10) are sketched as functions of ω in Figs. 8.2b and c with $A_m = 100$, $\omega_L = 10$ rad/s, and $\omega_H = 10^5$ rad/s. A logarithmic scale is used for ω because of the wide frequency range. Note that the amplification decreases in the low- and high-frequency extremes and that the phase is also variable in these ranges. Additionally, ω_H and ω_L are the high and low half-power frequencies, respectively, and the BW is $\omega_H - \omega_L = 10^5$ rad/s. The amplification over this region is essentially $A_m = 100$.

To obtain the curves shown in Figs. 8.2b and c, computations were carried out at various specific values of ω. For calculations such as these, a computer or programmable calculator provides considerable aid. However, an expeditious technique does exist for sketching the magnitude and phase angle of a complex quantity by inspection. This method only requires obtaining the complex function of frequency in standard form, as in (8.1–7). We examine this method next.

8.1.3   Bode Plots of Magnitude and Phase

*Bode plots** for a complex, frequency-dependent expression are straight-line, asymptotic sketches of *magnitude* and *phase* versus *frequency* that use special axes for magnitude and frequency. We will apply Bode's method to voltage amplification expressions for amplifiers, although it is applicable to general transfer functions that are complex functions of frequency.

In Bode plots, the frequency axis is compressed by using a log base 10 (log) scale. The magnitude axis is also compressed by using a log axis, where the magnitudes of voltage and current amplification are expressed in decibels (dB) and defined as

$$A_v \text{ (dB)} = 20 \log \left| \frac{V_{out}}{V_{in}} \right| \qquad (8.1–11)$$

* For further information see Bode, H. W., *Network Analysis and Feedback Design*, New York: D. Van Nostrand Co., 1945.

and

$$A_i \text{ (dB)} = 20 \log \left| \frac{I_{out}}{I_{in}} \right| \qquad (8.1–12)$$

Bode plots, then, consist of sketching the amplification in decibels versus log ω and the phase versus log ω, each of which can be approximated by straight-line segments.

To demonstrate the method, we use the amplification expression obtained in standard form in (8.1–7). While the method is examined here in great detail, application of the method for transistor amplifiers is done by inspection.

By definition, the magnitude of (8.1–7), expressed in decibels, is

$$A_v \text{ (dB)} = 20 \log A_m \left[\frac{1}{[1 + (\omega/\omega_H)^2]^{1/2}} \right]$$
$$\times \left[\frac{\omega/\omega_L}{[1 + (\omega/\omega_L)^2]^{1/2}} \right] \qquad (8.1–13)$$

Because of the properties of logarithms, we may write

$$A_v \text{ (dB)} = 20 \log A_m - 20 \log \left[1 + \left(\frac{\omega}{\omega_H} \right)^2 \right]^{1/2}$$
$$+ 20 \log \frac{\omega/\omega_L}{[1 + (\omega/\omega_L)^2]^{1/2}} \qquad (8.1–14)$$

Equation (8.1–14) indicates that A_v (dB) is the sum of three terms in which only the latter two are frequency dependent. We will consider each of these terms individually and then add them together.

The first term is constant with frequency and, as shown in Fig. 8.3a, is defined as $20 \log A_m = A_m$ (dB). The second term can be approximated at low frequency ($\omega \ll \omega_H$) by

$$- 20 \log \left[1 + \left(\frac{\omega}{\omega_H} \right)^2 \right]^{1/2} \cong 0 \qquad (8.1–15a)$$

and at high frequency ($\omega \gg \omega_H$) by

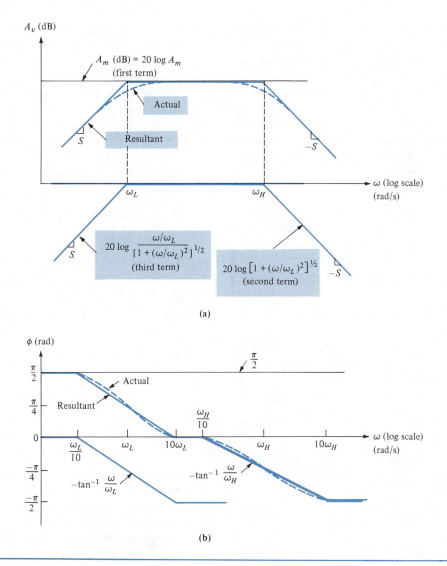

A_v (dB)

A_m (dB) = 20 log A_m
(first term)

Actual

Resultant

S

$-S$

ω_L ω_H ω (log scale)
(rad/s)

S

$20 \log \dfrac{\omega/\omega_L}{[1+(\omega/\omega_L)^2]^{1/2}}$
(third term)

$20 \log \left[1+(\omega/\omega_L)^2\right]^{1/2}$
(second term)

$-S$

(a)

ϕ (rad)

$\dfrac{\pi}{2}$

Actual

$\dfrac{\pi}{4}$

Resultant

$\dfrac{\omega_H}{10}$

0

$\dfrac{\omega_L}{10}$ ω_L $10\omega_L$ ω_H $10\omega_H$ ω (log scale)
(rad/s)

$-\dfrac{\pi}{4}$

$-\tan^{-1}\dfrac{\omega}{\omega_L}$

$-\tan^{-1}\dfrac{\omega}{\omega_H}$

$-\dfrac{\pi}{2}$

(b)

FIGURE 8.3   Bode Plots: (a) Magnitude, (b) Phase

$$-20 \log\left[1+\left(\frac{\omega}{\omega_H}\right)^2\right]^{1/2} \cong -20 \log \frac{\omega}{\omega_H}$$

(8.1–15b)

If we allow the entire range of ω for (8.1–15a) to be $\omega \leq \omega_H$ and that of (8.1–15b) to be $\omega \geq \omega_H$, then the variation in these regions can be represented by the two straight-line segments shown in Fig. 8.3a (labeled "second term"). Note that at

$\omega = \omega_H$, the straight-line approximate segments are in error with the original term, $20 \log[1 + (\omega/\omega_H)^2]^{1/2}$, by the maximum amount of $20 \log(2)^{1/2} \cong 3$ dB. Therefore, the characteristic frequency ω_H is sometimes referred to as a *3 dB* frequency.

The slope of the straight line for $\omega \geq \omega_H$ is defined as $-S$. The magnitude of S is obtained by considering two points on this line separated by a known frequency difference, which is quite often taken as a factor of 10 in frequency called a *decade*. Using a factor of 10 difference in frequency is extremely convenient because two frequencies differing by a factor of 10 are always separated by equal distances on a log scale (semi-log paper is thus not required for Bode plots). If the points selected are at frequencies $\omega = \omega_H$ and $\omega = 10\omega_H$, then the corresponding change in dB, $\Delta(\text{dB})$, is given by

$$\Delta(\text{dB}) = -20 \log \frac{\omega}{\omega_H}\Big|_{\omega = \omega_H}$$
$$- 20 \log \frac{\omega}{\omega_H}\Big|_{\omega = 10\omega_H}$$
$$= 0 - 20 \log 10 = -20 \text{ dB} \quad \textbf{(8.1–16)}$$

With this change for one decade, the slope is -20 dB/decade.

It is also sometimes convenient to use a frequency difference factor of 2 called an *octave*. The slope is then obtained as -6 dB/octave. We also denote the magnitude of this constant slope as S, where $S = 20 \text{ dB/decade} = 6 \text{ dB/octave}$.

The third term in (8.1–14) can be written as the difference of two terms, or

$$20 \log \frac{\omega/\omega_L}{[1 + (\omega/\omega_L)^2]^{1/2}} = 20 \log \frac{\omega}{\omega_L}$$
$$- 20 \log \left[1 + \left(\frac{\omega}{\omega_L}\right)^2\right]^{1/2} \quad \textbf{(8.1–17)}$$

Note that the first part of (8.1–17) varies linearly with $\log \omega$ for all ω with slope S. The second term of (8.1–17) is negative and behaves identically with (8.1–15a and b). Hence, it is approximately 0 for $\omega \leq \omega_L$ and decreases linearly for $\omega \geq \omega_L$ with slope $-S$. Adding the contributions on the right-hand side of (8.1–17) produces cancelation for $\omega \geq \omega_L$, as shown in Fig. 8.3a (labeled "third term").

Finally, we graphically add the contributions of the first, second, and third terms of (8.1–14) in Fig. 8.3a to obtain the resultant (straight-line)

curve. For comparison, the actual (computed) curve of Fig. 8.2b is also shown in Fig. 8.3a. The actual variation is observed to approach the Bode plot asymptotically, with maximum error at the 3 dB frequencies ω_L and ω_H. These frequencies are also called *corner*, or *break*, *frequencies* to indicate the abrupt change in the straight-line segment.

The Bode plot for phase angle versus frequency is obtained from the expression for phase (ϕ). In our example, ϕ is given from (8.1–10) as

$$\phi = \frac{\pi}{2} - \tan^{-1}\frac{\omega}{\omega_H} - \tan^{-1}\frac{\omega}{\omega_L} \quad \textbf{(8.1–18)}$$

As in the case of magnitudes, we sketch each individual term and add the results. They are shown in Fig. 8.3b, where the $\tan^{-1}$ terms are linearly approximated as follows:

$$\tan^{-1}\frac{\omega}{\omega_0} \cong 0 \text{ for } \omega \ll \omega_0 \text{ or } \omega \leq \frac{\omega_0}{10}$$
$$= \frac{\pi}{4} \text{ for } \omega = \omega_0$$
$$\cong \frac{\pi}{2} \text{ for } \omega \gg \omega_0 \text{ or } \omega \geq 10\omega_0$$

where ω_0 represents either ω_L or ω_H. In the region between $\omega_0/10 \leq \omega \leq 10\omega_0$, a straight line is constructed from the approximate values at the endpoints and the exact value in the middle. Again, for comparison, the actual plot of phase versus frequency is shown.

In summary, the Bode method for plotting the magnitude and phase of the complex function of frequency begins by obtaining the function in standard form, as indicated by (8.1–7). The characteristic frequencies, ω_L and ω_H in our example, are then obtained by inspection. These frequencies are marked on the log ω axis, and the linear variation of each term is sketched. As far as magnitudes are concerned, the term $1/[1 + (j\omega/\omega_H)]$ contributes a nonzero value to $A_v(\text{dB})$ for $\omega > \omega_H$, which reduces the amplification at a rate of

$S = 20\,\text{dB/decade}$ as ω decreases. Another important term that arises in some amplification expressions is $1 + (j\omega/\omega_0)$, which also contributes an increase in $A_v(\text{dB})$ at the rate of $S = 20\,\text{dB/decade}$ for $\omega > \omega_0$.

To sketch the phase of the complex function simply involves determining the linear representation of each $\tan^{-1}$ term and any constant term and adding the results. The values of the break frequencies as well as $1/10$ and 10 times these values are marked on the log ω axis. The corresponding straight lines are then drawn realizing that one point is at $\pm\pi/4$ for $\omega = \omega_0$ (considering a term involving $1 + (j\omega/\omega_0)$, where $+$ is appropriate for a numerator term and $-$ for a denominator term.

We will employ the Bode technique in determining the frequency behavior of electronic amplifiers. As we will see, this technique allows de-termination of the frequency response by inspection of the complex function in standard form.

─────────── EX. 8.1

Bode Plot Example

The complex current amplification equation for a capacitively coupled BJT amplifier is obtained as

$$\frac{I_o}{I_s} = 100\left(\frac{j\omega/\omega_1}{1 + j\omega/\omega_1}\right)\left(\frac{j\omega/\omega_2}{1 + j\omega/\omega_2}\right)$$
$$\times \left(\frac{1}{1 + j\omega/\omega_3}\right)\left(\frac{1}{1 + j\omega/\omega_4}\right)$$

Sketch the amplitude and phase Bode plots for $\omega_1 = 1$, $\omega_2 = 100$, $\omega_3 = 10^6$, and $\omega_4 = 10^8$ rad/s.

Solution: Since the current amplification expression is already in the standard form, the Bode plots are obtained by sketching the contribution of each term and adding the results. Figure E8.1a displays the magnitude Bode plot, and

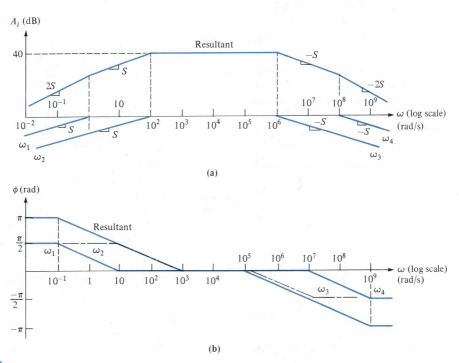

(a)

(b)

FIGURE E8.1

Fig. E8.1b displays the phase Bode plot. The separate curves are numbered according to their frequency in the given complex amplification expression.

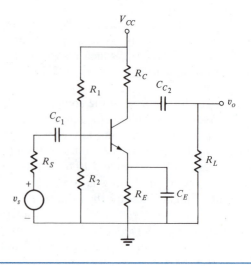

FIGURE 8.4   Common-Emitter Amplifier Circuit

8.2   LOW-FREQUENCY RESPONSE OF SINGLE-TRANSISTOR AMPLIFIERS

In considering actual transistor amplifier circuits, it is extremely convenient to separate the low- and high-frequency analyses from each other. This separation is valid since these frequency ranges in actual amplifier circuits are separated by the medium-frequency range, which by design is at least 20 kHz and, quite often, much more. The benefit from this practice is that the amount of complex algebra involved in determining a general amplification expression is considerably reduced.

The analysis at low frequencies must take into account the impedances of the capacitors and inductors in the amplifier circuits. Since these elements are used similarly in BJT and FET circuits, we expect to obtain similar results. However, for the FET, the results are simpler to obtain because of the open circuit at the gate terminal. For the BJT, in order to reduce the amount of complex algebra involved, we consider the effects of each frequency-dependent element separately. Furthermore, we will demonstrate the effects of these elements at low frequencies using only the common-emitter and common-source amplifier configurations. Note that similar effects are obtained for other amplifier configurations.

8.2.1   Common-Emitter Amplifier at Low Frequencies

Figure 8.4 shows the common-emitter amplifier circuit with coupling capacitors C_{C_1} and C_{C_2} and bypass capacitor C_E. Considering the ef-

fects of each capacitor separately is equivalent to assuming that the others have large values, which, in some instances, may be the case. For the general case, the individual capacitor results are still useful and provide a means of comparison. In general, we will see that the bypass capacitor C_E is the dominant factor in determining the low half-power frequency.

Effects of Input Coupling Capacitor.   Figure 8.5a displays the small-signal equivalent circuit for the common-emitter amplifier with only the capacitor C_{C_1}. The voltage amplification is obtained in the usual manner by starting at the output terminal and writing the expression for output phasor voltage as follows:

$$\mathbf{V}_o = -(h_{fe}\mathbf{I}_b)R_C \,\|\, R_L \qquad (8.2\text{–}1)$$

To determine $\mathbf{I}_b$, we use the current divider rule, where

$$\mathbf{I}_b = \frac{R_B}{R_B + h_{ie}}\mathbf{I}_t \qquad (8.2\text{–}2)$$

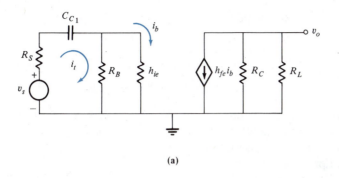

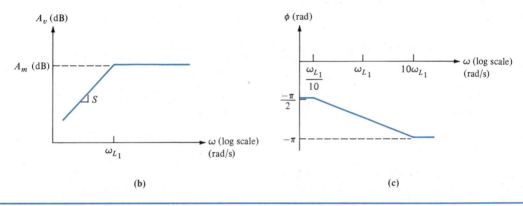

FIGURE 8.5 Small-Signal Model with C_{C_1}: (a) Equivalent circuit, (b) Magnitude Bode plot at low frequency, (c) Phase Bode plot at low frequency

I_t is obtained from KVL for the input loop as follows:

$$I_t = \frac{V_s}{R_S + (1/j\omega C_{C_1}) + R_B \| h_{ie}} \qquad (8.2-3)$$

By substitution of (8.2–3) into (8.2–2) into (8.2–1), we have

$$\frac{V_o}{V_s} = -h_{fe}(R_C \| R_L)\left(\frac{R_B}{R_B + h_{ie}}\right)$$
$$\times \left[\frac{1}{R_S + (1/j\omega C_{C_1}) + R_B \| h_{ie}}\right] \qquad (8.2-4)$$

By factoring the resistance $R_S + R_B \| h_{ie}$ out of the denominator of the term in brackets and multi-

plying numerator and denominator by $j\omega C_{C_1}$ $(R_S + R_B \| h_{ie})$, we obtain the standard form of the voltage amplification expression:

$$\frac{V_o}{V_s} = -\frac{h_{fe}(R_C \| R_L)(R_B)}{(R_S + R_B \| h_{ie})(R_B + h_{ie})}$$
$$\times \left[\frac{j\omega C_{C_1}(R_S + R_B \| h_{ie})}{1 + j\omega C_{C_1}(R_S + R_B \| h_{ie})}\right]$$
$$= -A_m\left[\frac{j\omega/\omega_{L_1}}{1 + j\omega/\omega_{L_1}}\right] \qquad (8.2-5)$$

Here, we recognize the medium-frequency amplification as

$$A_m = \frac{h_{fe}(R_C \| R_L)(R_B)}{(R_S + R_B \| h_{ie})(R_B + h_{ie})} \qquad (8.2-6)$$

and the characteristic low frequency (correspon-ding to C_{C_1}) as

$$\omega_{L_1} = \frac{1}{C_{C_1}(R_S + R_B \| h_{ie})} \qquad (8.2-7)$$

Note that (8.2–6) is equivalent to (5.2–38) ob-tained in Chapter 5. Corresponding Bode plots for magnitude and phase are obtained using the methods of the previous section. They are dis-played in Figs. 8.5b and c. The frequency ω_{L_1} is the half-power frequency.

Effects of Output Coupling Capacitor.   Figure 8.6 shows the small-signal circuit model for the common-emitter amplifier of Fig. 8.4 with only the capacitor C_{C_2}. The output voltage is written in terms of phasors as follows:

$$\mathbf{V}_o = -\mathbf{I}_o R_L \qquad (8.2-8)$$

where $\mathbf{I}_o$ is obtained by inspection of the output loop using the current divider rule. Substituting for $\mathbf{I}_o$ in (8.2–8) yields

$$\mathbf{V}_o = -\frac{h_{fe}\mathbf{I}_b R_C}{R_C + (1/j\omega C_{C_2}) + R_L} R_L \qquad (8.2-9)$$

The input base current is obtained from the input loop of Fig. 8.6 using a Thévenin equivalent cir-cuit. By inspection, we have

$$\mathbf{I}_b = \frac{\mathbf{V}_s R_B/(R_S + R_B)}{h_{ie} + R_S \| R_B} \qquad (8.2-10)$$

Substitution of (8.2–10) into (8.2–9) and solving for the voltage amplification yield

$$\frac{\mathbf{V}_o}{\mathbf{V}_s} = -\frac{h_{fe}R_C R_L R_B/(R_S + R_B)}{h_{ie} + R_S \| R_B}$$
$$\times \left[\frac{1}{R_C + (1/j\omega C_{C_2}) + R_L} \right] \qquad (8.2-11)$$

By multiplying numerator and denominator by $j\omega C_{C_2}(R_C + R_L)$ and rearranging, we obtain the standard form of the voltage amplification expression:

$$\frac{\mathbf{V}_o}{\mathbf{V}_s} = -\frac{h_{fe}(R_C \| R_L)[R_B/(R_S + R_B)]}{h_{ie} + R_S \| R_B}$$
$$\times \left[\frac{j\omega C_{C_2}(R_C + R_L)}{1 + j\omega C_{C_2}(R_C + R_L)} \right] \qquad (8.2-12)$$

For this case, we recognize that

$$A_m = \frac{h_{fe}(R_C \| R_L)[R_B/(R_S + R_B)]}{h_{ie} + R_S \| R_B} \qquad (8.2-13)$$

and (corresponding to C_{C_2}) that

$$\omega_{L_2} = \frac{1}{C_{C_2}(R_C + R_L)} \qquad (8.2-14)$$

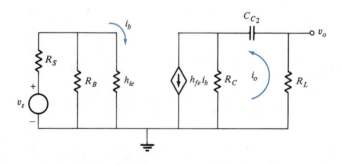

FIGURE 8.6   Small-Signal Model with C_{C_2}

Although it is not immediately obvious, the medium-frequency voltage amplification expression of (8.2–13) is actually equal to (8.2–6). The Bode plots are therefore identical to those of Figs. 8.5b and c with ω_{L_2} replacing ω_{L_1}. The effects of the input and output coupling capacitors are therefore also the same, although, in general, the half-power frequencies have different magnitudes.

Effects of Bypass Capacitor. Figure 8.7a shows the small-signal equivalent circuit for the com-mon-emitter amplifier of Fig. 8.4 with only the bypass capacitor C_E. To determine the voltage amplification, we first write the load voltage from Fig. 8.7a as follows:

$$\mathbf{V}_o = -(h_{fe}\mathbf{I}_b)R_C \| R_L \qquad (8.2–15)$$

The base current is then obtained from the input loop using a Thévenin equivalent circuit. Note that $i_e = (1 + h_{fe})i_b$ is the current through

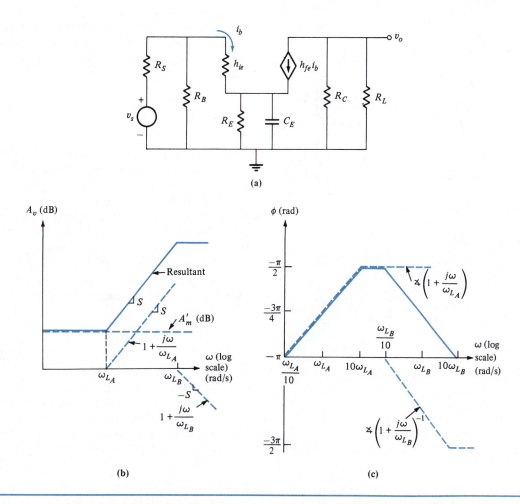

(a)

(b)

(c)

FIGURE 8.7   Small-Signal Model with C_E: (a) Equivalent circuit, (b) Magnitude Bode plot at low frequency, (c) Phase Bode plot at low frequency

$R_e \| Z_{C_E}$. By inspection, we have

$$\mathbf{I}_b = \frac{[R_B/(R_S + R_B)]\mathbf{V}_s}{R_S \| R_B + h_{ie} + (1 + h_{fe})R_E \| Z_{C_E}}$$

$$(8.2-16)$$

where $R_E \| Z_{C_E} = R_E/(1 + j\omega R_E C_E)$. Substitution of (8.2–16) into (8.2–15) and solving for the voltage amplification yield

$$\frac{\mathbf{V}_o}{\mathbf{V}_s} = -\frac{h_{fe}(R_C \| R_L)[R_B/(R_S + R_B)]}{R_S \| R_B + h_{ie} + (1 + h_{fe})R_E/(1 + j\omega R_E C_E)}$$

$$(8.2-17)$$

Multiplying numerator and denominator by $1 + j\omega C_E R_E$ yields

$$\frac{\mathbf{V}_o}{\mathbf{V}_s} = -\frac{(1 + j\omega R_E C_E)[h_{fe}(R_C \| R_L)R_B/(R_S + R_B)]}{(1 + j\omega R_E C_E)(R_S \| R_B + h_{ie}) + (1 + h_{fe})R_E}$$

$$(8.2-18)$$

Factoring the resistor factor, $R_S \| R_B + h_{ie} + (1 + h_{fe})R_E$, out of the denominator, we obtain the standard form as follows:

$$\frac{\mathbf{V}_o}{\mathbf{V}_s} = -\frac{h_{fe}(R_C \| R_L)[R_B/(R_S + R_B)]}{R_S \| R_B + h_{ie} + (1 + h_{fe})R_E}$$
$$\times \left[\frac{1 + j\omega R_E C_E}{1 + \dfrac{j\omega R_E C_E(R_S \| R_B + h_{ie})}{R_S \| R_B + h_{ie} + (1 + h_{fe})R_E}} \right]$$

$$(8.2-19)$$

Immediately, we may recognize that there are two characteristic low frequencies associated with C_E. They are defined as

$$\omega_{L_A} = \frac{1}{R_E C_E}$$

$$(8.2-20)$$

and

$$\omega_{L_B} = \frac{1}{R_E C_E} \left[\frac{R_S \| R_B + h_{ie} + (1 + h_{fe})R_E}{R_S \| R_B + h_{ie}} \right]$$

$$(8.2-21)$$

We note that $\omega_{L_B} > \omega_{L_A}$. Finally, the coefficient multiplying the frequency-dependent term in brackets in (8.2–19) is a constant (but not A_m), which we define as

$$A'_m = \frac{h_{fe}(R_C \| R_L)[R_B/(R_S + R_B)]}{R_S \| R_B + h_{ie} + (1 + h_{fe})R_E} \qquad (8.2-22)$$

With these definitions, (8.2–19) becomes

$$\frac{\mathbf{V}_o}{\mathbf{V}_s} = -A'_m \left(\frac{1 + j\omega/\omega_{L_A}}{1 + j\omega/\omega_{L_B}} \right) \qquad (8.2-23)$$

The complex form of (8.2–23) dictates that A'_m is not the medium-frequency amplification. We determine the medium-frequency amplification from this low-frequency expression by letting ω approach ∞ to yield

$$A_m = \left| \frac{\mathbf{V}_o}{\mathbf{V}_s} \right|_{\omega \to \infty} = A'_m \left(\frac{\omega_{L_B}}{\omega_{L_A}} \right) \qquad (8.2-24)$$

The Bode technique for magnitude and phase is used to obtain the corresponding plots shown in Figs. 8.7b and c. The individual terms are also sketched; dashed lines aid in our understanding the resultant for each case. Note that ω_{L_B} is the actual half-power frequency where the medium-frequency amplification is reduced by $\sqrt{2}$. Note also that the relative magnitudes chosen for ω_{L_A} and ω_{L_B} are such that overlapping in the phase plot does not occur. Nonoverlap is usually the case provided that $h_{fe} \gg 1$.

-- EX. 8.2

Low-Frequency Effects of C_E Example

Calculate the magnitudes of the medium-frequency gain and the break frequencies corresponding to the circuit of Fig. 8.7a. The external resistor values are $R_S = 5$ kΩ, $R_B = 5$ kΩ, $R_E = 0.5$ kΩ, $R_C = 5$ kΩ, and $R_L = 5$ kΩ. Additionally, $C_E = 5$ μF, and the transistor parameters are $1 + h_{fe} = 100$ and $h_{ie} = 3$ kΩ.

Solution: Substituting directly into (8.2–20) and (8.2–21) yields the break frequencies as follows:

$$\omega_{L_A} = \frac{1}{(0.5 \times 10^3)(5 \times 10^{-6})} = 400 \text{ rad/s or } 63.7 \text{ Hz}$$

and

$$\omega_{L_B} = 400 \left[\frac{2.5 + 3 + 100(0.5)}{2.5 + 3} \right]$$

$$= 4036.4 \text{ rad/s or } 642 \text{ Hz}$$

The medium-frequency gain is then obtained from (8.2−24), where A'_m is obtained from (8.2−22) as

$$A'_m = \frac{99(2.5)(1/2)}{2.5 + 3 + 100(0.5)} = 2.23$$

Thus, substituting into (8.2−23) yields

$$A_m = 2.23 \left(\frac{4036.4}{400} \right) = 22.5$$

Effects of Inductive Coupling.   To determine the low-frequency effects of inductive coupling, we consider the amplifier circuit of Fig. 8.8a. This circuit is identical to the previous amplifier circuit except that L has replaced R_C. The small-signal equivalent circuit is shown in Fig. 8.8b, where it

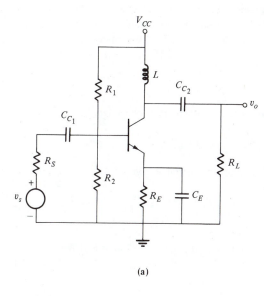

(a)

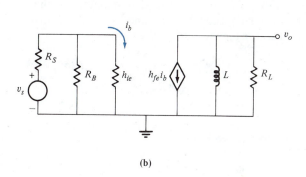

(b)

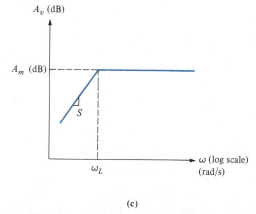

(c)

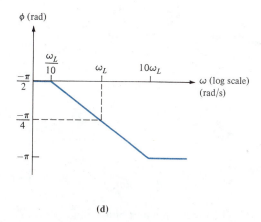

(d)

FIGURE 8.8   Common-Emitter Amplifier Circuit with Inductive Coupling: (a) Actual circuit, (b) Equivalent circuit, (c) Magnitude Bode plot at low frequency, (d) Phase Bode plot at low frequency

has been assumed that the capacitors are short circuits and L has been retained to show its effects at low frequencies. Note that the input portion of the circuit is unchanged from the medium-frequency case and that only the output loop is altered.

From the output loop of Fig. 8.8b, we have

$$\mathbf{V}_o = -(h_{fe}\mathbf{I}_b)Z_L \| R_L \qquad (8.2-25)$$

where $Z_L \| R_L = j\omega L R_L/(j\omega L + R_L)$. The base current is given by (8.2–10), and by substitution while we solve for $\mathbf{V}_o/\mathbf{V}_s$, we have

$$\frac{\mathbf{V}_o}{\mathbf{V}_s} = -\frac{h_{fe}R_B/(R_S + R_B)}{h_{ie} + R_S \| R_B}\left(\frac{j\omega L R_L}{j\omega L + R_L}\right) \qquad (8.2-26)$$

Factoring R_L out of the denominator of the frequency-dependent term and R_L^2 out of the numerator term yields

$$\frac{\mathbf{V}_o}{\mathbf{V}_s} = -\frac{h_{fe}[R_B/(R_S + R_B)](R_L)}{h_{ie} + R_S \| R_B}\left(\frac{j\omega L/R_L}{(j\omega L/R_L) + 1}\right) \qquad (8.2-27)$$

We now recognize that

$$A_m = \frac{h_{fe}[R_B/(R_S + R_B)]R_L}{h_{ie} + R_S \| R_B} \qquad (8.2-28)$$

and that

$$\omega_L = \frac{R_L}{L} \qquad (8.2-29)$$

Substituting (8.2–28) and (8.2–29) into (8.2–27) yields

$$\frac{\mathbf{V}_o}{\mathbf{V}_s} = -A_m\left(\frac{j\omega/\omega_L}{1 + j\omega/\omega_L}\right) \qquad (8.2-30)$$

The magnitude and phase Bode plots of (8.2–30) are displayed in Figs. 8.8c and d. The low-frequency effects of inductive coupling are observed to be the same as those associated with coupling and bypass capacitors.

EX. 8.3

Low-Frequency Effects of L

Consider the common-emitter amplifier circuit of Fig. 8.8a with $R_S = 0$, $L = 1$ H and $R_L = 1$ kΩ. Calculate the medium-frequency amplification A_m and the low half-power frequency ω_L. For the transistor, let $h_{fe} = 100$ and $h_{ie} = 2.5$ kΩ.

Solution: With $R_S = 0$, analysis of the small-signal equivalent circuit is considerably simplified at the input. From the circuit of Fig. 8.8a, by inspection,

$$\mathbf{I}_b = \frac{\mathbf{V}_s}{h_{ie}} \qquad (1)$$

Then, since the AC output voltage is

$$\mathbf{V}_o = -(h_{fe}\mathbf{I}_b)R_L \| Z_L \qquad (2)$$

substitution and rearranging yield

$$\frac{\mathbf{V}_o}{\mathbf{V}_s} = \frac{h_{fe}R_L}{h_{ie}}\left(\frac{j\omega L/R_L}{j\omega L/R_L + 1}\right) \qquad (3)$$

Thus, with $A_m = (h_{fe}R_L)/h_{ie}$ and $\omega_L = R_L/L$ and by substituting values, we have

$$A_m = \frac{100(1)}{2.5} = 40$$

and

$$\omega_L = \frac{10^3}{1} = 1000 \text{ rad/s}$$

8.2.2   Common-Source Amplifier at Low Frequencies

Figure 8.9a shows the common-source amplifier circuit with a JFET. The complete low-frequency small-signal equivalent circuit is shown

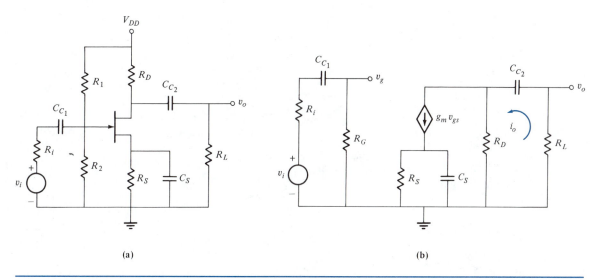

FIGURE 8.9   Common-Source Amplifier Circuit with JFET: (a) Complete circuit, (b) Small-signal equivalent circuit with C_{C_1}, C_{C_2}, and C_S

in Fig. 8.9b. Note that R_i has been used for the input source resistor symbol and R_S for the resistor attached to the source terminal. In this case, we retain all of the capacitors and determine the overall low-frequency voltage amplification. This method is practical algebraically in the FET case because the input of the FET is an open circuit.

The output voltage from Fig. 8.9b is given by

$$V_o = -I_o R_L$$
$$= -g_m V_{gs} \left[\frac{R_D}{R_D + R_L + (1/j\omega C_{C_2})} \right] R_L$$
$$(8.2\text{–}31)$$

where the current divider rule was used to substitute for I_o. Rearranging (8.2–31) to obtain the standard form yields

$$V_o = V_{gs} \left[\frac{-g_m R_L j\omega C_{C_2} R_D}{j\omega C_{C_2}(R_D + R_L) + 1} \right] \quad (8.2\text{–}32)$$

The gate-to-source voltage V_{gs} is just the difference in gate terminal voltage V_g and source ter-

minal voltage V_s, or

$$V_{gs} = V_g - V_s \quad (8.2\text{–}33)$$

The voltage at the source terminal is given by

$$V_s = g_m V_{gs}(R_S \| Z_{C_S}) = g_m V_{gs} \left(\frac{R_S}{j\omega R_S C_S + 1} \right)$$
$$(8.2\text{–}34)$$

By rearranging, we have

$$V_s = V_{gs} \left(\frac{g_m R_S}{j\omega R_S C_S + 1} \right) \quad (8.2\text{–}35)$$

Substitution of (8.2–35) into (8.2–33) yields

$$V_{gs} = V_g - V_{gs} \left(\frac{g_m R_S}{j\omega R_S C_S + 1} \right) \quad (8.2\text{–}36)$$

Solving for V_{gs}, we have

$$V_{gs} = \frac{V_g}{1 + g_m R_S/(j\omega R_S C_S + 1)} \quad (8.2\text{–}37)$$

Multiplying through by $j\omega R_S C_S + 1$ and rearranging into standard form yield

$$
\begin{aligned}
\mathbf{V}_{gs} &= \frac{\mathbf{V}_g(j\omega R_S C_S + 1)}{j\omega R_S C_S + 1 + g_m R_S} \\
&= \frac{\mathbf{V}_g(j\omega R_S C_S + 1)/(1 + g_m R_S)}{j\omega R_S C_S/(1 + g_m R_S) + 1}
\end{aligned}
\qquad (8.2\text{--}38)
$$

From the input loop, using the voltage divider rule, the voltage at the gate is

$$
\begin{aligned}
\mathbf{V}_g &= \frac{\mathbf{V}_i R_G}{R_i + (1/j\omega C_{C_1}) + R_G} \\
&= \frac{\mathbf{V}_i j\omega R_G C_{C_1}}{j\omega C_{C_1}(R_i + R_G) + 1}
\end{aligned}
\qquad (8.2\text{--}39)
$$

Finally, substitution of (8.2–39) into (8.2–38) into (8.2–32) and solving for the voltage amplification yield

$$
\begin{aligned}
\frac{\mathbf{V}_o}{\mathbf{V}_i} = -&\left[\frac{g_m R_L j\omega R_D C_{C_2}}{j\omega C_{C_2}(R_D + R_L) + 1}\right] \\
\times &\left[\frac{(j\omega R_S C_S + 1)/(1 + g_m R_S)}{j\omega R_S C_S/(1 + g_m R_S) + 1}\right] \\
\times &\left[\frac{j\omega R_G C_{C_1}}{j\omega C_{C_1}(R_i + R_G) + 1}\right]
\end{aligned}
\qquad (8.2\text{--}40)
$$

Note that (8.2–40) is in the standard form for complex voltage amplification. Greater insight is obtained by defining the characteristic frequencies somewhat arbitrarily as follows:

$$
\omega_1 = \frac{1}{(R_D + R_L)C_{C_2}}
\qquad (8.2\text{--}41)
$$

$$
\omega_2 = \frac{1}{R_D C_{C_2}}
\qquad (8.2\text{--}42)
$$

$$
\omega_3 = \frac{1}{R_S C_S}
\qquad (8.2\text{--}43)
$$

$$
\omega_4 = \frac{1 + g_m R_S}{R_S C_S}
\qquad (8.2\text{--}44)
$$

$$
\omega_5 = \frac{1}{(R_i + R_G)C_{C_1}}
\qquad (8.2\text{--}45)
$$

$$
\omega_6 = \frac{1}{R_G C_{C_1}}
\qquad (8.2\text{--}46)
$$

Note that $\omega_2 > \omega_1$, $\omega_4 > \omega_3$, and $\omega_6 > \omega_5$. Substituting these frequencies into (8.2–40) yields

$$
\begin{aligned}
\frac{\mathbf{V}_o}{\mathbf{V}_i} = -&\frac{g_m R_L}{1 + g_m R_S}\left[\frac{j\omega/\omega_2}{(j\omega/\omega_1) + 1}\right] \\
\times &\left[\frac{(j\omega/\omega_3) + 1}{(j\omega/\omega_4) + 1}\right]\left[\frac{j\omega/\omega_6}{(j\omega/\omega_5) + 1}\right]
\end{aligned}
\qquad (8.2\text{--}47)
$$

The medium-frequency voltage amplification, obtained by allowing ω to approach ∞, is given by

$$
A_m = \frac{g_m R_L}{1 + g_m R_S}\left(\frac{\omega_1}{\omega_2}\right)\left(\frac{\omega_4}{\omega_3}\right)\left(\frac{\omega_5}{\omega_6}\right)
\qquad (8.2\text{--}48)
$$

This equation may be written in terms of resistors, which is seen most easily from (8.2–40), to yield

$$
A_m = g_m R_L\left(\frac{R_D}{R_D + R_L}\right)\left(\frac{R_G}{R_i + R_G}\right)
\qquad (8.2\text{--}49)
$$

Equation (8.2–49) is identical to (5.3–11) given in Chapter 5 if we consider $R_{G_3} = 0$ and $1/g_o$ as an open circuit.

Note that each term in brackets in (8.2–47) corresponds to a different capacitor. That is, ω_1 and ω_2 correspond to C_{C_2}; ω_3 and ω_4, to C_S; ω_5 and ω_6, to C_{C_1}. Additionally, each of the terms in brackets represents a reduction in the magnitude of voltage amplification in the low-frequency region. Similar results were obtained for the common-emitter capacitors.

The Bode plot for the magnitude of voltage amplification corresponding to (8.2–47) is shown in Fig. 8.10, where it has been assumed that $\omega_5 > \omega_4$ and $\omega_3 > \omega_1$. The primary break frequency of interest is the largest of these, or ω_5 as has been assumed. At this frequency, the actual medium-

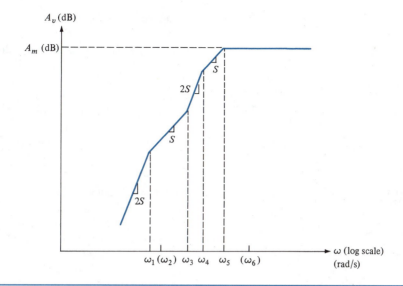

FIGURE 8.10   Magnitude Bode Plot at Low Frequency

frequency amplification is reduced by approximately 3 dB. Note that the actual amplification is reduced by more than this amount due to the other corner frequencies; therefore, this result is approximate. Thus, we let ω_5 be the approximate low half-power frequency. The smaller corner frequencies ω_4, ω_3, and ω_1, are just that; they are not half-power frequencies since a much greater reduction than 3 dB is present at these frequencies.

We should be aware that the assumption concerning the magnitudes of the break frequencies ($\omega_5 > \omega_4$ and $\omega_3 > \omega_1$) is not valid in general. This fact is not a problem, however, because we can easily calculate the magnitudes of each frequency and, by comparison, determine the largest. The largest break frequency will be the approximate low half-power frequency.

Because of the numerous characteristic frequencies, the phase Bode plot in this case is more of a challenge than it is useful. However, from (8.2–40), we can observe that the phase is modified from the usual 180° value. Each of the three terms in brackets causes deviation in the phase at low frequencies.

--- EX. 8.4

Common-Source Amplifier Example

Consider the common-source amplifier circuit of Fig. 8.9a and calculate the characteristic frequencies contained in the voltage amplification expression of (8.2–47). Use the following data: $R_S = 1$ kΩ, $R_L = 20$ kΩ, $R_G = 50$ kΩ, $R_D = 10$ kΩ, $R_i = 0$, and $C_S = C_{C_i} = C_{C_2} = 10$ μF. For the transistor, let $g_m = 5$ mS. Indicate which of these frequencies is the low half-power frequency and calculate the medium-frequency voltage amplification A_m from (8.2–48).

Solution: By direct substitution into (8.2–41) through (8.2–46), we have

$$\omega_1 = \frac{1}{(10 + 20)(10^3 \times 10^{-5})} = 3.33 \text{ rad/s}$$

$$\omega_2 = \frac{1}{(10^4)(10^{-5})} = 10 \text{ rad/s}$$

$$\omega_3 = \frac{1}{(10^3)(10^{-5})} = 100 \text{ rad/s}$$

$$\omega_4 = \frac{1 + (0.005)(10^3)}{(10^3)(10^{-5})} = 600 \text{ rad/s}$$

$$\omega_5 = \frac{1}{(5 \times 10^4)(10^{-5})} = 2 \text{ rad/s}$$

$$\omega_6 = \omega_5 = 2 \text{ rad/s}$$

Thus, since ω_4 is much greater than the other characteristic frequencies, we recognize ω_4 as the low half-power frequency.

The medium-frequency amplification is calculated from (8.2–42) as

$$A_m = \frac{(0.005)(20 \times 10^3)}{1 + (0.005)(10^3)} \left(\frac{3.33}{10}\right) \left(\frac{600}{100}\right) \left(\frac{2}{2}\right)$$

$$= 33.3$$

8.2.3 Resultant Low-Frequency Effects and Low Half-Power Frequency

In analyzing the common-emitter amplifier at low frequencies, we have observed that the major effect of each of the capacitors (when we treat them separately) as well as of inductive loading is to reduce the magnitude of amplification. The same result was obtained for the source follower (when we treat all of the capacitors collectively). Moreover, for other single-transistor amplifier circuits, the results are also similar, and a typical amplification response is like the one previously shown for the common-source amplifier in Fig. 8.10.

In general, a collective analysis of the common-emitter amplifier would not yield identical values as the break frequencies of this amplifier that were obtained by separate analyses. However, the most important of these characteristic frequencies is the largest one because this frequency determines (approximately) the value of the low half-power (or 3 dB) frequency, defined as ω_L (3 dB). The actual variation in amplifier amplification for frequencies less than this value is unimportant. Thus, a simple method of determining ω_L (3 dB) is to calculate the values of each of the break frequencies (determined separately) and

compare their magnitudes. The largest of the break frequencies is taken to be the low half-power frequency. As we will see in the following example, the bypass capacitor C_E in BJT amplifiers usually determines the value of ω_L (3 dB). Thus, a larger value of capacitance is usually used for C_E as compared to that of C_{C_1} and C_{C_2}.

EX. 8.5

Low-Frequency Common-Emitter Amplifier Example

For the common-emitter amplifier of Fig. 8.4, determine the low half-power frequency. Use the following values in the amplifier circuit: $R_S = 1 \text{ k}\Omega$, $R_B = 100 \text{ k}\Omega$, $h_{ie} = 2 \text{ k}\Omega$, $R_L = 5 \text{ k}\Omega$, $R_C = 2 \text{ k}\Omega$, $R_E = 1 \text{ k}\Omega$, $C_{C_1} = 10 \text{ }\mu\text{F}$, $C_{C_2} = 10 \text{ }\mu\text{F}$, $C_E = 200 \text{ }\mu\text{F}$, and $h_{fe} = 100$.

Solution: The break frequency for C_{C_1}, from (8.2–7), is

$$\omega_{L_1} = \frac{1}{C_{C_1}(R_S + R_B \| h_{ie})}$$

$$= \frac{1}{(10 \times 10^{-6})(1 \text{ k}\Omega + 100 \| 2 \text{ k}\Omega)}$$

$$= 33.75 \text{ rad/s or } 5.37 \text{ Hz}$$

For C_{C_2}, from (8.2–14), we have

$$\omega_{L_2} = \frac{1}{C_{C_2}(R_C + R_L)} = \frac{1}{(10 \times 10^{-6})(7 \text{ k}\Omega)}$$

$$= 14.3 \text{ rad/s or } 2.27 \text{ Hz}$$

For C_E, the break frequency is given by (8.2–21) as

$$\omega_{L_B} = \frac{1}{R_E C_E} \left[\frac{R_S \| R_B + h_{ie} + (1 + h_{fe})] R_E}{R_S \| R_B + h_{ie}} \right]$$

$$= \frac{1}{(10^3)(200 \times 10^{-6})} \left(\frac{0.99 + 2 + 101}{0.99 + 2} \right)$$

$$= 173.9 \text{ rad/s or } 27.7 \text{ Hz}$$

Hence, the half-power frequency is $\omega_{L_B} = 27.7 \text{ Hz}$ and depends upon C_E. Note that the magnitude of C_E is twenty times that of C_{C_1} and C_{C_2}, and yet this frequency turned out to be the largest.

8.3   HIGH-FREQUENCY RESPONSE OF SINGLE-TRANSISTOR AMPLIFIERS

In the medium- and high-frequency ranges, the coupling and bypass elements, as well as the inductors, behave as intended. However, parasitic capacitors associated with each transistor have impedances associated with them that diminish with increased frequency ($Z_C = 1/j\omega C$). As mentioned previously, parasitic capacitors are in parallel with PN junctions and with the MOS layers. Thus, as ω is increased, the parasitic capacitance will eventually act like a short. However, even before this extreme condition occurs, the gain is severely reduced.

8.3.1   High-Frequency Transistor Models

To determine the high-frequency effects that occur in transistor amplifier circuits, we will examine high-frequency small-signal transistor models. These models are actually valid for all frequencies, but since their use is mandatory at high frequencies, they are classed as high-frequency models. At low and medium frequencies, the high-frequency small-signal transistor models reduce to those of Chapter 5 since the transistor capacitors act like open circuits.

High-Frequency BJT Model. A high-frequency small-signal BJT model for the common-emitter configuration is shown in Fig. 8.11. A similar model can be constructed for the common-base configuration. Note that the three elements C_{bc}, C_{be}, and r_b have been added in Fig. 8.11 as compared to the small-signal model of Chapter 5. Another modification is that the current source is given by $h_{fe}i_b'$, where i_b' is the defined current through h_{ie}. Neglecting the resistance r_b relative to h_{ie}, by Ohm's law, the current source can also be written as $h_{fe}i_b' = h_{fe}v_{be}/h_{ie} = g_m v_{be}$, where we define $g_m = h_{fe}/h_{ie}$. This voltage dependent form

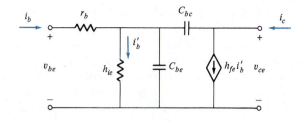

FIGURE 8.11   High-Frequency Small-Signal Common-Emitter BJT Model

of the current source is sometimes more convenient.

The capacitors C_{bc} and C_{be} are directly across the base–collector and base–emitter junctions, respectively. Physically, these capacitors represent the ratio of the variation in charge on either side of the junction to the variation in voltage across the junction. They are very small in magnitude, typically in the pF range (10^{-12} F). Since these capacitors are detrimental to device operation, they are referred to as *parasitic* elements.

Note that the current entering the base terminal of the model in Fig. 8.11 divides between h_{ie}, C_{be}, and C_{bc}. Hence, a reduction in current through h_{ie} (i_b') occurs, which causes the current source magnitude ($h_{fe}i_b'$) to be reduced. This mechanism allows the amplification to be reduced at high frequencies. At low and medium frequencies, the impedance magnitude of C_{bc} and C_{be} are quite large and force these elements to act like open circuits, which results in $i_b = i_b'$. Under these conditions and except for the resistor r_b, Fig. 8.11 reduces to the model of Chapter 5.

The resistor r_b is the *base-spreading resistor* and is due to lateral flow of majority charges through the base region from the base terminal to the emitter junction. Typically, the resistance value for r_b is on the order of 10 Ω, and it is definitely negligible at low and medium frequencies. However, in the range of high frequencies, this small resistor must be accounted for when $R_S = 0$ in order to obtain the correct high half-power frequency.

BJT Cutoff Frequency and Gain–Bandwidth Product. From the high-frequency BJT model of Fig. 8.11, we will see that the ideal current amplification reduces with frequency because of the parasitic capacitors C_{bc} and C_{be}. Under ideal conditions, a load resistor is represented by a short circuit, and if we neglect the current through C_{be}, we have

$$\mathbf{I}_c = h_{fe}\mathbf{I}_b' \qquad (8.3\text{–}1)$$

where, from the current divider rule,

$$\mathbf{I}_b' = \mathbf{I}_b\left[\frac{1/j\omega(C_{bc} + C_{be})}{h_{ie} + 1/j\omega(C_{bc} + C_{be})}\right]$$

$$= \mathbf{I}_b\left[\frac{1}{j\omega h_{ie}(C_{bc} + C_{be}) + 1}\right] \qquad (8.3\text{–}2)$$

By substituting (8.3–2) into (8.3–1) and solving for the complex current amplification, we have

$$\frac{\mathbf{I}_c}{\mathbf{I}_b} = \frac{h_{fe}}{j\omega h_{ie}(C_{bc} + C_{be}) + 1} \qquad (8.3\text{–}3)$$

This expression indicates that the medium-frequency current amplification is h_{fe} with a 3 dB half-power frequency of $1/h_{ie}(C_{bc} + C_{be})$, which is defined as ω_β and called the *beta cutoff frequency*. Thus, (8.3–3) can be written as

$$\frac{\mathbf{I}_c}{\mathbf{I}_b} = \frac{h_{fe}}{j\omega/\omega_\beta + 1} \qquad (8.3\text{–}4)$$

Another frequency of importance for the BJT is obtained by setting $|\mathbf{I}_c/\mathbf{I}_b| = 1$ using (8.3–4) and determining the corresponding value of ω, which is defined as ω_T. Thus,

$$\left|\frac{\mathbf{I}_c}{\mathbf{I}_b}\right| = 1 = \left.\frac{h_{fe}}{[(\omega/\omega_\beta)^2 + 1]^{1/2}}\right|_{\omega = \omega_T} \qquad (8.3\text{–}5)$$

Solving for ω_T yields

$$\omega_T = \omega_\beta(h_{fe}^2 - 1)^{1/2} \cong h_{fe}\omega_\beta \qquad (8.3\text{–}6)$$

We can now observe the real meaning of ω_T. It is the current gain–bandwidth product for the BJT. Often, the frequency $f_T = \omega_T/2\pi$ is specified by manufacturers on transistor data sheets, as was noted in Section 3.4.

In general, the *gain–bandwidth product* (*GBWP*) is regarded as a figure of merit for amplifiers. This quantity can be defined as the product of medium-frequency amplification A_m, which is the gain, and the bandwidth given approximately by the upper half-power frequency (since the lower half-power frequency is usually negligible in comparison).

High-Frequency FET Model. Figure 8.12 displays the high-frequency small-signal FET model for the common-source configuration that, like the BJT model, is valid at all frequencies. Here, the two elements C_{gs} and C_{gd} have been added, as compared to the FET model of Chapter 5. For JFETs, these capacitors represent the parasitic capacitance associated with the junctions. For MOSTs, C_{gs} and C_{gd} come about because of the overlap of the gate conductor and the source and drain regions, respectively. The MOS layers provide a natural parallel-plate capacitive structure, and the oxide provides the insulator region between the gate conductor top plate and semiconductor bottom plate.

Capacitors C_{gs} and C_{gd} have very small magnitudes, similar to parasitic capacitors of the BJT, and typical values are also in the pF range. However, as the frequency is increased, the impedance

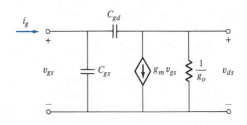

FIGURE 8.12 High-Frequency Small-Signal Common-Source FET Model

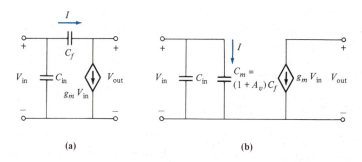

FIGURE 8.13   Parasitic Capacitance and Current Source Portion of High-Frequency Model: (a) Exact equivalent circuit, (b) Approximate equivalent circuit

associated with each of these capacitors reduces in magnitude. At high enough frequencies, the magnitude of these impedances allows a nonzero gate current, and the input gate terminal no longer acts like an open circuit.

Miller Effect and Simplified High-Frequency Transistor Models.   The high-frequency small-signal BJT and FET circuit models have identical forms as far as the two parasitic capacitors and the current source are concerned. Figure 8.13a shows the exact equivalent circuit that represents only these elements. Note that C_f is a feedback capacitor connecting the output back to the input and represents C_{bc} of the BJT or C_{gd} of the FET. The capacitor C_{in} represents C_{be} of the BJT or C_{gs} of the FET.

The difficulty with the circuit of Fig. 8.13a, in terms of transistor amplifier analysis, is that it directly connects the output back to the input and is therefore not susceptible to a solution for the amplification by inspection. However, we will develop an approximate equivalent circuit that is nearly exact and that provides the desired separation of input and output portions of the circuit. The approximate equivalent circuit also provides greater insight into which of the capacitors (C_f or C_{in}) is more important. As we will see, the feedback capacitor plays the dominant role.

One way of separating the input portion of the circuit from the output portion is simply to

remove C_f from the circuit of Fig. 8.13a. However, we would then have to assume that the current through C_f, defined as I in Fig. 8.13a, is negligible on both sides of the circuit. In fact, I is a small current relative to the output portion of the circuit because it is only a portion of the small input base current. The approximation that we will therefore make is that I is negligible on the output side of Fig. 8.13a. The effect of including this component of current on the output side is to add a small capacitor in parallel with $g_m\mathbf{V}_{in}$.

To account for **I** on the input side of the circuit in Fig. 8.13a, we begin by writing the current–voltage expression for the capacitor C_f as follows:

$$\mathbf{I} = j\omega C_f(\mathbf{V}_{in} - \mathbf{V}_{out}) \qquad (8.3–7)$$

Since the circuit of Fig. 8.13a is a portion of an amplifier circuit, we may also write

$$\mathbf{V}_{out} = -A_v\mathbf{V}_{in} \qquad (8.3–8)$$

where A_v is the voltage amplification and is known for the particular amplifier. The negative sign indicates that the amplifier inverts. Thus, by substitution into (8.3–7), we have

$$\mathbf{I} = j\omega C_f(1 + A_v)\mathbf{V}_{in} \qquad (8.3–9)$$

This equation may be represented by

$$\mathbf{I} = j\omega C_m\mathbf{V}_{in} \qquad (8.3–10)$$

where a magnified capacitance has been defined as $C_m = (1 + A_v)C_f$. This equivalent capacitance is often called the *Miller capacitance*.

From the form of (8.3–10), we realize that the current $\mathbf{I}$ can be represented by a capacitor of magnitude C_m with voltage V_{in} across its terminals. Hence, this equivalent element is used to replace C_f, as shown in Fig. 8.13b, where the input and output portions of the circuit are now separated. The new equivalent circuit of Fig. 8.13b is extremely valuable in analyzing transistor amplifiers at high frequencies. Note that the effects

of C_{in} and C_f are similar in that these elements equivalently shunt the input. However, the effects of the feedback capacitor C_f are amplified. This condition is referred to as the *Miller effect*. We will use this model in our analysis of high-frequency BJT and FET amplifiers.

_____ EX. 8.6

High-Frequency FET Nodal Analysis Example

Determine the voltage amplification expression in standard form for the amplifier circuit shown in Fig. E8.6a by writing

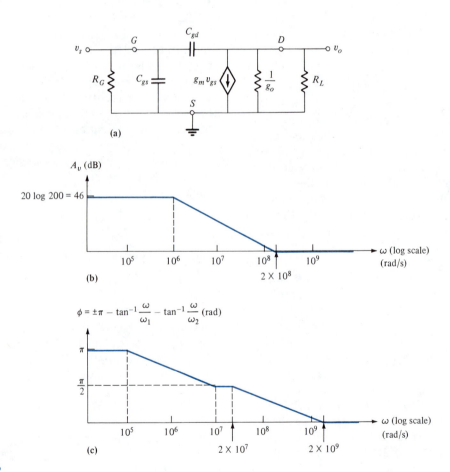

FIGURE E8.6

KCL for the output voltage terminal. We note that this equation is sometimes referred to as a *nodal* equation. Calculate the characteristic frequencies and the medium-frequency amplification. Sketch the Bode plots for amplitude and phase for element values as follows: $R_G = 100$ kΩ, $R_L = 20$ kΩ, $g_o = 0$, $g_m = 10$ mS and $C_{gd} = C_{gs} = 50$ pF.

Solution: Writing KCL for the output voltage terminal yields

$$\left(g_o + \frac{1}{R_L} + j\omega C_{gd}\right)\mathbf{V}_o - j\omega C_{gd}\mathbf{V}_{gs} = -g_m\mathbf{V}_{gs} \qquad (1)$$

However, $\mathbf{V}_{gs} = \mathbf{V}_s$. By substituting and solving for $\mathbf{V}_o/\mathbf{V}_s$, we have

$$\frac{\mathbf{V}_o}{\mathbf{V}_s} = \frac{-g_m + j\omega C_{gd}}{g_o + (1/R_L) + j\omega C_{gd}} = -A_m\left(\frac{1 - j\omega/\omega_2}{1 + j\omega/\omega_1}\right) \qquad (2)$$

where we have defined

$$A_m = \frac{g_m}{g_o + (1/R_L)} \qquad (3)$$

$$\omega_1 = \frac{1}{C_{gd}/[g_o + (1/R_L)]} \qquad (4)$$

$$\omega_2 = \frac{1}{C_{gd}/g_m} \qquad (5)$$

Substituting numerical values into (3), (4), and (5) yields

$$A_m = (0.01)(2 \times 10^4) = 200$$

$$\omega_1 = \frac{1}{(0.5 \times 10^{-10})(2 \times 10^4)} = 10^6 \text{ rad/s}$$

$$\omega_2 = \frac{1}{(0.5 \times 10^{-10})/0.01} = 2 \times 10^8 \text{ rad/s}$$

Thus, (2) can be written as

$$\frac{\mathbf{V}_o}{\mathbf{V}_s} = -200\left[\frac{1 - (j\omega/2 \times 10^8)}{1 + (j\omega/1 \times 10^6)}\right] \qquad (6)$$

The Bode plots corresponding to (6) are obtained by inspection and are displayed in Figs. E8.6b and c.

8.3.2   BJT Amplifiers at High Frequencies

To demonstrate the high-frequency behavior of BJT amplifiers, we will again use the common-emitter amplifier. Note that similar effects are observed for other amplifier configurations. The complete circuit for the common-emitter amplifier is shown in Fig. 8.4. At high frequencies, the coupling and bypass capacitors act like short circuits, but the transistor parasitic capacitors must now be included.

Figure 8.14a displays the small-signal equivalent circuit valid at high frequencies (and medium frequencies, as well). The approximate equivalent circuit for the BJT of the previous section has been used and explicitly indicates the Miller effect. The Miller capacitance is given by

$$C_m = (1 + A_v)C_{bc} \qquad (8.3\text{–}11)$$

where $A_v = \mathbf{V}_{out}/\mathbf{V}_{in}$. For the common-emitter amplifier circuit of Fig. 8.14a (where the voltages $\mathbf{V}_{out}$ and $\mathbf{V}_{in}$ are indicated), this amplification factor is obtained as

$$A_v = \frac{\mathbf{V}_{out}}{\mathbf{V}_{in}} = \frac{(h_{fe}\mathbf{I}_b')R_C \| R_L}{\mathbf{V}_{in}} = g_m(R_C \| R_L) \qquad (8.3\text{–}12)$$

where the substitution $h_{fe}\mathbf{I}_b' = h_{fe}\mathbf{V}_{in}/h_{ie} = g_m\mathbf{V}_{in}$ was used. Thus, the Miller capacitance for the BJT case is given by

$$C_m = C_{bc}[1 + g_m(R_C \| R_L)] \qquad (8.3\text{–}13)$$

Analysis of Fig. 8.14a to obtain the overall phasor voltage amplification $\mathbf{V}_o/\mathbf{V}_s$ is now straight-forward. From the output portion of the circuit, we have

$$\mathbf{V}_o = -(h_{fe}\mathbf{I}_b')R_C \| R_L \qquad (8.3\text{–}14)$$

Furthermore, by inspection of the input portion of the circuit, we observe that

$$\mathbf{I}_b' = \frac{\mathbf{V}_{in}}{h_{ie}} \qquad (8.3\text{–}15)$$

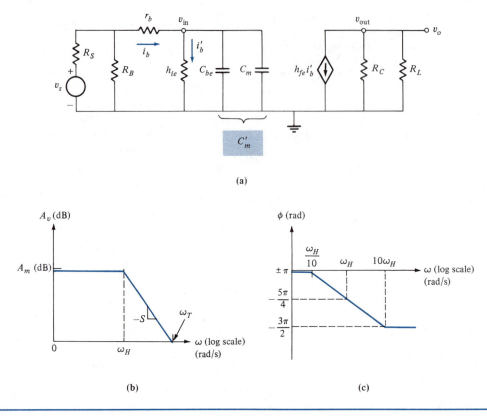

(a)

(b)

(c)

FIGURE 8.14   Medium- and High-Frequency Small-Signal Common-Emitter Amplifier Circuit: (a) Approximate equivalent circuit, (b) Magnitude Bode plot at high frequency, (c) Phase Bode plot at high frequency

Additionally, from the input loop of Fig. 8.14a and by using a Thévenin equivalent circuit for R_S, R_B, and the input signal voltage, we have

$$\mathbf{V}_{in} = \frac{h_{ie} \| Z_{C'_m}}{R_S \| R_B + r_b + h_{ie} \| Z_{C'_m}} \left(\frac{R_B}{R_S + R_B} \right) \mathbf{V}_s$$

$$(8.3\text{–}16)$$

Expanding the parallel combination yields

$$h_{ie} \| Z_{C'_m} = \frac{h_{ie}}{j\omega h_{ie} C'_m + 1} \qquad (8.3\text{–}17)$$

Substitution of (8.3–17) into (8.3–16) and rearranging yield

$$\mathbf{V}_{in} = \frac{h_{ie}[R_B/(R_S + R_B)]\mathbf{V}_s}{j\omega h_{ie} C'_m(R_S \| R_B + r_b) + R_S \| R_B + r_b + h_{ie}}$$

$$(8.3\text{–}18)$$

Substitution of (8.3–18) into (8.3–15) into (8.3–14) then yields

$$\mathbf{V}_o = -h_{fe}(R_C \| R_L)$$
$$\times \left[\frac{[R_B/(R_S + R_B)]\mathbf{V}_s}{j\omega h_{ie} C'_m(R_S \| R_B + r_b) + R_S \| R_B + r_b + h_{ie}} \right]$$

$$(8.3\text{–}19)$$

By solving (8.3–19) for $\mathbf{V}_o/\mathbf{V}_s$ and factoring the resistor $(R_S \| R_B + r_b + h_{ie})$ out of the denominator, we obtain the standard form of the voltage amplification as follows:

$$\frac{\mathbf{V}_o}{\mathbf{V}_s} = -\frac{h_{fe}(R_C \parallel R_L)[R_B/(R_S + R_B)]}{R_S \parallel R_B + r_b + h_{ie}}$$

$$\times \left[\frac{1}{\dfrac{j\omega h_{ie} C'_m (R_S \parallel R_B + r_b)}{R_S \parallel R_B + r_b + h_{ie}} + 1}\right] \qquad (8.3-20)$$

We now define the medium-frequency amplification and the high break frequency, respectively, as

$$A_m = \frac{h_{fe}(R_C \parallel R_L)[R_B/(R_S + R_B)]}{R_S \parallel R_B + r_b + h_{ie}} \qquad (8.3-21)$$

and

$$\omega_H = \frac{1}{C'_m[h_{ie} \parallel (R_S \parallel R_B + r_b)]} \qquad (8.3-22)$$

Therefore, (8.3–20) becomes

$$\frac{\mathbf{V}_o}{\mathbf{V}_s} = -A_m\left(\frac{1}{j\omega/\omega_H + 1}\right) \qquad (8.3-23)$$

Note that (8.3–21) is identical to (8.2–13) except for r_b. If r_b had also been considered at low frequencies, the expressions for A_m would be identical.

The form of (8.3–23) indicates that ω_H is the high half-power (3 dB) frequency. Because of the Miller effect, this frequency is reduced to a value considerably less than ω_β (the beta cutoff frequency). The corresponding magnitude and phase Bode plots are displayed in Figs. 8.14b and c. Note that in the calculation of values for A_m and ω_H from (8.3–21) and (8.3–22), the value of r_b plays an insignificant role because it is quite small. However, in instances where the input signal source resistance R_S is zero, the small value of r_b is extremely important in determining ω_H. This fact is observed by considering $R_S = 0$ in (8.3–22), which yields

$$\omega_H = \frac{1}{C'_m(h_{ie} \parallel r_b)} \qquad (8.3-24)$$

Note that if r_b were not in the original circuit ($r_b = 0$), then ω_H would be infinite; that is, the parasitic capacitors would have no effect. From the circuit of Fig. 8.14a, we see that if $R_S = r_b = 0$, then the input signal voltage is directly across h_{ie}, which results in $i'_b = v_s/h_{ie}$ regardless of the capacitors.

Note that in Fig. 8.14b, we have defined another characteristic frequency as ω_T. This frequency corresponds to $A_v(\text{dB}) = 0$ or $|A_v| = 1$ and is called the gain–bandwidth product of the amplifier, similar to that of the BJT by itself. We obtain an equation for ω_T from (8.3–23) by setting the magnitude of $\mathbf{V}_o/\mathbf{V}_s$ equal to unity for ω_T, or

$$\left|\frac{\mathbf{V}_o}{\mathbf{V}_s}\right| = \frac{A_m}{[(\omega_T/\omega_H)^2 + 1]^{1/2}} = 1 \qquad (8.3-25)$$

Solving for ω_T yields

$$\omega_T = (A_m^2 - 1)^{1/2}\omega_H \cong A_m\omega_H \qquad (8.3-26)$$

The quantity $\omega_T\ (=2\pi f_T)$ is also called the *upper cutoff frequency* of the amplifier and denotes the frequency at which the gain is reduced to unity.

———————————————————————— EX. 8.7

High-Frequency BJT Common-Emitter Miller Effect Example

Calculate the medium-frequency gain A_m and the high break frequency ω_H for the common-emitter amplifier circuit of Fig. 8.14a. Use the following values: $R_S = 0$, $R_B = R_C = R_L = 10\ \text{k}\Omega$, $h_{fe} = 100$, $h_{ie} = 2\ \text{k}\Omega$, $r_b = 50\ \Omega$, and $C'_m = 10^3\ \text{pF}$.

Solution: By direct substitution into (8.3–21) and (8.3–22), we have

$$A_m = \frac{100(5)(1)}{0 + 0.05 + 2} = 244$$

and

$$\omega_H = \frac{1}{(10^{-9})(2)(0.05)/2.05} = 20.5 \times 10^9 \ \text{rad/s}$$

BJT Upper Cutoff Frequency Example

A BJT common-emitter amplifier has a medium-frequency voltage amplification of $A_m = 100$ and a high half-power (3 dB) frequency of $\omega_H = 10^8$ rad/s. Calculate the upper cutoff frequency and sketch the magnitude Bode plot.

Solution: From (8.3−26), we have

$$\omega_T \cong (100)(10^8) \cong 10^{10} \text{ rad/s}$$

The corresponding magnitude Bode plot is displayed in Fig. E8.8.

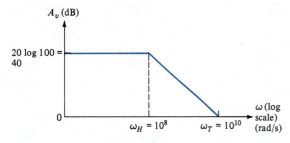

FIGURE E8.8

8.3.3   FET Amplifiers at High Frequencies

To demonstrate the high-frequency behavior of FET amplifiers, we will again use the JFET common-source amplifier of Fig. 8.9a. The small-signal equivalent circuit for this amplifier, valid at high (and medium) frequencies, is shown in Fig. 8.15. Note that the Miller effect equivalent circuit has been used to represent the FET, where $C_m = (1 + A_v)C_{gd}$ and $A_v = g_m(R_D \| R_L)$ here. Thus, the Miller capacitance is given by

$$C_m = [1 + g_m(R_D \| R_L)]C_{gd} \qquad (8.3-27)$$

We obtain the expression for overall voltage amplification by writing the equation for output voltage from the circuit of Fig. 8.15 as

$$V_o = -(g_m V_{gs})R_D \| R_L \qquad (8.3-28)$$

From the input loop and by using a Thévenin equivalent circuit, we have

$$V_{gs} = \frac{[R_G/(R_i + R_G)]V_s}{j\omega C_m'(R_i \| R_G) + 1} \qquad (8.3-29)$$

Substitution of (8.3−29) into (8.3−28) and solving for the voltage amplification yield

$$\frac{V_o}{V_s} = -\frac{g_m(R_D \| R_L)[R_G/(R_i + R_G)]}{j\omega C_m'(R_i \| R_G) + 1} \qquad (8.3-30)$$

Immediately, we recognize that

$$A_m = g_m(R_D \| R_L)\left(\frac{R_G}{R_i + R_G}\right) \qquad (8.3-31)$$

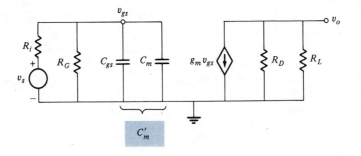

FIGURE 8.15   Medium- and High-Frequency Small-Signal Common-Source Amplifier Circuit

and that

$$\omega_H = \frac{1}{C'_m(R_i \| R_G)} \qquad (8.3-32)$$

where (8.3–31) is identical to (8.2–49). Substituting (8.3–32) and (8.3–31) into (8.3–30) yields

$$\frac{V_o}{V_s} = -A_m \left[\frac{1}{(j\omega/\omega_H) + 1} \right] \qquad (8.3-33)$$

Note that (8.3–33) is identical in form to (8.3–23) for the BJT common-emitter amplifier. Therefore, the magnitude and phase Bode plots are the same as those shown in Figs. 8.14b and c.

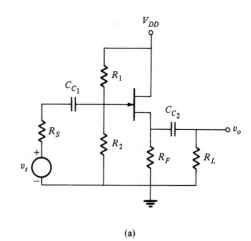

(a)

—————————————————————— EX. 8.9

High-Frequency Source Follower Example

Figure E8.9a displays the common-drain amplifier, or source follower, circuit. Using a small-signal equivalent circuit that includes the Miller capacitance, determine the voltage amplification V_o/V_s at medium and high frequencies. Use the following values: $R_S = 1\ \text{k}\Omega$, $R_G = 50\ \text{k}\Omega$, $R_F = R_L = 2\ \text{k}\Omega$, $g_o = 0$, $C'_m = 250\ \text{pF}$, and $g_m = 50\ \text{mS}$.

Solution: Figure E8.9b displays the corresponding small-signal equivalent circuit. From the output loop, the output voltage is given by

$$V_o = (R_F \| R_L) g_m V_{gs} \qquad (1)$$

where V_{gs} is seen from the circuit to be

$$V_{gs} = V_g - V_o \qquad (2)$$

Substituting (2) into (1) and solving for V_o yield

$$V_o = V_g \left[\frac{g_m(R_F \| R_L)}{1 + g_m(R_F \| R_L)} \right] \qquad (3)$$

Now, the gate voltage V_g is obtained from Fig. E8.9b by using a voltage divider in the impedance network. By inspection, we have

$$V_g = V_s \left(\frac{R_G \| Z_{C'_m}}{R_S + R_G \| Z_{C'_m}} \right) \qquad (4)$$

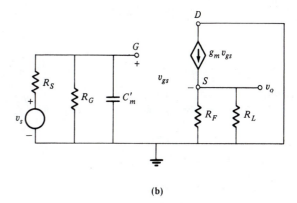

(b)

FIGURE E8.9

where

$$R_G \| Z_{C'_m} = \frac{R_G}{1 + j\omega C'_m R_G} \qquad (5)$$

Thus, substitution of (5) into (4) yields

$$V_g = V_s \left[\frac{R_G/(1 + j\omega C'_m R_G)}{R_S + [R_G/(1 + j\omega C'_m R_G)]} \right] \qquad (6)$$

Multiplying numerator and denominator by $1 + j\omega C'_m R_G$, we have

$$V_g = V_s \left[\frac{R_G}{R_S(1 + j\omega C'_m R_G) + R_G} \right] \qquad (7)$$

Substitution of (7) into (3) yields

$$
\mathbf{V}_o = \mathbf{V}_s\left[\frac{R_G}{R_S(1 + j\omega C'_m R_G) + R_G}\right]
$$
$$
\times\left[\frac{g_m(R_F \| R_L)}{1 + g_m(R_F \| R_L)}\right] \tag{8}
$$

Finally, by dividing by $\mathbf{V}_s$ and factoring $R_S + R_G$ out of the denominator of the first term in brackets, we have

$$
\frac{\mathbf{V}_o}{\mathbf{V}_s} = \left[\frac{R_G/(R_S + R_G)}{1 + j\omega C'_m(R_S \| R_G)}\right]\left[\frac{g_m(R_F \| R_L)}{1 + g_m(R_F \| R_L)}\right] \tag{9}
$$

Thus, by inspection of (8), we have

$$
A_m = \frac{R_G}{R_S + R_G}\left[\frac{g_m(R_F \| R_L)}{1 + g_m(R_F \| R_L)}\right] \tag{10}
$$

and

$$
\omega_H = \frac{1}{C'_m(R_S \| R_G)} \tag{11}
$$

Substituting numerical values into (9) and (10) yields

$$
A_m = \frac{50}{1 + 50}\left[\frac{(0.05)(10^3)}{1 + (0.05)(10^3)}\right] = 0.96
$$

and

$$
\omega_H = \frac{1}{(0.25 \times 10^{-9})[(1 \times 50/51)](10^3)}
$$
$$
= 4.08 \times 10^6 \text{ rad/s}
$$

8.4 OVERALL FREQUENCY RESPONSE OF TRANSISTOR AMPLIFIERS

In the previous section, we considered the effects of low and high frequencies separately. We now combine these effects and consider the overall frequency response of transistor amplifiers.

8.4.1 Single-Transistor Amplifiers

In the preceding analyses of the common-emitter BJT and the common-source FET amplifier circuits, we determined the same expression for A_m in the low- and high-frequency ranges. A similar result is obtained with other single-transistor amplifier circuits and indicates that the low- and high-frequency effects do not interact. The overall frequency response of single-transistor amplifiers can therefore be obtained simply by extending the results of the separate frequency ranges.

Figure 8.16 displays the resultant magnitude Bode plot for a typical overall frequency response of a single-transistor amplifier. Note that only the largest of the low corner frequencies is of importance in determining the low half-power frequency ω_L (3 dB) when it is assumed that ω_L (3 dB) is much greater than the other corner frequencies. Although this assumption is not always true, it turns out to be practical in almost all cases be-

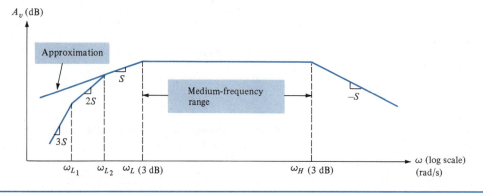

FIGURE 8.16   Magnitude Bode Plot Superimposing Low and High Frequency

cause $\omega_L(3\,\text{dB})$ is quite small (typically, $f_L <$ 100 Hz). Additionally, since the actual variation of amplification below $\omega_L(3\,\text{dB})$ is unimportant, we represent the low-frequency region with only a single corner frequency, as shown in Fig. 8.16. Of course, we cannot disregard $\omega_L(3\,\text{dB})$ entirely because it keeps us mindful of the fact that capacitively and inductively coupled amplifiers do not amplify at low frequency and, in particular, DC. Furthermore, in such multistage coupled amplifiers, $\omega_L(3\,\text{dB})$ increases with an increase in the number of stages and can become important, as we will see.

The bandwidth of the single-stage amplifier is defined as the difference in frequency between the low and high half-power frequencies, or

$$BW = \omega_H(3\,\text{dB}) - \omega_L(3\,\text{dB}) \qquad (8.4\text{--}1)$$

However, since $\omega_H(3\,\text{dB}) \gg \omega_L(3\,\text{dB})$, we have

$$BW \cong \omega_H(3\,\text{dB}) \qquad (8.4\text{--}2)$$

8.4.2   Multistage-Transistor Amplifiers

To simplify the description of multistage amplifiers, we assume that the stages cascaded together are identical. Additionally, we will represent each low-frequency range with a single corner frequency, ω_L, as was indicated in Fig. 8.16. We also assume that the input impedance of each stage is very large and that the output impedance of each stage is very small. Therefore, the overall voltage amplification is given by the product of the individual amplification expressions.

The voltage amplification expression for a single-stage is given by

$$A_{v_1} = A_m \left(\frac{j\omega/\omega_L}{1 + j\omega/\omega_L} \right) \left(\frac{1}{1 + j\omega/\omega_H} \right) \qquad (8.4\text{--}3)$$

For n such stages connected in cascade, the voltage amplification is just the product of each stage, or

$$A_{v_n} = A_{v_1}^n = A_m^n \left(\frac{j\omega/\omega_L}{1 + j\omega/\omega_L} \right)^n \left(\frac{1}{1 + j\omega/\omega_H} \right)^n \qquad (8.4\text{--}4)$$

Equation (8.4–4) indicates that the medium-frequency gain has been increased to a value corresponding to the product of the individual medium-frequency amplification factors (A_m^n). In terms of dB, the improvement in amplification is a factor of n.

However, it is not immediately obvious from (8.4–4) how the low and high 3 dB half-power frequencies vary with the number of transistor stages. An indication of the effect of multiple stages is obtained by studying the Bode amplitude plots of Fig 8.17, where $n = 1$, 2, and 3. Note that with

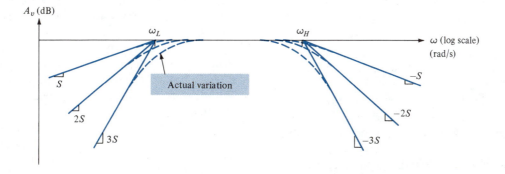

FIGURE 8.17   Normalized Magnitude Bode Plots

the increase of each additional stage, the original 3 dB half-power points $\omega_L(3\,\text{dB})$ and $\omega_H(3\,\text{dB})$ become frequencies with an additional 3 dB drop per added stage. Thus, the actual 3 dB frequencies shift toward one another and thus decrease the bandwidth.

An expression for the amount of the shift can be obtained quite easily for each half-power frequency by assuming $\omega_H \gg \omega_L$. At low frequencies, where ω is in the range of ω_L, only the first term in brackets in (8.4–4) is significant. Setting the magnitude of the first term in brackets equal to $1/\sqrt{2}$, we have

$$\left[\frac{\omega/\omega_L}{[(\omega/\omega_L)^2 + 1]^{1/2}}\right]^n = \frac{1}{2^{1/2}} \qquad (8.4–5)$$

where ω in (8.4–5) now represents the low half-power frequency, by definition. Solving (8.4–5) for ω yields

$$\omega = \omega_L(3\,\text{dB}) = \omega_L \frac{1}{(2^{1/n} - 1)^{1/2}} \qquad (8.4–6)$$

where we recall that ω_L is the single-stage low half-power frequency. We observe from (8.4–6) that as n increases, the low half-power 3 dB frequency increases.

At high frequencies, where ω is in the range of ω_H, only the second term in brackets in (8.4–4) varies with frequency. Equating the magnitude of this term equal to $1/\sqrt{2}$ results in

$$\omega = \omega_H(3\,\text{dB}) = \omega_H(2^{1/n} - 1)^{1/2} \qquad (8.4–7)$$

We note from (8.4–7) that as n increases, the high half-power 3 dB frequency reduces.

Thus, as the number of stages is increased, the bandwidth, $= \omega_H(3\,\text{dB}) - \omega_L(3\,\text{dB})$, is reduced. The medium-frequency amplification A_m, however, is increased with additional stages. Trade-offs between A_m and bandwidth are therefore possible. That is, if the gain of each individual stage is reduced, which increases the bandwidth, then connecting such individual stages into a

multistage amplifier can result in lower medium-frequency amplification but higher bandwidth.

8.5   FREQUENCY DEPENDENCE OF OP AMPS

Since capacitors and inductors are not used in the basic internal circuitry of an op amp, the amplification is not a function of frequency at low frequencies. Thus, for an op amp, $\omega_L(3\,\text{dB}) = 0$, and the amplification is a function of frequency at high frequencies only.

To account for this high-frequency dependence, we consider the example of the op amp given in Section 6.4. The circuit shown in Fig. 6.21 consists of two difference amplifiers, a DC level shifter, and an emitter follower output amplifier.

The high-frequency AC equivalent circuit corresponding to Fig. 6.21 is displayed in Fig. 8.18. A nonzero input v_s is applied to one input terminal of the op amp, and the other is grounded (consider C_{comp} to be an open circuit for the present analysis). Note that the current sources T_3, T_6, and T_8 are represented by large resistors corresponding to $1/h_{ob}$. Furthermore, the two difference amplifiers are represented at high frequencies by their corresponding Miller capacitors. However, in the level shifter and emitter follower stages, the capacitors between base and emitter are in series with very large resistors. Hence, only the base-to-collector capacitors are necessary in the AC equivalent circuit. Finally, to simplify the notation, h_{fe} is assumed to be equal for all transistors.

Using phasor representations for the frequency-dependent currents and voltages in Fig. 8.18, we begin the analysis at the output terminal. With the input impedance of the emitter follower (R_i) treated like an open circuit, the output load voltage is obtained (by using the voltage divider rule) in terms of the voltage at the collector of T_4 as

$$\mathbf{V}_l = \frac{\mathbf{V}_{c_4}}{j\omega R_7 C_{bc_9} + 1} \qquad (8.5–1)$$

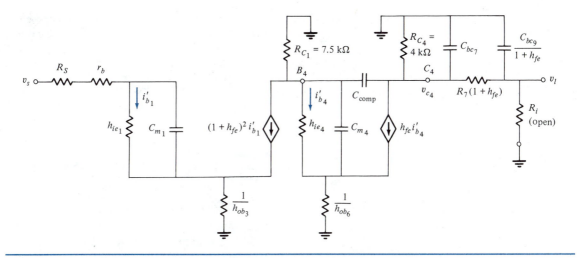

FIGURE 8.18   High-Frequency AC Equivalent Circuit for Op Amp of Fig. 6.21

The voltage $\mathbf{V}_{c_4}$ is then obtained (by using Ohm's law and treating C_{comp} as an open circuit) from

$$\mathbf{V}_{c_4} = -h_{fe}\mathbf{I}'_{b_4}\left(\frac{R_{C_4}}{j\omega R_{C_4}C_{bc_7} + 1}\right) \qquad (8.5\text{–}2)$$

where the current in $R_7(1 + h_{fe})$ is neglected in comparison to that in R_{C_4} and C_{bc_7}. The current $\mathbf{I}'_{b_4}$ is now obtained by realizing $V_{b_4} = (1 + h_{fe})^2 I'_{b_1} R_{C_1}$ and

$$I'_{b_4} = I_{b_4}/(j\omega C_{m_4}h_{ie_4} + 1)$$
$$= \frac{V_{b_4}/(j\omega C_{m_4}h_{ie_4} + 1)}{h_{ie_4}\|(1/j\omega C_{m_4}) + (1 + h_{fe})(1/h_{ob_6})}$$
$$\cong \frac{CI'_{b_1}}{1 + j\omega C_{m_4}h_{ie_4}} \qquad (8.5\text{–}3)$$

where C is a constant and $(1 + h_{fe})(1/h_{ob_6}) \gg h_{ie_4}$. Finally, for simplification, an additional frequency-dependent term results because of C_{m_1} and $R_S + r_b$ at the input. However, we will assume that $R_S + r_b$ is approximately zero so that from the circuit of Fig. 8.18, we have

$$\mathbf{I}'_{b_1} = \frac{\mathbf{V}_s}{h_{ie_1}} \qquad (8.5\text{–}4)$$

By combining (8.5–1) and (8.5–4), we obtain

$$\frac{\mathbf{V}_l}{\mathbf{V}_s} = \frac{A_m}{[(j\omega/\omega_1) + 1][(j\omega/\omega_2) + 1][(j\omega/\omega_3) + 1]} \qquad (8.5\text{–}5)$$

where A_m is the medium-frequency amplification and where

$$\omega_1 = \frac{1}{h_{ie_4}C_{m_4}} \qquad (8.5\text{–}6)$$

$$\omega_2 = \frac{1}{R_{C_4}C_{bc_7}} \qquad (8.5\text{–}7)$$

$$\omega_3 = \frac{1}{R_7C_{bc_9}} \qquad (8.5\text{–}8)$$

The Bode plots for the magnitude of the voltage amplification and the phase versus frequency are displayed in Fig. 8.19. Note that each stage behaves like an RC low-pass filter network. The following example indicates typical values for the half-power frequencies.

── EX. 8.10

Op Amp Half-Power Frequency Example

Calculate the half-power frequencies corresponding to (8.5–6) through (8.5–8). Use the following values for the

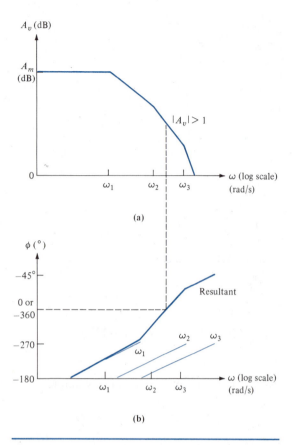

FIGURE 8.19   Bode Plots for Uncompensated Op Amps: (a) Magnitude, (b) Phase

resistors and capacitors: $R_7 = 4.6\,\text{k}\Omega$, $C_{bc9} = 2.5\,\text{pF}$, $R_{C_4} = 4\,\text{k}\Omega$, $C_{bc7} = 5\,\text{pF}$, $C_{m4} = 120\,\text{pF}$, $h_{ie4} = 2.5\,\text{k}\Omega$, $R_{C_1} = 7.5\,\text{k}\Omega$, and $h_{fe} = 100$.

Solution: Substituting these values into (8.5–6) through (8.5–8) yields

$$\omega_1 = \frac{10^{12}}{(2.5 \times 10^3)(120)}$$

$$= 3.3 \times 10^9 \text{ rad/s or } 0.53 \text{ GHz}$$

$$\omega_2 = \frac{10^{12}}{(4 \times 10^3)(5)} = 5 \times 10^7 \text{ rad/s or } 8 \text{ MHz}$$

$$\omega_3 = \frac{10^{12}}{(4.6 \times 10^3)(2.5)}$$

$$= 8.6 \times 10^7 \text{ rad/s or } 13.6 \text{ MHz}$$

From this example, we note that the lowest half-power frequency corresponds to the first stage following the input difference amplifier. Furthermore, this frequency is reduced from the other due to the Miller effect, which amplifies the collector-to-base capacitance.

8.6   FREQUENCY COMPENSATION OF OP AMPS

In the previous section, we observed that the magnitude of the voltage amplification for a typical op amp reduces at higher frequencies. This effect is caused by the inevitable low-pass filter behavior of each amplifier state. Of course, the reduction in magnitude is a detrimental effect, but even more detrimental is the 90° change in phase associated with each half-power frequency.

The problem that arises is that the negative feedback typically used with op amps will become positive feedback when the frequency is increased such that the phase of the output voltage is changed by 180°. Then, if the voltage amplification is still greater than 1, an unstable situation occurs, and the positive feedback will cause oscillation, which is the production of an output voltage with no input voltage.

This situation is depicted in Fig. 8.19, where the magnitude and phase Bode plots for an op amp without feedback are shown. Between ω_2 and ω_3 we note that the phase shift is zero, while the magnitude of the amplification is greater than unity. The output then reinforces the input, and oscillation occurs.

This oscillation can be a desired effect, as we will see in Chapter 14, which describes the design of basic oscillators. However, when the op amp is being used for other purposes (for example, as an amplifier), the oscillation produces instability and is undesirable.

To eliminate this effect, the upper half-power frequency, which is $\omega_H(3\,\text{dB}) = \omega_1$ in Fig. 8.19, is designed deliberately to be very low, typically in the range of 1 to 20 Hz. Thus, as the operating frequency is increased beyond this half-power frequency, the voltage amplification reduces at the rate of 20 dB/decade. This reduction with frequency is designed such that the voltage amplification is less than 1 near the first natural upper half-power frequency of the op amp. Hence, the problem of zero phase shift with amplification greater than 1 is removed.

8.6.1   Compensation Capacitors

Perhaps the easiest way to introduce a low half-power frequency is to place an additional capacitor in the circuit. This capacitor is shown in Fig. 8.18 labeled as C_{comp} for *compensation capacitor,* and its magnitude is amplified by the Miller effect. The amplified capacitance is then in parallel with C_{m_4}, and the corresponding half-power frequency of this stage is reduced to the low-frequency range.

Op amps that have a compensation capacitor such as C_{comp} in Fig. 8.18 built into the integrated circuit are said to be *internally compensated.* However, if C_{comp} is connected externally to special op amp pins, then the op amp is said to be *externally compensated.* Both types of op amps are commercially available, as we observed in Section 6.6.

8.6.2   Unity Gain Bandwidth and Gain–Bandwidth Product

Figure 8.20a shows the magnitude Bode plot for a frequency-compensated op amp with various amounts of negative feedback. The amount of feedback is controlled by varying R_F in the inverting amplifier circuit, as shown in Fig. 8.20b. As we have seen, the medium-frequency gain depends directly upon $-(R_F/R_S)$. Hence, as R_F is increased, the magnitude of the medium-frequency gain is increased, as indicated in Fig. 8.20a.

Note that the top curve represents the case without feedback ($R_F = \infty$, $R_S = 0$), whereas the lower curves indicate increased amounts of feedback. For each case, a different half-power frequency or bandwidth is obtained.

Note also that the frequency at which the gain is reduced to unity ($|A_v|_{dB} = 0$) is labeled ω_{unity} and is called the *unity gain bandwidth.* This frequency represents a range of frequencies or bandwidth over which the voltage gain is greater than unity.

Recall that the bandwidth (BW) corresponds to the medium-frequency range (the range of frequencies between the upper and lower half-power frequencies). Since the lower half-power frequency is zero for an op amp, BW is therefore equal to the upper half-power frequency. Hence, for the magnitude Bode plot of Fig. 8.20a, as R_F is increased from R_{F_1} to R_{F_2} to R_{F_3}, BW is decreased from ω_{H_3} to ω_{H_2} to ω_{H_1}. Furthermore, the amount of reduction in BW is inversely proportional to R_F/R_S, as we will now verify.

First, we consider an op amp without feedback. The signal voltage v_s is applied across the op amp input with $v_d = -v_s$. This signal condition corresponds to Fig. 8.20b with $R_F = \infty$ and $R_S = 0$. Hence, the output voltage is given by

$$v_o = +A_d v_d = -A_d v_s \qquad (8.6-1)$$

If we assume that A_d is frequency dependent and given by

$$A_d = \frac{A_m}{1 + j\omega/\omega_H} \qquad (8.6-2)$$

then the voltage amplification without feedback depends on frequency as follows:

$$\frac{V_o}{V_s} = \frac{-A_m}{1 + j\omega/\omega_H} \qquad (8.6-3)$$

Thus, the gain is A_m, and the upper half-power frequency or BW is ω_H without feedback. Furthermore, the gain–bandwidth product is

$$\text{GBWP} = A_m \omega_H \qquad (8.6-4)$$

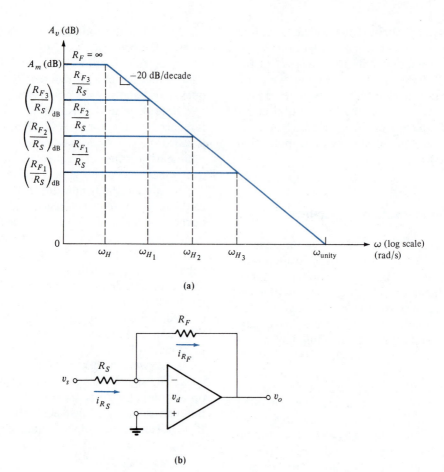

FIGURE 8.20   Frequency-Compensated Op Amps: (a) Magnitude Bode plot $(R_{F_3} > R_{F_2} > R_{F_1})$, (b) Inverting amplifier circuit

With negative feedback $(R_F \neq \infty$ and $R_S \neq 0)$, and by using KCL at the negative terminal and neglecting the input current to the op amp, we have

$$\frac{v_s - (-v_d)}{R_S} = \frac{-v_d - v_o}{R_F} \qquad (8.6\text{--}5)$$

Substituting $v_o = A_d v_d$ and solving for v_o yield

$$v_o = \frac{-A_d(R_F/R_S)v_s}{A_d + (R_F/R_S)} \qquad (8.6\text{--}6)$$

where we have assumed $R_F/R_S \gg 1$, which is appropriate for the amplifier example we are considering.

If we assume that the frequency dependence of A_d is given again by (8.6–2), then, by direct substitution into (8.6–6), using phasor notation we have

$$\mathbf{V}_o = \frac{-\left(\dfrac{A_m}{1 + j\omega/\omega_H}\right)\left(\dfrac{R_F}{R_S}\right)\mathbf{V}_s}{\left(\dfrac{A_m}{1 + j\omega/\omega_H}\right) + \dfrac{R_F}{R_S}} \qquad (8.6\text{--}7)$$

Multiplying numerator and denominator by $1 + j\omega/\omega_H$ and dividing by V_s yield

$$\frac{V_o}{V_s} = \frac{-A_m(R_F/R_S)}{A_m + (R_F/R_S)(1 + j\omega/\omega_H)} \qquad (8.6\text{–}8)$$

By rearranging (8.6–8), we obtain the standard form of the voltage amplification:

$$\frac{V_o}{V_s} = -\frac{A_m(R_F/R_S)}{A_m + (R_F/R_S)}$$
$$\times \left[\frac{1}{1 + j(\omega/\omega_H)(R_F/R_S)/[A_m + (R_F/R_S)]} \right]$$
$$(8.6\text{–}9)$$

Finally, if $A_m \gg (R_S + R_F)/R_S$, then (8.6–9) becomes

$$\frac{V_o}{V_s} = -\frac{R_F}{R_S}\left[\frac{1}{1 + j(\omega/\omega_H)(R_F/R_S)/A_m} \right]$$
$$(8.6\text{–}10)$$

Thus, from (8.6–10), we observe that the medium-frequency gain is R_F/R_S, while the upper half-power frequency or BW is $\omega_H A_m(R_S/R_F)$. The gain–bandwidth product with feedback is therefore

$$\text{GBWP} = \left(\frac{R_F}{R_S}\right)\left(\omega_H A_m \frac{R_S}{R_F}\right)$$
$$= \omega_H A_m \qquad (8.6\text{–}11)$$

which is identical to the gain–bandwidth product feedback.

Finally, we conclude that the gain–bandwidth product is constant and independent of the amount of feedback because we know that as the amount of feedback is increased, the gain reduces but the bandwidth is increased by the same factor. This important result is true provided that the feedback elements (R_F and R_S) are purely resistive and that $A_d \gg \dfrac{(R_F + R_S)}{R_S} \gg 1$.

CHAPTER 8 SUMMARY

- In general, voltage and current amplification are complex functions of frequency. At low frequencies, capacitors and inductors in the circuit external to the transistor are responsible for a reduction in the magnitude of the amplification, where the phase is also altered from that at medium frequencies. At high frequencies, parasitic capacitors associated with the junctions and MOS layers of transistors cause a reduction in the magnitude of amplification and a change in phase from that at medium frequencies.

- Bode plots are used to describe the magnitude and phase of complex amplification expressions. These plots are straight-line, asymptotic sketches of magnitude in decibels and phase in radians (or degrees) versus frequency in ra-

dians (or hertz) on a log axis. The plots are obtained by inspection of the complex amplification expression in standard (dimensionless) form.

- The impedances of external capacitors ($Z_C = 1/j\omega C$) and inductors ($Z_L = j\omega L$) are taken into account at low frequencies, and circuit analysis is used to derive complex expressions for voltage amplification that verify the detrimental low-frequency behavior. For BJT amplifier circuits, bypass and coupling capacitors are considered individually (to reduce the algebra involved); for FETs, all three capacitors are accounted for simultaneously.

- At high frequencies, transistor models must be modified to include parasitic capacitors. For BJTs, these capacitors are associated with the base–emitter junction C_{be} and the base–collector junction C_{bc}. For FETs, these capacitors represent the gate-to-source capacitance C_{gs} and the gate-to-drain capacitance C_{gd}.

■ For the common-emitter (or common-source) amplifier at high frequencies, the effects of the feedback capacitor $C_F = C_{bc}$ (or C_{gd}) can be accounted for by an input capacitor C_m. This Miller capacitance is represented by $C_m = (1 + A_v)C_F$, where $A_v = (h_{fe}/h_{ie})R_C \| R_L = g_m(R_C \| R_L)$ for the BJT (or $A_v = g_m(R_D \| R_L)$ for the FET).

■ In the high-frequency region ($\omega > \omega_H$), the magnitude of the amplification decreases as the frequency is increased. The phase also changes by a factor of 90° for each amplifier stage.

■ The bandwidth of a single-transistor amplifier stage is approximately $BW \cong \omega_H\,(3\,\text{dB})$. The bandwidth of a multistage amplifier is reduced

from that of a single stage. However, the medium-frequency amplification is increased.

■ Bode plots for an uncompensated op amp indicate that the magnitude of the amplification is reduced and that the phase angle is increased as the frequency is increased. This situation introduces the possibility of oscillation in which the feedback becomes positive. To eliminate the effect, capacitors are added to the op amp circuitry either internally or externally to deliberately produce a very low upper half-power frequency. These capacitors are said to provide internal or external compensation, respectively.

■ The gain–bandwidth product of an op amp with feedback is identical to the gain–bandwidth product without feedback. Thus, a reduction in gain due to the feedback results in an increase in bandwidth by the same factor.

CHAPTER 8 PROBLEMS

8.1
The current amplification for a BJT amplifier circuit is obtained as follows:

$$\frac{\mathbf{I}_o}{\mathbf{I}_s} = 25\left[\frac{j\omega/10}{(1 + j\omega/20)(1 + j\omega/50)}\right]$$

Sketch the amplitude and phase Bode plots.

8.2
Sketch the amplitude and phase Bode plots for the voltage amplification expression given by

$$\frac{\mathbf{V}_o}{\mathbf{V}_s} = 100\left[\frac{j\omega/100}{(1 + j\omega/100)(1 + j\omega/10^5)}\right]$$

8.3
Sketch the amplitude Bode plot for the voltage amplification expression given by

$$\frac{\mathbf{V}_o}{\mathbf{V}_s} = 10^5\left[\frac{(1 + j\omega/1)(1 + j\omega/10)}{(1 + j\omega/5)(1 + j\omega/20)(1 + j\omega/10^4)}\right]$$

8.4
Consider two identical amplifiers of the type in Problem 8.3 connected in cascade so that the overall voltage amplification is $|\mathbf{V}_o/\mathbf{V}_s|^2$. Sketch the resulting Bode plot for the amplitude of overall voltage amplification.

8.5
For the measured amplitude Bode plot of Fig. P8.5, determine the voltage amplification expression.

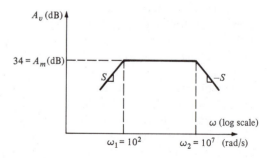

FIGURE P8.5

8.6
Repeat Problem 8.5 for the current amplification amplitude Bode plot of Fig. P8.6.

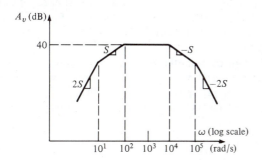

FIGURE P8.6

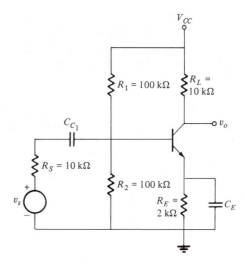

FIGURE P8.8

8.7
Analyze the circuit of Fig. P8.7 at low frequency where the impedance of C_{C_1} must be considered. Let $h_{fe} = 100$ and $h_{ie} = 0.1$ kΩ and determine the voltage amplification V_o/V_s. Sketch the amplitude and phase Bode plots. Let $C_{C_1} = 1$ μF.

V_o/V_s. Use $C_E = 200$ μF and sketch the amplitude and phase Bode plots.

8.9
Determine the voltage amplification at low frequency for the circuit of Fig. P8.9. Assume that C_E acts like a short circuit but that C_{C_1} and C_{C_2} do not ($C_{C_1} = C_{C_2} = 1$ μF). Let $h_{fe} = 100$ and $h_{ie} = 5$ kΩ and determine the voltage amplification V_o/V_s. Sketch the amplitude Bode plot.

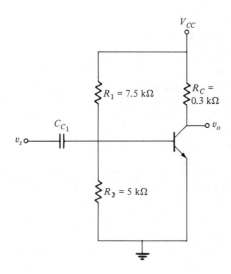

FIGURE P8.7

8.8
Analyze the circuit of Fig. P8.8 at low frequency where the impedance of C_E must be considered but C_{C_1} acts like a short circuit. Let $h_{fe} = 100$ and $h_{ie} = 0.5$ kΩ and determine the voltage amplification

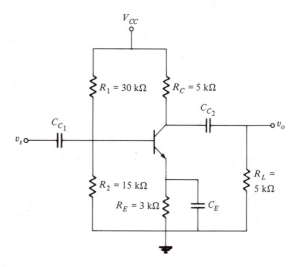

FIGURE P8.9

8.10

Suppose that C_E, in the circuit of Fig. P8.9, had inadvertently been left out. Determine the medium-frequency voltage amplification under these conditions. Compare the result with Problem 8.9.

8.11

Determine the current amplification $\mathbf{I}_o/\mathbf{I}_s$ at low frequency for the circuit of Fig. P8.11. Sketch the amplitude Bode plot. Let $h_{fe} = 100$ and $h_{ie} = 1\ \text{k}\Omega$.

8.12

Figure P8.12 displays a common-collector amplifier circuit. Determine the current amplification $\mathbf{I}_o/\mathbf{I}_s$ at low frequency taking into account the effect of C_{C_1}. Let $h_{fe} = 100$ and $h_{ie} = 2\ \text{k}\Omega$ with $C_{C_1} = 5.6\ \mu\text{F}$. Sketch the amplitude Bode plot.

8.13

For the common-base amplifier circuit of Fig. P8.13, determine the voltage amplification at low frequency accounting for C_{C_1} and C_{C_2} with C_B acting like a short circuit. Use $h_{fe} \gg 1$ and $h_{ib} = 5\ \Omega$, with $C_{C_1} = C_{C_2} = 0.5\ \mu\text{F}$.

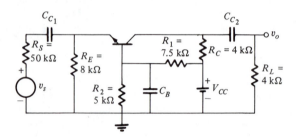

FIGURE P8.13

8.14

Repeat Problem 8.13 accounting for C_B (0.33 μF) and treating C_{C_1} and C_{C_2} like short circuits. Use $\alpha = 0.99$.

8.15

The common-collector amplifier of Fig. P8.12 has R_E replaced with the combination of elements shown in Fig. P8.15. Determine the current amplification $\mathbf{I}_o/\mathbf{I}_s$ accounting for the effects of capacitor C_{C_2} only. Let $h_{fe} = 100$, $h_{ie} = 2\ \text{k}\Omega$, and $C_{C_2} = 1\ \mu\text{F}$.

8.16

For the JFET amplifier of Fig. P8.16, determine the voltage amplification $\mathbf{V}_o/\mathbf{V}_s$ at low frequency. Assume

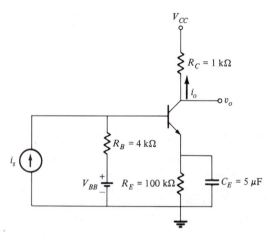

FIGURE P8.11

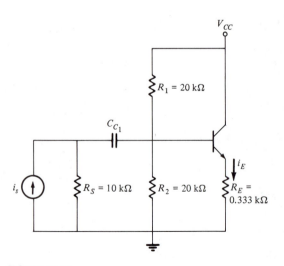

FIGURE P8.12

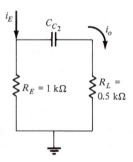

FIGURE P8.15

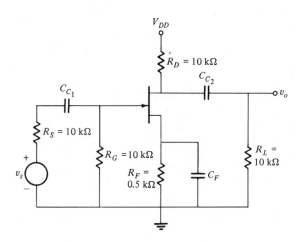

FIGURE P8.16

that C_F is a short circuit but that C_{C_1} and C_{C_2} are not ($C_{C_1} = C_{C_2} = 0.1\ \mu F$). Use $g_m = 8$ mS and $g_o = 0$. Sketch the magnitude Bode plot.

8.17
Repeat Probem 8.16 accounting only for $C_F = 2\ \mu F$ (treating C_{C_1} and C_{C_2} like short circuits).

8.18
For the E-D MOST amplifier of Fig. P8.18, determine the voltage amplification V_o/V_s at low frequency. Assume that C_F is a short circuit but that C_{C_1} and C_{C_2} are not ($C_{C_1} = C_{C_2} = 3.5\ \mu F$). Use $g_m = 0.01/\Omega$ and $g_o = 0$. Sketch the magnitude Bode plot.

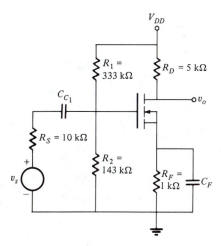

FIGURE P8.20

8.19
Repeat 8.18 accounting only for $C_F = 2\mu F$ and treating C_{C_1} and C_{C_2} like short circuits.

8.20
For the E-O MOST amplifier of Fig. P8.20, determine the voltage amplification V_o/V_s at low frequency accounting only for C_{C_1} ($C_{C_1} = 1\ \mu F$). Treat C_F as a short circuit, and use $g_m = 2.7$ mS and $1/g_o = 20\ k\Omega$.

8.21
Repeat Problem 8.20 to determine the effect of $C_F = 2\ \mu F$. Assume that C_{C_1} is negligible. For simplicity, let $g_o = 0$.

8.22
Figure P8.22 displays a BJT high-frequency small-signal equivalent circuit. Determine the current amplication I_o/I_s and sketch the amplitude and phase

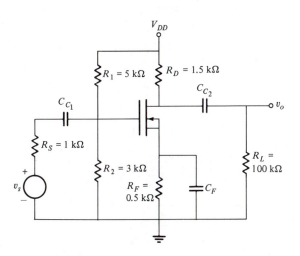

FIGURE P8.18

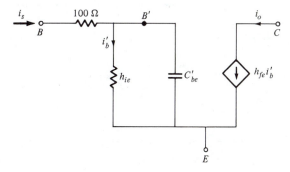

FIGURE P8.22

Bode plots. Use $h_{fe} = 100$, $h_{ie} = 2.5$ kΩ, and $C'_{be} = 4$ pF.

8.23

For the amplifier circuit of Fig. P8.7, determine the voltage amplification $\mathbf{V}_o/\mathbf{V}_s$ at high frequency. Use the high-frequency equivalent circuit with $r_b = 100$ Ω, $C'_m = 100$ pF, $h_{fe} = 100$, and $h_{ie} = 0.1$ kΩ. Sketch amplitude and phase Bode plots in the high-frequency range.

8.24

Combine the results of Problems 8.7 and 8.23 to sketch the amplitude Bode plot for all frequencies. What causes the discrepancy of medium-frequency amplification?

8.25

For the amplifier circuit of Fig. P8.8, determine the voltage amplification $\mathbf{V}_o/\mathbf{V}_s$ at high frequency. Use the high-frequency equivalent circuit with $r_b = 100$ Ω, $C'_m = 100$ pF, $h_{fe} = 100$, and $h_{ie} = 0.5$ kΩ. Sketch the amplitude and phase Bode plots.

8.26

For the amplifier circuit of Fig. P8.9, determine the voltage amplification $\mathbf{V}_o/\mathbf{V}_s$ at high frequency. Use the high-frequency equivalent circuit with $r_b = 100$ Ω, $C'_m = 100$ pF, $h_{fe} = 100$, and $h_{ie} = 0.5$ kΩ. Sketch the amplitude and phase Bode plots.

8.27

For the common-collector amplifier circuit of Fig. P8.12, determine the current amplification $\mathbf{I}_o/\mathbf{I}_s$ at high frequency with $h_{fe} = 100$, $h_{ie} = 2$ kΩ, $r_b = 0$ Ω, and $C'_m = 100$ pF.

8.28

Combine the results of Problems 8.12 and 8.27 to sketch the amplitude Bode plot for all frequencies.

8.29

For the common-base amplifier circuit of Fig. P8.13, determine the voltage amplification at high fre-

quency. Use the common-base high-frequency small-signal model for the BJT as displayed in Fig. P8.29. Let $h_{ib} = 5$ Ω, $h_{fe} \gg 1$, and $C'_m = 100$ pF. Sketch the amplitude and phase Bode plots.

8.30

Repeat Problem 8.29 for $R_S = 1$ kΩ.

8.31

Combine the results of Problems 8.13 and 8.29 to sketch the amplitude Bode plot for all frequencies.

8.32

Determine the voltage amplification $\mathbf{V}_o/\mathbf{V}_s$ for the circuit of Fig. P8.32. Sketch the amplitude and phase Bode plots. Hint: Write one nodal equation for the output terminal. Use $g_m = 50$ mS.

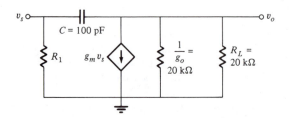

FIGURE P8.32

8.33

For the circuit of Fig. P8.33, determine the current amplification at high frequency. Use the high-frequency small-signal circuit model with $h_{fe} = 100$, $r_b = 0$, $h_{ie} = 0.5$ kΩ, $C'_{bc} = 2$ pF, and $C'_{be} = 40$ pF. Sketch the amplitude Bode plot.

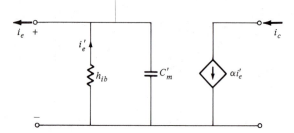

FIGURE P8.29

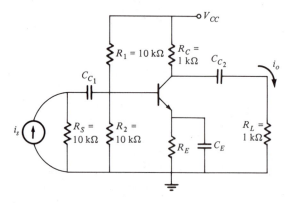

FIGURE P8.33

8.34

For the same circuit of Problem 8.33, carry out a low-frequency analysis with $C_{C_2} = 20$ μF and treat C_{C_1} as well as C_E like short circuits. Sketch the amplitude Bode plot and include the high-frequency results from Problem 8.33.

8.35

Determine the voltage amplification for the JFET amplifier of Fig. P8.16 at high frequency. Use the parameters of Problem 8.16 and $C'_m = 200$ pF. Sketch the magnitude Bode plot for high frequency and add the low-frequency results from Problem 8.16.

8.36

Repeat Problem 8.35 for the E-D MOST amplifier of Fig. P8.18. Use the parameters of Problem 8.18 (as well as the low-frequency results) and $C_m = 200$ pF.

8.37

Repeat Problem 8.35 for the E-O MOST amplifier of Fig. P8.20. Use the parameters of Problem 8.20 (as well as the low-frequency results) and $C'_m = 200$ pF.

8.38

For the multistage amplifier of Fig. P6.1, determine the low-frequency voltage amplification V_o/V_s accounting only for coupling capacitors $C_{C1} = C_{C2} = 1$ μF. Use a small-signal equivalent circuit along with the parameters of Problem 6.1. Sketch the amplitude Bode plot.

8.39

Repeat Problem 8.38 for high frequency and assume that $C'_m = 100$ pF and that $r_b = 10$ Ω for each BJT. What change results for $r_b = 0$?

8.40

For the multistage JFET amplifier of Fig. P6.8, determine the low-frequency voltage amplification expression for V_o/V_s accounting only for the bypass capacitors $C_F = 1$ μF. Use a small-signal equivalent circuit and the parameters of Problem 6.8 with $R_S = 0$ and $R_F = 1$ kΩ.

8.41

Repeat Problem 8.40 for high frequency and assume that $C'_m = 200$ pF for each JFET. What change results for $R_s \neq 0$?

8.42

Determine the gain–bandwidth product for an op amp with a magnitude Bode plot as shown in Fig. P8.42. What is the unity gain–bandwidth for this case?

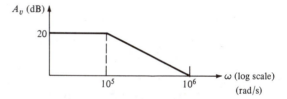

FIGURE P8.42

8.43

Consider an op amp with $\omega_{unity} = 10^7$ rad/s. For $|A_m|_{dB} = 40$ dB, determine the bandwidth. Sketch the magnitude Bode plot.

FUNDAMENTAL DIGITAL LOGIC OPERATIONS, GATES, AND DEFINITIONS

The first eight chapters of this text have been concerned primarily with amplifier analysis and design along with analog circuit applications. This chapter introduces the basic logic elements that are used in digital circuit applications. These applications employ discrete signal levels rather than continuous signals.

The five basic logic operations are defined as follows: NOT, AND, OR, NAND, and NOR. When each of these operations is individually carried out by an electronic circuit, the circuit that performs the particular logic operation is called a *gate*. Logic gates are simply an electronic implementation of designated logic functions. The logic gates for the five basic logic operations are referred to as *combinational* when their output depends upon the present value of the input. Logic gates are of the *sequential* type when their output depends upon the past value of the input, as well as the present value of the inputs.

After the basic logic operations and their sample gates are described, we will discuss the limitations of simple BJT gates. Basic definitions of important logic circuit parameters that define the behavior of the gate are fan-in, fan-out, logic swing, noise margin, switching speed, and power–delay product. These parameters are defined and described in this chapter and are used in comparing the logic families in the next two chapters. This chapter concludes by describing monostable and astable multivibrator circuits made up of logic gates. These circuits are basic to the production of a precisely timed train of pulses and serve as practical examples of logic circuit applications.

9.1   COMBINATIONAL LOGIC OPERATIONS AND GATES

The voltages or currents in digital circuits have only two possible states, which suggests that these variables are *binary*. We will consider the voltages to be the variables. The two possible states then correspond to a high voltage or a low voltage. Further, we will let these two states be represented by the binary system 1 and 0, with the high voltage corresponding to 1 and the low voltage corresponding to 0. This representation of states is referred to as a *positive logic* binary system of voltage variables and is entirely adequate for our purposes, although other choices of logic are also possible.

The definitions and conventional symbols of combinational (and sequential) logic operations are quite straightforward. Since the operation is binary, the method of describing each of the basic operations is indicated by what is called a *truth table*, or *function table*. This tabular description gives all possible combinations of inputs and their corresponding outputs. Quite often, truth tables use a 1 or a 0 to describe each state, but at times we will also use the high and low voltage levels (V_{HI} and V_{LO}, respectively). These tables are also called function tables because they indicate the logic function performed.

9.1.1   NOT Operation and Gate

The first and most important combinational logic function to be defined is the NOT operation, or logic inverse operation. This logic function is best introduced by considering the BJT common-emitter amplifier circuit shown in Fig. 9.1a. However, the input voltage in this digital logic network is not amplified. The voltage levels are either high or low (1 or 0), and the input voltage level A determines the state of the output voltage level B.

The circuit of Fig. 9.1a is called a NOT gate because it performs the operation of *logic inversion*. That is, if the input voltage is low, then very little base current will flow, the transistor will be essentially cut off, and the output voltage will be high. If the input voltage is high, however, then a large base current will flow, the transistor (by bias design) will be forced into saturation, and the output voltage will be low. The output voltage is therefore the logical inversion of the input voltage.

- **NOT Operation:** B is the inverse of A, or B is NOT A. Stated mathematically,

$$B = \bar{A} \qquad\qquad (9.1\text{–}1) \quad \blacksquare$$

The corresponding truth table in terms of the binary numbers 1 and 0 is shown in Fig. 9.1b. In

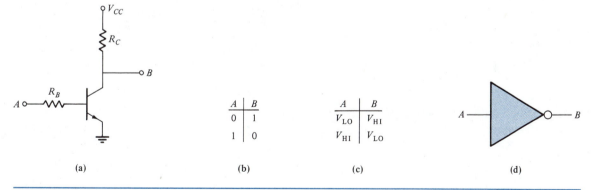

A	B
0	1
1	0

(b)

A	B
V_{LO}	V_{HI}
V_{HI}	V_{LO}

(c)

(a)   (d)

FIGURE 9.1   The BJT Logic Inverter NOT Gate: (a) Inverter circuit NOT gate, (b) 0, 1 truth table, (c) Voltage-level truth table, (d) NOT gate circuit symbol

this case, all possible combinations of inputs are $A = 0, 1$. The truth table in terms of voltage levels (for positive logic) is shown in Fig. 9.1c.

The conventional symbol for the NOT gate is shown in Fig. 9.1d. Note that there is only one input (A) and one output (B). The small circle at the output is called an *inverting circle* and indicates the NOT, or inverse, operation.

_____ EX. 9.1

BJT Logic Inverter Example

In the logic inverter circuit of Fig. 9.1, $R_B = R_C = 2$ kΩ and $V_{CC} = 5$ V. Determine the voltage at B for inputs at A given by 0 and 5 V. Let $\beta = 100$, $V_0 = 0.7$ V, and $v_{CE\,sat} = 0.2$ V.

Solution: When the input voltage at $A = 0$ (low), the base current $i_B = 0$, and the transistor is cut off. The output voltage at B is then V_{CC} (high) since there is no current through R_C.

When the input voltage at $A = 5$ V (high), then the base current is given by

$$i_B = \frac{5 - 0.7}{R_B} = \frac{4.3}{2} = 2.15 \text{ mA} \tag{1}$$

The corresponding collector current would be $i_C = \beta i_B = 216$ mA except that the maximum collector current for this case is $(V_{CC} - v_{CE\,sat})/R_C = 2.4$ mA. Thus, the transistor is in saturation, and

$$v_{CE} = v_{CE\,sat} = 0.2 \text{ V} \tag{2}$$

while

$$i_C = i_{C\,sat} = \frac{5 - 0.2}{2} = 2.4 \text{ mA} \tag{3}$$

_____

9.1.2   AND Operation and Gate

A BJT logic AND gate is shown in Fig. 9.2a. This gate has two BJTs with two inputs (A and B) and one output (C). It is, however, entirely possible to have more than two inputs by adding more BJTs.

This gate is designed such that the BJTs will operate either in cutoff or in saturation. When no input (or a low voltage) exists for either BJT, each BJT will be cut off; when the inputs are each at

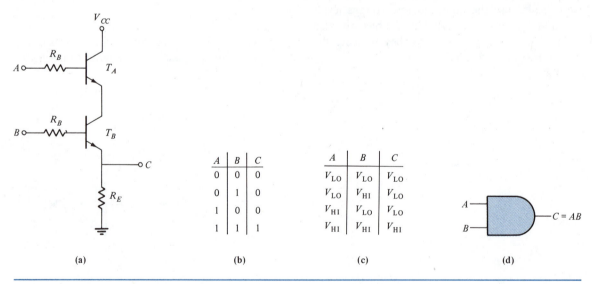

A	B	C
0	0	0
0	1	0
1	0	0
1	1	1

A	B	C
V_{LO}	V_{LO}	V_{LO}
V_{LO}	V_{HI}	V_{LO}
V_{HI}	V_{LO}	V_{LO}
V_{HI}	V_{HI}	V_{HI}

$C = AB$

(a)   (b)   (c)   (d)

FIGURE 9.2   BJT Logic AND Gate: (a) BJT AND gate, (b) 0, 1 truth table, (c) Voltage-level truth table, (d) AND gate circuit symbol (two inputs)

a high voltage, the BJTs will be in saturation. From the circuit of Fig. 9.2a, we see that the output voltage level will be low (0) unless both A and B are high because with no input the BJT behaves like a huge resistor between collector and emitter. For high inputs at A and B, T_A and T_B are in saturation, and the voltage at C is essentially V_{CC}.

- **AND Operation:** If A AND B are high (1), then C is high (1); otherwise, C is low (0). Stated mathematically,

$$C = A \cdot B = AB \qquad (9.1\text{--}2) \quad \blacksquare$$

The standard multiplication dot here indicates the logical AND operation. This dot is often omitted, however, as indicated in (9.1–2), but the meaning is still retained.

The corresponding truth table for this case in terms of the binary numbers 1 and 0 is displayed in Fig. 9.2b. Note that the two inputs have four different combinations. The number of possible input combinations is multiplied by 2 for each additional input so that for N inputs there are 2^N different combinations. Figure 9.2c shows the truth table in turns of high and low values of voltage.

The circuit symbol for the AND gate is shown in Fig. 9.2d. This symbol indicates a two-input AND gate, but additional inputs are possible.

———————————————————————————— EX. 9.2

BJT AND Gate Example

In the AND gate circuit of Fig. 9.2a, $R_B = R_E = 2$ kΩ and $V_{CC} = 5$ V. Verify the first three rows of the truth table of Fig. 9.2c by letting $\beta = 100$, $V_0 = 0.7$ V, and $V_{LO} = 0$. Then, verify the last row of this truth table by letting the input high voltages $= 6$ V and $v_{CE\,sat} = 0.2$ V. A larger input voltage is necessary in order to drive the BJTs into saturation.

Solution: For the cases in which an input at A or B (or both) is low ($V_{LO} = 0$), the base current and the collector

current are zero. Hence, the output at $C = 0$, or V_{LO}, and the first three rows of Fig. 9.2c are verified.

When the inputs at A and $B = 6$ V, then the base currents are given by

$$i_{B_A} = \frac{6 - (V_{CC} - 0.2 + 0.7)}{R_B} = \frac{6 - 5.5}{R_B} = \frac{0.5}{2}$$
$$= 0.25 \text{ mA}$$

and

$$i_{B_B} = \frac{6 - (V_{CC} - 0.4 + 0.7)}{R_B} = \frac{0.7}{2} = 0.35 \text{ mA}$$

Since the maximum collector current is $V_{CC}/R_E = 5/2 = 2.5$ mA, each of these base currents is sufficient to drive the BJTs into saturation because $\beta = 100$. Therefore, the output voltage at C is

$$V_C = V_{CC} - 2v_{CE\,sat} = 5 - 2(0.2) = 4.6 \text{ V}$$

which is a high voltage. Note that the input voltages must be higher than V_{CC}, while the output voltage is less than V_{CC}. This condition is referred to as *voltage degradation.*

————————————————————————————

——————
9.1.3   OR Operation and Gate
——————

Figure 9.3a shows a BJT logic OR gate. Like the AND gate, the OR gate displayed has two BJTs with two inputs and one output. More inputs are also possible in this case.

Like the AND gate, the OR gate of Fig. 9.3a is designed so that each BJT is in saturation when its input voltage is high and so that operation in cutoff prevails for a low input voltage.

- **OR Operation:** If either A OR B is high (1), then C is high (1); otherwise, C is low (0). Stated mathematically,

$$C = A + B \qquad (9.1\text{--}3) \quad \blacksquare$$

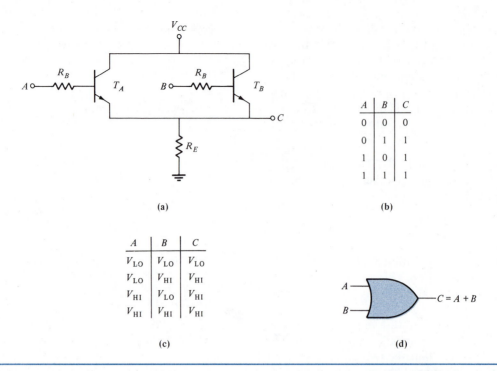

FIGURE 9.3 BJT Logic OR Gate: (a) BJT OR gate, (b) 0, 1 truth table, (c) Voltage-level truth table, (d) OR gate circuit symbol (two inputs)

The standard plus sign here represents the logical OR operation.

The two truth tables corresponding to the OR operation are shown in Figs. 9.3b and c. The corresponding circuit symbol for a two-input OR gate, which can also have more inputs, is shown in Fig. 9.3d.

EX. 9.3

BJT OR Gate Example

In the circuit of Fig. 9.3a, let $V_{CC} = 5$ V, $R_B = 2$ kΩ, and $R_E = 2$ kΩ. Verify the truth table of Fig. 9.3c by letting the input low voltages $= 0$ V and the input high voltages $= 6$ V. Let $v_{CE\,sat} = 0.2$ V, $\beta = 100$, and $V_0 = 0.7$ V.

Solution: The first row of the truth table is verified by realizing that with low (0) inputs the BJTs are cut off and no current passes through R_E. Hence, the output voltage $= 0$ V.

However, in each of the next three rows of the truth table, one or both of the inputs are high. For a high input of 6 V, the base current of the BJT is given by

$$i_B = \frac{6 - 0.7 - (V_{CC} - 0.2)}{R_B} = \frac{0.5}{2} = 0.25 \text{ mA}$$

Note that this current is sufficient to drive the transistor into saturation because $\beta = 100$. Hence, the output voltage is $V_{CC} - 0.2 = 4.8$ V, which corresponds to the output high voltage.

9.1.4 NAND Operation and Gate

The NAND operation is just the inverse of the AND operation. Figure 9.4a displays the

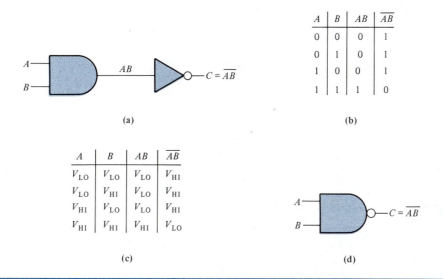

A	B	AB	$\overline{AB}$
0	0	0	1
0	1	0	1
1	0	0	1
1	1	1	0

(a)   (b)

A	B	AB	$\overline{AB}$
V_{LO}	V_{LO}	V_{LO}	V_{HI}
V_{LO}	V_{HI}	V_{LO}	V_{HI}
V_{HI}	V_{LO}	V_{LO}	V_{HI}
V_{HI}	V_{HI}	V_{HI}	V_{LO}

(c)   (d)

FIGURE 9.4   Logic NAND Gate: (a) Equivalent NAND Gate, (b) 0, 1 truth table, (c) Voltage-level truth table, (d) NAND gate circuit symbol (two inputs)

equivalent circuit that represents the NAND operation with two of the logic circuit symbols just defined. A BJT circuit using the two-input AND gate of Fig. 9.2a with the inverter of Fig. 9.1a connected to its output also performs this function.

- **NAND Operation:**   The result of A AND B is inverted (NOT A AND B). Stated mathematically,

$$C = \overline{AB} \qquad\qquad \textbf{(9.1–4)} \;\blacksquare$$

The truth tables for this two-input NAND gate are given in Figs. 9.4b and c. Also included in these tables is a column for the AND operation. This column clearly indicates that the NAND operation is the inverse of the AND operation.

The conventional symbol for a NAND gate is shown in Fig. 9.4d. Note that this symbol is just the AND gate symbol with an inverting circle added at the output of the symbol.

9.1.5   NOR Operation and Gate

The description of the five basic combinational logic gates is completed here with the definition of the NOR gate. Figure 9.5a shows an equivalent circuit that performs the NOR operation by using an OR gate and an inverter. This operation may also be implemented by connecting the previous BJT circuits for the OR and NOT operations in the manner indicated in Fig. 9.5a.

- **NOR Operation:**   The NOR operation is the inverse of the OR operation. Stated mathematically,

$$C = \overline{A + B} \qquad\qquad \textbf{(9.1–5)} \;\blacksquare$$

The truth tables and circuit symbol for this case are displayed in Figs. 9.5b, c, and d. Note that the symbol consists of the OR gate symbol with an inverting circle at the output.

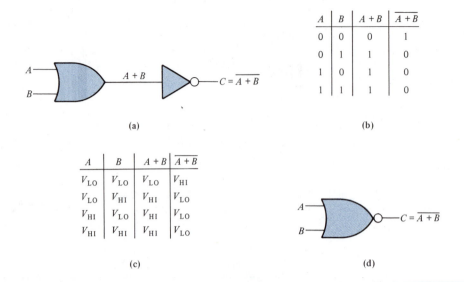

A	B	A + B	$\overline{A+B}$
0	0	0	1
0	1	1	0
1	0	1	0
1	1	1	0

(a)

(b)

A	B	A + B	$\overline{A+B}$
V_{LO}	V_{LO}	V_{LO}	V_{HI}
V_{LO}	V_{HI}	V_{HI}	V_{LO}
V_{HI}	V_{LO}	V_{HI}	V_{LO}
V_{HI}	V_{HI}	V_{HI}	V_{LO}

(c)

(d)

FIGURE 9.5   Logic NOR Gate: (a) Equivalent NOR gate, (b) 0, 1 truth table, (c) Voltage-level truth table, (d) NOR gate circuit symbol (two inputs)

9.2   SEQUENTIAL LOGIC OPERATIONS AND GATES

Sequential logic gates have output voltages that depend upon past and present inputs. That is, input voltages can be applied, and the output voltage is fixed according to these applied voltages. The output then remains unchanged when the inputs are removed. This quality is the essential characteristic of a memory, or sequential, element. Memory elements that are configured using basic combinational logic gates are called *flip-flops*. Since the output can have two stable states, flip-flops are also called *bistable memory elements*.

9.2.1   *RS* Flip-Flop

Figure 9.6a displays an implementation of a memory element that uses two NOR gates (*G*1 and *G*2). Each NOR gate output is fed back to the other NOR gate input. A similar arrange-

ment can also be implemented with NAND gates. Circuits such as these are primary examples of bistable memory elements. The applied inputs in Fig. 9.6a are *R* (for *reset*) and *S* (for *set*). Hence, this bistable multivibrator is known as an *RS flip-flop* (*RSFF*). The *RS* flip-flop is also sometimes called a *latch*.

Note that the outputs of the NOR gates (*Q* and *Q̄*, as shown in Fig. 9.6a) are defined a priori to be the inverses of each other, which is true only under certain conditions, as we will see. For the inputs shown in Fig. 9.6a, the outputs of *G*1 and *G*2 are given, respectively, by

$$Q = \overline{R + \bar{Q}} \tag{9.2-1}$$

and

$$\bar{Q} = \overline{S + Q} \tag{9.2-2}$$

To see the implications of these expressions, we will investigate the results of applying all possible combinations of inputs *R* and *S*. Each combina-

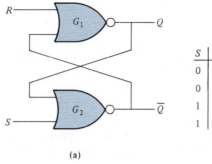

S	R	Q	$\bar{Q}$	State
0	0	Q	$\bar{Q}$	Unchanged
0	1	0	1	Reset
1	0	1	0	Set
1	1	0	0	Not used

(a)   (b)

FIGURE 9.6   *RS* Flip-Flop, or Latch: (a) Arrangement using two NOR gates, (b) Truth table

tion will result in a particular state or condition for the output of the *RSFF*. The truth table for the *RSFF* of Fig. 9.6a is shown in Fig. 9.6b and indicates all possible combinations of inputs with their corresponding outputs.

Reset State.   We define this state as $R = 1$ and $S = 0$. From (9.2–1), we obtain

$$Q = \overline{1 + \bar{Q}} = \bar{1} = 0$$

which implies that Q must either change to 0 or remain 0 depending upon its initial value (the value before $R = 1$ and $S = 0$). Thus, Q is fed back as a 0 to the input of G_2 in Fig. 9.6a, which, from (9.2–2), results in

$$\bar{Q} = \overline{0 + 0} = \bar{0} = 1$$

Thus, $\bar{Q}$ must change or remain the same, but it must be a 1 after these inputs are applied. In summary, then, we see that the results of applying inputs $R = 1$ and $S = 0$ are outputs $Q = 0$ and $\bar{Q} = 1$. This condition is called the *reset state*.

Set State.   We define this state as $R = 0$ and $S = 1$. From (9.2–2), $\bar{Q}$ is given by

$$\bar{Q} = \overline{1 + Q} = \bar{1} = 0$$

which means that $\bar{Q}$ is set at 0 regardless of its prior state. Thus, with $Q = 0$ fed back to the input

of G_2 in Fig. 9.6a, we obtain

$$Q = \overline{0 + 0} = \bar{0} = 1$$

In summary, we see that the results of inputs $R = 0$ and $S = 1$ are outputs $Q = 1$ and $\bar{Q} = 0$. This condition is called the *set state*.

Next, we consider the other two possible input states (which are not given specific names).

$R = S = 0$.   For this state, no input voltages are applied to the input terminals of the *RSFF*. From (9.2–1) and (9.2–2), we have (for $R = S = 0$)

$$Q = \overline{0 + \bar{Q}} = \bar{\bar{Q}} = Q$$

and

$$\bar{Q} = \overline{0 + Q} = \bar{Q}$$

Note that Q and $\bar{Q}$ are unchanged when $R = S = 0$. The importance of this state is that the set or reset condition can be applied to the *RSFF*, but the inputs can then be removed ($R = S = 0$) with no change to the output as a result. The gate "remembers" what was applied in the past and remains in that state.

$R = S = 1$.   For this state, (9.2–1) and (9.2–2) lead to the same result. That is, from (9.2–1),

$$Q = \overline{1 + \bar{Q}} = \bar{1} = 0$$

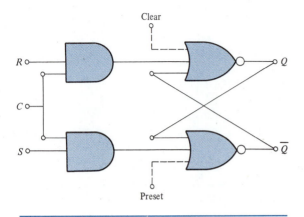

FIGURE 9.7   Gated, or Clocked, *RS* Flip-Flop

and from (9.2–2),

$$\bar{Q} = \overline{1 + Q} = \bar{1} = 0$$

Since these results are equal and since they are supposed to be inverses, we see that they are inconsistent with the original definition for outputs Q and $\bar{Q}$. This inconsistency is eliminated by never using or allowing the condition $R = S = 1$.

An *RSFF* with the additional input of C, incorporated by using two AND gates, is shown in Fig. 9.7. This additional terminal is called the *clock* terminal. When an input is present at this terminal ($C = 1$), inputs at R and S will affect outputs Q

and $\bar{Q}$. However, if $C = 0$, Q and $\bar{Q}$ will not be affected by any inputs at R and S. The reason for using such a setup is to allow synchronization of all the gates in an overall digital system. The voltage associated with the clock is actually a precise string of pulses alternating periodically between 1 and 0. With this method of timing, the output of each gate is independent of the precise arrival times of the voltages at R and S, which is quite important since gates have delay times associated with them that can result in erroneous voltage levels at certain instants and hence erroneous logic.

Two more terminals are shown in Fig. 9.7. Represented with dashed lines, they are labeled *clear* and *preset*. These terminals are often incorporated into the clocked *RSFF* (*CRSFF*) to force $Q = 0$ when clear $= 1$ or $\bar{Q} = 0$ when preset $= 1$.

Block diagrams that denote the *RSFF*, the *CRSFF*, and the *CRSFF* with preset and clear terminals are shown in Fig. 9.8. These diagrams are very convenient for the representation of flip-flops and allow generality.

————————————————————— **EX. 9.4**

RS Flip-Flop Example

The gated *RSFF* of Fig. 9.7 (or Fig. 9.8c) is to be analyzed. After the clear terminal is set equal to 1, a square wave is applied to the *S* terminal and its inverse to the *R* terminal, as shown in Fig. E9.4 along with the clock pulse train. De-

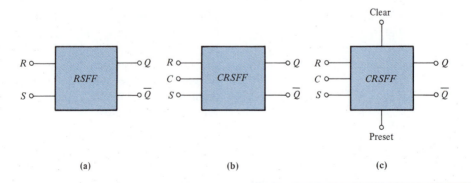

(a)   (b)   (c)

FIGURE 9.8   Block Diagrams of *RS* Flip-Flops: (a) *RSFF*, (b) Clocked *RSFF*, (c) Clocked *RSFF* with clear and preset terminals

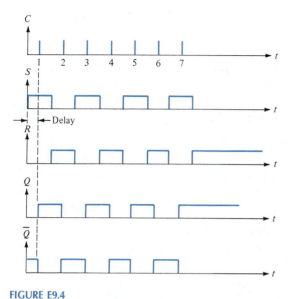

FIGURE E9.4

9.2.2   *JK* Flip-Flop

One of the disadvantages of the *RSFF* is that the state $R = S = 1$ is forbidden. Several techniques can be employed to eliminate this undesirable feature, however. One method is to use the *JK flip-flop* (*JKFF*) shown in Fig. 9.9a. This circuit consists of two three-input AND gates and an *RSFF*. The outputs of the *RSFF* are fed back to the additional terminal of each AND gate. The new input terminals are labeled *J* and *K* (instead of *R* and *S*). $J = 1$, $K = 0$, and $C = 1$ sets the *RSFF*, while $J = 0$, $K = 1$, and $C = 1$ resets it. The truth table for this *JKFF* is given in Fig. 9.9b, and a block diagram of the *JKFF* is shown in Fig. 9.9c. To verify the truth table, we will analyze each input case by considering that $C = 1$. If $C = 0$, then each input AND gate will be disabled (disallowing the effect of any other inputs), and $R = S = 0$, which is one condition of no change in the output of the *RSFF*.

termine the waveshapes for Q and $\bar{Q}$. This *RSFF* for which $S = \bar{R}$ is sometimes referred to as a *delay* flip-flop because the waveshapes for Q and $\bar{Q}$ are delayed in time.

Solution: Immediately after clear = 1, $Q = 0$ and $\bar{Q} = 1$. Then, after clock pulse 1 (see Fig. E9.4), $S = 1$ and $R = 0$, which forces $Q = 1$ and $\bar{Q} = 0$. At the instant of clock pulse 2, $S = 0$ and $R = 1$, which forces $Q = 0$ and $\bar{Q} = 1$. The same changes are then repeated after the succeeding clock pulses. Figure E9.4 shows the resulting square waves for Q and $\bar{Q}$. Note that the waves are delayed in time with respect to the inputs at S and R.

$J = 0, K = 0, C = 1$.   This is the other condition of no change in the output of the flip-flop. For this condition, both AND gates (G_1 and G_2 in Fig. 9.9a) are disabled, which forces their outputs to be 0. Thus, $R = S = 0$, and Q and $\bar{Q}$ are unchanged.

$J = 0, K = 1, C = 1$.   Since $J = 0$, G_1 is disabled, which forces $S = 0$. Then, since $K = 1$, $R = Q$. The value of R depends upon the value of the

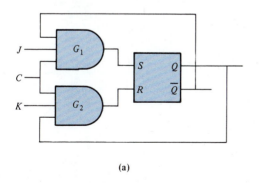

J	K	Q State
0	0	Unchanged
0	1	0; Reset
1	0	1; Set
1	1	Changed

| (a) | (b) | (c) |

FIGURE 9.9   *JK* Flip-Flop: (a) Arrangement using two AND gates and an *RSFF*, (b) Truth table ($C = 1$), (c) Block diagram

RSFF output. Suppose that $Q = 1$ initially. Then $R = 1$, the output will change to $Q = 0$ (since $S = 0$), and the reset condition results for the *RSFF*. However, if $Q = 0$ initially, then $R = 0$, and (with $S = 0$) there is no change in the output. Hence, in either case ($Q = 1$ or 0 initially), the output of the flip-flop becomes $Q = 0$ and $\bar{Q} = 1$. Therefore, the application of $J = 0$ and $K = 1$ results in the reset condition.

$J = 1$, $K = 0$, $C = 1$.   This case is just the opposite of the previous one. It results in the set condition for the flip-flop, or $Q = 1$ and $\bar{Q} = 0$.

$J = 1$, $K = 1$, $C = 1$.   For this case, the output of G_1 is $S = \bar{Q}$ and that of G_2 is $R = Q$. Thus, if $Q = 1$ ($\bar{Q} = 0$) initially, then $R = 1$ and $S = 0$, and the *RSFF* output is reset to $Q = 0$ ($\bar{Q} = 1$). However, if $Q = 0$ ($\bar{Q} = 1$) initially, then $S = 1$ and $R = 0$, and the *RSFF* is set to $Q = 1$ and $\bar{Q} = 0$. Hence, the state of the *JKFF* changes whenever $J = K = 1$.

An important application of the *JKFF* is *frequency division*, which is accomplished by applying a 1 to the J and K terminals and a periodic square wave or pulse train to the C terminal. Since J and K are 1, whenever C is 1, the output will change state. Thus, suppose that $Q = 0$ initially. Then, the first clock pulse will change Q to $Q = 1$, the second clock pulse will change Q to $Q = 0$,

and so on. Pulse trains for this case are shown in Fig. 9.10. Notice that the frequency of the output has been divided by 2. If the output of this *divide-by-two gate* is fed into another *JKFF* with $J = K = 1$ and the same square wave is applied to C, then the result will be a pulse train with frequency equal to 1/4 of that of the clock pulse train. This process can be repeated to generate a wide range of periodic pulse trains, and conveniently an input frequency can be divided to obtain the desired output frequency.

JK Flip-Flop Example

Two *JK* flip-flops that respond to downward high-to-low voltage transitions at the clock pulse terminal (negative edge-triggered) are connected together as shown in Fig. E9.5a. For the given square wave input, draw the waveshapes for Q_1 and Q_2. Assume that $J = K = 1$ and that the initial values of Q_1 and Q_2 are zero.

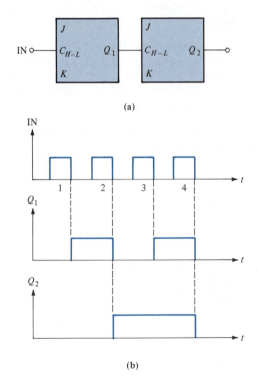

(a)

(b)

FIGURE E9.5

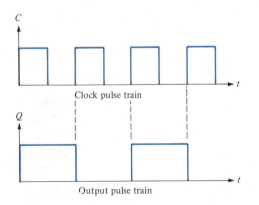

Clock pulse train

Output pulse train

FIGURE 9.10   Divide-by-Two Voltage Waveshapes

Solution: At the end of the first clock pulse, the downward transition causes Q_1 to change state and go high. However, there is no change in the state of the second flip-flop, since the flip-flops only respond to high-to-low changes at the clock pulse terminal, as specified. At the end of the second pulse, Q_1 again changes state. It changes from high to low, which forces the second flip-flop to change state. After the third clock pulse, Q_1 again goes high, while Q_2 is unchanged. Finally, after the fourth clock pulse, Q_1 goes high to low and forces Q_2 to also change state. The corresponding outputs for Q_1 and Q_2 are shown in Fig. E9.5b.

9.2.3   Edge-Triggered Master–Slave JK Flip-Flop

A difficulty can occur with the *JKFF* that is referred to as *toggling*. This effect is a switching back and forth (toggling) of the output voltage when $C = 1$ for a period of time that is longer than the propagation delay (amount of time for the flip-flop output to react to the input) of the flip-flop. That is, after the output switches, these new outputs become the inputs and, if the condition $C = 1$ is still present, the outputs will switch again. This toggling continues until C changes state. Thus, the clock signal must be in the form of a pulse with the pulse width less than the propagation delay time of the flip-flop. This restriction is not practical, however, because propagation delays of flip-flops are designed to be very small themselves ($\cong$ 10 ns). The difficulty is eliminated by using a *master–slave* flip-flop configuration, as shown in Fig. 9.11a.

To describe the operation of the master–slave *JKFF* of Fig. 9.11a, we begin by considering the clock transition from low to high (from $C = 0$ to

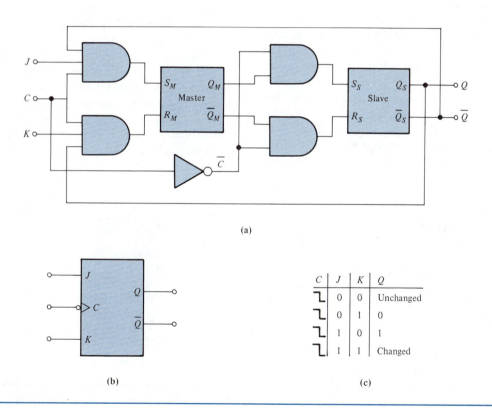

(a)

(b)

(c)

C	J	K	Q
⌐⌐	0	0	Unchanged
⌐⌐	0	1	0
⌐⌐	1	0	1
⌐⌐	1	1	Changed

FIGURE 9.11   Master–Slave *JK* Flip-Flop: (a) Negative edge-triggered configuration, (b) Block diagram, (c) Truth table

$C = 1$). This transition causes $\bar{C}$ to have the opposite transition, from high to low (1 to 0). Hence, the master $RSFF$ is enabled, and the slave $RSFF$ is disabled ($R_S = S_S = 0$) because any inputs at the J and K terminals will fix the outputs Q_M and $\bar{Q}_M$ of the master $RSFF$ through the input values of S_M and R_M but cannot affect those of the slave $RSFF$ since $R_S = S_S = 0$. Then, when C undergoes a transition back to 0, $\bar{C}$ becomes 1, which disables the master $RSFF$ and maintains the same outputs. However, $\bar{C} = 1$ immediately enables the slave $RSFF$, and its inputs from the master $RSFF$ cause the slave outputs to change to those of the master. Since the new slave outputs Q_S and $\bar{Q}_S$ are fed back to the input of the master $RSFF$, which is disabled, toggling does not occur. The only restriction on the length of the clock pulse is that it must be greater than the propagation delay time of the flip-flop. This restriction is easily accommodated.

Figure 9.11b displays the block diagram for the clocked master–slave $JKFF$. Since it is triggered on the high-to-low, or negative, edge of the clock pulse, the triangle and small inverting circle are included. The triangle is used to denote *edge-triggering*, while the inverting circle denotes the negative, or *trailing*, edge. The truth table is given in Fig. 9.11c. Flip-flops can also be configured to react to the *leading*, or low-to-high, edge of the clock pulse. For this type of flip-flop, the small inverting circle is omitted.

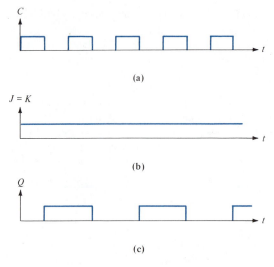

FIGURE E9.6

9.2.4 *D* and *T* Flip-Flops

The D, or *delay, flip-flop (DFF)* is displayed in Fig. 9.12a. Its corresponding truth table is given in Fig. 9.12b. An inverter is connected between the J and K terminals and forces these inputs to always be inverses. We observe that for $D = 0$, $J = 0$ and $K = 1$; hence, from the $JKFF$ truth table of Fig. 9.11c, $Q = 0$, which corresponds to the reset condition. Also, for $D = 1$, $J = 1$ and

---

EX. 9.6

Trailing Edge-Triggered *JK* Flip-Flop Example

An edge-triggered master–slave flip-flop, as shown in Fig. 9.11a, has a fixed high voltage applied to the J and K terminals (obtained by shorting these together) and a pulse train applied to C. The voltage wave shapes for these inputs are shown in Figs. E9.6a and b. Sketch the resulting wave train for Q.

Solution: At each downward transition of C, $J = K = 1$. Therefore, the output of the flip-flop changes. If we assume that $Q = 0$ initially, we obtain the resulting pulse train for Q as displayed in Fig. E9.6c. This type of flip-flop is referred to as a *toggle flip-flop*.

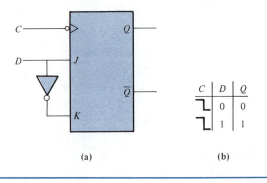

FIGURE 9.12 D Flip-Flop: (a) Block diagram, (b) Truth table

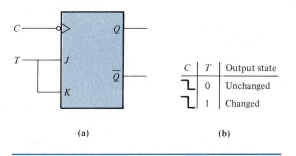

FIGURE 9.13   T Flip-Flop: (a) Block diagram, (b) Truth table

C	T	Output state
⌐_	0	Unchanged
⌐_	1	Changed

(a)   (b)

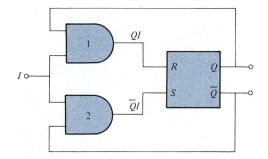

FIGURE E9.7

$K = 0$; thus, $Q = 1$, which corresponds to the set condition. The *DFF* truth table is thereby verified.

The *DFF* is used to provide a time delay. The input at D is read in and must remain present. It is transferred to the output after the desired time delay by the particular clock pulse.

The T, or *toggle, flip-flop* (*TFF*) changes state (toggles) with each clock pulse. A *JKFF* is easily converted into a *TFF* by connecting the J and K terminals directly together, as shown in Fig. 9.13a. The corresponding truth table is given in Fig. 9.13b and is verified by considering the truth table of the trailing edge-triggered *JKFF* of Fig. 9.11c.

The *TFF* can be used as a divide-by-two gate by maintaining a high voltage at T and applying a train of square wave pulses at C. The output will be a train of square wave pulses with half the frequency of the clock.

─────────────────────────────── **EX. 9.7**

T Flip-Flop Example

The memory element shown in Fig. E9.7 consists of two AND gates and an *RSFF*. Analyze the circuit by considering the *I* terminal to be 1 or 0, while the initial state of the *RSFF* is either $Q = 1$ or 0. What type of flip-flop is represented?

Solution: Since the input to each AND gate involves outputs from the *RSFF*, the final state of the *RSFF* depends on the initial state (Q_0 and $\bar{Q}_0$). Note that from the output of each AND gate, we have

$$R = QI \qquad (1)$$

and

$$S = \bar{Q}I \qquad (2)$$

Thus, for $Q_0 = 0$ or 1 and $I = 0$, then $R = 0$ and $S = 0$, and there is no change in the output. However, if $Q_0 = 0$ and $I = 1$, then $R = 0$ and $S = 1$, and the flip-flop changes state. Similarly for $Q_0 = 1$ and $I = 1$, we obtain from (1) and (2) $R = 1$ and $S = 0$, and the flip-flop again changes state. Thus, this flip-flop is a toggle, or T, flip-flop.

9.2.5   Binary Counter

The *binary counter* is a series of clocked T flip-flops (with $T = 1$) connected in cascade that counts incoming pulses in the binary system. Figure 9.14a shows the basic arrangement of clocked T flip-flops that trigger on the high-to-low changes of the pulse train applied at C_1. The corresponding input and output waveshapes are shown in Fig. 9.14b. Note that each flip-flop performs the divide-by-two operation.

In this instance, the flip-flops change state in synchronism with the clock transitions. When each C input undergoes a high-to-low transition, the output changes state. This type of counter is often called a *synchronous counter*. If a signal is applied that is not in synchronism with the clock, this signal is said to be *asynchronous*.

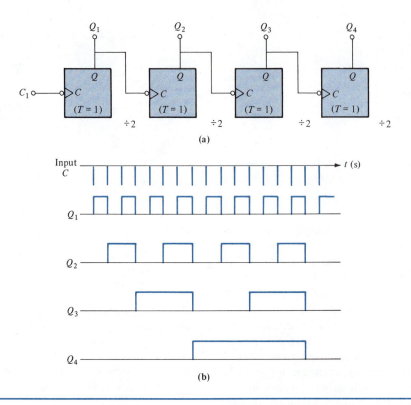

FIGURE 9.14   Binary Counter: (a) Arrangement using clocked *T* flip-flops in cascade, (b) Input and output wave-shapes

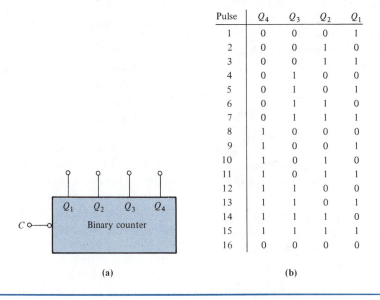

Pulse	Q_4	Q_3	Q_2	Q_1
1	0	0	0	1
2	0	0	1	0
3	0	0	1	1
4	0	1	0	0
5	0	1	0	1
6	0	1	1	0
7	0	1	1	1
8	1	0	0	0
9	1	0	0	1
10	1	0	1	0
11	1	0	1	1
12	1	1	0	0
13	1	1	0	1
14	1	1	1	0
15	1	1	1	1
16	0	0	0	0

(a)

(b)

FIGURE 9.15   Binary Counter: (a) Block diagram, (b) Truth table

Figure 9.15a displays the standard representation of a binary counter. The outputs are listed in reverse order in the truth table of Fig. 9.15b. Each of the outputs Q_1, Q_2, Q_3, and Q_4 represents one bit in a binary number from 1 to 15. The actual number (the count of pulses) is represented by the reverse of the outputs, or by $Q_4 Q_3 Q_2 Q_1$.

——————————————— EX. 9.8

Binary Counter Example

Design a binary counter that will count to 4 and then start counting over again. Assume that the flip-flops are cleared initially so that $Q_1 = Q_2 = Q_3 = 0$ before the first pulse enters.

Solution: To count to 4, which is represented by 100 in binary numbers, three flip-flops are needed. An arrangement of three edge-triggered T flip-flops is used, as shown in Fig. E9.8a (disregard the AND gate at present). The out-

puts Q_1, Q_2, and Q_3 after each pulse at C_1 (high-to-low transition) are displayed in Fig. E9.8b. The reason for the AND gate is that when the fifth pulse is received, Q_3 and $Q_1 = 1$, which forces the output of the AND gate to 1. Thus, Q_2 and Q_3 are cleared, and Q_1 is preset. The binary number $Q_3 Q_2 Q_1 = 001$ results with a renewed count.

—————

9.2.6 Decade Counter

As the number of decimal digits in a number increases, it becomes cumbersome to express this number in binary form since the numbers become unrecognizable. Therefore, a coding system is used in which each decimal digit (0 to 9) is represented by four bits (0000 to 1001). This system is called the *binary-coded decimal system* (*BCD*). A number such as 739 would be presented

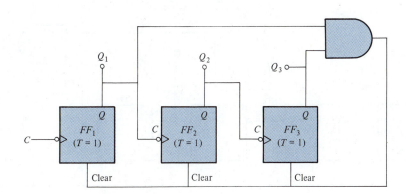

(a)

Pulse	Q_3	Q_2	Q_1
0	0	0	0
1	0	0	1
2	0	1	0
3	0	1	1
4	1	0	0
5	1	0	1
6	0	0	1

(b)

FIGURE E9.8

BCD as 0111 0011 1001. Note that, with twelve bits, such a system can represent decimal numbers from 0 to 999. However, note also that BCD is less efficient (requires more bits) than the plain binary representation. BCD uses twelve bits to represent numbers from 0 to 999, whereas twelve bits in binary are used to represent numbers from 0 to 2^{12}, or 4096.

Applications of BCD arise in instances where decimal data is an input or output of the logic system. For example, hand-held calculators use BCD, as do voltmeters and clocks that have digital displays. It should be mentioned, however, that large computers do not use BCD because more bits are required, which reduces the number of available memory locations.

A *decade counter* uses cascaded flip-flops along with AND gates to count pulses in the base 10 numbering system. Four flip-flops together with an AND gate are required for each digit.

Each flip-flop also has a clear terminal provided (as in Figs. 9.7 and 9.8) that forces $Q = 0$ when this input is 1. Figure 9.16a shows the specific arrangement for a decade counter using T flip-flops. Each flip-flop changes state when the input to C goes high to low. The binary number is represented by $Q_4 Q_3 Q_2 Q_1$. When ten pulses have entered, the decade counter should have 0 for all outputs. The AND gate is therefore connected as shown so that all flip-flops are cleared when $Q_4 Q_3 Q_2 Q_1 = 1010$ (binary 10). Thus, the decade counter begins counting pulses again, each time ten pulses are counted. Figure 9.16b shows a convenient representation of a decade counter. As indicated, additional decade counters may be interconnected to accommodate counting of large numbers.

Binary and decade counters are readily available in IC form. A popular IC binary counter is the TTL 7493. A widely used IC decade counter

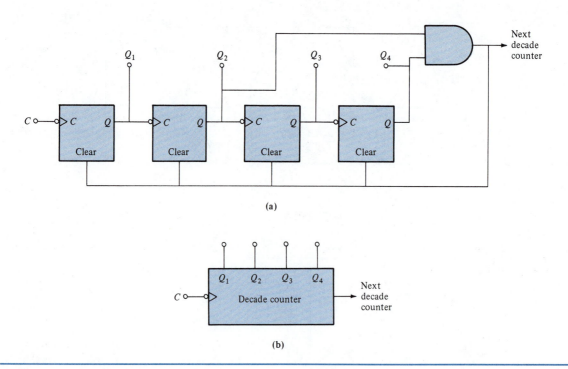

(a)

(b)

FIGURE 9.16   Decade Counter: (a) Arrangement using T flip-flops and an AND gate, (b) Block diagram

is the TTL 7490. These two chips are members of the transistor–transistor logic (TTL) family, which will be described in Chapter 10. More advanced versions will be described in Chapter 11.

9.3   OTHER LOGIC OPERATIONS AND IMPLEMENTATION

9.3.1   Half Adder (Exclusive OR)

Figure 9.17a shows a block diagram of a logic element that is called a *half adder*. This gate carries out half of the addition operation for a single digit (bit) of two binary numbers and omits the carry from the previous bit. With single-bit binary inputs A and B, the output S is their sum. The output C is the carry for the next bit obtained in adding two binary bits. Denoting these outputs in logic expressions yields

$$S = (A + B)\overline{AB} = \overline{A}B + A\overline{B} \qquad (9.3\text{–}1a)$$

and

$$C = AB \qquad (9.3\text{–}1b)$$

The carry operation of (9.3–2) is easily understood in that only if A AND B are nonzero will C be nonzero; otherwise, there will be no carry ($C = 0$), which is exactly the case in the addition of two binary bits. Similarly, (9.3–1) can be understood by realizing that it states that A OR B AND NOT A AND B, which is the operation of addition for two binary bits without a carry from the previous bit. This logic operation S is also widely known as the *exclusive OR operation*. Since the exclusive OR operation is used quite frequently, the special symbol $\oplus$ is used to represent it.

■ *Exclusive OR Operation:* Stated mathematically,

$$A \oplus B = (A + B)\overline{AB}$$
$$= \overline{A}B + A\overline{B} \qquad (9.3\text{–}2) \quad ■$$

A circuit to implement the half adder using basic logic gates is shown in Fig. 9.17b. This circuit is easily understood by checking the validity of the outputs of each gate that correspond to the given inputs.

9.3.2   Full Adder

To perform the operation of complete or full addition of two binary bits that includes adding a possible carry, two half adders along with an OR gate are connected as shown in Fig. 9.18. This circuit performs complete addition and is thus called a *full adder*. The subscript N refers to the Nth binary bit in overall binary numbers.

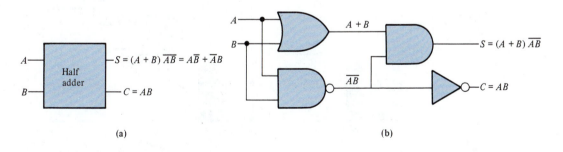

(a)   (b)

FIGURE 9.17   Half Adder: (a) Block diagram, (b) Implementation of half adder

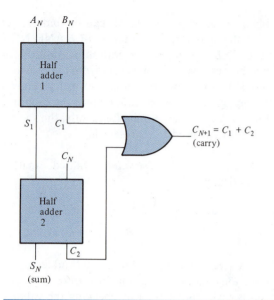

A_N and B_N are the bits added, and C_N is the carry from the previous column. The carry for the next column C_{N+1} is an output. Large numbers of these full adders are required in digital calculators just to carry out the process of addition.

Since addition of binary bits can be accomplished using full adder circuits, it should be realized that multiplication (which is addition repeated) as well as all other mathematical operations can be performed using similar gates. We should note, therefore, that the full adder circuit is a fundamental element in digital computational circuits. Both the half adder and full adder are available in IC form in various logic families, as described in Chapters 10 and 11.

───────────────────────────────────── EX. 9.9

Half Adder Example

Using half adders and OR gates, design a circuit that will add three-bit integer numbers.

Solution: Figure E9.9 indicates the basic connections required to add three-bit integer numbers. In each succeeding column after the first, the carry from the previous column must be added in.

───────────────────────────────────── EX. 9.10

Full Adder Example

Construct a truth table for the full adder of Fig. 9.18 and show that the correct outputs (S_N and C_{N+1}) are obtained

FIGURE 9.18   Full Adder

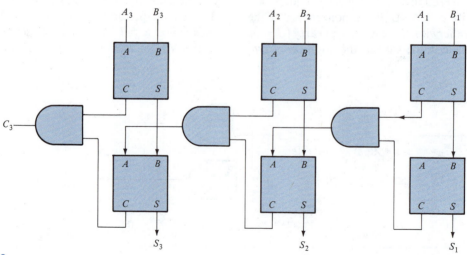

FIGURE E9.9

Row	A_N	B_N	C_N	$A_N \oplus B_N$ S_1	$A_N B_N$ C_1	$C_N S_1$ C_2	$S_1 \oplus C_N$ S_N	$C_1 + C_2$ C_{N+1}
1	0	0	0	0	0	0	0	0
2	0	0	1	0	0	0	1	0
3	0	1	0	1	0	0	1	0
4	0	1	1	1	0	1	0	1
5	1	0	0	1	0	0	1	0
6	1	0	1	1	0	1	0	1
7	1	1	0	0	1	0	0	1
8	1	1	1	0	1	0	1	1

FIGURE E9.10

and correspond to full addition of two single-bit binary numbers with the carry included.

Solution: Construction of the desired truth table, as shown in Figure E9.10, is initiated by first forming the columns A_N, B_N, and C_N for all possible combinations of these inputs. Then, succeeding columns are obtained by carrying out the function at the top of the column.

To show that these results are correct, we consider each row. In row 1, all three variables (A_N, B_N, and C_N) are 0, giving a sum (S_N) and a carry (C_{N+1}) = 0. In rows 2, 3, and 5, one variable is 1 while the others are 0, giving $S_N = 1$ and $C_{N+1} = 0$. In rows 4, 6, and 7, two variables are 1 while one variable is 0, giving $S_N = 0$ and $C_{N+1} = 1$. Finally, in row 8, all three variables are 1, giving $S_N = 1$ and $C_{N+1} = 1$.

9.3.3 Implementing Logic Operations

In Fig. 9.17b, we observed that a half adder could be implemented by using three basic gates: an OR gate, an AND gate, and a NAND gate. A full adder and more complex systems can be implemented in similar fashion. It is possible to connect various types of discrete gates in this way to carry out desired operations. However, for integrated circuits, combinations such as these with different types of gates are usually not used. Typical IC families provide conveniently either the NAND or the NOR operation. Fortunately, if an IC logic family provides the basic NOR

operation, all other logic operations can be implemented by using specific connections to the NOR gate and/or additional NOR gates. The same statement applies to NAND gates. However, it does not apply to the OR operation or to the AND operation because it is impossible to produce the NOT operation using only OR gates or AND gates.

Figure 9.19 shows the various connections and additional NOR gates that are necessary to implement all the other combinational logic operations using only NOR gates. Similarly, Fig. 9.20 shows the circuits that provide all the other operations using only NAND gates.

9.3.4 DeMorgan's Theorem

To verify that the outputs of Figs. 9.19c and 9.20c do provide the designated logical function, we will consider a Boolean algebra theorem that is extremely useful in logic variable algebra.

▪ **DeMorgan's Theorem:**   Stated mathematically in either of two ways,

$$\overline{ABC\cdots} = \bar{A} + \bar{B} + \bar{C} + \cdots \quad (9.3\text{–}3)$$

and

$$\overline{A + B + C + \cdots} = \bar{A}\bar{B}\bar{C}\cdots \quad (9.3\text{–}4) \quad ▪$$

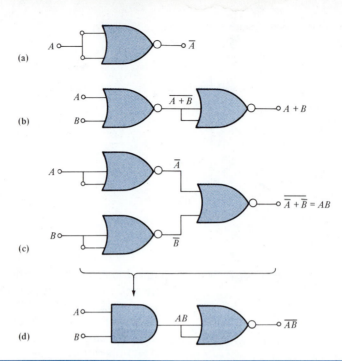

FIGURE 9.19   NOR Gate Implementation of Other Logic Gates: (a) NOT gate, (b) OR gate, (c) AND gate, (d) NAND gate

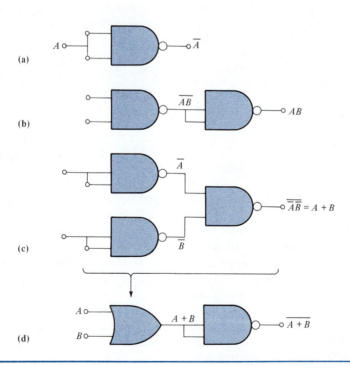

FIGURE 9.20   NAND Gate Implementation of Other Logic Gates: (a) NOT gate, (b) AND gate, (c) OR gate, (d) NOR gate

Column	1	2	3	4	5	6	7	8	9	10
Variable	A	B	$\bar{A}$	$\bar{B}$	AB	$\overline{AB}$	$\bar{A}+\bar{B}$	$A+B$	$\overline{A+B}$	$\overline{A}\overline{B}$
	0	0	1	1	0	1	1	0	1	1
	0	1	1	0	0	1	1	1	0	0
	1	0	0	1	0	1	1	1	0	0
	1	1	0	0	1	0	0	1	0	0

FIGURE 9.21   Truth Table for Verification of Logic Expressions

The proof of DeMorgan's theorem can best be understood if only two variables are considered. Thus, for two variables, we have

$$\overline{AB} = \bar{A} + \bar{B} \tag{9.3–5}$$

and

$$\overline{A + B} = \bar{A}\bar{B} \tag{9.3–6}$$

A brief examination of the truth table in Fig. 9.21 confirms the validity of these equations. (The proof with additional variables is left to the reader.) To understand this table, note that each column is constructed by carrying out the logic operation specified at the top of the column. We see that columns 6 and 7, along with columns 9 and 10, are equal and thus verify (9.3–5) and (9.3–6), respectively. Furthermore, the outputs of Figs. 9.19c and 9.20c are also thus verified.

───────────────────────────── EX. 9.11

DeMorgan's Theorem Example

Use DeMorgan's theorem to show that the exclusive OR operation is equivalent to an *inequality comparator*, which provides an output of 1 when A and B are unequal.

Solution: The exclusive OR operation is given by

$$A \oplus B = (A + B)\overline{AB} \tag{1}$$

Using DeMorgan's theorem on the last term yields

$$A \oplus B = (A + B)(\bar{A} + \bar{B}) \tag{2}$$

However, by using the rule of distribution (identical to that of ordinary algebra), (2) can be written as

$$A \oplus B = A\bar{A} + A\bar{B} + B\bar{A} + B\bar{B} \tag{3}$$

Now, note that the functions $A\bar{A}$ and $B\bar{B}$ must be zero. Therefore,

$$A \oplus B = A\bar{B} + B\bar{A} \tag{4}$$

From this expression, we observe that the exclusive OR operation is equivalent to A AND NOT B OR B AND NOT A and hence will be 1 whenever A and B are unequal.

9.4   PARAMETERS AND DEFINITIONS FOR LOGIC GATES

All families of integrated circuits to be considered in Chapters 10 and 11 have restrictions associated with them. In this section, we will examine the important restrictions of logic circuits and the general definitions of the essential parameters used to describe their performance. We also will consider the logic inverter and observe its nonideal operation.

As an aid in our description, we will consider the simple case of the resistor–transistor logic (RTL) multiple-input NAND gate shown in Fig. 9.22. This type of logic family is no longer used in practice, but it serves as a useful and simple example. We will assume in the description that identical gates are used to drive the inputs A, B, C, and so on of the gate under consideration and that the output is also driving identical gates.

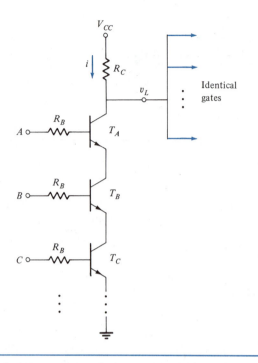

FIGURE 9.22   RTL Multiple-Input NAND Gate

9.4.1   Pull-Up Resistor

The collector resistor R_C in the circuit of Fig. 9.22 is sometimes called a *pull-up*, or *passive pull-up*, *resistor*. This terminology can be understood if we consider the identical output

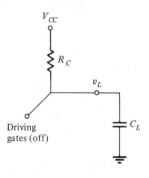

FIGURE 9.23   Equivalent Circuit with v_L High

gates to have a capacitance load C_L associated with them. Then, when one or more of the inputs goes low, the load voltage v_L will rise exponentially through the equivalent RC series network shown in Fig. 9.23. The output load voltage v_L is said to be "pulled up" toward V_{CC}. In practical ICs, active pull-up is employed to aid this process of switching.

9.4.2   Fan-In, Fan-Out, and Voltage Degradation

The first parameters to be introduced and defined are fan-in and fan-out. Their basic definitions are as follows:

- *Fan-in* is the number of inputs connected to a gate input terminal.
- *Fan-out* is the number of outputs connected to a gate output terminal.

The maximum fan-in and fan-out are limited by the current-carrying capability of the gate as well as by the voltage variation, or *voltage degradation*, that is permissible. Note that if the output voltage v_L in Fig. 9.22 is high corresponding to a logical 1 (when A, B, or any of the inputs are 0), then current must flow from V_{CC} through R_C and into the identical gates connected at the output. The more gates that are attached, the more current (indicated as i in Fig. 9.22) that must flow through R_C. Since $v_L = V_{CC} - iR_C$, we see that v_L decreases as the number of driven gates is increased. This condition defines the voltage degradation of the high voltage level (1). For a larger number of gates connected at the output, v_L can be reduced to a very small voltage that can no longer be considered to correspond to a logical 1. This problem is a serious one for the simple RTL logic circuit being considered.

The fan-in of the driving gate is restricted in a similar manner. Consider Fig. 9.22 again. If all of the inputs (A, B, C, and so on) are high, each of the BJTs will be in saturation and v_L should be small corresponding to logical 0. However, if we consider that $v_{CE\,sat}$ for each transistor is 0.2 V and that the number of inputs is N, then we must have $v_L = N(0.2)$ V. Hence, as N is increased, the output voltage will increase from 0 and approach V_{CC}, which results in an unacceptable voltage degradation of the low level (0).

The problem of voltage degradation can be quite serious, particularly when the logical 1 and 0 levels can no longer be distinguished from each other. To avoid this problem, ranges of voltage values are specified by the gate manufacturer for the logical 1 level (high voltage) and the logical 0 level (low voltage). These specifications include both minimum and maximum values for these ranges. The voltage ranges are then directly related to fan-in and fan-out, as we will see next.

9.4.3   Fundamental BJT Inverter

To further observe the effect of voltage degradation when output gates are connected, we first consider the basic BJT inverter shown in Fig. 9.24 with its output open circuited. We begin by generating the voltage transfer characteristic v_{OUT} versus v_{IN} by considering the overall range of v_{IN}, where $v_{IN} \geq 0$ for the NPN BJT.

In the low input voltage range, $0 \leq v_{IN} \leq V_0$, where V_0 is the diode forward voltage (and

$V_0 \cong 0.7$ V for a silicon junction), the BJT will be cut off since the collector junction is reversed biased. Thus, there will be no current through R_L and $v_{OUT} = V_{CC}$. This cutoff region is indicated in Fig. 9.25. When $v_{IN} \geq V_0$ but less than the voltage required to drive the BJT into saturation, the transistor operates in the normal active mode (as an amplifier) where $i_C = \beta i_B$, $i_B = (v_{IN} - V_0)/R_B$, and $v_{OUT} = V_{CC} - R_C i_C$. Thus, we have

$$v_{OUT} = \frac{V_{CC} - R_C \beta(v_{IN} - V_0)}{R_B} \qquad (9.4\text{--}1)$$

This equation is for a straight line with slope $-\beta R_C/R_B$, starting at $v_{IN} = V_0$ (where $v_{OUT} = V_{CC}$). This middle region is indicated in Fig. 9.25. However, as v_{IN} is further increased above V_0, the BJT will eventually go into saturation, and $v_{CE} = v_{CE\,sat}$. Then, v_{OUT} will remain approximately fixed at $v_{CE\,sat}$, which is often approximated as 0.2 V for a silicon BJT. An equation for the input voltage at which the BJT just saturates ($v_{I\,sat}$) can be obtained from (9.4–1) by substituting $v_{OUT} = v_{CE\,sat}$. For typical resistor, V_{CC}, and β values, the required $v_{I\,sat}$ will be approximately 0.1 V greater than V_0. Hence, the range of input voltage swing corresponding to switching the transistor from cutoff to saturation is quite narrow, but it is not as abrupt as it would be ideally.

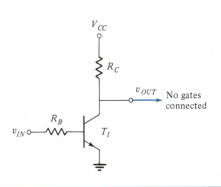

FIGURE 9.24   Basic RTL BJT Inverter

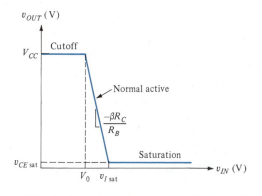

FIGURE 9.25   Voltage Transfer Characteristic for RTL Inverter of Fig. 9.24

Now suppose we connect identical gates to the output of the inverter under consideration. Then, when v_{OUT} is high, current will flow through R_C and into these connected gates. As mentioned previously, v_{OUT} will be reduced by the voltage across R_C, and the amount of reduction depends directly upon the number of attached gates. Thus, v_{OUT} will not be equal to V_{CC} in the cutoff region of the basic inverter, as shown in Fig. 9.25. When v_{OUT} is a high voltage corresponding to logical 1, it can have a range of values less than V_{CC} and is directly dependent upon the specific number of output gates. The output high voltage also depends upon the magnitude of V_{CC} and the specific type of circuitry (not necessarily BJT, as in this case).

— EX. 9.12

BJT Inverter Example

Consider the basic BJT inverter of Fig. 9.24 and determine the voltage transfer characteristic with no output gates connected. Let $V_{CC} = 5$ V, $R_C = 1$ kΩ, and $R_B = 1$ kΩ. Let $V_0 = 0.7$ V and $\beta = 100$, or if the BJT is in saturation, $v_{CE} = v_{CE\,sat} = 0.2$ V.

Solution: For $v_{IN} \leq V_0 = 0.7$ V, $i_B = 0$ and the transistor is cut off with $i_C = 0$ and $v_{OUT} = V_{CC} = 5$ V.

When v_{IN} is increased beyond V_0, i_B becomes greater than zero and the BJT turns on. For v_{IN} slightly above V_0, the base current is given by

$$i_B = \frac{v_{IN} - V_0}{R_B} \qquad (1)$$

while the collector current is $i_C = \beta i_B$. The output voltage is therefore

$$v_{OUT} = V_{CC} - i_C R_C = V_{CC} - \beta \left(\frac{v_{IN} - V_0}{R_B} \right) R_C \qquad (2)$$

By substituting values, we have

$$v_{OUT} = 5 - 100(v_{IN} - 0.7) = 75 - 100 v_{IN} \qquad (3)$$

From (3), we observe that as v_{IN} is increased, v_{OUT} decreases and the BJT saturates when v_{OUT} becomes equal to $v_{CE\,sat} = 0.2$ V. By substitution of $v_{OUT} = 0.2$ V into (3),

we obtain the corresponding input voltage from

$$0.2 = 75 - 100 v_{IN} \qquad (4)$$

Solving for v_{IN}, we have

$$v_{IN} = v_{I\,sat} = 0.748 \text{ V}$$

This result is the minimum value of v_{IN} that causes the BJT to saturate, or $v_{I\,sat} = V_{IH}$.

Finally, for $v_{IN} > 0.748$ V, the BJT remains in saturation. Note that $v_{I\,sat} - V_0 = 0.048$ V and that the input voltage swing from cutoff to saturation is indeed quite narrow.

9.4.4 High- and Low-Voltage Noise Margins

With variations in the high and low voltage levels occurring, it is convenient to standardize the terminology used to describe these fluctuations. Figure 9.26 displays the idealized voltage transfer characteristic of an inverter circuit. Indicated on the output voltage axis are the voltages V_{OH} and V_{OL}, which correspond to the output voltage high and low levels, respectively.

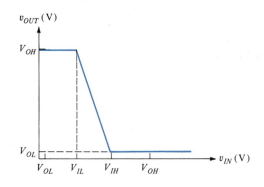

FIGURE 9.26 General Voltage Transfer Characteristic for an Inverter

For the RTL logic gates under consideration in this section, V_{OH} would be less than V_{CC} and $V_{OL} = v_{CE\,sat}$. On the input voltage axis, the voltages corresponding to the low and high input voltages are V_{IL} and V_{IH}. V_{IL} is the maximum input low voltage that will provide a logical 1 output, and V_{IH} is the minimum input high voltage that will provide a logical 0 output. Also indicated on the input voltage axis of Fig. 9.26 are the voltages V_{OL} and V_{OH}, which are outputs from this gate. They are also inputs to the next gate. It must be understood that we can never have

$$V_{OH} < V_{IH} \tag{9.4-2}$$

or

$$V_{OL} > V_{IL} \tag{9.4-3}$$

if the high and low logic levels are to be distinguishable from each other. The specific meaning of these inequalities is defined by the following equations:

$$V_{NMH} = V_{OH} - V_{IH} \tag{9.4-4}$$

and

$$V_{NML} = V_{IL} - V_{OL} \tag{9.4-5}$$

Equation (9.4–4) defines the *high-voltage noise margin* (V_{NMH}) for the logic 1 voltage level, and (9.4–5) defines the *low-voltage noise margin* (V_{NML}) for the logic 0 voltage level. The voltage noise margins represent a safety margin for the high and low levels. Any extraneous noise voltage must be less than that corresponding to the noise margin voltages.

Note that these voltage noise margins must always be greater than zero to distinguish the logical 1 voltage level from the logical 0 voltage level. The actual magnitude of the voltage level is of no concern. What is of concern is that the magnitudes remain in the voltage ranges that provide a positive noise margin.

Fan-Out and Voltage Degradation Example

Consider the basic BJT inverter of Fig. 9.24 with eight identical gates connected at the output (a fan-out of 8) and determine V_{OH} and V_{IH}. Let $V_{CC} = 5$ V, $R_C = R_B = 1$ kΩ, $V_0 = 0.7$ V, and $\beta = 100$, or for BJT operation in saturation, $v_{CE} = v_{CE\,sat} = 0.2$ V.

Solution: To determine V_{OH}, we consider $v_{IN} \leq 0.7$ V, and the BJT T_I will be cut off. The output voltage v_{OUT} will then be V_{OH}, but it is less than V_{CC} because current flows into the eight identical gates connected at the output. To determine V_{OH}, we realize that

$$v_{OUT} = V_{OH} = V_{CC} - i_{R_C}R_C \tag{1}$$

where $i_{R_C} = 8i_B$ and i_B is the base current to each output gate. Thus, by substitution,

$$V_{OH} = V_{CC} - 8i_B R_C \tag{2}$$

Also, for each output gate, we have

$$v_{OUT} = V_{OH} = i_B R_B + V_0 \tag{3}$$

By eliminating V_{OH}, we have

$$V_{CC} - 8i_B R_C = i_B R_B + V_0 \tag{4}$$

or

$$i_B = \frac{V_{CC} - V_0}{R_B + 8R_C} \tag{5}$$

Substituting (5) into (2) and (3) then yields

$$V_{OH} = V_{CC} - 8\left(\frac{V_{CC} - V_0}{R_B + 8R_C}\right)R_C$$

$$= \left(\frac{V_{CC} - V_0}{R_B + 8R_C}\right)R_B + V_0 \tag{6}$$

Finally, by substituting numerical values, we have

$$i_B = \frac{5 - 0.7}{1 + 8(1)} = 0.48 \text{ mA}$$

and

$$V_{OH} = 5 - 8(0.48) = 1.16 \text{ V}$$

We observe that the output high voltage is considerably degraded from the value $V_{CC} = 5$ V. Additional output gates (an increased fan-out) would degrade the output even more.

To calculate V_{IH}, note that $v_{OUT} = v_{CE\,sat}$ for this condition and that the output current to the connected gates is zero (they are operating in cutoff). Therefore, V_{IH} is obtained by using exactly the same formulation as in the previous example, and $V_{IH} = 0.748$ V.

—————————————————————— EX. 9.14

Noise Margin Example

Determine the low and high noise margins for the inverter circuit of Ex. 9.13.

Solution: The noise margin for the high voltage is the excess of V_{OH} over the minimum value required to barely saturate the output gates, or

$$V_{NMH} = V_{OH} - V_{IH} = 1.16 - 0.748 = 0.412 \text{ V}$$

Similarly, the noise margin for the low voltage is the difference between the maximum allowable input voltage for the low state and the actual output voltage for the low state, or

$$V_{NML} = V_{IL} - V_{OL} = 0.7 - 0.2 = 0.5 \text{ V}$$

9.4.5   Switching Speed and Propagation Delay

The RTL inverter of Fig. 9.24 has a finite *switching speed*. That is, as the input voltage is changed from high to low or vice versa, the output voltage response does not occur simultaneously but is delayed in time. This limitation holds true for all IC gate families and is encountered in digital logic circuits. To understand this limitation, we will first consider the mechanism for single-transistor gates and then extend this understanding to multiple-transistor gates (IC families).

For the BJT, the switching speed is primarily dependent upon the length of time necessary to either remove or replace the stored charge in the base region. (This charge movement is actually a one-junction effect and hence is true for PN diodes as well). Let us consider that the emitter junction of the BJT is unbiased (v_{IN} is low corresponding to logical 0) and that the collector junction is reverse biased by V_{CC}. Then, no currents flow and the BJT is in the cutoff state. If the input voltage is now switched high (corresponding to logical 1), carriers from the emitter are injected into the base. These carries flow primarily by diffusion (the flow of particles from regions of high concentration to regions of low concentration) across the base to the collector, which results in collector current. However, to support or allow this diffusion current flow, some of the injected charge must distribute itself in the base such that a higher concentration exists near the emitter junction than the collector junction. Once this distribution is established, carrier flow from the emitter to the collector is possible, and therefore so is current flow (the actual direction is dependent upon whether the charge is positive or negative). However, a finite time is required to establish this equilibrium distribution of stored charge in the base, and a time delay thus occurs in switching the BJT on (the on state occurs when the BJT is in saturation). A similar time delay occurs in switching the BJT off; however, in this case, the delay is due to stored-charge removal . In order for the collector current to cut off, the stored charge must be removed, which requires a finite amount of time that is even larger than the turn-on time.

Figure 9.27a shows an input voltage waveshape switching from $v_{IN} = 0$ to $v_{IN} = V_{CC}$ and back again after time τ, where τ is considered to be larger than the transient periods of the output. The corresponding collector current, shown in Fig. 9.27b, switches from 0 (cutoff) to $i_{C\,sat}$ (saturation) and back again. Note the time delays involved between switching the input voltage and the response of the output current. For consistency, manufacturers define the transient time periods in terms of 10% and 90% of the maximum value. The characteristic times from Fig. 9.27b are defined as

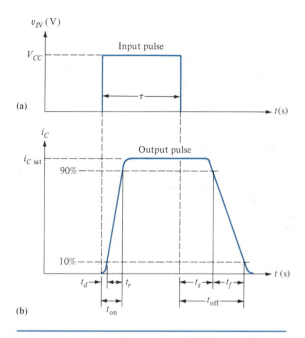

FIGURE 9.27   Time Delay Definitions: (a) Input wave-shape, (b) Output response

follows:

t_d = delay time

t_r = rise time

t_s = storage time

t_f = fall time

$t_{on} = t_d + t_r$

$t_{off} = t_s + t_f$

Note that t_f, the time required to switch i_C from 90% $i_{C\,sat}$ to 10% $i_{C\,sat}$, is larger than t_r, the time required to switch i_C from 10% $i_{C\,sat}$ to 90% $i_{C\,sat}$. This difference is due to the fact that more time is required to remove the stored charge than to establish the equilibrium distribution of this charge. In some BJT IC families, additional current paths are provided to remove the stored charge more rapidly and reduce t_{off}.

Time delays also occur in FET logic gates. However, the corresponding mechanism is en-tirely different since these transistors do not have stored charge. The switching speed of an FET is limited by the charge and discharge rate of load capacitors, which is the equivalent representation of the input gate terminal of each FET. Since the resistors used in FET gates are large in magnitude, the equivalent capacitors charge and discharge with a large time constant (RC), which is the primary limitation on the switching speed of FET logic gates.

The time delays that occur for single-transistor gates are increased in multiple-transistor gates. Fortunately, each IC family employs circuit techniques that reduce the overall time delays. However, these delays cannot be eliminated entirely, and manufacturers specify the rise and fall times associated with the input–output characteristics of a gate as well as the propagation delay times. The *propagation delay time* for gates is defined basically as the time required for the output voltage to respond to the input voltage. Figure 9.28 shows more precise definitions of propagation delay time used by most manufacturers of IC logic gate families. The input and output voltage waveforms are shown because the propagation delay depends upon both. Note that the propagation delay time for a high-to-low transition (t_{PHL}) and the propagation delay time for a low-to-high transition (t_{PLH}) are defined in terms of 50% levels. The overall propagation delay (t_P) is then defined as the average of these two, or

$$t_P = \frac{t_{PHL} + t_{PLH}}{2} \qquad (9.4\text{–}6)$$

Also shown in Fig. 9.28 are the general definitions of rise time and fall time.

9.4.6   Power–Delay Product

An important parameter for logic gates is the product of power dissipation (P_D) and propagation delay (t_P). Ideally, very high speed (low

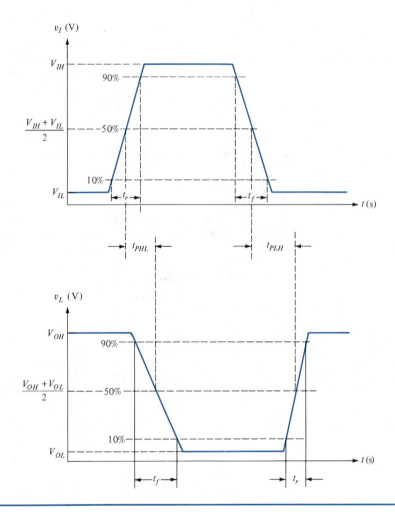

FIGURE 9.28   Propagation Delay Definitions

propagation delay) and very small power dissipation are desirable. However, these parameters are in direct conflict with each other. To achieve lower power dissipation, DC supply voltages and/or currents must be reduced, which causes an increase in the propagation delay. Thus, a figure of merit called the *power–delay product (PD)* is defined as follows:

$$PD = t_P P_D \qquad (9.4\text{--}7)$$

Note that the smaller the value of the power–delay product, the more ideal the gate.

9.5   MULTIVIBRATORS USING LOGIC GATES

In Chapter 7, we considered one type of multivibrator, namely, the astable multivibrator, that was configured by using a Schmitt trigger. In general, there are three classes of multivibrators: monostable, bistable, and astable.

Bistable multivibrators are circuits that have two stable states. Examples are the flip-flops of Section 9.2. *Monostable multivibrators* are devices that have one stable state and one quasi-stable state. *Astable multivibrators* possess two quasi-stable states. In this section, we will examine some monostable and astable multivibrators that are configured by using logic gates. The principle application of multivibrators is their use in timing circuits that provide precisely timed waveshapes.

9.5.1   Monostable Multivibrator

The monostable multivibrator, possessing one stable state and one quasi-stable state, is sometimes called a *one-shot* multivibrator. A trigger signal is used to cause a change in state of the monostable multivibrator. After the trigger signal is applied (and removed), the altered state lasts for only a short time because the circuit returns by itself to the original stable state.

As an initial example, we consider the discrete BJT monostable multivibrator shown in Fig. 9.29. The resistor values, V_{CC}, and the magnitude of the trigger pulse voltage v_I are chosen to force T_1 into saturation. Thus, upon application of a short-duration trigger pulse, T_1 saturates and v_{CE_1} drops low. Since the voltage across C cannot change instantaneously, v_{BE_2} also drops low and T_2 cuts off. However, C charges as current passes from V_{CC} through R, C, and the output of T_1 (see Chapter 1), which causes an exponential increase in voltage

across the capacitor with the polarity as indicated in Fig. 9.29. Hence, V_C increases (approaching $V_{CC} - v_{CE\,sat}$) until its magnitude is sufficient to switch T_2 into saturation. Then, T_1 turns off with $v_{CE_1} = V_{CC}$. Current then passes from V_{CC} through R_{C_1}, C, and T_2, which charges C to a voltage opposite in polarity to that shown in Fig. 9.29. T_1 is forced to remain off, with T_2 on (through V_{CC}, R, and R_{C_2}), which is the stable state of the circuit.

A difficulty with the discrete monostable multivibrator is that the output voltage pulse width cannot be accurately predetermined. It depends upon the specific value of the forward base–emitter voltage and $v_{CE\,sat}$, both of which are variable.

The monostable multivibrator shown in Fig. 9.30a uses NOR gates to overcome this difficulty. We assume that the NOR gates are of the CMOS family (to be discussed in Chapter 11) whose idealized voltage transfer characteristic is displayed in Fig. 9.30b. Note that the threshold voltage V_T is equal to $V_{SS}/2$ for this CMOS case. Note also that the input terminal of each NOR gate draws negligible current since a large input impedance is present (as in the case of op amps).

To describe the operation of Fig. 9.30a, we begin by considering $v_{I_1} = 0$, which forces the output of G_1 to be high because of the transfer characteristic (Fig. 9.30b). Since the input of G_2 does not conduct current, $v_{I_2} = V_{SS}$ and from the transfer characteristic, $v_{O_2} = 0$. Hence, both inputs to G_1 are low and C is initially uncharged.

If a voltage pulse of length τ (as shown in Fig. 9.31a) is applied at the input terminal of G_1, the output of G_1 switches to 0. So also does v_{I_2} because the capacitor voltage cannot change abruptly. Hence v_{O_2} switches high. These voltages are displayed in Figs. 9.31b, c, and d. Upon removal of the input pulse, since v_{O_2} is high, the output of G_1 is low.

Current then passes from V_{SS} through R, C, and the output of G_1 and causes C to charge. The expression for the exponentially decaying voltage (as described in Chapter 1) is

$$v_C = v_F + (v_I - v_F)e^{-t/RC} \qquad (9.5-1)$$

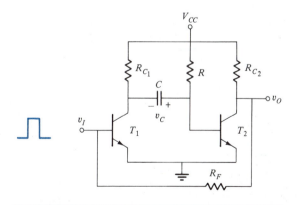

FIGURE 9.29   Discrete BJT Monostable Multivibrator

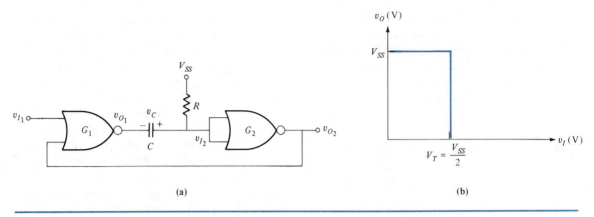

FIGURE 9.30   Monostable, or One-Shot, Multivibrator: (a) Configuration using NOR gates, (b) CMOS NOR gate transfer characteristic (inverter)

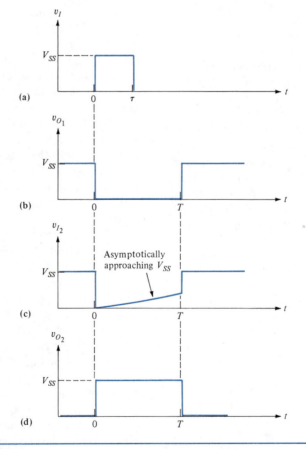

FIGURE 9.31   Voltage Waveshapes for CMOS Monostable Multivibrator: (a) Input voltage, (b) G_1 output voltage, (c) G_2 input voltage, (d) G_2 output voltage

where v_F and v_I are the final and initial voltages of the capacitor. Note that for this situation $v_I = 0$ and $v_F = V_{SS}$. However, with the output of G_1 low, $v_{I_2} = v_C$, and when v_C increases to $v_C = V_T$, G_2 changes state. From the waveshape shown in Fig. 9.31c, switching occurs at time T after application of the pulse. Thus, (9.5–1) becomes

$$V_T = V_{SS} - V_{SS}e^{-T/RC} \qquad \text{(9.5–2)}$$

Solving for T yields

$$T = RC \ln \frac{V_{SS}}{V_{SS} - V_T} \qquad \text{(9.5–3)}$$

For CMOS gates, $V_T = V_{SS}/2$, and thus

$$T = RC \ln 2 \cong 0.7RC \qquad \text{(9.5–4)}$$

Note that we have neglected the propagation delay times of the gates. However, this approximation is valid provided that τ in Fig. 9.31 is much greater than the sum of the propagation delays of each gate.

9.5.2   Astable Multivibrator

The astable multivibrator, as was observed in Chapter 7, possesses two quasi-stable states. Without external triggering, transitions between these states continually occur, and this operation is often called *free running*. Astable multivibrators are often used as pulse train generators and are ideal for generating the clock pulses required in digital circuits.

Figure 9.32 displays a discrete astable multivibrator using BJTs where T_1 and T_2 are operating in either cutoff or saturation. For this circuit, there is no stable state for v_L, so v_L toggles between V_{CC} and $v_{CE\,\text{sat}}$.

The circuit is analyzed by assuming that one transistor is off while the other is in saturation.

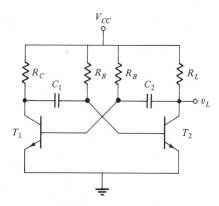

FIGURE 9.32   Discrete BJT Astable Multivibrator

This situation reverses as the capacitors C_1 and C_2 charge and discharge. For example, if T_1 is on ($v_{CE_1} = v_{CE\,\text{sat}}$) and T_2 is off ($v_L = V_{CC}$), then C_2 charges such that its voltage approaches $V_{CC} - V_0$. When this voltage becomes large enough, T_1 turns off (V_0 cannot be maintained for v_{BE_1}) and v_{CE_1} becomes V_{CC}, which causes T_2 to turn on. Then, C_1 mimics the behavior of C_2 until T_2 turns off, which causes T_1 to turn on again. This process repeats and repeats. Since the timing in this circuit is not very precise, IC astable multivibrators, as shown in Figs. 9.33a and b, can be used to alleviate this problem.

Note that we again neglect the propagation delays of the gates and that each of the gates have the same transfer characteristics of Fig. 9.30b. We begin the analysis by assuming that the input voltage to G_1 (v_{I_1}) has just increased slightly above V_T, which (from Fig. 9.33a) results in $v_{O_1} = v_{I_2} = 0$ and $v_{O_2} = V_{SS}$. But since the voltage across C cannot change abruptly, v_{I_1} jumps to $V_T + V_{SS}$, as shown in Fig. 9.34a. This voltage (which is across R in Fig. 9.33a) then decreases exponentially as C charges. This charging is due to the current supplied by $v_{O_2} = V_{SS}$ passing through C, R, and the output of G_1. When v_{I_1} becomes less than V_T, G_1 switches state with $v_{O_1} = v_{I_2}$ becoming V_{SS}. G_2 is caused to switch state with v_{O_2} dropping from

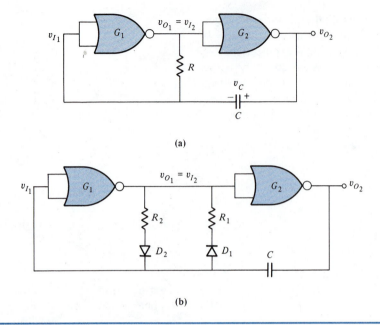

FIGURE 9.33   Astable Multivibrator: (a) Configuration using NOR gates, (b) Configuration with asymmetric square wave output

V_{SS} to 0. This abrupt change forces v_I to reduce further by an amount V_{SS} since the capacitor voltage cannot change abruptly. Capacitor C now charges (with opposite polarity from that shown in Fig. 9.33) as current passes from the output of G_1, through R and C, to the low output of G_2. The capacitor voltage v_{I_1} will increase until it becomes just greater than V_T. Then, the same process repeats and repeats. The voltage waveshapes are displayed in Fig. 9.34.

Qualitatively, in the first time period (T_1 of Fig. 9.34), the exponentially decaying voltage v_{I_1} is again given by (9.5–1) with $V_I = V_{SS} + V_T$ and $V_F = 0$. Thus,

$$v_{I_1} = 0 + (V_{SS} + V_T)e^{-t/RC} \qquad (9.5\text{–}5)$$

At $t = T_1$, $v_{I_1} = V_T$ to yield

$$V_T = (V_{SS} + V_T)e^{-T_1/RC} \qquad (9.5\text{–}6)$$

Solving for T_1 yields

$$T_1 = RC \ln \frac{V_{SS} + V_T}{V_T} \qquad (9.5\text{–}7)$$

For CMOS gates, where $V_T = V_{SS}/2$, $T_1 = RC \ln 3$.

To obtain the time T_2 corresponding to the second portion of one cycle, we again use (9.5–1) with $V_I = V_T - V_{SS}$ and $V_F = V_{SS}$:

$$v_{I_1} = V_{SS} + (V_T - V_{SS} - V_{SS})e^{-t/RC} \qquad (9.5\text{–}8)$$

Thus, when $v_{I_1} = V_T$, $t = T_2$ to yield

$$V_T = V_{SS} + (V_T - 2V_{SS})e^{-T_2/RC} \qquad (9.5\text{–}9)$$

Solving for T_2 yields

$$T_2 = RC \ln \left(\frac{2V_{SS} - V_T}{V_{SS} - V_T} \right) \qquad (9.5\text{–}10)$$

For CMOS gates, we obtain $T_2 = RC \ln 3$, which is equal to T_1.

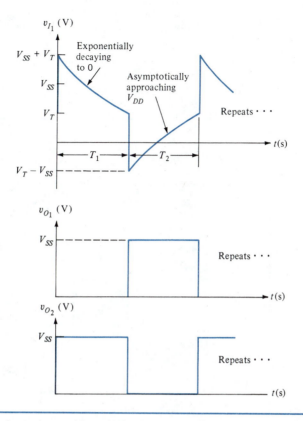

FIGURE 9.34   Voltage Waveshapes for Astable Multivibrator of Fig. 9.33

The overall period is then given by

$$T = T_1 + T_2$$

$$= RC \ln\left(\frac{V_{SS} + V_T}{V_T}\right)\left(\frac{2V_{SS} - V_T}{V_{SS} - V_T}\right)$$

$$\tag{9.5-11}$$

and, for CMOS,

$$T = 2RC \ln 3 \tag{9.5-12}$$

ent values for R_1 and R_2, the rate of charging for capacitor C is changed during each overall period $T_1 + T_2$ and $T_1 \neq T_2$. In this case, the time periods are given by

$$T_1 = R_1 C \ln \frac{V_{SS} + V_T}{V_T} \tag{9.5-13}$$

and

$$T_2 = R_2 C \ln \frac{2V_{SS} - V_T}{V_{SS} - V_T} \tag{9.5-14}$$

9.5.3   Pulse Train Generator

Figure 9.33b displays a circuit that produces an asymmetric square wave. By using differ-

Thus, for example, with $T_1 \ll T_2$ by selecting $R_1 \ll R_2$, a train of pulses is generated. The magnitude of each pulse is equal to the DC supply voltage, and the pulse width is dependent upon the product R_1C. By using circuits similar to this one, pulses of nanosecond duration can be produced.

Astable Multivibrator Example

A CMOS astable multivibrator of the type shown in Fig. 9.33a has $V_{SS} = 5$ V, $R = 100$ kΩ, and $C = 0.1$ μF. Calcu-

late the period of the square waves generated for v_{O_1} and v_{O_2}.

Solution: Substituting the values of R and C directly into (9.5–12) yields

$$T = 2(10^5)(10^{-7}) \ln 3 = 0.02 \ln 3 = 0.022 \text{ s}$$

EX. 9.15

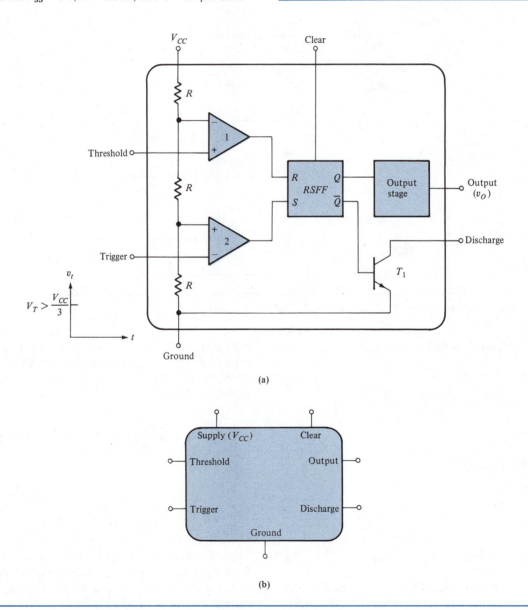

(a)

(b)

FIGURE 9.35   555 Timer Chip: (a) Basic configuration, (b) Block diagram

EX. 9.16

Pulse Train Generator Example

A train of pulses is to be generated using the modified astable multivibrator shown in Fig. 9.33b. Select values for R_1 and R_2 with $C = 0.1 \, \mu F$ that will provide $T_1 = 100$ ms while $T_2 = 1$ ms.

Solution: From (9.5–13) and (9.5–14), we have

$$T_1 = R_1(10^{-7}) \ln 3$$

and

$$T_2 = R_2(10^{-7}) \ln 3$$

where we have substituted $V_T = V_{SS}/2$. Thus, by equating T_1 and T_2 to the desired time lengths, we obtain

$$100 \text{ ms} = R_1(10^{-7}) \ln 3$$

and

$$1 \text{ ms} = R_2(10^{-7}) \ln 3$$

Thus,

$$R_1 = \frac{10^{-1}}{10^{-7} \ln 3} = 0.91 \text{ M}\Omega$$

and

$$R_2 = \frac{10^{-3}}{10^{-7} \ln 3} = 9.1 \text{ k}\Omega$$

9.6   PRECISION IC TIMERS

Precision timers in the form of IC chips are available from numerous semiconductor manufacturers. These devices can provide a precise time delay or produce an accurate train of pulses. The time delay mode is monostable operation, and the amount of delay is controlled by a single external resistor and capacitor. For astable operation, the frequency and duty cycle (length of on time relative to off time) of the pulse train are controlled by two external resistors and a capacitor.

One of the most widely used IC timers is the 555 timer chip. This chip, introduced by Signetics, originally used bipolar ICs. Presently, the 555 timer chip is available in bipolar or CMOS circuits from various manufacturers.

The basic components of the 555 timer chip are shown in Fig. 9.35a. Two comparators with fixed voltages at their positive terminals are used at the input and feed an *RSFF*. The output of the *RSFF* is connected to discharge transistor T_1 and an output buffer stage. Figure 9.35b shows the block diagram of the 555 timer. Seven external connections are provided for the basic timer, as indicated. Their function will be described shortly. Note the effect of the three equal resistors at the input. By the voltage divider rule, the voltages at the positive terminals of the comparators are set at $v_{p_1} = (2/3)V_{CC}$ and $v_{p_2} = (1/3)V_{CC}$.

9.6.1   Monostable Multivibrator Using the 555 Timer

Figure 9.36 displays the circuit connection of the 555 timer chip for monostable multivibrator operation. The external elements are R_E

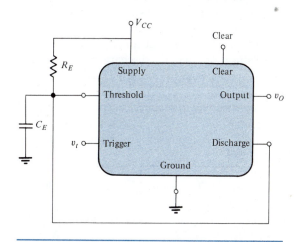

FIGURE 9.36   Monostable Multivibrator Circuit Using the 555 Timer

and C_E, and their precise values determine the output pulse duration. The magnitude of the output voltage depends upon the flip-flop output, with logic level $Q = 1$ corresponding to voltage V_{DC} and logic level $Q = 0$ corresponding to 0 V.

To initiate the operation, a voltage is applied to the clear terminal, which resets the *RSFF* so that Q is at logic level 0 and $\bar{Q}$ is at 1. Hence, from the circuit of Fig. 9.35a, since $Q = 0$, the output terminal is low. Furthermore, since $\bar{Q} = 1$, T_1 is in saturation with the discharge output low ($v_{CE\,sat_1}$). Thus, in Fig. 9.36, the initial voltage for the output (v_O) and the threshold terminals are each essentially 0 V. This input, along with the voltage divider, forces the outputs of both comparators to be low. Thus, the initial state of the *RSFF* is $R = S = 0$, with $Q = 0$ and $\bar{Q} = 1$.

When a negative-going trigger pulse of short duration is applied (as indicated in Fig. 9.35a) for which $v_I < v_{p_2} (= V_{CC}/3)$, the output of comparator 2 goes high and forces $S = 1$ while $R = 0$. This set condition forces $Q = 1$ and $\bar{Q} = 0$. Thus, v_O goes high, and T_1 switches from saturation to cutoff. Then, the voltage at the threshold terminal increases since C_E charges through V_{CC} and R_E. Meanwhile, the short-duration trigger pulse is removed, and $S = 0$ (with $R = 0$, as well), causing no change in the output of the *RSSF*. Hence, as the voltage across C_E increases to $(2/3)V_{CC}$, the output of comparator 1 switches high and forces $R = 1$ while $S = 0$. This reset condition forces $Q = 0$ and $\bar{Q} = 1$. Transistor T_1 is then driven into saturation, and C_E discharges very rapidly since T_1 has a small saturation resistance.

The width of the voltage pulse at the output terminal (v_O) is determined by the time required for the capacitor to charge to a final voltage v_F when the initial voltage is v_I. From Chapter 1, this expression is given by

$$v_{C_E} = v_F - (v_F - v_I)e^{-t/RC} \qquad (9.6-1)$$

or, since $v_F = v_{CC}$ and $v_I = 0$, we have

$$v_{C_E} = V_{CC} - V_{CC}e^{-t/RC} \qquad (9.6-2)$$

However, C_E discharges at time $t = T$ when $v_{C_E} = (2/3)V_{CC}$. By substitution,

$$\frac{2}{3}V_{CC} = V_{CC}(1 - e^{-T/R_EC_E}) \qquad (9.6-3)$$

Solving for T yields

$$T = R_EC_E \ln 3 = 1.1R_EC_E \qquad (9.6-4)$$

9.6.2 Astable Multivibrator Using the 555 Timer

Figure 9.37 indicates the circuit connection of the 555 timer chip for astable multivibrator operation. In this case, the two external resistors R_{E_1} and R_{E_2} and the one external capacitor C_E are required in order to control the on and off duration times.

Since this circuit will switch back and forth from one quasi-stable state to the other, we begin by assuming an initial state. If we let the voltage across C_E be high, $v_{C_E} \gtrsim (2/3)V_{CC}$, both the threshold and trigger inputs will be high. Comparator 1 is forced to have a high output ($R = 1$); comparator 2, to have a low output ($S = 0$). The *RSFF* is thus in the reset condition with $Q = 0$ and $\bar{Q} = 1$. Hence, v_O goes low, and transistor T_1 is in saturation. Then, C_E is allowed to discharge through R_{E_2} and the output of T_1, and the voltage across C_E is forced to reduce. When this voltage becomes less than $(2/3)V_{CC} = v_{n_1}$, the output of comparator 1 switches low and forces $R = 0$ while $S = 0$ also. No change occurs in the operation, and the voltage across C_E continues to reduce with continued discharge. However, when this voltage reaches $(1/3)V_{CC}$, the output of comparator 2 switches high and forces $S = 1$. The *RSFF* goes into the set condition with $Q = 1$ and $\bar{Q} = 0$. Hence, v_O goes high, and the discharge transistor is cut off. The capacitor C_E then begins to charge through V_{CC}, R_{E_1} and R_{E_2}.

The equations for this operation are as follows. From (9.6-1), with $V_I = (2/3)V_{CC}$ and $V_F = 0$,

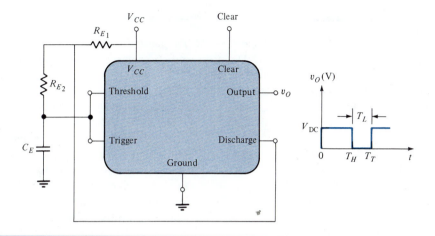

FIGURE 9.37   Astable Multivibrator Circuit Using the 555 Timer

the voltage across C_E is given by

$$v_{C_E} = \frac{2}{3} V_{CC} e^{-t/R_{E_2}C_E} \qquad (9.6-5)$$

The time duration for this low state is defined as T_L. Its value is obtained by setting $t = T_L$, while $v_{C_E} = (1/3)V_{CC}$ in (9.6–5):

$$\frac{1}{3} V_{CC} = \frac{2}{3} V_{CC} e^{-T_L/R_{E_2}C_E} \qquad (9.6-6)$$

Solving for T_L yields

$$T_L = R_{E_2}C_E \ln 2 = 0.693 R_{E_2}C_E \qquad (9.6-7)$$

Similarly, when the output is high from (9.6–1), the voltage across C_E is given by

$$v_{C_E} = V_{CC} - \left(V_{CC} - \frac{1}{3} V_{CC} \right) e^{-t/(R_{E_1} + R_{E_2})C_E}$$

$$= V_{CC} - \frac{2}{3} V_{CC} e^{-t/(R_{E_1} + R_{E_2})C_E} \qquad (9.6-8)$$

We now let T_H define the duration of time during which V_O is high and realize that switching out of this state occurs when v_{C_E} just exceeds $(2/3)V_{CC}$. Substituting $v_{C_E} = (2/3)V_{CC}$ at $t = T_H$ into (9.6–8)

thus yields

$$\frac{2}{3} V_{CC} = V_{CC} - \left(\frac{2}{3} V_{CC} \right) e^{-t_H/(R_{E_1} + R_{E_2})C_E}$$

$$(9.6-9)$$

Solving for T_H yields

$$T_H = (R_{E_1} + R_{E_2})C_E \ln 2$$
$$= 0.693(R_{E_1} + R_{E_2})C_E \qquad (9.6-10)$$

The overall period of the output voltage is thus

$$T_T = T_L + T_H = 0.693(R_{E_1} + 2R_{E_2})C_E$$
$$(9.6-11)$$

9.6.3   Selection Guide for Precision Timers

In the appendix to this chapter, Data Sheet A 9.1 lists precision timers that are commonly used in pulse waveshaping applications. Note that a wide variety of timers is available with time durations starting with 1 μs intervals on up to days. More detailed information on timer chips is given in Appendix 6 at the end of the book.

CHAPTER 9 SUMMARY

■ The five basic logic operations or functions are defined as follows: NOT, AND, OR, NAND, and NOR. A gate is the circuit implementation of a logic operation. The circuit symbols with corresponding outputs for one or two inputs are as follows:

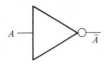

■ The truth, or function, table for a logic operation is a description of all possible combinations of inputs and their corresponding outputs in tabular form. The truth table for various logic operations is as follows:

A	B	$\bar{A}$	AB	$A + B$	$\overline{AB}$	$\overline{A + B}$
0	0	1	0	0	1	1
0	1	1	0	1	1	0
1	0	0	0	1	1	0
1	1	0	1	1	0	0

■ Sequential logic gates are memory elements that have output voltages that are dependent upon past and present input voltages. These memory elements are called flip-flops.

■ The basic reset–set flip-flop ($RSFF$) is reset by $R = 1$ and $Q = 0$. It is set by $R = 0$ and $Q = 1$. The $RSFF$ can also have a clock input but has one inconsistent state.

■ The basic JK flip-flop ($JKFF$) uses a clock input and eliminates the inconsistent state of the $RSFF$. The $JKFF$ is reset by $J = 0$ and $K = 1$ and is set by $J = 1$ and $K = 0$. The input $J = K = 1$ causes the output to change state, while $J = K = 0$ causes no change in output state. To avoid toggling, a master–slave arrangement is used with the $JKFF$ and results in edge-triggering by the clock terminal. Negative edge-triggering corresponds to a high-to-low voltage transition.

■ The delay flip-flop (DFF) is edge triggered, has $J = K$, and provides a time delay. The reset condition corresponds to $D = 0$, and the set condition corresponds to $D = 1$. The toggle flip-flop (TFF) has $J = K$ and is also edge triggered. For $J = K = 0$, the output is unchanged; for $J = K = 1$, the output is changed.

■ The binary counter, decade counter, half adder, and full adder are applications of combinational and sequential gates. One output of the half adder with inputs A and B is the exclusive OR function $(A + B)\overline{AB}$. The other output is the carry AB.

■ DeMorgan's theorem in two variables can be stated as $AB = \bar{A} + \bar{B}$ or as $A + B = \bar{A}\bar{B}$. This theorem is useful in the implementation of logic circuits.

■ Specific connections of only NOR gates can be used to implement all other logic operations. The same statement applies to NAND gates.

■ Fan-in, fan-out, voltage degradation, noise margins, switching speed, propagation delay, and power–delay product are parameters used to describe logic circuits.

■ The monostable, or one-shot, multivibrator possesses one stable state and one quasi-stable stable. A trigger signal changes the state for a short time, and the circuit returns by itself to the stable state. The astable multivibrator possesses two quasi-stable states, and transitions between these states continually occur. The bistable multivibrator is a circuit that, like a flip-flop, has two stable states. Multivibrator circuits consist of transistors or logic gates along with RC networks in which a capacitor charges and discharges. The time periods for stable operation depend upon the resistor and capacitor values, as well as the DC supply voltage and V_T for MOS logic gates. Multivibrators can be used to generate square waves.

■ Precision timers, such as the 555 timer chip, are used to precisely control the pulse duration or square wave period as well as the magnitude of the voltage. The specific time duration for the on and off states corresponds to external resistor and capacitor values, while the voltage magnitude depends upon the DC supply voltage.

CHAPTER 9 PROBLEMS

9.1
In the logic inverter circuit of Fig. 9.1, $R_B = R_C = 4\ k\Omega$ and $V_{CC} = 5$ V. Determine the voltage at B for inputs at A given by 0 and 5 V. Let $\beta = 100$, $V_0 = 0.7$ V, and $v_{CE\ sat} = 0.2$ V.

9.2
The circuit shown in Fig. P9.2 is a diode logic OR gate. Show that the high output voltage level is degraded by the forward voltage drop of a diode (V_0). Additionally, if this output is the input to another similar OR gate, show that the output is degraded by an additional V_0. Construct a truth table with voltages and consider the input high voltages to be equal to 5 V.

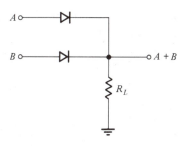

FIGURE P9.2

9.3
Using the truth tables, verify the distribution rules for three variables as given by the following:
  (a) $A(B + C) = AB + AC$
  (b) $A + AC + BA + BC = (A + B)(A + C)$

9.4
Construct a circuit for a half adder using basic logic gates.

9.5
Given the logic function $f = AB + \overline{AB} + \overline{A}B$, simplify this expression by applying DeMorgan's theorem to the first term. Construct a circuit corresponding to this expression using basic logic gates with inputs A and B.

9.6
Construct a circuit with three inputs to realize each of the following logic functions using NOR gates only: **(a)** AND, **(b)** OR, and **(c)** NAND.

9.7
Repeat Problem 9.6 using NAND gates only.

9.8
In the AND gate circuit of Fig. 9.2a, $R_B = R_E = 1\ k\Omega$ and $V_{CC} = 5$ V. Verify the first three rows of the truth table of Fig. 9.2c by letting $\beta = 100$, $V_0 = 0.7$ V, and $V_{LO} = 0$. To verify the last row of this truth table, what is the minimum input high voltage required?

9.9
Consider the circuit of Fig. 9.3a with $V_{CC} = 5$ V, $R_B = 1\ k\Omega$, and $R_E = 1\ k\Omega$. Verify the truth table of Fig. 9.3c by letting the input low voltages be 0 V, the input high voltages be 6 V, $v_{CE\ sat} = 0.2$ V, $\beta = 100$, and $V_0 = 0.7$ V.

9.10
Analyze the logic circuit of Fig. P9.10 and construct the truth table. Simplify the logic statement using

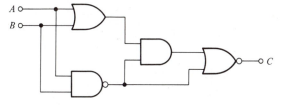

FIGURE P9.10

DeMorgan's theorem and the distribution rule from part a in Problem 9.3.

9.11

Synthesize a logic circuit to realize the following function as written: $AB(A + B) + \bar{A}B(A + \bar{B}) = C$. Using the distribution rule from part a in Problem 9.3, simplify the expression and synthesize again using only one gate.

9.12

The expression for the sum of two binary bits is given by $S = \bar{A}B + A\bar{B}$. Using DeMorgan's theorem, show that $S = (A + B)\overline{AB}$. Synthesize the latter expression using three gates, each with two inputs.

9.13

Construct an RS flip-flop using two NAND gates and feedback connections (similar to that in Fig. 9.6a). Construct a truth table for this circuit and verify the various states.

9.14

Synthesize the circuitry for a half adder, where $S = A\bar{B} + \bar{A}B$ and $C = AB$.

9.15

Obtain the inverse of $S = A\bar{B} + \bar{A}B$ and show by using truth tables that $\bar{S}$ is unity only when $A = B$ (equality comparator). Additionally, use DeMorgan's theorem to show that $\bar{S} = \bar{A}\bar{B} + AB$.

9.16

An $RSFF$ and a TFF are connected as shown in Fig. P9.16. The T flip-flop responds to downward transitions only. For the given square wave input, draw the corresponding waveshapes for Q_1 and Q_2 assuming that they are initially zero.

9.17

For the BJT inverter of Fig. 9.24, determine the values of R_B and R_C such that the transistor operates at the edge of saturation for $V_{IH} = 2$ V. Let $i_{C\,sat} = 10$ mA, $V_0 = 0.7$ V, $v_{CE\,sat} = 0.2$ V, $\beta = 100$, and $V_{CC} = 5$ V.

9.18

For the BJT inverter in Fig. 9.24, calculate the fan-out such that V_{OH} does not fall below 2 V. Assume identical driven gates and $V_{CC} = 5$ V, $R_C = 0.48$ kΩ, and $R_B = 6.25$ kΩ.

9.19

For the BJT inverter of Fig. 9.24, determine the transfer characteristic v_{OUT} versus v_{IN}. Let $R_C = 2$ kΩ, $R_B = 20$ kΩ, $V_{CC} = 6$ V, $\beta = 100$, $v_{CE\,sat} = 0.2$ V, and $V_0 = 0.7$ V. Additionally, determine the low and high noise margins.

9.20

For the circuit of Fig. P9.20, V_1 and V_2 are either 0 or 5 V. If $v_{CE\,sat} = 0.2$ V, what is the fan-in of this gate that will maintain the output low voltage less than $V_0 = 0.7$ V.

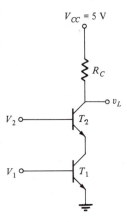

FIGURE P9.20

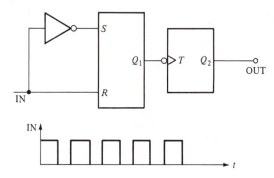

FIGURE P9.16

9.21

For the BJT inverter of Fig. 9.24, calculate the minimum input voltage required to drive the BJT into saturation (V_{IH}). Let $R_C = R_B = 5$ kΩ, $V_{CC} = 10$ V, $\beta = 100$, $V_0 = 0.7$ V, and $v_{CE\,sat} = 0.2$ V.

9.22
Repeat Problem 9.21 for $R_B = 2$ kΩ.

9.23
For the BJT inverter of Fig. 9.24, calculate the value of R_C that provides $V_{IH} = 0.8$ V. Let $V_{CC} = 5$ V, $R_B = 4.5$ kΩ, $\beta = 100$, $V_0 = 0.7$ V, and $v_{CE\,sat} = 0.2$ V.

9.24
Determine the maximum fan-out of the BJT inverter of Fig. 9.24. Assume that the output gates are identical as shown in Fig. P9.24. Let $R_B = 10$ kΩ, $R_C = 1$ kΩ, and $V_{CC} = 5$ V. For the BJT, let $\beta = 50$, $V_0 = 0.7$ V, and $v_{CE\,sat} = 0.2$ V.

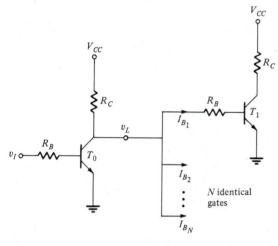

FIGURE P9.24

9.25
Repeat Problem 9.24 using $\beta = 100$. (Note that for reliable fan-out, the minimum value of β must be used.)

9.26
Draw a circuit that will count from 0 to 8 in binary. Use T flip-flops and an AND gate.

9.27
Draw the representation for a decade counter using T flip-flops and show the output $Q_4Q_3Q_2Q_1$ for clock pulses between 8 and 21.

9.28
Using half adders and OR gates, design a circuit that will add four bit integer numbers.

For the BJT monostable multivibrator of Fig. 9.29, consider v_I to be a pulse of voltage that switches T_1 into saturation, which forces T_2 to cut off. Determine the length of time that this state prevails using $R_{C_1} = R_{C_2} = R = 1$ kΩ, $C = 1$ μF, and $V_{CC} = 3$ V. Also, let $\beta = 100$, $V_0 = 0.7$ V, and $v_{CE\,sat} = 0.2$ V. Assume that the voltage across the capacitor is zero before the pulse is applied.

9.30
Repeat Problem 9.29 assuming $V_0 = 0.8$ V where the increase of 0.1 V is due to an increase in temperature.

9.31
The circuit of Fig. P9.31 is a monostable multivibrator. Assuming that one transistor is cut off, determine I_B, I_C, and V_{CE} for the other transistor. Let each transistor have parameters that are equal, with $V_0 = 0.7$ V, $v_{CE\,sat} = 0.2$ V, and $\beta = 100$.

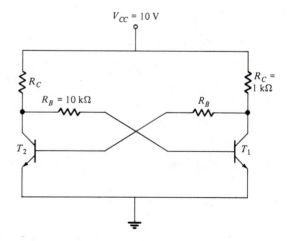

FIGURE P9.31

9.32

Construct a truth table for the CMOS circuit shown in Fig P9.32. Consider the inputs at *A* or *B* to be either high voltage corresponding to V_{CC} or low voltage corresponding to 0. Indicate the state of each MOST (on or off) for each combination of inputs. What function does this circuit perform? (This circuit is an example of CMOS as discussed in Chapter 11.)

9.33

Repeat Problem 9.32 for the CMOS circuit of Fig. P9.33.

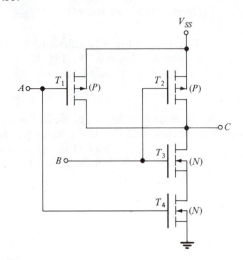

FIGURE P9.33

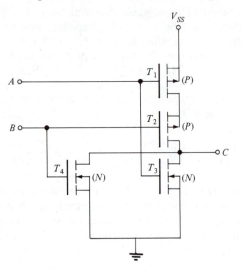

FIGURE P9.32

DATA SHEET A 9.1   Selection Guide for Precision Timers

SPECIAL FUNCTIONS

precision timers

commercial temperature range   (values specified for T_A = 25°C)

DEVICE NUMBER	DESCRIPTION	TIMING		OUTPUT CURRENT	PACKAGES	PAGE
		FROM	TO			
NE555	Single Timer	1 μs	1 s	± 200 mA	D,JG,P	5-21
NE556	Dual Timer	1 μs	1 s	± 200 mA	D,J,N	5-31
TLC551C	LinCMOS, Single High-Speed Timer	1 μs	1 s	100 mA − 10 mA	D,N	5-89
TLC552C	LinCMOS, Dual High-Speed Timer	1 μs	1 s	100 mA − 10 mA	D,N	5-93
TLC555C	LinCMOS, Single High-Speed Timer	1 μs	1 s	100 mA − 10 mA	D,JG,P	5-97
TLC556C	LinCMOS, Dual High-Speed Timer	1 μs	1 s	100 mA − 10 mA	D,N	5-97
uA2240C	Programmable Timer/Counter	10 μs	Days	4 mA	N	5-109

automotive temperature range   (values specified for T_A = 25°C)

DEVICE NUMBER	DESCRIPTION	TIMING		OUTPUT CURRENT	PACKAGES	PAGE
		FROM	TO			
SA555	Single Timer	1 μs	1 s	± 200 mA	D,JG,P	5-21
SA556	Single Timer	1 μs	1 s	± 200 mA	D,J,N	5-31

military temperature range   (values specified for T_A = 25°C)

DEVICE NUMBER	DESCRIPTION	TIMING		OUTPUT CURRENT	PACKAGES	PAGE
		FROM	TO			
SE555	Single Timer	1 μs	1 s	± 200 mA	FH,FK,JG	5-21
SE555C	Single Timer	1 μs	1 s	± 200 mA	FH,FK,JG	5-21
SE556	Single Timer	1 μs	1 s	± 200 mA	FH,FK,J	5-31
SE556C	Single Timer	1 μs	1 s	± 200 mA	D,J,N	5-31
TLC555M	LinCMOS, Single High-Speed Timer	1 μs	1 s	100 mA − 10 mA	JG	5-97
TLC556M	LinCMOS, Dual High-Speed Timer	1 μs	1 s	100 mA − 10 mA	J	5-97

10

SATURATING LOGIC FAMILIES

The logic families described in this chapter utilize resistors, diodes, and BJTs. Since the operation of the controlling BJTs (those that control the function of each gate) is limited by design to the saturation region, these logic families are known as *saturating* logic families. Operation in saturation, however, is disadvantageous because the propagation delay required to switch a BJT from saturation into cutoff (and vice versa) is considerably larger than that required to switch from the active region into cutoff (and vice versa).

As compared to the sample RTL gates of Chapter 9, the various logic families to be described here utilize circuit modifications that improve their performance. Each of the saturating logic families will be described in a chronological manner starting with a more complete description of RTL that initially employed discrete components. This family is no longer in use but is very convenient for developing reasons for seeking improved logic circuits. Such improvement in saturating logic families is provided by diode–transistor logic (DTL), which leads into transistor–transistor logic (TTL). DTL and TTL are also early IC families. Other families that offer much more improvement in operation are the three important nonsaturating logic families discussed in Chapter 11: Schottky transistor–transistor logic (STTL), which uses specialized BJTs; emitter-coupled logic (ECL), which also utilizes BJTs; and MOS logic, which uses either N-channel or P-channel MOSTs or both. All of these more advanced families are available only in IC form.

This chapter concludes by providing a description of the standard TTL 7400 series. This series of ICs was the original workhorse of the industry before the Schottky lines were introduced.

10.1   RESISTOR–TRANSISTOR LOGIC (RTL)

Resistor–transistor logic (RTL) circuits were introduced as basic logic gate examples in Chapter 9. This logic family was the first to become commercially available. Low-, medium-, and high-power configurations were manufactured that differed only in the magnitude of the collector and base resistors used in the circuitry. As we will see, RTL has several associated drawbacks that have been improved upon with the more advanced logic families.

10.1.1   Basic RTL NOR Gate and Inverter

Figure 10.1 displays the basic RTL NOR gate that uses the common-collector resistor R_C for all of the input gate BJTs. This gate is also called the *driving gate*. The circuit is designed such that each BJT operates in saturation for a high input voltage and in cutoff for a low input voltage. Since the collector and emitter terminals of the BJTs for this gate are in parallel, the magnitude of the output load voltage v_L is controlled by any transistor in saturation that forces v_L low. BJTs that are cut off have no effect unless all are cut off, which forces v_L high. Note that if the driv-

ing gate consisted of only a single transistor, the circuit of Fig. 10.1 would become the basic RTL inverter discussed in Chapter 9.

The NOR operation is provided at the output load terminal and is verified by determining v_L for all possible combinations of the inputs. From Fig. 10.1, we note that for the condition of all inputs low, the output is high. Additionally, for any (or all) inputs high, the output is low. Hence, this gate performs the NOR operation.

10.1.2   Direct-Coupled Transistor Logic (DCTL)

Historically, RTL was developed to improve another type of logic family called *direct-coupled transistor logic* (DCTL). This type of logic family is identical to RTL except that the gates are direct coupled without the base resistors R_B ($R_B = 0$). However, DCTL suffers from the difficulty referred to as *current hogging*. Due to slight differences in the base–emitter forward voltage (V_0) of the driven gates, it is possible that one of the BJTs (with the smallest V_0) can turn on before the others and therefore "sink" most (or all) of the output current. The rest of the driven gates would then be deprived of their share of current, and some of them would not turn on. By including base resistors, this difficulty is overcome.

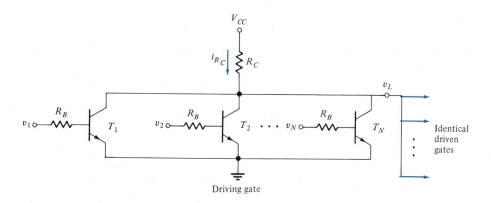

FIGURE 10.1   Basic RTL NOR Gate

DCTL Current Hogging Example

Consider the RTL gate of Fig. 10.1 with the resistors $R_B = 0$. Consider also that the output of this gate feeds three other single BJT gates (fan-out of 3) and that each BJT has a slightly different forward voltage given by 0.69, 0.7, and 0.7 V, with V_{CC} much greater than these values. Draw the equivalent circuit for the condition of v_L high (T_1 through T_N are cut off), and assume that each forward-biased base–emitter junction can be represented by an ideal diode and a DC voltage corresponding to the forward voltage V_0. Use the equivalent circuit to determine the base current for each transistor and the load voltage v_L.

Solution: The equivalent circuit is shown in Fig. E10.1. The branches for the base–emitter junction of each driven transistor have been replaced by an ideal diode and the given forward voltages. Since the transistors T_1 through T_N are cut off, they are represented by open circuits.

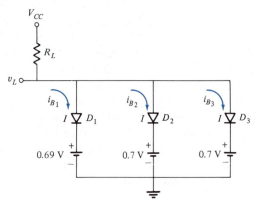

FIGURE E10.1

For V_{CC} greater than any of the forward voltages, D_1 shorts and v_L becomes equal to 0.69 V. The base current through this branch from Fig. E10.1 is given by KVL as

$$i_B = \frac{V_{CC} - 0.69}{R_C}$$

The other two branches will be open circuited because v_L is insufficient in magnitude to cause D_2 and D_3 to short. Hence, T_1 "hogs" all of the current, and only one of the driven BJTs is in saturation.

10.1.3 RTL Fan-Out

In the circuit of Fig. 10.1, if any or all of the inputs are high, then the corresponding transistors will be operating in saturation, which will force the load voltage v_L to be low. Under these conditions, the driven gates will be operating in cut off, and there is no restriction on the fan-out.

However, if all of the inputs to the driving gate are low, v_L will be high and each of the driven gates, as shown in Fig. 10.1, should be in saturation. The actual value of v_L in this case is very dependent upon the magnitude of the fan-out. If the fan-out is too large, the driven BJTs may not be in saturation, as we will see.

The equivalent circuit with v_L high is displayed in Fig. 10.2, where the fan-out is assumed to be K. We have also assumed that the forward voltage for the base–emitter junction of each driven BJT is identical and given by V_0. Hence, the base current of each driven transistor is identical ($i_B = i_{B_1} = i_{B_2} = \cdots = i_{B_K}$). From Fig. 10.2 (using KVL), we have

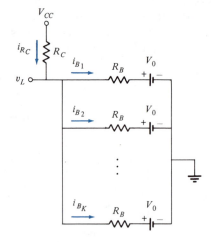

FIGURE 10.2   Equivalent Circuit for RTL NOR Gate with v_L High

$$i_B = \frac{v_L - V_0}{R_B} \qquad \text{(10.1–1)}$$

Furthermore, by rearranging this expression, we obtain the load voltage as

$$v_L = V_0 + i_B R_B \qquad \text{(10.1–2)}$$

Since the base currents for the driven gates must be supplied by V_{CC}, from Fig. 10.2, we observe that

$$i_{R_C} = K i_B \qquad \text{(10.1–3)}$$

where

$$i_{R_C} = \frac{V_{CC} - v_L}{R_C} \qquad \text{(10.1–4)}$$

Equating (10.1–3) and (10.1–4) yields

$$K i_B = \frac{V_{CC} - v_L}{R_C} \qquad \text{(10.1–5)}$$

By substituting for i_B from (10.1–1) and solving for K, we obtain

$$K = \frac{R_B}{R_C}\left(\frac{V_{CC} - v_L}{v_L - V_0}\right) \qquad \text{(10.1–6)}$$

The underlying restriction on K is brought out by realizing that in order to maintain the driven BJTs in saturation, the base current of each must satisfy

$$i_B \geq \frac{i_{C\,sat}}{\beta} \qquad \text{(10.1–7)}$$

and that for driven gates that are identical to the driving gate, the saturation collector current is given by

$$i_{C\,sat} = \frac{V_{CC} - v_{CE\,sat}}{R_C} \qquad \text{(10.1–8)}$$

Substituting (10.1–8) into (10.1–7) yields

$$i_B \geq \frac{V_{CC} - v_{CE\,sat}}{\beta R_C} \qquad \text{(10.1–9)}$$

Hence, the minimum base current required for saturation is

$$i_{B\,min} = \frac{V_{CC} - v_{CE\,sat}}{\beta R_C} \qquad \text{(10.1–10)}$$

Substitution of (10.1–10) into (10.1–2) provides the corresponding minimum load voltage as

$$v_{L\,min} = V_0 + \frac{R_B(V_{CC} - v_{CE\,sat})}{\beta R_C} \qquad \text{(10.1–11)}$$

The maximum fan-out is now obtained when v_L has this minimum value. By substituting $v_L = v_{L\,min}$ into (10.1–6), we have

$$K_{max} = \frac{R_B}{R_C}\left(\frac{V_{CC} - v_{L\,min}}{v_{L\,min} - V_0}\right) \qquad \text{(10.1–12)}$$

This expression can be simplified further by substituting for $v_{L\,min}$ in the denominator of (10.1–12) and rearranging to obtain

$$K_{max} = \beta \frac{V_{CC} - v_{L\,min}}{V_{CC} - v_{CE\,sat}} \qquad \text{(10.1–13)}$$

At first glance, (10.1–13) does not appear to be overly restrictive because for large V_{CC}, $K_{max} \cong \beta$. However, large V_{CC} creates unacceptable power dissipation. Furthermore, the transistors used in this logic family have a range of β values given by $\beta_{min} \leq \beta \leq \beta_{max}$. Thus, to ensure that all of the driven transistors are in saturation, we must use $\beta = \beta_{min}$ in (10.1–13), or

$$K_{max} = \beta_{min} \frac{V_{CC} - v_{L\,min}}{V_{CC} - v_{CE\,sat}} \qquad \text{(10.1–14)}$$

The following example illustrates a calculation of maximum fan-out for RTL.

───────────────────────────────────── EX. 10.2

Maximum RTL Fan-Out Example

Calculate the maximum fan-out K_{max} by using (10.1–14) and assuming that $R_B = 10$ kΩ, $R_C = 1$ kΩ, $V_{CC} = 5$ V, and $\beta_{min} = 25$. Let $V_0 = 0.7$ V, while $v_{CE\,sat} = 0.2$ V.

Solution: We begin by calculating $v_{L\,min}$ from (10.1–11) as follows:

$$V_{L\,min} = 0.7 + \frac{10(5 - 0.2)}{25} = 2.62 \text{ V}$$

Substitution into (10.1–14) then yields

$$K_{max} = 25 \frac{5 - 2.62}{5 - 0.2} = 12.4$$

───────────────────────────────────────

We note from this example that the value of K_{max} is low and that it might still be acceptable were it not for the fact that transistors in saturation are often in a state referred to as *deep saturation*. That is, the saturation base current is greater than $i_{B\,sat}$, which causes $v_{L\,min}$ to increase because of the increase in voltage across R_B. Furthermore, when a BJT is deeply saturated, V_0 increases from the usual value of 0.7 V to perhaps 0.8 V, which creates a further increase in $v_{L\,min}$. The following example demonstrates the seriousness of this dilemma.

───────────────────────────────────── EX. 10.3

Realistic RTL Fan-Out Example

Calculate the maximum fan-out for the RTL NOR gate when the BJTs are in deep saturation that results in $V_0 = 0.8$ V with a saturation base current that is twice the minimum value. Use the same circuit values of Ex. 10.2 and $\beta_{min} = 25$ with $v_{CE\,sat} = 0.2$ V.

Solution: Under these conditions, the minimum load voltage is obtained from (10.1–11) by doubling the R_B term to obtain

$$V_{L\,min} = \frac{0.8 + 2(10)(5 - 0.2)}{25} = 4.64 \text{ V}$$

Substitution into (10.1–14) then yields

$$K_{max} = 25 \frac{5 - 4.64}{5 - 0.2} = 1.875$$

───────────────────────────────────────

10.1.4   RTL NOR Gate with Active Pull-Up

To increase the fan-out of the RTL gates to an acceptable level, active *pull-up* is used. Figure 10.3a displays the modified RTL circuit for an inverter in which the pull-up transistor T_P provides the necessary increased output current. An improvement in fan-out by a factor of 3 to 4 is achieved, as we will see.

Transistor T_2 in Fig. 10.3a is the controlling transistor that provides logic inversion between v_I and v_L. Transistor T_1 operates in an identical manner to T_2 (since their inputs are the same), and therefore the inputs to T_2 and T_P are always opposite. The maximum fan-out for this case is determined, as in the previous section, for v_L high. Under these conditions v_I is low, T_1 and T_2 are cut off, and T_P is saturated.

The equivalent circuit for v_L high is shown in Fig. 10.3b, where K identical branches are connected at the output and represent the driven gates. Once again, we have assumed that the driven gates are identical to the driving gate. Note that each branch has resistor $R_B/2$ in series with V_0, which is the Thévenin equivalent of two branches in parallel, each with R_B in series with V_0. Figure 10.3c displays a simplified equivalent circuit where all of the output branches have been combined into a single Thévenin equivalent circuit.

For our analysis, we will assume that the base current for T_P is negligible and therefore that $i_{C_P} = i_{E_P}$. Writing KVL for the output loop of Fig. 10.3c thus yields

$$V_{CC} - V_0 - v_{CE\,sat} = i_{E_P}\left(R_{C_P} + \frac{R_B}{2K}\right)$$

$$(10.1–15)$$

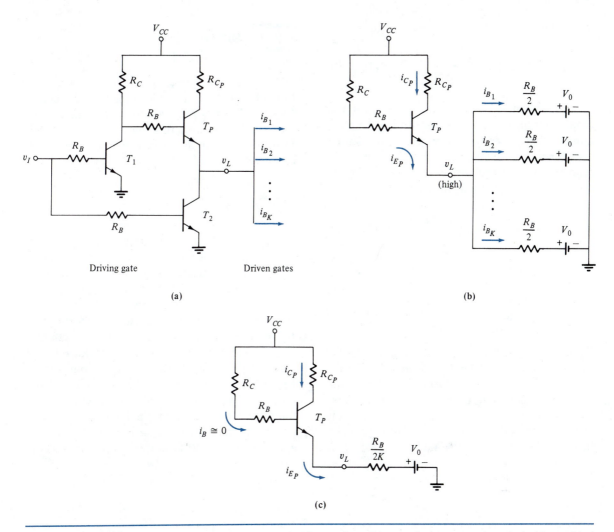

FIGURE 10.3   Modified RTL Circuit: (a) RTL inverter with active pull-up, (b) Equivalent circuit with v_L high, (c) Simplified equivalent circuit with v_L high

Solving for i_{E_P} gives

$$i_{E_P} = \frac{V_{CC} - V_0 - v_{CE\,\text{sat}}}{R_{C_P} + (R_B/2K)} \qquad (10.1\text{–}16)$$

Note that each driven gate consists of two transistors in saturation and that the emitter current of T_P must now equal

$$i_{E_P} = 2K i_B \qquad (10.1\text{–}17)$$

where i_B is the base current of each transistor. Equating (10.1–16) and (10.1–17) thus yields

$$2K i_B = \frac{V_{CC} - V_0 - v_{CE\,\text{sat}}}{R_{C_P} + (R_B/2K)} \qquad (10.1\text{–}18)$$

To simplify the analysis, we assume that K is large enough such that $R_B/2K \ll R_{C_P}$. Solving (10.1–18) for K thus yields

$$K = \frac{V_{CC} - V_0 - v_{CE\,sat}}{2i_B R_{C_P}} \qquad (10.1-19)$$

The maximum value of fan-out (K_{max}) is now obtained from (10.1–19) by substituting the minimum i_B that forces the driven gates into saturation. Since the driving gate of Fig. 10.3a is also a representation of each driven gate, we see by inspection of T_1 that we again have

$$I_{B\,min} = \frac{i_{C\,sat}}{\beta} = \frac{V_{CC} - v_{CE\,sat}}{\beta R_C} \qquad (10.1-20)$$

Substitution into (10.1–19) thus yields

$$K = \frac{\beta R_C}{2R_{C_P}} \left(\frac{V_{CC} - V_0 - v_{CE\,sat}}{V_{CC} - v_{CE\,sat}} \right) \qquad (10.1-21)$$

By comparing (10.1–21) to (10.1–13), we observe that improvement in fan-out is provided by the factor $R_C/2R_{C_P}$ and also through the larger numerator voltage difference of (10.1–21) as compared to (10.1–13). The value of R_C is typically five to six times that of R_{C_P}, which alone provides justification for using active pull-up.

10.2   DIODE–TRANSISTOR LOGIC (DTL)

Diode–transistor logic (DTL) is similar to RTL except that diodes are used to replace some of the resistors. An improvement in fan-out is achieved with this logic family; however, DTL also has an associated reduction in speed, as we will see.

10.2.1   DTL Inverters

The basic DTL inverter is displayed in Fig. 10.4a. We begin our discussion by verifying the transfer characteristic shown in Fig. 10.4b.

Note that for v_I low ($0 \le v_I < V_0$), D_1 is short circuited, with $v_{D_1} = V_0$. However, the voltage at P is then insufficient in magnitude to short circuit D_2 and forward bias the base–emitter junction of T_1. Therefore, D_2 is open and T_1 is cut off. The input current path for this case is, as indicated in Fig. 10.4a, v_I low. Under these conditions, the output voltage v_L is high ($v_L = V_{OH}$), and its magnitude is dependent upon the fan-out.

As v_I is increased to $v_I = V_0 = 0.7$ V, the voltage at P in Fig. 10.4a is then sufficient to short

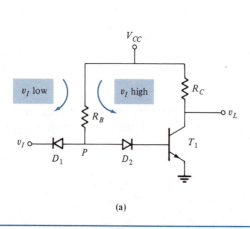

(a)

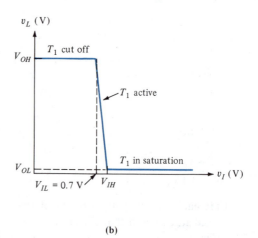

(b)

FIGURE 10.4   DTL Inverter: (a) Basic circuit, (b) Transfer characteristic

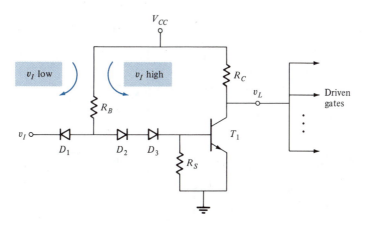

FIGURE 10.5   Modified DTL Inverter

circuit D_2 and turn on T_1. Thus, T_1 begins conduction with operation in the active region. Increasing v_I slightly causes the voltage at P to increase slightly, which is sufficient to drive T_1 into saturation and open circuit D_1. Then, the output voltage is forced to be $v_L = v_{CE\,sat}$. As v_I is increased further, D_1 remains open, and the voltage at P as well as the saturation condition of T_1 remain unchanged. The current path at the input is then through D_2 and T_1, as indicated in Fig. 10.4a by v_I high.

The corresponding transfer characteristic is shown in Fig. 10.4b. Note that V_{IL} ($= 0.7$ V) and V_{IH} are nearly equal and that the transfer characteristic is quite abrupt. The specific width of this region is not determinable except by measurement because of the assumption that the forward voltage is fixed at V_0 (0.7 V). Additionally, the output low voltage is $V_{OL} = v_{CE\,sat}$, and if no output gates are connected, the output high voltage is $V_{OH} = V_{CC}$.

The basic gate of Fig. 10.4a suffers from several drawbacks. One drawback is that there is no path for stored-charge removal from T_1 when T_1 is switched from saturation to cutoff. Unacceptably long transient periods and correspondingly large propagation delay result. This problem is reduced by providing an additional current

path, as shown in Fig. 10.5, through the addition of resistor R_S.

Another intolerable feature of the basic DTL inverter of Fig. 10.4a is that T_1 may begin to turn on before v_I reaches 0.7 V because of slight differences in forward voltage V_0 for D_1, D_2, and T_1. This problem is eliminated by including an additional diode (or two) in series with D_2 as also shown in Fig. 10.5. Sometimes, D_2 is replaced with a transistor.

--- EX. 10.4

DTL Inverter Example

For the basic DTL inverter of Fig. 10.4a, let $R_B = R_C = 1$ kΩ and $V_{CC} = 5$ V. For a fan-out of 0, determine V_{OH} and V_{OL} corresponding to Fig. 10.4b.

Solution: For v_I low (≤ 0.7 V), the voltage at point P is insufficient to allow D_2 and the base–emitter junction of T_1 to conduct. Hence, T_1 is cut off and $v_L = V_{OH} = V_{CC}$.

However, for v_I high ($\simeq V_{CC}$), D_1 is open, while D_2 and the base–emitter junction of T_1 become forward biased with voltage $V_0 = 0.7$ V. The base current for T_1 is then

$$I_{B_1} = \frac{V_{CC} - 2(0.7)}{R_B} = \frac{5 - 1.4}{1} = 3.6 \text{ mA} \qquad (1)$$

which is sufficient to saturate T_1. Hence, $V_{OL} = v_{CE\,sat} = 0.2$ V.

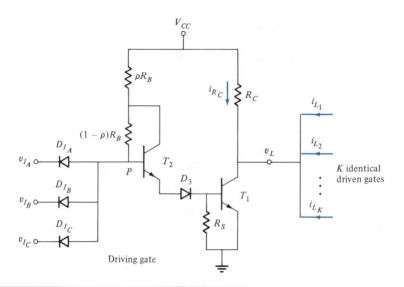

FIGURE 10.6   DTL NAND Gate

10.2.2   DTL NAND Gate

Figure 10.6 displays a three-input DTL NAND gate that has been further modified to improve the fan-out by replacing diode D_2 with transistor T_2 and an additional resistor, where ρ is a fraction. Note that when all inputs are high, the voltage at P is high and allows T_1 to saturate and v_L to go low. When any one or all inputs are low, the voltage at P is low and T_1 is cut off with v_L high. Hence, the DTL logic family provides the NAND operation. The circuit of Fig. 10.6 is fabricated quite conveniently in IC form and was used extensively prior to the introduction of TTL. Some manufacturers continue to supply DTL ICs. However, these ICs are used only for replacement purposes. Other logic families offer improvement in all aspects of operation.

10.2.3   DTL Fan-Out

To determine the maximum fan-out for the DTL logic gate, we proceed (as in the RTL cases) by assuming that K identical gates are connected at the output load terminal. However, a different procedure is inevitable in this case because fan-out is restricted when v_L is low. We will see that the maximum fan-out depends upon the degree of saturation of the controlling transistor.

With v_L low (for all v_I high), each of the load currents in Fig. 10.6 will be equal. Thus, $i_{L_1} = i_{L_2} = \cdots = i_{L_K}$. Additionally, an expression for these load currents can be written using KVL as follows:

$$i_{L_1} = \frac{V_{CC} - V_0 - v_{CE\,\text{sat}}}{R_B} \qquad (10.2\text{--}1)$$

which is obtained by realizing that each driven gate has identical circuitry to that of the driving gate. Hence, the driven gates have current flowing from V_{CC} through ρR_B and $(1 - \rho)R_B$ (the sum of which is R_B) and the input diode whose input voltage is given by $v_{CE\,\text{sat}}$. The collector current of the saturated transistor is given by

$$i_{C_1} = i_{R_C} + K i_{L_1} \qquad (10.2\text{--}2)$$

where, by substitution for i_{R_C} and i_{L_1}, we have

$$i_{C_1} = \frac{V_{CC} - v_{CE\,sat}}{R_C}$$
$$+ K\frac{V_{CC} - V_0 - v_{CE\,sat}}{R_B} \qquad \text{(10.2–3)}$$

Writing KVL between V_{CC} and point P yields

$$V_{CC} - V_P = \rho R_B i_{E_2} + (1 - \rho)R_B\left(\frac{i_{E_2}}{1 + \beta}\right)$$
$$\text{(10.2–4)}$$

where we have realized that i_{E_2} is the current through ρR_B and that $i_{E_2} = (1 + \beta)i_{B_2}$ (since T_2 is always active). By solving for i_{E_2}, we obtain

$$i_{E_2} = \frac{V_{CC} - V_P}{\rho R_B + (1 - \rho)R_B/(1 + \beta)} \qquad \text{(10.2–5)}$$

Finally, we note that when T_1 is in saturation, V_P is the sum of three forward voltage drops, or $V_P = 3V_0$. Additionally, when T_1 is in saturation, its base current is appreciable, and for typical values of R_S, $i_{E_2} \cong i_{B_1}$ (see Fig. 10.6).

We now define a *saturation parameter* as

$$\sigma = \frac{i_C}{\beta i_B} \qquad \text{(10.2–6)}$$

where if $\sigma = 1$, $i_C = \beta i_B$ and the transistor is active. However, when a transistor is in saturation, $\sigma < 1$, where a typical value of σ for a BJT in saturation is $\sigma = 0.85$. To determine a relationship for the fan-out, we substitute i_C and i_B for T_1 into (10.2–6), where we use i_C from (10.2–3) and i_B from (10.2–5). Thus, (10.2–6) becomes

$$\sigma = \frac{\dfrac{V_{CC} - v_{CE\,sat}}{R_C} + K\dfrac{V_{CC} - V_0 - v_{CE\,sat}}{R_B}}{\beta\dfrac{V_{CC} - 3V_0}{\rho R_B + (1 - \rho)R_B/(1 + \beta)}}$$
$$\text{(10.2–7)}$$

By solving for K, we obtain

$$K = \frac{\sigma\beta\left[\dfrac{V_{CC} - 3V_0}{\rho R_B + (1 - \rho)R_B/(1 + \beta)}\right] - \dfrac{V_{CC} - v_{CE\,SAT}}{R_c}}{\dfrac{V_{CC} - V_0 - v_{CE\,SAT}}{R_B}}$$
$$\text{(10.2–8)}$$

or

$$K = \sigma\beta\left[\frac{(V_{CC} - 3V_0)/(V_{CC} - V_0 - v_{CE\,SAT})}{\rho + (1 - \rho)/(1 + \beta)}\right]$$
$$- \frac{R_B}{R_C}\left(\frac{V_{CC} - v_{CE\,sat}}{V_{CC} - V_0 - v_{CE\,sat}}\right) \qquad \text{(10.2–9)}$$

To understand the implication of (10.2–9), we will make two assumptions. First, we let the voltage ratios in each term be approximated by unity, which is valid for $V_{CC} \gg V_0$ and $v_{CE\,sat}$. Second, for reasonably large β, the term $(1 - \rho)/(1 + \beta)$ may be neglected with respect to ρ. Thus, (10.2–9) is reduced to

$$K \cong \frac{\sigma\beta}{\rho} - \frac{R_B}{R_C} \qquad \text{(10.2–10)}$$

From (10.2–10), we realize that K can be much larger than in the previous cases as ρ (recall that ρ is a fraction) is reduced in magnitude. The ratio of R_B/R_C is typically 5 or less, and it is thus relatively unimportant. Note that for $\rho = \sigma/2$ and neglecting R_B/R_C, (10.2–10) becomes

$$K = 2\beta \qquad \text{(10.2–11)}$$

which indicates an improvement in fan-out by a factor of 2 over the corresponding maximum RTL fan-out of (10.1–13), without active pull-up.

Thus, we see that the fan-out of DTL is improved over that available with RTL. However, the controlling transistor (T_1 in the driving gate and similar transistors in other gates) becomes heavily saturated ($\sigma \ll 1$) as ρ is reduced.

Undesirable power dissipation in all controlling transistors results, as well as increased switching speed and propagation delay. Hence, further improvement is warranted.

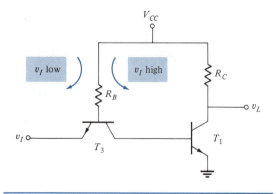

FIGURE 10.7   Basic TTL Inverter

--- EX. 10.5

Fan-Out Calculation for DTL Example

Calculate the fan-out for the DTL NAND gate of Fig. 10.6. Let $R_B = R_C = 2\ k\Omega$, $V_{CC} = 5$ V, and $\rho = 0.425$, while $\beta = 25$, $V_0 = 0.7$ V, $v_{CE\ sat} = 0.2$ V, and $\sigma = 0.85$.

Solution: Substituting these values directly into (10.2–9) yields

$$K = (0.85)(25) \left[\frac{(5 - 2.1)/(5 - 0.7 - 0.2)}{0.425 + (1 - 0.425)/26} \right]$$

$$- \frac{2}{2} \left(\frac{5 - 0.2}{5 - 0.7 - 0.2} \right)$$

$$= 21.25[1.582] - 1(1.171)$$

$$= 32.4$$

For comparison, note that the approximate expression given by (10.2–11) predicts $K = 50$.

10.3   TRANSISTOR–TRANSISTOR LOGIC (TTL)

The logic families RTL and DTL are limited in speed because of the slow mechanism associated with the removal of stored charge from saturated transistors. The specialized circuitry used with *transistor–transistor logic* (TTL) provides an order of magnitude improvement in speed and also allows greater fan-out, as we will see.

10.3.1   Basic TTL Inverter

The basic TTL inverter is shown in Fig. 10.7, where the numbering of transistors T_1 and T_3 is intentional because of the additional transistors to be included later. Note the similarity of

the TTL gate of Fig. 10.7 with the DTL gate of Fig. 10.4a. The circuits are identical except for T_3, which has replaced diodes D_1 and D_2.

To show that the TTL gate of Fig. 10.7 does indeed provide logical inversion, we consider high and low values of v_I and determine the corresponding outputs. For v_I high, the base–emitter junction of T_3 is reverse biased, while the base–collector junction is forward biased by V_{CC}. Therefore, T_3 is operating in what is referred to as the *inverse active mode* (the inverse of the active mode). Under these conditions, very little current flows at the input terminal. However, a significant current passes from V_{CC} through R_B, the base–collector junction of T_3, and the base–emitter junction of T_1. This current path is indicated in Fig. 10.7 as v_I high. The magnitude of this current is sufficient to saturate T_1, which forces v_L low ($v_L = v_{CE\ sat}$).

However, for v_I low, both junctions of T_3 in Fig. 10.7 are forward biased by V_{CC}. Therefore, T_3 is operating in saturation. If we consider that T_1 was in saturation just before v_I was switched low, then the stored charge from T_1 is rapidly removed through T_3 when v_I goes low. After the stored-charge removal, T_1 operates in cutoff because the voltage at the base of T_3 is less than $2V_0$. Current continues to flow from V_{CC} through R_B and the base–emitter junction of T_3 and out the input terminal. This current path is shown in Fig. 10.7 as

v_I low. Since T_1 is cut off, v_L goes high and provides the desired logical inversion operation.

10.3.2 Comparison of Stored-Charge Removal for DTL to TTL

To compare the speed of stored-charge removal for DTL to that of TTL, we consider the basic inverter gates of Figs. 10.5 and 10.7. In each case, the output transistor T_1 is considered to be operating in saturation due to a high input. The input is then switched low, and each T_1 must undergo stored-charge removal until cutoff is reached. We will see that the current level upon switching is enormous for TTL as compared to DTL. Since current is the rate of change of charge with time, much more rapid stored-charge removal results for TTL and, hence, much faster speed.

For the DTL inverter in Fig. 10.5, when v_I is high, T_1 is in saturation. Current then passes along the v_I high path. When v_I is switched low, the initial stored-charge removal current (I_{SCR}) from T_1 through R_S will be

$$I_{SCR} = \frac{V_0}{R_S} \qquad (10.3\text{–}1)$$

where V_0 is the forward voltage initially across the base–emitter junction of T_1.

For the case of TTL, from the basic inverter of Fig. 10.7, the initial stored-charge removal current is obtained by considering the state of T_3. When v_I switches low, current passes through the v_I low path, and $v_{E_3} = v_{CE\,sat}$ while $v_{BE_3} = V_0$, which results in $v_{B_3} = v_{CE\,sat} + V_0$. Meanwhile, v_{BE_1} remains at V_0 until the stored charge is removed. Thus, initially (upon switching), $v_{C_3} = V_0$, and $v_{CB_3} = v_{C_3} - v_{B_3} = -v_{CE\,sat}$. Although v_{CB_3} is a forward voltage, it is insufficient in magnitude to turn on the collector junction of T_3. Therefore, T_3 initially operates in the active mode. Thus, i_{C_3} is related to i_{B_3} through

$$i_{C_3} = \beta i_{B_3} \qquad (10.3\text{–}2)$$

Since $i_{B_3} = (V_{CC} - V_0 - v_{CE\,sat})/R_B$, by substitution, we have

$$i_{C_3} = \frac{\beta(V_{CC} - V_0 - v_{CE\,sat})}{R_B} \qquad (10.3\text{–}3)$$

Note that this current is the initial stored-charge removal current for TTL, or

$$I_{SCR} = \frac{\beta(V_{CC} - V_0 - v_{CE\,sat})}{R_B} \qquad (10.3\text{–}4)$$

By comparing (10.3–4) to (10.3–1), we observe an enormous difference. For TTL, a factor of β improvement is achieved, as well as a large increase in the voltage factor since $V_{CC} - V_0 - v_{CE\,sat} \gg V_0$.

10.3.3 Basic TTL NAND Gate

The basic gate of the TTL family provides the NAND operation. Figure 10.8a displays the basic TTL NAND gate, which utilizes a specialized transistor (T_3) that has multiple emitters. An alternative to this configuration would be to use multiple transistors with common-base and common-emitter terminals. However, the multiple-emitter transistor is the "heart" of TTL because it is easily fabricated using IC technology and requires much less chip area than multiple BJTs do. Figure 10.8b shows the cross section of a multiple-emitter transistor with five emitters. Note that it is no more difficult to fabricate this transistor than it is to fabricate an ordinary (single-emitter) transistor. This cross section can be compared to the one presented in Chapter 3. From the cross section in Fig. 10.8b, we note that the base and collector regions of each transistor are common because they are the same region. Therefore, the new circuit symbol for T_3 in Fig. 10.8a is used.

To show that the NAND operation is provided by the gate of Fig. 10.8a, we consider all

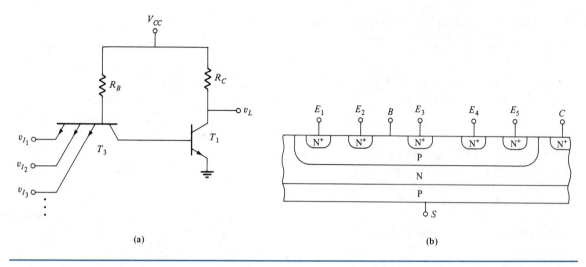

FIGURE 10.8   Basic TTL NAND Gate: (a) Multiple-emitter transistor configuration, (b) Multiple-emitter transistor cross section

combinations of input and their corresponding outputs. For any or all inputs low, T_3 operates in saturation with both junctions forward biased, and T_1 cuts off. Thus, the output load voltage is high. However, for all inputs high, T_3 is inverse active, T_1 saturates, and v_L goes low ($v_L = v_{CE\,sat}$). Hence, this gate provides the logic NAND operation.

10.3.4   TTL NAND Gate with Active Pull-UP

Although the basic TTL NAND gate drastically reduces the speed limitation due to stored-charge removal, another speed limitation is still present. This additional limitation is due to the rise time of the output voltage and amounts to the time required to charge an equivalent capacitor connected at the output as the output switches from low to high voltage. In the basic gate of Fig. 10.8a, the equivalent output capacitor must charge through the pull-up resistor R_C. The associated rise time for this case is quite large and is considerably improved using active pull-up.

Figure 10.9 displays the circuit for the TTL NAND gate with active pull-up as well as pull-down. Its operation will be described in detail in the next section. The BJT T_P and the pull-up resistor R_{C_P} are the primary additional elements for active pull-up. However, transistor T_2 must also be included to provide logic inversion to T_P such that T_P is off when T_1 is on and vice versa. The resistor R_E is then also required to provide a path for stored-charge removal, which is less severe in this case because T_1 does not become heavily saturated.

Moreover, diode D is included in Fig. 10.9 to ensure that T_P is operating in cutoff when T_1 is in saturation. Note that with T_1 in saturation, T_2 is also in saturation. Therefore, by KVL,

$$v_{BP} = v_{BE_1} + v_{CE_2} \qquad (10.3\text{–}5)$$

or

$$v_{BP} = V_0 + v_{CE\,sat} \qquad (10.3\text{–}6)$$

This voltage is insufficient to turn on T_P since D provides an additional forward voltage V_0. With-

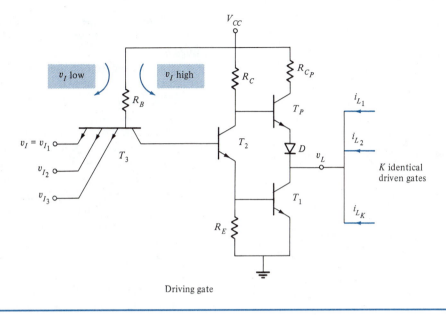

FIGURE 10.9   TTL NAND Gate with Active Pull-Up and Pull-Down Transistors

out D, however, T_P might turn on, which would result in erroneous operation. The output branch of the TTL NAND gate of Fig. 10.9 is often called a *totem pole* connection.

To observe the improvement in rise time when active pull-up is used as compared to passive pull-up, we consider each case separately. First, for the case of passive pull-up, the equivalent capacitor at the output (C_o) charges directly through R_C and V_{CC}, which results in a rise time given by

$$\tau_{\text{passive}} = R_C C_0 \qquad (10.3\text{--}7)$$

with an initial current given by $(V_{CC} - v_{CE\,\text{sat}})/R_C$.

For active pull-up, the equivalent capacitor charges through resisto: R_{C_P}. Thus, the associated active rise time is

$$\tau_{\text{active}} = R_{C_P} C_0 \qquad (10.3\text{--}8)$$

with a much larger initial current given by $(V_{CC} - V_0 - 2v_{CE\,\text{sat}})/R_{C_P}$ (which neglects i_{B_2}). The im-

provement in rise time with TTL is then provided by choosing $R_{C_P} \ll R_C$, where a factor of 10 is typical.

Note that the *pull-up transistor* (T_P in Fig. 10.9) is used to switch the output terminal voltage high when it turns on. Hence, this transistor is also called the *current-sourcing transistor* because, when it is turned on, it supplies current to the output. Similarly, the transistor T_1 is called the *pull-down transistor* because, when it turns on, it forces the output voltage to the low level. The pull-down transistor is also known as the *current-sinking transistor* because it conducts (sinks) the current from the output gates when it turns on.

10.3.5   TTL NAND Gate Transfer Characteristic

Figure 10.10 displays the TTL NAND gate transfer characteristic corresponding to the gate of Fig. 10.9 with active pull-up. We can verify this transfer characteristic by varying only one

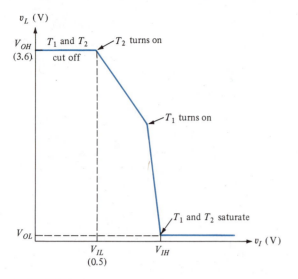

FIGURE 10.10　TTL NAND Gate Transfer Characteristic

input voltage ($v_{I_1} = v_I$) and assuming that the others are disconnected or high.

For the circuit of Fig. 10.9, for v_I low, current passes from V_{CC} through R_B and T_3 and out the input terminal, as shown by the path labeled v_I low. Furthermore, V_{CC} forces T_3 into saturation under these conditions, and the voltage at the base of T_2 is given by

$$v_{B_2} = v_I + v_{CE\,sat3} \qquad (10.3\text{–}9)$$

Hence, if v_I is low and $v_I + v_{CE\,sat3} < V_0$, T_2 will be cut off as will T_1. With T_1 and T_2 cut off, transistor T_P and diode D are both on. With the base current of T_P neglected, the output load voltage is

$$v_L = V_{CC} - 2V_0 = v_{OH} \qquad (10.3\text{–}10)$$

as shown in Fig. 10.10. Assuming silicon transistors where $V_0 = 0.7$ V and $V_{CC} = 5$ V, we have $V_{OH} = 3.6$ V. However, as v_I increases, v_{B_2} increases and T_2 will turn on and become active initially. Since T_3 is in saturation, T_2 will turn on and become active when $v_I + v_{CE\,sat3} = V_0$, or

$$v_I = V_0 - v_{CE\,sat3} = V_{IL} \qquad (10.3\text{–}11)$$

where we have indicated that this value of v_I corresponds to V_{IL} as shown in Fig. 10.10. By using values for silicon, we have $V_{IL} = 0.7 - 0.2 = 0.5$ V.

As v_I is increased further, the voltage across R_E increases and i_{E_2} increases. With T_2 being driven harder, v_{BE_1} increases and T_1 turns on when $v_{BE_1} = V_0$. Then, the input voltage is $v_I = 2V_0 - v_{CE\,sat3}$ (since T_3 is still saturated) and results in another break point in the transfer characteristic shown in Fig. 10.10. Transistor T_P also turns off in this region because the increase in i_{C_2} causes v_{B_P} to reduce as v_I increases.

Finally, when v_I increases to the level where T_1 and T_2 both become saturated, $v_I = v_{IH}$ where

$$v_{IH} \gtrsim 2V_0 \qquad (10.3\text{–}12)$$

For silicon transistors, this value is 1.4 V. The reason for indicating that v_{IH} is greater than but approximately equal to $2V_0$ in (10.3–12) is that at this high level of input voltage, T_2 and T_1 have both been driven into saturation, which requires a slightly larger value of forward voltage across the base–emitter junctions than active operation does. The output voltage when T_2 is in saturation is then given by

$$v_L = v_{CE\,sat} = v_{OL} \qquad (10.3\text{–}13)$$

where $v_{OL} \cong 0.2$ V for the case of silicon transistors.

It should be noted that after T_1 turns on and v_I is increased further, T_3 will eventually become inverse active. However, since T_1 and T_2 are both in saturation, the output remains unchanged.

10.3.6　TTL Fan-Out

The fan-out associated with TTL families is (as in the case of DTL) restricted by the low output level, $v_L = v_{CE\,sat}$, because this condition requires that T_1 sink a large current that is directly proportional to the fan-out. That is, for v_L high, the output gates conduct very little current because each of the connected multiple-emitter transistors operates in the inverse active mode. Hence, we determine the TTL fan-out by considering v_L to be low and therefore T_1 in saturation with $v_L = v_{CE\,sat}$. We will proceed here in a manner similar to the analysis of DTL and obtain expressions for the currents of the controlling transistor T_1 along with the degree of saturation σ.

For T_1 in saturation, T_P is cut off, and the collector current of T_1 is given by

$$i_{C_1} = K i_L \qquad (10.3-14)$$

where i_L is the load current for each gate with $v_L = v_{CE\,sat}$ and where we have assumed identical driven gates with a fan-out of K. Additionally, since the gates are identical,

$$i_{C_1} = K\left(\frac{V_{CC} - V_0 - v_{CE\,sat}}{R_B}\right) \qquad (10.3-15)$$

where i_{C_1} was obtained by observing the identical driving gate and writing KVL for the v_I low path.

The base current expression for T_1 is obtained by considering Fig. 10.9 and KCL at the base of T_1:

$$i_{B_1} = i_{E_2} - \frac{V_0}{R_E} \qquad (10.3-16)$$

However,

$$i_{E_2} = i_{C_2} + i_{B_2} \qquad (10.3-17)$$

By using KVL for the collector branch of T_2, we have (neglecting i_{B_P})

$$i_{C_2} = \frac{V_{CC} - v_{CE\,sat} - V_0}{R_C} \qquad (10.3-18)$$

and by using KVL for the base branch of T_3, we have

$$i_{B_2} = \frac{V_{CC} - 3V_0}{R_B} \qquad (10.3-19)$$

Substituting back into (10.3–16) yields

$$i_{B_1} = \frac{V_{CC} - v_{CE\,sat} - V_0}{R_C} + \frac{V_{CC} - 3V_0}{R_B} - \frac{V_0}{R_E} \qquad (10.3-20)$$

We again define the saturation parameter σ for TTL (in exactly the same manner as we did for DTL) as $\sigma = i_C / \beta i_B$ so that substitution for i_C and i_B yields

$$\sigma = \frac{[K(V_{CC} - V_0 - v_{CE\,sat})]/R_B}{\beta\left(\dfrac{V_{CE} - V_0 - v_{CE\,sat}}{R_C} + \dfrac{V_{CC} - 3V_0}{R_B} - \dfrac{V_0}{R_E}\right)} \qquad (10.3-21)$$

Solving for K yields

$$K = \sigma\beta\left[\frac{R_B}{R_C} + \frac{V_{CC} - 3V_0}{V_{CC} - V_0 - v_{CE\,sat}} - \frac{R_B}{R_E}\left(\frac{V_0}{V_{CC} - V_0 - v_{CE\,sat}}\right)\right] \qquad (10.3-22)$$

This equation shows some similarity to (10.2–9) for DTL. Again, we observe that K is proportional to the $\sigma\beta$ product. However, for TTL, R_B/R_C is a multiplying factor. This ratio can be made ~ 10 to provide very large fan-out. In (10.3–22), the term containing the factor $V_{CC} - 3V_0$ is approximately unity and adds to the R_B/R_C term to increase the fan-out. The last term in (10.3–22) reduces the fan-out because it is negative; however, its magnitude is also about unity because R_B/R_E is typically 10 at most, while the voltage ratio is typically of order 1/10. Thus, the fan-out for TTL is given approximately by

$$K \cong \sigma\beta \frac{R_B}{R_C} \qquad (10.3-23)$$

——————————————————— EX. 10.6

Fan-Out Calculation for TTL Example

Calculate the fan-out for the TTL NAND gate of Fig. 10.9. Let $R_B = 4\ \text{k}\Omega$, $R_C = 1.4\ \text{k}\Omega$, $R_E = 1\ \text{k}\Omega$, and $V_{CC} = 5\ \text{V}$, while $\beta = 25$, $V_0 = 0.7\ \text{V}$, $v_{CE\,sat} = 0.2\ \text{V}$, and $\sigma = 0.85$.

Solution: Substituting these values directly into (10.3–22) yields

$$K = (0.85)(25)\left[\frac{4}{1.4} + \frac{5 - 2.1}{5 - 0.7 - 0.2}\right.$$
$$\left. - \frac{4}{1}\left(\frac{0.7}{5 - 0.7 - 0.2}\right)\right]$$
$$= 21.25[2.86 + 0.707 - 0.68]$$
$$= 61.3$$

For comparison, note that substitution into the approximate expression given by (10.3–23) yields $K = 60.7$.

Note that the fan-out for TTL is larger than for any of the previous logic families. However, it should further be noted that manufacturers typically specify a maximum fan-out for particular logic families. For TTL, the maximum fan-out is usually 15.

Additional circuit modifications of TTL are often included to further improve operating performance. These modifications are also used with Schottky TTL, which will be discussed in Chapter 11.

10.3.7   Open-Collector Output

Many logic gates in the TTL family are available with an *open-collector output*, which is indicated in data sheets by specifying "open-collector." This specification means that the pull-up transistor (T_P in Fig. 10.9) is not in the circuit. The advantage of such an arrangement is that open-collector outputs can be wired together with one external pull-up resistor. This connection is referred to as a *wired-AND* because this output is high only when all pull-down transistors are off. A wired-AND connection results in the saving of one AND gate.

Wiring the outputs of normal TTL gates together leads to an intolerable situation. That is, one gate may have its pull-up transistor in saturation (supplying or sourcing current), while another gate may have its pull-down transistor in saturation (acting as a low-resistance or sinking current). The use of open-collector gates avoids the resultant excessive power dissipation.

10.4   STANDARD TTL INTEGRATED CIRCUITS

This section introduces logic gates in the 5400/7400 series of standard TTL ICs. The same numbering system is used for the Schottky TTL (STTL) ICs that are described in Chapter 11. The difference between the 5400 and 7400 TTL series is that the 5400 series is operable in the range of temperatures from $-55°$ C to 125° C, which satisfies military specifications, whereas the 7400 series is operable in the range of temperatures from 0° C to 70° C. Often, this line of ICs is referred to as the 54/74 series. It has for many years been the dominant logic family for general-purpose applications.

In completing this chapter, we will examine standard TTL parameters and then consider a few of the commonly used gates that are available in standard TTL form. We will observe that the packages used for TTL gates are similar to those used for the linear ICs described in Chapter 6.

The internal operation and exact circuitry of gates packaged as ICs are no longer of primary concern. Instead, the primary concerns are the input and output parameters and the pin layouts for each particular gate. Typical parameters for standard TTL are described next; their particular values are indicated in the data sheets of Appendix 6 at the end of the book.

10.4.1   Standard TTL Parameters

The supply voltages (V_{CC}) used with the 7400 series of TTL range from 4.75 V to 5.25 V for reliable operation. For the 5400 series, this range is widened to 4.5 V and 5.5 V. Voltage values greater than these can cause circuit damage (breakdown of the base–emitter junction of T_3), and voltages values that are less can cause erratic operation.

As far as standard TTL is concerned, typical values of propagation delay, power dissipation per gate, and power–delay product are 10 ns, 10 mW, and 100 pJ, respectively. These values can each be improved upon by approximately an order of magnitude by using the various Schottky lines as described in Chapter 11. Typical DC noise margins are approximately 1 V to 2 V, and a nominal fan-out is 10.

10.4.2   Combinational TTL Gates

7400 Quad Two-Input NAND Gate.   As described in Section 10.3, the primary combinational gate for TTL is the NAND gate. Hence, it is natural that the first logic gate in the 7400 series is the 7400 quad two-input NAND gate. This IC consists of four two-input NAND gates contained in a 14-pin dual in-line package (DIP). Seven pins (or terminals) are located in-line on either side of the chip, as shown in Fig. 10.11. In this type of diagram, referred to as a *pin layout,* or *pinout,* the pin numbers and the notch are positioned accord-

ing to a top view of the IC. Note that it is customary to indicate the particular gates and their circuit symbols (four NAND gates in this case) on the IC with specific connections to the pins. Pinouts are given on data sheets regardless of the type of logic family.

7402 Quad Two-Input NOR Gate.   Figure 10.12 displays the pinout for a 7402 quad two-input NOR gate. As shown, this IC is also packaged in a 14-pin DIP with the NOR gates and their pin connections clearly evident. Note that in comparing the 7400 and the 7402 14-pin DIPs, pin 7 is used for ground (*GND*), while pin 14 is used for the DC supply (V_{CC}). These two terminals are typically reserved for these purposes.

7404 Hex Inverter.   The pinout for the 7404 hex inverter is displayed in Fig. 10.13. Note that six (hex) inverters are presented on this IC.

7410 Triple Three-Input NAND Gate.   As another example of a 7400 series logic gate, the 7410 is shown in Fig. 10.14. This IC consists of three NAND gates and each gate has three inputs. This IC is also contained in a 14-pin DIP.

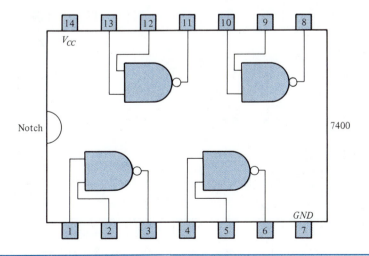

FIGURE 10.11   7400 Quad Two-Input NAND Gate

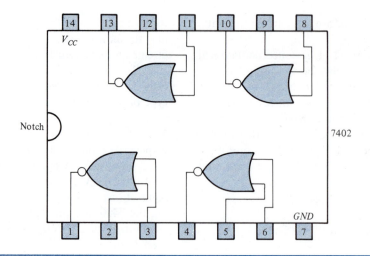

FIGURE 10.12   7402 Quad Two-Input NOR Gate

10.4.3   Sequential TTL Gates

Two examples of dual flip-flops in the 7400 series are shown in Figs. 10.15 and 10.16 along with their truth tables. Each chip has two flip-flops with preset ($Q = 1$) and clear ($Q = 0$) inputs. These flip-flops are edge triggered. The 7474

dual D type is positive-edge triggered, and the 7476 dual JK type is negative-edge triggered.

Numerous other standard TTL chips are available from manufacturers. Additional information on TTL gates is presented in Chapter 11 and in Appendix 6 at the end of the book.

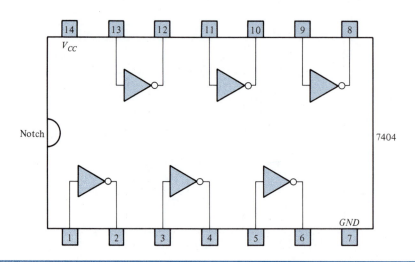

FIGURE 10.13   7404 Hex Inverter

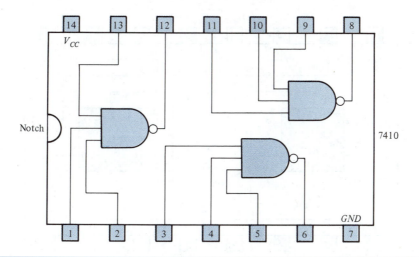

FIGURE 10.14   7410 Triple Three-Input NAND Gate

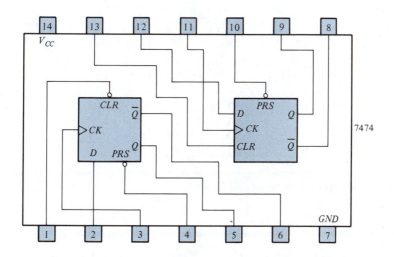

Truth Table

	Inputs			Outputs	
PRS	CLR	CK	D	Q	$\overline{Q}$
L	H	—	—	H	L
H	L	—	—	L	H
L	L	—	—	H*	H*
H	H	↑	H	H	L
H	H	↑	L	L	H
H	H	L	—	Q_0	$\overline{Q_0}$

FIGURE 10.15   7474 Dual D Flip-Flops with Preset and Clear

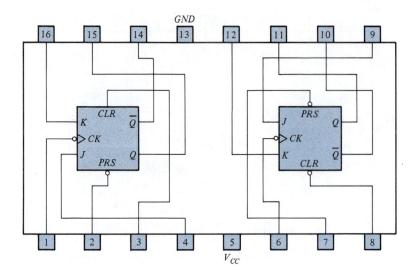

Truth Table

Inputs					Outputs	
PRS	CLR	CK	J	K	Q	$\overline{Q}$
L	H	—	—	—	H	L
H	L	—	—	—	L	H
L	L	—	—	—	H*	H*
H	H	⊓	L	L	Q_0	$\overline{Q_0}$
H	H	⊓	H	L	H	L
H	H	⊓	L	H	L	H
H	H	⊓	H	H	Toggle	

FIGURE 10.16   7476 Dual *JK* Flip-Flop with Preset and Clear

---

CHAPTER 10 SUMMARY

- Saturating logic families utilize BJTs that are designed to operate either in cutoff or in saturation. A serious drawback for these types of logic families is the time delay associated with switching the BJT from saturation to cutoff.

- RTL and DTL are two saturating logic families that exhibit undesirable voltage degradation and low fan-out, as well as poor speed characteristics. RTL stands for resistor–transistor logic, and DTL stands for diode–transistor logic. These families have been superceded by TTL, or transistor–transistor logic, and more advanced families.

- Voltage transfer characteristics for RTL and DTL inverters exhibit narrow active regions, which become narrower for larger β. The transfer characteristic for TTL displays a slightly wider active region.

- The basic logic circuit for DTL and TTL is the NAND gate.

- The TTL NAND gate transfer characteristic displays $V_{OH} = 3.6$ V, $V_{OL} = 0.2$ V, $V_{IL} = 0.5$ V, and $V_{IH} \cong 1.4$ V.

- TTL provides an order of magnitude improvement in speed and increased fan-out over that available with RTL and DTL because of its specialized circuitry, which includes the input multiple-emitter transistor and the output active pull-up.

■ The standard TTL logic family is numbered beginning at 5400 or 7400. The 5400 line is operable over a temperature range of $-55°$ C to $125°$ C, whereas the 7400 is operable over the range of $0°$ C to $70°$ C. Typical values of standard TTL parameters are propagation delay of 10 ns, power dissipation per gate of 10 mW, nominal fan-out of 10, and power–delay product of 100 pJ. Many various TTL gates are available from manufacturers.

CHAPTER 10 PROBLEMS

10.1
Analyze the circuit of Fig. P10.1 to determine the currents i_{R_1}, i_{R_2}, i_B, and i_C for $v_I = 0$, 0.5, 2 and 5 V, use $\beta = 100$ and $V_0 = 0.7$ V. Can the transistor ever become saturated?

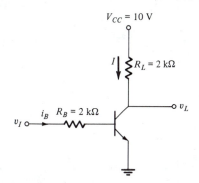

FIGURE P10.2

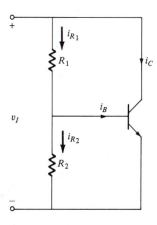

FIGURE P10.1

10.2
For the RTC circuit shown in Fig. P10.2, determine i_C, i_B, and v_L for $v_I = 10$ V. Let $\beta = 100$, $V_0 = 0.7$ V, and $v_{CE \, sat} = 0.2$ V. For a fan-out of 4 (assuming identical gates), let $v_I = 0$ and determine I and v_L. Draw an equivalent circuit and assume that the driven transistors are in saturation to obtain the solution.

10.3
Determine and sketch the voltage transfer characteristic (v_L versus v_I) for the inverter circuit of Fig. P10.2 (with no output gates). Let $\beta = 100$, $V_0 = 0.7$ V, and $v_{CE \, sat} = 0.2$V.

10.4
Repeat Problem 10.3 for a fan-out of 4. Assume identical output gates.

10.5
For the circuit of Fig P10.2, let $R_B = R_C = 1$ kΩ and $V_{CC} = 5$ V. Determine i_C, i_B, and v_L for $v_I = 5$ V. Let $\beta = 100$, $V_0 = 0.7$ V, and $v_{CE \, sat} = 0.2$ V. For a fan-out of four identical gates, let $v_I = 0$ and determine I, i_B, and v_L. Draw an equivalent circuit with the driven transistors in saturation to obtain the solution.

10.6
For the RTL inverter of Fig. P10.6, determine i_{Bp}, i_{Cp}, and v_L for $\beta = 100$, $V_0 = 0.7$, $v_{CE \, sat} = 0.2$ V, and $v_I = 3$ V (with $i_L = 0$). Use an equivalent circuit.

10.7
Repeat Problem 10.6 for $v_I = 0$ V.

10.8
Determine the voltage transfer characteristic for the circuit of Fig. P10.6 assuming a fan-out of two identical circuits. Use $\beta = 100$, $V_0 = 0.7$ V, $v_{CE \, sat} = 0.2$ V, and an equivalent circuit to obtain the solution.

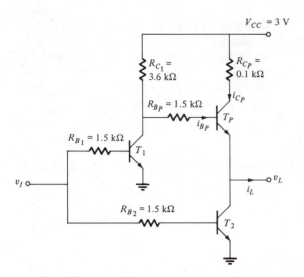

FIGURE P10.6

10.9

For the DTL circuit of Fig. P10.9, determine the transfer characteristic for v_L versus v_{I_1} (with $v_{I_2} = 10$ V). Use $\beta = 100$, $V_0 = 0.7$V, and $v_{CE\,sat} = 0.2$ V.

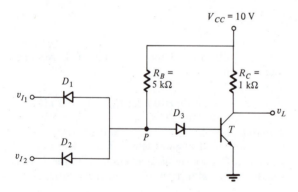

FIGURE P10.9

10.10

For the DTL gate of Fig. P10.9, determine the output voltage v_L for all possible combinations of input voltages (high and low). Let $\beta = 100$, $V_0 = 0.7$ V, and $v_{CE\,sat} = 0.2$ V. What logic operation does the circuit perform? Construct a truth table with voltages.

10.11

For the DTL gate of Fig. P10.11, determine the transistor base current, collector current, and output voltage v_L for all possible combinations of input (high and low). Let $\beta = 100$, $V_0 = 0.7$ V, and $v_{CE\,sat} = 0.2$ V. What logic operation does the circuit perform? Construct a truth table with voltages.

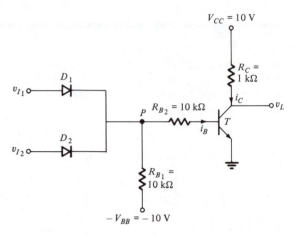

FIGURE P10.11

10.12

For the DTL gate of Fig P10.12, determine i_B, i_C, and v_L for all possible combinations of inputs. Let $\beta = 100$, $V_0 = 0.7$ V, and $v_{CE\,sat} = 0.2$ V. Construct a truth table with voltages.

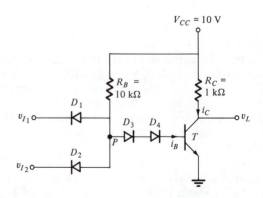

FIGURE P10.12

10.13

For the DTL gate of Fig. P10.13, determine i_B, i_C, and v_L for all possible combinations of inputs. Let $\beta = 100$, $V_0 = 0.7$ V, and $v_{CE \, \text{sat}} = 0.2$ V.

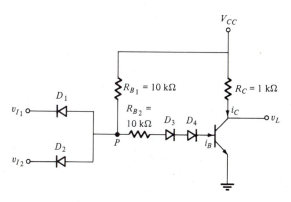

FIGURE P10.13

10.14

For the gate of Fig. P10.14, all BJTs have $\beta = 100$, $V_0 = 0.7$ V, and $v_{CE \, \text{sat}} = 0.2$ V. Determine the following (assuming that T_3 is in saturation when on):

(a) i_{D_3}, i_{B_3}, i_{C_3}, and v_L for both inputs low (0.2 V)

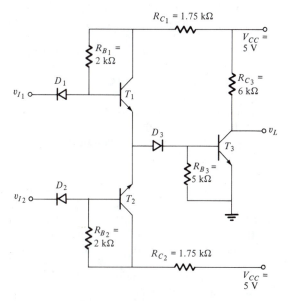

FIGURE P10.14

(b) Repeat for one input high (5 V). What logic operation does the circuit perform? Construct a truth table with voltages.

10.15

Note that the gate of Fig. P10.14 does not have a fan-out limitation for v_L high; however, for v_L low, a fan-out problem does exist. Determine the maximum fan-out under this condition if the maximum collector current for T_3 is 20 mA. Use an equivalent circuit and the data of Problem 10.14.

10.16

For the DTL NAND gate of Fig. 10.6, let $R_C = R_B = 1$ kΩ, $\rho = 0.4$, and $V_{CC} = 5$ V, while $\beta_{\text{min}} = 25$, $V_0 = 0.7$ V, $V_{CE \, \text{sat}} = 0.2$ V, and $\sigma = 0.85$. For a fan-out of 10, determine the high and low values of v_L. Also, determine the maximum fan-out.

10.17

Repeat Problem 10.16 for $\rho = 0.3$ and all other parameters unchanged.

10.18

For the basic TTL inverter of Fig. P10.18, assume that $v_I = 0$ (low) and determine i_I, i_{R_B}, i_{B_1}, i_{C_1}, and v_L. Under these conditions, realize that T_3 is in saturation, draw a diagram for T_3 and T_1, and indicate the specific value of all terminal voltages. Use $\beta = 100$, $V_0 = 0.7$ V, and $v_{CE \, \text{sat}} = 0.2$ V.

10.19

Repeat Problem 10.18 for $v_I = 3$ V (high). What type of altered operation does T_3 now display?

10.20

Sketch the transfer characteristic for the basic TTL gate of Fig. P10.18. Label all high and low voltages

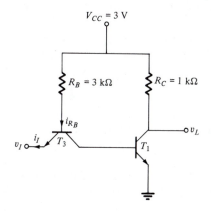

FIGURE P10.18

for v_I and v_L and assume that T_1 saturates when $v_{BC_3} = 0.7$ V.

10.21

For the TTL gate of Fig. P10.21, determine the high and low output voltage levels (V_{OH} and V_{OL}). Use $\beta = 100$, $V_0 = 0.7$ V, and $v_{CE\,sat} = 0.2$ V and indicate the operating state of T_3, T_2, and T_1 in each case.

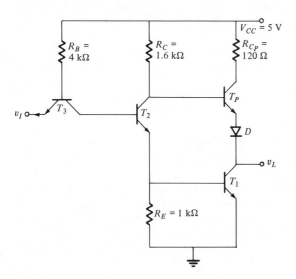

FIGURE P10.21

10.22

For the TTL inverter of Fig. P10.21, assume that T_1 and T_2 are in saturation and determine the currents i_{B_P} and i_{C_P}. What operating state is T_P in?

10.23

For the TTL invert of Fig. P10.21, assume that T_1 and T_2 are cut off and that V_L is measured as 3.5 V with output gates connected. Determine i_{B_P} and i_{C_P} for $\beta = 100$, $V_0 = 0.7$ V, and $v_{CE\,sat} = 0.2$ V.

10.24

Verify the voltage transfer characteristic of the TTL inverters of Fig. 10.10. Assume that T_1 saturates when $v_{B_3} = 2.1$ V.

10.25

For the TTL inverter of Fig P10.21, determine the power supplied by V_{CC} under the following conditions:

    (a) v_I is high (≥ 3.6 V).
    (b) v_I is low (≤ 0.2 V).

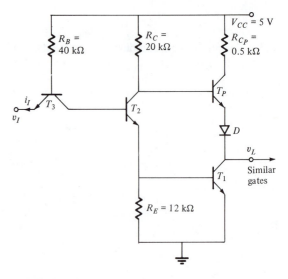

FIGURE P10.27

10.26

For the inverter of Fig. P10.21, T_1 and T_P can saturate simultaneously when T_1 is switching form saturation to cutoff. Under these conditions, determine i_{C_P} and the terminal voltages of T_P. Also, calculate the power supplied for this case.

10.27

For the TTL inverter of Fig. P10.27, assume that the output is connected to similar gates and determine the following:

    (a) for $v_I = 0.2$ V and a fan-out of 1 (assuming T_P is active), determine v_{B_3}, i_{B_3}, i_I, and v_L.
    (b) For v_I high, determine the same quantities.

10.28

Determine the maximum fan-out for the TTL gate of Fig. P10.21. Let $\sigma = 0.85$, $\beta_{\min} = 25$, $V_0 = 0.7$V, and $v_{CE\,sat} = 0.2$ V. Derive the expression and calculate the value.

10.29

Repeat Problem 10.28 for the TTL gate of Fig. P10.27.

10.30

For the TTL inverter of Fig. P10.27, determine the power supplied by V_{CC} under the following conditions:

    (a) v_I is high (≥ 3.6 V).
    (b) v_I is low (≤ 0.2 V).

Compare the results with those of Problem 10.25. Calculate the currents for T_P assuming $v_L = 3.5$ V.

11

NONSATURATING LOGIC FAMILIES

This chapter introduces *nonsaturating* logic families. The chapter begins with a description of the various versions of Schottky TTL (STTL). The emphasis is on the internal operation of these gates and the manner in which improved operation over standard TTL is provided. Following this description, the emitter-coupled logic (ECL) family is introduced. This type of logic is faster than TTL; however, more power dissipation is associated with this family of gates.

MOS logic circuits composed of N-channel MOSFETs, P-channel MOSFETs, or both (as in the case of CMOS, or complementary MOS) are introduced next. Logic families consisting of MOS transistors are far superior with regard to power dissipation because the gate currents are very small. As in the case of STTL and ECL, the emphasis in this chapter is initially on internal operation, and actual logic gates are presented in later sections.

In order to use different logic families in the same circuit, an interface circuit is necessary. Such interface circuits transform the output current and voltage levels from one family into compatible input current and voltage levels for the next logic family. A description of various interface circuits in given into Section 11.4. The chapter concludes by listing an assortment of functionally comparable STTL, ECL, and CMOS ICs.

11.1   SCHOTTKY TRANSISTOR–TRANSISTOR LOGIC (STTL)

Schottky transistor–transistor logic (STTL) utilizes the same circuitry as TTL as described in Chapter 10 except that specialized BJTs called *Schottky transistors* are used that do not operate in saturation. Hence, this logic family is of the nonsaturating type, and delay times are appreciably reduced from those achievable with TTL. The fan-out of STTL is high and is essentially the same as that of standard TTL.

11.1.1   Schottky Transistor

The speed limitation of TTL is primarily due to the time delay associated with the stored-charge removal from the base region of the BJT when it is switched from saturation into cutoff. The Schottky transistor is a modified BJT that does not permit operation in saturation. It is a simple and easily fabricated device, as we will see.

The Schottky transistor consists of a BJT with a Schottky diode connected directly in parallel with its base and collector terminals. The equivalent circuit for the Schottky transistor is shown in Fig. 11.1a.

Recall from Chapter 2 that a Schottky diode is a metal-semiconductor diode with a forward

voltage of $V_0 = 0.3$ V and reverse voltage behavior similar to that of a PN junction diode. Hence, as long as a reverse voltage is applied across the collector–base junction, the operation of the Schottky transistor is identical to that of the ordinary BJT.

When the base–collector junction is forward biased, however, the maximum forward voltage that can be applied across this junction (with the Schottky diode in parallel) is the forward voltage of the Schottky diode, or 0.3 V. This magnitude of forward voltage is insufficient to drive the BJT into saturation. Thus, the Schottky transistor never operates in saturation.

Figure 11.1b shows the circuit symbol used to represent the Schottky transistor. Also shown are the voltage values when both junctions of the BJT are forward biased to their extremes. Note that the collector–emitter voltage under these conditions is

$$v_{CE} = v_{BE} + v_{CB} \qquad (11.1–1)$$

or, with substitution of values,

$$v_{CE} = 0.7 - 0.3 = 0.4 \text{ V} \qquad (11.1–2)$$

This voltage is sufficiently larger than $v_{CE\,\text{sat}}$ ($\cong 0.2$ V) to maintain operation of the BJT in the active region.

Fabrication of the Schottky diode is easily accomplished, as indicated by the transistor cross section of Fig. 11.2. The only difference between the Schottky transistor cross section and that of the ordinary transistor is overlap of the base-metal contact with the collector region. Thus, the base metal is deposited in a larger surface region, and fabrication of a Schottky transistor requires no additional IC processing steps.

In cross sections of transistors shown previously, the metal layer was not indicated because the metal used (usually Al) makes an ohmic contact (one that exhibits a linear i–v relation) with N^+ and P-type silicon. However, in the Schottky

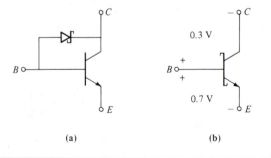

(a)   (b)

FIGURE 11.1   Schottky Transistor: (a) Equivalent circuit, (b) Circuit symbol and voltages corresponding to maximum base–collector and base–emitter voltages

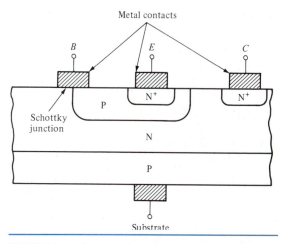

FIGURE 11.2   Cross Section of Integrated-Circuit Schottky Transistor

transistor case, the metal and lightly doped N region form a rectifying junction or a Schottky diode.

11.1.2 Basic STTL Gate

Figure 11.3 displays the basic STTL gate, which is essentially obtained from the basic circuit of TTL (Fig. 10.11a) by replacing the BJTs with Schottky transistors. However, two exceptions are present. First, the diode D and transistor T_P of Fig. 10.9 are replaced with a Darlington connection of transistors T_{P_1} and T_{P_2} and a resistor R_3. Note that the base–emitter voltages of T_{P_1} and T_{P_2} in Fig. 11.3 provide the necessary two

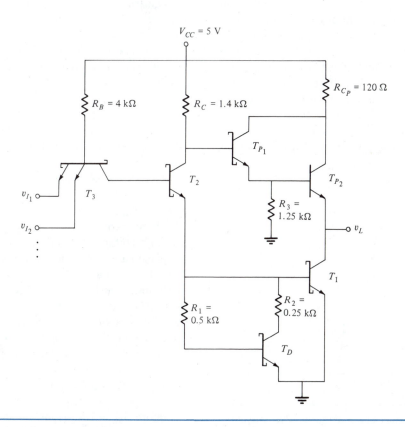

FIGURE 11.3   Basic High-Power STTL Gate

diode forward voltages between the collector of T_2 and the output terminal. Additionally, however, the Darlington-pair connection provides a much larger output current and fan-out improvement because of the larger current amplification available with this connection. Second, R_E in Fig. 10.9 is replaced with R_1, R_2, and T_D in Fig. 11.3. The transistor T_D provides active pull-down of T_1, which allows faster turnoff since a larger current passes through T_D than with R_E only, as T_1 switches from active operation to cutoff.

Note that the transistor T_{P_2} is not a Schottky transistor, because T_{P_2} can never operate in saturation, which requires the emitter and collector junctions to both be forward biased. However, the base–collector terminals of T_{P_2} are in parallel with the base–emitter terminals of T_{P_1}, and this voltage is 0.4 V when T_{P_1} is conducting. Hence, the collector junction of T_{P_2} cannot become sufficiently forward biased to drive T_{P_2} into saturation.

The transfer characteristic for the STTL gate of Fig. 11.3, with all inputs high except v_{I_1}, is displayed in Fig. 11.4. This characteristic is easily verified by considering the conditions in which T_2 is on (active) and off (cut off). Note that T_{P_1} is always on because of resistor R_3. Thus, for T_2 also on (v_I high), v_{C_2} is obtained using KVL as

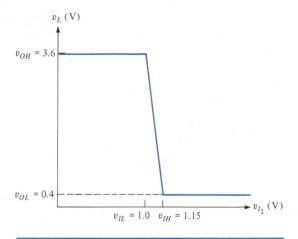

follows:

$$v_{C_2} = v_{BE_1} + v_{CE_2} \qquad \text{(11.1–3)}$$

or

$$v_{C_2} = 0.7 + 0.4 = 1.1 \text{ V}$$

Thus, the voltage at the base of T_{P_2} is

$$v_{BP_2} = v_{C_2} - v_{BP_1} \qquad \text{(11.1–4)}$$

or

$$v_{BP_2} = 1.1 - 0.7 = 0.4 \text{ V}$$

which results in T_{P_2} off and the lowest output voltage (since T_1 is a Schottky transistor). This voltage value is indicated in Fig. 11.4 as $V_{OL} = 0.4$ V.

For T_2 off (v_{I_1} low), both T_{P_1} and T_{P_2} are on, and the output voltage is obtained using KVL as follows:

$$v_L = V_{CC} - i_{BP_1} R_C - 2V_0 \qquad \text{(11.1–5)}$$

where i_{BP_1} is quite small. Thus,

$$v_L = 5.0 - 1.4 = 3.6 \text{ V}$$

which is V_{OH}, as also indicated in Fig. 11.4.

We now obtain the corresponding input low and high voltages V_{IL} and V_{IH}. As the input voltage is increased from the low range, where T_1 and T_2 in Fig. 11.3 are off, both T_1 and T_2 will turn on simultaneously when v_{B_2} becomes equal to $2V_0$. Both BJTs turn on simultaneously because no path for current exists from B_2 to ground until both T_1 and T_2 turn on. Since T_3 is a Schottky transistor, $v_{BC_3} = 0.4$ V at this point, and

$$v_{I_1} = 2V_0 - 0.4 = 1 \text{ V} \qquad \text{(11.1–6)}$$

FIGURE 11.4 Transfer Characteristic for STTL Gate of Fig. 11.3

Therefore, $V_{IL} = 1$ V, as indicated in Fig. 11.4. As v_I is increased above V_{IL}, both T_1 and T_2 draw additional current, and their base–emitter voltages must increase since they are being driven harder. The value of input voltage then required is about 0.1 V to 0.2 V greater than V_{IL}. Therefore $V_{IH} \cong 1.15$ V, as also indicated in Fig. 11.4.

11.1.3 Other Forms of STTL

The family of STTL described in the previous section is sometimes referred to as *high-power* STTL because each gate dissipates on the order of 20 mW of power, which is quite high. If the resistor values are increased, the power dissipation is reduced with, however, a corresponding reduction in speed. Nonetheless, *low-power*

STTL (LSTTL) uses resistor values that are approximately ten times larger than those of high-power STTL which results in a reduced power dissipation to approximately 2 mW/gate. The propagation delay for the low-power case is approximately 10 ns, whereas for the high-power case, it is approximately 2 ns.

Figure 11.5 displays the circuit configuration of the NAND gate for LSTTL. In addition to larger resistor values (which mean lower current), several circuit modifications can be observed by comparing Figs. 11.5 and 11.3.

Note that in the LSTTL circuit, the multiple-emitter transistor at the input has been replaced with diodes (D_{in_1}, D_{in_2}, ...). This replacement is possible for STTL since the original function of

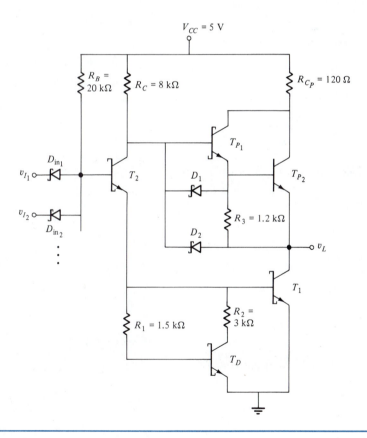

FIGURE 11.5 Basic Low-Power STTL Gate

the multiple-emitter transistor was to aid in the removal of stored charge from saturated transistors and Schottky transistors do not saturate.

Diodes D_1 and D_2 decrease the turnoff speed of T_{P_2} when v_L switches from high to low. Diode D_1 provides a direct discharge path for the base of T_{P_2} when T_2 turns on. Diode D_2 speeds the pull-down of the output voltage.

───────────────────────────── EX. 11.1

STTL and LSTTL Power Dissipation Example

Calculate the power dissipation for the high-power and low-power STTL gates of Figs. 11.3 and 11.5, respectively. Let $\beta = 30$ for each of the BJTs.

Solution for High-Power Case: For the circuit of Fig. 11.3, for v_L low (v_{I_1} high), current supplied by V_{CC} passes through R_B and R_C and is given by

$$i_{R_B} = \frac{V_{CC} - V_{BE_1} - V_{BE_2} - V_{BC_3}}{R_B} = \frac{5 - 1.4 - 0.3}{4}$$

$$= 0.825 \text{ mA}$$

and

$$i_{R_C} = \frac{V_{CC} - V_{CE_2} - V_{BE_1}}{R_C} = \frac{5 - 0.4 - 0.7}{1.4} = 2.79 \text{ mA}$$

and $i_{R_{CP}} = 0$. Total current supplied by V_{CC} is therefore

$$i_T = 0.825 + 2.79 = 3.615 \text{ mA}$$

The power dissipated for v_L low is therefore

$$P_D = (3.615)(5) = 18.075 \text{ mW}$$

For v_L high (v_{I_1} low), the currents supplied by V_{CC} are

$$i_{R_B} = \frac{V_{CC} - V_{BE_3} - V_{OL}}{R_B} = \frac{5 - 0.7 - 0.4}{4} = 0.975 \text{ mA}$$

and

$$i_{R_C} = \frac{V_{CC} - V_{BE_{P_1}} - i_{R_3}R_3}{R_C}$$

$$= \frac{5 - 0.7 - i_{R_3}(1.25)}{1.4} \tag{1}$$

where $i_{B_{P_1}}$ was neglected, and

$$i_{R_{C_P}} = \frac{V_{CC} - V_{CE_{P_1}} - i_{R_3}R_3}{R_{C_P}} = \frac{5 - 0.4 - i_{R_3}(1.25)}{0.12} \tag{2}$$

where i_{R_3} must be determined. An additional equation is obtained by writing KCL for the common-collector terminal of the Darlington:

$$i_{R_{C_P}} = i_{C_{P_1}} + i_{C_{P_2}} \tag{3}$$

By using transistor relations for active operation, we have

$$i_{R_{C_P}} = \beta i_{R_C} + \beta i_{B_{P_2}} \tag{4}$$

But,

$$i_{B_{P_2}} = i_{E_{P_1}} - i_{R_3} = (\beta + 1)i_{R_C} - i_{R_3} \tag{5}$$

where $i_{B_{P_1}} = i_{R_C}$ was used. Thus, by substituting into (3), we have

$$i_{R_{C_P}} = \beta i_{R_C} + \beta[(\beta + 1)i_{R_C} - i_{R_3}] \tag{6}$$

$$= [\beta + \beta(\beta + 1)]i_{R_C} - \beta i_{R_3}$$

$$= (\beta^2 + 2\beta)i_{R_C} - \beta i_{R_3}$$

or, with $\beta = 30$,

$$i_{R_{C_P}} = 960 i_{R_C} - 30 i_{R_3} \tag{7}$$

Equations (1), (2), and (7) are now solved for the unknowns to yield $i_{R_C} = 0.107$ mA, $i_{R_{C_P}} = 3.75$ mA, and $i_{R_3} = 3.32$ mA. Note that it is necessary to solve these three equations simultaneously because the magnitudes of $i_{B_{P_2}}$ and i_{R_3} are not known relative to each other. Thus, the total current for v_L high (v_{I_1} low) is

$$i_T = i_{R_B} + i_{R_C} + i_{R_{C_P}} = 0.975 + 0.107 + 3.75$$

$$= 4.832 \text{ mA}$$

The power dissipated is

$$P_D = (4.832)(5) = 24.16 \text{ mW}$$

The average power dissipated per gate is therefore

$$P_{D\text{avg}} = \frac{18.075 + 24.16}{2} = 21.12 \text{ mW}$$

Solution for Low-Power Case: For the circuit of Fig. 11.5, we first note that the resistor values for R_B and R_C have been increased from the previous case; hence, we expect lower power dissipation. For v_L low (v_{I_1} high), the currents are

$$i_{R_B} = \frac{V_{CC} - v_{BE_1} - v_{BE_2}}{R_B} = \frac{5 - 1.4}{20 \text{ k}\Omega} = 0.18 \text{ mA}$$

and

$$i_{R_C} = \frac{V_{CC} - V_{CE_2} - v_{BE_1}}{R_C} = \frac{5 - 0.4 - 0.7}{8 \text{ k}\Omega}$$
$$= 0.4875 \text{ mA}$$

and $i_{R_{C_P}} = 0$. The total current is therefore

$$i_T = 0.6675 \text{ mA}$$

The power dissipation is

$$P_D = (5)(0.6675) = 3.3375 \text{ mW}$$

For v_L high (v_{I_1} low), the currents are

$$i_{R_B} = \frac{V_{CC} - v_{D_{in_1}} - V_{OL}}{R_B} = \frac{5 - 0.7 - 0.4}{20 \text{ k}\Omega}$$
$$= 0.195 \text{ mA}$$

and, for no load, $i_{R_C} = i_{R_{C_P}} = 0$. With a load connected, these currents are increased; however, i_{R_B} remains the most significant. Thus, the total current for v_L low is the current through R_B, and the power dissipation is

$$P_D = (5)(0.195) = 0.975 \text{ mW}$$

The average power dissipated per gate is therefore

$$P_{D \text{ avg}} = \frac{3.3375 + 0.975}{2} = 2.16 \text{ mW}$$

Other forms of STTL besides the high-power and low-power forms are available from manufacturers. Advanced STTL families use advanced Schottky transistor structures that require less area and have reduced parasitic capacitance associated with them. The ASTTL family (advanced Schottky) utilizes small resistors that result in higher speed and higher power dissipation. The ALSTTL family (advanced low-power Schottky) also uses large resistors. The description of these specific STTL gates will be deferred until later in this chapter.

11.2   EMITTER-COUPLED LOGIC (ECL)

Emitter-coupled logic (ECL) is another nonsaturating logic family. Since the transistors operate in the active region or are cut off, propagation delays associated with saturation are avoided. This family is also referred to as *current-mode logic* (CML) because of the inherent current-switching mechanism of operation.

11.2.1   Basic ECL Inverter

The basic ECL inverter is displayed in Fig. 11.6a. This configuration is essentially the same as that of the difference amplifier introduced in Chapter 4. It also requires matched BJTs as well as collector resistors. The current through R_E is provided by either T_1 or the reference transistor T_R, depending upon whether T_1 is active (on) or cutoff (off). When $v_I < V_R$, T_1 is cutoff while T_R is active and the current through R_E is provided by T_R. When $v_I = V_R$, this current is provided by both transistors. Additionally, as v_I increases from low to high, T_R turns off while T_1 turns on and the current switches from T_R to T_1. Therefore, the term *current-mode logic* applies.

For purposes of analysis of the basic ECL inverter of Fig. 11.6a, we will initially take $V_{EE} = 0$ and assume that $V_R \geq V_0$. However, in actual ECL gates, V_{EE} is nonzero and $V_{CC} = 0$ so that negative voltages result throughout the entire circuit. However, this is not an amenable initial solution.

Figure 11.6b shows the voltage transfer characteristic for v_L versus v_I corresponding to the basic inverter circuit of Fig. 11.6a. Note that as

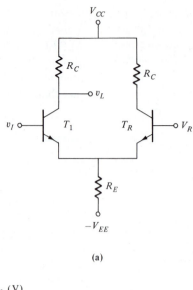

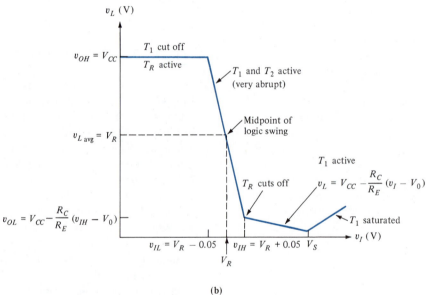

FIGURE 11.6   ECL Inverter: (a) Basic circuit, (b) Transfer characteristic for $V_{EE} = 0$

v_I increases from a low value $(v_I < V_R)$ to a high value $(v_I > V_R)$, the output switches from high to low and provides the logic inverter operation. When $v_I = V_R$, both T_1 and T_R are active; however, with only a slight increase (or decrease) in v_I from

V_R, the output becomes high (or low). Hence, the range of v_I over which both transistors are active is quite narrow.

We begin the analysis of Fig. 11.6a by considering the case in which $v_I = V_R$ and both BJTs are active. The base–emitter voltages of each transistor are then given by V_0. Under these conditions, the voltage at E is obtained from either the T_1 or T_R side as follows:

$$v_E = V_R - V_0 \qquad\qquad (11.2\text{--}1)$$

Note that this expression continues to be valid even for $v_I < V_R$ because T_R has a fixed voltage V_R applied at its base terminal and the reduction in v_I must therefore appear directly across the base–emitter terminals of T_1. Thus, for $v_I < V_R$,

$$v_{BE_1} = v_I - v_E = v_I - V_R + V_0 \qquad (11.2\text{--}2)$$

and since $v_I < V_R$, $v_{BE_1} < V_0$ so that T_1 turns off. Since T_1 is off, the output load voltage (v_L in Fig. 11.6a) is high and $v_L = V_{CC}$.

For $v_I > V_R$, the voltage at E must increase because T_1 is being driven harder by its increased base voltage. The voltage at E is then determined from the T_1 side of the circuit and is given by

$$v_E = v_I - V_0 \qquad\qquad (11.2\text{--}3)$$

which increases as v_I increases. The base–emitter voltage of T_R must therefore reduce because the base voltage of T_R is fixed at V_R. Thus,

$$v_{BE_R} = V_R - v_E = V_R - v_I + V_0 \qquad (11.2\text{--}4)$$

and since $v_I > V_R$, $v_{BE_R} < V_0$ so that T_R turns off. The load voltage is then given by

$$v_L = V_{CC} - i_{C_1} R_C \qquad\qquad (11.2\text{--}5)$$

where $i_{C_1} \cong i_{E_1}$ and

$$i_{E_1} = \frac{v_I - V_0}{R_E} \qquad\qquad (11.2\text{--}6)$$

By substitution of (11.2–6) into (11.2–5), we have

$$v_L = V_{CC} - \frac{R_C}{R_E}(v_I - V_0) \qquad (11.2\text{--}7)$$

Equation (11.2–7) shows that v_L is a linear function of v_I for $v_I > V_R$. This variation is indicated in Fig. 11.6b for $v_I > V_{IH}$ where we have defined the input voltage at which T_R is just cut off as V_{IH}. The corresponding output low voltage is obtained from (11.2–7) as

$$V_{OL} = V_{CC} - \frac{R_C}{R_E}(V_{IH} - V_0) \qquad (11.2\text{--}8)$$

The actual value of V_{IH} is very nearly equal to V_R, and it corresponds to the reduction in forward voltage across the base-emitter junction that is required to turn off T_R. Experimentally, this additional voltage is approximately 0.05 V. Similarly, when the input low voltage v_I is approximately 0.05 V less than V_R, T_1 is just cut off since circuit symmetry exists. We will therefore define the width of the transition region for v_I to be 0.1 V, centered about V_R. Thus, $V_{IL} = V_R - 0.05$ V and $V_{IH} = V_R + 0.05$ V, as shown in Fig. 11.6b.

Of course, as v_I is increased beyond V_{IH}, v_L reduces slightly and T_1 eventually saturates. When T_1 saturates, $v_I = V_S$ (as shown in Fig. 11.6b), beyond which v_L increases as v_I is increased still further. This region of operation with T_1 in saturation is avoided in usual operation.

——————————————————————————— **EX. 11.2**

ECL Transfer Characteristic Example

Determine and sketch the transfer characteristic v_L versus v_I for the circuit fo Fig. 11.6a with $R_C = 1\ \text{k}\Omega$, $R_E = 1\ \text{k}\Omega$, $V_{CC} = 5$ V, $V_R = 3$ V, and $V_{EE} = 0$ V.

Solution: For $v_I < V_R - 0.05$ V, T_1 is off while T_R is active. Thus, $v_L = V_{CC} = 5$ V. As v_I is increased into the range $V_R - 0.05 \le v_I \le V_R + 0.05$, both T_1 and T_R are active and v_L drops abruptly and linearly as v_I increases. Then, for $v_I > V_R + 0.05$ V, T_R is off while T_1 is active, with

$$v_L = V_{CC} - i_{C_1} R_C \qquad\qquad (1)$$

But,

$$i_{C_1} \cong i_{R_E} = \frac{v_I - 0.7}{R_E} \qquad\qquad (2)$$

where we let $v_I = V_R + 0.05 + \Delta v_I$, where Δv_I is defined as the increase in v_I beyond the active linear region. Thus, by substitution into (2),

$$i_{C_1} = \frac{V_R + 0.05 + \Delta v_I - 0.7}{R_E} \tag{3}$$

and, by substitution of (3) into (1),

$$v_L = V_{CC} - \left(\frac{V_R + \Delta v_I - 0.65}{R_E}\right) R_C \tag{4}$$

By substituting the element values, we obtain

$$v_L = 5 - \left(\frac{3 + \Delta v_I - 0.65}{2}\right)(1) = 3.825 - \frac{\Delta v_I}{2} \text{ V}$$

The overall variation is sketched in Fig. E11.2.

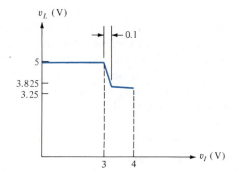

FIGURE E11.2

— EX. 11.3

ECL Noninverting Transfer Characteristic Example

Repeat Ex. 11.2 and let the collector voltage of T_R (v_{C_R}) be the output load voltage.

Solution: Under these conditions, when v_I is low, v_{C_R} is low; when v_I is high, v_{C_R} is high. The transfer characteristic is shown in Fig. E11.3 and is essentially the reverse of the one shown in Fig. E11.2. Note that for v_I low, T_1 is off and T_R is on, with

$$v_{C_R} = V_{CC} - i_{C_R} R_C \tag{1}$$

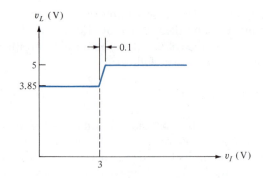

FIGURE E11.3

where

$$i_{C_R} \cong i_{E_R} = \frac{V_R - 0.7}{R_E} \tag{2}$$

Therefore,

$$v_{C_R} = V_{CC} - \left(\frac{V_R - 0.7}{R_E}\right) R_C \tag{3}$$

By substituting values, we have

$$v_{C_R} = 5 - \left(\frac{3 - 0.7}{2}\right)(1) = 3.85 \text{ V}$$

For v_I high, T_1 is on while T_R is off, with $v_{C_R} = 5$ V. As we will see, when additional inputs are present, this transfer characteristic represents the OR operation.

11.2.2 Basic ECL NOR/OR Gate

The basic ECL NOR/OR gate is displayed in Fig. 11.7a. Additional input transistors are connected with their collectors and emitters common to one another. This gate provides both NOR and OR logic operations at the terminals marked v_{NOR} and v_{OR}, respectively. The circuit symbol for this gate is shown in Fig. 11.7b.

The circuit of Fig. 11.7a behaves in essentially the same manner as the basic inverter described in the previous section. The reference transistor

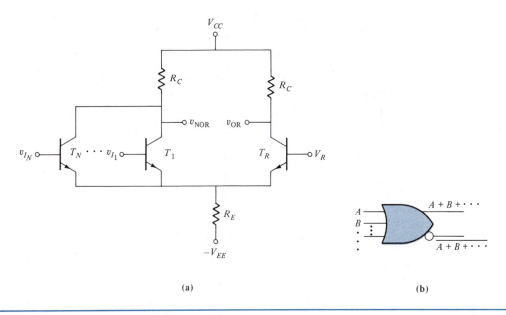

FIGURE 11.7   Basic ECL NOR/OR Gate: (a) ECL circuit, (b) Circuit symbol

T_R operates in the active mode if all of the input voltages are low since all of the input transistors are then cut off. The output v_{NOR} is then high, and v_{OR} is low.

If any or all of the inputs are high, the corresponding input transistors are active and T_R turns off. Under these conditions, v_{NOR} goes low and v_{OR} becomes high. Hence, the collector terminal of T_1 provides the NOR operation, and the collector terminal of T_R provides the OR operation.

11.2.3   ECL NOR/OR Gate with Output Buffers

The basic ECL NOR/OR gate of Fig. 11.7 has a major shortcoming in that when the output voltage (for v_{NOR} or v_{OR}) is high, insufficient current is available to drive even a few gates connected at the particular output. Thus, the fanout will be quite low. If the outputs of the basic ECL OR/NOR gate are fed into emitter followers as shown in Fig. 11.8a (disregard the particular element values for now), with the outputs of the overall gate taken from the emitters of T_3 and T_4

as shown, then the emitter followers provide the necessary output current. By acting as buffers between the gates connected at the output and the basic ECL NOR gate, they also provide isolation.

The improved ECL NOR/OR gate of Fig. 11.8a is also modified from the previous cases by grounding the top of the circuit (setting $V_{CC} = 0$) and using $V_{EE} > V_0$. As mentioned earlier, all of the voltages in the circuit are thereby forced to become negative. The applied reference voltage V_R must also be negative. The advantage of using a negative supply voltage is that the effects of a noise voltage coupled into the DC supply are essentially eliminated, which is observed by considering a noise voltage associated with the DC supply (in this case, V_{EE}) that appears in series with the DC supply voltage. This noise voltage will be almost entirely across the emitter follower resistor R_{EF} because the output impedance of the emitter follower is low and much less than R_{EF}. Then, since the reference terminal is at the top of the circuit, the output voltage will have a negligible noise component because of the small resistive voltage divider ratio.

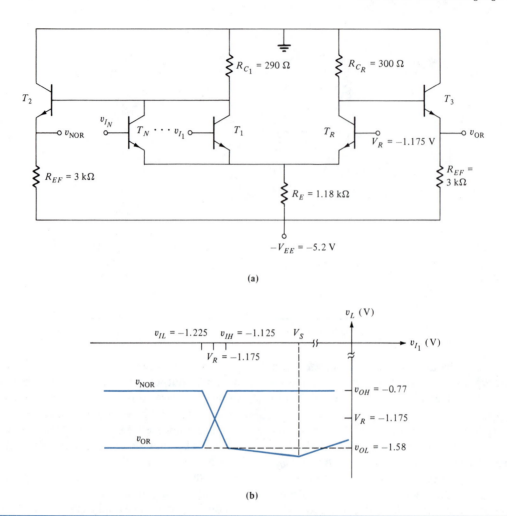

FIGURE 11.8   Improved ECL NOR/OR Gate: (a) ECL circuit with output buffers, (b) Transfer characteristics for the outputs

The voltage transfer characteristics for the outputs of the gate of Fig. 11.8a are displayed in Fig. 11.8b. The NOR output is essentially the same as that described earlier, except that the input and output voltages, since each is negative, have been translated to the third quadrant. The output v_{OR} is observed to essentially be the inverse of the output v_{NOR}.

We now consider the element values indicated in Fig. 11.8a and verify the transfer characteristics of Fig. 11.8b. A brief description indicating

the reason for using the particular element values is in order here. The voltage V_R is taken to have the peculiar value -1.175 V because it is obtained from a special temperature-compensating circuit to be described shortly. The specialized circuit places V_R directly at the midpoint of the logic swing regardless of the particular temperature. The special temperature-compensating circuit also requires a DC supply $-V_{EE}$ of -5.2 V; hence, this value is also used in the ECL circuit of Fig. 11.8a. Finally, note that the collector resistor for the input transistors is slightly less than that for the reference transistor. Thus, the voltage difference for the low level of the NOR output is reduced. Since multiple transistors are present at the input, additional current flows through this resistor, and its smaller value compensates for the reduction in low voltage level that occurs with the larger value of R_{C_1}.

In ECL circuit analysis, it is customary to use $V_0 = 0.75$ V instead of $V_0 = 0.7$ V for the forward-diode voltage operation of a silicon BJT to provide results that are in closer agreement with experiment. The larger value of V_0 is also warranted because the ECL transistors have smaller physical size than the BJTs of previously discussed logic families, which leads to larger V_0. Hence, in our ECL analysis, we will use $V_0 = 0.75$ V to represent the base–emitter voltage of a transistor operating in the active region.

We verify the transfer characteristics by considering the OR output in Fig. 11.8a. We determine the variation in v_{OR} versus v_{I_1}, with the assumption that the other inputs are all low.

For v_{I_1} high, the input transistor T_I is active and T_R is cut off. Therefore, the voltage at the collector of T_R is high and, by KVL, we have

$$v_{OR} = -i_{B_3} R_{C_R} - V_0 \qquad (11.2-9)$$

where i_{B_3} is obtained (from Chapter 4) for the emitter follower as follows:

$$i_{B_3} = \frac{V_{EE} - V_0}{R_{C_R} + (1 + \beta)R_{EF}} \qquad (11.2-10)$$

Substituting element values yields

$$i_{B_3} = \frac{5.2 - 0.75}{0.3 + 51(1.5)} = 0.058 \text{ mA}$$

where we have used $\beta = 50$, which is typically the minimum β value for ECL BJTs. Thus, the output voltage from (11.2–9) is

$$v_{OR} = -(0.3)(0.058) - 0.75 = -0.77 \text{ V}$$

Since this voltage is the output high voltage, $V_{OH} = -0.77$ V. Note that the inclusion of the base current of the emitter follower does not change v_{OR} significantly.

For v_{I_1} low, the input transistors are all cut off and T_R is active. The voltage at the collector of T_R is then

$$v_{C_R} = -R_{C_R} i_{C_R} \qquad (11.2-11)$$

where the base current of T_3 has been neglected in comparison to the much larger collector current of T_R. Then, since $i_{C_R} \cong i_{E_R}$, we have (using KVL)

$$i_{E_R} = \frac{V_{EE} + V_R - V_0}{R_E} \qquad (11.2-12)$$

Substituting element values yields

$$i_{E_R} = \frac{5.2 - 1.175 - 0.75}{1.18} = 2.78 \text{ mA}$$

Thus,

$$v_{C_R} = -(0.3)(2.78) = -0.83 \text{ V}$$

Since the output voltage of T_3 is

$$v_{OR} = v_{C_R} - V_0 \qquad (11.2-13)$$

subsituting values yields

$$v_{OR} = -0.83 - 0.75 = -1.58 \text{ V}$$

This voltage is V_{OL}, as indicated in Fig. 11.8b.

Note further that the middle of the transition region for the output voltage is given by the average of the high and low voltages, or

$$v_{L \text{ avg}} = \frac{V_{OL} + V_{OH}}{2} \qquad (11.2-14)$$

By substituting values, we have

$$v_{L \text{ avg}} = \frac{-0.77 - 1.58}{2} = -0.1175 \text{ V}$$

This value is equal to V_R, as indicated in Fig. 11.8b. This occurrence is not a coincidence; the DC voltage values and the resistances have been selected to provide this symmetry.

To obtain the corresponding input low and high voltages V_{IL} and V_{IH}, we consider the approximate details involved in switching one transistor off and the other on. For example, Fig. 11.9 indicates the situation in which the input transistor is just at cutoff and T_R is active. Slight variations in the base–emitter voltages then exist, as shown. For this particular situation, the corresponding point in the transfer characteristic is $v_{I_1} = V_{IL}$ and $v_{OR} = V_{OL}$. For the base–emitter loop of T_R in Fig. 11.9, using KVL yields

$$v_E = V_{EE} + V_R - 0.75 \qquad (11.2-15)$$

By substituting values, we have

$$v_E = 5.2 - 1.175 - 0.75 = 3.275 \text{ V}$$

Thus, writing KVL for the base–emitter loop of the input transistor yields

$$V_{IL} = -V_{EE} + v_E + 0.7 \qquad (11.2-16)$$

and by substituting values, we have

$$V_{IL} = -5.2 + 3.275 + 0.7 = -1.225 \text{ V}$$

which indicates that V_{IL} is 0.05 V less than $V_R = -1.175$ V.

A symmetrical result is obtained in a similar manner for V_{IH} such that $V_{IH} = -1.125$ V. These results are based upon the assumption that the base–emitter voltage varies by 0.05 V as a transistor is switched from cutoff to active operation, which approximates actual transistor behavior. The results obtained for V_{IH} and V_{IL} match quite closely those obtained experimentally.

For the ECL NOR output of Fig. 11.8a, the transition points for active operation of both transistors are positioned at approximately the same values of V_{IL} and V_{IH}. A slight difference exists because R_{C_1} and R_{C_R} are slightly different. The output levels are, however, reversed, but they also remain approximately equal in magnitude.

Moreover, as T_1 is driven harder for $v_{I_1} > V_{IH}$, the NOR output decreases linearly with v_{I_1}. This same variation was observed for the basic ECL inverter of Fig. 11.6a, and T_1 is active in this region up to $v_{I_1} = V_S$. Beyond $v_I = V_S$, T_1 saturates, and this region is avoided in actual operation.

For the ECL OR output of Fig. 11.8a, we have shown that V_R is essentially at the midpoint of the transition region of the transfer characteristic for v_{OR} versus v_{I_1}. The same is also true for the ECL NOR output. Hence, the high and low noise margins for the ECL OR and NOR gates must be equal, which provides symmetry for the acceptable range of low and high voltage levels. The following example indicates the magnitude of the noise margins.

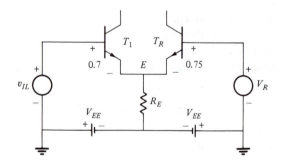

FIGURE 11.9   Base–Emitter Voltages That Exist When T_1 Is at Cutoff and T_R Is Active

——————————————————————— EX. 11.4

Noise Margin Example

Determine the high and low noise margins for the v_{NOR} transfer characteristic shown in Fig. 11.8b.

Solution: From the basic definitions of Chapter 9, we have

$$V_{NMH} = V_{OH} - V_{IH} = -0.77 + 1.125 = 0.365 \text{ V}$$

and

$$V_{NML} = V_{IL} - V_{OL} = -1.225 + 1.58 = 0.355 \text{ V}$$

11.2.4   ECL NOR/OR Gate Fan-Out

As we will see, the output buffers used with ECL families provide an entirely adequate drive current with corresponding large fan-out. To determine the maximum fan-out for the ECL gate of Fig. 11.8a, we will be concerned only with the NOR output (v_{NOR}). The results for the OR output are essentially the same. Figure 11.10 shows the ECL driving gate with the NOR output connected to K driven gates.

The maximum fan-out is determined for v_{NOR} high because when the output is low, the driven gates do not draw current since their transistors are cut off. Thus, for v_{NOR} high, we have $v_{NOR} = V_{OH} = -0.77$ V (from our previous analysis). This voltage must supply the input current for K identical driven gates. Each of these identical currents is determined by using the single driven gate ex-

plicitly indicated in Fig. 11.10. The input current to T_1' for v_{NOR} high is given by

$$i_{B_1}' = \frac{i_{E_1}'}{1 + \beta} \tag{11.2–17}$$

where

$$i_{E_1}' = \frac{V_{EE} + v_E'}{R_E} \tag{11.2–18}$$

where

$$v_E' = v_{NOR} - V_0 \tag{11.2–19}$$

By substituting values, we have

$$i_{E_1}' = \frac{5.2 - 0.77 - 0.75}{1.18} = 3.12 \text{ mA}$$

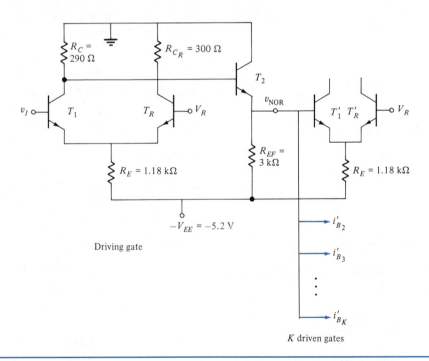

FIGURE 11.10   ECL Driving Gate with NOR Output Connected to K Driven Gates for Determining Maximum Fan-Out

Thus,

$$i'_{B_1} = \frac{3.12}{5} = 0.061 \text{ mA}$$

The driving gate must therefore supply K of these base currents, or

$$i_O = Ki'_{B_1} = K(0.061) \text{ mA} \qquad \textbf{(11.2–20)}$$

The current available from the driving gate is given by

$$i_O = i_{E_2} - i_{REF} \qquad \textbf{(11.2–21)}$$

where

$$i_{E_2} = (1 + \beta)i_{B_2}$$

$$i_{REF} = \frac{v_{\text{NOR}} + V_{EE}}{R_{EF}} \qquad \textbf{(11.2–22)}$$

$$i_{B_2} = \frac{-(v_{\text{NOR}} + V_0)}{R_C} \qquad \textbf{(11.2–23)}$$

By substituting values, we have

$$i_{B_2} = \frac{-(-0.77 + 0.75)}{0.290} = 0.069 \text{ mA}$$

$$i_{E_2} = 51(0.069) = 3.52 \text{ mA}$$

$$i_{REF} = \frac{-0.77 + 5.2}{3} = 1.48 \text{ mA}$$

Therefore,

$$i_O = 3.52 - 1.48 = 2.04 \text{ mA}$$

From (11.2–20), the fan-out is obtained as follows:

$$K = \frac{i_O}{i'_{B_1}} = \frac{2.04}{0.061} = 33.5$$

This magnitude of fan-out, however, is not recommended by manufacturers because of the capaci-

tive loading effect of such a large number of gates. A typical value of recommended gate fan-out is 15.

11.2.5   ECL Speed and Rise Time

The speed of ECL gates is adversely affected by the number of gates being driven due to the loading capacitances. However, for a fan-out of 15 or less, ECL is the fastest of the gates considered, with fall times on the order of a few nanoseconds and rise times even smaller.

The rise time of the output voltage is a lesser factor than the fall time because the effective load capacitance charges through the low-resistance path of the output of the emitter follower. However, the fall time requires the effective output capacitance to discharge through the emitter follower resistance R_{EF}, which is relatively large. For an improvement in speed of the fall time, we might consider decreasing the magnitude of R_{EF}. However, reduction in the magnitude of this resistor will reduce the fan-out.

───────────────────────────── **EX. 11.5**

Effective ECL Rise and Fall Times Example

The rise and fall times can be written in terms of an effective time constant defined as

$$\tau_E = R_E C_E \qquad \text{(1)}$$

where R_E is the effective resistance and C_E is the load capacitance that charges (rise time) or discharges (fall time) through R_E. Calculate the RC time constants for the ECL gate of Fig. 11.8a if $C_E = 1$ pF and $R_E = 300 \ \Omega$ while C_E is charging and $R_E = 3$ kΩ while C_E is discharging.

Solution: The time constant corresponding to C_E charging is

$$\tau_{E \text{ charge}} = (0.3 \times 10^3)(1 \times 10^{-12}) = 0.3 \text{ ns}$$

The time constant corresponding to C_E discharging is

$$\tau_{E \text{ discharge}} = (3 \times 10^3)(1 \times 10^{-12}) = 3 \text{ ns}$$

11.2.6   ECL Temperature-Compensated DC Supply for V_R

The ECL gate under discussion actually would not work under usual circumstances of temperature variation if it were not for the special temperature-compensating network that provides the reference voltage $V_R = -1.175$ V. Since the forward voltage of all junctions (V_0) is a function of temperature that changes at a rate corresponding to -2 mV/° C, corresponding changes in the logic levels are produced. Thus, V_R is actually designed to vary with temperature in the same manner as the forward voltage. The reference voltage is then maintained automatically in the middle of v_{IL} and v_{IH}, as well as v_{OH} and v_{OL}, regardless of temperature changes.

The temperature-compensated bias supply circuit for V_R is displayed in Fig. 11.11. Neglecting temperature variations, we first show that $V_R = -1.175$ V.

We assume that D_1 and D_2 have forward voltages corresponding to ECL with $V_0 = 0.75$ V, which is the assumed value at room temperature. With the transistor base current neglected, the diode current is given by

$$i_D = \frac{V_{EE} - 2V_0}{R_2 + R_3} \tag{11.2–24}$$

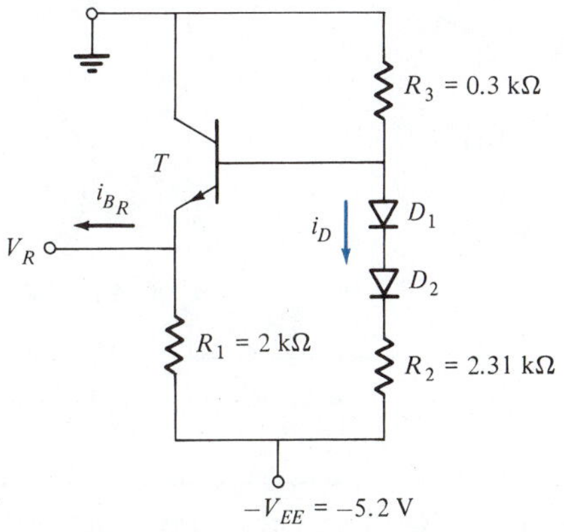

FIGURE 11.11   Temperature-Compensated Bias Supply Circuit for $V_R = -1.175$ V

Substituting element values yields

$$i_D = \frac{5.2 - 2(0.75)}{0.3 + 2.3} = 1.42 \text{ mA}$$

The voltage at the base of T in Fig. 11.11 is then given by

$$v_B = -i_D R_3 \tag{11.2–25}$$

and substituting for i_D and R_3 yields

$$v_B = -0.425 \text{ V}$$

Finally, the reference voltage is obtained from Fig. 11.11 as follows:

$$V_R = v_B - V_0 \tag{11.2–26}$$

or

$$V_R = -0.425 - 0.75 = -1.175 \text{ V}$$

To show that the circuit of Fig. 11.11 provides temperature compensation that places V_R in the middle of the logic swing for all temperatures, we begin by defining the change in base–emitter voltage with temperature for each forward-biased junction as

$$\Delta v_{BE}(T) = CT \tag{11.2–27}$$

where $C = -2$ mV/° C. From this change for each junction, we can obtain the corresponding change in V_R, or ΔV_R.

The voltage of (11.2–27) appears in series with each forward-biased junction. The circuit of Fig. 11.12 is now used to determine the corresponding change in V_R, or ΔV_R.

The change in input voltage at the base of the bias transistor is given by the resistive voltage divider ratio as

$$\Delta v_B = 2\Delta v_{BE}(T)\left(\frac{R_3}{R_2 + R_3}\right) \tag{11.2–28}$$

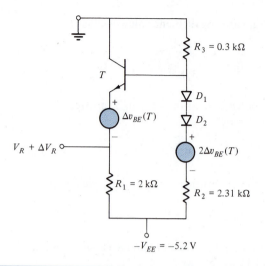

FIGURE 11.12   Equivalent Circuit Used to Account for Temperature Variation of V_R (ΔV_R)

Then, since the transistor acts as an emitter follower (unity voltage gain), we have

$$\Delta v_E = \Delta v_B \qquad (11.2-29)$$

Subtracting one temperature-dependent diode voltage yields

$$\Delta V_R = \Delta v_E - \Delta v_{BE}(T) \qquad (11.2-30)$$

With substitution from (11.2–28), we have

$$\Delta V_R = 2\Delta v_{BE}(T)\left(\frac{R_3}{R_2 + R_3}\right) - \Delta v_{BE}(T)$$

$$= \frac{R_3 - R_2}{R_2 + R_3}\,\Delta v_{BE}(T)$$

Substituting resistor values yields

$$\Delta V_R = \frac{0.3 - 2.31}{2.31 + 0.3}\,\Delta v_{BE}(T)$$

$$= -0.77\Delta v_{BE}(T) \qquad (11.2-31)$$

Equation (11.2–31) thus expresses the change in V_R with temperature, where at room temperature, $V_R = -1.175$ V.

Figure 11.13 is now used to obtain the changes in output high and low logic levels corresponding to changes in temperature. For simplicity, we consider only the OR output. The same results are obtained for the NOR output.

The two cases to consider are: T_R cut off and T_R active. For T_R cut off, the OR output is obtained by noting from Fig. 11.13 that a reduction in voltage by $\Delta v_{BE}(T)$ results, giving the high value of the OR output as

$$v_{\text{OR HI}} = V_{OH} - \Delta v_{BE}(T) \qquad (11.2-32)$$

where the change is just the change due to one diode drop.

For T_R active, the collector voltage of the reference transistor can be written as

$$v_{C_R} = V_{OL} + V_0 + \Delta v_{C_R}(T) \qquad (11.2-33)$$

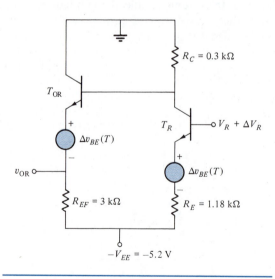

FIGURE 11.13   Equivalent Circuit for OR Output Emitter Follower and T_R Used to Obtain Changes in Output High and Low States Due to Changes in Temperature

where $\Delta v_{C_R}(T)$ is the change due to variation in T. For the circuit of Fig. 11.13, $\Delta v_{C_R}(T)$ can be written as the sum of two terms as follows:

$$\Delta v_{C_R}(T) = -\frac{R_C}{R_E}\Delta V_R + \frac{R_C}{R_E}\Delta v_{BE}(T)$$

$$(11.2-34)$$

where the first term is the "amplified" contribution for ΔV_R and the second term is the resistive divider result of the voltage $\Delta v_{BE}(T)$ corresponding to the base–emitter forward voltage of T_R. Substituting for ΔV_R and the resistor values yields

$$\Delta v_{C_R}(T) = -\frac{0.3}{1.18}(-0.77\Delta v_{BE}) + \frac{0.3}{1.18}\Delta v_{BE}$$

$$= 0.45\Delta v_{BE} \qquad (11.2-35)$$

Substitution into (11.2–33) yields

$$v_{C_R} = V_{OL} + V_0 + 0.45\Delta v_{BE} \qquad (11.2-36)$$

Thus, with T_R active, the output voltage at the OR output terminal of the emitter follower is, from Fig. 11.13,

$$v_{OR} = v_{C_R} - \Delta v_{BE}(T) - V_0$$

Substituting for v_{C_R} yields the low value of the OR output:

$$v_{OR\ LO} = V_{OL} + 0.45\Delta v_{BE} - \Delta v_{BE}$$

$$= V_{OL} - 0.55\Delta v_{BE} \qquad (11.2-37)$$

If we now consider the average value between the high and low states, we obtain

$$v_{avg} = \frac{v_{OR\ HI} + v_{OR\ LO}}{2}$$

$$= \frac{(V_{OH} - \Delta v_{BE}) + (v_{OL} - 0.55\Delta v_{BE})}{2}$$

or

$$v_{avg} = \frac{v_{OH} + v_{OL}}{2} - 0.775\Delta v_{BE} \qquad (11.2-38)$$

which is the midpoint of the voltage swing and is very nearly equal to

$$V_R + \Delta V_R = \frac{V_{OH} + V_{OL}}{2} - 0.77\Delta v_{BE} \qquad (11.2-39)$$

Thus, the temperature-compensating bias circuit maintains $V_R + \Delta V_R$ at the midpoint of the logic swing. Figure 11.14 displays the corresponding transfer characteristic. Note that the input voltage

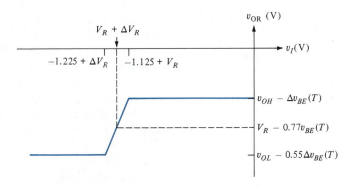

FIGURE 11.14   Transfer Characteristic for ECL OR Output Indicating Temperature Variation

levels of the transition region are centered about $V_R + \Delta V_R$. Finally, we observe that the symmetry of the transfer characteristic continues to provide essentially equal high and low noise margins.

ECL Power Dissipation Example

Calculate the power dissipation per gate in the circuit of Fig. 11.10 when the reference transistor T_R is active. Use $\beta = 50$, $V_R = -1.175$ V, and a fan-out of 0.

Solution: If T_R is active, the voltage across the emitter resistor R_E will be

$$v_{R_E} = (V_R - V_0) - (-V_{EE}) = -1.175 - 0.75 + 5.2$$
$$= 3.275 \text{ V}$$

Hence, the current through R_E is

$$i_{R_E} = \frac{3.275}{1.18} = 2.775 \text{ mA}$$

Also, when T_R is active, v_I is low and v_{NOR} is high, where $v_{NOR} = -0.77$ V. The voltage across the emitter follower resistor R_{EF} is

$$v_{R_{EF}} = v_{NOR} - (-V_{EE}) = 5.2 - 0.77 = 4.43 \text{ V}$$

Thus, the current is

$$i_{R_{EF}} = \frac{4.43}{3} = 1.48 \text{ mA}$$

Finally, the power dissipation for this ECL gate is

$$P_D = (i_{R_E} + i_{R_{EF}})V_{EE} = (2.775 + 1.48)(5.2) = 22.1 \text{ mW}$$

Note that this amount of dissipation is of the same order of magnitude as it is for high-power STTL.

11.3   MOS LOGIC

MOS logic families consist of N-channel or P-channel MOSTs and, in the case of CMOS (complementary MOS), pairs of N-channel and P-channel transistors. These families are another type of nonsaturating logic because MOSTs do not saturate (in the sense of BJT operation in saturation). However, all of the MOS families are still rather slow in speed as compared to the BJT families because of the inherent capacitance associated with the MOST input gate terminal.

The types of MOS families that are commercially available are PMOS, NMOS, and CMOS. Families comprised of JFETs are not used because precision fabrication of these devices in integrated-circuit form is much more difficult to achieve. The *PMOS family* utilizes only P-channel transistors; the *NMOS family*, only N-channel transistors. The *CMOS family* has matched pairs of N-channel and P-channel transistors that "complement" each other. The MOS logic families do not contain resistors; resistors are replaced with MOSTs to conserve chip area.

One of the primary advantages of the MOS logic families as compared to the BJT families is that larger packing densities (transistors/chip) are possible and therefore make MOS logic circuits more economical. The leader in this category is NMOS, with PMOS second, followed by CMOS. However, CMOS has the shortest propagation delay (with NMOS second) and, most importantly, has much lower power dissipation per gate. Thus, CMOS is highly advantageous for portable electronic applications (wristwatches, hand calculators, and so on) that require replaceable batteries.

The descriptions of MOS logic families that follow consider only NMOS and CMOS. PMOS is essentially the same as NMOS except for voltage polarities and current directions. However, PMOS is inherently slower because holes, instead of electrons, conduct the current.

It is of interest to note that PMOS was the first MOS logic family available on the commercial market since difficulty in fabricating N-channel enhancement-only (E-O) MOSTs was initially encountered. With the development of more advanced technology, this difficulty was eliminated, and NMOS became much more widely used than PMOS.

11.3.1 N-Channel MOST Behavior in Logic Circuits

In Chapter 3, the behavior of MOSTs was only partially described because the primary objective there was to explain the operation of a FET as an amplifier. Here, in Chapter 11, more detail concerning basic FET operation is necessary because the E-O and E-D MOSTs have several possible modes of operation in logic circuits.

In general, there are three regions of operation for the MOST in either the E-O or the E-D case. These regions of operation come about because of three possible channel conditions. The conditions for the N-channel E-O MOST are described first. Figure 11.15 displays the cross section of this device and indicates that voltages are applied to the gate (v_{GS}) and the drain (v_{DS}) relative to the source (and substrate). The three possible channel conditions for the N-channel E-O MOST are the following:

1. No channel exists at all, where $v_{GS} \leq V_T$. Operation under this condition is referred to as *operation in cutoff*, and $i_D = 0$.
2. A channel exists for the entire length from source to drain, where $v_{GS} \geq V_T$ and $v_{GS} - v_{DS} = v_{GD} \geq V_T$, which implies that $v_{DS} < v_{GS}$. This mode of operation is referred to as *operation in the linear region or triode region*, and $i_D \geq 0$.

3. A channel exists at the source end and is pinched off at the drain end, where $v_{GS} \geq V_T$ and $v_{GS} - v_{DS} = v_{GD} \leq V_T$. Operation under these conditions was referred to as *amplifier operation* in Chapter 3. However, it is also called *operation in saturation* (which is unfortunate terminology), which refers to the drain current reaching its maximum or saturated value as described in Chapter 3. The present discussion will refer to this state as *FET saturation*, and, for this case, $i_D > 0$.

Current–voltage equations describing E-O MOST operation in both the linear region and the saturation region have been developed. For the linear region, where $v_{GS} \geq V_T$ and $v_{GD} \geq V_T$, the drain-to-source current is given by

$$i_{DS} = K_N[2(v_{GS} - V_T)v_{DS} - v_{DS}^2] \quad (11.3\text{–}1)$$

where the subscript S for i_{DS} has been used to specifically indicate that this current flows from drain to source. This notation will be useful shortly. For the FET saturation region, where $v_{GS} \geq V_T$ and $v_{GD} \leq V_T$, the relationship is given by

$$i_{DS} = K_N(v_{GS} - V_T)^2 \quad (11.3\text{–}2)$$

which is the so-called *square-law relation* introduced in Chapter 3. The N-channel MOST proportionality constant K_N is given by

$$K_N = \frac{\mu_N \varepsilon}{2t}\left(\frac{W_N}{L_N}\right) \quad (11.3\text{–}3)$$

where ε and μ_N are the semiconductor material parameters for permittivity and electron mobility, respectively, and where $W_N, L_N,$ and t are geometry parameters for the MOST. W is the channel width, L is the channel length, and t is the thickness of the oxide layer. In device design, the width-to-length ratio W/L is an extremely important parameter.

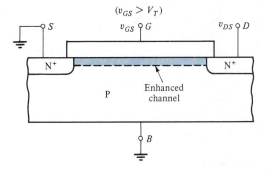

FIGURE 11.15 Cross Section of N-Channel E-O MOST

11.3.2   NMOS Inverter with Enhancement-Only Load

The basic N-channel MOST inverter shown in Fig. 11.16a consists of two enhancement-only (E-O) N-MOSTs. The upper transistor, with gate and drain interconnected, replaces the load resistor and is therefore called the *load transistor* (T_L). The load transistor can also be of the enhancement-depletion (E-D) type (this case is described in the next section). The bottom transistor is called the *driver transistor* (T_D) and is always an E-O MOST. A PMOS inverter is identical to this NMOS example except that P-channel transistors replace the N-channel transistors.

The geometries of T_L and T_D are quite different insofar as the width-to-length ratio W/L is concerned. A typical range of values for W/L begins approximately at 0.1 for load transistors and goes up to about 40 for driver transistors. As we will see, a small ratio for the load transistor and a large ratio for the driver transistor provide a very abrupt voltage transfer characteristic for the basic MOST inverter.

From Fig. 11.16a, note that T_L always operates in FET saturation since $v_{GS} = v_{DS}$ or $v_{GS} - v_{DS} = v_{GD} = 0$, which is always less than V_T. Thus, the load transistor current in this inverter circuit is obtained from (11.3-2) as

$$i_{DS_L} = K_L(v_{GS_L} - V_T)^2 = K_L(V_{DD} - v_L - V_T)^2 \tag{11.3-4}$$

where we have substituted $v_{GS_L} = V_{DD} - v_L$ from Fig. 11.16a and where K_N for the load transistor is defined as K_L.

The basic inverter voltage transfer characteristic for this case is shown in Fig. 11.16b. The specific form of this curve is determined by considering (11.3-1) through (11.3-4) and varying v_I. We will let the threshold voltage of the load and drive MOSTs be equal ($V_{T_L} = V_{T_D} = V_T$), which is typical in actual circuits.

For v_I low ($v_I \le V_T$), the driver transistor T_D operates in cutoff with $i_{DS_D} = 0$ since $v_{GS_D} \le V_T$. Thus, T_L must also have zero current, even if the output MOS gates are connected, since their input gate currents are quite small. The current for T_L

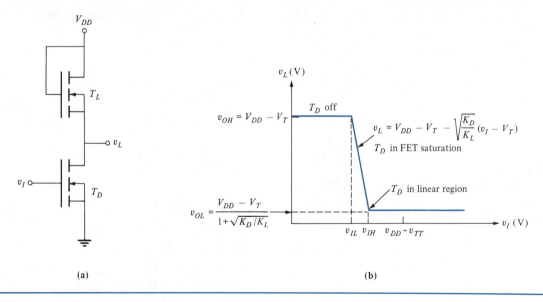

(a)   (b)

FIGURE 11.16   NMOS Inverter with Enhancement-Only Load: (a) Circuit design with T_L as an E-O N-MOST, (b) Basic inverter transfer characteristic corresponding to Fig. 11.16a

is given by (11.3–4). Thus, by setting (11.3–4) equal to zero, we have

$$i_{DS_L} = K_L(V_{DD} - v_L - V_T)^2 = 0 \qquad \text{(11.3–5)}$$

which yields

$$v_L = V_{DD} - V_T \qquad \text{(11.3–6)}$$

This portion of the transfer characteristic is labeled in Fig. 11.16b as T_D off, and we see that $V_{OH} = V_{DD} - V_T$.

As v_I is increased such that $v_I \geq V_T$, the driver transistor T_D begins to operate in FET saturation since $v_{GS_D} - v_{DS_D}$ is still less than V_T while $v_{GS_D} = v_I \geq V_T$. Hence, the current for T_D under these conditions is given by

$$i_{DS_D} = K_D(v_{GS_D} - V_T)^2 = K_D(v_I - V_T)^2 \qquad \text{(11.3–7)}$$

where K_D represents K_N for the driver transistor. Equating (11.3–7) to (11.3–5) then yields

$$K_D(v_I - v_T)^2 = K_L(V_{DD} - v_L - V_T)^2 \qquad \text{(11.3–8)}$$

By taking the square root of both sides of this equation and solving for the output voltage v_L, we obtain

$$v_L = V_{DD} - V_T - \sqrt{\frac{K_D}{K_L}}(v_I - V_T) \qquad \text{(11.3–9)}$$

This equation indicates a linear relationship between v_I and v_L and is indicated in Fig. 11.16b in the central region, where both T_L and T_D operate in FET saturation. Note that the parameter $\sqrt{K_D/K_L}$, which is the magnitude of the slope in this region, plays an important role in determining the width of the transition region. For larger values, the transition region is more abrupt, which is the reason for choosing widely different W/L ratios for T_L and T_D. Note also that (11.3–9) predicts the correct value for v_L, given by (11.3–6), by substituting $v_I = V_T$ into (11.3–9).

As v_I is increased still further, T_D begins to operate in the linear region. This operation occurs when $v_{GS_D} - v_{DS_D} = V_T$ or, since $v_{GS_D} = v_I$ and $V_{DS_D} = v_L$, in terms of v_I and v_L, we have

$$v_I - v_L = V_T \qquad \text{(11.3–10)}$$

By substituting into (11.3–9), eliminating v_I, and solving for v_L, we obtain

$$v_L = V_{OL} = \frac{V_{DD} - V_T}{1 + \sqrt{K_D/K_L}} \qquad \text{(11.3–11)}$$

Substitution of (11.3–11) into (11.3–10) then yields

$$V_{IH} = V_T + \frac{V_{DD} - V_T}{1 + \sqrt{K_D/K_L}} \qquad \text{(11.3–12)}$$

Equation (11.3–12) also indicates that for large K_D/K_L, V_{IH} is only slightly larger than $V_T(V_{IL})$ and therefore that the transition region is quite abrupt. The magnitude of the ratio K_D/K_L is controlled by the W/L ratio of the driver and load transistors.

Note that in Fig. 11.16a, as in Chapter 5, the terminal of T_2 is shown connected directly to the source of T_2 whereas the actual connection is to ground. However, the results obtained are not changed appreciably without this approximation. We may further note that this approximation involves using v_L as the substrate voltage instead of ground. Hence, the most serious error occurs when v_L is high.

EX. 11.7

NMOS Inverter with E-O Active Load Example

For the NMOS inverter with an E-O load as shown in Fig. 11.16a, calculate the transistor design ratio K_D/K_L and the W/L ratio for the driver and load transistors if $V_{DD} = 5.0$ V, $V_{T_D} = V_{T_L} = 1$ V, and $V_{IH} - V_{IL} = 0.5$ V.

Solution: From (11.3–12), since $V_{IL} = V_T$, we have

$$V_{IH} = V_{IL} + \frac{V_{DD} - V_T}{1 + \sqrt{K_D/K_L}} \qquad \text{(1)}$$

or

$$V_{IH} - V_{IL} = \frac{V_{DD} - V_T}{1 + \sqrt{K_D/K_L}} \qquad (2)$$

Substituting values yields

$$0.5 = \frac{5 - 1}{1 + \sqrt{K_D/K_L}} \qquad (3)$$

Rearranging gives

$$\sqrt{\frac{K_D}{K_L}} = 8 - 1 = 7 \qquad (4)$$

so that

$$\frac{K_D}{K_L} = 49 \qquad (5)$$

From (11.3–3), the W/L ratio for the driver and load transistors is therefore

$$\frac{(W/L)_D}{(W/L)_L} = 49$$

One way to accomplish this ratio is to use $W_D/W_L = 7$ and $L_L/L_D = 7$.

11.3.3 NMOS Inverter with Enhancement-Depletion Load

Figure 11.17a shows an alternate form of the NMOS inverter that uses an enhancement-depletion (E-D) N-MOST as the load transistor where the gate and source terminals are interconnected. This inverter circuit provides a larger V_{OH} and a more abrupt transition region but allows a smaller K_D/K_L ratio, which is advantageous because less chip area is then required. Hence, E-D loads are more widely used in NMOS logic circuits. However, the use of E-D loads requires an additional processing step to form the N-channel in these devices. The additional step is easily accomplished using advanced IC technology.

The voltage transfer characteristic of the NMOS inverter with E-D load as shown in Fig. 11.17a is given in Fig. 11.17b. The analytical development of this curve is more difficult than that of the previous case because the two N-channel MOSTs are different. In addition, the E-D load transistor has several modes of operation, which are generally described next in general.

First, the drain-to-source current for N-channel E-D MOST amplifier operation is given (from

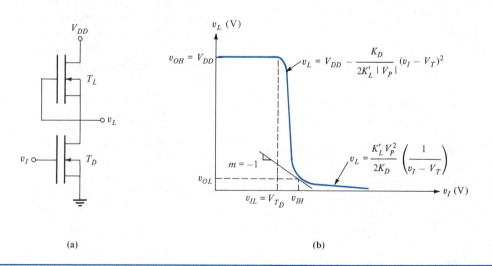

(a)   (b)

FIGURE 11.17   NMOS Inverter with Enhancement-Depletion Load: (a) Circuit design with T_L as an E-D N-MOST, (b) Basic inverter transfer characteristic corresponding to Fig. 11.17a

Chapter 3) by

$$i_{DS} = I_{DSS}\left(1 - \frac{v_{GS}}{V_P}\right)^2 \qquad (11.3\text{-}13)$$

for $v_{GS} - v_{DS} \leq V_P$, where for N-channel V_p is negative and $v_{GS} \geq V_P$. This type of operation is now referred to as *FET saturation* since this transistor is not being used as an amplifier.

A second mode of operation exists for the N-channel E-D MOST in which i_{DS} is given by

$$i_{DS} = \frac{I_{DSS}}{V_P}\left[2(v_{GS} - V_P)v_{DS} - v_{DS}^2\right] \quad (11.3\text{-}14)$$

for $v_{GS} - v_{DS} \geq V_P$, where, again, $v_{GS} \geq V_P$. This type of operation is called *linear operation* to denote linear dependence of i_{DS} upon v_{DS} for small v_{DS}. As in the case of E-O MOSTs, E-D MOSTs can also operate in cutoff, which occurs for $v_{GS} \leq V_P$.

Equations (11.3-13) and (11.3-14) are actually of the same form as corresponding equations (11.3-1) and (11.3-2) for the N-channel E-O MOST. By defining a new constant for the E-D MOST as

$$K_L' = \frac{I_{DSS}}{V_P^2} \qquad (11.3\text{-}15)$$

we can write (11.3-13) and (11.3-14) as follows:

$$i_{DS} = K_L'(v_{GS} - V_P)^2 \qquad (11.3\text{-}16)$$

for $v_{GS} - v_{DS} \leq V_P$, and

$$i_{DS} = K_L'[2(v_{GS} - V_P)v_{DS} - v_{DS}^2] \qquad (11.3\text{-}17)$$

for $v_{GS} - v_{DS} \geq V_P$.

For the specific case of $v_{GS} = 0$ and $v_{DS} = V_{DD} - v_L$ for the N-channel E-D MOST load transistor of Fig. 11.17a, (11.3-16) and (11.3-17) simplify to

$$i_{DS} = K_L'V_P^2 \qquad (11.3\text{-}18)$$

for $v_L - V_{DD} \leq V_P$ or $v_L \leq V_{DD} - V_P$, and

$$i_{DS} = K_L'[-2V_P(V_{DD} - v_L) \\ - (V_{DD} - v_L)^2] \qquad (11.3\text{-}19)$$

for $v_L - V_{DD} \geq V_P$ or $v_L \geq V_{DD} - V_P$.

For the N-channel E-D MOST driver transistor of Fig. 11.17a, where $v_{GS} = v_I$ and $v_{DS} = v_L$, (11.3-1) and (11.3-2) become

$$i_{DS} = K_D[2(v_I - V_T)v_L - v_L^2] \qquad (11.3\text{-}20)$$

for $v_I - v_L \geq V_T$, and

$$i_{DS} = K_D(v_I - V_T)^2 \qquad (11.3\text{-}21)$$

We now have the necessary "ammunition" to obtain the transfer characteristic of Fig. 11.17b.

With v_I low, the driver transistor is cut off and v_L is high. The E-D load transistor will then operate in the linear mode with $v_L \geq V_{DD} - V_P$. Since the driver transistor is cut off, the current in both transistors is zero. Setting (11.3-19) equal to zero yields

$$K_L'[-2V_P(V_{DD} - v_L) - (V_{DD} - v_L)^2] = 0 \\ (11.3\text{-}22)$$

with a valid solution for

$$v_L = V_{DD} \qquad (11.3\text{-}23)$$

This voltage corresponds to V_{OH}, as shown in Fig. 11.17b. Note that this voltage is higher than that possible with the E-O load transistor case of the previous section.

As v_I increases such that $v_I \gtrsim V_T$, the NMOS transistor turns on in FET saturation since $v_{GD} = v_I - v_L < V_T$ (with v_I low and v_L high). At the same time, the E-D load transistor continues to operate in the linear region. Therefore, by equating the currents of each transistor as given by (11.3-19) and (11.3-21), we have

$$K_D(v_I - V_T)^2 = K_L'[-2V_P(V_{DD} - v_L) \\ - V_{DD} - v_L)^2] \\ (11.3\text{-}24)$$

This expression is a nonlinear equation and is difficult to solve explicitly. However, in the region where $v_I \gtrsim V_T$, $V_{DD} - v_L$ is small. Therefore, we may neglect the last term in (11.3–24) to obtain a first-order solution. By carrying out this approximation and solving for v_L, we obtain

$$v_L = V_{DD} - \frac{K_D}{2K'_L|V_P|}(v_I - V_T)^2 \qquad \text{(11.3–25)}$$

where we have inserted the fact that V_P is negative for the N-channel MOST. Equation (11.3–25) thus provides the variation of v_L with v_I for $v_I \geq V_T$, as shown in Fig. 11.17b.

As v_I is increased to a larger value, v_L becomes low and forces the driver transistor to operate in the linear mode given by (11.3–20). Meanwhile, the E-D load transistor operates in the FET saturation mode given by (11.3–18). Equating these currents yields

$$K'_L V_P^2 = K_D[2(v_I - V_T)v_L - v_L^2] \qquad \text{(11.3–26)}$$

Again, a nonlinear relation for v_L versus v_I results. In this case, v_L is small, while v_I is large; v_L^2 can thus be neglected. Solving (11.3–26) under these conditions for v_L then yields

$$v_L = \frac{K'_L V_P^2}{2K_D}\left(\frac{1}{v_I - V_T}\right) \qquad \text{(11.3–27)}$$

Thus, as v_I increases, v_L reduces, as shown in Fig. 11.17b.

The transition region of the inverter characteristic will be more abrupt for larger values of K_D/K'_L. Additionally, this region is even more abrupt because v_L is inversely proportional to $v_I - V_T$.

In this case, the input high value of voltage (V_{IH}) is difficult to define because a distinct point at which the output reaches a particular state is not available. In cases such as this one, we may define V_{IH} corresponding to the input voltage at which the slope of the transfer curve is unity (as shown in Fig. 11.17b) or use the following approximate method.

To obtain the point V_{IH}, V_{OL}, we consider v_I high such that the driver transistor T_D is on with nonzero drain current. Under these conditions, the output voltage (v_{DS} for the driver transistor) is low and is assumed to be zero. The minimum input voltage for this condition (V_{IH}) is then determined by analyzing the rest of the circuit.

From Fig. 11.17a, for v_I high, T_D is on and $v_L \cong 0$. Thus, $V_{OL} = 0$ and $i_{DS} = I_{DSS}$ for the load transistor because $v_{GS} = 0$ for this transistor— since $i_{DS_L} = I_{DSS}(1 - v_{GS}/V_P)^2$. For the driver transistor, the drain current is given by

$$i_{DS} = K_N(v_{GS} - V_T)^2$$

By substituting $v_{GS} = V_{IH}$ and $i_{DS} = I_{DSS}$, we obtain

$$I_{DSS} = K_N(V_{IH} - V_T)^2$$

Solving for V_{IH} yields

$$V_{IH} = V_T + \left(\frac{I_{DSS}}{K_N}\right)^{1/2}$$

Numerical details are left to the problems at the end of the chapter.

_____ EX. 11.8

NMOS Inverter with E-D Active Load Example

For the NMOS inverter with an E-D load as shown in Fig. 11.17a, calculate the transistor design ratio K_D/K'_L. Compare this value with K_D/K_L obtained in Ex. 11.7. Let $V_{DD} = 5$ V, $V_T = 1$ V, $V_P = -4$ V, and $V_{IH} - V_{IL} = 0.5$ V. Also, let V_{IH} correspond to unity slope on the transfer curve.

Solution: To determine the slope of the transfer curve, we take the derivative of (11.3–27) and equate it to -1:

$$\frac{dv_L}{dv_I} = \frac{K'_L V_P^2}{2K_D}\left[\frac{-1}{(v_I - V_T)^2}\right] = -1 \qquad (1)$$

Thus, by solving for K_D/K'_L, we have

$$\frac{K_D}{K'_L} = \frac{V_P^2}{2(v_I - V_T)^2} \qquad (2)$$

Substituting values yields (using $v_I = V_{IH}$)

$$\frac{K_D}{K'_L} = \frac{4^2}{2(0.5)^2} = 32 \tag{3}$$

If we compare this value to that of Ex. 11.7, we observe that some improvement is achieved in the present case.

11.3.4   NMOS NAND and NOR Gates

The basic circuits for the two-input NAND and NOR gates using NMOS with an E-D load are displayed in Fig. 11.18. The NOR gate of Fig. 11.18a consists of a connection of two-input N-channel MOSTs in parallel, while the NAND gate of Fig. 11.18b has the input transistors connected in series. Each of these gates is

based upon the operation of the NMOS inverter. Additional input transistors can be added to increase the fan-in.

The operation of the NOR gate of Fig. 11.18a is understood by realizing that only if both inputs are low is the output high. Otherwise, for both or either of the inputs high, the output is low because a channel is enhanced. This operation is the basic NOR operation.

The operation of the NAND gate is similarly observed. Only if both inputs are high is the output low because both channels are enhanced. Thus, for both or either of the inputs low, the output is high. Hence, this gate performs the basic NAND operation.

Both NMOS NOR and NAND gates are used in NMOS logic circuits. However, NAND gates are usually avoided because they exhibit voltage degradation of the low voltage level. Since the

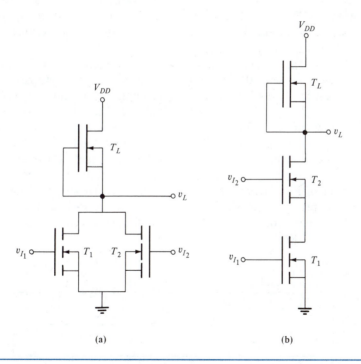

(a)   (b)

FIGURE 11.18   NMOS Two-Input NOR and NAND Gates: (a) NOR gate with input transistors connected in parallel, (b) NAND gate with input transistors connected in series

source–drain terminals of the input transistors are connected in series, the output voltage is the sum of these voltages. Since V_{OL} for each single transistor is greater than zero, the output low voltage with two input transistors will be doubled. For additional input transistors, this condition becomes worse. Elimination of this effect is possible by redesigning the input transistors such that their W/L ratio is increased as the number of input transitors is increased, which is accomplished by increasing the width of the input transitors. However, an additional problem is introduced because more chip area is then required. Thus, NMOS NOR gates are used in most NMOS logic circuits.

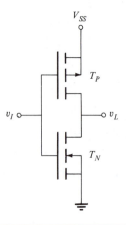

FIGURE 11.19   CMOS Inverter

11.3.5   CMOS Inverter

Complementary MOS, or CMOS, is a logic family that uses N-channel and P-channel MOSTs in matched or complementary pairs. Presently, CMOS is the most widely used digital circuit technology because it posseses the highest packing density and lowest power dissipation per gate of all the other families.

In the CMOS inverter circuit of Fig. 11.19, each of the transistors is an E-O MOST. The transistors are matched with identical threshold voltage values.

Note that the circuit of Fig. 11.19 indicates that the source and bulk region of each MOST are connected together. These connections are

physically possible in CMOS, as seen in Fig. 11.20, which shows in cross section the CMOS pair of transistors used in the basic inverter integrated circuits. Also note that the source of the P-channel transistor is connected directly to the DC supply; hence, this source is labeled V_{SS} in Fig. 11.19.

The voltage transfer characteristic for the CMOS inverter is displayed in Fig. 11.21. The labeling on this curve in various voltage ranges indicates the transistor states as v_I increases. To show the operation, we must first describe the behavior of the P-channel E-O MOST. Equations (11.3–1) through (11.3–3) are used to describe the current–voltage relations for the N-channel E-O transistor. However, these equations are also valid for the P-channel case when the subscripts for the device current and voltage are reversed.

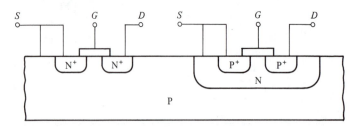

FIGURE 11.20   Cross Section of CMOS Complementary Pair

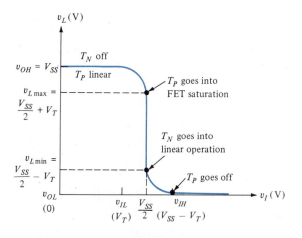

FIGURE 11.21   Transfer Characteristic for CMOS Inverter

Carrying out the subscript reversal yields the corresponding P-channel E-O MOST equations as follows:

$$i_{SD} = K[2(v_{SG} - V_T)v_{SD} - v_{SD}^2] \qquad (11.3\text{--}28)$$

for $v_{SG} \geq V_T$ and $v_{DG} \geq V_T$, and

$$i_{SD} = K(v_{SG} - V_T)^2 \qquad (11.3\text{--}29)$$

for $v_{SG} \geq V_T$ and $v_{DG} < V_T$, where

$$K = \frac{\mu_P \varepsilon}{2t}\left(\frac{W_P}{L_P}\right) \qquad (11.3\text{--}30)$$

Note that the threshold voltage of the P-channel MOST is denoted simply as V_T in (11.3–28) and (11.3–29) since, as stated previously, CMOS is designed to have identical threshold voltages for the complementary transistors. Additionally, the constant of proportionality is denoted simply as K since, in CMOS design, this constant is also equal to that of the N-channel transistor. Thus, in order that $K_P = K_N = K$, since μ_N and μ_P do not have equal values, the geometries of the two transistors must be different. This geometry is normally accomplished by making the width of the P-channel device approximately triple that of the N-channel device for silicon MOSTs because the mobilities differ approximately by a factor of 3.

The basic operation of the CMOS inverter of Fig. 11.19 now becomes exceedingly easy to understand. If the input voltage v_I is low, an N channel is not enhanced in T_N, whereas a P channel is enhanced in T_P. Thus, a resistive voltage divider is provided at the output, and v_L becomes a high voltage since this voltage is across the infinite resistance of the N channel. If v_I is high, an N channel is enhanced, while a P channel is not. Thus, the opposite resistive voltage divider at the output provides v_L low. This operation is the basic operation of the inverter shown in Fig. 11.19. The specific values shown in parentheses in Fig. 11.21 are now verified.

With v_I low ($v_I \leq V_T$), the N-channel transistor is cut off, while the P-channel transistor has $v_{SG} = V_{SS} - v_I > V_T$ and $v_{DG} = v_L - v_I > V_T$. Thus, the P-channel transistor current is given by (11.3–28), or

$$i_{SD} = K[2(v_{SG} - V_T)v_{SD} - v_{SD}^2] \qquad (11.3\text{--}31)$$

Furthermore, this current is zero since the N-channel current is zero. Hence, setting (11.3–31) equal to zero yields $v_{SD} = 0$. However, from Fig. 11.19, $v_{SD} = V_{SS} - v_L$. Therefore,

$$v_L = V_{SS} \qquad (11.3\text{--}32)$$

which is also the output high voltage V_{OH}.

As v_I is increased, T_N will turn on when $v_I = v_{GS} = V_T$. For this still low value of v_{GS}, $v_{GD_N} < V_T$ and the N-channel transistor turns on in FET saturation with $V_{IL} = V_T$.

As v_I is increased further, v_L reduces and T_p will also operate in FET saturation when v_{DG_P} just becomes equal to V_T. Substituting the voltages from the inverter circuit of Fig. 11.19 yields

$$v_{DG_P} = v_L - v_I = V_T \qquad (11.3\text{--}33)$$

with both T_N and T_P operating in FET saturation. Equating the currents of each transistor then yields

$$K(v_{GS_N} - V_T)^2 = K(v_{SG_P} - V_T)^2 \qquad \textbf{(11.3–34)}$$

By canceling K, taking the square root, and substituting for v_{GS_N} and v_{GS_P} from Fig. 11.19, we then obtain

$$v_I - V_T = V_{SS} - v_I - V_T \qquad \textbf{(11.3–35)}$$

or

$$v_I = \frac{V_{SS}}{2} \qquad \textbf{(11.3–36)}$$

Equation (11.3–36) corresponds to the central vertical region of the CMOS transfer characteristic of Fig. 11.21. This value of v_I can also be used to obtain the minimum and maximum values of v_L in this region.

At the bottom of the vertical region ($v_I = V_{SS}/2$), v_L has decreased sufficiently for T_N to just begin operating in the linear region, or $v_{GS} - v_{DS} = V_T$. By substituting corresponding circuit values from Fig. 11.19, we have

$$v_I - v_L = V_T \qquad \textbf{(11.3–37)}$$

Substituting for v_I from (11.3–36) then yields

$$v_{L\,\text{min}} = \frac{V_{SS}}{2} - V_T \qquad \textbf{(11.3–38)}$$

Similarly, at the top of the vertical region where T_P just becomes linear, we obtain

$$v_{L\,\text{max}} = \frac{V_{SS}}{2} + V_T \qquad \textbf{(11.3–39)}$$

Thus, we observe that the transition between the high and low output voltage levels is quite abrupt for CMOS. In fact, it is much more abrupt than that possible for NMOS or PMOS. More-over, the high and low values of output voltage are quite close to the DC supply voltage V_{SS} and to zero, respectively, which represents another improvement over NMOS and PMOS.

——————————————————————— EX. 11.9

CMOS Inverter Example

Consider the CMOS inverter of Fig. 11.19 and calculate the output low and high voltage levels. Let $V_{SS} = 5$ V, with the channel resistance of each transistor equal to 100 Ω and 10 kΩ for the on and off conditions, respectively.

Solution: Regardless of the input voltage condition, the load voltage for the CMOS inverter of Fig. 11.19 is given by a resistive voltage divider expression as follows:

$$v_L = \frac{R_N}{R_N + R_P} V_{SS} \qquad (1)$$

where R_N is the N-MOST channel resistance and R_P is the P-MOST channel resistance. Thus, for the case of v_L high,

$$V_{OH} = V_{SS} \frac{10}{10.1} = 4.95 \text{ V} \qquad (2)$$

For the case of v_L low,

$$V_{OL} = V_{SS} \frac{0.1}{10.1} = 0.0495 \text{ V} \qquad (3)$$

11.3.6   CMOS NAND and NOR Gates

Figure 11.22 displays the basic circuits for CMOS NOR and NAND gates. Neither of these logic gates is preferred over the other, and both are widely used in CMOS design. Note that in each case the P-channel and N-channel transistors are connected in pairs, and each input is connected to the gate of the complementary pair.

To observe that the CMOS NOR gate of Fig. 11.22a performs its operation, note that the output will be high only if v_{I_1} and v_{I_2} are low. Under these conditions, T_{N_1} and T_{N_2} are cut off, while T_{P_1} and T_{P_2} are turned on. Thus, v_L is high.

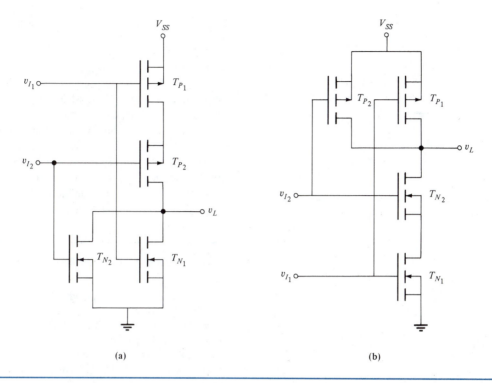

FIGURE 11.22   CMOS Two-Input NOR and NAND Gates: (a) NOR gate with P- and N-channel transistors connected in pairs, (b) NAND gate with P- and N-channel transistors also connected in pairs

For any other input state, T_{N_1} or T_{N_2} (or both) will be on, and v_L is low.

In a similar manner, the CMOS NAND gate of Fig. 11.22b is observed to carry out its logic function. Only if both v_{I_1} and v_{I_2} are high is the output voltage v_L low. For either v_{I_1} or v_{I_2} (or both) low, the output voltage v_L is high.

11.3.7   MOS Power Dissipation

As previously mentioned, MOS logic families have the lowest power dissipation per gate of any of the other families because of the relatively large values of resistance of the MOSTs that consequently draw only small amounts of current. We will see that, in particular, CMOS has the lowest power dissipation per gate by far.

An NMOS (PMOS) or a CMOS gate dissipates energy in three ways. Two ways are the usual cases involving either a high or low output and the power dissipation is the product of current and supply voltage. However, the third possibility for power dissipation occurs in switching from one logic state to another. In other logic families, this dissipation is negligible; for MOS families, this dissipation is appreciable.

We now formulate an expression for the power dissipation during the transient switching period, which is also called the *dynamic power dissipation*. We assume that the logic gate switches at a frequency f and that the gate has a load capacitance C_E as shown in Fig. 11.23 (or $2C_E$ for CMOS as show in Fig. 11.24). We also assume that the voltage across C_E changes by the amount V_{DD} (V_{SS} for CMOS) during each switching period. Thus, the amount of charge flow into or out of C_E is

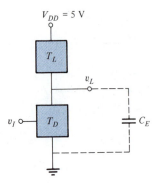

FIGURE 11.23   Block Diagram Representation of NMOS Inverter

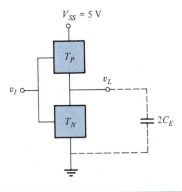

FIGURE 11.24   Block Diagram Representation of CMOS Inverter

given by

$$Q = C_E V_{DD} \qquad (11.3\text{--}40)$$

By relating this equation to the average current defined by

$$I_{avg} = \frac{Q}{T} \qquad (11.3\text{-}41)$$

we obtain, by substituting for Q,

$$I_{avg} = \frac{C_E V_{DD}}{T} = C_E V_{DD} f \qquad (11.3\text{--}42)$$

Where $f = 1/T$. Thus, the power dissipation due to switching is

$$P_D = I_{avg} V_{DD} = C_E V_{DD}^2 f \qquad (11.3\text{--}43)$$

The switching power dissipation given by (11.3–43) is not negligible, as we will see.

To demonstrate the low power dissipation per gate for MOS, we will consider an example that uses typical resistor values for the MOSTs and a supply voltage of 5 V. Calculations of specific amounts of power dissipation incurred in the basic MOS inverter circuits are then possible.

--EX. 11.10

NMOS and CMOS Power Dissipation Example

For the basic arrangement of the NMOS inverter of Fig. 11.23 with driver transistor T_D and load transistor T_L, assume that T_L has a resistance of 50 kΩ and T_D has resistance $R_{on} = 1$ kΩ and $R_{off} = 1$ MΩ. Calculate the average power dissipation if $V_{DD} = 5$ V, $f = 50$ kHz, and $C_E = 10$ pF. Repeat this calculation for the CMOS gate of Fig. 11.24 with $R_{on} = 5$ kΩ, $R_{off} = 1$ GΩ, $V_{SS} = 5$ V, and other parameters unchanged.

Solution for NMOS Case: We begin by calculating the power dissipation due to switching. From (11.3–43) by direct substitution, we have

$$P_D = (10 \times 10^{-12})(5^2)(50 \times 10^3) = 12.5 \ \mu W$$

For v_I low and v_L high (T_D off and T_L on), the drain–source current is

$$i_{DS} = \frac{5}{(50 \times 10^3) + 10^6} \cong 5 \ \mu A$$

where we have assumed that any gates connected at the output have much larger impedances than those under consideration. The power dissipation for this state is then

$$P_D = V_{DD} i_{DS} = (5)(5 \times 10^{-6}) = 25 \ \mu W$$

For the other state of the NMOS inverter, with v_I high and v_L low, the current is

$$i_{DS} = \frac{5}{(50 \times 10^3) + 10^3} = \frac{5}{51} \times 10^{-3} = 98 \ \mu A$$

The power dissipation for this state is then

$$P_D = V_{DD}i_{DS} = (5 \text{ V})(98 \text{ } \mu\text{A}) = 0.49 \text{ mW}$$

The average dissipation for the two states is thus 0.245 mW. With the addition of the switching power dissipation, the total average power dissipation is

$$P_{D \text{ avg}} = 257 \text{ } \mu\text{W}$$

Solution for CMOS Case: For the case of CMOS, the power dissipation is reduced still further from that of NMOS. In this case, the transistors are identical (since they are complementary) with an off resistance of 1 GΩ and an on resistance of 5 kΩ (R_{on} is usually larger for CMOS because of the P-channel transistor and making $T_P = T_N$). The two states of the CMOS inverter are then identical because the current in each case is the same and given by

$$i_{DS} = \frac{5}{10^9 + (5 \times 10^3)} = 5 \text{ nA}$$

Hence, the average power dissipation for these two states is

$$P_D = V_{DD}i_{DS} = (5)(5 \text{ nA}) = 25 \text{ nW}$$

The average power dissipation due to switching is twice that of the NMOS case ($2C_E$). Thus,

$$P_D = 25 \text{ } \mu\text{W}$$

The total average power dissipation is therefore

$$P_{D \text{avg}} = 25 \text{ nW} + 25 \text{ } \mu\text{W} \cong 25 \text{ } \mu\text{W}$$

These values indicate very strikingly that the power dissipation in MOS and, in particular, CMOS is exceedingly low.

11.3.8   MOS Fan-Out

Because of the extremely high input resistance of MOS gates, the fan-out achievable should be unlimited, based upon current considerations only. However, the input capacitance as-

sociated with MOS gates becomes the important factor as fan-out is increased. As additional gates are connected at the output, the loading capacitance will increase and is directly proportional to the fan-out. Also proportional to the fan-out is a corresponding increase in rise and fall times. The problem is even more serious in CMOS, where each output gate has a pair of transistors associated with it and where the effective capacitance is thereby doubled. Nonetheless, MOS logic families can still operate with a large fan-out (> 50) and are better than TTL circuits in this regard.

11.3.9   Rise and Fall Times in MOS Gates

Since MOSTs are majority-carrier devices, there are no minority-carrier storage time delays associated with the MOST families. The propagation delays are dependent only upon the rise and fall times that occur due to the charging and discharging of equivalent output capacitors.

The rise time of NMOS (and PMOS) gates is generally much longer than the rise time of bipolar devices because of the large resistance of the load MOST through which the output effective capacitance must charge. The fall time of NMOS (and PMOS) gates is, however, reduced since the resistance of the driver MOST through which the effective capacitor discharges is much smaller than the load MOST resistance. In CMOS, there is always a low-resistance path for charging and discharging the effective output capacitance. Therefore, the rise and fall times of CMOS are comparable in magnitude to the shorter fall time of NMOS gates.

To observe and compare the rise and fall times of NMOS and CMOS, we again consider the basic inverter for each case. We also include the effective output capacitance to account for the fan-out. The effective capacitor C_E is shown in dashed form in Figs. 11.23 and 11.24.

For the NMOS inverter of Fig. 11.23, as T_D goes off, v_L will rise as C_E charges through the current supplied by T_L. The resistance associated with $T_L (R_{T_L})$ is quite large (because of the small W/L ratio). The RC time constant, therefore also quite large, is given by

$$T_{\text{rise}} = R_{T_L} C_E \qquad (11.3\text{–}44)$$

For C_E in the picofarad range, T_{rise} is on the order of 100 ns. Thus, since the propagation delay is the time necessary for the output to attain 50% of its final value, a rough approximation indicates that t_{PLH} is also on the order of 100 ns, which is quite close to the value obtained with actual NMOS gates.

For the NMOS inverter as it switches from high to low, the capacitor C_E will discharge through the low-resistance path provided by $T_D (R_{\text{on}})$. The RC time constant for the fall time is then

$$T_{\text{fall}} = R_{\text{on}} C_E \qquad (11.3\text{–}45)$$

Since R_{on} is on the order of 1 kΩ and R_{T_L} is on the order of 50 kΩ, there is a difference of approximately a factor of 50 between the rise time and the fall time of NMOS gates. The corresponding high-to-low propagation delay is on the order of 1 ns and plays no role in determining the overall time delay.

For the case of the CMOS inverter, as depicted in Fig. 11.24, the rise time associated with v_L going from low to high is provided by T_P turning on. A low-resistance path is provided through which C_E charges. Similarly, as v_L switches from high to low, the fall time has a low-resistance path through T_N. Since T_N and T_P are complementary, the time constants corresponding to the rise and fall times are equal and given by

$$T_{\text{rise}} = T_{\text{fall}} = R_{\text{on}}(2C_E) \qquad (11.3\text{–}46)$$

where we have included a factor of 2 increase in effective capacitance. Based on the assumption that the fan-out is the same as that of the NMOS inverter and that R_{on} is larger for CMOS than for NMOS, the corresponding propagation delay for CMOS is on the order of 20 ns.

11.3.10   Input Diode Protection

MOS gates are notorious for self-destruction. Static charge can build up on a gate and develop a voltage across the oxide. Since the oxide is thin ($\cong 200$ Å in modern structures), the associated electric field (which is the ratio of the developed voltage to oxide thickness) can become sufficient in magnitude to cause oxide breakdown. When this breakdown occurs, the gate conductor becomes permanently shorted to the channel and creates a serious problem. Enough static charge buildup to create oxide breakdown is associated with merely rubbing the leads of the packaged device together or touching them without protective gloves. Furthermore, extraneous voltages applied to gates can also cause device destruction. Thus, protective diodes are often used at the inputs of MOS gates to remove any static charge buildup. For circuits in which these additional diodes are not used, specialized handling procedures are required. In addition, in shipping such ICs, special packaging that shorts all of the leads together is used to eliminate the possibility of static charge buildup.

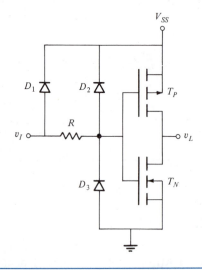

FIGURE 11.25   CMOS Inverter with Input Protective Diodes

A typical CMOS inverter with input protective diodes is shown in Fig. 11.25. The diodes have large breakdown voltages (much greater than any of the voltages in the circuit) and therefore do not operate in breakdown. Note that the diodes will have no effect on input voltages in the usual range of $0 \leq v_I \leq V_{SS}$ because the diodes are then open circuited. The diodes D_1 and D_2 do not allow v_I to exceed V_{SS}. The resistor R is used to prevent any large input voltage from being applied directly to the gate by providing a voltage drop. Finally, the diode D_3 prevents the possibility of large negative voltages appearing on the gate terminal.

11.4   INTERFACING LOGIC FAMILIES

Most often, digital logic systems consist of component circuits from a single logic family. However, in some instances, it is of benefit to use different logic families in an overall logic system. For example, high-speed ECL may be required at the "front end" of an overall logic system where speed is a necessity. In the memory portion of the system, however, slower-speed but lower-power CMOS logic with higher circuit density can be used more advantageously.

When different logic families are interconnected, we must be mindful of the fact that each family has different logic levels and that connecting the output of one family to the input of another requires that the voltage levels and current levels be compatible. Most often, a direct connection is not possible because of the different electrical characteristics of the two families. Under these conditions, an *interface circuit*, or *translator*, is inserted between the two IC families. Many of these translators are available commercially in IC form.

There are two requirements that the interface circuit must meet. First, its input electrical characteristics must match those of the driving family. Second, the interface output characteristics must match those of the driven family. We will consider

several examples of interfacing different families. The circuits used are by no means unique, as we will see.

11.4.1   Interfacing TTL and CMOS

The interface circuit in the case of TTL driving CMOS is merely a resistor, as we will verify. Actually, with both gates supplied by $V_{SS} = V_{CC} = 5$ V, a direct connection (without a resistor) is possible when the TTL output is low. No interface circuit is necessary because the CMOS input takes essentially zero current and V_{OL} for TTL is at most 0.5 V (for STTL), which is less than V_T.

However, there is an interface problem when the output of the TTL is in the high state (V_{OH}). Recall that we previously obtained $V_{OH} = 3.6$ V for TTL. This value is very nearly equal to the minimum value for the input high voltage for CMOS. If the TTL output is driving other TTL gates in addition to the CMOS gate, V_{OH} can be reduced even further. Hence, V_{OH} must be made higher by a voltage level shifter network. An external pull-up resistor is used, as indicated in Fig. 11.26. With the output of the TTL gate high, there will be very little current through R_P since both the TTL output impedance in this state and the CMOS input impedance are very large. Therefore, the TTL output voltage will rise to 5 V. Typically, R_P has a magnitude of 1 kΩ to 10 kΩ.

Now, consider the case of CMOS driving TTL, as indicated in Fig. 11.27a. When the CMOS

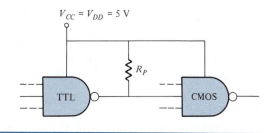

FIGURE 11.26   TTL Driving CMOS with External Pull-Up Resistor

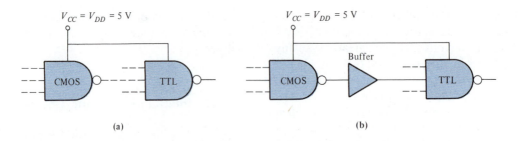

FIGURE 11.27   CMOS Driving TTL: (a) Without interface, (b) With buffer interface

output is high, the TTL gate draws very little current (since the input transistor is inverse active). For this state, a direct connection from CMOS to TTL is possible with no difficulty. However, when the CMOS output is low, it acts like a current sink for current to enter from the TTL input. Since this current can be quite large (1.3 mA for one gate), the voltage produced by this current and the output resistance can be larger than V_{IL} for the TTL gate. This situation is avoided by using a special interface circuit, or *buffer*, as indicated in Fig. 11.27b. This IC is commercially available and acts like an emitter follower. This element provides a fixed output voltage and a low-resistance path for the TTL input when its output is in the low state.

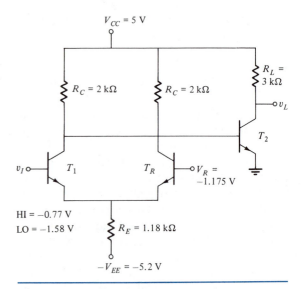

FIGURE 11.28   Basic ECL-to-TTL Translator

11.4.2   Interfacing ECL and TTL

Since ECL and TTL have completely different voltage levels, an interface circuit that shifts the voltage levels is an absolute necessity in translating from ECL to TTL or vice versa. Commercially available interface circuits to perform these tasks are almost always used. However, we will consider basic level translator circuits for each case.

ECL-to-TTL Translator.   Figure 11.28 displays a basic translator circuit to shift the voltage levels from ECL to TTL. Recall that logic level 1 cor-

responds to -0.77 V and logic level 0 corresponds to -1.58 V for ECL; the corresponding levels for TTL are 3.6 V and 0.2 V (or 0.4 V for STTL).

In the circuit of Fig. 11.28, when the input to T_1 is high, T_1 is on and T_R is off. The resultant emitter current for T_1 is given by

$$i_{E_1} = \frac{v_I - V_0 + V_{EE}}{R_E} \qquad \text{(11.4–1)}$$

By substituting values, we have

$$i_{E_1} = \frac{-0.77 - 0.7 - (-5.2)}{1.18 \text{ k}\Omega} = 3.16 \text{ mA}$$

Thus, the voltage at the collector of T_1 is

$$v_{C_1} \cong V_{CC} - R_C i_{E_1} \qquad (11.4\text{--}2)$$

Substituting values yields

$$v_{C_1} = 5 - (2)(3.16) = -1.32 \text{ V}$$

and T_2 is cut off. This value provides the high logic level for TTL at the output of the translator. When v_I is low, T_1 is off and T_2 is in saturation.

TTL-to-ECL Translator.   A basic voltage level shifter for translating from TTL to ECL is displayed in Fig. 11.29. For the high logic level input, $v_I = 3.6$ V and D is short circuited. Hence, the voltage at P is obtained as

$$V_P = V_{IH} + V_O = 3.6 + 0.7 = 4.2 \text{ V} \quad (11.4\text{--}3)$$

By using the superposition theorem for voltages V_P and V_{EE} (neglecting i_{B_1}), we have

$$v_{B_1} = V_P \left(\frac{R_{B_2}}{R_{B_1} + R_{B_2}} \right) - V_{EE} \left(\frac{R_{B_1}}{R_{B_1} + R_{B_2}} \right)$$
$$(11.4\text{--}4)$$

Substituting values yields

$$v_{B_1} = 4.2 \left(\frac{4}{3+4} \right) - 5.2 \left(\frac{3}{3+4} \right) \cong 0.17 \text{ V}$$

Note that since this voltage is larger than V_R, T_1 is on while T_R is off. Therefore, from the previous analysis of ECL, the translator output is -0.77 V.

For the low logic level input, $v_I = 0.2$ V and D is again short circuited. However, the voltage at P is now reduced to

$$V_P = V_{IL} + V_O = 0.2 + 0.7 = 0.9 \text{ V} \quad (11.4\text{--}5)$$

Superimposing V_P and V_{EE} again yields (11.4–4) and substituting values we have

$$v_{B_1} = 0.9 \left(\frac{4}{3+4} \right) - 5.2 \left(\frac{3}{3+4} \right) = -1.7 \text{ V}$$

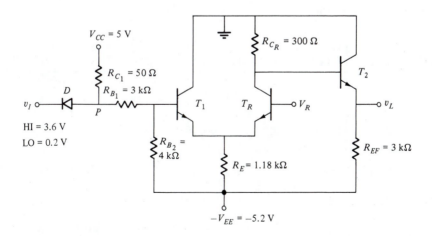

FIGURE 11.29   Basic TTL-to-ECL Translator

Note that since this voltage is less than $V_R = -1.175$ V, T_R is on. Thus, more current passes through R_{C_R} and v_L drops low to -1.58 V.

11.5   COMPARISON OF IC LOGIC FAMILIES

The various nonsaturating logic families discussed in this chapter have evolved through the years as state-of-the-art IC fabrication has advanced. As we have seen, each of these families has advantages as well as disadvantages.

In previous sections, analysis of the basic logic circuits for each family has led to the determination and comparison of logic parameters. In this section, we will verify the previously stated values by comparing these parameters with actual IC parameters obtainable from manufacturers' data sheets. The families considered are STTL, ECL, and CMOS.

Table 11.1 lists speed and power parameters for several lines of ICs, including STTL, ECL, and CMOS. These parameters are frequently used for determining the selection of one family over another. The values displayed are typical and are

specifically for Motorola, Inc. logic families. However, second-source vendors provide chips with similar parameters.

Note that three versions of Schottky TTL logic are listed in Table 11.1: LSTTL, or LS, for low-power Schottky; ALSTTL, or ALS, for advanced low-power Schottky; and FAST® (a Motorola trademark name), which offers an improvement in speed over the LS and ALS lines.

The LS line of ICs provides a current and power reduction improvement over standard TTL (described in Chapter 10) by about a factor of 5. Furthermore, the LS line is faster because of the Schottky transistors, which do not operate in saturation. Because of these improvements, LSTTL has forced standard TTL into obsolescence.

From Table 11.1 we can observe that the ALS line offers even lower power dissipation, as well as increased speed. Furthermore, the FAST STTL has additional speed improvement with increased power dissipation. The improvement in speed for these two lines is due to the use of advanced fabrication and processing of the IC chip.

The ECL families listed in Table 11.1 are labeled MECL, for Motorola ECL. Other manufacturers use different prefixes (Fairchild, for example, uses the letter F). The three versions

Table 11.1   Speed and Power Parameters for Nonsaturating Logic Families

Parameter	Units	STTL			MECL			CMOS	
		LS	ALS	FAST	10K	10KH	III	Standard	High Speed
Quiescent supply current per gate	mA	0.4	0.2	1.1	5.0	5.0	10	0.0001	0.0003
Power per gate (quiescent)	mW	2.0	1.0	5.5	26	26	54	0.0006	0.001
Propagation delay	ns	9.0	7.0	3.7	2.0	1.0	1.1	125	8.0
Speed–power product	pJ	18	7.0	19.2	52	26	59	0.075	0.01
Maximum clock frequency (*DFF*)	MHz	33	35	125	150	300	550	4.0	40
Maximum clock frequency (counter)	MHz	40	45	125	150	300	1200	5.0	40

From *Motorola Semiconductor Master Selection Guide and Catalog*, copyright of Motorola, Inc. Used by permission.

of ECL shown in Table 11.1 are MECL10K, MECL10KH, and MECL III. The 10K (10,000) stands for the numbering system used; logic gates are numbered from 10,000 and go up to higher numbers. For example, MC10104 is a Motorola ECL quad AND gate. The MECL10K has been the industry standard for high-speed applications for many years. Note that all three versions of MECL have the fastest speed by far of any logic family. However, the power dissipation per gate is also the largest.

The MECL10KH and MECL III are advanced versions of the MECL10K. The MECL10KH line offers high speed with an improvement of 100% in propagation delay from the standard MECL10K (see Table 11.1). The MECL III has a similarly small propagation delay; however, flip-flop toggle rates are improved by a factor of 2 to 4.

The last two columns in Table 11.1 provide parameter values for two versions of CMOS, standard CMOS and high-speed CMOS (abbreviated as HCMOS). Note that HCMOS has speed comparable to LSTTL. Furthermore, note that CMOS exhibits extremely low power dissipation per gate. Coupled with the fact that CMOS requires the least amount of chip area of any family for a given logic function, it is not surprising that portable electronic circuits (hand calculators, watches, and so on) all use CMOS circuitry.

11.6   SELECTION GUIDE FOR NONSATURATING LOGIC FAMILIES

Many functionally comparable nonsaturating logic integrated circuits are available from various manufacturers. In the appendix to this chapter, Data Sheet A11.1 lists the various logic circuits in TTL, CMOS, and MECL that are available from Motorola, Inc. Similar gates are also available from other semiconductor companies.

Note that the basic gates, such as the quad two-input AND, OR, and NOR gates, are avail-

able in all versions of all three families. Furthermore, some gates, such as ECL NAND gates, are not available in certain families. The basic ECL gate provides the OR/NOR output, and producing a NAND output with ECL is extremely inefficient. It should also be noted, however, that ECL is the only gate available for some of the OR and NOR complex gates.

The gates listed in Data Sheet A11.1 are packaged in standard 14-, 16-, or 20-pin cases. These cases are shown in Fig. 11.30. Recall from Chapter 6 that the 14-pin DIP is called the "N" package, and the 16-pin DIP is called the "J" package in the terminology of Texas Instruments, Inc.

Manufacturers' data sheets for several gates from each family are provided in Appendices 7, 8, and 9 at the end of the book. The pinout diagrams as well as the electrical parameters are given in these data sheets. The reader should note that

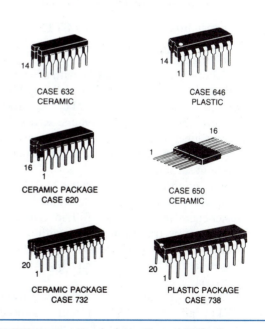

CASE 632
CERAMIC

CASE 646
PLASTIC

CERAMIC PACKAGE
CASE 620

CASE 650
CERAMIC

CERAMIC PACKAGE
CASE 732

PLASTIC PACKAGE
CASE 738

FIGURE 11.30   Standard 14-, 16-, and 20-Pin Cases

the pinouts given in detail for TTL in Chapter 10 are identical to those for the various lines of STTL.

Note that this value is the maximum power dissipated for all four gates on the quad chip.

EX. 11.11

Power Dissipation Calculation for Low-Power STTL Example

Calculate the average maximum power dissipated for the 7400LS (or the SN74LS00) quad two-input NAND gates by using the maximum DC supply currents from the data sheet in Appendix 7. These values are $I_{CC\,HI} = 1.6$ mA and $I_{CC\,LO} = 4.4$ mA at $V_{CC\,max} = 5.25$ V.

Solution: The average supply current for the 7400LS case is

$$I_{CC\,avg} = \frac{1.6 + 4.4}{2} = 3 \text{ mA}$$

The maximum average power dissipated in the 7400LS case is therefore

$$I_{CC\,avg}V_{CC} = (3)(5.25) = 15.75 \text{ mW}$$

EX. 11.12

Propagation Delay Calculation for Low-Power STTL Example

Calculate the average maximum propagation delay for the 7400LS quad two-input NAND gates by using the values from the data sheet in Appendix 7. These values are $t_{PLH} = 15$ ns and $t_{PHL} = 15$ ns. Comment on the power–delay product using the information from Ex. 11.11.

Solution: The average maximum propagation delay for the 7400LS is

$$t_{PD\,avg} = \frac{15 + 15}{2} = 15 \text{ ns}$$

Note that for low-power operation, a sacrifice in speed is unavoidable.

CHAPTER 11 SUMMARY

- The nonsaturating logic families STTL, ECL, MOS, and CMOS offer improvement over saturating logic families.
- The Schottky transistor is a BJT with a Schottky diode connected directly in parallel with its base and collector terminals. The Schottky diode prohibits the forward voltage across the collector junction from becoming greater than 0.3 V. Therefore, Schottky transistors never operate in saturation.
- STTL is essentially TTL with each BJT replaced with a Schottky transistor. The voltage transfer characteristic of the STTL inverter displays a narrow transition region, with $V_{IL} = 1$ V and $V_{IH} \cong 1.15$ V. The output high and low voltages are $V_{OH} = 3.6$ V and $V_{OL} = 0.4$ V, respectively.
- Additional forms of STTL are the low-power (LSTTL), advanced (ASTTL), and advanced low-power (ALSTTL) lines. For each of these lines of STTL, a trade-off exists between speed and power.
- Emitter-coupled logic (ECL) is another nonsaturating logic family in which BJTs are designed either to operate in the active region or to be cut off. The basic circuit for an ECL inverter is that of the two-transistor difference amplifier, where the input to the reference transistor is fixed at V_R and the input to the other transistor has v_I applied. ECL is also called CML (current-mode logic) because as v_I varies from low to high, the current switches from the reference transistor to the input transistor.
- The basic gate for ECL is the NOR/OR gate. Emitter follower circuits are used at the output of the basic ECL gate to increase the fan-out. ECL also utilizes negative voltage throughout the circuit to eliminate the effects of a noise voltage that could be associated with the DC supply V_{EE}.
- For the NOR and OR outputs, $V_{OH} = -0.77$ V

and $V_{OL} = -1.58$ V. For the inputs, $V_{IL} = -1.225$ V and $V_{IH} = -1.125$ V.

- The reference voltage V_R for ECL is temperature compensated. The usual design allows V_R to vary with temperature in the same manner as V_0. Thus, V_R is automatically maintained in the middle of the logic swing, regardless of temperature.

- The MOS logic families NMOS, PMOS, and CMOS offer a major advantage in that larger packing densities are possible because MOSTs can be made smaller than BJTs. The transfer characteristics of MOS logic families are non-linear in the transition region.

- NMOS (PMOS) families can be designed with either E-O loads or E-D loads. The family using E-D loads has a larger value for V_{OH}, and the transition region is more abrupt. $V_{OH} = V_{DD}$ for the case of E-D loads, and $V_{OH} = V_{DD} - V_T$ for the case of E-O loads. In both cases, $V_{IL} = V_{T_D}$. The voltages V_{IH} and V_{OL} depend upon the width-to-length ratio of the driver transistor to the load transistor.

- Two basic gates are natural applications of NMOS (PMOS) logic circuits. They are the NOR and NAND gates. NOR gates are used more frequently because NAND gates exhibit more low-level voltage degradation.

- The CMOS logic family uses N-channel and P-channel MOSTs in matched or complementary pairs. The CMOS inverter voltage transfer characteristic is more abrupt than that for NMOS (PMOS). For CMOS, $V_{OH} = V_{SS}$, $V_{OL} = 0$, $V_{IH} = V_{SS} - V_T$, and $V_{IL} = V_T$. Additionally, the abrupt portion of the transistion region occurs at $v_I = V_{SS}/2$.

- Similar to NMOS (PMOS), the basic circuits for CMOS are NAND and NOR gates. For CMOS, neither gate has preference over the other.

- MOS families require specialized handling procedures and/or gate protective circuitry to avoid static charge buildup on the gate, which can create an electric field large enough to cause oxide breakdown and permanent destruction of the IC.

- The interconnection of different logic families requires an interface or translator circuit that provides compatible voltage and current levels. Many translator circuits are commercially available in IC form.

- In comparing the nonsaturating logic families with one another, we see that CMOS (MOS) has the lowest power dissipation (by far), followed by STTL and ECL. For propagation delay, ECL is the fastest, followed by STTL and CMOS (MOS). CMOS (MOS) is first in packing density, followed by STTL and ECL.

CHAPTER 11 PROBLEMS

11.1

A Schottky transistor is used in the circuit of Fig. P11.1. Let the PN junction forward voltage be $V_0 = 0.7$ V and that of the Schottky diode be 0.3 V. Analyze the circuit to determine v_{CE}, draw the Schottky transistor symbol, and label the values of voltage between all terminals. Also, determine i_B, i_C, and i_D for $\beta = 100$. What is the operating state of the BJT?

11.2

Repeat Problem 11.1 for $V_{BB} = 20.7$ V.

11.3

Repeat Problems 11.1 and 11.2 with the Schottky diode removed from the circuit (leaving only the BJT). Let $v_{CE\,sat} = 0$ V.

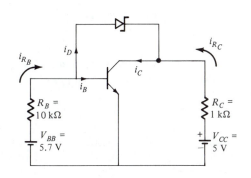

FIGURE P11.1

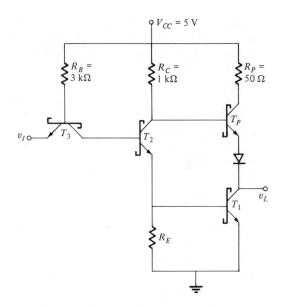

FIGURE P11.4

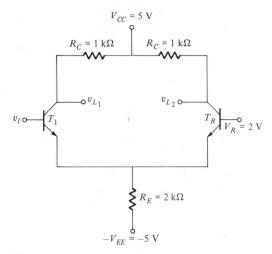

FIGURE P11.10

11.4
Sketch the transfer characteristic of the Schottky TTL inverter of Fig. P11.4 (v_L versus v_I). Explain the state of each transistor for the high and low states of v_I.

11.5
Calculate the power supplied for the Schottky TTL inverter of Fig. P11.4 for the high and low input voltages. Let v_L high = 3.5 V, $V_0 = 0.7$ V, and the voltage drop for the Schottky diodes = 0.3 V when forward biased.

11.6
Repeat Problem 11.4 for the STTL gate of Fig. 11.3. Assume that all inputs are high except v_{I_1}.

11.7
Repeat Problem 11.5 for the STTL gate of Fig. 11.3. Let $\beta = 50$.

11.8
Repeat Problem 11.4 for the STTL gate of Fig. 11.5. Assume that all inputs are high except v_{I_1}.

11.9
Repeat Problem 11.5 for the LSTTL gate of Fig. 11.5.

11.10
Sketch the transfer characteristics of v_{L_1} versus v_I and v_{L_2} versus v_I for the basic ECL gate of Fig. P11.10. Use $V_0 = 0.7$ V.

11.11
Repeat Problem 11.10 for $V_{CC} = 0$ V. Use $V_0 = 0.7$ V.

11.12
Repeat Problem 11.10 for $V_{CC} = 0$ and $V_R = -2$ V. Use $V_0 = 0.7$ V.

11.13
Determine the transfer characteristics v_{NOR} versus v_{I_1} for the ECL gate of Fig. 11.8a and treat all other inputs as low. Use $V_0 = 0.75$ V, which is more accurate for actual ECL gates. Neglect the base current of the emitter follower.

11.14
For the ECL NOR/OR gate of Fig. 11.8a, determine the transfer characteristic for $R_{EF} = 3$ kΩ, $R_{C_1} = R_{C_R} = 0.5$ kΩ, $R_E = 2$ kΩ, $V_R = -2$ V, and $V_{EE} = 5$ V. Use $V_0 = 0.75$ V, which is more accurate for actual ECL gates. Additionally, include the base current of the emitter followers for the high level of V_{NOR} and V_{OR}. Use $\beta = 50$.

11.15
For the ECL OR gate of Fig. P11.15, determine the values of V_{OH} and V_{OL}. Include the emitter follower base current and use $V_0 = 0.75$ V and $\beta = 50$.

11.16
For the gate of Problem 11.15, determine V_{IH} and V_{IL} using the approximate technique developed in the text. Draw the transfer characteristic using the solution of Problem 11.15.

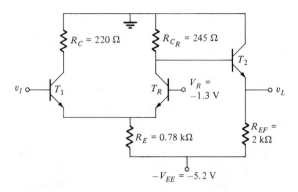

FIGURE P11.15

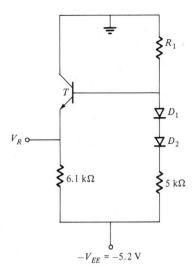

FIGURE P11.20

11.17
Determine the fan-out for the ECL gate of Fig. P11.15 and assume that the high-level noise margin is 0.2 V. For consistent answers, use $V_{IH} = -1.1$ V.

11.18
For the circuit of Fig. P11.18, determine V_{OH} and V_{OL} corresponding to $V_{IH} = -0.7$ V and $V_{IL} = -2$ V. Use $V_0 = 0.75$ V, $\beta = 50$, and $v_{CE \, sat} = 0.2$ V.

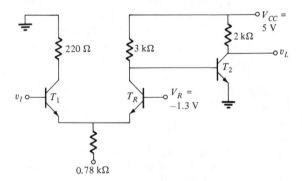

FIGURE P11.18

11.19
Determine the fan-out of the ECL NOR gate of Problem 11.14.

11.20
For the circuit of Fig. P11.20, determine the reference voltage V_R for $R_1 = 800$ Ω. Use $V_0 = 0.75$ V. Neglect the BJT base current.

11.21
Repeat Problem 11.20 for $R_1 = 5$ kΩ.

11.22
For the ECL gate of Fig. P11.15 (and Problem 11.15), determine the power supplied for the high and low states.

11.23
For the ECL NOR/OR gate of Problem 11.14, determine the power supplied for the high and low states.

11.24
Draw a circuit diagram for a CMOS two-input AND gate.

11.25
Draw a circuit diagram for a CMOS two-input OR gate.

11.26
Draw a circuit diagram for an *RSFF* using **(a)** basic NMOS NOR gates and **(b)** basic CMOS NOR gates.

11.27
Consider the NMOS inverter of Fig. 11.16a. Calculate the transistor design ratio K_D/K_L and the W/L ratio of the driver and load transistors. Let $V_{DD} = 5$ V, $V_{T_D} = V_{T_L} = 2$ V, and $V_{IH} - V_{IL} = 0.8$ V.

11.28
Consider the CMOS inverter of Fig. 11.19 for which $V_{DD} = 5$ V and the channel resistance of each transistor is either 1 kΩ or 1 MΩ for the on and off conditions, respectively. Calculate the output low and high voltage levels.

11.29

For the NMOS inverter of Fig. P11.29, determine V_{OH}, V_{OL}, V_{IH}, and V_{IL} for $v_{DD} = 5$ V, $R_D = 50$ kΩ, $V_T = 1$ V, and $K_N = 0.1$ mA/V². Use the approxi-

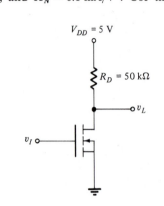

FIGURE P11.29

mate method and assume that $V_{OL} = 0$. Sketch the voltage transfer characteristic and also determine the noise margins V_{NML} and V_{NMH}.

11.30

For the NMOS inverter of Fig. 11.17a with $V_{DD} = 5$ V, determine V_{OH}, V_{OL}, V_{IH}, and V_{IL}. Data for the E-O MOST is $V_T = 2$ V and $K_N = 1$ mA/V²; for the E-D MOST, $I_{DSS} = 1$ mA and $V_P = -4$ V. Sketch the voltage transfer characteristic and also determine the noise margins V_{NML} and V_{NMH}.

11.31

For the CMOS inverter of Fig. 11.19, determine V_{OH}, V_{OL}, V_{IH}, and V_{IL} for $V_{SS} = 5$ V. For the inverter, $K = K_N = K_P = 1$ mA/V² and $V_T = V_{T_N} = V_{T_P} = 1$ V. Sketch the voltage transfer characteristic and also determine the noise margins V_{NML} and V_{NMH}.

11.32

Determine the maximum current for the CMOS inverter of Problem 11.31.

11.33

Calculate the total average power dissipated for a PMOS inverter that is the complement of the NMOS inverter of Fig. 11.23. Let $V_{DD} = -5$ V. T_D has

$R_{on} = 2$ kΩ and $R_{off} = 1$ MΩ, while T_L has a resistance value of 50 kΩ. Also let, $f = 100$ kHz and $C_E = 10$ pF.

11.34

Calculate the total average power dissipation for the CMOS inverter of Fig. 11.24. Let $V_{SS} = 5$ V, $R_{on} = 5$ kΩ, $R_{off} = 1$ kΩ, $f = 50$ kHz, and $C_E = 50$ pF (increased because of an increase in fan-out).

11.35

Consider a CMOS logic gate that is switched on and off every 10 μs. If $V_{SS} = 5$ V and $C_E = 150$ pF, determine the average transient current due to switching.

11.36

Consider a CMOS logic gate with $V_{DD} = 5$ V. If the total average power dissipation is 0.5 mW and the frequency of switching is 50 kHz, determine the load capacitance, which is $2C_E$.

11.37

Calculate the high and low voltage noise margins for a CMOS inverter with $V_{OL} = 0$, $V_{OH} = 5$ V, $V_{IH} = 4$ V, and $V_{IL} = 1$ V. Repeat this calculation for the case of two such inverters in cascade.

11.38

For the ECL-to-TTL translator of Fig. 11.28, determine the collector current of T_1 when **(a)** v_I is high $(-0.77$ V) and **(b)** v_I is low $(-1.58$ V). Also determine the corresponding output voltages. Assume that $v_{CE\,sat} = 0.2$ V.

11.39

Calculate the average maximum power dissipated for the 7402 LSTTL quad two-input NOR gate. Use the data sheet given in Appendix 7.

11.40

Calculate the average maximum propagation delay for the 7402 LSTTL quad two-input NOR gate. Use the data sheet given in Appendix 7.

11.41

Calculate the maximum power dissipation for the ECL quad two-input NOR gate MC10H102. Use the data sheet given in Appendix 8 at 25° C.

11.42

Repeat Problem 11.41 for average maximum propagation delay.

CHAPTER 11 APPENDIX

DATA SHEET A11.1   Functionally Comparable Logic Integrated Circuits

The following table offers a quick guide to logic circuits in the various families available from Motorola. Devices perform comparable functions within a specific logic family. Pinout configurations are generally identical when the numbers in the columns correspond, although there may be exceptions. Consult the data sheets for specific details.

The numbers in the various columns are suffixes that follow the basic line prefixes associated with each specific product line. These basic prefixes are as follows:

PREFIXES:	LS: SN54/74LS	Standard CMOS: MC14	MECL10K: MC10
	ALS: SN54/74ALS	Hi-Speed CMOS: MC54/74HC	MECL10KH: MC10H
	FAST: MC54/74F		MECL III: MC

Thus, the TTL LS Quad 2-input gate (item 1 in the table) can be ordered under the part number SN54LS06 (or SN74LS06). Similarly, the MECL10KH 2-input gate bears the part number MC10KH104.

Function†	TTL			CMOS		Package* TTL/CMOS	MECL			Package* MECL
	LS	ALS	FAST	STD	HC		10K	10KH	III	
AND Gates										
Quad 2-Input	06	06	06	061	06	632, 646	104	104		620, 648, 650
Quad 2-Input, Open-Collector	09	09								
Triple 3-Input	11	11	11	073	11					
Triple 3-Input, Open-Collector	15	15								
Dual 4-Input	21	21		062						
Hex							197			
NAND Gates										
Quad 2-Input	00	00	00	011	00	632, 646				
Quad 2-Input, Open-Collector	01	01								
Quad 2-Input, Open-Collector	03	03			03					
Quad 2-Input, High-Voltage	26									
Quad 2-Input Buffer	37	37								
Quad 2-Input Buffer, Open-Collector	36	38								
13-Input	133				133	620,648				
Triple 3-Input	10	10	10	023	10	632, 646				
Triple 3-Input, Open-Collector	12	12								
Dual 4-Input	20	20	20	012	20					
Dual 4-Input, Open-Collector	22	22								
Dual 4-Input Buffer	40									
8-Input	30			068	30					
OR Gates										
Quad 2-Input	32	32	32	071	32	632, 646	103	103	1664	620, 648, 650
Dual 3-Input 3-Output							110			
High-Speed Dual 3-Input 3-Output							210	210		
Triple 3-Input				075	4075					
Dual 4-Input				072						
NOR Gates										
Quad 2-Input	02	02	02	001	02	632, 646	102	102	1662	620, 648, 650
Quad 2-Input Buffer	28	28								
Quad 2-Input Buffer, Open-Collector	33	33								
Dual 5-Input	260									
Triple 3-Input	27	27		025	27					

†For Temperature Characteristics of the various Logic Lines, see end of table on page 83.
*Refer to page 84 for additional information on packaging, number of pins.

continues

DATA SHEET A11.1   Continued

Function†	TTL			CMOS		Package*	MECL			Package*
	LS	ALS	FAST	STD	HC	TTL/CMOS	10K	10KH	III	MECL
NOR Gates (continued)										
Quad 2-Input with Strobe						632,	100	100		620,
Triple 4-3-3 Input						646	106	106		648, 650
Dual 3-Input 3-Output							111			620,648
High-Speed Dual 3-Input 3-Output							211	211		620,
Dual 3-Input, plus Inverter				000						648,
Dual 4-Input				002	4002					650
8-Input				078	4078					
Exclusive OR Gates										
Quad	86		86	070	86	632,	113	113		620,
2-Input Quad	386					646				648,
Quad, Open-Collector	136									650
Triple 2-Input									1672	620
Exclusive NOR Gates										
Quad	266			077	266	632,				
Triple 2-Input						646			1674	620
Complex Gates										
Quad OR/NOR						632,	101	101		620,
Triple 2-3-2 Input OR/NOR						646	105	105	1688	648,
Triple 2-Input Exclusive OR/Exclusive NOR							107	107		650
Dual 4-5 Input OR/NOR							109	109		
Dual 4-5 Input OR/NOR								209		
Dual 2-Wide 2-3 Input OR-AND/ OR-AND-Invert							117	117		
Dual 2-Wide 3-Input OR-AND							118	118		
4-Wide 4-3-3-3 Input OR-AND Gate							119	119		
OR-AND/OR-AND-INVERT Gate							121	121		
High-Speed Dual 3-Input 3-Output OR/NOR							212			
Dual 4-Input OR/NOR									1660	
Dual AND-OR-INVERT Gate	51	51								
Dual AND-OR-INVERT Gate				506		620,648				
3-2-2-3 Input AND-OR-INVERT Gate	54					632,				
2-Wide and 4-Input AND-OR-INVERT Gate	55	55				646				
4-2-2-3 Input AND-OR-INVERT Gate			64							

†For Temperature Characteristics of the various Logic Lines, see end of table on page 83.
*Refer to page 84 for additional information on packaging, number of pins.

continues

DATA SHEET A11.1   Continued

Function†	TTL			CMOS		Package*	MECL			Package*
	LS	ALS	FAST	STD	HC	TTL/CMOS	10K	10KH	III	MECL
Complex Gates (continued)										
Triple Gate (Dual 4-Input NAND Gate and 2-Input NOR/OR Gate or 8-Input AND/NAND Gate)				501		620, 648				
4-Bit AND/OR Selector (Quad 2-Channel Data Selector or Quad Exclusive NOR Gate)				519						
Dual 5-Input Majority Logic Gate				530						
Hex Gate (Quad Inverter plus 2-Input NOR Gate plus 2-Input NAND Gate)				572						
Inverters/Buffers (Non 3-State)										
Hex Inverter	04	04	04	069	04	632, 646				620, 648
Hex Inverter, Open-Collector	05	05								
Dual Complementary Pair plus Inverter				007						
Hex Buffer				050	4050	620, 648				
Strobed Hex Inverter/Buffer				502						
Hex Buffer with Enable							188	188		
Hex Inverter with Enable							189	189		
Hex Inverter/Buffer				049	4049		195			
Translators										
Quad MTTL to MECL, ECL Strobe						620, 648	124	124		620, 648
Quad TTL to MECL, TTL Strobe								424		
Quad MECL to MTTL							125	125		
Quad MECL to TTL, Single Supply								350		
Triple MECL to NMOS							177			
TTL or CMOS to CMOS Hex Level Shifter				504						
Quad MST-to-MECL 10,000							190			
Hex MECL 10,000 to MST							191			

†For Temperature Characteristics of the various Logic Lines, see end of table on page 83.
*Refer to page 84 for additional information on packaging, number of pins.

continues

DATA SHEET A11.1   Continued

Flip-Flops/Registers

Dual JK				027		620,648	135	135		620,648
Dual JK	73A				73	632,				
Dual D	74A	74	74	013	74	646	131	131		
Dual JK	76A				76	620,648				
Dual JK with Preset	78A					632,646				
Dual JK with Preset	109A	109	109		109	620,648				
Dual JK with Clear	107A				107	632,646				
Dual JK Edge-Triggered	112A		112		112	620,648				
Dual JK Edge-Triggered	113A		113		113	632,646				
Dual JK Edge-Triggered	114A		114							
4-Bit D Register, 3-State	173			076	173	620,648				
Hex D with Clear	174		174	174	174					
Hex D with Enable	378		378			732,738				
Quad D with Clear	175		175	175	175	620,648				
Octal D with Clear	273	273			273	732,738				
Octal D, 3-State	374	374	374		374					
Octal D with Enable	377	377								
4-Bit D with Enable	379		379			620,648				
Hex D							176	176		
Hex "D" Master-Slave/with Reset							186	186		

†For Temperature Characteristics of the various Logic Lines, see end of table on page 83.
*Refer to page 84 for additional information on packaging, number of pins.

DATA SHEET A11.1   Continued

Function†	TTL			CMOS		Package*	MECL			Package*
	LS	ALS	FAST	STD	HC	TTL/CMOS	10K	10KH	III	MECL
Flip-Flops/Registers (continued)										
Octal D, Inverting, 3-State					564	732,738				620, 648
Octal D, 3-State		574			574					
High-Speed Dual Type D Master-Slave							231			
Dual Clocked R-S									1666	
Dual Clocked Latch									1668	
Master-Slave Type D									1670	
UHF Prescaler Type D									1690	
Octal D Flip-Flop, 3-State			534		534					
Counters										
Decade	90					632, 646				620, 648
Divide-By-12	92									
4-Bit Binary	93						154			
Decade, Asynchronously Presettable	196									
4-Bit Binary, Asynchronously Presettable	197									
BCD Decade, Asynchronously Reset	160A	160	160	160	160	620, 648				
4-Bit Binary, Asynchronous Reset	161A	161	161	161	161		178	016	1654	
BCD Decade, Synchronous Reset	162A	162	162	162	162					
4-Bit Binary, Synchronous Reset	163A	163	163	163	163					
Up/Down Decade, with Clear	192	192	192	510	192					
Up/Down Binary, with Clear	193	193	193	516	193					
Up/Down Decade	190	190	190				137			
Up/Down Binary	191	191	191	029			136	136		
Decade (Divide By 2 and 5)	290					632, 646	138		1678	
4-Bit Binary	293									
Dual Decade	390			518	390	620, 648				
Dual 4-Bit Binary				520						
Dual 4-Bit Binary	393				393	632,646				
Dual Decade	490					620,648				
Decade Up/Down, 3-State	568	568				732, 738				
Binary Up/Down, 3-State	569	569								
Synchronous 4-Bit Up/Down Decade	668					620, 648				
Synchronous 4-Bit Up/Down Binary	669									
Up/Down Decade	168	168	168							
Up/Down Binary	169	169	169							

†For Temperature Characteristics of the various Logic Lines, see end of table on page 83.
*Refer to page 84 for additional information on packaging, number of pins.

continues

12

SENSING OR DETECTING TRANSDUCERS

A *transducer* is a device that converts one form of energy into another. In this chapter, devices that convert one form of energy into electrical energy or light energy and that also act as sensors are described. The description of electromechanical transducers in the form of rotational machinery is deferred until Chapter 13.

The word *sensor* is often used synonymously for *transducer*. In general, however, a *sensor* is a device that receives and responds to a stimulus, whereas a transducer responds to one form of energy and converts this energy into another form. For example, a photoelectric diode is a transducer that converts light energy into electrical energy. Additionally, this device acts as a sensor of light by producing a current whose magnitude is dependent on the intensity of the light.

The types of sensing transducers described first are those that provide a variable R, L, or C (impedance) as an output quantity for various forms of inputs, except light. Then, voltage-generating sensors, again without light, that utilize the piezoelectric, thermocouple, or Hall effect for their operation are described. Finally, light-sensing and -producing devices are introduced. These devices are modernly known as optoelectronic transducers when light energy and electrical energy are coupled between each other.

12.1   VARIABLE-IMPEDANCE DEVICES

Variable-impedance devices are sensors that produce a change in impedance for a change in force, pressure, temperature, displacement, or voltage. For the particular device, the variable impedance may correspond to a resistance, an inductance, or a capacitance.

12.1.1   Strain Gages

A *strain gage* is a specialized device that creates a change in resistance for a change in force or pressure. Typical construction of a strain gage consists of a fine wire conductor firmly bonded or cemented to a mounting plane, as shown in Fig. 12.1. Often, instead of wire, a meandering pattern of metallic foil is used because a very accurate surface pattern can be obtained using photolithography techniques.

The operation of a strain gage involves a force or pressure applied to the device that creates a change in the resistance of the conductor. When a tensile (elongating) stress is applied as shown in Fig. 12.1, the conductor elongates and the cross-sectional area reduces. Each of these dimensional changes in conductor size causes an increase in conductor resistance. Hence, the magnitude of the tensile force or pressure is indicated by the amount of increase in the resistance of the con-

ductor. The resistance change associated with the change in area, however, is usually insignificant.

The sensitivity of the strain gage is defined by

$$S = \frac{\Delta R/R}{\Delta L/L} \tag{12.1–1}$$

where R and L are the unstrained resistance and length, respectively, while ΔR and ΔL are the corresponding changes in these quantities with applied elongating stress. The ratio $\Delta L/L$ is often referred to as the *strain* with units of distance/distance.

Typical metallic strain gages using fine wire or etched metal foil have a sensitivity S in the range of 2 to 4. Recently, however, semiconductor strain gages have become available with S on the order of 100 times larger. Unfortunately, semiconductor strain gages are more sensitive to temperature variations.

--- EX. 12.1

Strain Gage Example

A metallic strain gage with sensitivity $S = 2$ has an unstrained resistance $R = 100\ \Omega$. If the gage is subjected to a strain of 10^{-3} in./in., determine the change in resistance.

Solution: Direct substitution into (12.1–1) yields

$$2 = \frac{\Delta R/100}{10^{-3}}$$

Thus,

$$\Delta R = 2(100)(10^{-3}) = 0.2\ \Omega$$

To provide accuracy in measurements using strain gages, a bridge circuit is often used. As shown in Fig. 12.2, the measurement strain gage is in one arm of the bridge with resistance R_4. Additionally, to provide temperature compensation, R_2 is an identical strain gage with zero applied strain. If R_1 and R_4 are temperature-compensated resistors, V_{out} will be proportional to the applied strain and will be independent of temperature.

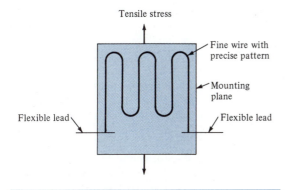

Tensile stress

Fine wire with precise pattern

Mounting plane

Flexible lead

Flexible lead

FIGURE 12.1   Typical Strain Gage

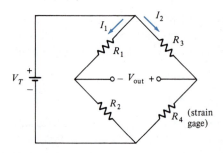

FIGURE 12.2   Strain Gage Used in Bridge Circuit

To obtain an expression for V_{out} in the bridge circuit of Fig. 12.2, we first realize that

$$V_{out} = I_2 R_4 - I_1 R_2 \qquad (12.1\text{--}2)$$

where

$$I_1 = \frac{V_T}{R_1 + R_2} \qquad (12.1\text{--}3)$$

$$I_2 = \frac{V_T}{R_3 + R_4} \qquad (12.1\text{--}4)$$

Thus, by substitution, we have

$$V_{out} = \frac{V_T R_4}{R_3 + R_4} - \frac{V_T R_2}{R_1 + R_2} \qquad (12.1\text{--}5)$$

Combining the two terms in (12.1–5) using the lowest common denominator yields

$$V_{out} = V_T \left[\frac{R_1 R_4 - R_2 R_3}{(R_1 + R_2)(R_3 + R_4)} \right] \qquad (12.1\text{--}6)$$

If we now let the resistors in branches 1 and 3 be equal in magnitude to the unstrained resistance R_2, we have $R_1 = R_2 = R_3 = R$, and substitution yields

$$V_{out} = V_T \left[\frac{R(R_4 - R)}{2R(R + R_4)} \right] = \frac{V_T}{2} \left(\frac{R_4 - R}{R + R_4} \right)$$

$$(12.1\text{--}7)$$

Note that if R_4 is under zero strain, R_4 is also equal to R and $V_{out} = 0$. Furthermore, if we define the resistance with strain as $R_4 = R + \Delta R$, (12.1–7) can be written as

$$V_{out} = \frac{V_T}{2} \left(\frac{\Delta R}{2R + \Delta R} \right) \qquad (12.1\text{--}8)$$

———————————————————————— EX. 12.2

Bridge Circuit Example

In the bridge circuit arrangement of Fig. 12.2, identical metallic strain gages used for the elements R_3 and R_4 have an unstrained resistance of $R = 100\ \Omega$ and a sensitivity of $S = 2$. If stress is applied only to strain gage R_4, calculate the approximate value of V_{out} for $V_T = 10$ V and $R_1 = R_2 = 1\ \text{k}\Omega$. Consider three different stress measurements where $\Delta R = 10\%$, 20%, and 30% of R.

Solution: In this example, we must use the general expression given by (12.1–6). By direct substitution,

$$V_{out} = 10 \left[\frac{10^3 (R + \Delta R) - 10^3 R}{(2 \times 10^3)(2R + \Delta R)} \right] = 5 \left(\frac{\Delta R}{2R + \Delta R} \right)$$

where $R = 100\ \Omega$ and $\Delta R = 10\%$, 20%, and 30% of R. Note that using a large resistor in branches 1 and 3 does not provide a numerical advantage; however, it is convenient at times. By substituting $\Delta R = 10\%(R) = 10\ \Omega$, we obtain

$$V_{out} = 5 \left(\frac{10}{200 + 10} \right) \cong 0.238 \text{ V}$$

Similarly, for $\Delta R = 20\%(R) = 20\ \Omega$, we have

$$V_{out} = 5 \left(\frac{20}{200 + 20} \right) \cong 0.45 \text{ V}$$

and for $\Delta R = 30\%(R) = 30\ \Omega$, we have

$$V_{out} = 5 \left(\frac{30}{200 + 30} \right) \cong 0.65 \text{ V}$$

We observe that the variation of V_{out} with stress is not linear in this example. However, the variation of V_{out} with stress will be essentially linear provided that $\Delta R \ll R$, which usually requires $\Delta R \leq 10\%(R)$.

12.1.2   Thermistors and Sensistors

Thermistors and *sensistors* are temperature-sensing variable resistors. Specifically, these devices are resistors whose resistance varies in a precise manner with temperature. The specific variation depends upon the material composition of the thermistor, as well as its geometry (length, width, and cross-sectional area).

Two classes of temperature-sensing variable resistors are available commercially. One class possesses a positive temperature coefficient, which means that the resistance increases as the temperature increases. The variable resistors in this class are sensistors. The other class has a negative temperature coefficient for which the resistance decreases as the temperature increases. The variable resistors in this class are thermistors. They are usually composed primarily of semiconductor material that characteristically exhibits a negative temperature coefficient.

Thermistors and sensistors are available in all sorts of shapes and small sizes, from beads, rods, and disks to washer shapes. They are also available with a broad range of resistances and temperatures. Resistance values at room temperature can range from 100 Ω to 10 MΩ. Furthermore, for a 300° C change in temperature, a typical change in resistor value is approximately three orders of magnitude for thermistors and one order of magnitude for sensistors.

One of the primary applications of a thermistor is to measure temperature. However, to provide an accurate measurement, a bridge network such as that of Fig. 12.2 is used. When a temperature-variable resistor is used in place of the strain gage, (12.1–6) again provides the expression for output voltage for the general case, where R_4 is now the resistance of the temperature sensor. Under the conditions where $R_1 = R_2 = R_3 = R$, (12.1–7) provides the simplified expression for V_{out}.

Thermistors are also used for temperature compensation. When one of these temperature-sensing resistors is placed in series with a resistor possessing the opposite temperature coefficient, the total resistance becomes independent of temperature. Quite often, this type of temperature compensation is used in ammeters and voltmeters that utilize the d'Arsonval movement (to be described in Chapter 14). The resistance associated with the wire coil of the meter has a positive temperature coefficient (metallic), and placing a thermistor (negative temperature coefficient) in series with the meter compensates for the change with temperature. Meters with this type of temperature compensation can be used over a wide range of temperatures with little error.

12.1.3   Variable-Inductance Transducers

Variable-inductance transducers are devices that provide an AC electrical output that varies in proportion to a mechanical displacement. When the device is specifically designed to provide an inductively coupled AC output voltage that is linearly proportional to the input displacement, it is called a *linear variable differential transformer* (LVDT). (Refer to Chapter 1 for a general discussion of transformers.)

Figure 12.3a displays a schematic diagram of the LVDT, and Fig. 12.3b shows its corresponding circuit symbol. As indicated, the LVDT consists of a primary coil and two identical secondary coils connected in series opposition. The movement of the core changes the inductive coupling between the primary and secondary windings and thereby provides corresponding changes in the AC secondary voltage V_{out}. This voltage can be either positive or negative depending upon the displacement direction.

In the starting position, where the movable core is centered and hence couples half of each secondary coil, the output voltage is zero. The secondary coil voltages are equal and thus cancel each other since V_{out} is the difference in these voltages. Then, as the core is displaced from the zero position, one secondary voltage increases while the other reduces. An increase in the magnitude of

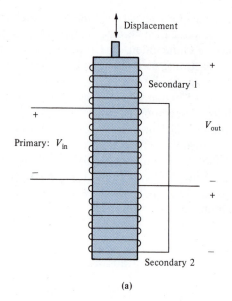

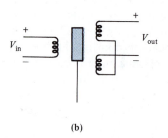

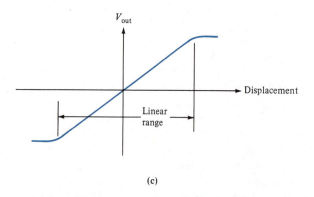

(c)

FIGURE 12.3   Linear Variable Differential Transformer (LVDT): (a) Physical schematic, (b) Circuit symbol, (c) Transfer characteristic

output voltage V_{out} results. Hence, for off-center displacements, the output voltage of the LVDT is a linear function of displacement.

Figure 12.3c shows the transfer characteristic of the LVDT. Typical ranges of displacement can be a few microinches or several inches, while typical source voltages are sinusoidal with $f = 60$ Hz to 20 kHz and amplitude on the order of 10 V rms. The output voltage undergoes a 180° phase reversal as the displacement goes through zero.

── **EX. 12.3**

LVDT Example

Consider an LVDT with an rms input voltage of 10 V, maximum rms output voltage of 5 V, and maximum displacement of $+6$ cm. Calculate the output voltage for displacements of 6 cm and 2.3 cm.

Solution: For a displacement of 6 cm, the output voltage is 5 V. For a displacement of 2.3 cm, the output voltage is (2.3)(5)/6, or 1.92 V.

12.1.4   Mechanically Variable-Capacitance Transducers

The operation of *mechanically variable-capacitance transducers* is based upon the parallel-plate capacitance expression, which is given by (see Chapter 1)

$$C = \frac{\varepsilon_r \varepsilon_0 A}{d} \qquad (12.1\text{–}9)$$

where

ε_r = relative dielectric constant of the material separating the plates

ε_0 = dielectric permittivity of free space (8.854×10^{-12} F/m)

d = spacing between the plates in meters (m)

A = area of the plates in m².

Several possibilities exist for capacitive transducers that involve varying the capacitance through geometric variations of either A or d (or both).

Three practical capacitive transducers are shown in Fig. 12.4. In Fig. 12.4a, one plate of the capacitor is movable and results in capacitance that is proportional to the amount of insertion or overlap area A of the plates. In Fig. 12.4b, similar capacitive variation exists except that one plate is rotated to produce a change in the overlap area A of the two plates. A third example is shown in Fig. 12.4c, where a conductive diaphragm is deflected due to variation in pressure. The deflection causes a decrease in the spacing between the plates (d) as well as an increase in the plate area A. Both of these variations result in an increase in capacitance.

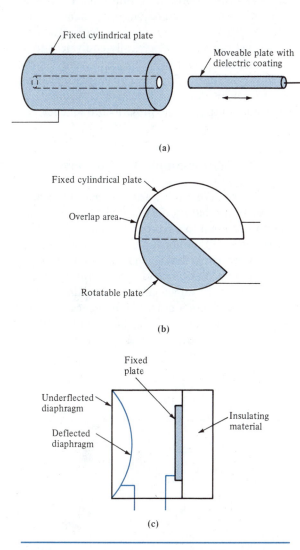

(a)

(b)

(c)

FIGURE 12.4   Variable-Capacitance Transducers: (a) Variable area example, (b) Another variable area example, (c) Variable dielectric thickness example

In figure (a): Fixed cylindrical plate; Moveable plate with dielectric coating

In figure (b): Fixed cylindrical plate; Overlap area; Rotatable plate

In figure (c): Fixed plate; Underflected diaphragm; Deflected diaphragm; Insulating material

---

EX. 12.4

Variable-Capacitance Transducer Example

Consider a capacitive transducer as shown in Fig. 12.4a. Calculate the magnitude of capacitance if the inner cylinder is inserted 2 cm. Let the diameter of the inner cylinder be 1 cm and the spacing between plates be 0.01 cm. The dielectric coating on the inner cylinder has $\varepsilon_r = 10$.

Solution: For this example, we use the parallel-plate capacitance expression of (12.1–9) with the average plate area given by

$$A = (\pi D_{avg})(L) = \pi(1.005)(2) = 6.31 \text{ cm}^2$$

where $L = 2$ cm is the length of insertion. Thus, the capacitance is

$$C = \frac{(10)(8.854 \times 10^{-14})(6.31)}{2} = 2.8 \text{ pF}$$

12.1.5   Voltage Variable-Capacitance Devices

The operation of *voltage variable-capacitance devices* is based upon the capacitance of a PN junction diode and its dependence upon reverse voltage across its terminals. These devices are also called *varactors*, from the term *variable reactors*.

Under reverse-bias conditions, the capacitance of a PN junction is given by the expression

$$C = K V_R^{-m}$$

where

V_R = applied reverse voltage

K = constant of proportionality

m = constant dependent upon the doping variation perpendicular to the junction

Varactor diodes are widely used in electronic applications. They are appropriate whenever voltage variable tuning is desired, for example, in detecting circuits or oscillator circuits where a variable C is required.

12.2   VOLTAGE-GENERATING SENSORS

Various transducers can be used to produce or generate voltage. In this section, we consider sensors that utilize the piezoelectric, thermocouple, or Hall effect to generate voltage. Electromechanical voltage generators will be considered in Chapter 13.

12.2.1   Piezoelectric Transducers

Mechanical pressure applied to particular crystals such as quartz or rochelle salt (and some ceramics) produces mechanical deformation and a small potential difference or voltage across the crystal. This process is referred to as the *piezoelectric effect*. Upon removal of the pressure, the original shape returns and the corresponding voltage reduces to zero. The opposite energy conversion process also occurs such that applying a voltage (and electric field) across the crystal produces mechanical deformation.

Figure 12.5a displays the basic constituents of a *piezoelectric transducer*. Pressure applied to the crystal exerts a force that slightly deforms the crystal and generates a potential difference proportional to the magnitude of the force or pressure.

One of the primary applications of a piezoelectric transducer is its use in crystal oscillator circuits (to be described in Chapter 14). Such oscillators are used to produce a sinusoidal voltage at a precise frequency and require a resonant circuit, as we will see. However, a quartz crystal under prescribed pressure is electrically equivalent to a resonant *RLC* circuit with specific element values, as shown in Fig. 12.5b. Subsequent placement of the crystal into an oscillator circuit results in a sinusoidal output at the prescribed frequency.

12.2.2   Thermocouple Transducers

A *thermocouple transducer* is a thermoelectric device that produces a difference in potential proportional to a temperature difference. The operation of a thermocouple transducer is based upon the principle that when two dissimilar metal wires are joined at one end, a temperature difference between the joined end and the other end of the wires produces a potential difference between the wires. The magnitude of this potential difference is very dependent upon the specific materials

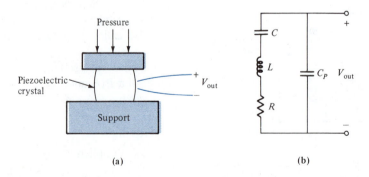

(a)   (b)

FIGURE 12.5   Piezoelectric Transducer: (a) Basic constituents, (b) Electrical equivalent

used for the wires as well as the temperature difference.

In applications of the thermocouple transducer for temperature measurements, two junctions of the dissimilar metal wires are used. The first junction is placed in the vicinity of the location under measurement and is usually referred to as the *hot junction,* as indicated in Fig. 12.6. Another junction, called the *cold junction,* is placed in a controlled environment at a fixed reference temperature, as also shown in Fig. 12.6. The output voltage V_{out} between the two points of contact is then directly proportional to the difference in temperature between these two junctions.

To accurately measure the prevailing temperature difference, specific materials are used for the metal wires such that significant voltage is developed even for small temperature differences. A commonly used combination for temperatures up to 300° C is copper to constantan (an alloy of nickel and copper), which provides a voltage change of approximately 43 $\mu V/^\circ$ C. Another combination that operates to temperatures of 1600° C is platinum to platinum-rhodium, which produces approximately 6 $\mu V/^\circ$ C. After specific wire materials are selected, each system is carefully calibrated, quite often with another method of temperature measurement. Finally, the generated output voltage V_{out} is measured with a sensitive voltmeter (see Fig. 12.6), which provides a direct reading of the temperature difference between the two junctions.

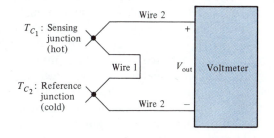

FIGURE 12.6   Thermocouple Transducer Used for Measuring a Temperature Difference

---

EX. 12.5

Thermocouple Transducer Example

Consider a thermocouple that uses the metals chrome (alloy of nickel and chromium) and constantan. For this combination, the circuit of Fig. 12.6 is used to measure the temperature of a heated electrode, where the reference thermocouple is immersed in an ice bath (0° C). The voltage is measured to vary linearly over the range of 0° C to 980° C with corresponding voltage change of 0 V to 75 mV. What is the temperature for voltage readings of 35.2 mV and 68.5 mV?

Solution: For the reading of 35.2 mV, the temperature is (35.2)(980)/75, or 460° C. Similarly, for 68.5 mV, the temperature is 895° C.

12.2.3   Hall Effect Generator

The operation of a *Hall effect generator* is based upon a phenomenon that is most pronounced in semiconductor materials. This phenomenon, known as the *Hall effect,* is the production of a transverse voltage due to the application of a longitudinal current and a perpendicular magnetic field. The Hall effect occurs because charged particles moving in a magnetic field are deflected by a magnetic field force (to be described in Chapter 13). This force is directly dependent upon the magnitude of the perpendicular magnetic field and the magnitude of the velocity of the charges (or the magnitude of the current). The deflection of charges then produces a voltage that is proportional to the amount of deflection. Hence, the generated voltage, which is transverse to the motion of charges, is proportional to the magnitude of the magnetic field and the magnitude of the current. The polarity of the generated voltage depends upon the type of majority-charge carrier.

Figure 12.7 indicates a typical geometry for the Hall effect device. To maximize the generated transverse Hall (*y*-directed) voltage V_H, the longitudinal (*x*-directed) current I_{LL} and the magnetic

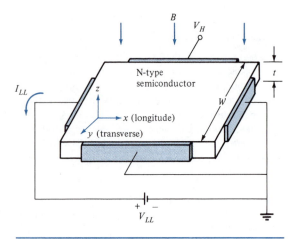

FIGURE 12.7   Hall Effect Device

field intensity (z-directed) B are mutually perpendicular. The transverse voltage produced is obtained as follows:

$$V_H = \frac{R_H I_{LL} B}{t} \qquad \text{(12.2–1)}$$

where t is the sample thickness and R_H is the Hall constant, which is a known quantity for a particular material. For most semiconductors, $|R_H| = |1/qN|$, where N is the majority-carrier concentration and $q = 1.6 \times 10^{-19}$. The basic unit for B in the Mks system is the Tesla (T) and R_H has derived units Vm/TA. See Chapter 13 for a more complete description of magnetic fields.

A primary application of the Hall effect generator is as a switch that has two discrete output levels for high and low (zero) voltage corresponding to on and off. In this application, the current is usually held constant, and a magnet is brought into close proximity to produce a nonzero V_H. This voltage can then be amplified to produce the output high voltage. When the magnet is removed ($B = 0$), the voltage is switched to zero.

───────────────────────────────────── EX. 12.6

Hall Effect Example

Calculate the value of V_H that would be produced in the Hall generator of Fig. 12.7 for $I_{LL} = 1$ μA, $t = 100$ μm, $B = 10^{-3}$ T, and $R_H = 10$ Vm/TA.

Solution: Substituting values directly into (12.2–1) yields

$$V_H = \frac{(10)(10^{-6})(10^{-3})}{10^{-4}} = 0.1 \text{ mV}$$

12.3   OPTOELECTRONIC TRANSDUCERS AND SENSORS

The word *optoelectronic* refers to light and electrical energy interaction. We will consider transducers that generate a voltage due to the absorption of light, as well as the opposite case of transducers that produce light through the application of a voltage. In addition to these transducer examples, we will also consider sensor applications such as the photoconductive cell. When this device absorbs light, its electrical resistance is changed, as we will see. However, to more fully understand and appreciate the operation of these devices, we begin by describing the basic characteristics of light.

12.3.1   The Electromagnetic and Optical Spectrum

Light is a form of electromagnetic radiation similar to radiowaves that also has frequency as well as wavelength associated with it. Figure 12.8 displays the complete electromagnetic spectrum and indicates the very small range of frequencies and wavelength that visible light occupies. As we will see, optoelectronic devices operate in the infrared, visible, and ultraviolet regions. The combination of these regions is referred to as the *optical spectrum*, as indicated in Fig. 12.8. The visible region has a wavelength range between 0.4 μm and 0.7 μm and separates the infrared and ultraviolet regions from each other.

The relationship between frequency and wavelength for any form of electromagnetic radiation

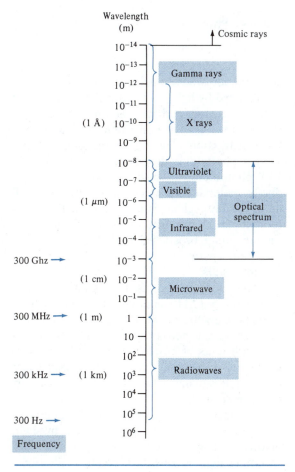

FIGURE 12.8   Electromagnetic Spectrum

TABLE 12.1	Visible Region		
Wavelength λ		**Frequency f**	
(μm)	**(Å)**	**(GHz)**	**Color**
0.41	4100	731,700	Violet
0.42	4200	714,290	Indigo
0.47	4700	638,300	Blue
0.52	5200	576,920	Green
0.58	5800	517,240	Yellow
0.60	6000	500,000	Orange
0.65	6500	461,540	Red

the visible light range are exceedingly large in comparison to usual electronic signal frequencies.

Table 12.1 also shows the various colors associated with the visible wavelength region. The longest wavelengths produce the sensation of red to the human eye, whereas the shortest wavelengths produce the sensation of violet. Figure 12.9 displays the visual response of the human eye to light as a function of wavelength. Also indicated are the ultraviolet and infrared (invisible) regions, to which the human eye does not respond. Note that by comparing Fig. 12.9 and Table 12.1, we see that the human eye responds best to green and yellow.

is given by the expression

$$f = \frac{c}{\lambda} \qquad\qquad (12.3\text{-}1)$$

where

c = velocity of light (3×10^8 m/s)

λ = wavelength

Table 12.1 indicates values of λ and f for the visible region. From this table, we observe that expressing visible light in terms of wavelength in μm is much more convenient than using frequency. Note also that the frequencies corresponding to

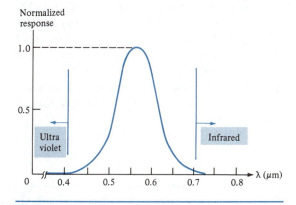

FIGURE 12.9   Relative Response of Human Eye to Light as a Function of Wavelength

12.3.2 Photoconductive Cells (Photoresistors)

A *photoconductive cell,* or *photoresistor,* is a variable resistor made of semiconductor material whose magnitude of resistance is dependent upon the magnitude of the incident light intensity absorbed by the semiconductor. The circuit symbol commonly used for such a light-dependent resistor is displayed in Fig. 12.10. The arrows indicate incoming light absorption. The photoresistor is a two-terminal nonjunction semiconductor device.

The mechanism by which the resistance is altered is simply that the absorbed light energy creates more electron-hole pairs in the semiconductor. With more free carriers present, the resistance is reduced. Additionally, as the light intensity is increased, even more electron-hole pairs are created, and the resistance is further decreased.

The most commonly used semiconductor materials for photoresistors are cadmium sulfide (CdS) and cadmium selenide (CdSe). These materials have a very low intrinsic concentration and hence a very large resistance without light. Thus, when light is absorbed, a much larger change in resistance is possible.

An important parameter for the photoresistor is its resistance without light. This parameter is referred to as the *dark resistance.* Typical variation in resistance from dark to light is 10 MΩ to 10 kΩ.

Photoresistors are used in intrusion detector and automatic door opener applications. Such applications involve the blocking of light, which causes a circuit to be triggered to the on condition. Photodiodes and phototransistors, however, have become more popular in these applications.

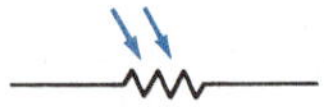

FIGURE 12.10   Photoresistor (Photoconductive Cell) Circuit Symbol

12.3.3 Photodiodes

A *photodiode* is a PN junction semiconductor diode whose current-voltage characteristic is altered by light absorption. The photodiode is constructed such that light can be focused on the vicinity of the junction. Figure 12.11a displays the construction of a typical metal-can encapsulated photodiode. For improved operation, a PIN photodiode, in which an intrinsic semiconductor material (I) is sandwiched between the P and N diode regions, is used. With a PIN photodiode, the electric field region in the vicinity of the junction is broadened so that more electron-hole pair separation is allowed and hence larger current results.

The circuit symbol used for a photodiode is shown in Fig. 12.11b, and typical current-voltage characteristics are shown in Fig. 12.11c. Incident light radiance, denoted as R_I, is used as the parameter with units of power/area or mW/cm^2. This parameter is the input light energy. Observe that as the light radiance is increased, the reverse current is increased.

The mechanism by which the PN diode current-voltage characteristic is altered by light involves the generation of additional carriers. Incoming light is absorbed, creating additional electron-hole pairs. Since additional charges are produced in the vicinity of the junction (or in the I region), they are separated by the electric field in this region and are swept across the junction. Because of the electric field direction, the charges move to their majority sides (electrons to N and holes to P). This charge flow constitutes additional current in the reverse direction, which is usually much larger than the diode reverse saturation current I_S (from Chapter 2). The reverse saturation current I_S is referred to as the *dark current* I_D in photodiode terminology.

For photodiode applications, when a reverse voltage is applied to the diode, a reverse current flows, which is the sum of the dark current I_D and the light-generated current I_G. The magnitude of

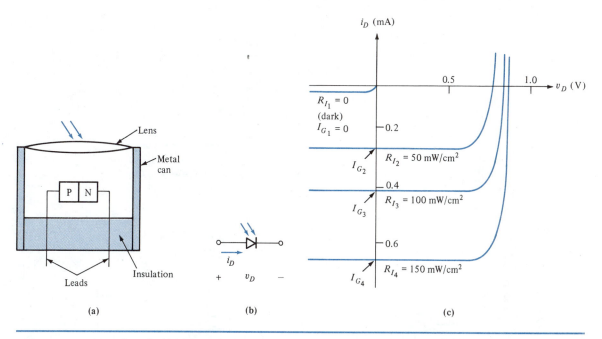

FIGURE 12.11   Basic Photodiode: (a) Construction, (b) Circuit Symbol, (c) $i_D - v_D$ characteristics (R_I = incident light radiance)

the current is then very dependent upon the amount of incoming light that is absorbed.

One of the primary applications of a photodiode is as a light detector. Figure 12.12a displays the simplest arrangement that can be used in this application. The output voltage V_L is inversely proportional to the incident light radiance of the incoming light, and the current is the sum of I_D and I_G. This relationship is observed in Fig. 12.12b by constructing a load line (from Chapter 2) and considering various incoming light radiances such that $R_{I_1} < R_{I_2} < R_{I_3}$. The corresponding output voltages are V_{L_1}, V_{L_2}, and V_{L_3}.

An op amp amplifier can be used, as shown in Fig. 12.12c, to invert as well as amplify the light-generated voltage V_L. From Chapter 7, recall that the input impedance of this inverting amplifier circuit is essentially R_S and that R_S must have a large value relative to R_L for V_L to be unaffected by the op amp connection. Additionally, for $R_F \gg$

R_S, the output voltage is given by

$$V'_L = -\frac{R_F}{R_S} V_L \qquad (12.3\text{--}2)$$

Another method of increasing the sensitivity of the photodiode detector is to use the op amp circuit shown in Fig. 12.12d. Recall from Chapter 7 that this circuit is a current-to-voltage converter. Since the n terminal is a virtual ground and essentially zero current enters the input of the op amp, the output voltage is given by

$$V'_L = R_F(I_G + I_D) \qquad (12.3\text{--}3)$$

and the light-generated current is converted into a voltage. The function of resistor R_0 is to balance the output voltage due to the op amp input offset current. The use of R_0 is important when the magnitude of the light-generated current is on the order of the input offset current.

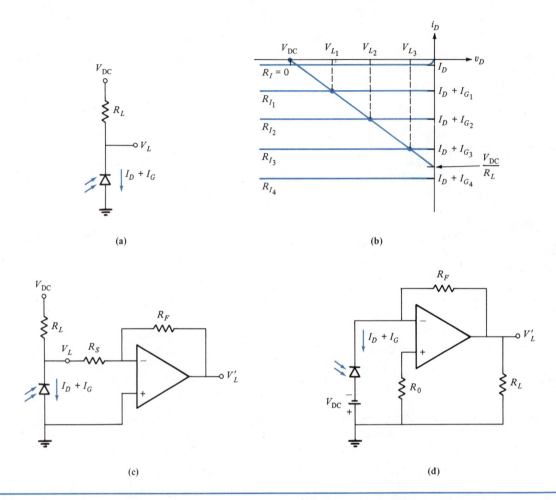

FIGURE 12.12   Photodiode Detector Circuits and Analysis: (a) Basic circuit, (b) Load-line analysis, (c) Basic circuit with amplification, (d) Current-to-voltage converter

---

EX. 12.7

Photodiode Example

A PIN photodiode whose i_D–v_D characteristics are displayed in Fig. E12.7 is used in the op amp circuit of Fig. 12.12d with $V_{DC} = 10$ V, $R_F = R_0 = 100$ kΩ, and $R_L = 1$ kΩ. If the incident light radiance is 2.0 μW/mm^2, determine the voltage across the load resistor. Repeat this calculation for a circuit without the op amp—that is, a circuit with only V_{DC}, R_L, and the photodiode.

Solution: We begin by determining the photodiode current in the op amp circuit. Since the input loop consists of V_{DC}, the photodiode, and essentially zero resistance, the photodiode current is obtained graphically from Fig. E12.7 (as shown) to be $i_D \cong 0.51$ μA. Thus, from (12.3–3), we have

$$V'_L = (100 \times 10^3)(0.51 \times 10^{-6}) = 51 \text{ mV}$$

Without the op amp, the photodiode current is essentially the same (visualize the load line with $V_{DC} = -10$ V and

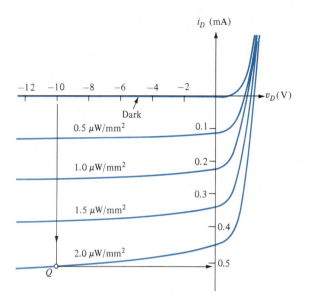

FIGURE E12.7

$R_L = 1 \text{ k}\Omega$. Thus, the output voltage is

$$V'_L = (1 \times 10^3)(0.51 \times 10^{-6}) = 0.51 \text{ mV}$$

12.3.4 Phototransistors

A *phototransistor* is a BJT whose current–voltage characteristics are modified by light absorption. As in the photodiode case, these characteristics are highly dependent upon the amount of light absorption.

Figure 12.13a shows the typical construction of a phototransistor, and Figs. 12.13b and c show the circuit symbols used when a base terminal is unnecessary (Fig. 12.13b) and when a base terminal is provided (Fig. 12.13c). In either case, the light is allowed to impinge on the emitter junction, which is forward biased for proper phototransistor operation. A bias voltage V_{DC} is applied as shown in Fig. 12.13d, which not only forward biases the emitter junction but also reverse biases the collector junction. Under these conditions, the phototransistor is operating in the amplifier

region (see Chapter 3). Hence, the light-generated current (which can be considered as input base current) is amplified by the common-emitter amplification factor β. For the general case in which a base connection is provided (Fig. 12.13c), the collector current is given by

$$I_C = \beta(I_B + I_G) \tag{12.3–4}$$

where I_B is the base current (supplied by the external circuit) and I_G is the light-generated current. For the case in which there is no base connection (Fig. 12.13b), $I_B = 0$, while I_G is still amplified by the factor β. Since the collector current I_C is proportional to βI_G in both phototransistor cases, the sensitivity of a phototransistor is much greater than that of a photodiode.

――――――――――――――――――― EX. 12.8

Phototransistor Example

Typical current–voltage characteristics for a phototransistor are displayed in Fig. E12.8. For an incident light radiance $R_I = 5 \text{ mW/cm}^2$, determine v_{CE} and i_C for the

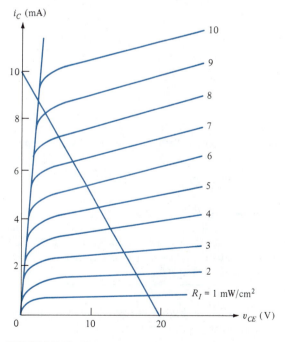

FIGURE E12.8

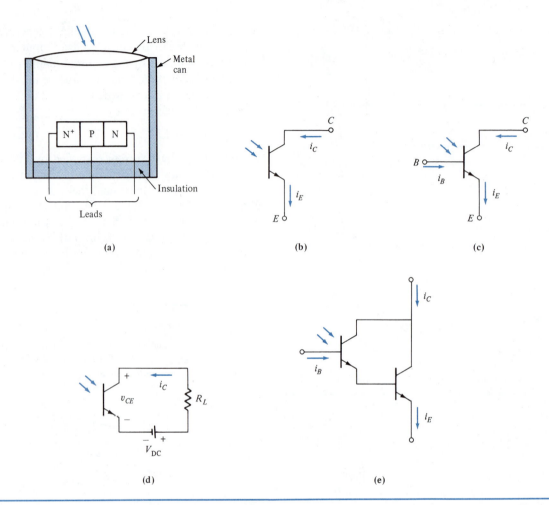

FIGURE 12.13   Basic Phototransistor: (a) Construction, (b) Circuit symbol without base connection, (c) Circuit symbol with base connection, (d) Basic circuit, (e) Darlington-pair phototransistor

basic phototransistor circuit of Fig. 12.13d with $R_L = 2$ kΩ and $V_{DC} = 20$ V.

Solution: The load line corresponding to $V_{DC} = 20$ V and $R_L = 2$ kΩ is shown in Fig. E12.8. The intersection of the load line with the transistor curve for $R_I = 5$ mW/cm^2 then provides the values of current and voltage as $v_{CE} \cong$ 11 V and $i_C \cong 4$ mA.

To provide higher current gain, Darlington-pair phototransistors are often used. Figure 12.13e shows the circuit symbol for this higher-gain con-

figuration with base connection. The collector current in this case (with proper bias applied) is given approximately by

$$I_C = \beta^2(I_B + I_G) \qquad (12.3\text{--}5)$$

Like photodiodes, phototransistors can also be used as light detectors. In these applications, the circuits of Fig. 12.12 are also appropriate and provide additional sensitivity using phototransistors.

12.3.5 Photovoltaic Cells (Solar Cells)

Photovoltaic cells are light-absorbing photodiodes that are used as transducers to produce or generate output electrical energy from input light energy. These cells are often referred to as *solar cells* to imply simply that the sun is the source of light.

The basic solar cell circuit is a photodiode without an applied reverse voltage. When a resistor is connected directly across the photodiode terminals, the absorbed light then creates a current and voltage.

The current–voltage characteristics of the solar cell are the same as those of the photodiode as shown in Fig. 12.11c. However, connecting a resistor across the diode terminals results in operation in the fourth quadrant, which is the region of operation for the solar cell, as we will now see.

Figure 12.14a shows a circuit diagram for solar cell operation. The resistor R_L is the load resistor to which electrical power is delivered through conversion of the incoming light energy. This process is identical to that of the photodiode.

The generated current and voltage for the solar cell are most easily determined graphically (this technique was presented in Chapters 3 and 4). Figure 12.14b shows the current–voltage characteristic of the solar cell assuming a particular incoming light radiance (R_{I_0}). Also shown is the current–voltage characteristic of the load resistor R_L. Of course, the operating point is located at the intersection of the two curves and is labeled Q. Different resistor values are possible that lead to larger generated voltage or larger generated current, whichever is desirable. Note that the intersections of the solar cell characteristic with the current and voltage axes are denoted as the short-circuit current I_{SC} and the open-circuit voltage V_{OC}, respectively. Furthermore, at these extremes, the output power is zero. The maximum output power for a solar cell occurs in the "knee" region of the current–voltage characteristic, as the next example indicates.

––––––––––––––––––––––––––––––––––––– EX. 12.9

Solar Cell Maximum Power Example

Figure E12.9 shows a family of curves for a silicon photovoltaic cell with the current axis inverted, which is the usual representation on data sheets. The characteristic

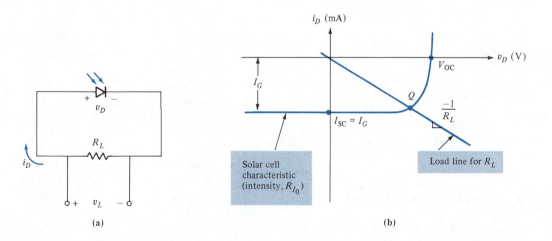

(a)   **(b)**

FIGURE 12.14   Solar Cell: (a) Circuit diagram, (b) Determination of Q for the solar cell

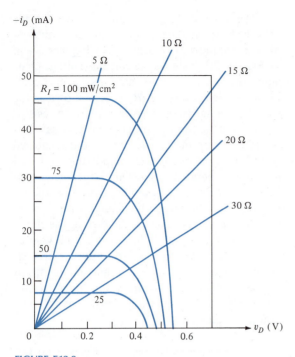

FIGURE E12.9

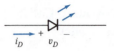

FIGURE 12.15   LED Circuit Symbol

curve labeled $R_I = 100$ mW/cm^2 is regarded as the maximum value provided by sunlight (on a clear day at noon). Sketch load lines for resistor values of $R_L = 5, 10, 15, 20$, and 30 Ω and determine which of these cases would provide maximum power output. What is the approximate maximum power output for the case in which $R_I = 50$ mW/cm^2?

Solution: Sketching the load lines for $R_L = 5, 10, 15, 20$, and 30 Ω yields respective intersection points with the $R_I = 100$ mW/cm^2 curve approximately at (0.22,45), (0.4,41), (0.47,31), (0.5,25), and (0.51,17). The respective power outputs corresponding to the product of i_0 and v_0 are then 9.9, 16.4, 14.57, 12.5, and 8.67 W. The maximum power output for $R_I = 100$ mW/cm^2 is thus $R_L = 10$ Ω.

12.3.6   Light-Emitting Diodes (LEDs)

Light-emitting diodes (LEDs) are PN junction semiconductor diodes that are specially designed to emit (generate) light when they are

sufficiently forward biased. The circuit symbol for an LED is shown in Fig. 12.15 and indicates the opposite function from that of a photodiode. The light emitted is very dependent upon the type of semiconductor material, as we will now see.

LED Materials and Operation.   Light-emitting diodes require special construction for visible light emission. Particular semiconductor materials and a special junction arrangement that allows the emitted light from the junction to be observed are used.

Operation of LEDs involves forward biasing the PN junction to cause electron and hole carrier recombination that results in light emission. Light emission from the diode, as indicated by the arrows in Fig. 12.15, requires the use of semiconductor materials that emit radiation upon carrier recombination.

The primary light-emitting materials are gallium arsenide (GaAs) and gallium phosphide (GaP) in combination with various constituents. Semiconductors such as silicon and germanium are not suitable for LEDs because carrier recombination in these materials results only in heating the semiconductor.

When the semiconductor material used for the LED is pure GaAs, the forward-biased diode emits radiation in the infrared region with wavelength given by $\lambda \cong 0.905$ μm. Since this light is not visible to the human eye, GaAs LEDs have no display applications. However, they can be used in other applications where invisible emission is useful. One possibility is as an optical coupler, as shown in Fig. 12.16. This circuit acts as an isolator and uses an LED to excite a photodiode in an enclosed package. Optical couplers will be described in more detail in the next section.

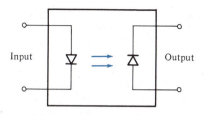

FIGURE 12.16 Optical Coupler

Another material that is often used in LED infrared applications is a combination of GaAs with Si possessing $\lambda = 0.94\ \mu m$. This combination emits more efficiently than pure GaAs, which means that more light emission occurs for a given forward bias.

When the semiconductor material used for the LED is GaP, the emitted light is in the visible range and is red with wavelength given by $\lambda = 0.55\ \mu m$. In this case, impurities—oxides of cadmium (CdO) or Zinc (ZnO)—are also added to GaP that provide more efficient light emission at this wavelength. Furthermore, by adding only nitrogen or sulfur to GaP, the LED will emit green light; by adding mixtures of these two sets of impurities, the LED will emit yellow and orange light.

As indicated in Fig. 12.17, light is emitted in the vicinity of the junction, and the top region (P, in this case) is made thin in order to improve emission. Light emission is achieved by forward biasing the diode, which creates a significant forward current due to majority carriers crossing the junction. A small fraction of these carriers recombine as they attempt to cross the junction. For radiating-type semiconductors, the energy of this recombination is given off in the form of light at a particular wavelength (or frequency). The intensity of the light emitted is directly dependent upon the amount of recombination and therefore the magnitude of the forward current I_F and forward voltage V_F. The wavelength of the emitted light is directly dependent upon the semiconductor material and its constituents. Table 12.2 indicates various materials that are used for LEDs and the corresponding wavelength and color of the emitted light for each material.

Except for the light emission, the LED behaves like an ordinary PN diode with i_D–v_D characteristics similar to those described in Chapter 2. However, when piecewise linear characteristics are used to approximate the LED, the value of forward voltage for the GaP and GaAs diode is about twice that of the Si diode, and the forward voltage is approximated as $V_{0\ LED} \cong 1.5$ V. Furthermore, the breakdown voltage of LEDs is typically quite low, on the order of 10 V. However, LEDs are not normally operated under reverse-bias conditions and are easily protected from this occurrence by placing an ordinary diode in parallel with the LED, as shown in Fig. 12.18.

As is the case for other electronic devices, the operation of LEDs must not exceed maximum voltage and current ratings. Frequently, a resistor

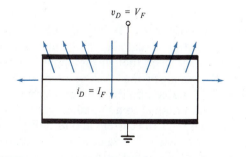

FIGURE 12.17 LED Light Emission

TABLE 12.2 Commonly Used LED Materials and Characteristics of the Light Emission

Semiconductor Material	Wavelength (μm)	Color
ZnS	0.35	Violet
CdS	0.52	Green
GaP (CdO)	0.7	Red–infrared
GaP (N)	0.55	Green
GaAs	0.88	Infrared

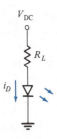

V_{DC}

R_L

i_D

FIGURE 12.18   LED Bias Circuit with Reverse Voltage Protection

is placed in series with an LED to limit the forward current. Figure 12.19 displays the basic arrangement for limiting the forward current of an LED. From Ohm's law, the expression for R_L is

$$R_L = \frac{V_{DC} - V_{0\,LED}}{I} \qquad (12.3\text{–}6)$$

where $V_{0\,LED}$ is the forward drop across the LED.

———————————————————————— **EX. 12.10**

LED Logic Indicator Example

An LED is used in the inverter circuit of Fig. E12.10 and is to be turned on when the BJT is operating in saturation. Select the value of R_C that will provide an LED current of 20 mA. Use $V_{0\,LED} = 1.5$ V for the LED and $v_{CE\,sat} = 0.2$ V.

Solution: When v_I corresponds to a high voltage (logic level 1), the inverter provides an output load voltage that corresponds to a low voltage (logic level 0). Additionally,

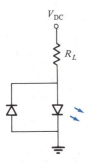

V_{DC}

R_L

FIGURE 12.19   Basic LED Bias Circuit

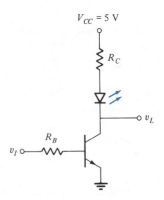

$V_{CC} = 5$ V

R_C

v_L

R_B

v_I

FIGURE E12.10

the transistor is in saturation and $v_L = v_{CE\,sat} = 0.2$ V. Thus, by writing KVL for the output branch, we obtain the LED current as

$$i_C = \frac{V_{CC} - 1.5 - 0.2}{R_C} = \frac{5 - 1.5 - 0.2}{R_C} \qquad (1)$$

Solving for R_C yields

$$R_C = \frac{3.3}{i_C} = \frac{3.3}{20} = 165\ \Omega \qquad (2)$$

LEDs in Numerical Displays.   LEDs are used frequently in arrays for numerical displays. In particular, in digital-display instruments, a seven-segment format is used to delineate any integer from 0 to 9. The discussion of this important application is deferred until Chapter 14, where digital-readout instruments are described in detail.

LED Current Limiter.   To limit the LED current I_{LED} to a fixed value that maintains constant emission brightness even though the applied voltage varies, the circuit of Fig. 12.20 is used. Furthermore, the variable applied voltage V_{DC} must be greater than or equal to a minimum voltage V_{min} in order that the LED current and resulting emission be so restricted. The value of V_{min} is obtained by considering T_2 to just turn on (become active) when $R_E I_{LED} = V_0$ and $V_{DC} = V_{min}$. Then, by writ-

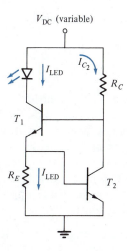

V_{DC} (variable)

FIGURE 12.20   LED Current Limiter

ing KVL for the left-hand branch of Fig. 12.20, we have

$$V_{DC} = V_{min} = R_E I_{LED} + v_{BE_1} + v_{CB_1} + V_{0\,LED}$$
$$(12.3-7)$$

Also, since $v_{BE_1} = V_0$ and the minimum value of $v_{CB_1} = 0$, we have

$$V_{min} = 2V_0 + V_{0\,LED} \qquad (12.3-8)$$

The value predicted by (12.3–8) is very nearly equal to 3 V (assuming $V_0 = 0.7$ V and $V_{0\,LED} \cong 1.6$ V).

The design equation for the value of R_E is obtained by considering T_2 to become active, or

$$v_{BE_2} = I_{LED} R_E = V_0 \qquad (12.3-9)$$

Rearranging this equation then yields

$$R_E = \frac{V_0}{I_{LED}} = \frac{0.7}{I_{LED}} \qquad (12.3-10)$$

where I_{LED} is the constant LED current desired and the BJTs are assumed to be made of silicon.

The resistor R_C is required in order to support the additional voltage of V_{DC} when its value is larger than $2V_0$. The magnitude of R_C is selected as $R_C = 100 R_E$, which avoids driving T_2 into saturation.

—————————————————— **EX. 12.11**

LED Constant-Current Regulator Example

For the circuit of Fig. 12.20, determine the value of R_E required to produce an LED current of 35 mA provided that $V_{DC} \geq V_{min}$. Also, calculate R_C and the value of V_{min}. Finally, determine I_{LED} and I_{C_2} for $V_{DC} = 5$ V and 10 V. Neglect the base currents for the silicon BJTs and let $V_{0\,LED} = 1.8$ V.

Solution: From (12.3–10) and substituting values, we have

$$R_E = \frac{0.7}{35} = 20\ \Omega$$

Therefore,

$$R_C = 100 R_E = 2\ \text{k}\Omega$$

From (12.3–8), the minimum value of V_{DC} is obtained as

$$V_{min} = 2(0.7) + 1.8 = 3.2\ \text{V}$$

Thus, for $V_{DC} > V_{min}$, we have (from Fig. 12.20)

$$I_{C_2} = \frac{V_{DC} - 2(0.7)}{R_C}$$

For $V_{DC} = 5$ V and 10 V, the respective values of I_{C_2} are

$$I_{C_2} = \frac{5 - 1.4}{2} = 1.8\ \text{mA}$$

and

$$I_{C_2} = \frac{10 - 1.4}{2} = 4.3\ \text{mA}$$

The LED current is unchanged for $V_{DC} > V_{min}$ because $R_E I_{LED}$ remains essentially fixed; therefore, $I_{LED} = 35$ mA for $V_{DC} = 5$ V and 10 V.

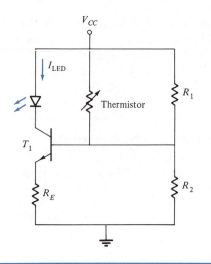

FIGURE 12.21   Constant Emission with Temperature Variation

Temperature Regulator Using a Thermistor.
LEDs become less efficient light emitters as the temperature is increased. Hence, to obtain essentially constant emission over a certain range of temperatures, a BJT using a thermistor can be employed, as shown in Fig. 12.21. The resistor R_E is selected to limit the LED current, with the

maximum LED current given by

$$I_{\text{LED max}} = \frac{V_{CC} - V_{0\,\text{LED}} - V_0}{R_E} \qquad \textbf{(12.3–11)}$$

Thus, since the thermistor resistance decreases as the temperature rises, the base current of T_1 will increase and I_{LED} will also increase, offsetting the original decrease due to the increased temperature. Similarly, if the temperature is reduced, the thermistor resistance increases and causes I_{LED} to change in the opposite manner, thus maintaining constant I_{LED}.

DC Polarity Indicator.   Connecting two LEDs in parallel as shown in Fig. 12.22a provides a simple means of indicating voltage polarity. Manufacturers offer packaged LEDs in this configuration that utilize two LEDs with different colors. Since each LED only emits light when it is forward biased, D_1 will emit for positive V_{DC} and D_2 will emit for negative. Figure 12.22b shows a polarity indicator that is much more sensitive because of the amplification provided by the noninverting amplifier. Another form of polarity indicator is shown in Fig. 12.22c, where the output

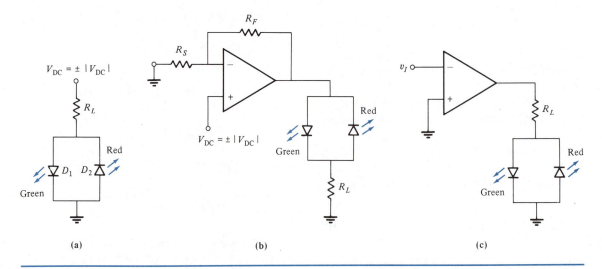

(a)   (b)   (c)

FIGURE 12.22   DC Polarity Indicators: (a) With two LEDs in parallel, (b) With noninverting amplifier, (c) With comparator

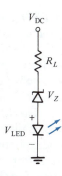

FIGURE 12.23 Voltage Level Indicator

of a comparator is driven into positive or negative saturation corresponding respectively to red and green emission.

Voltage Level Indicator. An LED used in combination with a zener diode and a limiting resistor, as shown in Fig. 12.23, can be used as a voltage level indicator. When V_{DC} becomes greater than V_{LED} and V_Z, the LED will begin to emit. Additionally, as V_{DC} is increased, the emission becomes brighter.

LED Smoke Detector. An LED in combination with a light detector can be used to sense smoke. The emitted light from the LED is allowed to pass through an air-filled dark tube. If smoke is present, less light reaches the photodetector and its output is reduced.

LED Intrusion Alarm. The basic LED intrusion alarm consists of an LED biased to emit infrared (and invisible) radiation at all times. This radiation is directed (aimed) toward a solar cell receiver that detects this particular emission. If the infrared emission is blocked, the detector circuitry sounds an alarm.

Figure 12.24 displays the basic circuit for an intrusion alarm. As long as radiation from the LED is detected by the solar cell, the amplified output signal maintains the relay in a closed position with no signal reaching the alarm. However, when the radiation is blocked, the relay opens to sound the alarm.

In actual systems, battery-operated multivibrator pulsating circuits are used for the transmitter portion of the system to reduce power consumption but maintain the required LED forward current. Additionally, capacitively coupled receivers that also operate in a pulsed mode are employed. These receivers utilize high-gain amplifiers for use with long-range applications.

Various types of intrusion systems are possible. For example, a simple doorway protection system requires a transmitter and receiver located on opposite sides of a doorway. When an object enters the doorway, the signal is blocked and the alarm is sounded. Frequently, the transmitter and receiver are on the same side of the doorway and a mirror is placed on the opposite side.

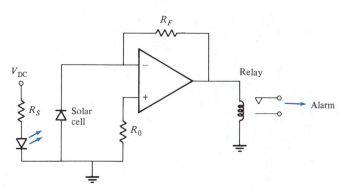

FIGURE 12.24 Basic Intrusion Alarm

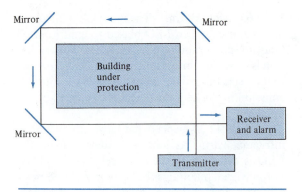

FIGURE 12.25   Building Intrusion Alarm System

An intrusion alarm system that protects an entire building is depicted in Fig. 12.25. The alignment of such a system must be carried out carefully so that each mirror is positioned rigidly and individually through measurements with a portable optical receiver.

12.3.7   Optical Couplers

Optoelectronic isolators, or *optical couplers,* are devices that provide total electrical isolation between an incoming and outgoing elec-

trical signal via an LED and a light-detecting device. An electrical circuit for such a device was shown earlier in Fig. 12.16. Generally speaking, such a coupler provides a noise-free interface in a data transmission system.

Typical construction of an optical coupler is shown in Fig. 12.26a. The LED can be either a visible or an infrared emitter, and the detector is usually a photodiode, phototransistor, or Darlington-pair phototransistor. Figure 12.26b shows the schematic diagram for this optical coupler where a phototransistor is used as the detector.

The most important parameter for these devices is the current transfer ratio I_{OUT}/I_{LED}. These currents are indicated in Fig. 12.26b. This ratio is the measure of the efficiency of the device in coupling the applied signal from the input terminals to the output terminals. This ratio depends upon the efficiency of the LED in transducing current into light emission, the amount of this light that enters the detector, the sensitivity of the detector, and the amplification properties of the detector. For the optical coupler itself, the current transfer ratio is always less than 1, with a typical value on the order of 0.1 when a Darlington-pair phototransistor is used. For a single-phototransistor

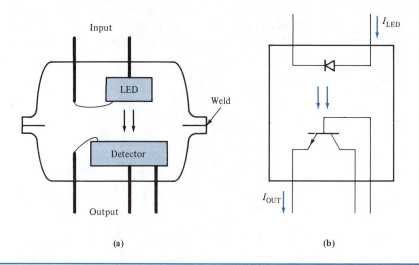

(a)   (b)

FIGURE 12.26   Optoelectronic Isolator or Coupler: (a) Typical construction, (b) Schematic diagram

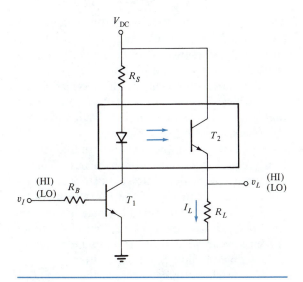

FIGURE 12.27   Digital Logic Optoelectronic Coupler

$$I_{LED} = \frac{V_{DC} - V_{0\,LED} - v_{CE\,sat}}{R_S} = \frac{5 - 1.8 - 0.2}{1.5}$$

$$= 2 \text{ mA} \qquad \qquad (1)$$

Thus, with a current transfer ratio of 0.4, the output current is $I_L = 2(0.4) = 0.8$ mA, and the output voltage is

$$v_L = I_L R_L = 0.8(3.75) = 3 \text{ V} \qquad (2)$$

detector, the current transfer ratio is typically ten times less. The output current, however, can always be amplified further, if desired.

A primary application of optoelectronic isolators is in digital logic coupling. In this application, the device is used to transmit a logic level in a logic circuit with complete electrical isolation, as indicated in Fig. 12.27. Visual inspection of this circuit reveals that for v_I high, v_L is high and that for v_I low, v_L is low. However, complete isolation is provided, and delicate computers can be interfaced to control the operation of large machines without the danger of feedback occurring (which could be destructive).

––––––––––––––––––––––––––––––––––––––– EX. 12.12

Optical Coupler Example

Consider the digital logic optical coupler of Fig. 12.27. Let $V_{DC} = 5$ V, $R_S = 1.5$ kΩ, $R_L = 3.75$ kΩ, and the current transfer ratio $I_0/I_{LED} = 0.4$. For v_I equal to a high input, determine v_L. Let $V_{0\,LED} = 1.8$ V and $v_{CE\,sat} = 0.2$ V.

Solution: For v_I high, T_1 is in saturation and $v_{CE_1} = v_{CE\,sat} = 0.2$ V. The LED is then turned on with a forward current given by

12.4   DETECTING TRANSDUCER EXAMPLES: RADIO AND TELEVISION

Two of the most significant detecting transducer examples are radio and television. Each of these systems requires an antenna at the receiving end that intercepts a portion of the radio or TV signals in the form of electromagnetic waves. The receiving antenna, or *receiver,* then performs the function of converting the electromagnetic waves into time-varying electrical currents, which vary according to the changes in the waves striking the antenna.

An antenna is also required to generate and transmit the electromagnetic waves at their originating position. The transmitting antenna in the *transmitter,* performs the opposite function of the receiver and converts time-varying currents into electromagnetic waves, which are then broadcast through the air via this antenna.

The following discussion of radio and television involves the description of the basic constituents of receivers and transmitters as well as their operation.

12.4.1   Radio

Radio can be defined as one-way communication of encoded electrical signals via electromagnetic waves. Although many different radio systems are available, all utilize the same basic processes.

Figure 12.28 displays a complete radio system with both the transmitter and receiver components indicated in block diagram form. In the transmitter (Fig. 12.28a), the first process is the conversion of sound into an electrical signal by the microphone. In the receiver (Fig. 12.28b), the last process is the conversion of the electrical signal back into sound by the speaker. The operation of these components involves specific translational transducers, which will be described in Chapter 13.

The main function of the transmitter is to alter the af (audio frequency) signal (5 Hz to 15 kHz) to be transmitted through a process of modulation. This process, to be described in the next section, involves "mixing" a high-frequency rf (radio frequency) signal with the low-frequency af signal. The new signal (the carrier signal) is then amplified and fed to the antenna for transmission through air to the receiver. The electromagnetic waves radiating from the transmitting antenna are then picked up by a receiving antenna with reduced amplitudes (which are smaller for increased transmitter-to-receiver distances).

The next component in the receiver is the tuner, which consists of a resonant LC filter circuit that is used to select the desired signal and reject all others. The resonant frequency is given by $\omega_0 = (1/LC)^{1/2}$, and only signals with frequencies nearly equal to ω_0 are allowed to pass. Tuning a radio to a particular station consists of adjusting the resonant frequency of the tuner (usually by varying the capacitance) until it matches the desired rf signal frequency. The output of the tuner is then a weak version of the modulated voltage from a single radio station transmitter.

The weak incoming signal is then amplified and demodulated, as shown in Fig. 12.28b. Demodulation, also to be described shortly, is the method by which a replica of the original audio

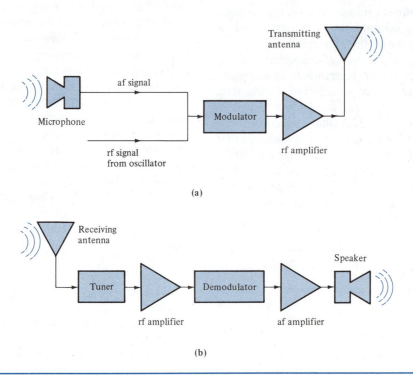

(a)

(b)

FIGURE 12.28   Block Diagram of Radio System: (a) Radio transmitter, (b) Radio receiver

signal is reproduced. After demodulation, the af signal is then amplified again and fed into a speaker that converts the electrical signal into sound waves.

Now that we have examined the basic operation of a radio system, we will consider the individual processes in more detail.

Modulation.   The first signal process performed in the transmitter is *modulation*. This frequency-shifting operation is necessary in order to allow efficient transmission of the relatively low-frequency af signal via electromagnetic waves. Low-frequency signals converted directly by the antenna into electromagnetic waves experience much more attenuation through transmission in air than high-frequency signals do.

Figure 12.29 shows rf signal waveshapes, or *carrier signals,* for two commonly used types of modulation for radio (as well as television): *amplitude modulation* (AM) and *frequency modulation* (FM). Amplitude modulation, as shown in Fig. 12.29a, is characterized by a signal whose amplitude varies in proportion to the low-frequency modulating signal variation. For frequency modulation, however, as shown in Fig. 12.29b, the frequency of the carrier signal is proportional to the modulating signal frequency. The actual mod-

ulating signals are created by speech, music, a TV picture, or even digital signals.

Amplitude modulation can be described mathematically by considering the unmodulated (rf) signal to be sinusoidal, as follows:

$$v_c = A_c \sin \omega_c t \qquad (12.4-1)$$

where

A_c = amplitude of the carrier signal

ω_c = frequency of the carrier signal

Our discussion here will concern only the process of amplitude modulation (and demodulation).

For amplitude modulation, the amplitude is a function of time and can be written as

$$A_c(t) = A(1 + m \sin \omega_m t) \qquad (12.4-2)$$

where

ω_m = modulation frequency

m = modulation index ($0 < m < 1$)

A = constant

Substitution of (12.4–2) into (12.4–1) yields the expression of the amplitude-modulated signal as

$$v_c = A(1 + m \sin \omega_m t) \sin \omega_c t$$
$$= A[\sin \omega_c t + m \sin \omega_m t \sin \omega_c t] \quad (12.4-3)$$

Figure 12.30 shows a plot of (12.4–3) for the case in which $A = 1$. Substituting the trigonometric identity for the product of two sine functions yields an alternative expression to (12.4–3) given by

$$v_c = A\left[\sin \omega_c t + \frac{m}{2} \cos(\omega_c - \omega_m)t \right.$$
$$\left. - \frac{m}{2} \cos(\omega_c + \omega_m)t \right] \qquad (12.4-4)$$

Note that this amplitude-modulated signal is the summation of three sinusoids, the frequencies of which are ω_c, $\omega_c + \omega_m$, and $\omega_c - \omega_m$. The latter two frequencies, called *side-band frequencies,* are

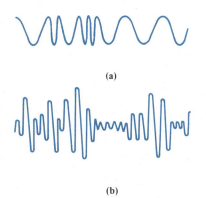

(a)

(b)

FIGURE 12.29   Radio Frequency Signal Waveshapes: (a) Amplitude modulation (AM), (b) Frequency modulation (FM)

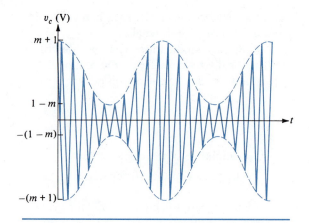

FIGURE 12.30 Amplitude-Modulated Sine Wave

Amplitude Modulation Circuit. Figure 12.31a shows one example of a circuit that performs the process of amplitude modulation along with amplification and filtering. The source voltages v_m and v_c are the modulating and carrier signals, respectively. The resistors R_1, R_2, and R_E along with V_{CC} are chosen to provide a suitable Q point for the BJT. The operating point then shifts as the signals v_m and v_c are applied and change with time. Since $\omega_m \ll \omega_c$, the af signal v_m can be regarded as a DC voltage over small ranges of time. However, as v_m changes, the Q point shifts and creates an amplitude-modulated output. The capacitors in the circuit are either coupling or bypass capacitors, except for C_2. Proper choice of capacitor C_2 and the inductor L provides an LC circuit that resonates at ω_c. That is, only the resonant frequency signal, given by $\omega_0 = (1/LC_2)^{1/2}$, reaches R_L. The low-frequency signals are shunted through L, while the high-frequency signals are shunted through C_2.

An amplitude-modulated output voltage is produced in this circuit due to the nonlinearity of the BJT. Thus, the output voltage has the same form as (12.4–3) or (12.4–4). To verify this fact,

nearly equal in magnitude since $\omega_c \gg \omega_m$. In general, the amplitude-modulated wave will contain more frequencies than just these three. However, the magnitude of all of the signal frequencies produced is approximately the same and is in the high-frequency rf range, which allows the required more efficient transmission of the signal information.

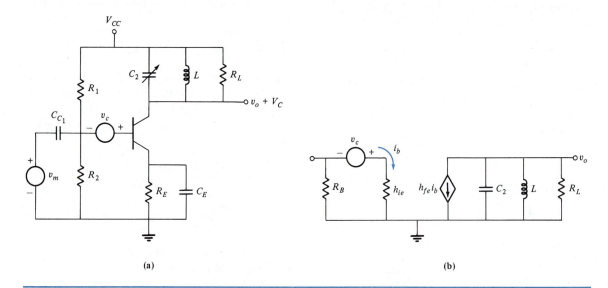

(a)

(b)

FIGURE 12.31 Amplitude Modulation Amplifier: (a) Actual circuit, (b) Small signal equivalent circuit

we first consider $v_c = A \sin \omega_c t$ and assume that v_m is a constant. Under these conditions, the AC output voltage v_o is an amplified version of v_c when L and C have also been selected to resonate at $\omega_0 = \omega_c = (1/LC)^{1/2}$. Therefore,

$$v_o = -|A_V|v_c \qquad (12.4\text{–}5)$$

where $|A_V|$ is the magnitude of the voltage amplification.

By using a small-signal model and the techniques of Chapter 5, we develop an expression for $|A_V|$. First, the AC base current is obtained from Fig. 12.31b as follows:

$$i_b = \frac{v_c}{h_{ie} + R_B} \qquad (12.4\text{–}6)$$

Additionally, the output voltage is given by

$$v_o = -i_c R_L = -(h_{fe} i_b) R_L \qquad (12.4\text{–}7)$$

provided that the frequency is $\omega_0 = (1/LC)^{1/2}$, or nearly so. By substituting for i_b, we have

$$v_o = \frac{-h_{fe} R_L v_c}{h_{ie} + R_B} = -|A_V|v_c \qquad (12.4\text{–}8)$$

We note that $|A_V|$ depends linearly upon h_{fe}, or

$$|A_V| = C h_{fe} \qquad (12.4\text{–}9)$$

where C is a constant.

We now take into consideration the slow variation in the modulating voltage v_m (at frequency ω_m) and the corresponding change in the Q point. Since the BJT behaves nonlinearly to some degree, we realize that this nonlinearity will cause h_{fe} to change. By expanding h_{fe} in a Taylor series and retaining only the first two terms, we have

$$h_{fe} = h_{feo} + \left.\frac{dh_{fe}}{di_B}\right|_Q \Delta i_B \qquad (12.4\text{–}10)$$

where Δi_B is the change in base current due to the modulating signal v_m. Since v_m is directly in parallel with $R_B = R_1 \| R_2$ (see Figure 12.31), we have

$$\Delta i_B = \frac{v_m}{h_{ie}} \qquad (12.4\text{–}11)$$

Substituting (12.4–11) into (12.4–10) yields

$$h_{fe} = h_{feo} + \left.\frac{dh_{fe}}{di_B}\right|_Q \frac{v_m}{h_{ie}}$$
$$= h_{feo}(1 + Kv_m) \qquad (12.4\text{–}12)$$

where K is another constant. Finally, by substituting (12.4–12) back into (12.4–9), we obtain the voltage amplification dependence upon modulation voltage v_m as follows:

$$|A_V| = C h_{feo}(1 + Kv_m) \qquad (12.4\text{–}13)$$

Substitution into (12.4–5) then yields

$$v_o = -C h_{feo}(1 + Kv_m)v_c \qquad (12.4\text{–}14)$$

Thus, we observe that (12.4–14) has the same form as (12.4–3).

— EX. 12.13

Amplitude Modulation Example

For the AM amplifier circuit of Fig. 12.31a, determine the voltage amplification (without v_m) and the amplitude-modulated output voltage corresponding to (12.4–14). Let $h_{fe} = 100$, $dh_{fe}/di_B = K = 0.1$ $(mA)^{-1}$, $R_L = 5$ kΩ, $h_{ie} = 1$ kΩ, and $R_B = 3$ kΩ. Treat L and C_2 as open circuits, assuming that $\omega_0 = \omega_c$.

Solution: The voltage amplification without v_m is given by (12.4–8). Substituting values yields

$$|A_V| = \frac{h_{fe} R_L}{h_{ie} + R_B} = \frac{100(5)}{1 + 3} = 125 \qquad (1)$$

or $|A_V| = 1.25 h_{fe}$. If we take v_m into consideration, then $|A_V|$ is given by (12.4–13) and by substituting values, we have

$$|A_V| = 125(1 + 0.1 v_m) \qquad (2)$$

Finally, the amplitude-modulated output voltage is obtained from (12.4–14) as

$$v_o = -125(1 + 0.1v_m)v_c \tag{3}$$

Demodulation or Detection.   The process by which the coded signal information in the modulated wave is converted back to its original form is called *demodulation* or *detection*. For example, a waveshape similar to that shown in Fig. 12.30 is to be reconstructed in its original form, which is the envelope of the AM signal. By realizing that only the top or bottom waveshape variation in Fig. 12.30 is needed for reconstruction of the af signal, we can electronically eliminate half of the AM waveshape. This elimination is most easily accomplished by using a peak rectifier circuit (in the form of a diode rectifier from Chapter 2), as shown in Fig. 12.32. Positive portions of the AM signal voltage pass through the diode detector, while negative portions are blocked. Furthermore, the slow variation of the envelope is recovered, while the rapid variation at ω_c is ignored for $R_L C \gg 1/\omega_c$. A reasonable design rule is $R_L C = 1/\omega_m$. The af signal component then appears across the load resistor R_L.

Automatic Gain Control.   A circuit that used op amps to provide automatic gain control (AGC) was described in Chapter 7. In radio receiver systems, AGC is an absolute necessity because the strength of the signal received from a particular station is variable depending upon each of the

following:

- Distance to the station
- Magnitude of the power output of the station
- Weather conditions

Hence, in a particular location under constant weather conditions, each radio station provides a certain signal intensity. However, this intensity varies due to changing weather conditions as well as location when the radio is mobile (for example, in an automobile). Even when the radio is immobile with fixed weather conditions, closer stations provide a stronger signal, while distant stations provide only a weak signal. Thus, since the signal from any particular station changes with time, without AGC, continual adjustment of the volume control would be necessary to provide a constant level of sound.

Op Amp AGC Amplifier Circuit.   Figure 12.33 displays an op amp AGC audio amplifier circuit in which v_i is the input audio signal that is widely variable. The connection of the feedback elements $R_1, R_2, C_1, C_2,$ and T_1 provides the control voltage

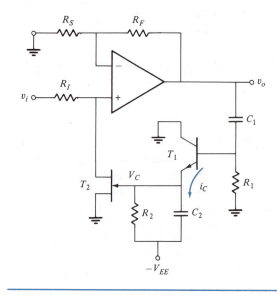

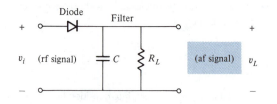

FIGURE 12.32   Amplitude Demodulation or Detection Circuitry (Diode Detector)

FIGURE 12.33   Op Amp AGC Audio Amplifier Circuit

V_C to the gate of the JFET T_2. We can verify that as the magnitude of v_i varies, v_o remains essentially constant.

To analyze the circuit of Fig. 12.33, recall from Chapter 7 that variation of the control voltage V_C causes the channel resistance R_{CH} of T_2 to vary. Additionally, R_I and R_{CH} form a resistive divider network at the input. Thus, the voltage at the positive terminal of the op amp is given by

$$v_p = \frac{R_{CH}}{R_{CH} + R_I} v_i \qquad (12.4\text{--}15)$$

and the output voltage v_o is given by the noninverting amplifier expression as

$$v_o = \left(\frac{R_F}{R_S} + 1\right) v_p \qquad (12.4\text{--}16)$$

Hence, by substituting (12.4–16) into (12.4–15), we obtain

$$v_o = \left(\frac{R_F}{R_S} + 1\right)\left(\frac{R_{CH}}{R_{CH} + R_I}\right) v_i \qquad (12.4\text{--}17)$$

To show that v_o remains essentially constant as v_i varies, we now consider the effects of the feedback network. Capacitor C_1 acts like a coupling capacitor to the AC output signal v_o, and the operation of T_1 depends upon the magnitude and sign of v_o. However, when the magnitude of v_i is small, such that the magnitude of v_o is less than 0.7 V, then T_1 is cut off and $V_C = -V_{EE}$. The value of V_{EE} is selected such that under these conditions R_{CH} is large, with $R_{CH} \gg R_I$. Hence, from (12.4–5), $v_p = v_i$, with T_2 acting like an open circuit and

$$v_o = \left(\frac{R_F}{R_S} + 1\right) v_i \qquad (12.4\text{--}18)$$

Now, we consider the magnitude of v_i to be increased such that the magnitude of v_o becomes large enough to forward bias the base–emitter

junction of T_1. T_1 thus operates in the active region during this time, with $i_c > 0$ in the direction shown in Fig. 12.33. Thus, C_2 begins to charge, increasing V_C from its minimum value of $-V_{EE}$. R_{CH} is forced to reduce, and the voltage at p is given by

$$v_p = \frac{R_{CH}}{R_{CH} + R_I} v_i < v_i \qquad (12.4\text{--}19)$$

Thus, v_p is also reduced. Furthermore, as v_i continues to increase, V_C increases, causing R_{CH} to reduce along with v_p and thus keeping v_o essentially constant as desired.

AGC audio amplifiers are available in IC form from various manufacturers. One popular example is the LM170 AGC audio amplifier.

Basic Receiver Circuit.   Figure 12.34 displays a basic radio receiver circuit that includes an antenna, tuner, automatic gain control, and speaker. The antenna receives the rf electromagnetic radiation from all local radio stations and converts it into a signal current. The *LC* tuning network at the input is then adjusted (by varying C_V) to select only the desired station signal and reject all others. If we disregard the AGC feedback connection for the moment, we see that the BJT common-emitter amplifier amplifies the rf signal current and that the diode demodulator and filter reconstruct the original signal. This signal is then fed to the speaker, and the corresponding sound is radiated.

If we now take the AGC connection into account, we note that the demodulated voltage to the speaker is negative and that the larger the magnitude of the output signal, the louder the sound. Furthermore, the voltage across C_F (with the polarity as indicated in Fig. 12.34) is also larger. Thus, more DC current is forced through R_{F_2}, which reduces the DC base current of the BJT. In turn, the amplification of the BJT amplifier is reduced (since $v_o/v_s \cong (h_{fe}R_L)/h_{ie}$ and $h_{ie} = 0.025/I_B$ from Chapter 5). Hence, the BJT output and the

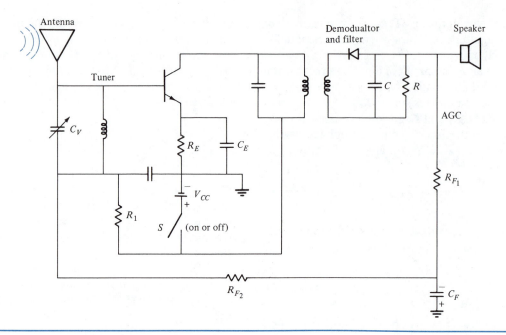

FIGURE 12.34   Basic Radio Receiver Circuit

resulting sound are reduced for larger output signals. For smaller output signals, the automatic gain control has little effect.

12.4.2   Television

Although radio and television systems use many of the same processes, converting a visual image into an electrical signal is unique to a television system. *Television* is the science of transmitting rapidly changing pictures via electrical signals. Similar to what happens to a radio signal, the picture is electronically disassembled at the transmitter and sent via electromagnetic waves through air (or a coaxial cable) to the receiver where it is electronically reassembled.

The process begins at the TV camera where the scene to be televised is converted into an electrical (video) signal. Synchronizing and blanking pulses are added to this signal to form a composite video signal that is then used to amplitude

modulate a carrier signal. After transmission occurs, the receiver uses this signal to reconstruct the scene.

TV Camera.   The beginning point in the process of television is the camera. The camera that is described here is the *imaging vidicon camera tube*. This camera, shown in cross section in Fig. 12.35, consists of a cathode, beam-focusing electrodes, light-producing lenses, a photoconductive (light-sensitive) target, and electrical target connections. The basic operation of the camera is as follows. The cathode is heated to emit electrons in all directions. Voltages applied to the focusing and accelerating electrodes then provide an electric field that accelerates the electrons toward the target. Additionally, these electrodes focus the electrons into a beam, and, at any particular instant in time, the electron beam converges toward a single desired spot on the target. At the same time, a set of lenses focuses the image (as in an ordinary camera) on the target. Since the photoconductive

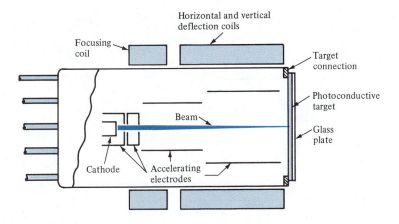

FIGURE 12.35   Imaging Vidicon Camera Tube

target is illuminated by the light image, the conductivity of the spot where the beam impinges is directly proportional to the amount of illumination at this position. Hence, areas where the light image is bright will have a higher voltage than areas of lower intensity. Thus, these higher-voltage areas allow more electrons to be absorbed by the target at these positions. The absorbed electrons for a particular spot then flow through the target connection and a load resistor and thus produce the video signal that represents the light image at one position on the target.

To produce a charge density pattern directly related to the light image over the entire target, the beam is scanned over the surface. As the beam scans the pattern, electrons are absorbed in varying amounts on the target surface depending upon the voltage of the region.

Very importantly, the camera and transmitter are connected to a *sync generator,* a device that adds pulses to the video signal and maintains the camera and receivers in perfect scanning step. Without this synchronizing signal, the image would constantly roll or jumble. That is, the source and receiver would tend not to scan the same portion of the scene at the same time. Actually, this roll or jumble is used by cable TV stations to scramble the picture for nonpaying customers.

Hence, because of this synchronizing signal, the electron beams of the camera tube and picture tube are ensured of being at corresponding points of their respective screens.

For transmission, the video and audio signals are mixed together into a composite signal. The video portion of the signal comprises 4 MHz of the typical 6 MHz wide TV signal. This band of frequencies contains all of the video light level and control signals needed to produce a full picture. The audio signal is included with the video signal. Although there is a slight time lag between video and audio signals as they enter a receiver, the net time difference is too small to be noticeable.

TV Picture Tube.   At the receiver, the picture tube uses the video signal to reproduce the image focused by the camera. Figure 12.36 displays a typical TV picture tube, which is a cathode-ray tube (to be described in detail in Chapter 14) that consists of cathode and control electrodes along with a phosphor screen. The uniformly coated phosphor screen emits light when electrons are absorbed. The intensity of the emitted light depends upon the strength of the electron beam, and an image is produced by varying the beam intensity over the screen surface. The video signal contains the information required to vary the strength and

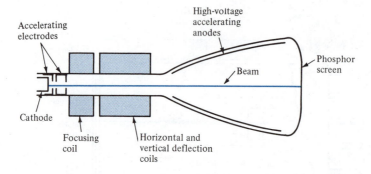

FIGURE 12.36   Television Picture Tube

position of the electron beam such that the original camera image is produced on the TV screen.

Station Selection.   Figure 12.37 shows a typical AM receiver for video and audio. After the incoming signal has been received by the TV antenna, it is amplified from a few millivolts to 1 V peak-to-peak, and the much stronger signal is then sent to the channel selector or tuner. Since the antenna signal represents all available local TV frequencies and hence all available TV stations, the selector is needed to attenuate undesired frequencies, while it passes those that contain the information for a single TV station. The channel selection uses a band-pass filter that consists of combinations of variable capacitors. The

variable frequency response of this circuit allows the band of desired signals passed to be adjustable in discrete steps corresponding to the allowable range of each station.

After the desired station is selected, the audio and video signals are demodulated and separated. The audio signal is then sent to amplifiers that drive the speaker(s), and the video and associated control signals are routed to the picture tube.

Scanning.   The television picture is produced by scanning the scene line by line according to the video information. The scanning process provides the method of electronically disassembling the picture at the camera tube and reassembling it at the picture tube. The relative intensity of the vi-

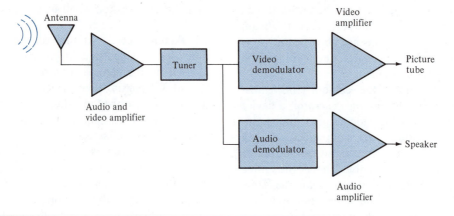

FIGURE 12.37   Receiver for Audio and Video

sual image is produced one line at a time. In the scanning process, the electron beam can be considered to start from the top left of the screen and to proceed horizontally to the right as it produces light and dark regions across the screen. The next line to be projected starts slightly lower and again at the left of the screen. However, in order for the beam to jump from the right of the screen to the left (and from the bottom to the top to repeat the sequence) without interfering with the picture, synchronized pulses in the video signal are used to blank the beam. Thus, the electron gun scans the entire screen back and forth, but only on a left-to-right scan is the video signal projected on the screen.

The scanning motion is controlled by deflection plates (as in the case of the oscilloscope, to be described in detail in Chapter 14). A periodic saw-tooth (sweep) signal is applied to the deflection plates that causes the beam to sweep horizontally as well as vertically, with the period of the vertical sweep signal much larger than that of the horizontal sweep signal.

The actual scanning process in a modern television is slightly more complicated. Instead of scanning each line directly after the next, cameras first scan every other line, then go back and scan the skipped lines. This process is called *interlace scanning*, which implies that each complete frame of a moving scene is projected as two separate pictures that have half the definition of the finished product. Hence, if a moving picture is made of 30 still pictures per second, and if every picture is made of two interlaced images, then the electron gun has to scan the tube 60 times per second. Interlace scanning greatly decreases flicker and dramatically increases the quality of the image.

Color Television. Color television utilizes the technique of mixing the three primary colors to produce all other colors. The three primary colors are red, green, and blue. Hence, in a color TV system, three imaging elements are used, one for each primary color. The receiver then recreates the color image by combining the individual primary color images.

Three separate tubes are used to collect the information corresponding to the intensity of each primary color in a scene. All of this information is contained in the 6 MHz TV signal corresponding to a particular station. Collecting all of the information is absolutely necessary in order that black and white television still receive the same signals without interference from the color information. In fact, it was this need to keep black and white television compatible with the new color signals that contributed to keeping color television out of the market for 20 years after it was first demonstrated. To reproduce the color image accurately, the face of the picture tube in a color television is made up of thousands of groups of fluorescent dots. Each group of dots consists of the three primary color elements. Mixing these colors allows a receiver to simulate all the colors of the rainbow. To prevent the electron beam from hitting an incorrectly colored dot, most color TV sets employ a shadow mask, a flat panel that has thousand of small holes lithographed into it. Each hole is positioned so that a beam from an electron gun is allowed to hit only the fluorescent dots assigned to it; all others are shaded.

CHAPTER 12 SUMMARY

- A sensor receives and responds to a stimulus, whereas a transducer responds to one form of energy and converts this energy into another form.
- Variable-impedance sensors produce a change in impedance for a change in force, pressure, temperature, displacement, or voltage.
- A strain gage produces a change in resistance for a change in force or pressure. The sensitivity S of a strain gage is defined as $(\Delta R/R)/(\Delta L/L)$,

which is in the range of 2 to 4 for metal strain gages and is 100 times larger for semiconductor gages. For improved accuracy, strain gages are usually used in bridge circuits.

- Thermistors and sensistors are temperature-sensing variable resistors. Sensistors are metallic resistors that have a positive temperature coefficient (the change in resistance is positive for a positive change in T). Thermistors are semiconductor resistors with a negative temperature coefficient. These devices can be used for temperature compensation.

- The LVDT (linear variable differential transformer) is a variable-inductance transducer that provides an inductively coupled AC electrical output voltage that is linearly proportional to an input displacement.

- Variable-capacitance transducers are available with mechanical or voltage control. For a mechanically variable capacitance, $C = (\varepsilon_r \varepsilon_0 A)/d$, where the magnitude of C is controlled by varying A or d. For a voltage variable capacitance (varactor diode), the magnitude of C is dependent upon the applied reverse voltage.

- Thermocouple transducers produce a difference in potential proportional to a temperature difference. Two junctions of dissimilar material are required. Thermocouples have been widely used to measure temperature.

- The Hall effect is the production of a transverse voltage due to the application of a longitudinal current and a perpendicular magnetic field. The Hall effect device is used as a switch with two discrete output levels corresponding to $B = 0$ and $B > 0$.

- The operation of optoelectronic devices is based upon light and electrical energy interaction. Light is a form of electromagnetic radiation with an associated frequency, wavelength, and color. Table 12.1 associates various colors with wavelength and frequency, where $f = c/\lambda$ and $c = 3 \times 10^8$ m/s.

- Photoconductive cells, or photoresistors, are variable resistors made of semiconductor material whose magnitude of resistance decreases as the amount of light absorbed by the semiconductor increases. The largest value of resistance is the dark resistance.

- A photodiode is a PN junction semiconductor diode or PIN diode whose current–voltage characteristic is dependent upon light absorption. The photodiode is reverse biased, and the current is a reverse current equal to the sum of the light-generated current I_G and the diode reverse saturation current I_S. Op amps can be used with photodiodes for amplification or current-to-voltage conversion.

- A phototransistor is a BJT that is biased into the amplifier region with V_{DC} applied between collector and emitter. Incoming light is absorbed near the emitter junction, which produces I_G that adds to I_B (although I_B may be zero). This current is then amplified by β for the BJT. For higher current amplification, Darlington-pair phototransistors are used and provide amplification proportional to β^2.

- A photovoltaic cell is a PN junction diode transducer that absorbs light energy to produce current and/or voltage depending upon the load resistor R_L connected. When sunlight is the source of light, the diode is referred to as a solar cell. Maximum power output occurs for a specific value of R_L that depends upon the diode i_D–v_D characteristics and the magnitude of the incoming light radiance.

- A light-emitting diode (LED) is a PN junction diode made of special light-emitting semiconductor materials. Under forward-bias conditions, carriers crossing the junction of an LED recombine and produce light emission. The emitted light may be in the visible spectrum or infrared region depending upon the diode semiconductor material. The value of forward voltage for an LED under light-emitting conditions is $V_{0\,LED}$ and is approximately 1.5 V.

- Applications for LEDs include numerical displays, light sources, voltage polarity indicators, voltage level indicators, smoke detectors, intrusion alarms, and optical couplers.

- Optoelectronic isolators, or optical couplers,

utilize an input LED whose light energizes an output detector that can be a photodiode, phototransistor, or Darlington-pair phototransistor. Darlington phototransistors provide the largest current transfer ratio of I_{OUT}/I_{LED} on the order of 0.1.

■ Radio and television are examples of detecting transducers. Each of these systems requires a transmitter and a receiver.

■ For a radio, the microphone converts the audio signal (sound) into an electrical signal that is modulated, amplified, and transmitted by an antenna. The receiver reverses this process by using an antenna to receive the signal, which is then demodulated and converted into sound via a speaker. Modulation is a frequency-shifting process that converts the signal to much higher frequencies that are transmitted more efficiently through air. Demodulation is the conversion of the modulated signal back to its original form.

■ Television is the science of transmitting rapidly changing pictures via video electrical signals with accompanying audio signals (the audio signals are transmitted as in radio). For the TV transmitter, a camera converts the light image into a synchronized video signal that is modulated and transmitted via an antenna. Scanning a scene line by line is the method used to electronically disassemble the picture. At the receiver, the antenna converts the electromagnetic radiation into a synchronized electrical signal. This signal is demodulated, amplified, and scanned across the phosphor screen of the picture tube, which reproduces the image. A color picture is produced by mixing red, green, and blue at thousands of dots on the face of the picture tube

CHAPTER 12 PROBLEMS

12.1

The wire in a strain gage is 15 cm in length and has an initial resistance of 100 Ω. An applied force causes a change in length of 0.15 mm and a change in resistance of 0.2 Ω. Calculate (a) the sensitivity of the strain gage and (b) the strain caused by the force.

12.2

Hooke's law states that the strain $\sigma = \Delta L/L$ is related to the stress s by the relation $\sigma = s/E$, where E is the modulus of elasticity (also known as Young's modulus). A strain gage with a sensitivity of 3 is fastened to a steel member subjected to a stress of 840 kg/cm^2. The modulus of elasticity of steel is 2.1×10^6 kg/cm^2. If the initial resistance is 100 Ω, determine the change in resistance due to the applied stress.

12.3

The resistance of wire with length L and cross-sectional area A is given by $R = \rho L/A$, where ρ is the resistivity of the material. Consider a gage wire with circular cross section of diameter D. Derive an expression for the sensitivity $S = (\Delta R/R)/(\Delta L/L)$ in terms of the fractional change in resistivity $\Delta\rho/\rho$ and Poisson's ratio μ. Poisson's ratio is defined by

$\Delta D/D = -\mu\Delta L/L$, where $\Delta D/D$ is the strain in the lateral direction and $\Delta L/L$ is the strain in the axial direction.

12.4

In the bridge circuit arrangement of Fig. 12.2, all four resistors have the same initial resistance of 100 Ω. Resistor R_4 is now mounted on a steel specimen. Calculate the minimum stress that can be detected if the maximum sensitivity of the monitoring equipment is 5 μV. The bridge supply is $V_T = 10$ V, and Young's modulus (see Problem 12.2) for steel is $E = 2.1 \times 10^6$ kg/cm^2. Additionally, the sensitivity $S = 2$. Neglect the fractional change of resistance due to the stress.

12.5

A strain gage possessing an unstrained resistance of 250 Ω is connected as one arm in the bridge circuit of Fig. 12.2. For $V_T = 10$ V, $S = 2.5$, and identical unstressed resistors in the other three arms of the bridge, calculate the output voltage for a strain applied to R_4 of magnitude 100 in./in.

12.6

The bridge circuit of Fig. 12.2 has two identical metallic strain gages for R_2 and R_4. Each has an initial resistance of 200 Ω, and sensitivity $S = 2$. If stresses are applied to both strain gages, calculate the approximate value of V_{out} for $V_T = 5$ V and $R_1 = R_3 = 1$ kΩ for the following cases:

(a) $\Delta R_2 = \Delta R_4 = 10\,\Omega$
(b) $\Delta R_2 = 10\,\Omega$, $\Delta R_4 = 5\,\Omega$
(c) $\Delta R_2 = 5\,\Omega$, $\Delta R_4 = 10\,\Omega$

12.7

Consider the bridge circuit of Fig. 12.2. Assume that $R_1 = R_2 = R_3 = R$ and that R_4 is a strain gage resistor with unstrained resistance R. Show that the variation of V_{out} with stress is linear provided that $\Delta R/R \ll 1$. Use information from Problem 12.2, where $\sigma = S/E$ and $E =$ Young's modulus.

12.8

The resistance of a thermistor can be expressed as $R_T = R_0 e^{\beta(1/T - 1/T_0)}$, where R_T and R_0 are the resistances of the thermistor at temperatures T and T_0 ($^\circ$K), respectively, and β depends upon the thermistor material. Consider a thermistor R_T used in the bridge circuit of Fig. P12.8 with $R_0 = 0.008\,\Omega$. If $R_1 = 250\,\Omega$, $R_2 = 1.5$ kΩ, $R_3 = 3.0$ kΩ, and $V_{in} = 5$ V, determine the coefficient β if the output voltage changes from 0.004 V to 2.0 V as the temperature varies from 0° C to 85° C.

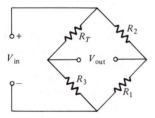

FIGURE P12.8

12.9

An LVDT has the following characteristics:

Core displacement (mm):
    0.025 0.05 0.075 0.1 0.125 0.15 0.175
Output (V): 0.05   0.1   0.15   0.2 0.25   0.25 0.25

Determine the linear range of the LVDT.

12.10

Consider the circuit of Fig. 12.3b. The output voltage of the LVDT is given by $V_{out} = (M_1 - M_2)di_P/dt$,

where M_1 and M_2 are mutual inductances of the secondary windings and i_P is the instantaneous primary current. The linear range of the LVDT is determined by the range over which the mutual inductance varies linearly with core displacement. Fig. P12.10 displays the variation of $M_1 - M_2$ as a function of displacement for a particular LVDT. If $i_P = 5 \sin \omega t$ mA, where $\omega = 2\pi f = 2\pi(1000)$ rad/s, sketch the rms output voltage versus displacement in the linear region.

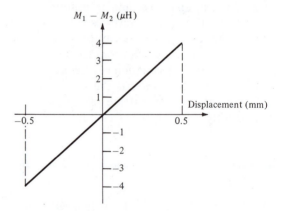

FIGURE P12.10

12.11

Consider the variable-capacitance transducer of Fig. 12.4a. The diameter of the inner cylinder is 2 cm, the spacing between the plates is 0.02 cm, and the dielectric coating on the inner cylinder has a relative permittivity of 10. Determine the capacitance if the inner cylinder is inserted a distance of 5 cm.

12.12

Consider a capacitive pressure transducer that consists of two parallel conducting plates supported by air. Without pressure, the spacing of the plates is 0.5 cm and the magnitude of capacitance is 500 pF. With 500 psi applied, the plate separation is reduced by 0.1 cm. If the plate area is 5 cm^2, determine the capacitance with this pressure applied.

12.13

A variable-capacitance transducer with air as the dielectric has a peak overlap area of 5 cm^2 and plate separation of 0.05 cm under air pressure conditions. Calculate the maximum capacitance.

12.14

The variable-capacitance transducer of Fig. 12.4c is used as the feedback element in Fig. P12.14. If the initial spacing d is 0.01 cm and the input and output voltages are 5 V and -2 V rms, respectively, determine the transducer capacitance and the diaphragm displacement. Let $C_1 = 12$ pF, $\varepsilon_r = 10$, and $A = 4$ mm^2. Neglect the variation of A with the displacement of the diaphragm.

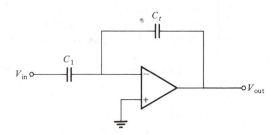

FIGURE P12.14

12.15

Consider the bridge circuit of Fig. P12.15. Let d be the equilibrium plate separation and x be the upward displacement from this separation. If $V_{in} = 5$ V and $V_{out} = 2$ V, determine the x displacement of the

plate. Calculate C_1 if $C_2 = 12$ pF. Let $C_3 = C_4 = 5$ pF and $d = 0.005$ cm.

12.16

The electrical equivalent circuit of a piezoelectric transducer is given in Fig. 12.5b. Determine the equivalent reactance of the piezoelectric crystal if the resistance R is neglected. Find the range of frequencies for which the reactance is inductive. Note: Use $\omega_s = \sqrt{1/LC}$ and $\omega_p = \sqrt{(1/LC)(C_P + C/C_P)}$ to simplify the results, where ω_s is the series resonant frequency and ω_p is the parallel resonant frequency.

12.17

Consider the circuit of Fig. 12.6, which is used to measure the temperature of a heated electrode. The thermocouple is made of chrome and constantan. Previous experiments with the thermocouple show that the voltage varies linearly with temperature for $0 \le T \le 450°$ C and varies exponentially as $V = V_0 e^{\beta(1/T_0 - 1/T)}$ for $T > 450°$ C. Calculate the voltage readings for 90° C and 500° C if $V_0 = 100$ mV, $\beta = 2500°$ K^{-1}, $T_0 = 723°$ K, and the voltage at $T = 0°$ C is 5 mV.

12.18

A thermocouple thermometer is displayed in Fig. P12.18. The internal resistance of the millivoltmeter is R_i. Assume that the thermocouple has a resistance $R_t = 3\ \Omega$ and that the resistance of the connected wire is 5 Ω. Additionally, the voltage V_{xy} increases linearly with temperature by 45 μV/° C in the temperature range 0° C to 900° C. Determine the voltage applied to the millivoltmeter at $T = 200°$ C if its internal resistance is 500 Ω.

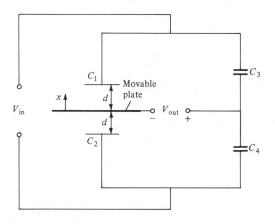

FIGURE P12.15

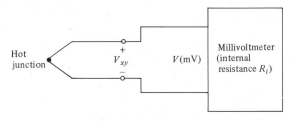

FIGURE P12.18

12.19

A semiconductor sample is contacted and oriented in a magnetic field with $B = 0.4$ T, as shown in

Fig. 12.7. The thickness of the sample is $t = 60\ \mu m$. If the measured current is 3 mA and $V_H = -1.5$ mV, determine the type and concentration of the majority carrier.

12.20
A sample of silicon is doped with 10^{18} phosphorus atoms/cm^3. Determine the Hall voltage if the longitudinal current is 2 mA and the magnetic field intensity is $B = 0.3$ T. The sample thickness is $t = 50\ \mu m$.

12.21
Consider the circuit of Fig. 12.12c with $R_F = 100$ kΩ, $R_S = 1$ kΩ, $V_{DC} = 0.5$ V, and $R_L = 100$ kΩ. Determine the photodiode current if the dark current is negligible. Let the difference-mode amplification for the op amp be $A_d = 10^5$.

12.22
For the circuit of Fig. 12.12d, a PIN photodiode is used with i_D–v_D characteristics as indicated in Fig. E12.7. For $V_{DC} = 6$ V, $R_F = R_0 = 100$ kΩ, and $R_L = 500$ Ω, determine the radiance of the incident light if the measured output voltage is 15 mV.

12.23
Consider the basic phototransistor circuit of Fig. 12.13d with $R_L = 5$ kΩ and $V_{DC} = 10$ V. The current–voltage characteristics of the phototransistor are displayed in Fig. E12.8. If the light radiance is $R_I = 2$ mW/cm^2, determine v_{CE} and i_C.

12.24
The current–voltage characteristics of a solar cell are given by $i_D = I_S(e^{v_D/v_T} - 1) - I_G$, where $I_S = 10^{-14}$ A, $v_T = 0.025$ V, and I_G is the photogenerated current $= 1\ \mu A$. This solar cell is used in the circuit of Fig. P12.24 with $R_S = 1$ kΩ and $R_F = 10$ kΩ. If the measured output voltage is $V_{out} = 5$ V, determine the diode voltage v_D and current i_D.

12.25
Show that the current–voltage expression for a solar cell can be written as $v_D = v_T \ln[1 + (i_D + I_G)/I_S]$, where I_G is also the short-circuit current I_{SC}. Consider a silicon solar cell with $I_S = 2$ nA and $I_{SC} = 2$ mA. Plot the current–voltage characteristic as in Fig. E12.9. Indicate the values of open-circuit voltage and short-circuit current. Remember that i_D is negative and is plotted as $-i_D$ in Fig. E12.9.

12.26
The electrical power of a solar cell is the product of i_D and v_D. Show that this power is a maximum when the following equation is satisfied: $1 + I_G/I_S = (1 + V_{mp}/V_T)e^{V_{mp}/v_T}$, where V_{mp} is the voltage for maximum power and the other parameters are as usual. For $I_S = 5$ mA and $I_G = 1$ mA, determine V_{mp}.

12.27
Fig. P12.27 is an equivalent circuit for a solar cell that includes the parallel or shunt resistance R_{SH} of the cell. If I_S is the reverse saturation current for the diode, write the relation for i_D versus v_D from the circuit. How are I_{SC} and V_{OC} obtained? Write equations for these quantities.

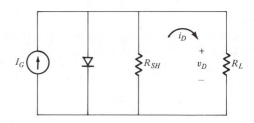

FIGURE P12.27

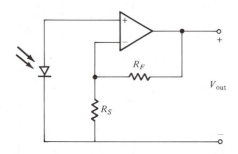

FIGURE P12.24

12.28
A GaAs red LED has a forward voltage $V_{0\ LED} = 1.8$ V. Determine the value of the bias resistor R_L in the circuit of Fig. 12.18 if $V_{DC} = 9$ V and the forward LED current is 10 mA.

12.29
Consider the circuit of Fig. E12.10 with $R_C = 1$ kΩ, $R_B = 10$ kΩ, and $v_I = V_{CC} = 5$ V. Determine the LED current i_C for $V_{0\ LED} = 1.5$ V, $V_0 = 0.7$ V, $\beta = 100$, and $v_{CE\ sat} = 0.2$ V.

12.30

Design an LED current limiter (see Fig. 12.20) such that the LED current does not exceed 20 mA. Specify the values of R_E and R_C (where $R_C = 100R_E$). Also, determine the minimum voltage of the battery required to ensure proper operation. Let $V_{0\,LED} = 1.5$ V and $V_0 = 0.7$ V.

12.31

The digital logic optoelectronic coupler circuit shown in Fig. 12.27 uses a GaAs LED with $V_{0\,LED} = 1.6$ V. For $V_{DC} = 5$ V, $R_S = 2$ kΩ, and $R_L = 4$ kΩ, determine the current transfer ratio I_{OUT}/I_{LED} such that $v_L = 4$ V when the input voltage is high. Use $v_{CE\,sat} = 0.2$ V.

12.32

Consider the circuit of Fig. 12.31a. Determine the amplitude-modulated voltage for the following parameters: $h_{fe} = 120$, $dh_{fe}/di_B = K = 0.2$, $R_L = 5$ kΩ, $h_{ie} = 1$ kΩ, and $R_B = 3$ kΩ.

12.33

An amplitude-modulated signal can be described as a function of time in the following form: $v(t) = A_c[1 + Km(t)]\cos \omega_c t = v_e(t)\cos \omega_c t$, where $m(t)$ is the modulating signal (information) and $v_e(t)$ is the envelope of the modulating signal. If $m(t) = A_m \cos \omega_m t$, the amplitude-modulated signal becomes $v(t) = A_c(1 + \mu \cos \omega_m t)\cos \omega_c t$, where $\mu = KA_m$. For $A_c = 5$ V, $A_m = 2$ V, $\omega_m = 2\pi(10\text{ Hz})$ and $\omega_c = 2\pi(100\text{ Hz})$, sketch $v(t)$ versus time for the following values of K:

    (a) 0.1/V

    (b) 0.5/V

    (c) 1/V (referred to as overmodulated)

Indicate the envelope on each sketch.

12.34

To avoid the condition of overmodulation indicated in Problem 12.33, the quantity μ (the modulation factor) should be less than 1. Using the equations of Problem 12.33, derive an expression and determine the minimum values for the envelope of the modulated signal.

12.35

Consider the amplitude demodulation circuit of Fig. 12.32 and let $v_i = A \sin \omega t$ as indicated in Fig. P12.35. For an ideal diode, derive the expression for v_L between times t_1 and t_3. Sketch v_L on the same graph as v_i and determine the peak-to-peak ripple or variation in output voltage. Assume $R_L C \gg t_2 - t_1$, where the diode turns on at t_2.

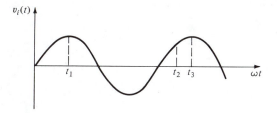

FIGURE P12.35

12.36

The AGC amplifier circuit of Fig. 12.33 is used to control the audio intensity of a mobile radio. Let $R_F = 100$ kΩ, $R_S = 1$ kΩ, and $R_I = 50$ Ω. Determine the value of the channel resistance R_{CH} if the output voltage is 0.6 V and the input voltage is 50 mV.

13

ELECTROMECHANICAL TRANSDUCERS

Electromechanical transducers are an extremely broad class of devices that convert mechanical energy into electrical energy or vice versa. Since these devices require a magnetic field to couple the energy from one form to the other, they might more appropriately be called *electromagnetic mechanical transducers*. This terminology, however, is seldom used.

In this chapter, the basic relations corresponding to translational and rotational motion of charged particles in a magnetic field are developed. The relations represent either an interaction between a current and a magnetic field that produces a mechanical force (motor action) or a mechanical force that causes motion of a conductor in a magnetic field to induce a current in the conductor (generator action). The developed relations are fundamental to the operation of rotating machines, as we will see. The chapter begins by describing properties of magnets and magnetic fields, from which basic translational and rotational transducer relations are developed. From these relations, the mechanics of electrical machinery are described.

13.1 PROPERTIES OF MAGNETS AND MAGNETIC FIELDS

13.1.1 Magnets

A *magnet* is a body that exerts an attractive magnetic force on another body made of iron and certain other materials without any physical connection between the two bodies. Magnets are usually composed of *ferromagnetic* materials such as steel, iron, and alloys of iron containing cobalt, nickel, or aluminum. Ferromagnetic materials have a specialized internal atomic structure that provides very large magnetic fields. The parameter that distinguishes the magnetic property of the particular medium is known as the *permeability* μ, where the permeability of free space is given by $\mu_0 = 4\pi \times 10^{-7}$, with units of henrys/meter (H/m). Ordinary materials have $\mu \cong \mu_0$; however, ferromagnetic materials have very large permeability, with $\mu \geq 1000\mu_0$. The permeability of the medium is usually written as $\mu = \mu_r\mu_0$, where μ_r is the relative permeability.

The attractive force that a magnet provides is described and visualized in terms of the magnetic field associated with the magnet. Magnetic fields are produced by *permanent magnets,* which are those made of ferromagnetic material possessing a built-in magnetization. However, magnetic fields are also created by the motion of charged particles, or current, as we will see.

Figure 13.1a displays a diagram of a permanent magnet called a *bar magnet*. Magnetic field lines that emanate from the north pole (N) and return to the south pole (S) are shown and represent the direction of the magnetic field at various positions in space. Such field-line directions are determined by placing a compass in the vicinity of the magnetic field. The needle of the compass aligns in the field direction at each position of measurement. Furthermore, the magnitude of the force tending to align the needle is strongest close to the magnet and decreases as the distance from the magnet is increased. This variation in the magnitude of the magnetic field is depicted in Fig.

13.1a where denser field lines appear in close proximity to the magnet and sparser field lines appear farther from it.

Another permanent magnet that has important practical applications is the *horseshoe magnet* shown in Fig. 13.1b. The horseshoe shape may be thought of as being obtained by bending a bar magnet until its poles face each other. The separation between the poles is referred to as the *air gap*. The magnetic field is essentially constant in magnitude in the air gap, is reduced at the edges, and is reduced further beyond the edges. As shown in Fig. 13.1b, the field lines bend in the edge region, which is referred to as the *fringe field region*.

Figure 13.1c shows another form of magnet called an *electromagnet,* which is composed of a coil of wire with N turns spread out as shown. This arrangement is called a *solenoidal coil,* or simply a *solenoid.* Observe that the magnetic field lines for the solenoid are very similar to those of the permanent bar magnet of Fig. 13.1a. Hence, we begin to realize that magnetic fields are associated with currents, as well as magnetized materials.

Figure 13.1d displays another form of an electromagnet in which a wire is wrapped around a ferromagnetic core that has an air gap in it. Application of current i produces an associated magnetic field that is confined to the high-permeability material. This highly intensified field is proportional to the applied current i. Furthermore, the magnetic field in the air gap is also proportional to i. Hence, the air gap magnetic field is controllable by varying the applied current. The electromagnets of Figs. 13.1c and d can be considered to be transducers of electrical input energy into magnetic output energy.

13.1.2 Magnetic Flux Density, Magnetic Flux, and Magnetic Field Intensity

In general, the magnitude and direction of a magnetic field are described by a vector possessing magnitude and direction, called the *magnetic flux density,* which is denoted by $\bar{B}$. The

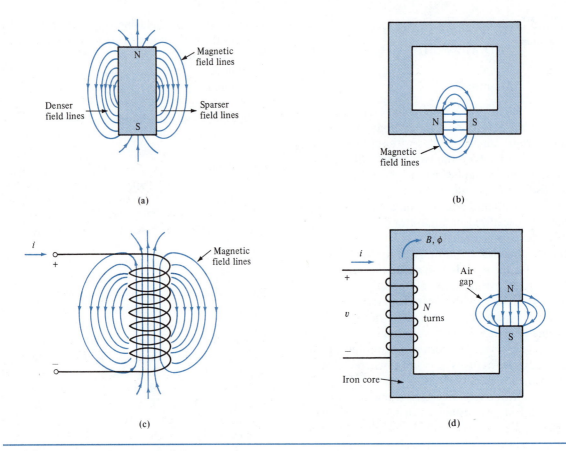

FIGURE 13.1　Different Types of Magnets: (a) Permanent bar magnet, (b) Permanent horseshoe magnet, (c) Electromagnet (solenoid), (d) Horseshoe electromagnet

magnitude of magnetic flux density is denoted by B, and its defined unit is the *tesla* (T). On the atomic level, the physical effect of a magnetic field and its associated flux density is the exertion of a force on a moving charged particle. From the Lorentz force law of physics, the magnitude and direction of the magnetic field force are given by the vector relation

$$\bar{F} = Q(\bar{u} \times \bar{B}) \qquad (13.1-1)$$

where

　　$\bar{u}$ = velocity of charge in units of m/s
　　Q = magnitude of charge in units of C

Note that since $\bar{F}$ is in newtons (N), the derived units for $\bar{B}$ are taken directly from (13.1–1) as T = N/(mC/s).

　　Observe from (13.1–1) that if $\bar{u}$ and $\bar{B}$ are parallel, the magnetic force is zero because the cross product is zero. Therefore, only the perpendicular components of $\bar{u}$ and $\bar{B}$ contribute to a magnetic force on a charged particle. Additionally, the three directions, consisting of the normal components of $\bar{u}$ and $\bar{B}$ along with the direction of the force, form a right-hand coordinate system $(\bar{u}, \bar{B}, \bar{F})$.

　　The magnetic field force that is produced creates, in general, an additional component of velocity that interacts with $\bar{B}$. The next example indicates the type of motion that a moving charged particle undergoes when unconfined in a magnetic field medium.

───────────────────────────── EX. 13.1

General Motion of a Charged Particle in a Magnetic Field Example

Figure E13.1 displays an unconfined positively charged particle with charge Q moving initially with constant velocity u_0 in the x direction. A uniform magnetic flux density is directed into the paper, as indicated. Determine the initial magnetic force on Q and describe the resulting motion.

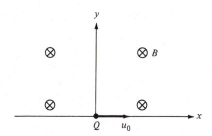

FIGURE E13.1

Solution: The initial magnetic force on the charge Q is given by (13.1–1), and since B and u_0 are perpendicular, we have

$$\bar{F} = Q(\bar{u} \times \bar{B}) = Qu_0 B\hat{y}$$

where $\hat{y}$ is a unit vector in the y direction. This force will produce a y component of velocity that creates a magnetic field force in the x direction because of the direction of $\bar{B}$, and the resulting trajectory remaining is the x–y plane. Solution of Newton's law of motion ($\bar{F} = m\bar{a}$) shows the trajectory to be a circle.

─────────────────────────────

Another important magnetic field quantity is the *magnetic flux*, which is denoted by Φ. The relationship between magnetic flux and the magnitude of the magnetic flux density is given mathematically by an integral over the area under consideration. However, for constant flux density

of magnitude B, the flux Φ is the product of B and the perpendicular area A_P through which the magnetic field passes, or

$$\Phi = BA_P \qquad (13.1–2)$$

where Φ is given in units of webers (Wb). From (13.1–2), Wb = Tm².

A final important quantity associated with a magnetic field is the *magnetic field intensity*, which is defined as a vector proportional to $\bar{B}$ as follows:

$$\bar{H} = \frac{\bar{B}}{\mu} \qquad (13.1–3)$$

The units of magnitude field intensity are T/(H/m), or A/m, as we will see. Also, H is independent of the permeability of the medium. In most instances, we will deal with only the magnitudes B and H.

Return to the horseshoe electromagnet of Fig. 13.1d and note that a coil with N turns is supplied with current i in amperes (A). This arrangement produces flux density B and corresponding magnetic flux Φ as indicated. The flux Φ in the iron core is analogous to electric current in an electric circuit. Similarly, the source of magnetic flux, or Ni, referred to as the *magnetomotive force* (mmf), is analogous to the source in an electric circuit (which is the voltage, or electromotive force, emf). As Ni is increased, Φ increases correspondingly; a typical measured curve is shown in Fig. 13.2.

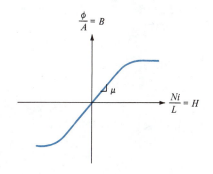

FIGURE 13.2   Typical Measured *B–H* Curve

Often, the axes are replaced with B and H, as indicated, where $B = \Phi/A$ and $H = Ni/L$ (L is the length of the flux path).

13.1.3   Energy Stored in a Magnetic Field

We now develop an expression for the energy stored in a magnetic field. This is an important concept in electromechanical transducers because the magnetic field acts as the coupling mechanism between electrical and mechanical forms of energy.

From the horseshoe electromagnet example of Fig. 13.1d, we write the expression for the supplied electrical energy as the time integral of the power, or

$$E = \int_0^t vi\,dt = \int_0^t N\left(\frac{d\Phi}{dt}\right)i\,dt = \int_0^\Phi Ni\,d\Phi$$

$$(13.1-4)$$

where we have substituted $v = N(d\Phi/dt)$, which is known as Faraday's law. This important law of physics relates the induced voltage or emf induced in a coil of N turns to the time rate of change of magnetic flux. In effect, this law states that the flux must be changing with time for a voltage to be induced. Substituting $Ni = HL$ and (for constant area) $d\Phi = A\,dB$ into (13.1–4) yields

$$E = \int_0^B (HL)(A\,dB)$$

$$(13.1-5)$$

If we now extract the volume (LA), the energy/unit volume is given by

$$\frac{E}{LA} = \int_0^B H\,dB$$

$$(13.1-6)$$

However, if we substitute $H = B/\mu$ and assume that μ is constant, then by integration

$$\frac{E}{LA} = \frac{B^2}{2\mu}$$

$$(13.1-7)$$

────────────────────────────────────── EX. 13.2

Magnetic Energy Example

For the electromagnet with applied current i in Fig. 13.1.d, calculate the ratio of energy stored in the iron to that in the air gap. Let $\mu_r = 3000$. The ratio of air gap length L_A to iron length L_I is $L_A/L_I = 0.01$.

Solution: The ratio of energy/unit volume is set up as

$$\frac{E_A/L_A A_A}{E_I/L_I A_I} = \frac{B_A^2/2\mu_0}{B_I^2/2\mu_r\mu_0} = \mu_r$$

where μ_r is the relative permeability of the iron and the fringe fields have been neglected so that $B_A = B_I$. Solving for the ratio of energies yields

$$\frac{E_A}{E_I} = \frac{\mu_r L_A A_A}{L_I A_I} = 3000\left(\frac{1}{100}\right) = 30$$

This result indicates that most of the magnetic energy supplied by the applied current is stored in the air gap.

13.2   ELECTROMECHANICAL TRANSLATIONAL TRANSDUCERS

To develop the basic relations for electromechanical *translational* transducers, we consider the highly idealized situation depicted in Fig. 13.3. A conductor of length d is moving with constant velocity u perpendicular to a uniform magnetic flux intensity B (into the page). Thus, the free charges in the conductor also move with velocity

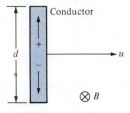

FIGURE 13.3   Moving Conductor in Uniform Magnetic Field

u in the magnetic field B, and a vertical magnetic field force, as given by (13.1–1), develops on each charge. We observe that the free charges in this case are confined to vertical movement in the conductor. Additionally, the free charges in the conductor are electrons and holes with charge $Q = \pm q$, where q is the magnitude of the charge on an electron (1.6×10^{-19} C). The magnetic force will therefore act in the vertical direction, as indicated in Fig. 13.3, on each free charge in the conductor, causing the positive holes to move upward and leave negative charge behind, while the electrons move downward and leave positive charge behind. Because of this charge movement, an electric field and corresponding voltage develop in the conductor. This electric field, in turn, creates an electric field force on each free charge that is given by

$$F = Q\mathscr{E} \qquad (13.2–1)$$

For the coordinate system of Fig. 13.3, the electric field force is also vertical, but opposite in direction to the magnetic field force. As additional charges separate, the electric field force increases until it becomes equal in magnitude to the magnetic field force. Charge separation then ceases, and an equilibrium condition results. Under these conditions, (13.1–1) and (13.1–2) are equal, and equating them yields

$$\bar{\mathscr{E}} = \bar{u} \times \bar{B} \qquad (13.2–2)$$

Since $\bar{u}$ and $\bar{B}$ are perpendicular,

$$\mathscr{E} = uB \qquad (13.2–3)$$

13.2.1 Induced Electromotive Force, or Voltage

Equations (13.2–2) and (13.2–3) represent an electric field that is developed by the motion of a conductor in a magnetic field. The electric field is related to a corresponding devel-oped voltage through $\mathscr{E} = v/d$, where the voltage v is often called the *electromotive force* (emf). By substitution for $\mathscr{E}$ in (13.2–3), we obtain the developed voltage as

$$v = Bud \qquad (13.2–4a)$$

Equation (13.2–4a) is a fundamental electromechanical relationship. It relates the developed voltage v across a moving conductor with velocity u and length d in a magnetic field having flux density B. Furthermore, it is a specialized relation in that u and B are perpendicular to each other and are also perpendicular to the conductor length direction. In the general case,

$$v = Bud \sin\theta \cos\alpha \qquad (13.2–4b)$$

where θ is the angle between d and u, while α is the angle between B and the perpendicular direction of the d–u plane.

———————————————————— EX. 13.3

Induced Voltage Example

A conducting rod is attached to an iron cylinder as shown in Fig. E13.3a. This cylinder is inserted in the magnet arrangement of Fig. E13.3b and rotated at a speed of 1800 rpm. Calculate the magnitude of the induced emf in the conductor during each half revolution. Assume that the flux from each pole is constant with $\Phi = 0.05$ Wb directed as shown in Fig. E13.3b. Sketch the resulting emf (v) versus time (t).

Solution: The magnetic flux density for each pole is given by

$$B = \frac{\Phi}{A} = \frac{0.05}{d(\pi R)} \qquad (1)$$

The rotational speed of 1800 rpm corresponds to a velocity given by

$$u = \frac{1800}{60}(2\pi R) = 60\pi R \qquad (2)$$

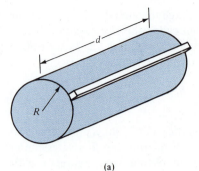

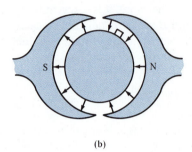

(a)

(b)

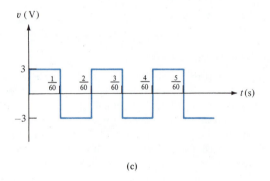

(c)

FIGURE E13.3

Thus, for each half revolution, the induced voltage is obtained from (13.2−4a) as

$$v = Bud = \frac{0.05}{d(\pi R)}(60\pi R)d = 3 \text{ V}$$

where this voltage changes sign each half revolution. The corresponding waveshape of v versus t is shown in Fig. E13.3c.

13.2.2 Magnetic Force Due to Current and Magnetic Field Interaction

As an extension of the moving conductor example of Fig. 13.3, consider connecting the ends of the moving conductor as shown in Fig. 13.4. Current is now possible in this arrangement because the conductor slides along conducting support rails that form a closed loop. We analyze this situation by considering the conductor to be moving with constant velocity u due to an applied force F_A. This motion is perpendicular to the magnetic flux density B, and, as in the previous moving conductor example, a magnetic field force causes separation of positive and negative charge in the conductor and hence charge motion in the vertical direction. This movement produces an emf (v) in the conductor, as well as a current (i), since the free charges may now exit and reenter the moving conductor via the connected loop. If we denote the average vertical velocity of the free charges in the conductor as u' (see Fig. 13.4), then we see that this vertical motion interacts with the magnetic field to create another magnetic field force (opposing $\bar{F}_A$) that is given by

$$\bar{F}_M = Q(\bar{u}' \times \bar{B}) \qquad \text{(13.2–5a)}$$

Since u' and B are perpendicular, the magnitude of this force is

$$F_M = Qu'B \qquad \text{(13.2–5b)}$$

Under equilibrium conditions, F_M is equal and

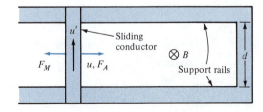

FIGURE 13.4   Moving Conductor with Support Rails

opposite to the applied force F_A, which is necessary for the conductor to move with constant velocity u. Furthermore, Qu' can be identically replaced with id to yield

$$F_M = Bid \qquad (13.2\text{--}6)$$

where i, B, and F_M are mutually perpendicular and form a right-hand coordinate system. Substitution of id for Qu' is based upon the following substitutions:

$$id = \frac{Q}{t}\,d = Q\,\frac{d}{t} = Qu' \qquad (13.2\text{--}7)$$

13.2.3   Use of Faraday's Law

In addition to the force F_M produced on the conductor, there is also a voltage developed along length d that is given by (13.2–4a). It is instructive to develop this expression for induced voltage from Faraday's law, introduced in Section 13.1.3 and given by

$$v = N\left(\frac{d\Phi}{dt}\right) \qquad (13.2\text{--}8)$$

where

N = number of turns of a general coil
Φ = magnetic flux through the coil

In the present example, $N = 1$, and since $\Phi = BA$, the induced voltage can be written as

$$v = \frac{d\Phi}{dt} = \frac{d(BA)}{dt} = B\left(\frac{dA}{dt}\right) \qquad (13.2\text{--}9)$$

since B is constant with time. However, from Fig. 13.4, we observe that $dA = (udt)d$; thus, $dA/dt = ud$. Therefore, by substitution, (13.2–9) becomes

$$v = Bud \qquad (13.2\text{--}10)$$

which is exactly the same relation as (13.2–4a).

13.2.4   Basic Translational Transducer

The electromechanical example of Fig. 13.4 can be regarded as an energy transducer in which mechanical energy is converted into electrical energy or vice versa. This property is immediately observable by considering the electrical and mechanical power contained in the system. The energy is then simply the integral of the power over time.

The electrical power is given by the product of current and voltage, or

$$P_E = iv = i(Bud) \qquad (13.2\text{--}11)$$

The mechanical power is the product of force and velocity, or

$$P_M = Fu = (Bid)u \qquad (13.2\text{--}12)$$

Note that the two equations for power are identical, which is the case in a lossless system. If losses such as frictional and resistive were considered, they would be subtracted from the mechanical and electrical power, respectively.

To obtain a better understanding of the energy transfer, consider Fig. 13.5, which shows a moving conductor system with its loop attached to a resistance R_L and a DC voltage V_{DC}. Under these conditions, two distinctly different possibilities exist: $v > V_{DC}$ and $v < V_{DC}$. We note that, depending upon which inequality prevails, the current will change direction and so also will the

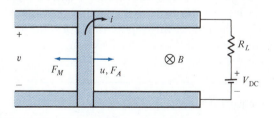

FIGURE 13.5   Moving Conductor with Support Rails Connected to External Resistor and DC Voltage

function of the transducer. Thus, if $v > V_{DC}$, then i will be in the direction shown in Fig. 13.5 and

$$i = \frac{v - V_{DC}}{R_L} \tag{13.2-13}$$

In this case, mechanical energy is supplied in the form of a force F_A and velocity u and produces electrical energy in the form of a current i, a voltage v, and associated power $iv = iBud$. Operation in this manner is an example of electrical power generation, and the device is usually referred to as a *generator*.

But, if $v < V_{DC}$, then

$$i = \frac{-V_{DC} + v}{R_L} < 0 \tag{13.2-14}$$

and the current reverses direction. In this case, the current is supplied by the battery V_{DC}. A mechanical force is thus produced with corresponding velocity, and input electrical energy is converted into output mechanical energy. Operation in this manner is an example of electrical power converted into mechanical power with resulting *motor* action.

The translational transducer of Fig. 13.5 is quite useful in showing that electrical and mechanical energy can be transformed into each other. The example, however, is not practical in general because there are limits as to how far the translational motion can continue. The system becomes practical, however, if rotational motion is used as a substitute for translational motion, which is the basis for all rotating machines. The translational transducer is practical in cases in which the movement is quite small or vibratory, as in the example of the next section.

EX. 13.4

Translational Transducer Motor Example

Consider the moving conductor system of Fig. 13.5. Calculate the induced emf for the system in which $B = 10$ T, $u = 5$ m/s, and $d = 4$ cm. Also calculate the electrical

power supplied to the battery $V_{DC} = 1$ V and the power dissipated in the resistor $R = 1$ kΩ. Finally, disregard frictional losses and calculate the required mechanical power and the applied force.

Solution: The induced emf is calculated from (13.2−4a) as

$$v = Bud = (10)(5)(0.04) = 2 \text{ V} \tag{1}$$

Thus, the current i in Fig. 13.5 is given by (13.2−13), or

$$i = \frac{v - V_{DC}}{R_L} = \frac{2 - 1}{1} = 1 \text{ mA} \tag{2}$$

Hence, the power supplied to the battery is

$$P_B = iV_{DC} = (1)(1) = 1 \text{ mw} \tag{3}$$

and the power dissipated in R_L is

$$P_R = i^2 R_L = 1 \text{ mW} \tag{4}$$

Finally, the mechanical power required is just the sum of the electrical powers, or $P_M = 1 + 1 = 2$ mW. Thus, the required force is obtained from

$$P_M = Fu \tag{5}$$

or

$$F = \frac{P_M}{u} = \frac{2 \text{ mW}}{5 \text{ m/s}} = 0.4 \times 10^{-3} \text{ N} \tag{6}$$

13.2.5 Translational Transducer Example: Loudspeaker

One example of a practical translational transducer is the loudspeaker. Figure 13.6 displays the basic geometry of a loudspeaker with its basic components of a permanent magnet, a speaker cone, and fine wire wrapped into a coil and attached to the cone support. The permanent magnet produces a constant radial magnetic field (in the annular air gap) between the north and

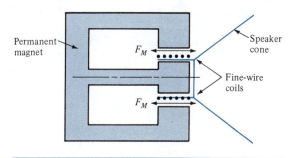

FIGURE 13.6   Basic Geometry of a Loudspeaker

south poles of the magnet. This magnetic field is perpendicular to the wire.

The mechanism of operation of the loudspeaker is as follows. An AC current that is proportional to the sound to be carried over the loudspeaker is applied to the fine-wire coil. This current produces a corresponding time-dependent magnetic field force F_M on the wire. The magnitude of the force is simply Bid, where d is the overall length of the wire coil. The direction of the force is horizontal, as indicated in Fig. 13.6. Since the coil is attached to the speaker cone, the varying magnetic force F_M causes horizontal vibratory motion of the speaker cone, which, in turn, creates waves of varying air pressure that result in sound being radiated.

It is of interest to realize that a microphone produces the opposite effect of converting sound into a time-varying signal. However, the microphone uses the same basic principles as the loudspeaker. In this case, sound waves strike a miniaturized speaker cone and cause vibration and hence movement of the fine wires in a direction perpendicular to a constant magnetic field. This translational movement induces a voltage and a current that are proportional to the sound.

––––––––––––––––––––––––––––––––––– **EX. 13.5**

Loudspeaker Example

Consider a loudspeaker as shown in Fig. 13.6 that consists of a 50-turn wire coil of 4 cm in diameter. If the permanent magnetic flux density B is 1 T and the coil current

varies as $i(t) = 10 \sin \omega t$ mA, calculate the magnetic force F_M on the speaker cone.

Solution: The force is given by (13.2–6) multiplied by N. Substituting values yields

$$F_M = BidN = (1)(10 \sin \omega t)[2\pi(0.04)](50)$$
$$= 0.126 \sin \omega t \text{ N}$$

Note that the force varies in direct proportion to the current.

13.3 ELECTROMECHANICAL ROTATIONAL TRANSDUCERS (ELECTRIC MACHINES)

In this section, the basic concepts of translational transducers are extended to *rotational* systems. Practical electromechanical transducers that are based upon rotating systems are known as *electric machines*. This section begins with the development of basic relations for an ideal rotational transducer and concludes with descriptions of the most important types of electric machines and their principles of operation. We will see that, in all cases, machine operation is based upon the interaction of electric currents and magnetic fields. Since electric currents produce magnetic fields, alternate views of machine operation are sometimes based upon the interaction of two magnetic fields or the interaction of two currents.

13.3.1 Electromechanical Rotational Relations

As in the case of translational transducers, the analysis of rotational systems is initiated by considering an idealized rotational transducer. Figure 13.7 displays a rectangular coil with N turns that is constructed from rigid wire. The coil is rotatable on a shaft, and continuous electrical

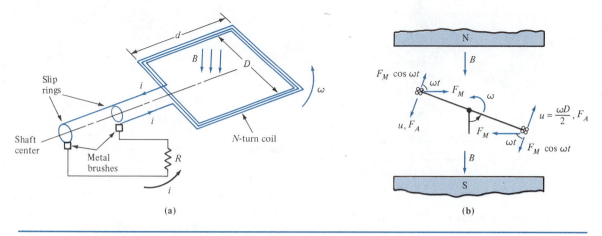

FIGURE 13.7   Idealized Rotational Transducer: (a) Three-dimensional view, (b) Cross-sectional view

contact to the coil ends is achieved through metal brushes resting on slip rings attached to the coil ends. A uniform magnetic field is applied in the downward direction, as indicated in Fig. 13.7.

A voltage is induced across the slip rings as the coil rotates about its shaft. The expression for this induced voltage is most easily obtainable from Faraday's law as given by (13.2–8), or

$$v = N\left(\frac{d\Phi}{dt}\right) = N\left(\frac{d}{dt}\right)(BA) = NB\left(\frac{dA}{dt}\right)$$

$$(13.3-1)$$

since B is constant. If we now consider the coil to be rotating with constant angular velocity ω in rad/s, then by defining angle ωt as in Fig. 13.7b, we obtain a time-varying flux area as $A = Dd \sin \omega t$. By substitution into (13.3–1), the induced voltage is obtained as

$$v = NBDd\omega \cos \omega t \qquad (13.3-2)$$

Equation (13.3-2) is also obtainable from (13.2–4b), where $\theta = \pi/2$ (angle between d and u), while $\alpha = \omega t$ (angle between B and normal to the d–u plane) and $u = \omega D$. Additionally, (13.3–2) indicates that the induced voltage across the ends

of a coil with N turns and area A that rotates with constant angular velocity ω in a constant magnetic field B varies sinusoidally with time. The polarity of the generated voltage is determined from the direction of B and the direction of rotation. For counterclockwise rotation and a downward magnetic field direction, as indicated in Fig. 13.7a, the current direction and polarity of induced voltage are also as indicated and are produced by the magnetic field force on the free charges in the top and bottom conductors (of length d). The side conductors of length D have no effect (except to act like short circuits).

With the current-carrying conductors of length d moving in the perpendicular magnetic field, a magnetic field force on each conductor is obtained from (13.2–6) and repeated here as

$$F_M = Bid \qquad (13.3-3)$$

Furthermore, the magnetic field force direction is perpendicular to the direction of B and i, as shown in Fig. 13.7b.

The magnetic field force on each conductor of length d contributes to the mechanical torque that provides rotation of the coil about its axis. An expression for the overall torque is obtained by considering the contribution from each conductor. This contribution is given by the product of the normal force and distance, or

$$T = F_M \cos \omega t \left(\frac{D}{2} \right) \qquad (13.3\text{--}4)$$

where F_M is multiplied by $\cos \omega t$ to obtain the normal component of the magnetic force and $D/2$ is the distance from the center of rotation to the force position. Substitution of (13.3–3) into (13.3–4) then yields

$$T = \frac{BiDd}{2} \cos \omega t \qquad (13.3\text{--}5)$$

which is valid for each individual conductor. Thus, for all $2N$ conductors, the total torque is given by

$$T_T = NBDdi \cos \omega t \qquad (13.3\text{--}6)$$

Observe that the torque varies sinusoidally in the same manner as the induced voltage.

   We now consider the electrical and mechanical aspects of the ideal rotational transducer as was done previously for the ideal translational transducer. The electrical power is given by

$$P_E = vi = (NBDd\omega \cos \omega t)i \qquad (13.3\text{--}7)$$

Similarly, the mechanical power is given by

$$P_M = T_T \omega = (NBDdi \cos \omega t)\omega \qquad (13.3\text{--}8)$$

where $P_E = P_M$, just as in the ideal translational system. If losses were taken into consideration, P_E and P_M would be reduced and, in general, would not be equal. However, a power balance equation would be valid with the electrical power minus the electrical losses equal to the mechanical power minus the mechanical losses.

   The rotational example under consideration is the basis for electric machines. When electrical energy is applied and converted into mechanical energy via a rotating electromechanical trans-

ducer, this type of machine is called a *motor*. When a mechanical input is converted into an electrical output via a rotational electromechanical transducer, the machine is called a *generator*.

13.3.2   Types of Rotating Machines

   Three primary classes of rotating machines are described briefly in this section and in detail in the sections that follow. These primary classes include DC, synchronous, and induction machines. The latter two classes operate with AC sinusoidal power. Each of these classes of machines can be used in the generator or motor mode of operation. However, some are more practical in one mode as opposed to the other. All rotating machines have basic constituents called the *stator* and the *rotor*. The stator is the stationary outer frame of the machine, and the rotor is the rotating inner portion.

   DC machines have DC currents and voltages associated with both the stator and rotor. The windings on the stator are called the *field windings*, whereas those on the rotor are called the *armature windings*. The armature windings of the rotor must be connected to slip rings and brushes in order to allow electrical contact to these windings. Additionally, the use of a switching device, usually in the form of a commutator, is required. As we will see, the commutator consists of a rotating mechanical switch, which is necessary to reverse the direction of current at particular instants and maintain unidirectional torque in the DC machine.

   Synchronous machines have AC current (and voltage) in the stator, along with DC applied to the rotor that produces a rotor electromagnet. The field windings are on the rotor for these machines, and the DC electrical connections are accomplished using slip rings and brushes. Rotation of the rotor then creates a rotating magnetic field that interacts with the generated or applied sinusoidal current in the stator coils. These machines are synchronous because they operate at constant speed. Usually, for efficient machine use, three separate sets of conductors, referred to as *three-phase conductors,* are used in the stator. When more than one phase is present, the machine is said to be a *polyphase machine.*

Induction machines consist primarily of motors because the induction generator is much less effective than the synchronous generator. The induction motor is operated by applying polyphase AC to the stator windings (which are often very similar to those of a synchronous machine). As we will see, the application of polyphase AC to the stator windings produces a rotating magnetic field that induces another rotating magnetic field in the specially constructed rotor. The two magnetic fields then interact to produce rotation and torque.

The interaction between two rotating magnetic fields is best understood by considering an analogous example of two bar magnets separated from each other by a thin piece of cardboard but mutually attracted by the magnetic force between opposite poles (N and S). If one magnet is then rotated, the other follows. Similarly, the interacting magnetic fields of the rotor and stator exert an attractive force upon each other and cause rotation of the rotor and motor action.

13.3.3   Losses and Efficiency

All rotating machines have power losses associated with them such that the average output power P_O is less than the average input power P_I by the amount of the average power losses P_L. The *efficiency* η is defined as

$$\eta = \frac{P_O}{P_I} \qquad\qquad (13.3\text{–}9)$$

Since $P_O = P_I - P_L$, we have alternative forms for efficiency given by

$$\eta = \frac{P_I - P_L}{P_I} = \frac{P_O}{P_O + P_L} \qquad (13.3\text{–}10)$$

The losses may be determined by testing the machine under load conditions. In general, the various power losses consist of electrical, mechanical, and magnetic losses, with each of these obtainable in watts (W). However, it is customary to use the unit of horsepower (hp) when we deal with mechanical power, where 1 hp = 746 W.

Electrical power loss is due primarily to the resistance of the stator and rotor circuits. These losses are sometimes referred to as *copper losses* or i^2R *losses.* Additionally, in DC machines, electrical power loss is associated with the contact resistance of the commutator brushes.

Mechanical power loss is due to frictional losses created by the rotational motion of the rotor. These losses include brush-friction, bearing-friction, and windage (wind-friction) losses.

Magnetic power loss is sometimes called *core loss* or *iron loss.* This loss is due to the existence of time-varying magnetic fields that create small, circulatory currents called *eddy currents.* In AC machines, this loss is confined primarily to the stator iron; in DC machines, this loss is mainly in the rotor iron.

--- **EX. 13.6**

Losses and Efficiency Example

Consider a 10 hp motor with losses consisting of an electrical loss of 800 W, a mechanical loss of 600 W, and a magnetic loss of 100 W. Calculate the efficiency for full-load operation—that is, when the motor output is 10 hp.

Solution: The efficiency is calculated from (13.3–10) as follows:

$$\eta = \frac{P_O}{P_O + P_L} = \frac{(10)(746)}{(10)(746) + 800 + 600 + 100}$$

$$= 83.3\%$$

DC Machines.   Depending upon whether the machine is a generator or a motor, the direction of power flow is reversed. In the case of a DC generator, electrical power is produced in the rotor windings by applying mechanical power P_M and DC field power P_F to the stator coils. The total input power is the sum of these two, or

$$P_I = P_M + P_F \qquad (13.3\text{–}11)$$

Furthermore, P_M is the product of the applied torque T_A to the rotor in newton meters $(\text{N}\cdot\text{m})$ and the angular velocity ω in rad/s, or

$$P_M = T_A\omega \qquad (13.3\text{–}12)$$

Similarly, the DC field power is the product of field current I_F and voltage V_F, or

$$P_F = I_F V_F \qquad (13.3\text{–}13)$$

The input power is then reduced by the electrical, mechanical, and magnetic losses (P_L), and the output power is given by

$$
\begin{aligned}
P_O &= P_I - P_L \\
&= T_A\omega + I_F V_F - P_L
\end{aligned} \qquad (13.3\text{–}14)
$$

In the case of a DC motor, the input power is the applied DC power to both the field windings $(I_F V_F)$ and the armature windings $(I_A V_A)$. The output power is the mechanical power $(T_A\omega)$, and the power flow equation for this case is given by

$$P_O = I_F V_F + I_A V_A - P_L \qquad (13.3\text{–}15)$$

AC Machines. AC machines often have more than one phase of windings associated with the stator. Since these machines have polyphase windings, all phases must be accounted for in considering the power flow.

For the case of a generator (or alternator in this AC case), the input power is the sum of the mechanical power $(T_A\omega)$ and the field power $(I_F V_F)$, the same as it is for the DC generator case. The output power, however, is due to sinusoidally varying voltage and current, which are typically polyphase. The average power for one phase is given by

$$P_O = |I_{ph}|\,|V_{ph}|\cos\theta \qquad (13.3\text{–}16)$$

where $|I_{ph}|$ and $|V_{ph}|$ are the current and voltage magnitudes per phase and $\cos\theta$ is the power factor. For typical three-phase machines, the total output power is $3|I_{ph}|\,|V_{ph}|\cos\theta$. Another expression for the output power can be written in terms of the input power minus the losses, or

$$P_O = T_A\omega + I_F V_F - P_L \qquad (13.3\text{–}17)$$

For the case of an AC motor, the input power consists of the DC field power $(P_F = I_F V_F)$ and the AC power applied to the polyphase stator coils, where (as in the alternator case) the per phase power is given by $P_{\text{stator}} = |I_{ph}|\,|V_{ph}|\cos\theta$. The output mechanical power $(T\omega)$ is then

$$P_O = I_F V_F + 3|I_{ph}|\,|V_{ph}|\cos\theta - P_L \quad (13.3\text{–}18)$$

where (13.3–18) represents the power developed for a three-phase motor.

In the case of an induction machine, no field power is supplied. However, there is additional electrical loss due to the resistance associated with the rotor windings.

13.4 DC MACHINES

DC motors and generators are used in many industrial applications. As we will see, DC motors exhibit good starting torque and excellent speed control. Specific motor applications range from miniature fractional-horsepower motors used in electronic watches to very large motors (10,000 hp) used in automotive factories. DC generators powered by gasoline motors are sometimes used in remote areas.

13.4.1 Basic Construction and Relations

Figure 13.8 displays the basic geometry of a DC machine. This machine consists of an iron rotor and stator with DC currents in windings associated with each.

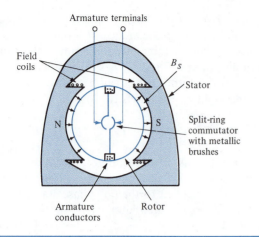

FIGURE 13.8   Basic Geometry of a DC Machine

The stator has relatively fine wire wrapped around its salient (protruding) poles. The field windings provide an essentially constant radial magnetic field (B_S) in the air gaps, as shown in Fig. 13.8. By varying the field current, the magnitude of B_S is varied.

The rotor has current-carrying conductors embedded in slots in its surface, as also shown in Fig. 13.8. These conductors, interconnected into a multi-turn coil, are the armature conductors. Relatively heavy wire is used for these conductors in order to support the relatively large armature current.

Electrical connections to the armature are made through a device called a *split-ring commutator*. A simple double split-ring commutator is shown in Fig. 13.8, and each end of the armature coil is connected to opposite sides of the split ring. During rotation, stationary metallic brushes ride on the split ring, providing electrical connection to the rotor. For the double split ring of Fig. 13.8, each half revolution reverses the connections to the armature coil. Thus, for example, for a DC motor, if a voltage and current are applied to the armature coils, the commutator reverses the current direction each half revolution. Then, since the stator magnetic field also reverses direction

each half revolution, design of the commutator to reverse the current direction simultaneously produces a constant torque on the shaft of the motor and corresponding emf associated with the armature conductors.

The emf and torque associated with the rotor are now obtained from relations derived previously. We consider each armature conductor to have length d, current I_A, and velocity $u = (D/2)\omega$, where D is the rotor diameter. The induced voltage v for each separate conductor is then, from (13.2–4a), given by

$$v = Bud = \frac{BDd\omega}{2} \qquad \text{(13.4–1)}$$

Additionally, the force on each conductor is, from (13.2–6), $F = Bid$, and the corresponding torque on each conductor is then

$$T = Bid\left(\frac{D}{2}\right) = \frac{BDdI_A}{2} \qquad \text{(13.4–2)}$$

Thus, for all $2N$ armature conductors, the total generated emf and torque, respectively, are obtained as follows:

$$V_G = 2Nv = NBDd\omega = KB\omega \qquad \text{(13.4–3)}$$

and

$$T_T = 2NT = NBDdI_A = KBI_A \qquad \text{(13.4–4)}$$

where the latter portion of (13.4–3) and (13.4–4) denote the functional dependence upon B, ω, and I_A, with K representing a constant.

It is of interest to show that energy is conserved in the ideal lossless system. For small field current, the electrical power is given essentially by the armature power, or

$$P_E = V_G I_A = (KB\omega)I_A \qquad \text{(13.4–5)}$$

where (13.4–3) was used to substitute for V_G. Furthermore, the mechanical energy is given by

$$P_M = T_T\omega = KBI_A\omega \qquad (13.4\text{–}6)$$

where (13.4–4) was used to substitute for T_T. Since losses have been neglected, we observe that $P_E = P_M$, just as in the previous lossless electromechanical transducers. Furthermore, the DC machine can be used to convert electrical energy into mechanical energy or vice versa. The basic operation of nonideal DC generators and motors is presented in the next two sections.

13.4.2   DC Generators

For the DC generator, the primary energy supplied is mechanical in that a constant torque applied to the shaft causes rotation of the rotor. A small amount of energy is also used to supply the field current to the field windings. The combination of these energies provides DC power generation in the armature.

The two important types of DC generators are the *separately excited generator* and the *self-excited generator*. These generators are named after the manner in which the field windings are electrically connected. Figure 13.9a shows the equivalent circuit for the DC machine with positive directions of current and polarity of voltage defined for the generator case. Figure 13.9b displays the specific electrical connections for the separately excited case, and Fig. 13.9c shows the self-excited case. The direction of I_A in each case indicates generator action. The additional variable resistor R_S is used to limit the field current.

Separately Excited Generator. For the separately excited generator of Fig. 13.9b, the output load voltage is given by

$$V_L = KB\omega - I_A R_A \qquad (13.4\text{–}7)$$

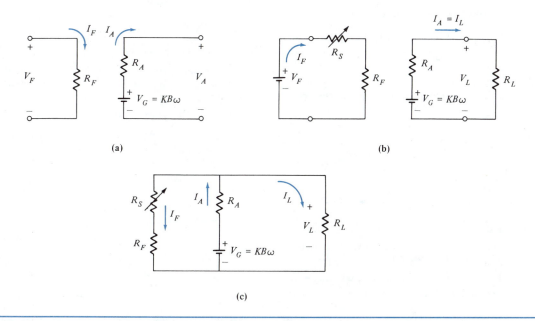

(a)

(b)

(c)

FIGURE 13.9   DC Generator Equivalent Circuits: (a) Basic circuits for the field and armature windings, (b) Separately excited connection, (c) Self-excited connection, or shunt generator

where $B = B(I_F)$. If a linear relation between B and I_F is assumed, an alternate expression for V_L can be written as

$$V_L = K'I_F\omega - I_A R_A \qquad (13.4\text{--}8)$$

Equations (13.4–7) and (13.4–8) are valid regardless of the value of R_L. An important special case for the generator is that of *no-load* conditions, occurring when R_L is not connected (or $R_L = \infty$). Under these conditions, the output voltage is maximum since $I_A = 0$ and is equal to the generated voltage under no-load conditions; $V_{NL} = V_G = K'I_F\omega$.

EX. 13.7

Separately Excited DC Generator Example

A separately excited DC generator has total losses given by $P_L = 500$ W. If the generator is operating with a load voltage $V_L = 230$ V and armature current $I_A = 100$ A, calculate the efficiency for $I_F = 1$ A. Let $R_A = 0.1\ \Omega$ and $R_F + R_S = 60\ \Omega$.

Solution: The output power is calculated as

$$P_O = V_L I_A = (230)(100) = 23\text{ kW}$$

whereas the input power is

$$P_I = P_O + P_L = 23{,}000 + 500 + (1)(60) + (100)^2(0.1)$$
$$= 24.56\text{ kW}$$

Thus, the efficiency is

$$\eta = \frac{23}{24.56} \cong 94\%$$

Self-Excited Generator. The self-excited generator, also called the *shunt generator*, as shown in Fig. 13.9c, has field current I_F supplied from the generated emf V_G. From the circuit, writing KVL for the armature branch and the field branch yields

$$V_L = I_F(R_F + R_S) = V_G - I_A R_A \qquad (13.4\text{--}9)$$

where (13.4–9) holds under all load conditions. In

particular, under no-load conditions, $I_L = 0$. Then, $I_A = I_F$; however, since this current is small, we have

$$V_{NL} = I_F(R_F + R_S) \cong V_G \qquad (13.4\text{--}10)$$

since R_A is quite small ($R_A \ll R_F + R_S$).

The output current–voltage characteristics for DC generators are shown in Fig. 13.10. For both types of DC generators,

$$V_L = V_G - I_A R_A \qquad (13.4\text{--}11)$$

However, V_G varies differently for each type because of different variations of I_A due to the different circuit connections.

For the separately excited case of Fig. 13.9b, V_L and I_L are linearly dependent upon R_L. For the self-excited case of Fig. 13.9c, as I_L is increased, I_A increases, which causes $I_A R_A$ to increase and V_L to reduce. Thus, I_F is reduced and therefore also B and V_G, with additional reduction in V_L. This variation is shown in Fig. 13.10.

EX. 13.8

DC Shunt Generator Example

A DC shunt generator has full-load values of 50 kW and 230 V with armature resistance of $R_A = 0.1\ \Omega$ and field resistance of $R_S + R_F = 60\ \Omega$. Calculate the induced voltage at full load if the generator is operating at rated voltage.

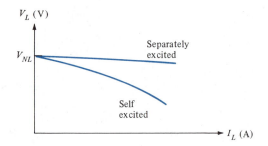

FIGURE 13.10 Generated Voltage Versus Load Current Characteristics

Solution: Since this device is a shunt generator, Fig. 13.9c applies, and

$$I_F = \frac{230}{60} = 3.83 \text{ A}$$

Additionally,

$$I_L = \frac{50 \times 10^3}{230} = 217.4 \text{ A}$$

where

$$I_A = I_L + I_F = 217.4 + 3.83 \cong 221 \text{ A}$$

Thus, writing KVL for the armature and load branches of Fig. 13.9c yields

$$V_L = V_G - I_A R_A$$

or

$$V_G = V_L + I_A R_A = 230 + (221)(0.1) = 252.1 \text{ V}$$

———————————————————————————————— EX. 13.9

DC Shunt Generator Efficiency Example

For the generator of Ex. 13.8, consider that the mechanical and magnetic losses are 2 kW. Calculate the electrical losses due to armature and field windings and determine the generator efficiency at full load.

Solution: Calculation of the electrical losses is carried out using the data from Ex. 13.8, where

$$I_A^2 R_A = (221.23)^2(0.1) = 4.89 \text{ kW}$$

and

$$I_F^2 R_F = (3.83)^2(60) = 0.88 \text{ kW}$$

Thus, the total losses are

$$P_L = 2 + 4.89 + 0.88 = 7.77 \text{ kW}$$

and, since $P_O = 50$ kw, the input power is

$$P_I = P_O + P_L = 57.77 \text{ kW}$$

and the efficiency is

$$\eta = \frac{P_O}{P_I} = \frac{50}{57.77} \cong 87\%$$

13.4.3   DC Motors

For the DC motor, DC voltages are applied to both the stator and rotor, thus producing field current and corresponding stator flux B_S as well as armature current I_A. Commutator interaction of B_S and I_A then produces torque and mechanical energy output.

The three important types of DC motors are the *shunt motor*, the *series motor*, and the *compound motor*. These motor types are also named, as in the case of DC generators, according to the manner in which the field and armature are electrically connected. Equivalent circuits for the various types of DC motors are displayed in Figs. 13.11a, b, and c, where I_A is shown in the direction

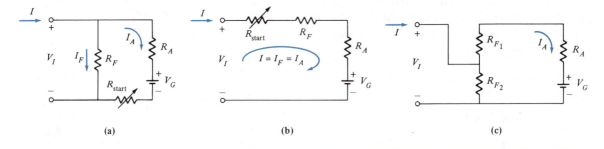

FIGURE 13.11   DC Motor Equivalent Circuits: (a) Shunt motor, (b) Series motor, (c) Compound motor

for motor operation. We observe that the compound motor of Fig. 13.11c has a portion of the field windings in series with the armature and a portion in shunt.

Shunt Motor. For the shunt motor of Fig. 13.11a, note that an additional variable resistor R_{start} has been inserted into the armature branch. This resistor is necessary in order to limit the armature current on starting. Because the generated voltage $V_G = 0$ as the motor is started ($\omega = 0$), without R_{start}, I_A would be enormous since R_A is small. After starting occurs, the resistor R_{start} is reduced to zero (usually automatically by a centrifugal switch).

The important equations for shunt motor operation are obtained by writing KVL for each branch of Fig. 13.11a to yield

$$V_I = V_G + I_A(R_A + R_{\text{start}}) \qquad (13.4\text{--}12)$$

and

$$V_I = I_F R_F \qquad (13.4\text{--}13)$$

where the generated emf is given by

$$V_G = KB\omega \qquad (13.4\text{--}14)$$

Additionally, the relationship for generated torque is

$$T = KBI_A \qquad (13.4\text{--}15)$$

An important dependence for a motor is the relationship between speed (ω) and torque (T). We can determine this relation for the shunt motor by substituting (13.4–14) and (13.4–15) into (13.4–12):

$$V_I = KB\omega + \left(\frac{T}{KB}\right) R_A \qquad (13.4\text{--}16)$$

Solving for ω then yields

$$\omega = \frac{V_I}{KB} - \left[\frac{T}{(KB)^2}\right] R_A \qquad (13.4\text{--}17)$$

Note that ω is often expressed in revolutions per minute (rpm) and denoted by n, where $n = 60\omega/2\pi$.

Thus, we observe a linear dependence of speed versus torque for a shunt motor. Additionally, the last term in (13.4–17) is sometimes small, and $\omega \cong V_I/KB$. Then, the DC shunt motor is essentially a constant-speed motor independent of the load torque T.

Figure 13.12 displays speed versus torque for a DC shunt motor using input voltage as the parameter. Note that increasing the input voltage increases the speed for the same range of torque. Thus, the input voltage can be used to change the motor speed.

Another important point from (13.4–17) is that B appears in the denominator of both terms. Thus, if $B \to 0$, $\omega \to \infty$. Thus, the field branch must never be opened during operation of a shunt motor; otherwise, an unstable, runaway, situation will occur.

Series Motor. For the series motor, the field and armature coils are connected in series as shown in Fig. 13.11b. Thus, $I_A = I_F$, the stator magnetic field is dependent upon I_A, and a much different speed–torque dependence is produced.

By writing KVL for the series motor, we have

$$V_I = V_G + I_A R_T \qquad (13.4\text{--}18)$$

where $R_T = R_{\text{start}} + R_F + R_A$. From (13.4–14), we note that $V_G = KB\omega$, where $B = B(I_A)$. If we as-

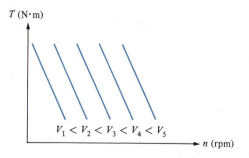

FIGURE 13.12   Speed Versus Torque for DC Shunt Motor

sume a linear relation between B and I_A ($B = kI_A$), then the generated voltage varies linearly with I_A, or

$$V_G = K(kI_A)\omega = CI_A\omega \qquad (13.4\text{--}19)$$

where $C = Kk$. Substituting into (13.4–18) and solving for I_A then yield

$$I_A = \frac{V_I}{C\omega + R_T} \qquad (13.4\text{--}20)$$

Additionally, from (13.4–15), $T = KBI_A$, and again assuming that B varies linearly with I_A, we have

$$T = C(I_A^2) \qquad (13.4\text{--}21)$$

Finally, substituting (13.3–20) into (13.3–21) yields

$$T = C\left(\frac{V_I}{C\omega + R_T}\right)^2 \qquad (13.4\text{--}22)$$

which indicates that the torque for a series motor varies inversely as the square of the speed.

Speed–Torque Curves. Normalized curves for speed versus torque using the machine-rated values (subscript R) are displayed in Fig. 13.13 for shunt, series, and compound motors. Note that since the field windings of the compound motor are partially in series and shunt with the armature, the speed–torque characteristic behavior is intermediate between the series and shunt cases.

From the speed–torque curve of the series motor, we observe that as the torque is reduced, the speed increases and vice versa. In practical motors, this speed without torque can be excessive. Therefore, a series DC motor is never operated without load torque (shaft-restoring torque). However, at low speed and, in particular, at starting ($\omega = 0$), the DC series motor has an advantage in that it possesses large starting torque.

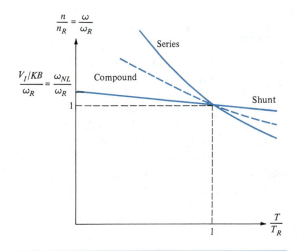

FIGURE 13.13   Normalized Speed Versus Torque for Typical DC Motors

─── EX. 13.10

DC Shunt Motor Example

Consider a 10 hp, 230 V shunt motor with armature resistance $R_A = 0.15\ \Omega$. If the armature current $I_A = 10$ A at no load with speed at 1800 rpm, determine the speed and torque for full-load conditions with $I_A = 40$ A. Consider $R_{\text{start}} = 0$ at this speed.

Solution: At no load, from the armature branch of Fig. 13.11a, the generated voltage is obtained as

$$V_G = V_I - I_A R_A = 230 - 10(0.15) = 228.5\ \text{V}$$

Similarly, under full-load conditions,

$$V_G = 230 - 40(0.15) = 224\ \text{V}$$

Thus, since I_F is unchanged, the field flux is unchanged, and the new speed is obtained through the proportionality

$$n = \frac{224}{228.5}\,(1800) = 1765\ \text{rpm}$$

At this speed, the mechanical torque is given by

$$T = \frac{I_A V_G}{\omega} = \frac{(40)(224)}{2\pi(n)/60} = \frac{(40)(224)}{2\pi(1765)/60} = 48.5\ \text{N·m}$$

13.5   SYNCHRONOUS MACHINES

Synchronous machines are opposite to DC machines in the sense that the field windings are on the rotor and the armature coils are on the stator. Furthermore, AC current and voltage in the armature or stator coils are either generated in the case of a generator or applied in the case of a motor. In the case of a generator, mechanical energy in the form of constant torque and rotation is applied to the rotor, and as we will see, induces a sinusoidal emf in the stator coils. For motor operation, however, sinusoidal voltages applied to the stator coils create a rotating magnetic field that produces torque and output mechanical energy.

13.5.1   Synchronous Generator Operation

Figure 13.14 shows in cross section a simplified synchronous machine that we will consider to be operating as a generator or alternator. The term *alternator* is used to represent an elec-

tric generator that produces alternating current. The rotor is an electromagnet with field windings that are excited by applying DC current through brushes and slip rings and thus produce north and south magnetic poles. Placing the field coils on the rotor is quite practical because much less power is associated with the field than with the armature. Additionally, in some instances, a permanent bar magnet is used to replace this electromagnetic rotor and thus eliminates brushes and slip rings entirely.

In the operation of this machine as an alternator, constant torque is applied to the rotor to create uniform rotation at constant angular velocity or speed of rotation. As we will see, motor operation also involves operation at constant rotational speed. This speed is referred to as the *synchronous speed* and is defined as n_s in rpm.

Sinusoidal Induced Voltage.   As shown in Fig. 13.14, the stator has N conductors embedded in slots on opposite sides of the rotor. These conductors are connected on end to form an N-turn rectangular coil. As the rotor rotates, its magnetic poles alternately pass the current-carrying conductors of the stator and thus induce an emf in the N-turn coil.

To produce sinusoidal AC in the stator windings of this basic synchronous alternator, the magnetic field perpendicular to the stationary conductors must vary sinusoidally, which is accomplished in the example of Fig. 13.14 by shaping the rotor appropriately with maximum flux density at the middle of each pole and reduced flux density to either side. Under these conditions, the induced emf for the synchronous machine is sinusoidal. By analogy with (13.3−2), the generated voltage for the basic alternator is

$$v_G = NB_R Dd\omega \cos \omega t \qquad (13.5-1)$$

where B_R is the peak magnetic flux density due to the rotor magnetic field.

Thus, we have used the same generated voltage expression as that for the rotating electrome-

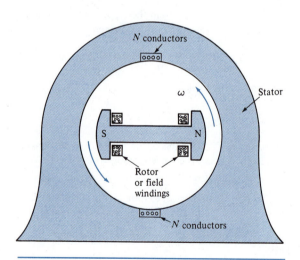

FIGURE 13.14   Synchronous Machine with Two Rotor Poles

chanical transducer. For that case, the magnetic flux density was fixed in space while the coil was rotating. In the present case, the coil is stationary while the magnetic flux density rotates. Since the relative motion for each case is identical, the same expression for induced voltage is appropriate.

Synchronous Speed.   Note that the electrical frequency of the induced emf is equal to that of the mechanical rotational frequency. In the case of a two-pole rotor, the frequency of the emf in terms of radian frequency ω_s is related to synchronous speed n_s in rpm as follows:

$$\omega = 2\pi \left(\frac{n_s}{60} \right) \tag{13.5-2}$$

Additionally, if we express frequency in Hz, we have

$$f = \frac{n_s}{60} \tag{13.5-3}$$

Rotors with multiple-pole pairs are often used because they have the advantage of operating at a lower mechanical rotating speed (synchronous speed) for the same electrical (AC) frequency. Examples of synchronous machines with four poles

(two-pole pairs) and six poles (three-pole pairs) are shown in Figs. 13.15a and b. With the number of pole pairs doubled, the rotational speed is halved. With the number of pole pairs increased by a factor of 3, speed is reduced by a factor of 3. To determine the frequency in the general case, recall that for a single pair of rotor poles, one entire revolution corresponds to a complete cycle of generated emf. Hence, the frequency is given by (13.5-2) or (13.5-3). Then, for more than one pole pair, this frequency is multiplied by $p/2$, where $p/2$ is the number of pole pairs. For a rotor with $p/2$ pole pairs, we then have

$$\omega = 2\pi \left(\frac{n_s}{60} \right) \left(\frac{p}{2} \right) = \frac{\pi(n_s p)}{60} \tag{13.5-4}$$

and

$$f = \left(\frac{n_s}{60} \right) \left(\frac{p}{2} \right) \tag{13.5-5}$$

Note that for $f = 60$ Hz, the highest possible synchronous speed occurs for a two-pole machine, which is 3600 rpm.

Three-Phase Generator.   Instead of the single-coil alternator arrangement of Fig. 13.14, a much

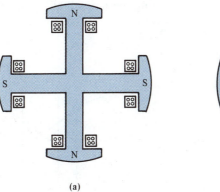
(a)

(b)

FIGURE 13.15   Rotors with Multiple-Pole Pairs: (a) Four poles, (b) Six poles

more efficient system uses three separate sets of conductors placed in slots and arranged symmetrically around the stator such that each coil takes up one third of the stator. Figure 13.16 shows such an arrangement, with the sets labeled $A–A'$, $B–B'$, and $C–C'$. If the windings are identical, three balanced sinusoidal voltages are produced that are equal in magnitude but that differ in phase by $120°$.

Expressions for the induced voltage of each coil are then

$$v_{G_A} = NB_R D\omega \cos \omega t$$
$$= \sqrt{2}V_g \cos \omega t \tag{13.5–6}$$

$$v_{G_B} = NB_R D\omega \cos(\omega t + 120°)$$
$$= \sqrt{2}V_g \cos(\omega t + 120°) \tag{13.5–7}$$

$$v_{G_C} = NB_R D\omega \cos(\omega t - 120°)$$
$$= \sqrt{2}V_g \cos(\omega t - 120°) \tag{13.5–8}$$

where the basic variation given by (13.5–1) has been used and where we have defined $\sqrt{2}V_g = NB_R D\omega$, with V_g as the rms value of the generated voltage. A machine with three such windings is referred to as a *balanced three-phase* synchronous machine.

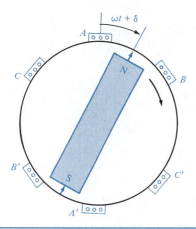

FIGURE 13.16   Three-Phase Synchronous Machine

13.5.2   Rotating Stator Magnetic Field

When AC voltages are induced (or applied) in the three-phase AC windings of the stator, an associated magnetic field is produced. This stator magnetic field has fixed amplitude and rotates at the synchronous speed, as we will see. This effect is not confined to synchronous machines but occurs in all machines with polyphase stator windings.

To demonstrate the rotating stator magnetic field phenomenon, we consider the balanced three-phase arrangement of Fig. 13.16. Since the voltages vary according to (13.5–6) through (13.5–8), the associated currents also have identical phase relationships. These balanced three-phase currents are sketched in Fig. 13.17. Each sinusoidal current produces a corresponding magnetic flux density that varies proportionately. Hence, since the currents vary sinusoidally, the magnetic flux densities also vary sinusoidally. Figure 13.18 indicates the direction of magnetic flux density for each coil when each current is positive. Note that each direction is perpendicular to the plane of the coil. Thus, each direction of magnetic flux density is $120°$ out of phase with the other flux density directions. Additionally, when the current is negative, the flux density reverses direction.

The resultant magnetic flux density associated with the three stator windings is now obtained by adding the individual vector components at spe-

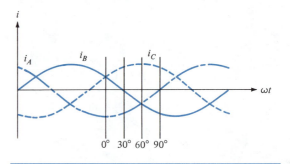

FIGURE 13.17   Balanced   Three-Phase   Instantaneous Currents

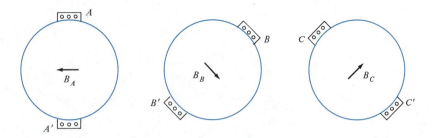

FIGURE 13.18 Direction of Magnetic Flux Density for Separate Coils

cific instants of time. The components and resultant magnetic flux densities are shown in Figs. 13.19a through d for instants of time $\omega t = 0°$, $30°$, $60°$, and $90°$, respectively, as indicated in Fig. 13.17. Note that at $\omega t = 0°$, current i_A has negative maximum value, while currents i_B and i_C are positive with half the maximum value. Adding the corresponding magnitudes of the magnetic flux densities in their proper directions (shown in Fig. 13.19a) yields the resultant flux density. The corresponding diagrams for later instants in time, $\omega t = 30°$, $60°$, and $90°$, are also shown in Fig. 13.19. Note that the magnitude of the resultant magnetic field remains constant at each instant and that uniform counterclockwise rotation is occurring. Other instants of time can be selected to further verify this fact. Hence, we observe that the stator magnetic field produced by balanced three-phase windings rotates counterclockwise at a uni-

form speed relative to the stator. This constant speed of rotation is the synchronous speed n_s.

13.5.3 Three-Phase Systems

In general, three-phase systems are used not only in the generation of AC but also in the transmission and distribution of electrical power. Such systems are more efficient than single-phase systems, particularly when a special circuit topology is used.

Figure 13.20a displays a symbolic representation for the three branches of a three-phase system. The two special circuit topologies that can be used with balanced three-phase systems are shown in Figs. 13.20b and c. For obvious reasons, these circuits are referred to as the *wye* (Y) and *delta* (Δ) connections, respectively. Observe that

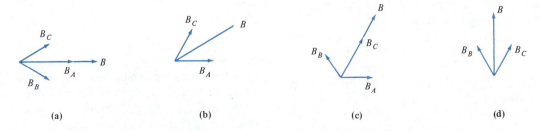

FIGURE 13.19 Vector Components and Resultant Flux Densities Indicating Magnitude and Angle: (a) $\omega t = 0°$, (b) $\omega t = 30°$, (c) $\omega t = 60°$, (d) $\omega t = 90°$

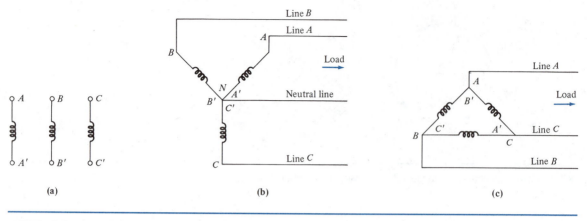

FIGURE 13.20   Three-Phase Systems: (a) Three-phase equivalent representation, (b) Wye (Y) connection, (c) Delta (Δ) connection

the delta configuration requires three lines to connect all three-phase voltages to a load, while the wye connection requires four. Both topologies therefore amount to a considerable savings in wire from the six lines that would be necessary if separate phase connections were used. Furthermore, the delta configuration has a slight wire savings advantage but is seldomly used because for a small imbalance in any of the three phases, large circular currents can develop that may be damaging to the windings. Note that if the three voltages are exactly balanced (with equal magnitudes and 120° phase difference), the sum of the voltages around the loop in the delta configuration is zero. However, if an imbalance in any of the voltages occurs, a net voltage around the loop will exist and a very large current can be created. This situation is similar to connecting two voltage sources in parallel (without a resistor in the loop). If the voltage sources are unequal, an infinite current (limited only by the wire resistance) will pass.

When the balanced three-phase voltages are applied to a load, the load is also connected in the wye configuration. Figure 13.21 displays the overall circuit for a Y-connected, balanced three-phase load with identical impedances in each branch. Note that the line currents ($\mathbf{I}_L$) and phase currents ($\mathbf{I}_{ph}$) for this configuration are identical

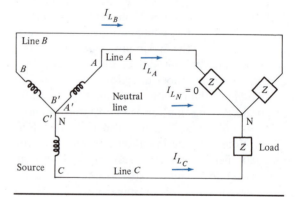

FIGURE 13.21   Y-Connected Source and Load

so that

$$\mathbf{I}_{ph} = \mathbf{I}_L \qquad (13.5\text{--}9)$$

However, the line-to-line voltages are equal to the phasor difference in two of the phase voltages. Defining the magnitude of the voltage of one phase as $|\mathbf{V}_{ph}|$ and the magnitude of the voltage between two lines as $|\mathbf{V}_L|$, we can show that

$$|\mathbf{V}_L| = \sqrt{3}|\mathbf{V}_{ph}| \qquad (13.5\text{--}10)$$

To obtain (13.5–10), consider the voltages of two of the balanced phases to be given by

$$v_A = |\mathbf{V}_{ph}| \cos \omega t \tag{13.5–11}$$

and

$$v_B = |\mathbf{V}_{ph}| \cos(\omega t + 120°) \tag{13.5–12}$$

or, using phasor notation (from Chapter 1),

$$\mathbf{V}_a = \mathbf{V}_{ph} \, \measuredangle \, 0° \tag{13.5–13}$$

and

$$\mathbf{V}_b = \mathbf{V}_{ph} \, \measuredangle \, 120° \tag{13.5–14}$$

The line-to-line phasor voltage $\mathbf{V}_{l_{ab}}$ is given by the difference between the two phase voltages, or

$$\mathbf{V}_{l_{ab}} = \mathbf{V}_a - \mathbf{V}_b \tag{13.5–15}$$

where Fig. 13.22 shows the component phasor voltages and their resultant. By using complex algebra, we can write $\mathbf{V}_a$ and $-\mathbf{V}_b$ as

$$\mathbf{V}_a = \mathbf{V}_{ph} \tag{13.5–16}$$

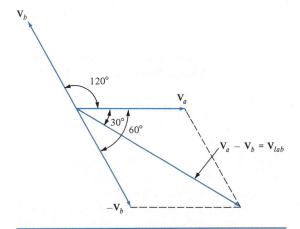

$\mathbf{V}_b$

120°

30° 60°

$\mathbf{V}_a$

$\mathbf{V}_a - \mathbf{V}_b = \mathbf{V}_{lab}$

$-\mathbf{V}_b$

FIGURE 13.22   Graphical Phasor Addition of Two Phases of Voltage

and

$$-\mathbf{V}_b = \mathbf{V}_{ph} \cos 60° - j\mathbf{V}_{ph} \sin 60°$$
$$= \mathbf{V}_{ph}\left(\frac{1}{2} - \frac{j\sqrt{3}}{2}\right) \tag{13.5–17}$$

Thus, by forming the difference $\mathbf{V}_a - \mathbf{V}_b$, we have

$$\mathbf{V}_{l_{ab}} = \mathbf{V}_a - \mathbf{V}_b = \mathbf{V}_{ph}\left(\frac{3}{2} - \frac{j\sqrt{3}}{2}\right)$$
$$= \sqrt{3}\mathbf{V}_{ph} \, \measuredangle \, -30° \tag{13.5–18}$$

and hence (13.5–10) is verified. Additionally, we observe that the line-to-line voltage lags the phase voltage by 30°. An alternate form of (13.5–18) is given by

$$\mathbf{V}_{ph} = \frac{\mathbf{V}_{l_{ab}}}{\sqrt{3}} \, \measuredangle \, +30° \tag{13.5–19}$$

If we now consider the power in each phase, the instantaneous power delivered to each individual load will differ; however, the average power delivered to each load is identical. With the average power delivered to each load denoted as P_{ph}, the total average power delivered P_T is given by

$$P_T = 3P_{ph} \tag{13.5–20}$$

The total average power may be written in terms of phase currents and phase voltage by recalling from Chapter 1 that

$$P_{ph} = |\mathbf{I}_{ph}| \, |\mathbf{V}_{ph}| \cos \theta \tag{13.5–21}$$

where $|\mathbf{I}_{ph}|$ and $|\mathbf{V}_{ph}|$ are the magnitudes of the phase current and voltage, respectively, and θ is the phase angle of the impedance (recall also that $\cos \theta$ is the power factor).

In some instances, it is advantageous to express the power per phase in terms of line current and voltage. Substituting the relations between line and phase quantities for the wye connection from (13.5–9) and (13.5–10), we obtain the average power per phase as

$$P_{ph} = \frac{|\mathbf{I}_L||\mathbf{V}_L|}{\sqrt{3}} \cos \theta \qquad (13.5–22)$$

The total average power is therefore

$$P_T = 3P_{ph} = \sqrt{3}|\mathbf{I}_L||\mathbf{V}_L| \cos \theta \qquad (13.5–23)$$

It is also of interest to note that another advantage of using a balanced three-phase system is that the total instantaneous power delivered to the three loads is a constant, independent of time. This fact can be shown by adding the instantaneous power in each phase, or

$$P_T = v_A i_A + v_B i_B + v_C i_C \qquad (13.5–24)$$

By substitution for the voltages and currents of the Y-connected load and by using trigonometric relations, it can be shown that the instantaneous power is given by

$$P_T = 3|\mathbf{I}_{ph}||\mathbf{V}_{ph}| \cos \theta$$
$$= \sqrt{3}|\mathbf{I}_L||\mathbf{V}_L| \cos \theta \qquad (13.5–25)$$

which is a constant. Verification of (13.5–25) is left to an end-of-chapter problem.

——————————————————————————— EX. 13.11

Three-Phase Wye Connection Example

A balanced set of load impedances are connected in the wye configuration. The value of each load impedance is $Z_L = 15 \angle 60° \, \Omega$, and the line-to-line voltage is 200 V. Calculate the phase voltages, the line currents, and the total power delivered to the loads.

Solution: The phase voltages and line voltages are related by (13.5–19). Therefore, for one phase,

$$\mathbf{V}_{ph} = \frac{\mathbf{V}_{lab}}{\sqrt{3}} \angle +30° = \frac{220}{\sqrt{3}} \angle 30° = 127 \angle 30° \text{V}$$

The other phases are displaced by 120° and −120°. The phase current is given by

$$\mathbf{I}_{ph} = \frac{\mathbf{V}_{ph}}{Z_L} = \frac{127 \angle 30°}{15 \angle 60°} = 8.5 \angle -30° \text{ A}$$

The other two phases are also displaced 120° and −120°. Thus, the power per phase is given by (13.5–21), or

$$P_{ph} = |\mathbf{I}_{ph}||\mathbf{V}_{ph}| \cos \theta = 8.5(127) \cos 60° = 539.75 \text{ W}$$

and the total power is

$$P_T = 3P_{ph} = 1619.25 \text{ W}$$

—————————————————————

13.5.4 Synchronous Generator Equivalent Circuit and Phasor Representation

We will consider synchronous generators with balanced three-phase stator conductors. The analysis of only one phase is necessary since the other phases behave identically except with different phase angles. The magnitude of the generated phase voltage in one phase is obtained from (13.5–1) as

$$|\mathbf{V}_g| = NB_R Dd\omega \qquad (13.5–26)$$

A more useful form of this equation is obtained by substituting for ω from (13.5–4) to yield

$$|\mathbf{V}_g| = NB_R Dd\left[\frac{\pi(n_s p)}{60}\right] \qquad (13.5–27)$$

Equation (13.5–27) shows the dependence of the generated voltage on the indicated parameters. For a given machine, if we let N, D, d, n_s, and p be constants, the generated voltage magnitude can be written as a function of the peak rotor flux density B_R, or

$$|\mathbf{V}_g| = KB_R \qquad (13.5–28)$$

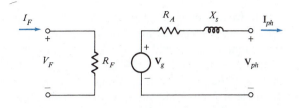

FIGURE 13.23   Synchronous Generator Equivalent Circuit

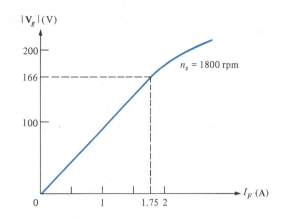

FIGURE 13.24   Magnetization Curve of $|V_g|$ Versus I_F

Furthermore, since B_R depends upon the field current I_F, assuming that a linear dependence exists yields

$$|\mathbf{V}_g| = K'I_F \qquad (13.5\text{–}29)$$

Thus, we conclude that the generated voltage magnitude can be varied by varying the field current.

Figure 13.23 displays the electrical circuit model for a synchronous generator. Note that this model is very similar to the separately excited DC generator, except that the stator coil reactance X_s must be included because AC is present. In fact, the armature resistance R_A is usually negligible in comparison to X_s, the *synchronous reactance*.

The magnitude of the generated voltage in Fig. 13.23 is given by (13.5–26) or (13.5–27). However, a more accurate description of this voltage is obtained from measurements and displayed graphically as $|V_g|$ versus I_F. Such a curve is called a *magnetization curve*, an example of which is shown in Fig. 13.24 for $n_s = 1800$ rpm. Since $|V_g|$ is directly dependent on ω (and therefore on n_s), by changing the frequency ω, the speed can be changed as well as the magnetization curve. Additional magnetization curves can be constructed for various speeds by using the given curve and changing the value of $|V_g|$ at a given I_F in proportion to the ratio of the speeds.

To analyze the circuit of Fig. 13.23, we neglect R_A and by writing KVL obtain

$$\mathbf{V}_{ph} = \mathbf{V}_g - \mathbf{I}_{ph}(jX_s) \qquad (13.5\text{–}30)$$

Rearranging yields

$$\mathbf{V}_g = \mathbf{V}_{ph} + \mathbf{I}_{ph}(jX_s) \qquad (13.5\text{–}31)$$

The phasor diagram for this equation is shown in Fig. 13.25 for the usual inductive load case with $\mathbf{I}_{ph}$ lagging $\mathbf{V}_{ph}$ by angle θ. We select $\mathbf{V}_{ph}$ to have zero phase angle for convenience. Note that the product of $\mathbf{I}_{ph}$ and jX_s is another phasor that is perpendicular to $\mathbf{I}_{ph}$.

Synchronous generators are usually rated in terms of the maximum kilovolt-ampere (KVA) load at a specific voltage, frequency, and power factor.

─────────────────────────────────────── **EX. 13.12**

Three-Phase Generator Example

Consider a three-phase generator with a four-pole rotor that has rated values of 5 kVA, 220 V at 60 Hz, and a synchronous reactance X_s of 5 Ω. For full-load operation with a power factor of 0.9 (current lagging voltage), calculate the required magnitude of generated voltage. Also, from the no-load magnetization curve of Fig. 13.24, determine the corresponding field current.

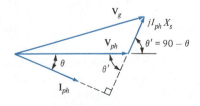

FIGURE 13.25   Phasor Diagram for Generator Operation

Solution: The magnitude of the phase voltage for the wye connection is given by

$$|\mathbf{V}_{ph}| = \frac{220}{\sqrt{3}} \cong 127 \text{ V}$$

Thus, since the power per phase is just 1/3 of the total power, the magnitude of phase current is given by

$$|\mathbf{I}_{ph}| = \frac{\mathbf{P}_T/3}{|\mathbf{V}_{ph}|} = \frac{5000/3}{127} = 13.1 \text{ A}$$

The phasor diagram is the same as that shown in Fig. 13.25, where $\theta \cong 26°$ (from $\cos \theta = 0.9$) and $\theta' = 90 - 26 = 64°$. Thus, the generated voltage is

$$\mathbf{V}_g = \mathbf{V}_{ph} + j\mathbf{I}_{ph}X_s = \mathbf{V}_{ph} + (\mathbf{I}_{ph})X_s \cos \theta'$$
$$+ j(\mathbf{I}_{ph})X_s \sin \theta'$$
$$= 127 + (13.1)(5)(0.438) + j(13.1)(5)(0.899)$$
$$= 155.7 + j59 = [155.7^2 + 59^2]^{1/2} \; \angle \; \tan^{-1}\frac{59}{155.7}$$
$$= 166 \; \angle \; 20.8°$$

Finally, from the magnetization curve of Fig. 13.24, we have $I_F \cong 1.75$ A.

13.5.5   Synchronous Motor Operation and Start-Up

In synchronous motors, DC is again applied to the field windings to provide control of the magnitude of the rotor magnetic field. Furthermore, in large industrial motors, three-phase AC is applied to the stator coils to produce a rotating magnetic field with angular frequency corresponding to the AC frequency (or speed corresponding to the synchronous speed). The rotating stator magnetic field then produces torque on the rotor shaft because of the magnetic attractive forces between the rotating stator magnetic field and the magnetic field of the rotor. The north and south poles of the rotor are attracted to the opposite poles of the rotating stator magnetic field, and uniform rotation is maintained.

If we consider the rotor to be rotating at synchronous speed along with the stator field, the magnetic forces will be totally radial, and no torque can be developed. However, as a load is placed on the shaft, the rotor will fall back through an angle α. A circumferential magnetic field force is produced (because of the attraction between opposite poles of the rotor and stator), and this force produces torque. As the angle α is increased, the torque is increased, while the rotor continues to rotate at synchronous speed. When this angle becomes equal to 90°, the maximum possible torque is developed. This torque is called the *pull-out torque* because if this value is exceeded (because of additional shaft torque), the rotor will slow down and stop.

Upon starting the motor (when the rotor is stationary) with the stator magnetic field rotating at synchronous speed, the attractive magnetic field forces create a sinusoidally varying torque. Thus, the net torque developed is zero, and a synchronous motor therefore does not self-start. To start this motor, special conductors are built into the rotor that allow AC current to be induced in them. The machine then operates an an induction motor (to be described in the next section) until the rotor is brought up to synchronous speed.

13.5.6   Synchronous Motor Equivalent Circuit and Phasor Representation

The equivalent circuit for the synchronous motor is shown in Fig. 13.26a. Note that the current direction is reversed from that corresponding to the synchronous alternator of Fig. 13.23.

By writing KVL for the stator portion of the circuit, we have

$$\mathbf{V}_{ph} = \mathbf{V}_g + j\mathbf{I}_{ph}X_s \qquad (13.5\text{--}32)$$

where (as in the generator case) $R_A \ll X_s$. The corresponding phasor diagram for the synchronous motor is displayed in Fig. 13.26b.

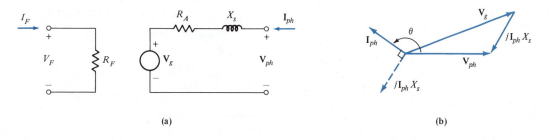

(a)   **(b)**

FIGURE 13.26   Synchronous Motor: (a) Equivalent circuit, (b) Phasor diagram for motor operation

---

─────────────────────────── **EX. 13.13**

Synchronous Motor Example

Consider a synchronous motor with specifications given as follows: 50 hp, 1200 rpm, 220 V, three-phase Y-connected, 60 Hz, and 0.85 leading power factor. Additionally, the stator resistance is 0.1 Ω/phase, the synchronous reactance is 1 Ω/phase, the field resistance is 50 Ω, the stator input is 50 kVA, and the sum of the magnetic and mechanical losses is 2 kW. Calculate the power output in hp and construct a phasor diagram to determine $\mathbf{V}_g$. From the magnitude of $\mathbf{V}_g$, determine the field current from the magnetization curve of Fig. E13.13a. Finally, calculate the efficiency.

Solution: The stator current magnitude per phase is given by

$$|\mathbf{I}_{ph}| = \frac{50,000}{\sqrt{3}\,(200)} \cong 131 \text{ A}$$

which is needed in order to calculate the copper losses of the stator. To determine the output mechanical power, the losses are subtracted from the stator input power, or

$$P_o = 50,000(0.85) - 3(131)^2(0.1) - 2000 = 35,352 \text{ W}$$

Converting this value to hp, where 1 hp = 746 W, yields

$$P_o = 47.4 \text{ hp}$$

By substituting values into (13.5–32), we have the phasor relationship

$$\mathbf{V}_g = \mathbf{V}_{ph} - j\mathbf{I}_{ph}X_s = \frac{220}{\sqrt{3}} - j(131\,\angle\,32°)(1)$$

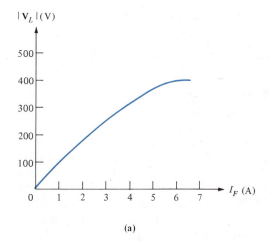

$|\mathbf{V}_L|$ (V)

(a)

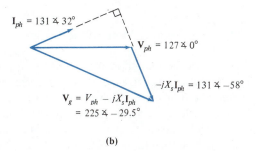

$\mathbf{I}_{ph} = 131\,\angle\,32°$

$\mathbf{V}_{ph} = 127\,\angle\,0°$

$-jX_s\mathbf{I}_{ph} = 131\,\angle\,-58°$

$\mathbf{V}_g = V_{ph} - jX_s\mathbf{I}_{ph}$
$= 225\,\angle\,-29.5°$

(b)

FIGURE E13.13

Writing the result in phasor form yields

$$\mathbf{V}_g \cong 225\,\angle\,-29.5°$$

The corresponding phasor diagram is shown in Fig. E13.13b. Multiplying the phase voltage by $\sqrt{3}$ yields the magnitude of the line voltage as

$$|\mathbf{V}_L| = \sqrt{3}(225) = 390 \text{ V}$$

Thus, from Fig. E13.13a, we observe that $\mathbf{I}_F \cong 5$ A.

Finally, to determine the efficiency, the total power input is obtained by adding the stator power to that of the field, or

$$P_I = 50,000(0.85) + (5)^2(50) \cong 43,750 \text{ W}$$

Thus, the efficiency is

$$\eta = \frac{P_O}{P_I} = \frac{35,352}{43,750} = 81\%$$

13.6   INDUCTION MOTORS

The induction motor is the most frequently used of all motors. Operation is such that AC is applied to the stator windings, which induces currents in the rotor windings. Although capable of generator action, induction machines are used primarily as motors because other generator types are more efficient and less costly.

3.6.1   Induction Motor Operation

In the induction motor, AC is applied to each stator winding exactly as it is in the synchronous motor case, which produces a rotating magnetic field associated with the stator that possesses synchronous speed n_s. The rotating stator magnetic field then induces AC currents and emfs in a specially constructed rotor. The special rotor consists of either of two types.

The first and most often used type of rotor is the *squirrel-cage rotor*. As shown schematically in Fig. 13.27, this rotor is constructed of aluminum or copper conductors placed in slots in the iron rotor and shorted together by circular conductors on the ends. This type of rotor construction does

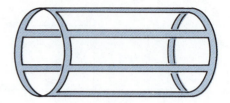

FIGURE 13.27   Squirrel-Cage Rotor

not require brushes or slip rings. The second type of rotor is a *wound rotor* and consists of polyphase windings similar to those of the stator. The ends of these windings are connected to slip rings so that a variable resistance can be connected into the rotor circuit that provides speed and torque variation. The wound rotor motor allows speed and torque adjustment by varying this external resistance. In both rotor cases, the induction mechanism is analogous to transformer action.

The rotating stator magnetic field induces *Bud* voltages in the rotor conductors that produce rotor current. These currents then interact with the rotating magnetic field to produce *Bid* forces and resulting torque. Rotation then begins in the same direction as the stator field. Thus, we observe that the induction motor is self-starting. This mechanism of starting the synchronous motor was referred to in the previous section.

13.6.2   Basic Relations for the Induction Motor

As the rotor begins to turn, its speed increases and begins to approach (but always remains less than) the synchronous speed of the rotating stator magnetic field. The difference in rotor speed n and synchronous speed n_s is defined as the *slip s*. Slip is the percentage difference in these speeds and is given specifically by the defining equation

$$s = \frac{n_s - n}{n_s} \qquad\qquad \textbf{(13.6–1)}$$

The magnitude of the slip s is typically in the range of $2\% \leq s \leq 5\%$ under full-load operation. Rearranging (13.6–1) and solving for n yield

$$n = n_s - sn_s = (1 - s)n_s \qquad (13.6\text{–}2)$$

From (13.6–2), we observe that at standstill ($s = 1$), the rotating stator magnetic field speed is n_s relative to the stationary rotor windings (and relative to the stationary stator windings). Under these conditions, the frequency of the induced rotor currents (f_R) must be the same as the frequency of the stator currents (f). However, if the rotor is turning at synchronous speed ($s = 0$), the stator and rotor magnetic fields are stationary relative to each other, and the frequency of the rotor currents is zero, as well as the rotor current, because the flux is unchanging. For intermediate rotor speeds, the frequency of the rotor currents is given by

$$f_R = sf \qquad (13.6\text{–}3)$$

Thus, the rotor turns at slightly less than synchronous speed. However, the rotor magnetic field itself revolves faster and at exactly synchronous speed, as the next example shows.

————————————————— EX. 13.14

Induction Motor Speed Example

A six-pole induction motor has three-phase AC supplied with $f = 60$ Hz. If the motor is running under load with $s = 0.03$, determine the synchronous speed, the rotor speed, the induced rotor current frequency, and the speed of the rotor magnetic field relative to the stationary stator.

Solution: The synchronous speed is obtained from (13.5–5), and by substitution,

$$n_s = \frac{120f}{p} = \frac{120(60)}{6} = 1200 \text{ rpm} \qquad (1)$$

Thus, the speed of the rotor is obtained from (13.6–2) as

$$n = (1 - s)n_s = (0.97)(1200) = 1164 \text{ rpm} \qquad (2)$$

and the frequency of the rotor current from (13.6–3) is

$$f_R = (0.03)(60) = 1.8 \text{ Hz} \qquad (3)$$

Next, since the stator poles induce the same number of poles in the rotor, the rotating rotor magnetic field has speed relative to the rotor given by

$$n_R = \frac{120f_R}{p} = \frac{(120)(1.8)}{6} = 36 \text{ rpm} \qquad (4)$$

Thus, since the speed of the rotor relative to the stationary stator from (2) is $n = 1164$ rpm, the speed of the rotor magnetic field relative to the stator is

$$n'_R = n_R + n = 1200 \text{ rpm} \qquad (5)$$

Hence, we observe that the two rotating magnetic fields have the same speed of rotation.

————————————————— EX. 13.15

Induction Motor Output Power Example

Consider a six-pole squirrel-cage induction motor with rated values of 10 hp, 230 V, and three-phase stator voltages at 60 Hz. If $s = 0.05$ under rated conditions, calculate the synchronous speed and the rotor speed. Also calculate the torque in N·m corresponding to full-load operation.

Solution: The synchronous speed is obtained from (13.5–5) as

$$n_s = \frac{120f}{p} = \frac{120(60)}{6} = 1200 \text{ rpm}$$

and the rotor speed is obtained from (13.6–2) as

$$n = (1 - s)n_s = (0.95)(1200) = 1140 \text{ rpm}$$

Thus, since the output power is the product of torque in N·m with angular velocity of the rotor in rad/s,

$$T = \frac{P_o}{\omega} = \frac{P_o}{2\pi(n)/60} = \frac{(10)(746)}{2\pi(1140)/60} = 62.5 \text{ N·m}$$

13.6.3   Rotor Equivalent Circuit

To develop an equivalent circuit for the rotor, we first consider the rotor under standstill conditions, where $s = 1$. This mode of operation is achieved by holding the rotor stationary using a brake. The induced rotor voltage in one phase $\mathbf{V}_r$ will then have the same frequency as the stator. Under this condition of $s = 1$, the induced rotor current I_r can be wirtten in terms of the rotor voltage (for one phase) as

$$\mathbf{I}_r = \left(\frac{\mathbf{V}_r}{R_R + j2\pi f L_R} \right)_{s=1} \qquad (13.6\text{–}4)$$

where R_R and L_R are the resistance and inductance of the rotor (also for one phase).

If we now consider a nonzero rotor speed, where $s < 1$, the induced rotor voltage becomes $s\mathbf{V}_r$, and the corresponding rotor frequency becomes sf. Substituting these values into (13.6–4) yields the general expression for the rotor current, or

$$\mathbf{I}_r = \frac{s\mathbf{V}_r}{R_R + j2\pi(sf)L_R} = \frac{s\mathbf{V}_r}{R_R + jsX_R} \qquad (13.6\text{–}5)$$

where $X_R = 2\pi f L_R$ is the rotor reactance for one phase. Dividing the numerator and denominator of (13.6–5) by s yields

$$\mathbf{I}_r = \frac{\mathbf{V}_r}{(R_R/s) + jX_R} \qquad (13.6\text{–}6)$$

Equation (13.6–6) can be rearranged into a more recognizable form by adding and substracting R_R from the denominator to yield

$$\mathbf{I}_r = \frac{\mathbf{V}_r}{R_R + jX_R + (R_R/s) - R_R}$$

$$= \frac{\mathbf{V}_r}{R_R + jX_R + [R_R(1 - s)/s]} \qquad (13.6\text{–}7)$$

An alternative form of (13.6–7) is now obtained by solving for the induced rotor voltage:

$$\mathbf{V}_r = \mathbf{I}_r R_R + jX_R \mathbf{I}_r + \mathbf{I}_r R_R \left(\frac{1 - s}{s} \right) \qquad (13.6\text{–}8)$$

This expression is very useful in describing the operation of induction motors.

13.6.4   Mechanical Power and Torque

The physical significance of (13.6–8) is now described. Note that the first two terms on the right side represent the induced voltage $\mathbf{V}_r$ under standstill conditions, where $s = 1$. Hence, the last term is due to slip when $s < 1$. Furthermore, the last term can be regarded as a voltage generation term $(\mathbf{V}_g)$ that, as we will see, represents electrical to mechanical energy conversion where

$$\mathbf{V}_g = \mathbf{I}_r R_R \left(\frac{1 - s}{s} \right) \qquad (13.6\text{–}9)$$

A modified equivalent circuit with this idea is shown in Fig. 13.28.

To delve further into the assumption that $\mathbf{V}_g$ represents a generated voltage, consider Fig. 13.29a, which displays a phasor diagram corresponding to (13.6–8). If we mutiply (13.6–8) by $\mathbf{I}_r$, a power equation is obtained as follows:

$$\mathbf{I}_r\mathbf{V}_r = \mathbf{I}_r^2 R_R + j\mathbf{I}_r^2 X_R + \mathbf{I}_r\mathbf{V}_g \qquad (13.6\text{–}10)$$

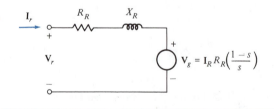

FIGURE 13.28   Rotor Equivalent Circuit for One Phase

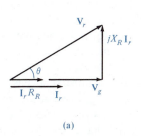

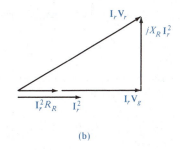

(a)   (b)

FIGURE 13.29   Phasor and Power Triangles: (a) Phasor diagram for (13.6–8), (b) Phasor diagram converted to power triangle

This equation is displayed graphically in Fig. 13.29b. From this figure, we realize that the real power is the sum of the power loss term $I_r^2 R_R$ and the additional term $I_r V_g$. Since it is also real, this latter term is recognized as the electrical power that is converted into mechanical power.

Since mechanical power (P_M) is the product of torque and angular velocity of the rotor (ω) in rad/s, the torque per phase can be written as

$$T = \frac{P_M}{\omega} = \frac{P_M}{2\pi n/60} \tag{13.6–11}$$

where we have substituted the speed of the rotor in rpm. By substituting for $P_M = I_r V_g$ and eliminating V_g using (13.6–9) and n using (13.6–2), we obtain

$$T = \frac{I_r I_r R_R (1-s)/s}{(2\pi/60)(1-s)n_s} = \frac{I_r^2 R_R}{\omega_s s} \tag{13.6–12}$$

where we have defined a synchronous angular velocity in rad/s as $\omega_s = 2\pi n_s/60$. Equation (13.6–12) represents the torque provided by one phase. For three phases, the total developed torque is

$$T_T = \frac{3 I_r^2 R_R}{\omega_s s} \tag{13.6–13}$$

From this expression, we observe that the starting torque (when $s = 1$) of an induction motor is directly dependent upon R_R. Thus, adding an external resistor to the terminals of a wound rotor will increase the starting torque.

To observe the dependence of torque upon s, for $s < 1$, we eliminate I_r using (13.6–6) to obtain

$$T_T = \frac{3 R_R}{\omega_s s} \left[\frac{V_r}{(R_R/s) + jX_R} \right]^2$$

$$= \frac{3|V_r|^2}{\omega_s} \left[\frac{R_R/s}{(R_R/s)^2 + X_R^2} \right] \tag{13.6–14}$$

where only the magnitude has been retained. Multiplying through by s^2 yields the general expression for developed torque as follows:

$$T_T = \frac{3|V_r|^2}{\omega_s} \left(\frac{s R_R}{R_R^2 + s^2 X_R^2} \right) \tag{13.6–15}$$

From (13.6–15), we observe that for small s (nearly synchronous speed), the torque is quite small (for $s = 0$, $T = 0$) and is given approximately by

$$T_T = \frac{3|V_r|^2}{\omega_s} \left(\frac{s}{R_R} \right) \tag{13.6–16}$$

Hence, for small s or nearly synchronous speed, T varies as $1/R_R$.

Another interesting result is obtained if we determine the maximum torque from (13.6–15). From calculus, by taking the derivative of T with *s* and setting this value equal to zero, we obtain

$$(R_R^2 + s^2 X_R) - s(2sX_R^2) = 0 \qquad (13.6\text{--}17)$$

where we have treated $\mathbf{V}_r$ as being independent of *s*. Solving (13.6–17) for *s* yields

$$s = s_{\text{max}} = \frac{R_R}{X_R} \qquad (13.6\text{--}18)$$

where s_{max} is the value of *s* that provides maximum torque. Substitution back into (13.6–15) yields the maximum total torque as

$$T_{\text{max}} = \frac{3|\mathbf{V}_r|^2}{\omega_s} \left[\frac{(R_R/X_R)(R_R)}{R_R^2 + (R_R^2/X_R^2)X_R^2} \right]$$

$$= \frac{3|\mathbf{V}_r|^2}{2\omega_s} \left(\frac{1}{X_R} \right) \qquad (13.6\text{--}19)$$

Finally, note that the maximum torque is independent of the rotor resistance R_R.

The typical speed–torque curves (obtained through measurements) for an induction motor are displayed in Fig. 13.30. Note that the effects of increasing the rotor resistance R_R can essen-

tially be considered to cause a reduction in the speed of the rotor. Observe also that excellent starting torque is available in all cases. Furthermore, the maximum torque is independent of the magnitude of R_R.

Thus, at normal speeds (small *s*), the torque varies as $1/R_R$, while at low speed (upon starting), the torque is proportional to R_R. It is therefore desirable to have large R_R upon starting and small R_R under normal operation. Obviously, this situation is accomplished using a wound rotor motor by varying the external resistance connected through slip rings to the rotor windings. Not as obviously, this situation can also be accomplished using a squirrel-cage rotor by connecting a resistor into the winding circuit through a centrifugal switch. Due to centrifugal force, the switch then shorts out the added resistance as the rotor increases speed.

— EX. 13.16

Induction Motor Example

Consider a six-pole, 208 V (line-to-line), three-phase 60 Hz induction motor with rated torque (at full load) of 120 N·m. Assume that the induced voltage per phase of the rotor is equal to the per phase voltage of the stator and that measurements indicate that $R_R = 0.2\ \Omega$ and $X_R = 0.5\ \Omega$. Calculate the synchronous speed, the slip at rated torque, the rated speed, the rated mechanical output power, the starting torque, and the maximum torque.

Solution: The synchronous speed is given by

$$n_s = \frac{120f}{p} = \frac{120(60)}{6} = 1200 \text{ rpm} \qquad (1)$$

From (13.6–16), we have

$$T_T = \frac{3|\mathbf{V}_r|^2}{\omega_s} \left(\frac{s}{R_R} \right) = \frac{3|\mathbf{V}_r|^2}{2\pi(n_s)/60} \left(\frac{s}{R_R} \right) \qquad (2)$$

By substitution,

$$120 = \frac{3|208/\sqrt{3}|^2}{2\pi(1200)/60} \left(\frac{s}{0.2} \right) \qquad (3)$$

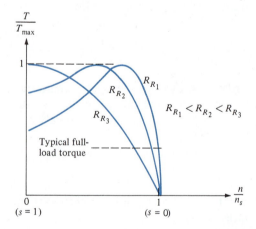

FIGURE 13.30   Normalized Speed Versus Torque for Induction Motors

Solving for s yields

$$s \cong 0.07 \qquad (4)$$

The rated speed is then obtained from

$$n = (1 - s)n_s = (1 - 0.07)(1200) = 1116 \text{ rpm} \qquad (5)$$

The rated mechanical output power is now obtained from

$$P_M = T\omega = 120 \left[\frac{2\pi(1116)}{60} \right] \cong 14 \text{ kW} = 18.8 \text{ hp} \qquad (6)$$

Setting $s = 1$ in (13.6–15) yields the starting torque as

$$T_{\text{start}} = \frac{3|\mathbf{V}_r|^2}{\omega_s} \left(\frac{R_R}{R_R^2 + X_R^2} \right) \qquad (7)$$

Substituting numerical values yields

$$T_{\text{start}} = \frac{3(208)^2/3}{2\pi(1200)/60} \left(\frac{0.2}{0.2^2 + 0.5^2} \right) \cong 237 \text{ N·m} \qquad (8)$$

Finally, the maximum torque is obtained by substituting numerical values into (13.6–19) to yield

$$T_{\text{max}} = \frac{3|\mathbf{V}_r|^2}{2\omega_s X_R} = \frac{3(208)^2/3}{2[2\pi(1200)/60](0.5)} \cong 344 \text{ N·m}$$

13.7   SINGLE-PHASE AND FRACTIONAL-HORSEPOWER MOTORS

For small-motor applications of less than 10 hp, single-phase motors are used almost exclusively. In many locations, only single-phase power is available, and through special design, single-phase motors can operate nearly as well as polyphase motors.

In single-phase induction and synchronous motors, special methods are required in order to produce starting torque because the magnetic field associated with the single stator winding is fixed in space. A rotating magnetic field is produced by including an auxiliary stator coil that is displaced by 90° from the main stator winding. This motor is called a *split-phase motor*.

13.7.1   Single-Phase Induction Motors

Figure 13.31a displays the circuit diagram for a split-phase (but still single-phase) squirrel-cage induction motor. This motor has two windings, the *main winding* and the *start-up winding*, that have their axes displaced by 90° in space. The main winding is primarily inductive, and thus the current lags the voltage by approximately 90° in time. The start-up winding is highly resistive, and hence the current and voltage for this winding are approximately in phase. The two currents are therefore about 90° out of time phase with each other. Hence, the stator field of the main winding reaches a maximum, while that of the start-up winding is small. This situation is reversed 90° later. In combination, the two windings produce a two-phase motor with stator flux that rotates in a similar manner to that of the three-phase motor. Finally, as the rotor comes up to approximately 80% of synchronous speed, the start-up winding is disconnected by a centrifugal switch. Since the start-up winding is essentially resistive, the motor is referred to as *resistance-start*.

Similar operation can be accomplished using a start-up winding that is reactive or capacitive. Figure 13.31b displays a *reactance-start* motor that utilizes a reactive branch in series with the start-up coil. The reactance is also switched out centrifugally as the motor comes up to speed. Finally, Fig. 13.31c shows a *capacitance-start* and *capacitance-run* induction motor. Note that two capacitors are used. When both capacitors are in the circuit, the magnitude of capacitance is largest and provides good starting torque. After starting occurs, the effective capacitance is reduced in value

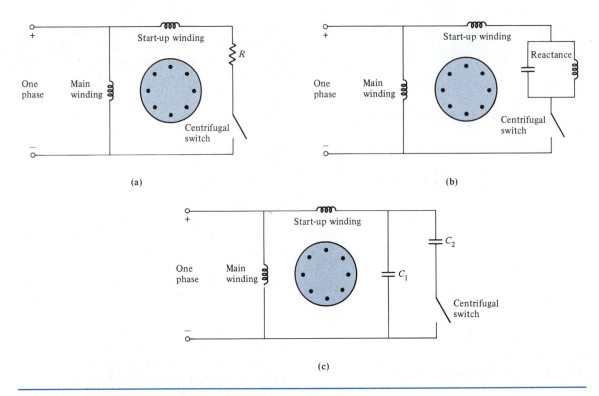

FIGURE 13.31   Split-Phase Squirrel-Cage Induction Motors: (a) Resistance-start, (b) Reactance-start, (c) Capacitance-start

as C_2 (in Fig. 13.31c) is disconnected by the centrifugal switch. Capacitor C_2 is chosen to be large in magnitude in order to provide good starting, while capacitor C_1 is smaller and provides optimum running performance with relatively high efficiency and power factor.

13.7.2   Single-Phase Synchronous Motors

Figures 13.32a and b show schematic diagrams of two single-phase synchronous motors. As indicated, a split phase with either

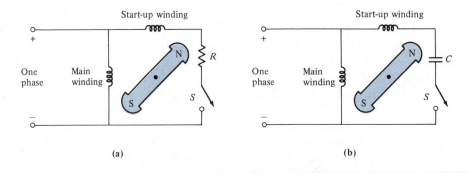

FIGURE 13.32   Single-Phase Synchronous Motors: (a) Resistance-start, (b) Capacitance-start

resistance-start or capacitance-start is used in order to produce nonzero starting torque. The rotor is usually a permanent magnet with north and south poles. After synchronous speed is achieved, as the stator current alternates, interaction with the rotor magnetic poles creates torque that continually causes rotation at the synchronous speed.

13.7.3   The Universal Motor

The universal motor is essentially a DC series motor that may be operated using DC or AC—hence, the name *universal*. By applying single-phase AC directly to the series-connected armature and field windings of this motor, a pulsating but unidirectional torque is produced. The stator and rotor, however, must be laminated to reduce eddy current losses when the motor is operated with AC. The universal motor has the disadvantage of using commutators and brushes. However, like the series DC motor, the universal motor has excellent speed–torque characteristics.

13.7.4   Stepper Motors

Stepper motors are fractional-horse-power motors that provide partial rotation in precise angular steps. These motors provide very precise position control and have many applications. Two important examples are control of the hands of an electrical analog watch and control of the magnetic head of a tape recorder. Precise clock pulses are used to control the stepper motor movement.

13.8   MAGNETIC SYSTEM EXAMPLE: TAPE RECORDERS

Tape recorders are instruments that are used to record, store, and replay audio, video, and general data information. Different constituents are required, depending upon the frequency of the electrical information to be processed. The description of tape recorders is confined here to the magnetic types. Audio recorders are described first.

13.8.1   Audio Tape Recorders

Basic Operation.   Figure 13.33 displays a schematic diagram that indicates in block diagram form the basic operation of an audio recorder. Note that three magnetic heads are used to record, play back, and erase the audio information on

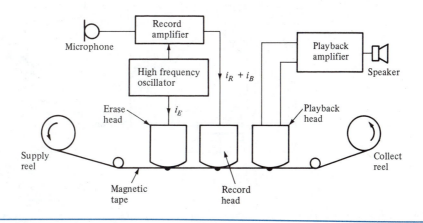

FIGURE 13.33   Basic Components in a Magnetic Tape Recorder

magnetic tape. Each magnetic head is engaged separately to perform its particular function. The supply and collect reels are motor controlled and allow the magnetic tape to pass over and interact with the desired head.

The basic process of recording is initiated as the incoming sound wave is converted by the microphone into a current. As shown in Fig. 13.33, this current is then amplified, producing the recording current i_R, which is added to a current i_B provided by the high-frequency oscillator. The resultant time-varying current $i_R + i_B$ is transformed into a proportional magnetic field by the recorder head. As the tape passes through the recorder head, it becomes magnetized and stores the information in magnetic form. During playback, the magnetic information on the tape is transformed by the playback head into a signal voltage proportional to the original sound. As shown in Fig. 13.33, this signal is then amplified in the playback amplifier and converted to sound waves via a speaker.

Recording Mode.   When the audio recorder is in the recording mode, fluctuations in sound are converted to variations in magnetic field intensity that are stored on the magnetic tape. This conversion

is accomplished through the use of a magnetic-sensitive tape material and the magnetic recording head.

A typical magnetic recording head is shown in Fig. 13.34 and consists of a special magnetic core with turns of fine wire wrapped around it. When an electric current containing the sound information is passed through the coil, a directly proportional magnetic field is produced in the magnetic head. Thus, as the audio current varies, the magnetic field varies in direct proportion. Additionally, since the core is constructed with a narrow air gap, a fringing magnetic field (as shown in Fig. 13.34) is produced that is also proportional to the sound current. The fringing field region provides the means for the magnetic field to interact with the tape. Thus, as the tape moves at constant speed over the fringing field region, the magnetic character of the tape is altered according to the magnitude of the fringing field. In this manner, the audio current information is transferred to the tape. The tape is then said to be *magnetized* with the desired information stored on the tape in readiness for the playback mode.

Audio signals range from approximately 15 Hz to 15 kHz. However, all audio recorders do not accommodate this entire range. To record

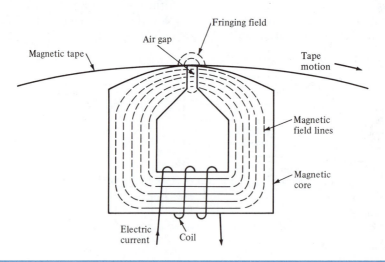

FIGURE 13.34   Magnetic Recording Head

higher frequency ranges, the tape speed is either increased or the head gap is made narrower. These modifications are extremely important in video signal recording.

Playback and Erase Modes.   In the playback process, the magnetic information encoded on the tape is transformed back into an electrical signal. During playback, the opposite process from record occurs as the tape passes over the playback head. The magnetic field lines encoded on the tape penetrate the surface of the head and induce a voltage across the terminals of the playback coil. The electrical signal produced is proportional to the original signal used to record on the tape. This signal is then amplified using audio amplifiers similar to the amplifiers described in Chapter 6.

The erasure process is accomplished by subjecting the magnetic tape to a decaying high-frequency magnetic field. This field is produced by the high-frequency oscillator current (i_E in Fig. 13.33) and the erase head. In the erase head, i_E is transduced into a magnetic field via the coil and magnetic core. As the tape passes in the vicinity of the fringing field region, each element of the tape is subjected to many cycles of the magnetic field, which is ensured by providing a large air gap for the erase head (larger than that of the record head). Additionally, as each tape element reaches the end of the fringing field region, the magnitude of the magnetic field reduces to zero, and the induced magnetic field is therefore also reduced to zero for each tape element. Thus, as the tape passes over the engaged erase head, the tape is said to be *demagnetized*.

13.8.2   Video Tape Recorders

The first video tape recorders were designed using audio recorders as a basis. However, modifications in the recording process were necessary in order to adjust to the 6 MHz video frequency range. Since there is a limit to how small the head gap can be made, it became necessary

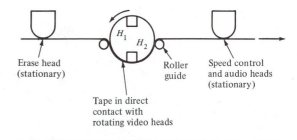

FIGURE 13.35   Video Recorder Heads

to greatly increase the tape speed of the video recorder. However, the required high speed is much too high to be achieved by the simple supply and collect reels of the audio recorder.

Rotating Drum System.   High tape speed in video recorders is accomplished by using multiple recording heads mounted on a rapidly rotating drum, as shown in Fig. 13.35. The tape is moved relatively slowly across the rapidly rotating record heads to achieve a high head-to-tape speed ratio. Thus, the length of tape necessary is greatly reduced, and practical tape lengths are possible.

As shown in Fig. 13.35, two recording video heads on the drum are separated by 180°. Extra heads are used for special modes of operation such as freeze-frame and speed control.

VHS and Beta.   Although the Video Home System (VHS) and Betamax (Beta) methods are very similar, they are completely incompatible for several reasons. Not only are cassette sizes different, but the VHS and Beta systems thread video tapes differently. Figures 13.36a and b show the threading systems as well as the component layouts for typical VHS and Beta units, respectively. The VHS threading system is referred to as an *M-type* systems, whereas the Beta system is an *omega-type* system (the names comes from the shapes of the tape path).

The two systems also differ in the size of the video head gaps and the speed at which the tape moves. Unfortunately, these differences in design between Beta and VHS (and other systems under development) are not likely to be reconciled.

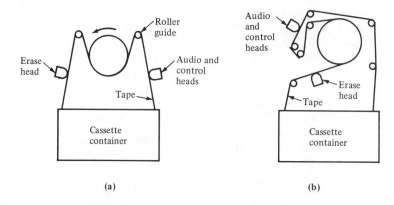

FIGURE 13.36   Tape Paths: (a) VHS, (b) Beta

CHAPTER 13 SUMMARY

■ Magnets are made of ferromagnetic materials that possess a built-in magnetization and an associated magnetic field whose magnitude decreases as the distance from the magnet increases. A magnet exerts an attractive force on iron and certain other materials.

■ Associated with a magnetic field are the density B, magnetic flux Φ ($= BA$), and magnetic field intensity H ($= B/\mu$). A charged particle with charge Q moving with velocity $\bar{u}$ in a magnetic field of intensity $\bar{B}$ has a force exerted upon it given by $F = Q\,(\bar{u} \times \bar{B})$. The energy per unit volume stored in a magnetic field is given by $E/LA = B^2/2\mu$, where μ, the permeability of the medium, is a magnetic parameter. For free space, $\mu = \mu_0 = 4\pi \times 10^{-7}$ H/m. For other materials, $\mu = \mu_r\mu_0$, where μ_r is the relative permeability.

■ Electromechanical translational transducers convert mechanical energy to electrical energy or vice versa via a magnetic field. Basic relations for a conductor of length d and velocity u perpendicular to a magnetic field B involve the induced voltage v and magnetic field force F_m given by $v = Bud$ and $F_M = Bid$. For a lossless transducer, $P_E = P_M = iBud$. A loudspeaker is an example of a translational transducer.

■ Electromechanical rotational transducers are an extension of translational transducers. Basic relations for a transducer consisting of an N-turn coil having dimensions D by d and rotating with angular velocity ω in rad/s are $v = NBD d\omega \cos \omega t$ and $T_T = NBDdi \cos \omega t$. For a lossless system, $P_E = P_M = iv = T\omega = NBDd\omega i \cos \omega t$.

■ Rotating machines are primary examples of rotational transducers. Three classes of rotating machines include DC, synchronous, and induction machines, each of which may be operated as a motor or as a generator.

■ All rotating machines have associated power losses. For an average input power P_I, output power P_O, and power loss P_L, the efficiency is given by

$$\eta = \frac{P_O}{P_I} = \frac{P_I - P_L}{P_I} = \frac{P_O}{P_O + P_L}$$

For a DC generator with applied torque T_A,

$$P_O = T_A\omega + I_F V_F - P_L$$

where $P_I = T_A\omega + I_F V_F$. For a DC motor,

$$P_O = I_F V_F + I_A V_A - P_L$$

where $P_I = I_F V_F + I_A V_A$. For an AC generator, the power output per phase is

$$P_O = |\mathbf{I}_{ph}||\mathbf{V}_{ph}| \cos\theta = T_A\omega + I_F V_F - P_L$$

For a three-phase AC motor,

$$P_O = T\omega = I_F V_F + 3|\mathbf{I}_{ph}||\mathbf{V}_{ph}| \cos\theta - P_L$$

- Basic relations for DC machines are $V_G = KB\omega$ and $T = KBI_A$. For the lossless case, $P_E = P_M = KBI_A\omega$.
- The two types of DC generators are the separately excited generator and the self-excited generator. In the self-excited case, or shunt generator, the field windings are in parallel with the armature windings and I_F is supplied by V_G. For the separately excited case, $V_L = KB\omega - I_A R_A \cong K'I_F\omega - I_A R_A$. For the self-excited case, $V_L = I_F(R_F + R_S) = KB\omega - I_A R_A$.
- The three types of DC motors are the shunt motor, the series motor, and the compound motor. They are named according to the type of electrical connection for the field and armature windings. For the shunt motor,

$$\omega = \frac{V_I}{KB} - \left[\frac{T}{(KB)^2}\right] R_A$$

We observe a linear speed (ω) versus torque (T) relation for the shunt motor. For the series motor,

$$T = C\left(\frac{V_I}{C\omega + R_T}\right)^2$$

and T varies inversely with the square of ω. The compound motor speed–torque variation is intermediate between the shunt and series cases.
- DC series motors exhibit high starting torque, and speed varies with load (torque). Shunt motors possess low starting torque but constant speed for a varying load. Compound motors exhibit speed variation that is much less

than series motors but greater than shunt motors.
- Synchronous machines are AC machines with field windings on the rotor (usually three-phase) and armature coils on the stator. For the generator case, mechanical energy is applied to the rotor and induces sinusoidal voltages in the stator coils. For the motor case, AC is applied to the stator coils and produces a rotating magnetic field with angular speed corresponding to the AC frequency. DC applied to the rotor provides a rotor magnetic field that locks in step with the rotating stator magnetic field. For a synchronous machine with $p/2$ pole pairs on the rotor, the synchronous speed is $n_s = 120f/p$, and the speed is said to be synchronous with the frequency.
- A balanced three-phase synchronous generator produces sinusoidal voltages in each phase that are 120° out of phase with one another. The generated voltage for one phase is given by $V_g = NB_R Dd\omega \cos\omega t$, where B_R is the rotor peak magnetic flux density. The magnitude of the generated voltage in one phase is given by

$$|\mathbf{V}_g| = NB_R Dd\omega = NB_R Dd\left[\frac{\pi(n_s p)}{60}\right]$$

where B_R depends upon I_F.
- Three-phase AC systems are used in the generation, transmission, and distribution of electrical power. The instantaneous power in a balanced three-phase system is a constant given by

$$p_t = 3|\mathbf{I}_{ph}||\mathbf{V}_{ph}| \cos\theta = \sqrt{3}|\mathbf{I}_L||\mathbf{V}_L| \cos\theta$$

where $\mathbf{V}_L = \sqrt{3}\mathbf{V}_{ph} \angle -30°$ for a Y-connected three-phase balanced load.
- The equivalent circuit for a synchronous generator consists of $\mathbf{V}_g$ in series with R_A and X_s, with R_F separately representing the field winding. The voltage for one phase is given by

$$\mathbf{V}_{ph} = \mathbf{V}_g - \mathbf{I}_{ph}(jX_s)$$

where R_A is neglected.

■ The equivalent circuit for a synchronous motor is the same as that of the generator except that the current is reversed. The voltage for one phase is given by

$$V_{ph} = V_g + jI_{ph}X_s$$

Since synchronous motors are not self-starting, induction motor starting is employed.

■ The two types of induction motor rotors are the squirrel-cage rotor and the wound rotor. When AC is applied to each stator winding, a rotating magnetic field is produced that induces currents and emfs in the rotor analogous to transformer action. The induced currents interact with the rotating stator magnetic field to produce torque. Slip s is defined as the difference between the synchronous speed and the actual speed of the rotor.

■ Squirrel-cage induction motors have copper or aluminum conductors set into the rotor periphery and connected on end to form a cage. Wound-rotor induction motors are commutator machines similar to DC motors that provide slip rings to which external resistance can be connected for speed adjustment.

■ The developed torque for a three-phase induction motor with rotor resistance R_R and reactance X_R is given by

$$T_T = \frac{3|I_r|^2 R_R}{\omega_s s} = \frac{3|V_r|^2}{\omega_s} \left(\frac{sR_R}{R_R^2 + s^2 X_R^2} \right)$$

where I_r and V_r are the induced rotor current and voltage per phase and $\omega_s = 2\pi n_s/60$. For motor operation near synchronous speed,

$$T_T = \frac{3|V_r|^2}{\omega_s} \left(\frac{s}{R_R} \right)$$

and the maximum total torque is

$$T_{max} = \frac{3|V_r|^2}{2\omega_s} \left(\frac{1}{X_R} \right)$$

■ For small-motor applications, single-phase motors are used almost exclusively. These motors require special methods to produce starting torque.

■ A split-phase induction motor has two windings, the main winding and the start-up winding. The main winding is inductive, and the current lags the voltage by approximately 90°. By providing a start-up winding with current and voltage in phase, a two-phase motor is produced. This situation is achieved by a resistive circuit (resistance-start) or a capacitive circuit (capacitance-start).

■ A universal motor is a DC motor that may be operated using DC or AC.

■ Magnetic tape recorders are used to record, store, and replay audio, video, and general data information. Magnetic heads magnetically encode the signal onto the magnetic tape. Playback is achieved by converting the magnetic signal back into an electrical signal.

CHAPTER 13 PROBLEMS

13.1

A coil is wrapped around a donut-shaped core as shown in Fig. P13.1. Such a coil is called a *toroid*. From basic laws of physics, the magnetic flux is $\Phi = \mu NiA/l$, where $\mu = \mu_r \mu_0$, N is the number of turns, i is the current in the coil, A is the cross-sectional area of the coil, and l is the average length in the toroid ($l = 2\pi R$). For a flux of 0.1 mWb, calculate the required current. Let the dimensions of the core

be $r = 1$ cm, $R = 10$ cm, and $N = 500$ turns. The core material is iron with $\mu_r = 1000$.

13.2

Repeat Problem 13.1 for the case in which the core is nonmagnetic ($\mu_r = 1$).

13.3

Consider the toroid core of Problem 13.1 to have an air gap of width 0.2 cm. Calculate the energy stored in the air gap and in the iron.

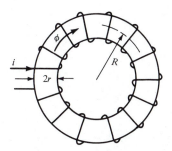

FIGURE P13.1

13.4
Consider a toroidal coil (as in Problem 13.1) with N turns and a ferromagnetic core material with $\mu_r = 2000$. Write the expression for emf and assuming that B is uniform throughout the core, determine expressions for H, B, and Φ. For a coil with $N = 1000$ and $r = 1$ cm wrapped tightly around the core of radius 5 cm, determine the current required to produce a magnetic flux of 0.4 mWb.

13.5
Consider a toroidal core with rectangular cross section. Let the height be 0.25 in., the inside diameter 1 in., the outside diameter 1.25 in., and $\mu_r = 1000$. If $B = 0.2$ Wb/m^2, determine Φ and i for $N = 1000$.

13.6
A moving conductor transducer, as shown in Fig. 13.5, has $d = 0.5$ m, $B = 1.5$ T, and $u = 5$ m/s. Let $V_{DC} = 0$ and $R = 50\ \Omega$. Calculate the emf induced in the conductor (v), the current (i), and the power delivered to R.

13.7
For the moving conductor transducer of Fig. 13.5, let $d = 1$ m, $u = 10$ m/s, and $B = 1$ T. For $V_{DC} = 5$ V and $R = 10\ \Omega$, determine the induced emf and the current. Also, calculate the force on the conductor, the mechanical power, and the electrical power delivered to R_L.

13.8
Repeat Problem 13.7 for the case in which the conductor has a resistance of 1 Ω. What is the power loss in the 1 Ω resistor? Calculate the efficiency of power transfer (P_{R_L}/P_M).

13.9
A loudspeaker with the basic geometry as shown in Fig. 13.6 has dimensions given as length = 3 mm,

mean radius = 5 cm, and width = 2 cm. If $B = 0.1$ T, calculate the number of coil turns for a magnetic field force of 0.1 N to correspond to a current of 20 mA. Assume that all of the magnetic field energy is in the air gap.

13.10
For the rotational transducer shown in Fig. P13.10, determine and indicate the following: **(a)** the direction of current flow from $F = Q(u \times B)$, **(b)** the polarity of the induced emf across R, **(c)** the expression for emf from $v = Nd\Phi/dt$, and **(d)** the developed torque expression from $T_d = rf_n$, where f_n is the normal force to r.

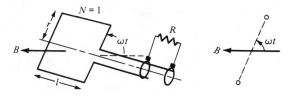

FIGURE P13.10

13.11
For Problem 13.10, let $B = 0.1$ T, $r = 5$ cm, $l = 10$ cm, and $\omega = 2\pi f$, where $f = 60$ Hz. Determine the time-varying expressions and calculate the peak values of the induced emf and the developed torque for $R = 10\ \Omega$.

13.12
The rotational transducer of Fig. 13.7 has $N = 20$, $B = 0.5$ T, $d = 10$ cm, and $D = 5$ cm. If a gasoline motor drives the shaft such that $\omega = 100$ rad/s, determine the induced voltage, the current, the average electrical power, and the average mechanical power. Let $R = 100\ \Omega$.

13.13
Repeat Problem 13.12 with $R = 10\ \Omega$.

13.14
Sketch the magnetization curve (V_G versus I_F) corresponding to the following measured data at 1000 rpm for a separately excited DC generator: $I_F = 1.5, 1.25, 1.0$, and 0.5 A for corresponding voltages 250, 230, 200, 100 V. If the field current I_F is fixed at 1.25 A, how fast must the generator be driven to generate 250 V at no load?

13.15

Sketch the magnetization curve for Problem 13.14 for a separately excited DC generator. What is the field current required to generate 200 V at 800 rpm under no-load conditions? Sketch the magnetization curve for 800 rpm.

13.16

A separately excited DC generator is rated at 1 kW, 100 V, and 20 A at 100 rpm. Let $R_A = 1\,\Omega$ and $R_F = 100\,\Omega$. Predict the no-load voltage at rated speed. Draw the circuit model to aid in the solution.

13.17

A separately excited DC generator is rated at 250 V and 100 A at a speed of 2000 rpm. The armature resistance $R_A = 0.2\,\Omega$ and the field resistance $R_F = 125\,\Omega$. Draw the circuit model and from it predict the no-load terminal voltage if the field current is reduced by 10%.

13.18

A separately excited DC generator is running at a speed of 1000 rpm with an applied field voltage of 10 V. For this generator, $V_G = 0.05\Phi n$, where $\Phi = 10I_F$ (with MKS units) and the total field resistance is 100 Ω. Draw the circuit model for this machine and determine the field current and flux. Also, determine the full-load voltage for $I_A = 5$ A and $R_A = 1\,\Omega$, as well as the no-load voltage.

13.19

A DC generator is rated at 10 kW, 200 V, and 50 A at 1000 rpm with $R_A = 0.4\,\Omega$ and $R_F = 80\,\Omega$. What is the full-load voltage at 1000 rpm, the no-load voltage at 1000 rpm?

13.20

A 10 kW, 50 V shunt generator has $R_A = 0.1\,\Omega$ and $R_F = 50\,\Omega$. Draw the circuit diagram and determine the generated voltage at rated load.

13.21

A 1 kW, 100 V shunt generator has $R_A = 0.3\,\Omega$ and $R_F = 150\,\Omega$. Determine the field current, the armature current, the load current, and the generated voltage.

13.22

The magnetization curve for a 1200 rpm DC shunt generator is shown in Fig. P13.22. If the total field resistance is 100 Ω and R_A can be neglected, determine the no-load terminal voltage. Use a circuit diagram to aid in the solution.

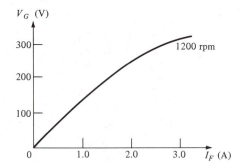

FIGURE P13.22

13.23

The DC generator of Problem 13.21 is driven at 1500 rpm with a separately supplied field current of 1.5 A. If a 50 Ω load is connected, determine the load current and voltage. Let $R_A = 0.5\,\Omega$.

13.24

A 230 V shunt motor has an armature resistance of 0.2 Ω and is operating at 1150 rpm with armature current of 100 A. An external resistance of 1.0 Ω is now inserted in series with the armature; the field resistance is unchanged. Use a circuit model as an aid to determine the new value of V_G and the speed (recall that $V_G = K\Phi n$).

13.25

A DC shunt motor has rated values of 10 hp, 250 V, 20 A, and 1200 rpm with $R_A = 0.5\,\Omega$ and $R_F = 100\,\Omega$. Determine the speed under no-load conditions. Also determine the speed when the motor is operated at 100 V and with a line current of 10 A, while I_F is maintained at a rated value of 1 A (by adjusting R_F).

13.26

Consider a series DC motor operating with 100 V and $R_A = R_F = 0.5\,\Omega$. An additional resistor $R_{start} = 100\,\Omega$ is used to limit the starting current. Draw the circuit model for this motor and from it determine the armature current under starting conditions. Calculate the starting current without R_{start}.

13.27

A 50 hp series motor is rated at 50 A at 1200 rpm and has $R_A = R_F = 0.5\,\Omega$. For $I_A = 25$ A, determine the speed and mechanical power output. Assume that the flux density varies linearly with I_A ($B = kI_A$).

13.28

Sketch the circuit diagrams for a DC series motor and a DC shunt motor. From these circuits, write the relationships for terminal voltage in terms of the induced voltage and the armature current. Then, use the relations $V_G = k_1 \Phi n$, $T = k_2 \Phi IA$, and $\Phi = k_3 I_F$ to show that **(a)** speed n varies linearly with torque T for a shunt motor and **(b)** torque T varies inversely with speed n^2 for a series motor.

13.29

A DC shunt motor rated at 120 V and 50 A has $R_A = 0.2 \, \Omega$ and $R_F = 100 \, \Omega$. If the mechanical loss is 357 W and the magnetic loss is negligible, calculate the total losses, the output power, and the efficiency.

13.30

Consider a 15 hp, 150 V DC shunt motor with $R_A = 0.5 \, \Omega$. If $I_A = 5$ A under no-load conditions with $n = 1800$ rpm, determine the speed and torque for full-load conditions, where $I_A = 20$ A.

13.31

A 12 hp, 150 V DC shunt motor has a full-load line current of 50 A. Also, $R_A = 0.3 \, \Omega$ and $R_F = 120 \, \Omega$. If the mechanical and magnetic losses are 300 W, calculate the efficiency of the motor.

13.32

Draw the electrical circuit model for a synchronous generator. Indicate the direction of power flow. How would this direction change for a motor?

13.33

A three-phase six-pole synchronous generator is rated at 100 kVA, 1200 V, and 60 Hz. The synchronous reactance is $X_s = 3 \, \Omega$. Calculate the rated speed and the generated emf for full-load operation with 0.8 leading power factor (I leading V). Assume that the load is Y connected and draw the phasor diagram.

13.34

Verify equation (13.5–25) using the following voltages and currents for a Y-connected balanced load:
$V_{AA'} = \sqrt{2} V_{ph} \sin \omega t$, $i_{A_L} = \sqrt{2} I_L \sin(\omega t - \theta)$; $V_{BB'} = \sqrt{2} V_{ph} \sin(\omega t - 120)$, $i_{B_L} = \sqrt{2} I_L \sin(\omega t - \theta - 120)$; $V_{CC'} = \sqrt{2} V_{ph} \sin(\omega t - 120)$, $i_{C_L} = \sqrt{2} I_L \sin(\omega t -$

$\theta + 120)$. Also use the relations $\sin \alpha \sin \beta = 1/2[\cos(\alpha - \beta) - \cos(\alpha + \beta)]$ and $\cos(\gamma + \beta) = \cos \gamma \cos \beta + \sin \gamma \sin \beta$.

13.35

Determine the synchronous speed for various synchronous generators having the following number of poles: $n = 2, 4, 6, 8$, or 10.

13.36

Consider a three-phase synchronous generator rated at 1 kVA and 120 V. Also, $X_s = 2 \, \Omega$ and $R_A = 0.5 \, \Omega$ per phase. Construct the phasor voltage diagram for this machine for a lagging power factor of 0.8. What is the magnitude and phase of V_g?

13.37

A synchronous generator has $X_s = 5 \, \Omega$ with a terminal voltage that is held constant at 50 kV. Draw the phase voltage diagram and calculate the magnitude of the generated voltage for a load that has 100 MW of power per phase with a lagging power factor of 0.6.

13.38

A synchronous motor has a rated input of 220 V and 100 kVA at 1800 rpm and 60 Hz. The motor is three-phase Y connected with a leading power factor of 0.8. $R_A = 0.1 \, \Omega$ per phase and $X_s = 5 \, \Omega$ per phase. The sum of mechanical and magnetic losses is 3000 W, and the field current loss is negligible. Calculate the power output and the efficiency.

13.39

Consider a 4-pole induction motor with rated values of 20 hp, 120 V, and three-phase voltage at 60 Hz. If $s = 0.04$ under rated conditions, calculate the synchronous speed and the rotor speed. Also, calculate the torque corresponding to full-load operation.

13.40

Consider a 220 V, 60 Hz, three-phase induction motor with four poles and rated torque of 100 N·m. If the induced voltage per phase of the rotor is equal to the per phase voltage of the stator and $R_R = 0.1 \, \Omega$ and $X_R = 0.3 \, \Omega$, calculate the following: **(a)** synchronous speed, **(b)** slip, **(c)** n, **(d)** mechanical output power, and **(e)** the starting torque and the maximum torque.

14

OPERATING PRINCIPLES OF LABORATORY INSTRUMENTS

This chapter describes the basic mechanisms of operation for instruments and laboratory equipment used in testing electronic circuits. Specific operating procedures for any particular instrument may be obtained from the instruction manual for that instrument.

The operating principles of basic electrical instruments are described first. These instruments include ammeters (for measuring current), voltmeters (for measuring voltage), and multimeters (for measuring either current, voltage, or resistance). These older electrical instruments form the basis of modern electronic instruments that employ electronic devices. Ammeters and voltmeters that measure DC as well as AC will be described.

The difference between electrical instruments and electronic instruments is that electronic instruments utilize electronic devices (tubes or transistors) that permit (among other advantages) higher sensitivity because of electronic amplification. Electronic instruments to be described include the electronic voltmeter and multimeter along with the cathode-ray oscilloscope (CRO), the curve tracer, and digital instruments. As we will see, the CRO is the most versatile electronic measuring device available because of the multitude of electrical quantities that are measurable with this instrument.

This chapter also includes a description of the sources of excitation of electronic circuits. These sources include DC power supplies and signal generators.

14.1   DC AND AC METERS

DC and AC meters have been in wide use for many years. The basic instruments that are used to measure DC current or voltage (and, quite often, resistance) are called *DC electrical meters,* or simply *DC meters*. The operating principle of DC meters is based upon a moving coil system that involves the d'Arsonval meter movement.

14.1.1   The d'Arsonval Meter

The *d'Arsonval meter* involves the principle of DC motor operation in which two interacting magnetic fields are arranged to produce a force that induces continual rotation of the rotor. In the moving coil system, the fields interact to produce a partial rotation that is counteracted upon by a spring.

Figure 14.1 displays the components of the basic d'Arsonval meter. A permanent horseshoe magnet is used to produce a fixed magnetic field. This field is concentrated between the gaps of the north and south pole pieces and the iron core as shown in Fig. 14.1. The iron core is mounted in a jewel setting that allows rotation, and a coil of fine wire is wrapped around the rotatable core. As a small current from the system under measurement passes through this coil, a second magnetic field is created. The magnitude of the coil magnetic field is directly proportional to the amount of current that passes through the coil. Interaction of the two magnetic fields then causes the coil to rotate, and the magnitude of the current determines the amount of rotation. When the measurable element is disconnected so that the current is zero, the spring returns the pointer to its original position. Note that the windings can deflect the pointer in only one direction and that polarity markings are placed on DC meters to avoid the application of current in the wrong direction.

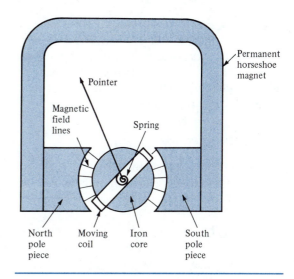

FIGURE 14.1   Basic d'Arsonval Meter

14.1.2   DC Ammeters

The *DC ammeter* is used to measure the DC current in some branch of a circuit. This instrument is placed directly in series with the branch. Since the fine wires of the coil have only a small resistance associated with them, only a limited range of current is measurable, and additional circuitry is required for a general measurement.

To increase the measurable current range, a small-valued resistor is placed in parallel (or shunt) with the windings of the coil. Thus, a portion (usually most) of the current is diverted through this small-valued resistance. Figure 14.2

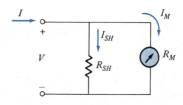

FIGURE 14.2   Ammeter Circuit with Shunt Resistor

indicates the circuit arrangement used to increase the range of measurable current. The resistance of the meter is denoted as R_M; that of the shunt resistor as R_{SH}. From KCL, the current to be measured, denoted as I, is the sum of the shunt resistor current I_{SH} and the meter current I_M, or

$$I = I_{SH} + I_M \qquad (14.1-1)$$

Furthermore, by defining the voltage across each of these resistors as V and using Ohm's law, we have

$$V = R_{SH}I_{SH} = R_M I_M \qquad (14.1-2)$$

Thus, the current that the meter measures is

$$I_M = I - I_{SH} \qquad (14.1-3)$$

By substituting for I_{SH} from (14.1–2), we have

$$I_M = I - \frac{R_M I_M}{R_{SH}} \qquad (14.1-4)$$

Solving for I_M yields the reduced value of meter current as

$$I_M = \frac{I}{1 + (R_M/R_{SH})} \qquad (14.1-5)$$

The magnitude of the shunt resistance required can be determined from the known value of R_M along with the measured current (I_M) and an assumed value of input current (I). From Ohm's law, $R_{SH} = V/I_{SH}$, and substituting for V and I_{SH} yields

$$R_{SH} = \frac{I_M R_M}{I - I_M} \qquad (14.1-6)$$

However, since we do not know the specific value of I (recall that we are trying to measure this current), (14.1–6) is not really useful in its present form.

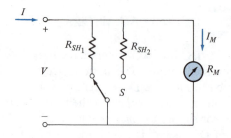

FIGURE 14.3   Switching Arrangement to Measure Two Current Ranges

Equation (14.1–6) can be rearranged into a more useful form by assuming that the current to be measured is a multiple of the meter current, or

$$I = NI_M \qquad (14.1-7)$$

where N is an integer. Substituting (14.1–7) into (14.1–6) and canceling I_M yields

$$R_{SH} = \frac{R_M}{N - 1} \qquad (14.1-8)$$

The operation of DC ammeters is based upon (14.1–8) and the use of several precision shunt resistors. Figure 14.3 displays an arrangement that might be used except for the fact that upon switching, all of the current can pass through the meter coil, which could cause permanent damage to the coil.

––––––––––––––––––––––––––––––––––––– EX. 14.1

Shunt Resistance Design Example

The meter coil in Fig. 14.3 has a resistance $R_M = 100\ \Omega$. Without any shunt resistance, currents in the range 0–1 mA can be measured. Determine the values of R_{SH_1} and R_{SH_2} that will allow the range of measurable currents to be increased to 0–10 mA and 0–100 mA, respectively.

Solution: The current ranges change by multiples of 10, or $N = 10$ and 100. Thus, to achieve the 0–10 mA range, the shunt resistance is

$$R_{SH_1} = \frac{R_M}{N-1} = \frac{100}{9} = 11.11 \ \Omega$$

For the 0–100 mA range, $N = 100$, and

$$R_{SH_2} = \frac{R_M}{N-1} = \frac{100}{99} = 1.01 \ \Omega$$

An improved arrangement for switching the shunt resistance into the measuring circuit is employed in Fig. 14.4. We note that no matter what position the switch is in, there is a shunt resistor in the circuit to limit the current through the meter coil. This arrangement is called the *Ayrton*, or *universal, shunt* and often uses more than two shunt resistors.

To analyze the Ayrton shunt ammeter, note that with the switch in position 1, the meter is the most sensitive because the shunt resistance is the largest. The total shunt resistance is given by the sum of the two resistors, or

$$R_{SH_T} = R_{SH_1} + R_{SH_2} = \frac{R_M}{N_1 - 1} \qquad (14.1–9)$$

where R_M and N_1 are known.

With the switch in position 2, R_{SH_2} is in parallel with $R_{SH_1} + R_M$. Under these conditions,

$$R_{SH_2}I_{SH} = (R_{SH_1} + R_M)I_M \qquad (14.1–10)$$

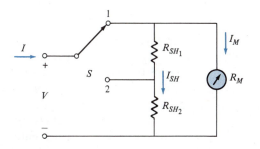

FIGURE 14.4   Ayrton Shunt Ammeter

where

$$I = I_{SH} + I_M = N_2 I_M \qquad (14.1–11)$$

with I, I_{SH}, and I_M in this case having different values than they had previously. Solving (14.1–10) for R_{SH_2} yields

$$R_{SH_2} = (R_{SH_1} + R_M)\frac{I_M}{I_{SH}} \qquad (14.1–12)$$

By substituting for I_{SH} from (14.1–11), we have

$$R_{SH_2} = \frac{R_{SH_1} + R_M}{N_2 - 1} \qquad (14.1–13)$$

Knowing R_M and the desired current ranges (N_1 and N_2) then allows the determination of R_{SH_1} and R_{SH_2} from (14.1–9) and (14.1–13) by solving these equations simultaneously. These results are demonstrated through the following example.

————————————————————————— EX. 14.2

Ayrton Shunt Ammeter Design Example

Determine the values of shunt resistance to be used with the meter coil of Ex. 14.1 ($R_M = 100 \ \Omega$, 0–1 mA range of current) that extend the range of measurable currents to 0–10 mA and 0–100 mA.

Solution: The total resistance, $R_{SH_1} + R_{SH_2}$, is first calculated from (14.1–9) as follows:

$$R_{SH_T} = R_{SH_1} + R_{SH_2} = \frac{R_M}{N_1 - 1}$$

$$= \frac{100}{9} = 11.11 \ \Omega \qquad (1)$$

Then, from (14.1–13),

$$R_{SH_2} = \frac{R_{SH_1} + R_M}{N_2 - 1} = \frac{R_{SH_1} + 100}{99} \qquad (2)$$

or

$$99R_{SH_2} - R_{SH_1} = 100 \qquad (3)$$

Adding (1) and (3) yields

$$100R_{SH_2} = 111.11 \qquad (4)$$

or

$$R_{SH_2} \cong 1.11 \; \Omega \qquad (5)$$

Therefore, substitution into (1) yields

$$R_{SH_1} = 10 \; \Omega \qquad (6)$$

The use of small precision resistors in the Ayrton shunt arrangement thus allows extension of the measurable current range. DC ammeters based upon this principle are constructed with more than two switch positions. It is advisable to always initially measure the current with the switch in the highest position to avoid damaging the meter coil. Note that some error is introduced by inserting the ammeters into the circuit under measurement because of the resistance of the ammeters. However, this error is tolerable provided that the ammeter resistance is small in comparison to the resistors in series with the instrument. Switching to a higher current range (less ammeter resistance) with the reading remaining the same allows a check on the accuracy of the measured current.

14.1.3   DC Voltmeters

The *DC voltmeter* is used to measure the DC voltage across a branch in a circuit. Hence, this measurement is obtained by placing the instrument in parallel with the branch. To obtain accurate results in this case, the resistance of the voltmeter must be much larger than that of the branch being measured.

The principle of the d'Arsonval meter movement is also used in the DC voltmeter. The essential quantity that must be known is the current for the coil that causes full-scale deflection of the meter. This current is denoted here as I_{FS}.

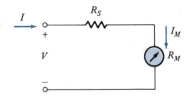

FIGURE 14.5   Simplest DC Voltmeter

If a resistor R_S is now placed in series with the coil, as shown in Fig. 14.5, we at first see that R_S has the desirable feature of limiting the current. However, as we will soon see, R_S has an additional outstanding feature in that it acts like a "multiplier" resistor.

If we assume that $I = I_{FS}$ such that the ammeter is reading full scale, then the corresponding full-scale voltage V_{FS} must satisfy the following relationship (from Ohm's law):

$$V_{FS} = (R_M + R_S)I_{FS} \qquad (14.1–14)$$

Additionally, if R_S is a multiple of R_M, where

$$R_S = KR_M \qquad (14.1–15)$$

then

$$V_{FS} = (1 + K)R_M I_{FS} \qquad (14.1–16)$$

and the current for full-scale deflection is simply

$$I_{FS} = \frac{V_{FS}}{(1 + K)R_M} \qquad (14.1–17)$$

The voltmeter thus constructed will accept any voltage from 0 to V_{FS}, while the current ranges correspondingly from 0 to I_{FS}. To increase the range of measurable voltages, the value of R_S is simply increased. When R_S is a precise multiple of R_M, this circuit can be used to precisely measure the voltage across a resistor. Great accuracy is obtained provided that the resistance of the voltmeter is much larger than the resistance of the

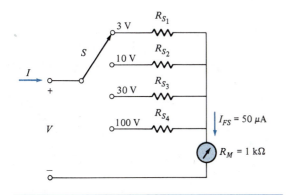

FIGURE 14.6   DC Voltmeter with Multiple Voltage Ranges

branch under measurement. Figure 14.6 displays a DC voltmeter in which various voltage ranges are measurable as indicated.

———————————————— EX. 14.3

DC Voltmeter Multiplier Resistance Design Example

Determine values for the series multiplier resistances R_{S_1}, R_{S_2}, R_{S_3}, and R_{S_4} in the circuit of Fig. 14.6.

Solution: From Ohm's law, the resistor values are calculated from

$$R_S = \frac{V_{FS} - I_{FS} R_M}{I_{FS}}$$

Substituting values from the circuit of Fig. 14.6 yields the specific values of resistance as follows:

$$R_{S_1} = \frac{3 - (50\ \mu A)(1\ k\Omega)}{50\ \mu A} = 59\ k\Omega$$

$$R_{S_2} = \frac{10 - (50\ \mu A)(1\ k\Omega)}{50\ \mu A} = 199\ k\Omega$$

$$R_{S_3} = \frac{30 - (50\ \mu A)(1\ k\Omega)}{50\ \mu A} = 599\ k\Omega$$

$$R_{S_4} = \frac{100 - (50\ \mu A)(1\ k\Omega)}{50\ \mu A} = 1.999\ M\Omega$$

As the resistor values are increased, the error introduced by the resistance of the voltmeter is

reduced. Often, as in the ammeter case, an indication of the amount of error is obtained by switching to a higher voltage range. Then, if the new reading is essentially unchanged, very little error is incurred.

The sensitivity S of a voltmeter is defined by

$$S = \frac{1}{I_{FS}} \qquad (14.1\text{–}18)$$

where the units of S are the inverse of current, or Ω/V. DC voltmeters are rated according to the magnitude of their sensitivity. This value is the resistance of the instrument for a 1 V range. A typical value of voltmeter sensitivity for use in electronic circuits should be no less than 20 kΩ/V.

14.1.4 Ohmmeters

The *ohmmeter* is a device that measures resistance, and the construction of this instrument is also based upon the d'Arsonval meter movement. Two additional elements required in the ohmmeter circuit, in addition to the d'Arsonval meter, are a DC voltage (often a battery) and a variable resistor. These elements are connected as shown in Fig. 14.7. The pointer of the meter provides a direct measure of the resistance value provided that a special calibrated scale is used. However, the circuit of Fig. 14.7 provides only a single range of resistance values, which is often inadequate for measuring resistance in general.

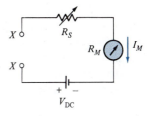

FIGURE 14.7   Basic Ohmmeter

As we will see, multi-range ohmmeters are constructed in a similar manner to ammeters and
voltmeters by incorporating a switch and additional resistors.

The basic operation of the ohmmeter involves
zeroing the instrument initially, which is carried
out by shorting the measurement terminals (X–X
in Fig. 14.7) together. By adjusting the variable
resistor, the full-scale current I_{FS} passes around
the loop, and this condition corresponds to 0 Ω
on the special resistance scale. Furthermore, under this shorted condition,

$$I_{FS} = \frac{V_{DC}}{R_S + R_M} \tag{14.1-19}$$

When the additional unknown resistor R_X is connected, the reduced current is given by

$$I = \frac{V_{DC}}{R_X + R_S + R_M} \tag{14.1-20}$$

The ratio of (14.1–20) to (14.1–19) is then obtained as

$$F = \frac{I}{I_{FS}} = \frac{R_S + R_M}{R_X + R_S + R_M} \tag{14.1-21}$$

where F denotes the fraction of full-scale current
that passes through the meter when the unknown
resistor is in the loop. Equation (14.1–21) can be
written as

$$F = \frac{R_T}{R_X + R_T} \tag{14.1-22}$$

where $R_T = R_S + R_M$ is the total resistance of the
loop without an unknown resistor R_X.

The special resistance scale is calibrated using (14.1–22) and noting that F is essentially
a resistive divider ratio. Thus, if $R_X = R_T$, $F =$
1/2, and the meter will read half scale. Similarly,
if $R_X = 2R_T$, $F = 1/3$; if $R_X = R_T/2$, $F = 2/3$. For
these examples, the meter pointer then indicates
1/3 and 2/3 full scale, respectively. Thus, since R_X

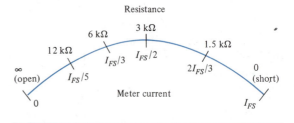

FIGURE 14.8   Nonlinear Ohmmeter Scale ($R_T = 3$ kΩ)

is directly dependent upon R_T (the total meter
loop resistance), the resistance scale is also directly
dependent upon the value of R_T.

Figure 14.8 displays an ohmmeter scale for
the special case in which the total loop resistance
$R_T = 3$ kΩ. The resistance values are indicated
on the upper scale corresponding to the fraction
of full-scale current values on the lower scale.
Note that the resistance scale is highly nonlinear.
Note further that for this example, a resistance
value between 12 kΩ and an open circuit ($R_X =$
∞) is represented by only 20% of the scale. Hence,
this single-resistance scale cannot provide accurate resistance readings in this range.

More accurate resistance readings are obtained by constructing multi-range ohmmeters
that include additional precision resistors and a
switch. An example of a multi-range ohmmeter
with three settings is displayed in Fig. 14.9. The

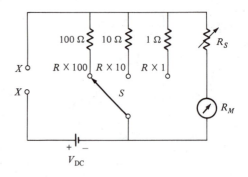

FIGURE 14.9   Multi-Range Ohmmeter with Three Settings ($R_T = R_S + R_M \gg 100$ Ω)

switch and additional resistors (which must have small values in comparison to $R_T = R_S + R_M$) allow accurate measurement of resistor values on the order of 1 Ω, 10 Ω, or 100 Ω in this case. The value of R_T is made large by selecting the value of the variable resistor R_S to be large since R_M is fixed in value. Furthermore, R_S usually consists of two resistors in series, one with a large fixed value of resistance and the other being a small-valued variable resistor for zero adjust. Thus, when the unknown terminals (X–X in Fig. 14.9) are shorted, the DC voltage (V_{DC}) is directly across a particular small-valued precision resistor as well as R_T. Under this condition, the value of V_{DC} and R_T must provide the full-scale deflection current I_{FS}, and a slight zero adjust of R_S is usually necessary for each switch position. After the zero adjust and with the switch in any particular setting, half-scale deflection will occur for an unknown resistor equal in magnitude to the particular small-valued resistance value. Therefore, for the $R \times 1$ setting, a mid-scale reading corresponds to 1 Ω. For the $R \times 10$ and $R \times 100$ settings, mid-scale readings correspond to 10 Ω and 100 Ω, respectively. Resistors with larger values can also be used provided that the value of $R_T = R_S + R_M$ remains much greater.

The principle involved in the multi-range ohmmeter is to divert most of the current through the small-valued resistor. By using additional resistors with precise values in multiples of 10, the resistance is read on the same calibrated scale and multiplied by the corresponding multiple of 10. Therefore, these resistors are also sometimes called *multiplier resistors*.

EX. 14.4

Ohmmeter Example

Consider an ohmmeter that uses a d'Arsonval meter with $I_{FS} = 1$ mA and a battery with $V_{DC} = 3$ V. If the battery voltage reduces to 2.75 V because of aging, determine the error in a measurement at mid range.

Solution: With terminals X–X shorted, the resistance of the meter before aging is (3 V)/(1 mA) = 3 kΩ, and the

meter would read mid range for an unknown resistance of the same value. However, when V_{DC} reduces to 2.75 V, the meter will indicate mid range for an unknown resistance of 2.75 kΩ. Thus, the error in the measurement is $(3 - 2.75)/3$, or 8.3%.

14.1.5   Sinusoidal AC Voltmeters

If a sinusoidally time-varying voltage is applied directly to a d'Arsonval meter, the pointer will rotate up and down the scale corresponding to the particular sinusoidal time variation, with pinning below zero occurring for the negative half periods. Relevant measurements in this case would be extremely tedious, if not impossible. However, if the sinusoidal voltage is first rectified, a DC voltage is produced. Applying this DC voltage to the d'Arsonval meter then allows the measurement of the constant value. Thus, the basic principle of the *AC voltmeter* is that the input sinusoidal AC signal is rectified and then measured with a DC meter. Specific calibration of the measurement is then also required in the form of either rms or peak voltage of the sinusoid.

Figure 14.10a displays a basic circuit that permits the measurement of the peak or rms value of a sinusoidal voltage. This circuit consists of a single diode, a series resistor, and a d'Arsonval meter. A sinusoidal input voltage is applied, and half-wave rectification (as described in Chapters 2 and 7) is provided as indicated in the figure, assuming that the diode is ideal. The voltage to which the meter responds is the average of the rectified voltage. Since the average of a sine wave over a half period is $0.636V_P$, where V_P is the peak value, the average of the rectified half sine wave over one period is one half this value, or $0.318V_P$. In comparing this magnitude to that of a DC voltage, we substitute $V_P = \sqrt{2}V_{rms}$, and the corresponding voltage in terms of the rms input voltage is thus $0.45V_{rms}$. Thus, we observe that for the half-wave rectifier case under consideration, the meter sensitivity for AC has been reduced to 45%

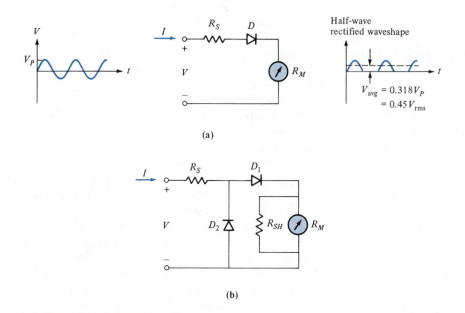

(a)

(b)

FIGURE 14.10   Half-Wave Rectifier Meter: (a) Basic circuit, (b) Practical implementation

of its DC value. The meter current is given by the ratio of the rms voltage to the total loop resistance, or

$$I_M = \frac{0.45 V_{rms}}{R_T} \qquad (14.1\text{–}23)$$

where again $R_T = R_S + R_M$. A meter of this type is adjusted to read I_{FS} for a particular maximum V_{rms} by setting the value of R_T (through R_S).

The circuit arrangement displayed in Fig. 14.10b is a practical implementation of the half-wave rectifier meter. The shunt resistor R_{SH} has a relatively small value in comparison to R_M and therefore allows a larger current to flow through diode D_1. The larger current allows D_1 to operate in a more linear manner. Diode D_2 is included to provide an alternate path for current flow in the reverse direction so that the reverse current of D_1 is eliminated and does not flow through the meter. Note that D_2 has no effect when D_1 is conducting because it is reverse biased.

To improve the sensitivity of AC meters, full-wave rectification instead of half-wave rectification is usually employed. Figure 14.11 shows the basic circuit for a full-wave rectifier meter that uses the bridge circuit described in Chapter 2. Since the average value of the full-wave rectifier output is twice that of the half-wave rectifier, the sensitivity is doubled. The basic design formula for meter current is then given by twice that of

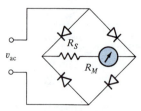

FIGURE 14.11   AC Voltmeter Using a Full-Wave Bridge Rectifier

(14.1–23), or

$$I_M = \frac{0.9 V_{\text{rms}}}{R_T} \qquad (14.1\text{--}24)$$

This equation is then used along with I_{FS} to set the value of R_T that corresponds to the maximum V_{rms} desired.

Resistor networks consisting basically of multiplier resistors are then employed to change the possible range of measurements. These circuits are similar to those previously described for the DC case and will be omitted here.

However, one additional type of AC voltmeter is of sufficient importance to be described. This type enables measurement of a peak-to-peak value of voltage where the waveforms need not be sinusoidal. A specific circuit to demonstrate the principle involved is shown in Fig. 14.12.

Note that D_1, C_1, and the AC input voltage v_{AC} form a clamping circuit (as described in Chapter 2). The voltage at P in Fig. 14.12 (V_P) is the capacitor voltage V_{C_1} added to v_{AC}, where the magnitude of V_{C_1} is equal to the negative peak value of v_{AC}. Capacitor C_2 will then charge through D_2 to the peak value at P, which is the sum of the positive and negative peaks of v_{AC}. A voltmeter consisting of R_S and a d'Arsonval meter as shown in Fig. 14.12 are used to measure the peak-to-peak voltage resulting across C_2. The

total resistance $R_T = R_S + R_M$ is assumed large enough so that C_2 does not discharge appreciably.

————————————————————————— EX. 14.5

AC Voltmeter Example

Consider the AC voltmeter of Fig. 14.12. For $v_{\text{AC}} = V_S \sin \omega t$, sketch the waveshapes of the voltages across C_1 and C_2 versus t. Treat D_1 and D_2 as ideal, with $R_S + R_M$ as an open circuit. Let $C_1 = C_2$ and $V_S = 1$ V.

Solution: From the methods of Chapter 2, the waveshapes for v_{C_1} and v_{C_2} are determined as shown in Fig. E14.5. During the first quarter-cycle of v_{AC}, D_2 conducts while D_1 is open, and $|v_{C_1}| = |v_{C_2}| = (1/2)v_{\text{AC}}$. During the second quarter-cycle of v_{AC}, both v_{C_1} and v_{C_2} remain fixed until v_{AC} decreases to one half of its peak value, at which time D_1 shorts and v_{C_1} begins to discharge, with $v_{C_1} = v_{\text{AC}}$ and v_{C_2} unchanged. During the third quarter-cycle, D_1 remains shorted and D_2 open, and C_1 charges to the peak of v_{AC} with polarity as indicated. During the fourth quarter-cycle, C_2 charges to V_S with D_2 shorted. Finally, during the

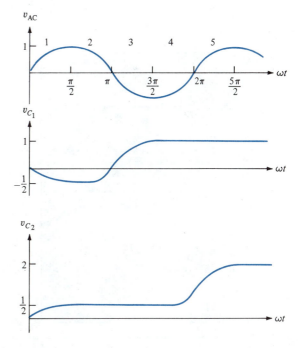

FIGURE E14.5

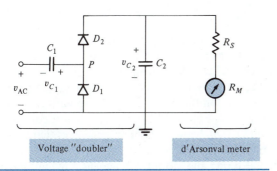

Voltage "doubler"   d'Arsonval meter

FIGURE 14.12   Peak-to-Peak AC Voltmeter

fifth quarter-cycle, D_2 remains shorted, and C_2 charges to the peak of $v_{AC} + V_S$, which is $2V_S$. After this time period, D_1 and D_2 remain open.

14.1.6 Multimeters

Each of the instruments described thus far utilizes the basic d'Arsonval movement. We have observed that DC ammeters and voltmeters consist of resistors interconnected in a particular way with the d'Arsonval winding. We have also seen that ohmmeters are constructed with a particular circuit arrangement of resistors along with a DC voltage source. Furthermore, sinusoidal AC voltmeters use a rectifier circuit in combination with a DC voltmeter. Quite simply, a *multimeter* is a measuring instrument that contains the associated circuitry to perform all three of the basic DC measurements, as well as the sinusoidal AC voltage measurement. Such an in-

strument is called a *volt-ohm-milliampere* (VOM) measuring apparatus.

Figure 14.13 displays the front panel of a typical multimeter. The function switch is initially set to the specified measurement to be made, with the range switch in the highest setting. To obtain an accurate reading, the range switch setting is then carefully changed to lower ranges that allow readings that do not pin the meter.

14.2 ELECTRONIC VOLTMETERS (EVMs)

Although the meters discussed thus far have a wide variety of uses, they also have inherent disadvantages. One of the primary disadvantages is lack of sensitivity. Another is that, as in the case of an electrical voltmeter, the impedance of the instrument varies with the particular switch setting and in certain instances may not be much greater than that of the circuit under measurement. Hence, error will be introduced into the measurement. These disadvantages can be eliminated through the use of the *electronic voltmeter* (EVM). This instrument includes AC and DC amplification of the voltage to be measured and is designed to provide a very large input impedance (on the order of $M\Omega$).

14.2.1 EVM Multimeters

In general, EVMs are instruments that are used to measure DC and AC voltages. However, DC and AC currents can also be measured directly by using a current-to-voltage converter based upon the principle of the op amp circuit of Fig. 7.14. Figure 14.14 displays such a converter with a switch and resistor arrangement that multiplies the current by factors of 10. For $R_F = 1, 10,$ and $100 \, \Omega$, the output voltage has a magnitude of i_s, $10i_s$, and $100 \, i_s$, respectively.

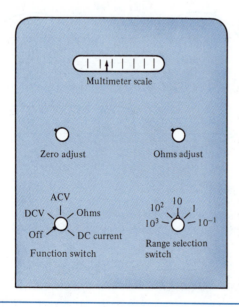

FIGURE 14.13   Front Panel of a Typical Multimeter

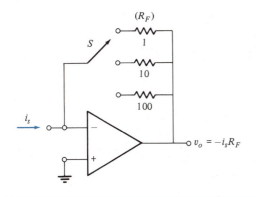

FIGURE 14.14   Current-to-Voltage Converter with Three Ranges of Current

Furthermore, with additional circuitry, the meter can also be used to measure resistance. Hence, we observe that EVMs can also be constructed in multimeter form and that the primary advantage is greater sensitivity.

14.2.2   Types of EVMs

Many different types of EVMs are presently available for use. Originally, vacuum tubes were used in these instruments, and they were known as *vacuum tube voltmeters* (VTVMs). Some VTVMs are still in use; however, the more modern solid-state models are far superior.

EVMs are categorized as either *analog* or *digital,* depending on the type of readout. Both analog and digital EVMs operate in the same analog manner with the incoming voltage being linearly amplified. However, the digital EVM must convert the analog output voltage to digital form, which is appropriate for the input of a digital display. This conversion is carried out by a special digital circuit called an analog-to-digital (A to D) converter. This type of converter, as well as instruments whose entire operation is based solely on digital logic circuits, are described in a later section of this chapter.

14.2.3   DC Electronic Voltmeters

The operation of EVMs is most easily described by initially considering the simplest possible case, which is the single-function DC EVM. This type of instrument is available commercially; however, multimeter EVMs are more prevalent as well as economical.

As a specific example, we consider the JFET difference amplifier DC EVM of Fig. 14.15. The use of JFETs provides a large input impedance (as described in Chapter 5) that we assume here to be 1 MΩ. The EVM of Fig. 14.15 contains the necessary components to measure DC voltages on three ranges as indicated, where the total resistance of the input attenuator branch is selected to match the input impedance of the JFET. The input attenuator permits a reduction (when required) in the magnitude of input voltage to be amplified. Note that the difference amplifier circuit is the same as described in Chapter 6 and displayed in Fig. 6.18 except for the adjustment potentiometer (R_A) that is necessary to balance the circuit. We will neglect these elements in our analysis and assume a balanced arrangement.

To analyze the DC EVM of Fig. 14.15, recall from Chapter 6 that the voltages at the drain terminals, without the meter connected in the circuit, are given by (6.2–83) and (6.2–84), where $A_c \ll A_d$. Thus, by rewriting these equations, we have under open-circuit output conditions,

$$v_{o_1} = \frac{g_m R_d v_d}{2} \qquad (14.2-1)$$

and

$$v_{o_2} = \frac{-g_m R_d v_d}{2} \qquad (14.2-2)$$

where v_d is the difference in input gate voltages $(v_{g_2} - v_{g_1})$.

For the EVM of Fig. 14.15, $v_d = -v_I$ (since $v_{g_2} = 0$), and thus

$$v_{o_1} = -v_{o_2} = \frac{-g_m R_d v_I}{2} \qquad (14.2-3)$$

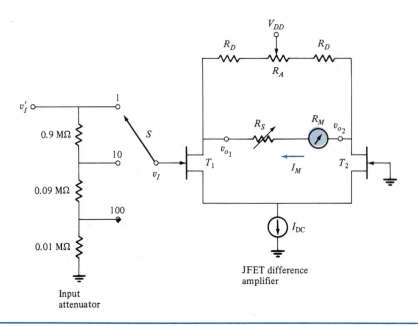

FIGURE 14.15   JFET Difference Amplifier DC EVM

Therefore, the voltage from drain to drain without the meter connection present is simply the difference in these voltages, or

$$v_{o_2} - v_{o_1} = g_m R_d v_I = V_{OC} \qquad (14.2\text{–}4)$$

where the open-circuit output voltage has been explicitly denoted as V_{OC}.

To determine the meter current with the value of V_{OC} known, we next determine the output impedance of the amplifier, which is the Thévenin equivalent resistance (R_{TH}) obtained by setting the sources of the amplifier equal to zero. By inspection of the circuit of Fig. 14.15, $R_{TH} = 2R_D$ since the output drain resistors of the JFETs along with the AC resistance of the current source are assumed to be large. Thus, the current through the meter is given by

$$I_M = \frac{V_{OC}}{R_{TH} + R_S + R_M} \qquad (14.2\text{–}5)$$

By substituting for V_{OC} and R_{TH}, we have

$$I_M = \frac{g_m R_D v_I}{2R_D + R_S + R_M} \qquad (14.2\text{–}6)$$

Hence, the meter current is directly proportional to the input DC voltage. Proper calibration of the meter scale then allows DC voltage measurement.

_____ EX. 14.6

DC EVM Example

Consider the JFET DC EVM of Fig. 14.15 with identical transistors and parameters $g_o = 0$ and $g_m = 0.01$ S. Let $R_D = 1$ kΩ, $v_I = 100$ mV, and $R_M + R_S = 1$ kΩ. Calculate the meter current I_M.

Solution: From (14.2–6), we have

$$I_M = \frac{g_m R_D v_I}{2R_D + R_S + R_M} = \frac{(0.01)(10^3)(0.1)}{3 \times 10^3} = 0.33 \text{ mA}$$

14.2.4   AC Electronic Voltmeters

A rectifier in combination with a voltage amplifier and meter provides the necessary components for AC voltage measurement. Two circuit arrangements are possible in that rectification can either take place before or after the amplification stage.

For the rectifier stage, either half-wave or full-wave rectification can be used. Additionally, active rectifiers using op amps are often used to replace the ordinary diode rectifier circuits. The active diode rectifier circuits of Section 7.3.1 are entirely appropriate for this AC voltmeter application.

14.2.5   Overall Operation of an EVM

Figure 14.16 displays in block diagram form the essential elements of an EVM that measures AC and DC voltages. The input attenuator is similar to the one in Fig. 14.15. The range selection switch on the instrument panel provides various alternative settings.

The DC amplifier stage amplifies AC as well as DC and has a large input impedance. Modern EVMs use difference amplifiers or an op amp to provide the necessary high gain and input impedance.

The last component is the rectifier stage. As previously mentioned, this portion of the circuit consists of diodes along with op amps. Additional amplification can also be provided by this stage.

Note that the EVM indicates a voltage proportional to the input voltage. The voltage reading is either DC or the rms value, depending upon whether the voltage being measured is DC or sinusoidal, respectively. In the case of a digital readout, the meter voltage is converted to digital form and displayed digitally, as will be described in detail in Section 14.7.

14.3   DC POWER SUPPLIES

As we have seen, all electronic circuits require a DC supply voltage. In portable circuits (such as in a hand calculator or portable radio), the required DC supply voltage is provided by a battery that converts electrochemical energy into electrical energy. In the laboratory, for testing purposes, a *DC power supply* is used to convert the AC power available at the 115 V, 60 Hz power outlet into DC. Similar power conversion circuitry is contained in nonportable electronic circuits (such as in a stereo or large television).

The DC power supply is an instrument whose output voltage can be varied within a specific range. Usually, this range begins at 0 V and can be increased to a certain maximum with coarse and fine tuning knobs. Many power supplies have both positive and negative voltage terminals available, which are also adjustable and are very con-

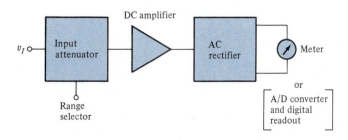

FIGURE 14.16   Block Diagram of an EVM

venient for use in circuits that require a positive and negative supply voltage.

A maximum current limit is associated with each DC power supply. When additional current is required, two power supplies can be paralleled; when additional voltage is required, two power supplies can be connected in series. For series operation, one of the power supplies must be operated in a floating mode (with its ground disconnected).

14.3.1   Regulated DC Power Supplies

Special circuitry is used to stabilize the DC output voltage of a power supply to variations in load current, AC input (line) voltage, and temperature. The output voltage of the power supply is then referred to as being *regulated*. Such regulation is totally impossible using simple rectifiers and filters and even the zener diode regulator example described in Chapter 2.

More elaborate circuitry utilizing negative feedback provides the improvement in regulation. A portion of the output voltage is fed back to the input and is compared to a desired voltage. If these voltages differ, the negative feedback automatically adjusts the output voltage to the correct value. The regulated DC power supply then provides a fixed output voltage independent of load variation, AC line variation, and temperature.

14.3.2   Series-Regulated DC Power Supply

Figure 14.17 displays one example of the basic circuitry for a series voltage regulator. This circuit is called a *series regulator* because the entire load current passes through the power transistor T_1. Shunt regulators are also possible where the power transistor is in parallel with the load. However, this arrangement has the disadvantage of yielding larger power loss.

The input portion of the regulator circuit consists of a full-wave bridge rectifier being supplied

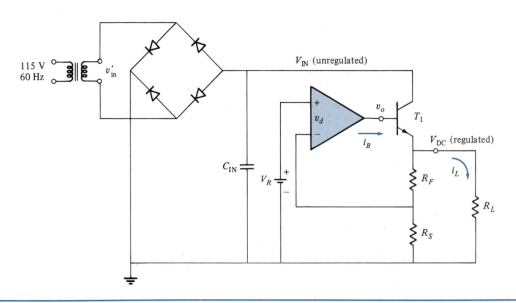

FIGURE 14.17   Series-Regulated DC Power Supply

with AC line power. This voltage is usually stepped down from the input voltage so that $|v'_{in}| <$ 115 V. The rectifier portion of the circuit then produces an unregulated DC voltage V_{IN}. As observed in Chapter 2, V_{IN} is essentially DC with a ripple voltage superimposed upon it.

To analyze the output portion of the circuit of Fig. 14.17, we first consider the operation of the op amp. The op amp is connected as a difference amplifier and amplifies the difference in voltage between the reference voltage V_{ref} and the fraction of output voltage, or $[R_S/(R_F + R_S)]V_{DC}$. The output voltage of the op amp is therefore

$$v_O = A_d(v_p - v_n) = A_d\left(V_{ref} - \frac{R_S}{R_F + R_S}V_{DC}\right)$$

$$(14.3\text{--}1)$$

where we have assumed that $R_L \gg R_o$, the output impedance of the op amp (we remove this restriction in the next example). Additionally, the transistor acts like an emitter follower, and thus

$$v_O = V_{DC} + 0.7 \qquad (14.3\text{--}2)$$

Equating (14.3–1) with (14.3–2) and solving for V_{DC} yield

$$V_{DC} = \frac{A_d V_{ref}}{1 + A_d\left(\dfrac{R_S}{R_F + R_S}\right)} - \frac{0.7}{1 + A_d\left(\dfrac{R_S}{R_F + R_S}\right)}$$

$$(14.3\text{--}3)$$

Finally, for large A_d, the last term is negligible, and

$$V_{DC} = \frac{R_F + R_S}{R_S} V_{ref} \qquad (14.3\text{--}4)$$

which is the regulated voltage output expression. Note that this voltage is adjustable by varying either R_F or R_S. Furthermore, V_{DC} is temperature compensated provided that the reference voltage is temperature compensated since the resistors

have the same temperature variation. Finally, note that V_{IN} must always be greater than V_{DC} in order for the transistor to operate as an emitter follower.

It should be noted that the DC supply voltage for the op amp is obtained from the unregulated voltage V_{IN}. Furthermore, the reference voltage V_{ref} is obtained by using a special temperature-compensation circuit (similar to that described in Chapter 11, Fig. 11.12) that is supplied by feedback from V_{DC}.

——————————————————————— EX. 14.7

Series-Regulated DC Power Supply Example

Determine the variation in V_{DC} with i_L for the regulated power supply of Fig. 14.17. Let the op amp output parameters be $R_o = 50\ \Omega$ and $A_d = 10^3$, while the BJT has $\beta = 100$ and $V_{BE} = 0.7$ V. Also let $R_S/(R_F + R_S) = 0.03$, $R_L = 1\ k\Omega$, and $V_{ref} = 0.31$ V. Neglect the current through the feedback path, assuming that R_S and R_F are in the MΩ range and R_L is much smaller.

Solution: In this example, the output impedance of the op amp is taken into account. To obtain V_{DC} as a function of i_L, the circuit shown in Fig. E14.7 is used (where the current through R_F and R_S is neglected). The magnitude of the controlled voltage source is given by

$$A_d v_d = A_d(V_{ref} - k V_{DC}) \qquad (1)$$

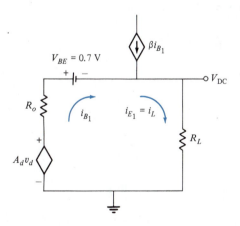

FIGURE E14.7

where $k = R_S/(R_F + R_S)$. Writing KVL for the output loop of Fig. E14.7 yields

$$V_{DC} + 0.7 = A_d v_d - R_o i_B$$

$$= A_d(V_{ref} - kV_{DC}) - R_o\left(\frac{i_L}{\beta + 1}\right) \tag{2}$$

By solving for V_{DC}, we have

$$V_{DC} = \frac{A_d V_{ref}}{1 + kA_d} - \frac{R_o i_L}{(\beta + 1)(1 + kA_d)} - \frac{0.7}{1 + kA_d} \tag{3}$$

Substituting values yields

$$V_{DC} = \frac{10^3(0.31)}{1 + 0.03(10^3)} - \frac{50i_L}{101(1 + 30)} - \frac{0.7}{1 + 30} \tag{4}$$

or

$$V_{DC} = 10 - \frac{i_L}{62.6} \tag{5}$$

Finally, if $R_L = 1\ k\Omega$, then $V_{DC} = i_L(1000)$. By substituting into (5), we obtain

$$V_{DC} = 10 - \frac{V_{DC}}{62.6 \times 10^3} \tag{6}$$

or by solving for V_{DC}, we have

$$V_{DC} = \frac{10}{1 + [1/(62.6 \times 10^3)]} = 9.9998\ V \tag{7}$$

Note that even for $R_L = 10\ \Omega$, we have

$$V_{DC} = 10 - \frac{V_{DC}}{626} \tag{8}$$

Solving for V_{DC} yields

$$V_{DC} = \frac{10}{1 + (1/626)} = 9.984\ V \tag{9}$$

14.3.3　IC Voltage Regulators

Very accurate voltage regulators in the form of IC chips are available from numerous manufacturers. These devices can provide up to several amps of current with typical output voltages from 2.6 V to 24 V. Data Sheet A14.1 in the appendix to this chapter indicates various voltage regulators with both positive and negative voltage outputs. The KC and LP packages are three-terminal ceramic and plastic, respectively. The reader is referred to the appendix at the back of the book for more detailed information and data sheets for several representative voltage regulators.

14.4　SIGNAL GENERATORS

Signal generators are instruments that provide signals for general electronic test purposes. Various different output signals are available with these instruments including sine, square, triangular, ramp, and pulse waveshapes. Several of these functions are available at the output of a function generator. Each particular waveshape can be achieved by some form of an oscillator and associated signal-conditioning circuitry. An oscillator is a circuit in which DC power delivered by the supply voltage is converted into a purely sinusoidal voltage.

We begin our discussion in this section by first describing the principle of operation of basic oscillators and laboratory-type oscillators. We then consider the basic principles involved in function generators.

14.4.1　Basic Oscillator Circuits

As stated previously, an *oscillator* is a circuit in which DC power delivered by the supply voltage is converted into a purely sinusoidal voltage. Hence, oscillators do not have an AC input; however, the AC sinusoidal output is generated through the use of positive feedback. Fundamentally, an oscillator can be considered to be an amplifier whose AC input consists of the positive feedback signal.

Figure 14.18 displays the four basic oscillator configurations using op amps to provide amplifi-

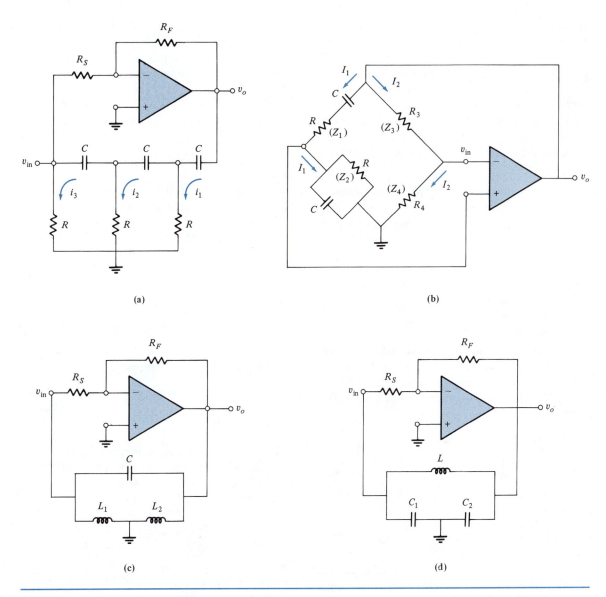

FIGURE 14.18   Basic Oscillators: (a) Phase shift oscillator, (b) Wien bridge oscillator, (c) Hartley oscillator, (d) Colpitts oscillator

cation. Earlier versions of these basic oscillators used transistors, and many other configurations are possible. The *phase shift oscillator* and the *Wien bridge oscillator* both employ *RC* networks in their feedback branches. These oscillators are

useful for generating signals with frequencies up to approximately 100 kHz.

The other two circuits of Fig. 14.18 are the *Hartley oscillator* and *Colpitts oscillator*, which use *LC* phase shift networks. These oscillators are useful for generating high frequency signals in the rf range (100 kHz to 100 MHz).

For these circuits to be effective oscillators, each must provide zero phase shift between the output and input terminals. Under these conditions, the output voltage v_o reinforces the feedback input voltage v_{in}, and sustained oscillation results at one particular frequency provided that the gain of the amplifier is sufficient to overcome losses.

Phase Shift Oscillator.　The phase shift oscillator of Fig. 14.18a provides a 180° phase shift due to the inverting op amp configuration, for which $v_o = -(R_F/R_S) v_{in}$ (assumed to be independent of frequency). At the same time, the RC phase shift network provides an additional 180° phase rotation at one particular frequency. Note that three capacitors are necessary because each provides a maximum of 90° phase shift, and hence only two would barely be sufficient.

To determine the voltage transfer function v_{in}/v_o and the phase shift of the RC network, we carry out a loop analysis (as introduced in Chapter 1). Writing KVL for each loop of Fig. 14.18a thus yields (assuming $R_S \gg R$)

$$v_o = \left(\frac{1}{j\omega C} + R\right) i_1 - R i_2 \tag{14.4-1}$$

$$0 = -R i_1 + \left(2R + \frac{1}{j\omega C}\right) i_2 - R i_3 \tag{14.4-2}$$

$$0 = -R i_2 + \left(2R + \frac{1}{j\omega C}\right) i_3 \tag{14.4-3}$$

This set of equations is easily solved for i_1, i_2, and i_3. However, we are interested only in determining the voltage transfer function v_{in}/v_o, where $v_{in} = R i_3$. Solving for i_3, multiplying the result by R_1, dividing by v_o, and grouping real and imaginary parts thus yield the desired transfer function as

$$\frac{v_{in}}{v_o} = \frac{1}{1 - \dfrac{5}{(\omega RC)^2} + j\left[\dfrac{1}{(\omega RC)^3} - \dfrac{6}{\omega RC}\right]} \tag{14.4-4}$$

where verification of (14.4–4) is left as an exercise for the reader.

To achieve 180° phase shift, the imaginary part in the denominator must be zero. Under these conditions, (14.4–4) becomes

$$\frac{v_{in}}{v_o} = \frac{1}{1 - \dfrac{5}{(\omega_o RC)^2}} \tag{14.4-5}$$

where ω_o is the particular oscillator frequency. Note also that the second term in the denominator of (14.4–5) must be greater than 1 in order to obtain the desired 180° phase shift between v_{in} and v_o.

By setting the imaginary part of (14.4–4) equal to zero, we obtain

$$\frac{1}{(\omega_o RC)^3} - \frac{6}{\omega_o RC} = 0 \tag{14.4-6}$$

Solving for ω_o thus yields

$$\omega_o = \frac{1}{\sqrt{6}RC} \tag{14.4-7}$$

which is the particular oscillator frequency.

At this frequency, the voltage transfer function of (14.4–5) is then given by

$$\frac{v_{in}}{v_o} = \frac{1}{1 - (5 \times 6)} = \frac{-1}{29} \tag{14.4-8}$$

which results in a reduction in amplitude from output to input by a factor of 29. The amplification of the inverting amplifier must at least recover this loss, where $R_F/R_S > 29$ in order to sustain the oscillation.

The phase shift oscillator is useful primarily because of its simplicity. However, adjusting the frequency of oscillation is inconvenient because all three capacitors (or resistors) must be varied. As we will see, the Wien bridge is a more suitable oscillator alternative and, in addition, provides frequency stability.

EX. 14.8

Phase Shift Oscillator Design Example

For the phase shift oscillator of Fig. 14.18a, determine values for R and R_F such that the frequency of oscillation is 15 kHz, $C = 0.001 \ \mu F$, and $R_S = 100 \ k\Omega$.

Solution: Solving for R from (14.4–7), we have

$$R = \frac{1}{\sqrt{6}\omega_o C} = \frac{-1}{2.45(2\pi)(15 \times 10^3)(10^{-9})}$$

$$= 4.33 \ k\Omega$$

Additionally, from (14.4–8), $v_{in}/v_o = -1/29$, and thus the amplification necessary from R_F/R_S is

$$\frac{R_F}{R_S} = 29$$

By substituting $R_S = 100 \ k\Omega$, we obtain $R_F = 2.9 \ M\Omega$. Usually, R_F is chosen slightly greater than 2.9 MΩ to ensure oscillation.

Wien Bridge Oscillator.   The circuit of Fig. 14.18b displays a phase shift network in the form of an impedance bridge, where the impedances Z_1, Z_2, Z_3, and Z_4 are connected end to end and the op amp is bridged diametrically across opposite ends. This circuit is a Wien bridge when the impedance elements are selected as shown in the figure.

To analyze the circuit, we first recall from Chapter 7 that the op amp draws negligible current. Additionally, for amplifier operation, the difference voltage v_d is approximately zero. Under these conditions, the bridge circuit is said to be balanced with

$$I_1 Z_1 = I_2 Z_3 \tag{14.4–9}$$

and

$$I_1 Z_2 = I_2 Z_4 \tag{14.4–10}$$

By taking the ratio of (14.4–9) to (14.4–10), we have

$$\frac{Z_1}{Z_2} = \frac{Z_3}{Z_4} \tag{14.4–11}$$

Then, since the specific element values for the four impedances are

$$Z_1 = R + \frac{1}{j\omega C} \tag{14.4–12}$$

$$Z_2 = \frac{R}{1 + j\omega RC} \tag{14.4–13}$$

$$Z_3 = R_3 \tag{14.4–14}$$

$$Z_4 = R_4 \tag{14.4–15}$$

substitution into (14.4–11) gives

$$\frac{R + \dfrac{1}{j\omega C}}{\dfrac{R}{1 + j\omega RC}} = \frac{R_3}{R_4} \tag{14.4–16}$$

Rearranging yields

$$2 + j\left(\omega RC - \frac{1}{\omega RC}\right) = \frac{R_3}{R_4} \tag{14.4–17}$$

Equating the real and imaginary parts then yields

$$\frac{R_3}{R_4} = 2 \tag{14.4–18}$$

and

$$\omega RC - \frac{1}{\omega RC} = 0 \tag{14.4–19}$$

or

$$\omega_o = \frac{1}{RC} \tag{14.4–20}$$

where $\omega = \omega_o$ is the solution of (14.4–19).

Thus, we see that for balanced bridge operation, (14.4–18) and (14.4–20) must be satisfied with oscillation only occurring at a particular frequency, where $\omega_o = 1/RC$. Under these conditions, the voltage at the output of Fig. 14.18b is in phase with the voltage at the left side of the bridge circuit, which is in phase with v_{in}.

Adjustable resistors and capacitors are used for the R and C elements, which then provide a variable oscillation frequency. Selection of R_3 and R_4 provides the negative feedback gain (as can be seen from the circuit of Fig. 14.18b). Often, a variable resistor with a positive temperature coefficient is used for R_4 to provide automatic gain control and to stabilize the oscillation frequency.

Hartley and Colpitts Oscillators. The phase shift networks for the Hartley and Colpitts oscillators displayed in Figs. 14.18c and d are LC combinations that are often called *tank circuits* because they store energy. These oscillators are used for the generation of rf signals and act like band-pass filters to allow transmission of the oscillation frequency and block all other frequencies. A 180° phase change is provided by the amplifier for which $v_o = -(R_F/R_S)v_{in}$, and the tank circuit provides another 180° phase change. Hence, v_{in} is reinforced by v_o, and oscillation can occur at a particular frequency.

To determine the oscillation frequency for either the Hartley or the Colpitts oscillators, we simply determine the tank circuit resonant frequency. This frequency is obtained by equating the capacitive and inductive reactances, or

$$X_C = X_L \qquad (14.4-21)$$

Thus, for the Hartley oscillator, the solution for the particular frequency is obtained from

$$\frac{1}{\omega_o C} = \omega_o(L_1 + L_2) \qquad (14.4-22)$$

or

$$\omega_o = \frac{1}{\sqrt{C(L_1 + L_2)}} \qquad (14.4-23)$$

Similarly, for the Colpitts oscillator,

$$\frac{1}{\omega_o\left(\dfrac{C_1 C_2}{C_1 + C_2}\right)} = \omega_o L \qquad (14.4-24)$$

Solving for ω_o yields

$$\omega_o = \frac{1}{\sqrt{L[C_1 C_2/(C_1 + C_2)]}} \qquad (14.4-25)$$

where the result of capacitors in series similar to resistors in parallel is observed.

To show that the tank circuit does indeed provide an additional 180° phase change, we first check this result for the Hartley oscillator. From Fig. 14.18c, we observe from the tank circuit that v_o divides across C and L_1 to yield (assuming that R_S for the op amp is large)

$$\begin{aligned} v_{in} &= v_o\left[\frac{j\omega L_1}{j\omega L_1 + (1/j\omega C)}\right] \\ &= v_o\left[\frac{-\omega^2 L_1 C}{-\omega^2 L_1 C + 1}\right] \end{aligned} \qquad (14.4-26)$$

Substituting for ω from (14.4–23) yields

$$v_{in} = v_o\left[\frac{-\dfrac{1}{C(L_1 + L_2)}L_1 C}{\dfrac{-1}{C(L_1 + L_2)}L_1 C + 1}\right] \qquad (14.4-27)$$

Rearranging yields

$$v_{in} = v_o\left(-\frac{L_1}{L_2}\right) \qquad (14.4-28)$$

The negative sign of (14.4–28) indicates the desired result.

  EX. 14.9

Hartley Oscillator Example

Determine the frequency of oscillation for the Hartley oscillator with $L_1 = 0.1$ mH, $L_2 = 0.4$ mH, and $C = 1$ pF.

Solution: Substituting directly into (14.4–23) yields

$$\omega_o = \frac{1}{\sqrt{10^{-12}(0.5 \times 10^{-3})}} = 0.45 \times 10^8 \text{ rad/s}$$

or

$$f_o = \frac{\omega_o}{2\pi} = 7.1 \text{ MHz}$$

Similarly, for the Colpitts oscillator, we have

$$v_{\text{in}} = v_o \left[\frac{1/j\omega C_1}{j\omega L + (1/j\omega C_1)} \right]$$

$$= v_o \left[\frac{1}{-\omega^2 LC_1 + 1} \right] \qquad (14.4-29)$$

Substituting for ω from (14.4-25) yields

$$v_{\text{in}} = v_o \left[\frac{1}{\dfrac{-LC_1}{L[C_1 C_2/(C_1 + C_2)]} + 1} \right]$$

$$= v_o \left(-\frac{C_2}{C_1} \right) \qquad (14.4-30)$$

In the Hartley oscillator, the frequency is adjustable by using a variable capacitor for C. A tapped inductor is used for L_1 and L_2. For the Colpitts oscillator, either a variable inductor or a variable capacitor is used.

Quartz Crystal Oscillators.   Some oscillator applications (electronic watches, for example) require extremely stable and precise control of the oscillator frequency. A *quartz crystal oscillator* meets this requirement and is often referred to as a *quartz resonator,* while the oscillator frequency is called the *resonant frequency.*

The quartz crystal oscillator operates on the principle of the piezoelectric effect as described in Chapter 12. The crystal is fabricated very precisely by vacuum depositing thin conducting electrodes onto the flat faces of a thin quartz substrate. The resulting structure is a parallel-plate capacitor with the quartz as the insulator sandwiched between the two plates. When an alternating voltage is applied between the plates, the quartz vibrates proportionally with the applied voltage. The mechanical resonant frequency is

directly dependent upon the frequency of the applied voltage and the dimensions of the quartz slab. Very precise control of the dimensions of the quartz resonator produces an ultrastable and accurate oscillator frequency.

The circuit symbol for the quartz crystal is shown in Fig. 14.19a. The mechanical resonance can be represented in an electrical analog circuit consisting of series elements R_S, C_S, and L_S in parallel with a parallel-plate capacitance C_P, as shown in Fig. 14.19b. The capacitance C_P is used to model the parallel-plate capacitance associated with the two electrodes separated by the thin slab of quartz.

A quartz crystal oscillator circuit is displayed in Fig. 14.19c. Defining v_{in} as the voltage at the positive terminal of the op amp, we have

$$\frac{v_o}{v_{\text{in}}} = \left(1 + \frac{R_F}{R_S} \right) \left(\frac{Z_C}{Z_C + R} \right) \qquad (14.4-31)$$

where Z_C is the parallel impedance of the crystal. At the resonant frequency, Z_C is large and zero phase shift results between v_o and v_{in}. Under these conditions, oscillation occurs and v_{in} is reinforced by v_o. At all other frequencies, Z_C is reduced and the zero phase shift oscillation requirement is not satisfied.

Voltage-Controlled Oscillators.   Oscillators are available in IC form with output frequency dependent upon the magnitude of an input control voltage. Such ICs are called *voltage-controlled oscillators* (VCOs). Variation in output frequency over a range of about three orders of magnitude is typical. A primary example is the Schottky TTL 74124.

14.4.2   Radio-Frequency Signal Generators

Commonly available *radio-frequency signal generators* provide an output voltage whose frequency is variable from approximately 100 kHz

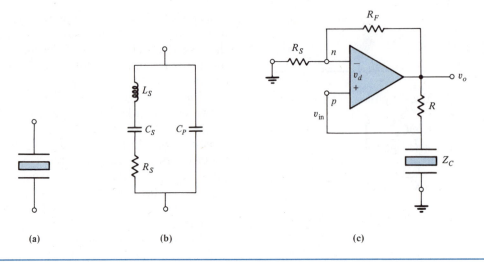

(a)   (b)   (c)

FIGURE 14.19   Quartz Crystal Oscillator: (a) Circuit symbol, (b) Equivalent circuit, (c) Crystal oscillator circuit

to the 100 MHz region. Typical output voltage magnitudes are also variable from the μV to 10 V range. These instruments use a combination of circuitry based upon one of the rf oscillators (using an *LC* tank circuit) described in the previous section.

A block diagram showing the basic constituents of an rf signal generator is displayed in Fig. 14.20. The rf oscillator is designed to have a very stable frequency that is roughly adjusted by tuning to a band of frequencies. The frequency is then finely adjusted to a particular value in the selected band with a vernier frequency selector. The broadband amplifier is designed to operate over a wide range of frequencies and has a small output im-

pedance. Generally, this value is designed to be approximately 50 Ω.

The attenuator network at the output of Fig. 14.20 consists of a resistive voltage divider network with small values of resistance so that the output impedance of the generator itself is also about 50 Ω. The selector switch of the attenuator network reduces the output voltage amplitude to the desired level.

14.4.3   Function Generators

As previously mentioned, a *function generator* is an instrument that can provide several different types of output voltage waveforms. The commonly generated output waveforms from function generators are sinusoidal, square, triangular, ramp, and pulse.

Function generators have three basic control features associated with them. First, a switch is provided to select the particular waveshape desired. Second, the frequency is adjustable and directly determines the period of the waveshape.

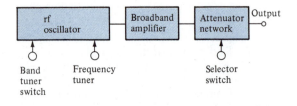

FIGURE 14.20   Radio-Frequency Signal Generator

Finally, the amplitude is adjustable by means of an output attenuator network.

Figure 14.21 displays a simple configuration for the generation of sine, pulse, square, triangular, and ramp voltage waveforms. Because of the use of the comparator for square wave generation, some distortion is introduced in the rise and fall times of the output. However, this simple example is used to demonstrate the basic principles involved in function generators.

With a sine wave at the positive input of the comparator, the output is driven into either positive or negative saturation. The peak values of the square wave then correspond to the positive and negative saturation voltages of the comparator. Choosing these voltages to be equal in magnitude then results in a symmetrical square wave output, as displayed in Fig. 14.21.

As also shown in Fig. 14.21, a pulse waveshape is produced by rectifying the square wave using a half-wave rectifier. Additionally, one

particular form of a triangular waveshape is generated by using an op amp integrator. Other triangular waveshapes may be generated by similar means, as we will see. The ramp function is provided by using full-wave rectification of the square wave and an op amp integrator.

In this simplified example, the frequency (or period) of each waveshape is adjustable by providing a variable element in the oscillator that can be set to a desired value as indicated by the frequency adjust in Fig. 14.21. Although the amplitudes of the generated waveshapes are not adjustable, using an output attenuator and corresponding switch could easily accomplish this adjustment.

Square and Triangular Wave Generators.   As we have seen, a simple square wave generator is composed of a comparator with a sine wave input with peak values equal to the saturation voltages of the comparator. These values are normally the

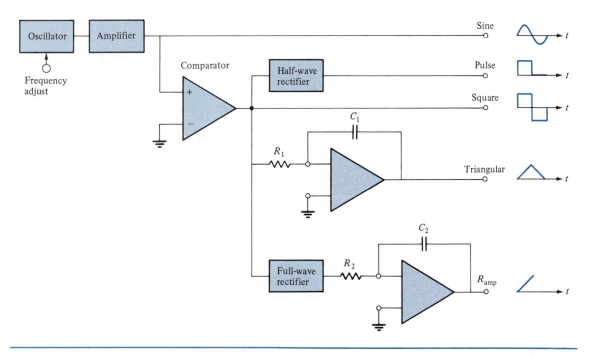

FIGURE 14.21   Simple Function Generator

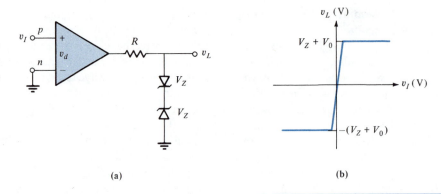

(a)   (b)

FIGURE 14.22   Voltage Comparator with Reduced Saturation Output Voltage: (a) Circuit, (b) Transfer characteristic

positive and negative DC supply voltages. However, Fig. 14.22a displays a comparator circuit in
which the peak values of the square wave are reduced to another value by using two zener diodes
in a back-to-back series configuration. To obtain
a symmetrical square wave, the diodes are chosen
to be identical where the sum of $V_Z + V_0$ is less
than the op amp output saturation voltage. We
observe from Fig. 14.22a that as the input voltage
v_I is increased, the output voltage will increase
and that when $v_L = V_Z + V_0$, where V_0 is the diode
forward voltage, the zener diodes will conduct.
For further increases in v_I, the output voltage then
remains fixed at $V_Z + V_0$. Similar variation occurs
for negative v_I.

───────────────────────── EX. 14.10

Voltage Comparator Design Example

Choose element values for a voltage comparator such that
the positive and negative saturation voltages are ± 5.7 V
and the input voltage range for linear (amplifier) operation
is less than 10 mV. Use a high-gain op amp with negative
feedback and silicon zener diodes.

Solution: Figure E14.10a displays a circuit that can provide the desired operation if the element values are chosen
properly. Note that for $|(R_F/R_S)v_I| \leq V_Z + 0.7$, $v_L = -(R_F/R_S)v_I$. Also note that for $-(R_F/R_S)v_I > V_Z + 0.7$, $v_L = 0.7 +$
V_Z; for $-(R_F/R_S)v_I < -(V_Z + 0.7)$, $v_L = -(V_Z + 0.7)$. Thus,

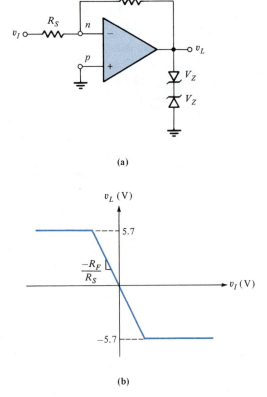

(a)

(b)

FIGURE E14.10

choosing $V_Z = 5$ V yields a saturation output voltage of ± 5.7 V.

To determine the required values for R_F and R_S, we consider the voltage transfer characteristic shown in Fig. E14.10b. Since the input voltage range for amplification is to be less than 10 mV, we have

$$\frac{V_Z + V_0}{R_F/R_S} \le 5 \text{ mV}$$

Solving for R_F/R_S yields

$$\frac{R_F}{R_S} \ge \frac{5.7}{5 \times 10^{-3}} = 1.14 \times 10^3$$

Thus, choosing $R_S = 1$ kΩ and $R_F = 1.2$ MΩ yields the desired characteristic.

The voltage transfer characteristic for the circuit of Fig. 14.22a is displayed in Fig. 14.22b. The resistor R in Fig. 14.22a is necessary in order to support the excess voltage from the output of the op amp. Hence, when v_I is a sinusoid, a square wave is generated with peak values equal to $v_L = \pm(V_Z + V_0)$.

We now emphasize, however, that the generation of a square wave (or triangular wave) does not require a sinusoidal input. Figure 14.23a displays a basic square as well as a triangular wave generator that is based upon the principle of astable multivibrator action. The comparator is used in a Schmitt trigger configuration with positive feedback provided by R_1 and R_2. The zener diodes determine the maximum and minimum values of the symmetrical square wave voltage v_L. This voltage charges and discharges the input capacitor, thus varying the voltage at the negative (n) terminal.

To explain the operation of this circuit, we first note that the difference voltage at the input is given by

$$v_d = v_p - v_n = \frac{R_2}{R_1 + R_2} v_L - v_C \qquad \textbf{(14.4–32)}$$

where v_C is either a rising or a decaying exponential. For the periods of time where $v_L = V_Z + V_0$,

v_C rises exponentially approaching $V_Z + V_0$. However, switching occurs when $v_d = 0$ so that from (14.4–32), we have

$$v_d = 0 = \frac{R_2}{R_1 + R_2}(V_Z + V_0) - v_{C_{\text{peak}}}$$

$$\textbf{(14.4–33)}$$

Therefore,

$$v_{C_{\text{peak}}} = \frac{R_2}{R_1 + R_2}(V_Z + V_0) \qquad \textbf{(14.4–34)}$$

Similarly, the negative peak value has the same magnitude.

The waveshapes for v_L and v_C are displayed in Fig. 14.23b. Note that v_C is only an approximate triangular wave because the capacitor current is not constant; v_C varies exponentially as the capacitor charges and discharges. This variation can be linearized, as we will now see.

Figure 14.23c shows another square and triangular wave generator that uses a different circuit arrangement with an additional op amp (labeled 2) connected as an integrator. The output voltage of the comparator (v_{o_1}) is a symmetrical square wave, and the output of the integrator (v_{o_2}) is a symmetrical triangular wave.

To analyze the circuit of Fig. 14.23c, we begin by writing (by superposition) the general equation for the voltage at the positive terminal for the comparator (labeled 1) as follows:

$$v_{p_1} = \frac{R_2}{R_1 + R_2} v_{o_1} + \frac{R_1}{R_1 + R_2} v_{o_2} \qquad \textbf{(14.4–35)}$$

Since the reference voltage at n_1 is zero ($V_{\text{ref}} = 0$) in this circuit, we note that the comparator switches output states whenever $v_{p_1} = 0$, which occurs when v_{o_1} is either $\pm(V_Z + V_0)$, while v_{o_2} is either decreasing or increasing linearly.

If we initially assume that v_{o_1} is high (because v_{p_1} is positive) corresponding to $v_{o_1} = +(V_Z + V_0)$, then the output of the integrator (v_{o_2}) is a negative ramp. This waveshape is generated because C_I (in Fig. 14.23c) has a constant current supplied from v_{o_1} through R_I, and thus the voltage across C_I (v_{C_I})

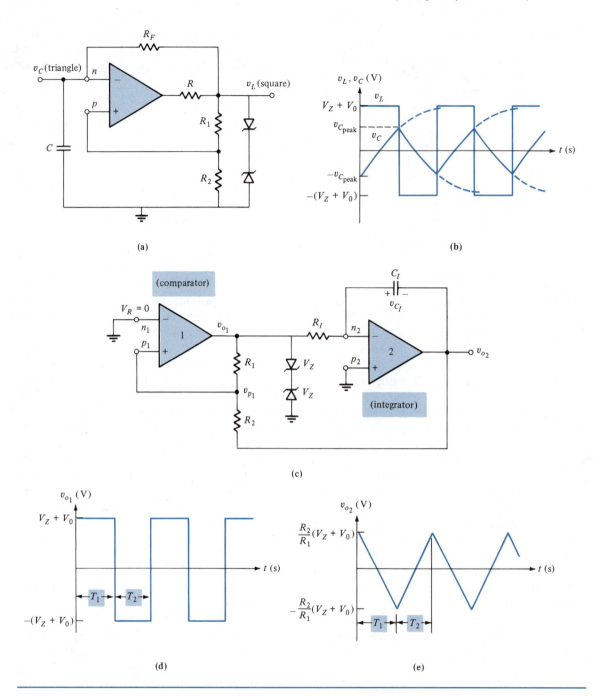

FIGURE 14.23   Square and Triangular Wave Generator: (a) Basic circuit, (b) Square (v_L) and triangular (v_C) waveforms, (c) More precise square and triangular wave generator, (d) Square wave output (v_{o_1}), (e) Triangular wave output (v_{o_2})

increases linearly causing v_{o_2} and also v_{p_1} to decrease linearly. As v_{p_1} becomes zero and goes slightly negative, v_{o_1} switches low to the negative value of $-(V_Z + V_0)$. At this point in time, the output voltage of the integrator (v_{o_2}) is at its negative peak value. Then, v_{o_2} becomes a positive ramp increasing linearly toward its peak value, where switching again occurs.

The positive and negative peak values of v_{o_2} are obtained by setting $v_{p_1} = 0$ in (14.4–35), while $v_{o_1} = \pm(V_Z + V_0)$. Solving for v_{o_2} yields the magnitude of the peak voltage as

$$\left|v_{o_{2\text{peak}}}\right| = \frac{R_2}{R_1}(V_Z + V_0) \qquad (14.4–36)$$

The corresponding waveshapes are displayed in Figs. 14.23d and e.

A more general triangular wave can be obtained as follows. If instead of grounding n_1 (in Fig. 14.23c), we apply a reference voltage V_{ref}, then the triangular wave can be shifted vertically. Furthermore, if the integrator resistor R_I is replaced with an element possessing a variable resistance that changes value corresponding to the two current directions, we can individually control the time length of the positive and negative ramps. The effect of a variable resistance dependent on current direction is easily accomplished by replacing R_I with two diodes and two resistors as displayed in Fig. 14.24a.

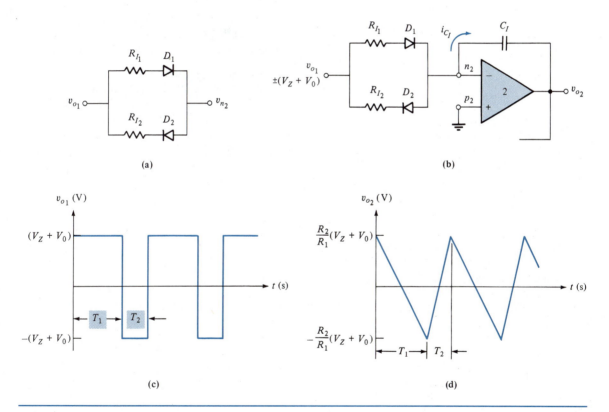

FIGURE 14.24   Modifications to Wave Generator and Resulting Waveforms: (a) Replacement for R_I in Fig. 14.23c, (b) Integrator portion of modified wave generator, (c) Square wave output (v_{o_1}) (d) Triangular wave output (v_{o_2})

To analyze this situation, consider the integrator circuit of Fig. 14.24b in which the input voltage is the square wave v_{o_1} with peaks at $\pm(V_Z + V_0)$. As C_I charges, since n_2 is a virtual ground, the current through C_I is given by

$$i_{C_I} = -C_I\left(\frac{dv_{o_2}}{dt}\right) \tag{14.4-37}$$

where the current is either positive and given by

$$i_{C_I} = +\frac{V_0 + V_Z}{R_{I_1}} \text{ for } v_{o_1} = (V_Z + V_0) \tag{14.4-38}$$

or negative and given by

$$i_{C_I} = -\frac{V_0 + V_Z}{R_{I_1}} \text{ for } v_{o_1} = -(V_Z + V_0) \tag{14.4-39}$$

Thus, for the positive current case, equating (14.4-37) and (14.4-38) yields

$$\frac{dv_{o_2}}{dt} = \frac{V_Z + V_0}{R_{I_1}C_I} \text{ for } v_{o_1} = +(V_Z + V_0) \tag{14.4-40}$$

We recognize (14.4-40) as being the slope for the negative ramp case, and we define the negative ramp time as T_1, as shown in Figs. 14.24c and d. This slope may also be written as the difference in voltage divided by the time period, or from Fig. 14.24d,

$$\frac{dv_{o_2}}{dt} = \frac{\frac{R_2}{R_1}(V_Z + V_0) - \left[-\frac{R_2}{R_1}(V_Z + V_0)\right]}{-T_1}$$

$$= -\frac{2\frac{R_2}{R_1}(V_Z + V_0)}{T_1} \tag{14.4-41}$$

Thus, equating (14.4-40) and (14.4-41) yields

$$\frac{V_Z + V_0}{R_{I_1}C_I} = \frac{2\frac{R_2}{R_1}(V_Z + V_0)}{T_1} \tag{14.4-42}$$

or

$$T_1 = 2\frac{R_2}{R_1}C_I R_{I_1} \tag{14.4-43}$$

In a similar manner, we obtain

$$T_2 = 2\frac{R_2}{R_1}C_I R_{I_2} \tag{14.4-44}$$

Very interestingly, we note that it is possible to force either of these times to be zero by choosing the resistor R_{I_1} or R_{I_2} to be zero. In this manner, a vertically shifted sawtooth waveform is obtained.

Sawtooth Wave Generators. Figure 14.25 displays a vertically shifted sawtooth waveshape that is generated by the circuit of Fig. 14.24b with $R_{I_1} = 0$. Under these conditions, $T_1 = 0$. Actually, T_1 will be slightly greater than zero, as indicated in Fig. 14.25, and will depend upon the diode forward resistance R_D, which is usually negligible, assuming that $R_{I_2} \gg R_D$.

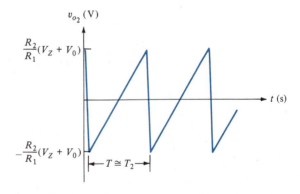

FIGURE 14.25   Shifted Sawtooth Waveshape with Period $T \cong T_2$

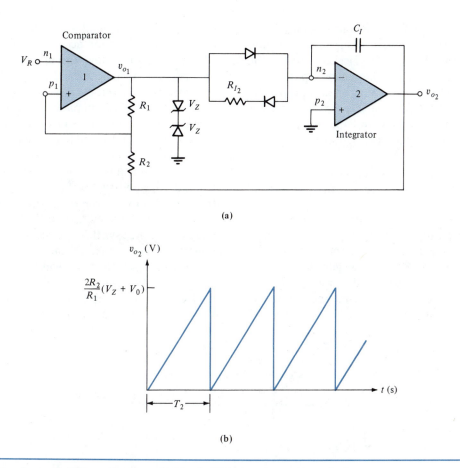

(a)

(b)

FIGURE 14.26   Sawtooth Voltage Generator: (a) Circuit, (b) Sawtooth waveshape (v_{o_2})

The sawtooth generator circuit shown in Fig. 14.26a shifts the waveshape vertically so that its minimum is at 0 V, as shown in Fig. 14.26b. We apply a particular reference voltage to n_1 as indicated in Fig. 14.26a. The value required must make the peak of v_{o_2} become twice that of the previous value, while the minimum becomes zero. The total amount of shift vertically is therefore $2(R_2/R_1)$ $(V_Z + V_0)$. Since switching of the comparator now occurs for $v_{p_1} = V_{ref}$, we determine the value of V_{ref} from (14.4–35) with $v_{o_2} = 0$ as

$$V_{ref} = \frac{R_2}{R_1 + R_2}(V_Z + V_0) \qquad (14.4-45)$$

The resulting sawtooth waveshape is displayed in Fig. 14.26b. This waveshape is extremely important in cathode-ray oscilloscopes, as we will see.

14.5   CATHODE-RAY OSCILLOSCOPE (CRO)

The *cathode-ray oscilloscope* (CRO) is a widely used instrument for displaying periodic voltage waveshapes. The primary function of this instrument is to display the particular waveshape as a function of time on a screen. This function is achieved through voltage control of a focused electron beam that impinges upon a phosphorescent screen. The screen surface emits light upon electron impact, and an illuminated spot occurs. For the usual mode of operation and proper adjustment of the CRO, the beam traces out the vertical input voltage waveshape on the screen and thus allows visual observation.

14.5.1   Basic Components and Operation

Figure 14.27 displays the basic elements of a CRO: a cathode-ray tube (CRT) along with its internal components and associated circuitry. The CRT consists of an electron gun with focusing plates, deflection plates, and a phosphorescent screen. These components are mounted inside an evacuated tube to prolong the life of the electron gun. The deflection plates are used to alter the electron beam direction according to the applied voltages and thus move the illuminated pattern on the screen. As indicated in Fig. 14.27, an amplified version of the vertical input voltage is applied to the vertical deflection plates. An external horizontal input voltage or a sawtooth voltage (similar to that of Fig. 14.26b) is also applied through an amplifier to the horizontal deflection plates. The sawtooth voltage is usually generated by the internal triggered sweep (S_1 set to INT SYNC) that provides a sawtooth wave period that is identical to or an exact multiple of the input voltage. Only under these synchronous conditions is a stationary pattern of the vertical input voltage presented on the screen, as we will see.

The vertical deflection plates are oriented horizontally in the tube. The vertical input signal to be viewed on the screen is amplified and ap-

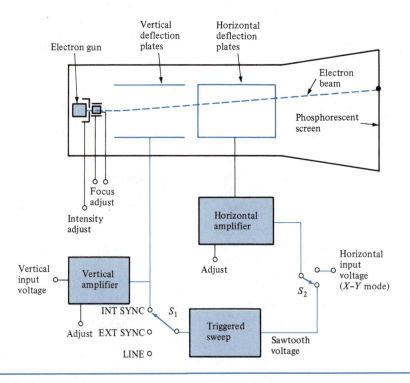

FIGURE 14.27   Cathode-Ray Oscilloscope

plied to the top plate relative to the bottom, which is grounded. The input vertical voltage can be either positive or negative and thus attracts or repels the beam of electrons. The beam is deflected upward or downward (as the vertical input voltage changes polarity), striking the screen in a higher or lower position with corresponding illumination.

The horizontal deflection plates focus the beam in the same manner but are oriented vertically in the tube. Horizontal "sweep" motion of the beam results when a sawtooth voltage is applied to one vertically oriented plate relative to the other, which is also grounded. During each period of the sawtooth voltage, the light spot on the screen moves horizontally at a constant speed (sweeps) across the screen. When the sawtooth voltage drops to zero, the spot returns to its original position at the left of the screen.

14.5.2   Producing a Fixed Pattern on the Screen

The simplest possible operating conditions for the CRO occur at warm-up, when the scope is turned on and there are no voltages applied to the vertical or horizontal deflection plates. The display upon the screen is then a dot whose position and intensity may be varied by additional electron gun focusing controls on the scope, as indicated in Fig. 14.27. Upon application of a time-varying vertical input voltage with no horizontal input, the dot is viewed as a vertical line. Similarly, with no vertical input and the triggered sweep of Fig. 14.27 activated to provide a sawtooth voltage to the horizontal plates, a horizontal line is observed. These lines occur because the screen continues to emit light for several tenths of a second after the beam has passed. We call this time the *screen emission time*. Thus, except at very low frequencies, the spot moves so rapidly that a stationary line is observed. At very low frequencies (on the order of 1 Hz), the actual movement of the spot is discernable.

We now consider the case of a periodic waveshape voltage applied to the vertical input and a synchronized sawtooth voltage applied to the horizontal deflection plates. This situation is the usual mode of operation of a CRO. We define the period of the vertical voltage as T_V and that of the horizontal sawtooth as T_H. For synchronism, these periods are related by

$$T_H = nT_V \qquad\qquad (14.5-1)$$

where n is an integer. Under these conditions, a replica of the vertical input voltage will appear on the screen, as we will now observe.

Two different synchronized examples with sinusoidal vertical input voltage are displayed in Fig. 14.28. The horizontal sawtooth voltage sweeps the beam across a fixed width of the screen during each period T_H. The vertical voltage deflects the beam sinusoidally in the vertical direction for one period (Fig. 14.28a) or two periods (Fig. 14.28b) of the sinusoid (T_V) during each sweep period (T_H). Hence, one sweep results in the exact waveshape on the screen as shown for v_V in each case. Since the vertical and horizontal voltages are synchronized, succeeding periods of the sawtooth will retrace the same pattern on the screen. In fact, for T_H corresponding to 10 Hz ($T_H = 0.1$ s), the vertical voltage waveshape corresponding to T_H will be retraced 10 times in one second. Since this time is much greater than the screen emission time, the vertical input voltage appears as a stationary sine wave on the screen. By increasing T_H to a larger integer multiple of T_V, additional periods of the vertical input voltage are displayed.

However, consider the case where the vertical and horizontal waveshapes are not synchronized. Figure 14.29 shows an example for which the periods (or frequencies) are slightly different. Under these conditions, each sweep will illuminate a different sinusoidal pattern on the screen. The waveshape then appears to be moving or "walking" across the screen. Figure 14.29 depicts this movement for two periods of a sawtooth

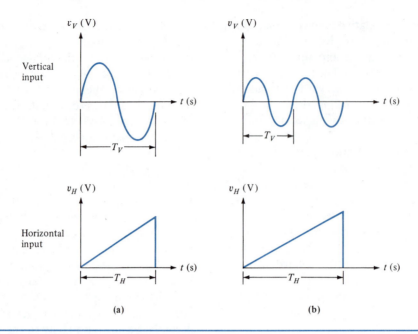

FIGURE 14.28   Vertical and Horizontal Voltage Waveforms (Synchronized): (a) Sinusoidal period equal to sawtooth period ($T_H = T_V$), (b) Sinusoidal period twice the sawtooth period ($T_H = 2T_V$)

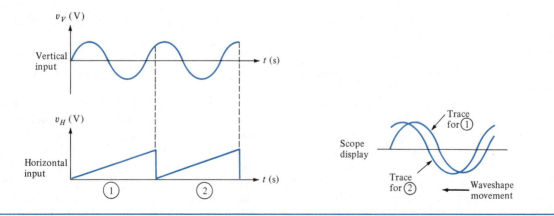

FIGURE 14.29   Vertical and Horizontal Voltages Waveforms (Not Synchronized)

waveshape. Hence, we observe that it is extremely important to synchronize the vertical and horizontal waveshapes with each other.

14.5.3   Triggered Sweep Circuit

As indicated in Fig. 14.27, a triggered sweep circuit can be used to produce a sawtooth voltage that is synchronized with the vertical in-put voltage. The triggered circuit provides precise timing of the output sawtooth voltage with the vertical input voltage. The required components of the sweep circuit are a special (triggered) saw-tooth generator preceded by a pulse generator. The pulse generator must produce a train of pulses that have the same period (or exact multiple) of the input trigger signal. Each pulse then triggers one sweep (or period) of the sawtooth voltage.

Figure 14.30a displays a circuit that provides the required periodic pulse generation. This pulse

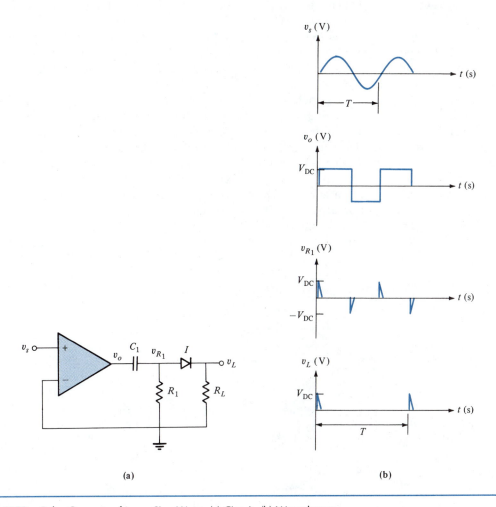

(a)   (b)

FIGURE 14.30   Pulse Generator from a Sine Wave: (a) Circuit, (b) Waveshapes

generator is sometimes called a *zero-crossing de-tector*. We assume that the input signal is a sine wave, as shown in Fig. 14.30b, and that the op amp saturation voltages are given by $V_{o\,max} = -V_{o\,min} = V_{DC}$. Also shown in Fig. 14.30b are the resulting voltage waveshapes for v_o, v_{R_1}, and v_L.

To explain the operation of the pulse generator, we note that as the input sine wave becomes slightly positive, the output voltage of the op amp becomes positively saturated and v_o switches to the positive saturation voltage ($V_{o\,max}$) of the op amp. Then, v_{R_1} also switches to this large voltage since the voltage across C_1 cannot change instantaneously. The ideal diode then acts like a short, and v_L is momentarily the large voltage ($V_{o\,max} = V_{DC}$). Immediately upon v_o switching high, C_1 begins to charge through R_1 and R_L and v_{R_1} and v_L decrease (decay exponentially). In order that these voltages correspond to a train of pulses, C_1 and R_1 and R_L are chosen to be small in magnitude such that C_1 charges in a small fraction of the input sine wave period. Thus, v_{R_1} and v_L become zero rapidly. When the sine wave goes slightly negative, the output of the op amp switches to the negative saturation voltage ($-V_{o\,max} = -V_{DC}$), producing a negative pulse of voltage for v_{R_1}. The diode in this case blocks the negative pulse from R_L, and v_L remains zero.

This pulse generation continues for each period of the sine wave. The corresponding waveshapes are displayed in Fig. 14.30b. The waveshape for v_L is observed to be a train of positive pulses with the identical period of the input signal. Often, for stability, some positive feedback is provided to eliminate erroneous switching due to noise at the input when v_s is small (just crossing zero). Additionally, the period of the pulse train can be multiplied if desired by passing this train through a frequency divider circuit.

A circuit that provides generation of the triggered sawtooth voltage is shown in Fig. 14.31. Each incoming positive pulse turns T_1 on, providing a discharge path for the capacitor in the feedback path of the op amp. At the end of the short pulse, T_1 cuts off and the capacitor charges with constant current (supplied from V_{CC}), given by $i_{C_I} = V_{CC}/R_I$. Then, the output voltage decreases linearly with time, with slope $dv_o/dt = -V_{CC}/R_I C_I$. Thus, the output voltage of the integrator varies linearly with time, given by $v_{o_1} = -(V_{CC}/R_I C_I)t$. This voltage is then inverted by the second op amp circuit (with $R_F = R_S$), giving $v_{o_2} = +(V_{CC}/R_I C_I)t$ and thus producing a sawtooth voltage for each input pulse of voltage.

The magnitude of the sweep voltage is roughly controlled by switching different capacitors into

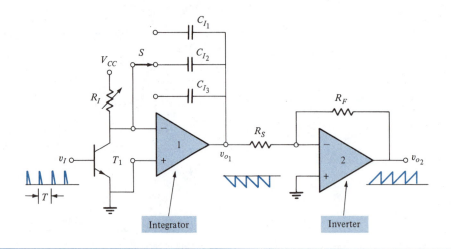

FIGURE 14.31   Triggered Sawtooth Generator

the circuit. Additionally, R_1 is usually a variable resistor that provides finer control of the sweep voltage magnitude.

Hence, the pulse circuit of Fig. 14.30 is used to excite the triggered sawtooth generator of Fig. 14.31. Combined, they provide the triggered sweep circuit of Fig. 14.27.

─────────────────────────── EX. 14.11

Triggered Sawtooth Generator Design Example

Choose element values for the sawtooth generator of Fig. 14.31 that provide a sawtooth peak voltage of 1 V and a period $T = 0.01$ s. Let $R_F = R_S$.

Solution: With $R_F = R_S$ in the circuit of Fig. 14.31, the peak output voltage is given by

$$V_P = \frac{V_{CC}}{R_I C_I} T$$

Element values that will provide $V_P = 1$ V are $V_{CC} = 5$ V, $R_I = 50$ MΩ, and $C_I = 10^3$ pF, with $T = 0.01$ s. The period $T = 0.01$ s is provided by using a pulse generator of the type shown in Fig. 14.30 with a sinusoidal input with frequency $f = 1/T = 1/0.01 = 100$ Hz.

─────────────────────────────

14.5.4   Different Modes of Triggering

Note that Fig. 14.27 indicates several alternatives for inputs that can be used to trigger the sweep voltage (controlled by switch S_1). When the triggered sweep circuit is connected to INT SYNC (for internal synchronization), the sawtooth voltage is automatically synchronized with the vertical input voltage. Setting S_1 to EXT SYNC (for external synchronization) allows triggering from any applied external voltage source. Similarly, if S_1 is set to LINE, the sawtooth is triggered by the 60 Hz AC line voltage and hence has a corresponding frequency.

14.5.5   X–Y Mode of Operation

The previous section described the usual mode of operation for an oscilloscope,

which is referred to as the *trigger-sweep mode*. This type of operation utilizes the internal triggered sweep generator of the scope (with several alternative switch S_1 positions) to display the input vertical voltage on the screen. Another mode of operation of an oscilloscope is to apply an independent voltage to the horizontal plates with the internal sweep circuitry disconnected through switch S_2 as shown in Fig. 14.27. This type of CRO operation is referred to as the *X–Y mode*. Voltages are applied externally to both the horizontal (X) and vertical (Y) plates.

Perhaps the most useful application of the scope in the $X–Y$ mode of operation is to determine the frequency and phase of an unknown sinusoidal voltage. This application is accomplished by applying the unknown sinusoidal voltage to the vertical input and a known sinusoidal voltage to the horizontal input and carefully observing the resulting patterns, called *Lissajou patterns*, on the screen.

By varying the frequency of the known signal source, Lissajou patterns, such as those displayed in Fig. 14.32, appear on the screen when the two signal frequencies are equal. The resulting pattern is then used to determine the phase angle difference between the two inputs. As shown in Figs. 14.32a, b, and c, the Lissajou patterns for equal frequencies are either a straight line, a tilted ellipse, or a circle, which indicate phase angle differences of 0°, 45°, and 90°, respectively. If the slope of the straight line or tilt of the ellipse is negative, the phase angle difference is between 90° and 180°. However, the phase angle difference observed in this manner can be either positive or negative (between 180° and 360°). Distinction between these two possibilities is provided by introducing an additional known phase angle. If the slope (or tilt) increases, the phase angle is between 0° and 180°, with the opposite occurrence for negative phase angles.

To describe the method of determining the phase angle, consider the known horizontal signal to be given by

$$v_H = V_H \sin \omega t \tag{14.5–2}$$

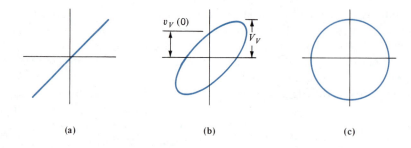

FIGURE 14.32   Lissajous Patterns for Phase Angle Differences of 0°, 45°, and 90°: (a) 0° phase angle (straight line), (b) 45° phase angle (tilted ellipse), (c) 90° phase angle (circle)

while the unknown vertical input voltage is given as

$$v_V = V_V \sin(\omega t + \phi) \qquad \text{(14.5–3)}$$

For $t = 0$, $v_H = 0$ and

$$v_V(0) = V_V \sin \phi \qquad \text{(14.5–4)}$$

Thus, the angle ϕ is given by

$$\sin \phi = \frac{v_V(0)}{V_V} \qquad \text{(14.5–5)}$$

The value of ϕ is then determined directly from the particular pattern on the scope screen. Figure 14.32b defines the values measured for the case of $\phi = 45°$, where an accurate measurement of $v_V(0)/V_V$ yields this ratio as 0.707.

Often, in varying the known signal frequency of the horizontal voltage, wildly different stationary patterns (different from those of Fig. 14.32) appear on the screen at particular values of frequency. Two such patterns are shown in Fig. 14.33. Such Lissajou patterns appear when the vertical and horizontal voltage frequencies are related by

$$n\omega_H = m\omega_V \qquad \text{(14.5–6)}$$

where m and n are integers. The integer values are

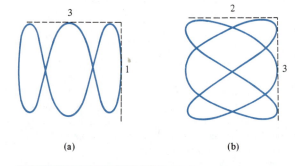

FIGURE 14.33   Lissajous Patterns for Sine Waves with Equal Phase and Different Frequency Ratios (ω_V/ω_H as Indicated): (a) $\omega_V/\omega_H = 3/1$, (b) $\omega_V/\omega_H = 2/3$

obtained by counting the peaks of the horizontal and vertical voltages, which is accomplished by counting the points of tangency on a portion of a rectangle drawn touching the peak horizontal and vertical voltage extremes. Note that Fig. 14.33a displays a frequency ratio $\omega_V/\omega_H = 3/1$, while Fig. 14.33b shows a 2/3 frequency ratio.

14.5.6   Dual-Beam and Dual-Trace Scopes

Dual-beam and dual-trace scopes are just what their names imply. Each scope provides a means of displaying two time-varying signals on the CRO screen. In a *dual-beam scope,* two electron guns along with separate focusing and deflection components are used to provide the two traces. Since dual systems are required, this type of instrument is quite costly.

A *dual-trace scope* provides the same function with a single electron beam and additional switching circuitry. Figure 14.34 displays a block diagram of the basic components for the two vertical input signals of a dual-trace CRO. Four modes of operation are usually provided. In two modes, only one of the inputs (single channel) is displayed. The other two modes provide a dual trace based upon either the *alternate* or the *chopped* mode of operation. These names imply that the beam is subjected to each incoming signal in an alternating or a chopped fashion.

In the alternate mode, each input is displayed on alternate sweeps and the switching between signals is synchronized with the sweep generator. Thus, one complete trace for each channel is produced on each alternate sweep. This mode is preferable for relatively high input frequencies. In the chopped mode, the signals are "chopped" by sampling each of them alternately many times during each sweep cycle. Portions of each waveshape are then omitted during each sweep but are produced later on succeeding sweeps. For low sweep rates, the display obtained using the alternate mode "flickers," and the chopped mode is therefore preferable.

14.6   CURVE TRACERS

A *curve tracer* is an instrument used to display current–voltage (*i* versus *v*) characteristics of two-terminal and three-terminal electronic devices. It consists of an oscilloscope and additional circuitry to provide the desired display on the scope screen. Self-contained instruments that perform only the function of curve tracing are one form of this instrument. Furthermore, the additional circuitry that converts an oscilloscope to perform this function is also a curve tracer and is available commercially. To describe the basic operation of a curve tracer, we first consider the principle involved in displaying the current–voltage characteristic of a two-terminal device.

14.6.1   Two-Terminal Device Display

To use an oscilloscope to display the current–voltage characteristic of a two-terminal device, the following conditions are required:

1. A voltage proportional to the device current must be applied to the vertical input of the scope.
2. A voltage proportional to the device voltage must be applied to the horizontal input of the scope.
3. The horizontal and vertical inputs must be synchronized.

The simple circuit of Fig. 14.35 satisfies the necessary requirements. For example purposes, a diode is used as the two-terminal element. The voltage source v_s is time varying (perhaps sinusoidal) and possesses both positive and negative values. The voltage applied to the vertical terminal of the scope is proportional to i_D and equal to $R_s i_D$. The voltage applied to the horizontal terminal is proportional to v_D and equal to $-v_D$. Thus, the voltage axis will be reversed on the scope.

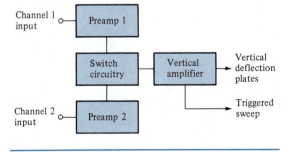

FIGURE 14.34   Basic Components of a Dual-Trace CRO

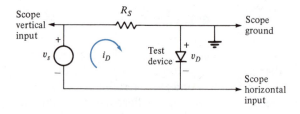

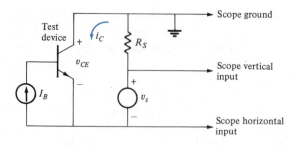

FIGURE 14.35   Circuit for Displaying i_D Versus v_D on a Scope Screen

FIGURE 14.36   Circuit for Displaying i_C Versus v_{CE} for a BJT

A similar problem occurs if the diode is reversed in the circuit correcting the voltage polarity but reversing the current. The vertical input to the scope then depends upon $-i_D$, and the current axis is reversed. Because of the grounding of the scope and the simple circuitry used, this problem is unavoidable (but certainly tolerable). In actual curve tracer circuitry, a polarity reversal switch allows reversal of the backward trace.

eration of this waveshape is described shortly. Furthermore, since only one polarity of v_s is necessary, a linearly varying sawtooth voltage is conveniently used for v_s (shown in Fig. 14.37 by dashed lines).

With these waveshapes applied to the BJT in the circuit of Fig. 14.36, the scope sweeps out characteristic curves corresponding to the parameter value of each step. If the frequency of the staircase is greater than 10 Hz, the entire family of curves will be individually swept across the screen in less than 1/10 of a second. Due to the screen emission time, a stationary pattern for each curve is thus observed.

14.6.2   Three-Terminal Device Display

Three-terminal device characteristics were first observed in Chapter 3. The output characteristics consist of a family of curves of current versus voltage for the output terminals, with an additional input current or voltage as the parameter. To obtain a display of such a family on a CRO, the three requirements as listed for a two-terminal device must also be satisfied here. Additionally, a means of changing the input parameter (current or voltage) must be provided.

Figure 14.36 displays a BJT connected as in the previous two-terminal device circuit. For a fixed value of I_B, the corresponding common-emitter $i–v$ characteristic will be traced on the scope screen. Then, by changing I_B to other fixed values, additional curves can be obtained.

A much more sophisticated approach, however, is to use a staircase current for I_B, as shown in Fig. 14.37. The actual circuitry for the gen-

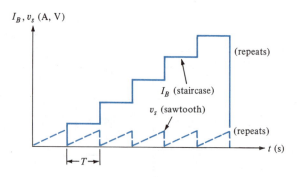

FIGURE 14.37   Staircase Input Parameter Waveform and Corresponding Sawtooth Voltage

The characteristics of FETs can be displayed in a similar manner by simply changing the input parameter. A staircase voltage function is now required. The current staircase is easily converted to a voltage staircase through use of a current-to-voltage op amp converter. The opposite conversion is also accomplished using a voltage-to-current op amp converter. These converters were described in Section 7.2.3.

14.6.3   Staircase Generator

Figure 14.38 shows a basic circuit for the generation of a voltage staircase with equal step heights for voltage changes and equal step widths for time changes. The arrangement of diodes and capacitors is sometimes called a *diode pump*. The op amps are included to provide more abrupt changes in voltage. Using a decade counter (as described in Chapter 9), the staircase will have ten steps, as we will see.

The input voltage to the staircase generator of Fig. 14.38 is a train of negative pulses with period T and amplitude denoted by V_{DC}. In general, a positive pulse train with fixed amplitude and period (or frequency) is provided by an astable multivibrator (using a circuit similar to that of Fig. 9.33). The required negative pulse train can then be produced by passing the positive pulse train through an op amp inverting amplifier. In curve tracer applications, the positive pulse train is also applied to the sawtooth generator of Fig. 14.31, providing exact synchronism of the sawtooth staircase functions.

The operation of the diode pump and associated op amps in Fig. 14.38 is as follows. Upon application of an input negative pulse, the output voltage of op amp 1 (v_{o_1}) and the voltage v both increase to the positive saturation voltage of op amp 1, which is $V_{o\,max}$. D_1 is forced to short while D_2 is open, and C_1 charges rapidly due to the current from the output of op amp 1 through C_1 and the small resistance of D_1. At the end of this charging period, the voltage of C_1 is then $V_{o\,max}$ with polarity as indicated in Fig. 14.38. The amount of charge for capacitor C_1 is then $C_1 V_{o\,max}$. Because of this voltage across C_1, D_2 becomes forward biased and hence current passes through D_2, C_1, and the output of op amp 1. This current is supplied through C_2 since op

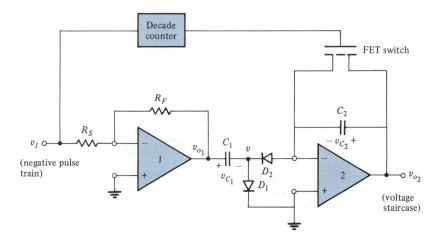

FIGURE 14.38   Staircase Voltage Generator with Ten Steps

amp 2 has negligible input current. Thus, the charge of capacitor C_1 is transferred to C_2, with $Q = C_1 V_{o\,max} = C_2 v_{o_2}$ since $v_{o_2} = v_{C_2}$. Thus, solving for v_{o_2} yields

$$v_{o_2} = \frac{C_1}{C_2} V_{o\,max} \qquad (14.6\text{--}1)$$

Note that in order for essentially all of the charge to transfer, we must have $C_2 \gg C_1$.

Hence, after one pulse, the output of the staircase is given by (14.6–1). After a second pulse, the output is doubled, and so on. During the time that these pulses are increasing v_{o_2} in this step-by-step manner, the decade counter is counting the pulses. When the count reaches 10, the output of the decade counter goes high, which causes the FET to turn on. The capacitor C_2 shorts and causes v_{o_2} to decrease to zero. The decade counter then starts counting again, and the staircase waveform with ten steps is repeated and repeated.

14.6.4　Basic Components of a Curve Tracer

Figure 14.39 displays the basic components of a curve tracer in block diagram form. The amplitude and frequency of the pulse generator are separately adjustable. The sweep generator provides a sawtooth voltage between the top and bottom terminals of the test transistor. The synchronized staircase generator provides the step changes of the input parameter of the device. Switch S allows selection of the input parameter to be either a voltage or a current staircase function. In the case of a BJT, S is set to the output of the voltage-to-current converter, which provides a staircase current. For FET measurements, S is set to the output of the voltage staircase. The voltage and current sampling elements also provide amplification or attenuation, as desired.

The terminals indicated in Fig. 14.39 are for a BJT (C, B, E) and a FET (D, G, S) in the com-

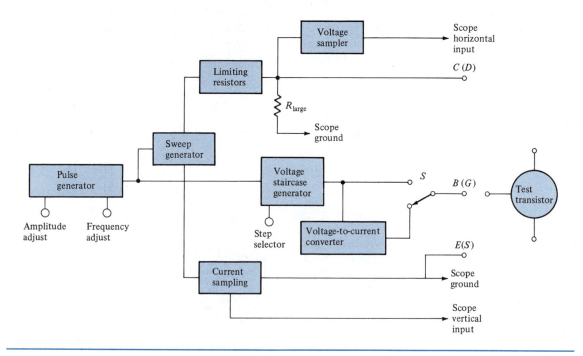

FIGURE 14.39　Basic Components of a Curve Tracer

mon-emitter or common-source configuration, respectively. The curve tracer can also be used with other transistor configurations by interchanging the terminals, as desired.

14.7   DIGITAL INSTRUMENTS

Digital instruments use digital logic circuits to carry out their intended function. In some instances, the input quantity is processed using analog circuitry and only the display is digital. In other instances, digital logic circuits are also used to process the input quantity. Such elements as basic logic gates, flip-flops, and counters (as described in Chapter 9) are essential components of both of these types of instruments. Furthermore, all digital instruments include some form of analog-to-digital (A/D) conversion. The incoming analog quantity to be measured is converted into digital form in which a train of voltage pulses represent the analog input. The element that performs this function is called an *analog-to-digital (A/D) converter*.

14.7.1   Analog-to-Digital (A/D) Converters

In making measurements with an analog instrument, the individual reading the meter scale functions as an analog-to-digital (A/D) converter. However, we are interested in describing circuits that perform this conversion, and there are many different types based upon various techniques. One of the most important, however, is the *voltage-to-frequency (V/F) converter*. We will consider this case to demonstrate the principle of A/D conversion. The V/F converter transforms a voltage into a train of pulses whose frequency depends linearly upon the magnitude of the input voltage (V_I), as we will see.

Figure 14.40a displays the basic circuit of a V/F converter. It consists of an integrator, a com-

parator, and a FET switch controlled by the output of the comparator. The operation of this circuit is as follows. At the instant that the input DC voltage (V_I) is applied (with $v_{o_2} = 0$), a constant current begins passing through C_I. The output of the integrator (v_{o_1}) begins decreasing linearly from zero and approaches $-V_{ref}$, which is the applied voltage to the positive terminal of the comparator. Since $v_{o_1} > -V_{ref}$ during this time period, v_{o_2} is at negative saturation, which is zero in this case (the negative DC supply terminal of the comparator is grounded). As v_{o_1} continues to decrease, it will eventually become equal to $-V_{ref}$ and become slightly more negative, at which time the comparator output (v_{o_2}) switches to positive saturation. This high output then turns the FET on and places a small resistance directly across the capacitor that allows discharge in a short time period (T_d). Then, v_{o_1} drops abruptly to zero, and v_{o_2} switches back to zero. The entire process is then repeated and repeated.

Figures 14.40b and c show the corresponding waveshapes for v_{o_1} and v_{o_2}. Note that v_{o_2} is a train of pulses with period very nearly equal to T ($T \gg T_d$) that corresponds to the time period when $v_{o_1} > -V_{ref}$. We will now see how the frequency corresponding to this period ($f = 1/T$) varies linearly with the magnitude of V_I.

If we let $t = 0$ correspond to the time that V_I is initially applied, the output of the integrator is given by

$$v_{o_1} = -\frac{1}{R_I C_I} \int_0^T V_I \, dt \qquad (14.7{-}1)$$

Integrating over one period yields

$$v_{o_1} = -\frac{1}{R_I C_I} V_I T \qquad (14.7{-}2)$$

However, when $t = T$, $v_{o_1} = -V_{ref}$. Substituting these values into (14.7–2) yields

$$T = \frac{R_I C_I V_{ref}}{V_I} \qquad (14.7{-}3)$$

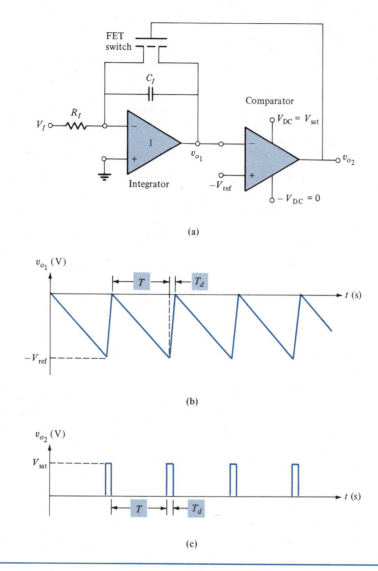

FIGURE 14.40   Voltage-to-Frequency Converter: (a) Basic circuit, (b) Output voltage v_{o_1} waveshape, (c) Output voltage v_{o_2} waveshape

Thus, the frequency is given by

$$f = \frac{1}{R_I C_I V_{ref}} V_I \qquad (14.7\text{--}4)$$

Note that for larger magnitudes of V_I, the number of pulses per unit time increases. Simi-larly, for a smaller magnitude of V_I, the number of pulses per unit time decreases. Therefore, if the pulses per unit time are counted, this number is directly proportional to the magnitude of voltage, and the process of analog-to-digital conversion is completed.

Various types of A/D converters are available in IC form. Typically, these converters can handle from seven to fourteen bits. The speed of conversion varies widely from one type to another depending upon the particular technology used. One IC example of a voltage-to-frequency converter is the LM331.

EX. 14.12

Voltage-to-Frequency Converter Design Example

For the V/F converter of Fig. 14.40a, let $V_I = 5$ V, $V_{ref} = 10$ V, and the frequency of the pulse train = 250 kHz. If $R_I = 10$ kΩ, determine the value of C_I.

Solution: By direct substitution into (14.7−4), we have

$$250 \times 10^3 = \frac{5}{(10^4)(C_I)(10)}$$

Solving for C_I yields

$$C_I = \frac{1}{(10^5)(50 \times 10^3)} = 200 \text{ pF}$$

14.7.2   Digital-to-Analog (D/A) Converters

A digital-to-analog (D/A) converter is used to reverse the process of A/D conversion. This process is inherently simpler than A/D conversion and is accomplished primarily by resistive networks.

Weighted-Resistor D/A Converter.   The D/A converter shown in Figure 14.41a uses weighted resistors of specific values as indicated. This circuit converts an N-bit binary number $(S_{N-1}S_{N-2}\cdots S_1S_0)$ into a voltage whose magnitude is proportional to the magnitude of the binary number represented. Each bit is either a high voltage (logic 1) or zero voltage (logic 0). Each bit is applied through a resistor of different magnitude (weighted) such that the voltage of the least significant bit is applied to the largest resistor, while that of the most significant bit is applied to the smallest resistor.

To show that the output load voltage v_L is proportional to the numerical value of the binary

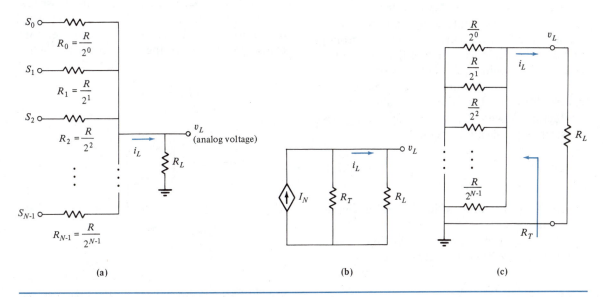

(a)   (b)   (c)

FIGURE 14.41   D/A Converter: (a) Circuit with weighted resistors, (b) Norton equivalent circuit, (c) Equivalent circuit for R_T

number, we first determine the equivalent circuit for the converter, as shown in Fig. 14.41b. We will treat each input as being 1 or 0 multiplied by a voltage V_{ref} to represent logic level 1 and 0, respectively. Then, the Norton equivalent current is obtained by replacing R_L with a short to yield

$$I_N = V_{\text{ref}}\left(\frac{S_0}{R_0} + \frac{S_1}{R_1} + \cdots + \frac{S_{N-1}}{R_{N-1}}\right)$$

$$(14.7\text{--}5)$$

Substituting the weighted resistor values, we have

$$I_N = V_{\text{ref}}\left(\frac{S_0}{R/2^0} + \frac{S_1}{R/2^1} + \cdots + \frac{S_{N-1}}{R/2^{N-1}}\right)$$

$$(14.7\text{--}6)$$

Rearranging yields

$$I_N = \frac{V_{\text{ref}}}{R}\left(S_0 2^0 + S_1 2^1 + \cdots + S_{N-1}2^{N-1}\right)$$

$$(14.7\text{--}7)$$

We note that $S_0, S_1, \ldots, S_{N-1}$ are each multiplied by a weighted power of 2, which increases by a factor of 2 with each integer increase in N. Thus, the magnitude of the current is directly proportional to the numerical value of the binary number. The Thévenin equivalent resistor R_T is obtained by setting all sources equal to zero, which places all weighted resistors in parallel with one another as shown in Fig. 14.41c. By adding these resistors in parallel, we obtain

$$R_T = \frac{R}{2^N - 1}$$

$$(14.7\text{--}8)$$

which is to be verified in an end-of-chapter problem.

The output load current i_L (in Fig. 14.41) is obtained using the current divider rule as

$$i_L = I_N\left(\frac{R_T}{R_T + R_L}\right)$$

$$(14.7\text{--}9)$$

and the output load voltage v_L is obtained as

$$v_L = R_L i_L = R_L\left(I_N \frac{R_T}{R_T + R_L}\right)$$

$$= I_N(R_L \| R_T)$$

$$(14.7\text{--}10)$$

Substituting for I_N and R_T then yields

$$v_L = \frac{V_{\text{ref}}}{R}\left(S_0 2^0 + S_1 2^1 + \cdots + S_{N-1}2^{N-1}\right)$$

$$\times \left(\frac{R_L \dfrac{R}{2^N - 1}}{R_L + \dfrac{R}{2^N - 1}}\right)$$

$$(14.7\text{--}11)$$

Rearranging yields

$$v_L = \frac{V_{\text{ref}}R_L}{R_L(2^N - 1) + R}$$

$$\times \left(S_0 2^0 + S_1 2^1 + \cdots + S_{N-1}2^{N-1}\right) \quad (14.7\text{--}12)$$

In usual applications, the weighted-resistor D/A converter is used with an op amp, as indicated in Fig. 14.42. To analyze this circuit, we first recall from Chapter 7 that the negative terminal of the op amp is a virtual ground. Furthermore, the output voltage of the op amp corresponding

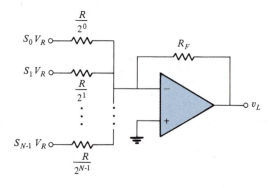

FIGURE 14.42   Weighted-Resistor D/A Converter with Op Amp

to each input voltage is the product of $-R_F/R_S$ (where R_S is the weighted resistor) and the corresponding input. If each input is again represented by $S_N V_{\text{ref}}$ (where $S_N = 0$ or 1), using superposition to add the voltages together thus yields

$$v_L = -R_F\left(\frac{V_{\text{ref}}S_0}{R/2^0} + \frac{V_{\text{ref}}S_1}{R/2^1} + \cdots + \frac{V_{\text{ref}}S_{N-1}}{R/2^{N-1}}\right)$$

$$(14.7\text{–}13)$$

Rearranging yields

$$v_L = \frac{-R_F}{R}\,V_{\text{ref}}(S_0 2^0 + S_1 2^1 + \cdots + S_{N-1}2^{N-1})$$

$$(14.7\text{–}14)$$

Note that (14.7–14) is identical in form to (14.7–12).

A major disadvantage of the weighted-resistor D/A converter is the wide range of resistor values that are required, as indicated in the following example.

─────────────────────────────────────── EX. 14.13

Weighted-Resistor D/A Converter Example

Calculate the maximum value of resistance required for a 10-bit D/A converter if the minimum resistor value is $R = 1\ \text{k}\Omega$.

Solution: The smallest resistor for the D/A converter is $R/2^{N-1}$; therefore, for a 10-bit binary number,

$$\frac{R}{2^{N-1}} = \frac{R}{2^9} = 1\ \text{k}\Omega$$

and $R = 512\ \text{k}\Omega$. Thus, we note that a very wide range of resistor values is required, beginning with $R_0 = 1\ \text{k}\Omega$ through $R_{N-1} = 512\ \text{k}\Omega$.

─────────────────────────────────────

Ladder Network D/A Converter.   A ladder network that provides D/A conversion using only two sizes of resistors (R and $2R$) is shown in Fig. 14.43a. To analyze this network, we again let each

voltage correspond to $S_N V_{\text{ref}}$, where S_N is either 0 or 1. Additionally, we initially consider that S_0 is high (1) and that all other S_N are low (0). This situation is depicted in Fig. 14.43b. Combining the top two branches into a Thévenin equivalent circuit then yields the network of Fig. 14.43c. Combining the top two branches into a Thévenin equivalent circuit $N-1$ more times then yields the circuit shown in Fig. 14.43d.

If we now similarly consider each of the other S_N inputs, the overall equivalent circuit of Fig. 14.43e is obtained by superposition. Therefore, the output load voltage is

$$v_L = \frac{S_0 V_{\text{ref}}}{2^N} + \frac{S_1 V_{\text{ref}}}{2^{N-1}} + \cdots + \frac{S_{N-1} V_{\text{ref}}}{2^1}$$

$$(14.7\text{–}15)$$

or

$$v_L = \frac{V_{\text{ref}}}{2^N}\,(S_0 + S_1 2^1 + \cdots + S_{N-1}2^{N-1})$$

$$(14.7\text{–}16)$$

Equation (14.7–16) represents the output load voltage without a load resistor connected. With a load resistor R_L connected between the output terminal and ground, the output voltage is reduced by the voltage divider ratio $R_L/(R + R_L)$; however, the relative weight of each input will remain the same.

In usual applications, the R–$2R$ ladder network is used with an op amp connected at the output as shown in Fig. 14.44a. The equivalent circuit for this D/A converter is indicated in Fig. 14.44b, from which we immediately observe that the op amp is connected as an inverting amplifier with amplification $-k$. The output voltage v_L of Fig. 14.44b is then given by

$$v_L = -k\left(\frac{S_0}{2^N} + \frac{S_1}{2^{N-1}} + \cdots + \frac{S_N}{2}\right)V_{\text{ref}}$$

$$(14.7\text{–}17)$$

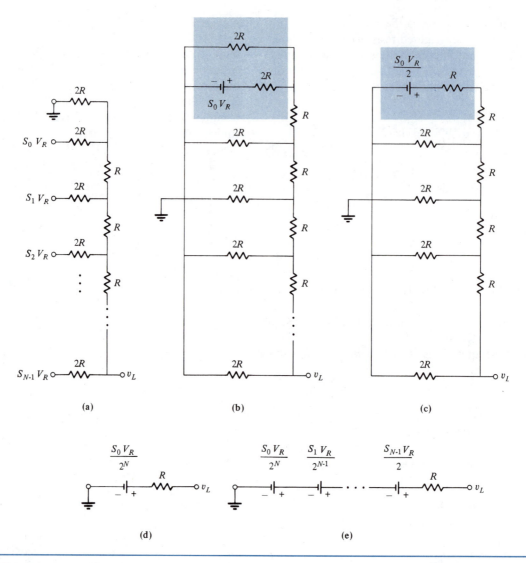

FIGURE 14.43 R–$2R$ Ladder Network D/A Converter: (a) Actual circuit, (b) Circuit with $S_1 = S_2 = \cdots = S_{N-1} = 0$, (c) Circuit with top two branches combined, (d) Thévenin equivalent circuit, (e) Overall equivalent circuit with non-zero S_N

14.7.3   Digital DC Voltmeter

Figure 14.45 displays the basic components of a *digital DC voltmeter*. The A/D converter uses an alternate form of conversion that requires a clocked input train of pulses with precise period or frequency. This input provides a means of accurately counting the number of pulses per unit time produced by the particular magnitude of V_I. An integrator is also used in this A/D converter and is contained in the clocked staircase generator, which has its maximum output voltage controlled by the output of the *RSFF*. The *RSFF*

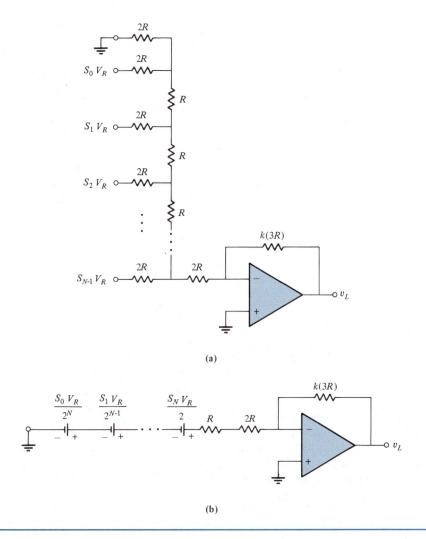

(a)

(b)

FIGURE 14.44   R–2R Ladder Network D/A Converter with Op Amp: (a) Actual circuit, (b) Equivalent circuit

also controls the transfer of the data into memory and ensuing display. The operation of the decade counter and the subsequent digital display are described in the next section.

The operation of the basic DC voltmeter of Fig. 14.45 is really quite simple. A DC voltage V_I is applied to the positive terminal of the comparator. Meanwhile, a staircase waveform (as described previously in Section 14.6.3) is applied to the negative terminal. As the staircase voltage increases from zero, in this case with each clock

pulse, the output of the comparator v_{o1} will be high until the staircase voltage becomes equal to V_I. While v_{o1} is high, the positive clock pulses entering the AND gate transmit an identical train of pulses to the decade counter. This train ceases when the staircase output becomes equal to V_I. Then, v_{o1} drops low, disabling the AND gate and ending the train of pulses for v_{o2}. Thus, the number of pulses produced for v_{o2} during this time is directly proportional to the magnitude of the input DC voltage.

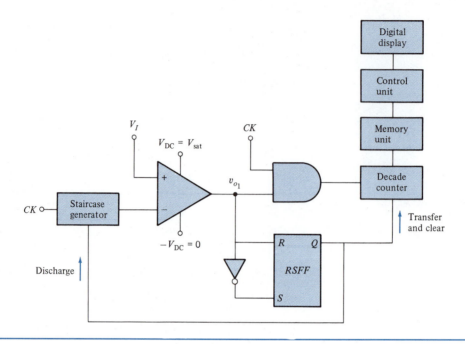

FIGURE 14.45   Basic Components of a Digital DC Voltmeter

When v_{o1} drops low, the output of the *RSFF* also changes, going high and discharging the capacitor in the staircase generator to force its output back to zero. The output of the *RSFF* also transfers the data into memory, resulting in a digital display that corresponds to the number of pulses that were counted. Additionally, the *RSFF* output clears the decade counter and sets its output back to zero. This process then repeats and repeats, and many new readings are transferred to display in a short time period.

Decimal Count	a_1	a_2	a_3	a_4
0	0	0	0	0
1	1	0	0	0
2	0	1	0	0
3	1	1	0	0
4	0	0	1	0
5	1	0	1	0
6	0	1	1	0
7	1	1	1	0
8	0	0	0	1
9	1	0	0	1

FIGURE 14.46   Truth Table for a Decade Counter

14.7.4   Decade Counter and Memory Unit

The basic operation of a *decade counter* was described in Chapter 9. The associated truth table is shown in Fig. 14.46. The output of this counter could be used along with controlled logic to digitally display the "running" pulse count. However, since this count may vary rapidly with time, a fixed reading would then be impossible. Hence, a method of storage or memory (such as that shown in Fig. 14.45) of the final count is a major requirement for a digital voltmeter.

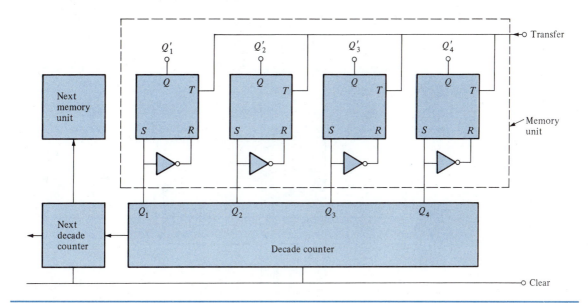

FIGURE 14.47   Decade Counters and Memory Units

Figure 14.47 displays an arrangement of four additional *RSFF*s, each connected to the outputs of a single decade counter that provides the desired memory unit for one digit. Additional decade counters and memory units are connected as indicated for additional digits. When the transfer terminal voltage goes high, the inputs from the decade counter are enabled and transferred to the additional *RSFF*s. The additional *RSFF*s form what is called a *memory unit,* or *register.* The output of the memory unit is applied through an interface circuit to a digital display, and the decade counter is then cleared and prepared for a new train of incoming pulses.

14.7.5   Seven-Segment Display

The most common technique for displaying a single digital readout is to use a *seven-segment display* as shown in Fig. 14.48a. Each segment consists of an LED (light-emitting diode) or an LCD (liquid-crystal display). The segments

turn on in specific combinations to form a decimal number from 0 to 9. Figure 14.48b displays the truth table required for each decimal number, with a 1 indicating a high voltage and the fact that the segment for that particular decimal count is on.

For the memory unit to turn on the correct segments, an interface or logic control circuit is required between the memory unit and the seven-segment display. This interface circuit consists of basic logic gates that provide a high voltage (1) to the segment when it is supposed to be on. For example, from the truth table of Fig. 14.48b, segment e is on for decimal counts of 0, 2, 6, and 8. The logic necessary for this display is obtained from the decade counter truth table in Fig. 14.46 as

$$e = \bar{Q}_1\bar{Q}_2\bar{Q}_3\bar{Q}_4 + \bar{Q}_1Q_2\bar{Q}_3\bar{Q}_4 + \bar{Q}_1Q_2Q_3\bar{Q}_4 + \bar{Q}_1\bar{Q}_2\bar{Q}_3Q_4 \qquad (14.7\text{--}18)$$

where each term respectively corresponds to 0, 2, 6, and 8 in binary. Thus, if any of these logic terms is 1, e will be 1, which is the desired result.

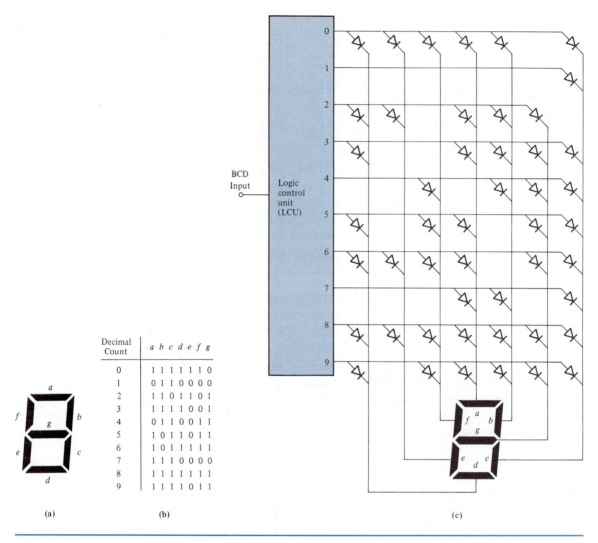

Decimal Count	a b c d e f g
0	1 1 1 1 1 1 0
1	0 1 1 0 0 0 0
2	1 1 0 1 1 0 1
3	1 1 1 1 0 0 1
4	0 1 1 0 0 1 1
5	1 0 1 1 0 1 1
6	1 0 1 1 1 1 1
7	1 1 1 0 0 0 0
8	1 1 1 1 1 1 1
9	1 1 1 1 0 1 1

(a)   (b)   (c)

FIGURE 14.48   Technique for Displaying a Single Digital Readout: (a) Seven-segment digital display, (b) Associated truth table, (c) Diode array (gate array) for seven-segment digital display

Similar expressions can also be written for the other six segments.

After logic expressions for the on condition of each segment are obtained, considerable simplification using Boolean algebra is usually in order. However, each logic expression without simplification can be implemented using basic logic gates. These logic circuits are then used as the interface between the memory unit and the seven-segment display.

Logic circuits that decode a binary number (in binary-coded decimal form, from Chapter 9) into the corresponding digital number are called *BCD-to-seven-segment decoder/driver circuits* and are available in IC form. A commonly used IC logic chip is the 74LS47.

An alternate method of excitation of the seven-segment display is shown in Fig. 14.48c. The logic control unit in this case consists of a A/D converter that provides an output voltage at one of the decimal (0 to 9) terminals, while the other outputs are zero. The diode array (gate array) turns on all the correct segments for each decimal number when it is high, as the reader may verify.

14.7.6   Digital Multimeters (DMMs)

A *digital multimeter* (DMM) is an instrument that is used to measure the same basic quantities as the analog multimeters previously described. As in the case of analog meters, a switch is available for selecting the desired quantity to be measured. Internally, however, there are two basically different forms of DMMs, as displayed in block diagrams form in Fig. 14.49.

Figure 14.49a shows the case in which analog processing is carried out as in the previously described analog instruments. The digital part of this type of instrument consists only of the A/D converter and the digital circuitry required to digitally display the output. This type of instrument is analog in nature with a digital readout.

Figure 14.49b indicates the components of a DMM that uses digital signal conditioning and processing to measure the analog input. When the input is time varying, the sample and hold (S/H) circuit samples the analog signal and holds the value fixed for A/D conversion. These samples are then averaged using digital processing and are subsequently also digitally displayed.

The primary advantages of DMMs over analog multimeters are convenience in reading and improved accuracy. Fundamentally, the improvement in accuracy results because digital data can be stored accurately as well as indefinitely. Referring to the type of instrument that is completely digital, the analog circuit difficulties of drift and stability are then eliminated because of the basic nature of the digital circuits. Manufacturers of digital instruments boast of an improvement in accuracy by a factor of 10 over counterpart analog instruments, providing accuracy on the order of 0.1% as compared to 1%.

A unique feature that digital display instruments possess is that the range of the readout can be extended to an additional "1/2 digit." For example, a 3 1/2 digit meter has a four-digit display, but the fourth digit can only be 1. Thus, the maximum reading of 999 can be extended to 1999. The additional half digit is referred to as an *over-range*

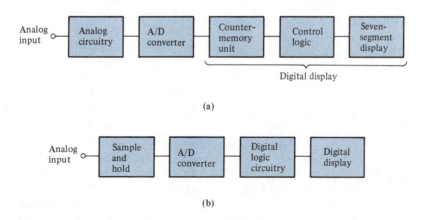

(a)

(b)

FIGURE 14.49   Digital Multimeters: (a) Analog processing with digital display, (b) Digital processing with digital display

digit. Additionally, the amount of over-ranging for a particular instrument is usually specified, with 50% being typical. Hence, for this case, a maximum reading of 999 can be extended to 1500.

14.8   TIMEPIECES

Perhaps the most important and practical measurement device ever invented is the timepiece in the form of a watch or clock. For nearly four centuries, civilized man has used the common timepiece called a clock to measure time. Furthermore, during these four hundred years, the operation of clocks was entirely mechanical except for approximately the last twenty years. During this recent time span, electronic clocks and wristwatches, in particular, have become a "boom" to the electronic industry. The advent of electronic watches was made possible only through improvements in semiconductor device fabrication techniques that permitted small enough circuitry (in the form of integrated circuits) for such timepieces to adorn a person's wrist.

14.8.1   Mechanical, Electrical, and Electronic Timepieces

We begin our description of the various types of clocks and watches by first considering mechanical timepieces. A description of these timepieces is essential in order to obtain an understanding of the required components in electronic clocks and watches.

Mechanical Timepieces.   The heart of the mechanical clock system is a wound spring and a balance wheel. The spring acts like an energy storage unit, and the balance wheel is a mechanical device that regulates the rate of rotation of the hands of the clock. Operation of this type of timepiece consists of compressing the spring (or "winding" the clock) and thus storing potential energy in the spring. The energy is then used with proper gearing arrangements to drive the balance wheel. This wheel, in turn, precisely regulates the movement of the hour, minute, and second hands of the clock. However, when the spring begins to show signs of wear, the timepiece will begin to lose or gain a certain number of minutes each day. The spring must then be replaced.

One of the major advances in mechanical watches was the development of a practical self-winding watch. The mechanism for storing energy in this case involved a cleverly designed sliding weight and spring arrangement. Motion of the watch provides kinetic energy, shifts the weights, and compresses the spring, thus storing the required energy.

Electrical Timepieces.   An additional improvement in mechanical watches was achieved when a miniaturized battery-driven motor was used to replace the spring and balance wheel. Although it was mostly mechanical, this timepiece was the first wristwatch that was partly electrical. Furthermore, improved timing accuracy was provided with a reduction in mechanical components.

Electronic Timepieces.   Electronic timepieces utilize a battery as the energy storage element that drives specialized electronic circuits. In the form of a watch, these timepieces possess either electronically driven contactless spring and balance movements or miniaturized oscillator-driven movements. Additionally, either a solid-state digital display or stepper motor is used to control the readout. However, the use of the oscillator movement instead of the spring balance increases the circuit components by about three orders of magnitude. This circuit requirement, along with small size for watch applications ($3-4$ cm^3), necessitates using integrated circuits in electronic watches.

Furthermore, since digital circuits have fewer components and superior component tolerances, these circuits are preferred over the more complex analog circuits. Another factor that influences the circuit designer in the selection of the particular circuitry is power consumption, which must be minimized in order that battery replacement be minimized. Thus, the integrated circuits that are most commonly used in electronic watch circuits are CMOS because of their characteristic low power consumption.

14.8.2   Electronic Watch Components and Operation

The five primary elements in an electronic watch are the battery, oscillator, counter (or frequency divider), driver or decoder circuitry, and output display unit. Figure 14.50 shows the connection of these basic elements in block diagram form, as well as the interface network that is often required to match the output of the counter with the input of the driver or decoder network.

The battery is the power source of the electronic watch. As shown in Fig. 14.50, the battery is the DC supply voltage for each of the electronic circuits as well as for the driver motor or decoder network. The batteries used in electronic wristwatches are by necessity very compact and have been designed to possess long life. After a great deal of research and development, the commonly used batteries that have been developed to possess these qualities are lithium, silver oxide, and mercuric oxide batteries. These same types of batteries are also used in hand calculators. Typical ratings for these batteries are 1.2–1.5 V and 100–250 mA-h, with a lifetime of 1 to 2 years.

The oscillator portion of the watch circuit directly replaces the spring and balance of the mechanical watch. The function of the oscillator is to generate a sinusoidal voltage at a particular frequency. The freqency is stabilized and precisely controlled by using a quartz crystal oscillator (as described in Section 14.4). A stabilized voltage pulse train is then generated from the sinusoidal voltage. This pulse train is summed and frequency divided to yield output pulse trains proportional to the time in seconds, minutes, hours, days, months, and even years.

The output of the counter is connected to an interface network and then to the input of the driver motor or decoder network, as shown in Fig. 14.50, because the output of the counter is typically incompatible with the input of the motor or decoder networks. For example, the output current may not be adequate to drive the subsequent stages, such as a stepper motor in the case of an analog watch. Furthermore, for digital display watches, an interface network is also required to make the oscillator output compatible with the digital display unit. Therefore, an additional interface network is required and is designed as a part of the oscillator and counter integrated circuit.

Various display units are presently available with electronic watches. Commonly used display units are the mechanical hands movement, light-emitting diodes, or the liquid-crystal display. The latter two of these three displays use seven segments for each numeral, as described in the previous section. Currently, however, LED displays are not used often because of their relatively high power consumption. When they are used, a push button is employed to display the time only when desired. Power is thus consumed by the display only during the viewing time. LCDs consume much less power and are therefore predominant.

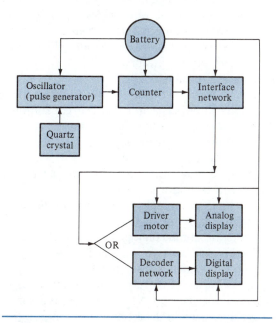

FIGURE 14.50   Block Diagram of an Electronic Wristwatch

- The operating principle of DC electrical meters is based upon the d'Arsonval meter movement, which involves the interaction of two magnetic fields that cause rotation of a coil and pointer mounted on a rotatable core. The rotation is counteracted upon by a spring, and the amount of rotation is directly dependent upon the applied current to the coil, which controls the magitude of the coil magnetic field.

- A DC ammeter measures DC current, and the range of measurements is extended by using the shunt resistor R_{SH}. The value of R_{SH} in terms of the meter resistance R_M is $R_{SH} = R_M/(N-1)$, where N is the ratio of incoming current to meter current. In practical ammeters, the Ayrton shunt circuit arrangement is used for switching to different current ranges.

- To measure DC voltage using the d'Arsonval meter movement, a series multiplier resistor is used ($R_S = KR_M$) and then provides the full-scale relationship $I_{FS} = V_{FS}/[(1 + K)R_M]$. Different voltage ranges are provided by switching to different multiplier resistances. The sensitivity of a voltmeter is $S = 1/I_{FS}$.

- An ohmmeter is an instrument used to measure an unknown resistance R_X and consists of R_M, R_S, and V_{DC} connected in series. Initially, the resistor R_S is adjusted with $R_X = 0$ to provide I_{FS}. Connecting R_X in series then produces a reduced current, and the value of R_X is obtained directly from the specially calibrated scale.

- AC voltmeters are used to measure the peak or rms value of a sinusoidal voltage. The voltage is rectified and then read on a meter scale that is calibrated for peak or rms values.

- EVM multimeters are instruments used to measure DC and AC voltages and currents. Current measurements are obtained by internally providing a current-to-voltage converter. EVMs can amplify the input voltage and have either analog or digital readout.

- A DC power supply converts the available AC power (115 V, 60 Hz) into DC. Regulated DC power supplies possess an adjustable output that is temperature compensated and relatively insensitive to the load.

- Oscillator circuits convert a DC supply voltage into a purely sinusoidal voltage. Four basic oscillator circuits are the phase shift oscillator, Wien bridge oscillator, Hartley oscillator, and Colpitts oscillator. The phase shift and Wien bridge oscillators employ RC networks and provide adjustable output frequencies up to 100 kHz. The Hartley and Colpitts oscillators use LC networks and provide adjustable output frequencies in the 100 kHz to 100 MHz range.

- Crystal oscillators provide an extremely stable and precise output frequency. A quartz crystal is used with precise dimensions and associated impedance that resonates at one frequency and thus produces zero phase shift between output and input and resultant oscillation.

- Signal or funtion generators are instruments that provide various periodic voltages such as sine, square, triangular, ramp, and pulse waveshapes. All of these waveshapes can be generated from a sinusoidal voltage using op amp circuitry consisting of a comparator along with rectifiers and integrators.

- Square and triangular voltage waveshapes can also be generated using astable multivibrator circuits with a capacitor that charges and discharges. More precise waveshapes are attainable if the capacitor is supplied with constant current as in the feedback branch of the integrator. By level shifting the triangular waveshape, a sawtooth waveshape is obtained.

- A cathode-ray oscilloscope (CRO) is used to visually display periodic voltage waveshapes as a function of time on a phosphorescent screen. The periodic voltage is applied to the vertical deflection plates, causing the electron beam to deflect vertically corresponding to the voltage magnitude and sign. A sweep voltage is applied to the horizontal deflection plates, causing the beam to move horizontally at constant speed across the screen. When the sweep voltage period is an integer multiple of that of the periodic voltage being displayed, a fixed pattern is produced on the screen. The sweep voltage can be internally synchronized, externally synchronized, or triggered by 60 Hz line voltage.

■ The frequency and phase of an unknown sinusoidal voltage can be determined using a CRO by applying the unknown sinusoidal voltage to the vertical input and a known sinusoidal voltage to the horizontal input and carefully observing the resulting Lissajou patterns. The frequency of the known sinusoid is varied until a recognizable pattern is displayed, from which the magnitude of the unknown frequency is then obtained. The unknown phase is given by $\phi = \sin^{-1} v_V(0)/V_V$, where ϕ can be between 0° to 180° or 180° to 360°. The specific value of ϕ is determined by introducing an additional known phase angle; if the tilt of the pattern increases, ϕ is between 0° and 180°.

■ A curve tracer is an instrument used to display i–v characteristics of two- and three-terminal devices. For two-terminal devices, an oscilloscope is used, and a voltage proportional to the device current is applied to the vertical input while a voltage proportional to the device voltage is applied to the horizontal input. These voltages are time varying and synchronized because they are produced by the same sweep voltage. For three-terminal devices, the parameter is generated using a staircase voltage waveform, and an oscilliscope displays the curves.

■ An A/D converter converts an analog voltage that depends upon a physical entity (displacement, pressure, voltage, and so on) into digital form. A voltage-to-frequency (V/F) converter transforms a voltage into a train of pulses whose frequency depends linearly on the magnitude of the input voltage. The basic V/F converter consists of an integrator, a comparator, and a FET switch controlled by the output of the comparator.

■ D/A converters convert a digital number into a voltage whose magnitude is proportional to the magnitude of the number. Weighted-resistor networks and R–$2R$ ladder networks are forms of D/A converters.

■ A digital DC voltmeter converts the voltage to be measured into digital form using a V/F converter. The pulses are then counted and are displayed digitally using seven-segment displays.

■ The electronic timepiece is a measurement device whose miniaturization was made possible through the advent of ICs. In the operation of an electronic timepiece, a battery energizes a crystal oscillator to produce a precise sine wave from which voltage pulse trains proportional to time in seconds, minutes, hours, days, months, and years are produced. The voltage pulses are counted and displayed digitally or used to drive a stepper motor that in turn drives the hands of an analog output watch.

CHAPTER 14 PROBLEMS

14.1

A d'Arsonval coil and associated meter scale indicate full-scale deflection for a current of 5 mA and a voltage of 30 mA. Specify the shunt resistance required to provide a current range of 0 to 5 A.

14.2

For the d'Arsonval coil of Problem 14.1, specify the multiplier resistance that will allow voltage measurements in the range of 0 to 300 V.

14.3

Repeat Problems 14.1 and 14.2 for current ranges 0 to 20 A and 0 to 100 A along with voltage ranges 0 to 10 V and 0 to 100 V.

14.4

Consider the ammeter arrangement of Fig. 14.3. The meter coil has $R_M = 5$ kΩ and $I_{FS} = 1$ mA. Determine R_{SH_1}, and R_{SH_2} that will extend the current range to:

(a) 0 to 10 mA   **(b)** 0 to 100 mA

14.5

Consider the Ayrton shunt ammeter circuit of Fig. 14.4. Obtain the expressions for R_{SH_1} and R_{SH_2} in terms of the meter resistance R_M and the current ratios N_1 and N_2, where $N_2 > N_1$. Determine the values of shunt resistances to extend the measureable current range to:

(a) 0 to 10 μA   **(b)** 0 to 100 μA

The meter resistance is 150 Ω, and the full-scale current without R_{SH} is 1 μA.

14.6

Repeat Problem 14.5 for $R_M = 100$ Ω and $I_{FS} = 1$ mA with current ranges 0 to 10 mA and 0 to 100 mA.

14.7

Consider the multi-range DC voltmeter shown in Fig. P14.7. For the meter $R_M = 1$ kΩ and $I_{FS} = 50$ A. Calculate values for the multiplier resistors R_{S_1}, R_{S_2}, R_{S_3}, and R_{S_4}.

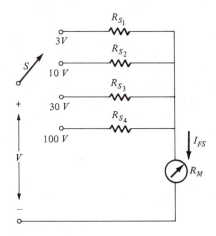

FIGURE P14.7

14.8

A multi-range DC voltmeter similar to that in Fig. P14.7 has voltage ranges of 0.1, 1.0, and 10. If the full-scale deflection current is 0.1 mA and $R_M = 100$ Ω, calculate values for the multiplier resistors R_{S_1}, R_{S_2}, and R_{S_3}.

14.9

For the circuit shown in Fig. P14.9, a voltmeter is used to measure the voltage across R_2. Calculate this

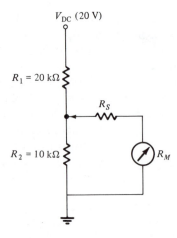

FIGURE P14.9

voltage for $R_S + R_M$ infinite and $R_S + R_M = 40$ kΩ. What is the error (%)?

14.10

An ohmmeter is used to measure an unknown resistance R_X. The ohmmeter has $V_{DC} = 3$ V, and when the ohmmeter is adjusted for zero resistance, $R_S + R_M = R_T = 4$ kΩ. Calculate R_X for a current of 0.2 mA.

14.11

Consider an ohmmeter that uses a d'Arsonval meter with $I_{FS} = 1$ mA and a battery with $V_{DC} = 9$ V. If the battery voltage reduces to 8 V after six months of use, determine the error in a measurement at mid range.

14.12

Figure P14.12 displays a multimeter configuration using an Ayrton shunt ammeter circuit, a multi-range voltmeter circuit, and a multi-range ohmmeter circuit. For $R_M = 1$ kΩ and a full-scale meter current of 50 μA, determine values for R_{SH_1}, R_{SH_2}, R_{S_1}, R_{S_2}, R_{S_3}, and R_{S_4}.

14.13

For the ohmmeter portion of the multimeter circuit of Fig. P14.12, determine the value of R_S for zero adjust if $V_{DC} = 3$ V and the meter has $R_M = 1$ kΩ and $I_{FS} = 50$ μA. Also determine the value of R_S if the voltage V_{DC} has decreased from 3 V to 2 V after repeated use.

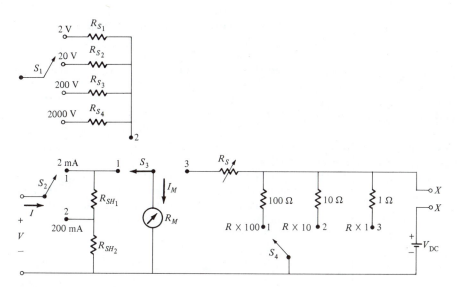

FIGURE P14.12

14.14

Figure P14.14 is a DC ammeter that uses two op amps. Determine the maximum input current for full-scale deflection of the meter. Let $I_{FS} = 1$ mA and note the direction of meter movement.

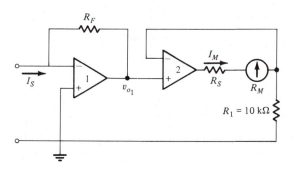

FIGURE P14.14

14.15

Determine the percentage of deflection the meter will have in the circuit shown in Fig. P14.15 for an input current $I_S = 2.2$ mA. Let $R_F = 1$ kΩ, $R_S = 5$ kΩ, $R_M = 1$ kΩ, and $I_{FS} = 1$ mA. What value of I_S causes 100% deflection?

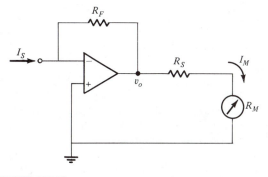

FIGURE P14.15

14.16

For the JFET EVM of Fig. 14.15, $T_1 = T_2$, $g_o = 0$, and $g_m = 10$ mS with $R_D = 5$ kΩ, $R_M + R_S = 100$ Ω, and $v_I = 1$ V. Determine the meter current.

14.17

For the JFET EVM of Problem 14.16, let $g_o = 1/(50\ \text{k}\Omega)$ and calculate the modified value for I_M. Begin by determining the modified expression for I_M.

14.18

Repeat Problem 14.17 for the following:
(a) $g_o = 1/(20\ \text{k}\Omega)$ (b) $g_o = 1/(10\ \text{k}\Omega)$
(c) $g_o = 1/(5\ \text{k}\Omega)$

What conclusion is reached by comparing these calculations?

14.19

For the circuit of Fig. 14.15, derive the expression for I_M without the use of the difference-mode analysis. Use a small-signal model and assume that the AC resistance of the DC current source is infinite.

14.20

Figure P14.20 displays a series-regulated DC power supply modified from that of Fig. 14.17. For this new circuit, determine V_{ref} for $R_F = 40$ kΩ, $R_S = 10$ kΩ, $R_3 = 20$ kΩ, $R_4 = 10$ kΩ, and $V_{DC} = 5$ V.

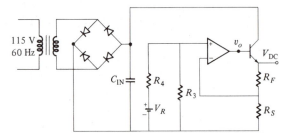

FIGURE P14.20

14.21

In connecting the circuit of Fig. P14.20, the only available reference source voltage is a 9 V battery. Calculate the resulting output voltage for $V_{ref} = 9$ V. If the output voltage is to be 5 V, what value of R_3 will provide this requirement?

14.22

Figure P14.22 displays another series DC regulator circuit that uses a BJT and a zener diode. Analyze this circuit to determine V_{DC}. Assume that the BJT is

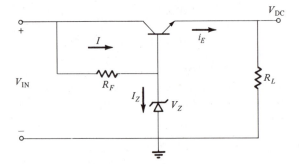

FIGURE P14.22

operating in the amplifier region with $\beta = 100$ and $V_{BE} = 0.7$ V. Additionally, $V_Z = 5.7$ V, $R_L = 1$ kΩ, $R_F = 10$ kΩ, and $V_{IN} = 15$ V. Note that $i_E = V_{DC}/R_L$ and that the minimum value of R_L (and the maximum value of V_{DC}) will depend upon the maximum current ratings of the transistor. Also, calculate I_Z.

14.23

For the phase shift oscillator of Fig. 14.18a, carry out the analysis to verify equation (14.4–4).

14.24

For the phase shift oscillator of Fig. 14.18a, determine the oscillator frequency for $R = 200\ \Omega$ and $C = 65$ nF. Also determine the magnitude of the amplification. Will this amount allow the phase shift oscillator to operate?

14.25

Figure P14.25 displays a JFET phase shift oscillator. Derive the expression for the oscillation frequency. Also, determine the voltage amplification v_0/v_1 provided by the JFET amplifier using a small-signal equivalent circuit.

14.26

Determine the oscillator frequency for the Wien bridge oscillator of Fig. 14.18b for $R = 1$ kΩ and $C = 10^{-8}$ F.

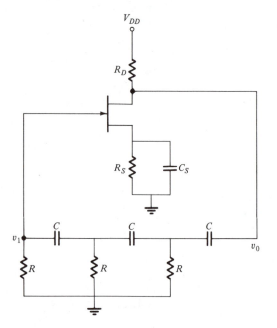

FIGURE P14.25

14.27
Carry out the analysis for the Wien bridge oscillator of Fig. 14.18b with general elements $Z_1 = R_1 + (1/j\omega C_1)$ and $Z_2 = R_2 \| (1/j\omega C_2)$ to determine the oscillator frequency. Calculate this frequency for $R_1 = R_2 = 1$ kΩ and $C_1 = C_2 = 10^{-8}$ F.

14.28
Redraw the circuit for the Wien bridge oscillator (Fig. 14.18b) showing the negative and positive feedback explicitly.

14.29
For a Wien bridge oscillator (Fig. 14.18b), $C = 1$ F and $R_4 = 1$ kΩ. Determine R and R_3 such that the oscillation frequency is 50 kHz.

14.30
For the Hartley oscillator of Fig. 14.18c, determine the operating frequency for $R_F = 10$ kΩ, $R_S = 2$ kΩ, $L_1 = 1$ mH, $L_2 = 2$ mH, and $C = 10$ pF. Check to ensure that the amplification criterion is satisfied.

14.31
For a Hartley oscillator (Fig. 14.18c), determine the value for C if the desired operating frequency is 50 kHz and the inductors have magnitudes $L_1 = 1$ mH and $L_2 = 3$ mH.

14.32
For the Colpitts oscillator of Fig. 14.18d, determine the operating frequency for $R_F = 31$ kΩ, $R_S = 3$ kΩ, $L = 300$ mH, $C_1 = 100$ pF, and $C_2 = 10$ pF.

14.33
For a Colpitts oscillator (Fig. 14.18d) with $C_1 = 1$ pF and $C_2 = 1$ pF, determine the value of L if the desired operating frequency is 42 MHz.

14.34
A voltage comparator with reduced saturation output voltage is shown in Fig. 14.22a. For $V_Z = 5$ V, $V_0 = 0.7$ V, and $A_d = 10^3$, sketch the voltage transfer characteristic. Assume that $V_{o\,max} = -V_{o\,min} = 10$ V.

14.35
A square and triangular wave generator as shown in Fig. 14.23c is to be designed with a frequency of 10 kHz. If $V_Z = 5$ V and $V_0 = 0.7$ V, determine the time T_1 of the negative ramp and the time T_2 of the positive ramp. Sketch the square and triangular wave outputs for one period. Let $R_I = 5$ kΩ, $R_1 =$

20 kΩ, and $R_2 = 10$ kΩ. Also determine the required value of C_I.

14.36
Repeat Problem 14.35 replacing R_I with the circuit given by Fig. 14.24a. Use $C_I = 10$ nF and determine values for R_{I_1}, R_{I_2}, T_1, and T_2 given that $R_{I2} = 3R_{I1}$.

14.37
A cathode-ray oscilloscope uses an external horizontal sweep signal generated by the sawtooth voltage generator of Fig. 14.26a. The horizontal deflection produced is equivalent to the horizontal switch set at 20 s/div, where the oscilloscope screen has 10 divisions. If $V_Z = 5$ V, $V_0 = 0.7$ V, $R_{I_2} = 10$ kΩ, and $C_I = 10$ nF, determine V_{ref} and draw the sawtooth waveshape for two periods.

14.38
For the triggered sawtooth generator of Fig. 14.31, determine C_{I_1}, C_{I_2}, and C_{I_3} for corresponding maximum values of $v_{o_2} = 1$ V, 5 V, and 9 V. Let $V_{CC} = 15$ V, $R_I = 100$ Ω, $T = 1$ μs, $R_F = 2$ kΩ, and $R_S = 2$ kΩ.

14.39
For the triggered sawtooth generator of Fig. 14.31, sketch v_{o1} and v_{o2} for $C_I = 1$ μF, $R_I = 5$ kΩ, $R_F = 2$ kΩ, $R_S = 4$ kΩ, $T = 1$ ms, and $V_{CC} = 10$ V.

14.40
Verify equation (14.7–8).

14.41
Consider the weighted-resistor D/A converter of Fig. 14.42. For $V_{ref} = 5$ V, $R_F = 1$ kΩ, and $R = 20$ kΩ, determine the analog output voltage corresponding to the following:
(a) 1001000   **(b)** 0000111   **(c)** 0000001

14.42
Repeat Problem 14.41 for $V_{ref} = 1$ V, $R_F = 1$ kΩ, and $R = 10$ kΩ.

14.43
Consider the R–$2R$ ladder D/A converter of Fig. 14.44a. For $V_{ref} = 5$ V and $k = 10$, determine the analog output voltage corresponding to the following:
(a) 1001000   **(b)** 0000111   **(c)** 0000001

14.44
Repeat Problem 14.43 for $V_{ref} = 10$ V and $k = 1$.

CHAPTER 14 APPENDIX

DATA SHEET A14.1   Voltage Regulators

positive-voltage regulators

DEVICE SERIES	OUTPUT VOLTAGE TOLERANCE	MINIMUM DIFFERENTIAL VOLTAGE	OUTPUT CURRENT RATING	PACKAGES	PAGE
LM2930-0	±10%	0.6 V	150 mA	KC	6-45
LM2931-0	±10%	0.6 V	150 mA	KC	6-51
LM330-0	±4%	0.6 V	150 mA	KC	6-27
LM340-00	+4%	2 V	1.5 A	KC	6-33
TL780-00C	±1%	2 V	1.5 A	KC	6-137
uA7800C	+4%	2 V – 3 V	1.5 A	KC	6-175
uA78L00AC	+5%	2 V	100 mA	LP	6-183
uA78L00C	±10%	2 V – 2.5 V	100 mA	LP	6-183
uA78M00C	±5%	2 V – 3 V	500 mA	KC	6-189
uA78M00M	±5%	2 V – 3 V	500 mA	KC	6-189

negative-voltage regulators

DEVICE SERIES	OUTPUT VOLTAGE TOLERANCE	MINIMUM DIFFERENTIAL VOLTAGE	OUTPUT CURRENT RATING	PACKAGES	PAGE
LM320-00	±4%	2 V	1.5 A	KC	6-21
MC79L00AC	±5%	1.7 V	100 mA	LP	6-57
MC79L00C	±10%	1.7 V	100 mA	LP	6-57
uA7900C	±5%	2 V – 3 V	1.5 A	KC	6-201
uA79M00C	±5%	2 V – 3 V	1.5 A	KC	6-207
uA79M00M	±5%	2 V – 3 V	1.5 A	KC	6-207

available output voltage for above regulator series

DEVICE SERIES	VOLTAGE SELECTIONS													
	2.6	5.0	5.2	6.0	6.2	8.0	8.5	9.0	10.0	12.0	15.0	18.0	20.0	24.0
LM2930-0		X				X								
LM2931-0		X												
LM320-00		X								X	X			
LM330-0		X												
LM340-00		X								X	X			
MC7900AC		X								X	X			
MC79L00C		X								X	X			
TL780-00C		X								X	X			
uA7800C		X		X		X	X		X	X	X	X		X
uA78L00AC	X	X		X		X		X	X	X	X			
UA78L00C	X	X			X	X		X	X	X	X			
uA78M00C		X		X		X			X	X	X		X	X
uA78M00M		X		X		X			X	X	X			
uA7900C		X	X	X		X				X	X	X		X
uA7900M		X	X	X		X				X	X	X		X
uA79M00C		X		X		X				X	X		X	X
uA79M00M		X		X		X				X	X			

Reprinted by permission of copyright holder. © 1984 Texas Instruments Incorporated.

DATA SHEET A14.1   Continued

positive-voltage series regulators

DEVICE NUMBER	OUTPUT VOLTAGE		MAXIMUM DIFFERENTIAL VOLTAGE	OUTPUT CURRENT RATING	PACKAGES	PAGE
	MIN	MAX				
LM217	1.2 V	37 V	$V_I - 1.2$ V	1.5 A	KC	6-11
LM317	1.2 V	37 V	$V_I - 1.2$ V	1.5 A	KC	6-11
LM350	1.2 V	33 V	$V_I - 1.2$ V	3 A	KC	6-41
TL317C	1.2 V	32 V	$V_I - 1.2$ V	100 mA	LP	6-91
TL317M	1.2 V	32 V	$V_I - 1.2$ V	100 mA	LP	6-91
TL783AC	5 V	200 V	200 V	700 mA	KC	6-141
TL783C	10 V	125 V	37 V	700 mA	KC	6-141
uA723C	3 V	38 V	37 V	25 mA	J, N	6-169
uA723M	3 V	38 V	37 V	25 mA	J, N	6-169

negative-voltage series regulators

DEVICE NUMBER	OUTPUT VOLTAGE		MAXIMUM DIFFERENTIAL VOLTAGE	OUTPUT CURRENT RATING	PACKAGES	PAGE
	MIN	MAX				
LM237	1.2 V	37 V	$V_I - 1.2$ V	1.5 A	KC	6-17
LM337	1.2 V	37 V	$V_I - 1.2$ V	1.5 A	KC	6-17

positive-shunt regulators

DEVICE NUMBER	SHUNT VOLTAGE		SHUNT CURRENT		TEMP COEFFICIENT RATING	PACKAGES	PAGE
	MIN	MAX	MIN	MAX			
TL430C	3 V	30 V	2 mA	100 mA	200 ppm/°C	LP	6-95
TL430I	3 V	30 V	2 mA	100 mA	200 ppm/°C	LP, P	6-95
TL431C	3 V	30 V	0.5 mA	100 mA	100 ppm/°C	LP, P	6-99
TL431I	2.55 V	36 V	1 mA	100 mA	100 ppm/°C	LP	6-99
TL431M	2.55 V	36 V	1 mA	100 mA	100 ppn/°C	JG	6-99

Adler, E.L., P.R. Belanger, and N.C. Rumin, *Introduction to Circuits with Electronics, An Integrated Approach.* New York: Holt, Rinehart and Winston, 1985.

Allen, P.E., and D.R. Holberg, *CMOS Analog Circuit Design.* New York: Holt, Rinehart and Winston, 1987.

Boctor, S.A., *Electric Circuit Analysis.* Englewood Cliffs, N.J.: Prentice-Hall, 1987.

Boylestad, R., and L. Nashelsky, *Electronic Devices and Circuit Theory,* 4th edition. Englewood Cliffs, N.J.: Prentice-Hall, 1987.

Burns, S.G., and P.R. Bond, *Principles of Electronic Circuits.* St. Paul, Minn.: West Publishing Company, 1987.

Carlson A.B., and D.G. Gisser, *Electrical Engineering Concepts and Applications.* Cambridge, Mass.: Addison-Wesley Publishing Company, 1981.

Castellucis, R.L., and S.D. Prensky, *Electronic Instrumentation,* 3rd ed. Englewood Cliffs, N.J.: Prentice-Hall, 1982.

Cathey, J.J., and S.A. Nasar, *Basic Electrical Engineering.* New York: McGraw-Hill Book Company, 1984.

Chin, A.F., and L. Jones, *Electronic Instruments and Measurements.* New York: John Wiley & Sons, 1983.

Comer, D.J., *Electronic Design with Integrated Circuits.* Reading, Mass.: Addison-Wesley Publishing Company, 1981.

Daugherty, D.G., and H.E. Talley, *Physical Principles of Semiconductor Devices.* Ames: Iowa State University Press, 1976.

DelToro, V. *Engineering Circuits.* Englewood Cliffs, N.J.: Prentice-Hall, 1987.

Douglas, M., and J.F. Passafiume, *Digital Logic Design, Tutorials and Laboratory Exercises.* New York: Harper and Row, Publishers, 1985.

Fitchen, F.C., *Electronic Integrated Circuits and Systems.* New York: Van Nostrand Reinhold Company, 1970.

Fitzgerald, A.E., D.E. Higginbotham, and A. Grabel, *Basic Electrical Engineering,* 4th ed. New York: McGraw-Hill Book Company, 1975.

Ghausi, M.S., *Electronic Devices and Circuits: Discrete and Integrated.* New York: Holt, Rinehart and Winston, 1985.

Grabel, A., and J. Millman, *Microelectronics,* 2nd ed. New York: McGraw-Hill Book Company, 1987.

Green, M.A., *Solar Cells, Operating Principles, Technology, and System Applications.* Englewood Cliffs, N.J.: Prentice-Hall, 1982.

Halkias, C.C., and J. Millman, *Electronic Devices and Circuits.* New York: McGraw-Hill Book Company, 1967.

Hayt, W.H. Jr., and J.E. Kemmerly, *Engineering Circuit Analysis,* 4th ed. New York: McGraw-Hill Book Company, 1986.

Horowitz P., and W. Hill, *The Art of Electronics.* Cambridge, Mass.: Cambridge University Press, 1980.

Karni, S., *Analysis of Electrical Networks.* New York: John Wiley & Sons, 1986.

Malvino, A.P., *Semiconductor Circuit Approximations,* 4th ed. New York: McGraw-Hill Book Company, 1985.

Margolis, S., and R. Mayne, *Introduction to Engineering.* New York: McGraw-Hill Book Company, 1982.

Miller, G.M., *Modern Electronic Communication,* 2nd ed. Englewood Cliffs, N.J.: Prentice-Hall, 1983.

Mitchell, F.H. Jr., and F.H. Sr., *Introduction to Electronics Design.* Englewood Cliffs, N.J.: Prentice-Hall, 1988.

Mitchell, F.H. Jr., and F.H. Sr., *Solutions Manual, Introduction to Electronics Design.* Englewood Cliffs, N.J.: Prentice-Hall, 1988.

Nasar, S.A., and L.E. Unnewehr, *Electromechanics and Electric Machines.* New York: John Wiley & Sons, 1979.

Pearman, R.A., *Power Electronics–Solid State Motor Control.* Reston, Va.: Reston Publishing Company, 1980.

Reinhard, D.K., *Introduction to Integrated Circuit Engineering.* Boston: Houghton Mifflin Company, 1987.

Roadstrum, W.H., and D.H. Wolaver, *Electrical Engineering for All Engineers.* New York: Harper and Row Publishers, 1987.

Roth, C.H. Jr., *Fundamentals of Logic Design,* 3rd ed. St. Paul, Minn.: West Publishing Company, 1985.

Schilling, D., and H. Taub, *Digital Integrated Electronics.* New York: McGraw-Hill Book Company, 1977.

Schilling, D.L., and C. Belove, *Electronic Circuits, Discrete and Integrated.* New York: McGraw-Hill Book Company, 1979.

Sedra, A.S., and K.C. Smith, *Microelectronic Circuits.* New York: Holt, Rinehart and Winston, 1987.

Smith, R.J., *Circuits, Devices and Systems,* 4th ed. New York: John Wiley & Sons, 1984.

Soclof, S., *Applications of Analog Integrated Circuits.* Englewood Cliffs, N.J.: Prentice-Hall, 1985.

Stout, M.B., *Basic Electrical Measurements.* Englewood Cliffs, N.J.: Prentice-Hall, 1950.

Strum, R.D., and J.R. Ward, *Electric Circuits and Networks,* 2nd ed. Englewood Cliffs, N.J.: Prentice-Hall, 1985.

Suprynowicz, V.A., *Electrical and Electronics Fundamentals.* St. Paul, Minn.: West Publishing Company, 1987.

Tocci, R.J., *Digital Systems, Principles and Applications.* Englewood Cliffs, N.J.: Prentice-Hall, 1985.

Triebel, W.A., *Integrated Digital Electronics.* Englewood Cliffs, N.J.: Prentice-Hall, 1985.

Van der Ziel, A., *Introduction to Electronic Circuits.* Boston: Allyn and Bacon, 1969.

Weick, C.B., *Applied Electronics.* New York: McGraw-Hall Book Company, 1976.

Wojslaw, C.F., *Electronic Concepts, Principles, and Circuits.* Reston. Va.: Reston Publishing Company, 1980.

APPENDIXES

APPENDIX 1   Diode Data Sheets

1N4001–1N4007

**LEAD MOUNTED
SILICON RECTIFIERS**

**50-1000 VOLTS
DIFFUSED JUNCTION**

"SURMETIC"▲ RECTIFIERS

. . . subminiature size, axial lead mounted rectifiers for general-purpose low-power applications.

Designers Data for "Worst Case" Conditions

The Designers▲ Data Sheets permit the design of most circuits entirely from the information presented. Limit curves — representing boundaries on device characteristics — are given to facilitate "worst case" design.

*MAXIMUM RATINGS

Rating	Symbol	1N4001	1N4002	1N4003	1N4004	1N4005	1N4006	1N4007	Unit
Peak Repetitive Reverse Voltage Working Peak Reverse Voltage DC Blocking Voltage	V_{RRM} V_{RWM} V_R	50	100	200	400	600	800	1000	Volts
Non-Repetitive Peak Reverse Voltage (halfwave, single phase, 60 Hz)	V_{RSM}	60	120	240	480	720	1000	1200	Volts
RMS Reverse Voltage	$V_{R(RMS)}$	35	70	140	280	420	560	700	Volts
Average Rectified Forward Current (single phase, resistive load, 60 Hz, see Figure 8, $T_A = 75^oC$)	I_O	1.0							Amp
Non-Repetitive Peak Surge Current (surge applied at rated load conditions, see Figure 2)	I_{FSM}	30 (for 1 cycle)							Amp
Operating and Storage Junction Temperature Range	T_J, T_{stg}	–65 to +175							^oC

*ELECTRICAL CHARACTERISTICS

Characteristic and Conditions	Symbol	Typ	Max	Unit
Maximum Instantaneous Forward Voltage Drop ($i_F = 1.0$ Amp, $T_J = 25^oC$) Figure 1	v_F	0.93	1.1	Volts
Maximum Full-Cycle Average Forward Voltage Drop ($I_O = 1.0$ Amp, $T_L = 75^oC$, 1 inch leads)	$V_{F(AV)}$	–	0.8	Volts
Maximum Reverse Current (rated dc voltage) $T_J = 25^oC$ $T_J = 100^oC$	I_R	0.05 1.0	10 50	μA
Maximum Full-Cycle Average Reverse Current ($I_O = 1.0$ Amp, $T_L = 75^oC$, 1 inch leads	$I_{R(AV)}$	–	30	μA

*Indicates JEDEC Registered Data.

MECHANICAL CHARACTERISTICS

CASE: Void free, Transfer Molded
MAXIMUM LEAD TEMPERATURE FOR SOLDERING PURPOSES: 350^oC, 3/8'' from case for 10 seconds at 5 lbs. tension
FINISH: All external surfaces are corrosion-resistant, leads are readily solderable
POLARITY: Cathode indicated by color band
WEIGHT: 0.40 Grams (approximately)

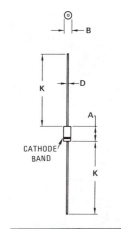

CATHODE
BAND

DIM	MILLIMETERS		INCHES	
	MIN	MAX	MIN	MAX
A	5.97	6.60	0.235	0.260
B	2.79	3.05	0.110	0.120
D	0.76	0.86	0.030	0.034
K	27.94	–	1.100	–

CASE 59-04
Does Not Conform to DO-41 Outline.

(continues)

APPENDIX 1   Diode Data Sheets   (Continued)

1N4001–1N4007

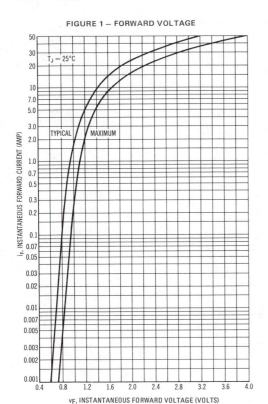

FIGURE 1 — FORWARD VOLTAGE

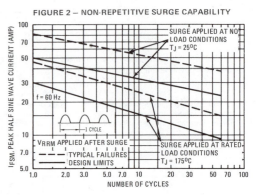

FIGURE 2 — NON-REPETITIVE SURGE CAPABILITY

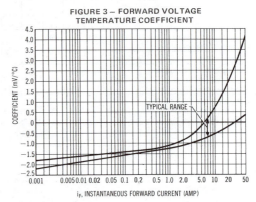

FIGURE 3 — FORWARD VOLTAGE
TEMPERATURE COEFFICIENT

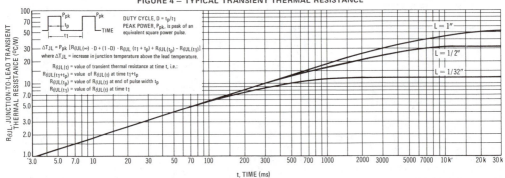

FIGURE 4 — TYPICAL TRANSIENT THERMAL RESISTANCE

The temperature of the lead should be measured using a thermocouple placed on the lead as close as possible to the tie point. The thermal mass connected to the tie point is normally large enough so that it will not significantly respond to heat surges generated in the diode as a result of pulsed operation once steady-state conditions are achieved. Using the measured value of T_L, the junction temperature may be determined by:

$$T_J = T_L + \triangle T_{JL}.$$

APPENDIX 1   Diode Data Sheets   (Continued)

1N4001–1N4007

CURRENT DERATING DATA

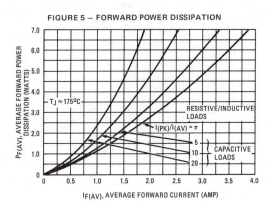

FIGURE 5 — FORWARD POWER DISSIPATION

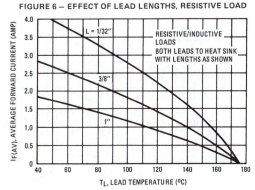

FIGURE 6 — EFFECT OF LEAD LENGTHS, RESISTIVE LOAD

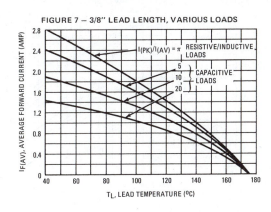

FIGURE 7 — 3/8″ LEAD LENGTH, VARIOUS LOADS

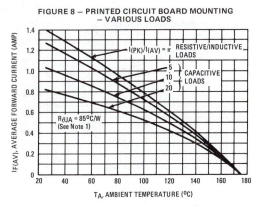

FIGURE 8 — PRINTED CIRCUIT BOARD MOUNTING — VARIOUS LOADS

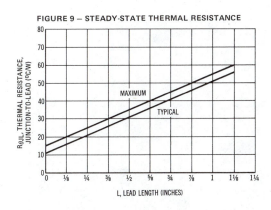

FIGURE 9 — STEADY-STATE THERMAL RESISTANCE

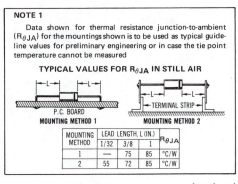

NOTE 1

Data shown for thermal resistance junction-to-ambient ($R_{\theta JA}$) for the mountings shown is to be used as typical guideline values for preliminary engineering or in case the tie point temperature cannot be measured

TYPICAL VALUES FOR $R_{\theta JA}$ IN STILL AIR

MOUNTING METHOD	LEAD LENGTH, L (IN.)			$R_{\theta JA}$
	1/32	3/8	1	
1	—	75	85	°C/W
2	55	72	85	°C/W

(continues)

APPENDIX 1   Diode Data Sheets   (Continued)

1N4001–1N4007

TYPICAL DYNAMIC CHARACTERISTICS

FIGURE 10 – FORWARD RECOVERY TIME

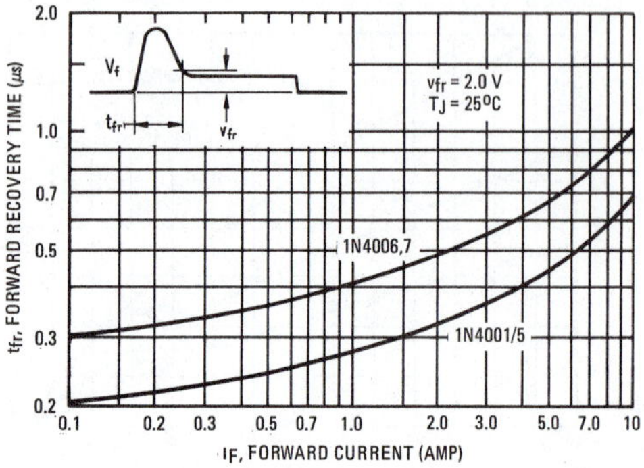

$v_{fr} = 2.0$ V
$T_J = 25^oC$

1N4006,7

1N4001/5

t_{fr}, FORWARD RECOVERY TIME (μs)

I_F, FORWARD CURRENT (AMP)

FIGURE 11 – REVERSE RECOVERY TIME

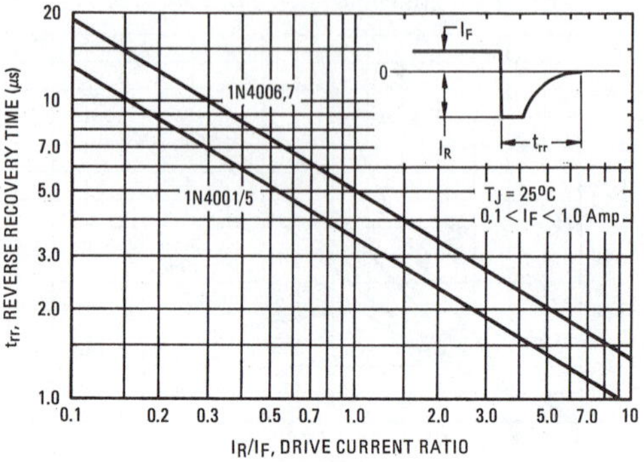

1N4006,7

1N4001/5

$T_J = 25^oC$
$0.1 < I_F < 1.0$ Amp

t_{rr}, REVERSE RECOVERY TIME (μs)

I_R/I_F, DRIVE CURRENT RATIO

FIGURE 12 – JUNCTION CAPACITANCE

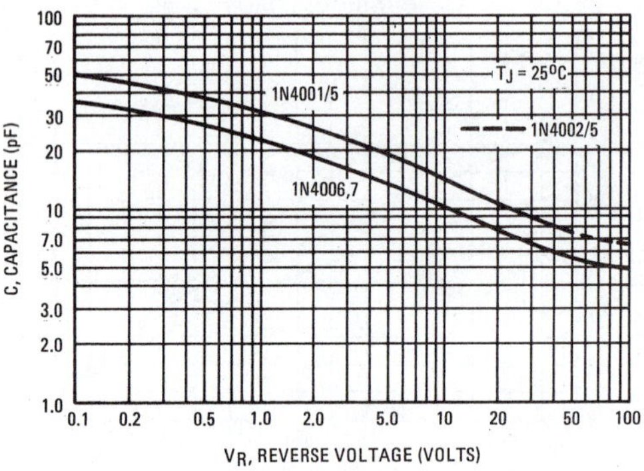

$T_J = 25^oC$

1N4001/5

1N4002/5

1N4006,7

C, CAPACITANCE (pF)

V_R, REVERSE VOLTAGE (VOLTS)

FIGURE 13 – RECTIFICATION WAVEFORM EFFICIENCY FOR SINE WAVE

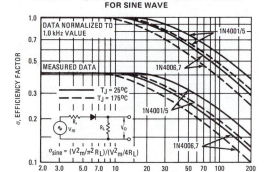

DATA NORMALIZED TO 1.0 kHz VALUE

1N4001/5

MEASURED DATA

1N4006,7

$T_J = 25^oC$
$T_J = 175^oC$

1N4001/5

1N4006,7

$\sigma_{sine} = (V^2m/\pi^2 R_L)/(V^2m/4R_L)$

σ, EFFICIENCY FACTOR

REPETITION FREQUENCY (kHz)

FIGURE 14 – RECTIFICATION WAVEFORM EFFICIENCY FOR SQUARE WAVE

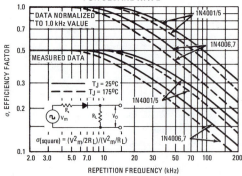

DATA NORMALIZED TO 1.0 kHz VALUE

1N4001/5

1N4006,7

MEASURED DATA

$T_J = 25^oC$
$T_J = 175^oC$

1N4001/5

1N4006,7

$\sigma_{(square)} = (V^2m/2R_L)/(V^2m/R_L)$

σ, EFFICIENCY FACTOR

REPETITION FREQUENCY (kHz)

RECTIFIER EFFICIENCY NOTE

The rectification efficiency factor σ shown in Figures 13 and 14 was calculated using the formula:

$$\sigma = \frac{P_{dc}}{P_{rms}} = \frac{\dfrac{V^2_O(dc)}{R_L}}{\dfrac{V^2_O(rms)}{R_L}} \cdot 100\% = \frac{V^2_O(dc)}{V^2_O(ac) + V^2_O(dc)} \cdot 100\% \quad (1)$$

For a sine wave input $V_m\sin(\omega t)$ to the diode, assumed lossless, the maximum theoretical efficiency factor becomes 40%; for a square wave input of amplitude V_m, the efficiency factor becomes 50%. (A full wave circuit has twice these efficiencies).

As the frequency of the input signal is increased, the reverse recovery time of the diode (Figure 11) becomes significant, resulting in an increasing ac voltage component across R_L which is opposite in polarity to the forward current thereby reducing the value of the efficiency factor σ, as shown in Figures 13 and 14.

It should be emphasized that Figures 13 and 14 show waveform efficiency only; they do not account for diode losses. Data was obtained by measuring the ac component of V_O with a true rms voltmeter and the dc component with a dc voltmeter. The data was used in Equation 1 to obtain points for the Figures.

1N4099–1N4135, 1N4614–1N4627

LOW-LEVEL SILICON PASSIVATED ZENER DIODES

. . . designed for 250 mW applications requiring low leakage, low impedance, and low noise.

- Voltage Range from 1.8 to 100 Volts
- First Zener Diode Series to Specify Noise — 50% Lower than Conventional Diffused Zeners
- Zener Impedance and Zener Voltage Specified for Low-Level Operation at I_{ZT} = 250 μA
- Low Leakage Current — I_R from 0.01 to 10 μA over Voltage Range

SILICON ZENER DIODES

(±5.0% TOLERANCE)

**250 MILLIWATTS
1.8–100 VOLTS**

SILICON OXIDE
PASSIVATED JUNCTION

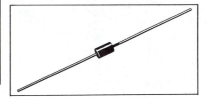

MAXIMUM RATINGS

Rating	Symbol	Value	Unit
DC Power Dissipation @ T_A = 25°C	P_D	250	mW
Derate above 25°C		1.43	mW/°C
Junction and Storage Temperature Range	T_J, T_{stg}	–65 to +200	°C

MECHANICAL CHARACTERISTICS

CASE: Hermetically sealed, all-glass.

DIMENSIONS: See outline drawing.

FINISH: All external surfaces are corrosion resistant and leads are readily solderable and weldable.

POLARITY: Cathode indicated by polarity band.

WEIGHT: 0.2 gram (approx.)

MOUNTING POSITION: Any

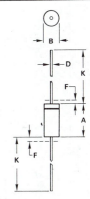

NOTES:
1. PACKAGE CONTOUR OPTIONAL WITHIN A AND B. HEAT SLUGS, IF ANY, SHALL BE INCLUDED WITHIN THIS CYLINDER, BUT NOT SUBJECT TO THE MINIMUM LIMIT OF B.
2. LEAD DIAMETER NOT CONTROLLED IN ZONE F TO ALLOW FOR FLASH, LEAD FINISH BUILDUP AND MINOR IRREGULARITIES OTHER THAN HEAT SLUGS.
3. POLARITY DENOTED BY CATHODE BAND.
4. DIMENSIONING AND TOLERANCING PER ANSI Y14.5, 1973.

DIM	MILLIMETERS		INCHES	
	MIN	MAX	MIN	MAX
A	3.05	5.08	0.120	0.200
B	1.52	2.29	0.060	0.090
D	0.46	0.56	0.018	0.022
F	—	1.27	—	0.050
K	25.40	38.10	1.000	1.500

All JEDEC dimensions and notes apply.

**CASE 299-02
DO-204AH**

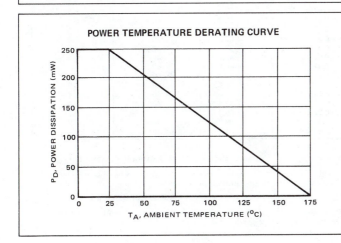

POWER TEMPERATURE DERATING CURVE

(*continues*)

APPENDIX 1   Diode Data Sheets   (Continued)

1N4099–1N4135, 1N4614–1N4627

ELECTRICAL CHARACTERISTICS

(At 25°C Ambient temperature unless otherwise specified) $I_{ZT} = 250\ \mu A$ and $V_F = 1.0$ V max @ $I_F = 200$ mA on all Types

Type Number (Note 1)	Nominal Zener Voltage V_Z (Note 1) (Volts)	Max Zener Impedance Z_{ZT} (Note 2) (Ohms)	Max Reverse Current I_R (μA)	@ (Note 4)	Test Voltage V_R (Volts)	Max Noise Density At $I_{ZT} = 250\ \mu A$ N_D (Fig 1) (micro-volts per Square Root Cycle)	Max Zener Current I_{ZM} (Note 3) (mA)
1N4614	1.8	1200	7.5		1.0	1.0	120
1N4615	2.0	1250	5.0		1.0	1.0	110
1N4616	2.2	1300	4.0		1.0	1.0	100
1N4617	2.4	1400	2.0		1.0	1.0	95
1N4618	2.7	1500	1.0		1.0	1.0	90
1N4619	3.0	1600	0.8		1.0	1.0	85
1N4620	3.3	1650	7.5		1.5	1.0	80
1N4621	3.6	1700	7.5		2.0	1.0	75
1N4622	3.9	1650	5.0		2.0	1.0	70
1N4623	4.3	1600	4.0		2.0	1.0	65
1N4624	4.7	1550	10		3.0	1.0	60
1N4625	5.1	1500	10		3.0	2.0	55
1N4626	5.6	1400	10		4.0	4.0	50
1N4627	6.2	1200	10		5.0	5.0	45
1N4099	6.8	200	10		5.2	40	35
1N4100	7.5	200	10		5.7	40	31.8
1N4101	8.2	200	1.0		6.3	40	29.0
1N4102	8.7	200	1.0		6.7	40	27.4
1N4103	9.1	200	1.0		7.0	40	26.2
1N4104	10	200	1.0		7.6	40	24.8
1N4105	11	200	0.05		8.5	40	21.6
1N4106	12	200	0.05		9.2	40	20.4
1N4107	13	200	0.05		9.9	40	19.0
1N4108	14	200	0.05		10.7	40	17.5
1N4109	15	100	0.05		11.4	40	16.3
1N4110	16	100	0.05		12.2	40	15.4
1N4111	17	100	0.05		13.0	40	14.5
1N4112	18	100	0.05		13.7	40	13.2
1N4113	19	150	0.05		14.5	40	12.5
1N4114	20	150	0.01		15.2	40	11.9
1N4115	22	150	0.01		16.8	40	10.8
1N4116	24	150	0.01		18.3	40	9.9
1N4117	25	150	0.01		19.0	40	9.5
1N4118	27	150	0.01		20.5	40	8.8
1N4119	28	200	0.01		21.3	40	8.5
1N4120	30	200	0.01		22.8	40	7.9
1N4121	33	200	0.01		25.1	40	7.2
1N4122	36	200	0.01		27.4	40	6.6
1N4123	39	200	0.01		29.7	40	6.1
1N4124	43	250	0.01		32.7	40	5.5
1N4125	47	250	0.01		35.8	40	5.1
1N4126	51	300	0.01		38.8	40	4.6
1N4127	56	300	0.01		42.6	40	4.2
1N4128	60	400	0.01		45.6	40	4.0
1N4129	62	500	0.01		47.1	40	3.8
1N4130	68	700	0.01		51.7	40	3.5
1N4131	75	700	0.01		57.0	40	3.1
1N4132	82	800	0.01		62.4	40	2.9
1N4133	87	1000	0.01		66.2	40	2.7
1N4134	91	1200	0.01		69.2	40	2.6
1N4135	100	1500	0.01		76.0	40	2.3

NOTE 1: TOLERANCE AND VOLTAGE DESIGNATION

The type numbers shown have a standard tolerance of ±5.0% on the nominal zener voltage.

NOTE 2: ZENER IMPEDANCE (Z_{ZT}) DERIVATION

The zener impedance is derived from the 60 cycle ac voltage, which results when an ac current having an rms value equal to 10% of the dc zener current (I_{ZT}) is superimposed on I_{ZT}.

NOTE 3: MAXIMUM ZENER CURRENT RATINGS (I_{ZM})

Maximum zener current ratings are based on maximum zener voltage of the individual units.

NOTE 4: REVERSE LEAKAGE CURRENT I_R

Reverse leakage currents are guaranteed and are measured at V_R as shown on the table.

APPENDIX 1   Diode Data Sheets   (Continued)

1N4099–1N4135, 1N4614–1N4627

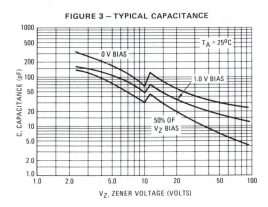

FIGURE 3 – TYPICAL CAPACITANCE

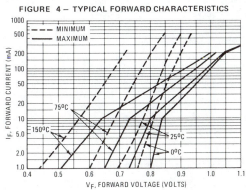

FIGURE 4 – TYPICAL FORWARD CHARACTERISTICS

MAXIMUM RATINGS

Rating	Symbol	2N4237	2N4238	2N4239	Unit
Collector-Emitter Voltage	V_{CEO}	40	60	80	Vdc
Collector-Base Voltage	V_{CBO}	50	80	100	Vdc
Emitter-Base Voltage	V_{EBO}		6.0		Vdc
Base Current	I_B		500		Vdc
Collector Current — Continuous	I_C		1.0 3.0*		Adc
Total Device Dissipation @ $T_A = 25°C$ Derate above 25°C	P_D		1.0 5.3		Watt mW/°C
Total Device Dissipation @ $T_C = 25°C$ Derate above 25°C	P_D		6.0 34		Watts mW/°C
Operating and Storage Junction Temperature Range	T_J, T_{stg}		-65 to $+200$		°C

THERMAL CHARACTERISTICS

Characteristic	Symbol	Max	Unit
*Thermal Resistance, Junction to Case	$R_{\theta JC}$	29	°C/W

2N4237
2N4238
2N4239

CASE 79-02, STYLE 1
TO-39 (TO-205AD)

3 Collector

2
Base

3   2 1

1 Emitter

GENERAL PURPOSE
TRANSISTOR

NPN SILICON

ELECTRICAL CHARACTERISTICS ($T_A = 25°C$ unless otherwise noted.)

Characteristic		Symbol	Min	Max	Unit
OFF CHARACTERISTICS					
Collector-Emitter Sustaining Voltage(1) ($I_C = 100$ mAdc, $I_B = 0$)	2N4237 2N4238 2N4239	$V_{CEO}(sus)$	40 60 80	— — —	Vdc
Collector Cutoff Current ($V_{CE} = 50$ Vdc, $V_{EB} = 1.5$ Vdc) ($V_{CE} = 80$ Vdc, $V_{EB} = 1.5$ Vdc) ($V_{CE} = 100$ Vdc, $V_{EB} = 1.5$ Vdc) ($V_{CE} = 30$ Vdc, $V_{EB} = 1.5$ Vdc, $T_C = 150°C$) ($V_{CE} = 50$ Vdc, $V_{EB} = 1.5$ Vdc, $T_C = 150°C$) ($V_{CE} = 70$ Vdc, $V_{EB} = 1.5$ Vdc, $T_C = 150°C$)	2N4237 2N4238 2N4239 2N4237 2N4238 2N4239	I_{CEX}	— — — — 	0.1 0.1 0.1 1.0 1.0 1.0	mAdc
Collector Cutoff Current ($V_{CB} = $ Rated V_{CBO}, $I_E = 0$) ($V_{CE} = $ Rated V_{CEO}, $I_B = 0$)		I_{CBO}	— —	0.1 .07	mAdc
Emitter Cutoff Current ($V_{EB} = 6.0$ Vdc, $I_C = 0$)		I_{EBO}	—	0.5	mAdc
ON CHARACTERISTICS					
DC Current Gain(1) ($I_C = 50$ mAdc, $V_{CE} = 1.0$ Vdc) ($I_C = 250$ mAdc, $V_{CE} = 1.0$ Vdc) ($I_C = 500$ mAdc, $V_{CE} = 1.0$ Vdc) ($I_C = 1.0$ Adc, $V_{CE} = 1.0$ Vdc)		h_{FE}	30 30 30 15	— 150 — —	—
Collector-Emitter Saturation Voltage(1) ($I_C = 500$ mAdc, $I_B = 50$ mAdc) ($I_C = 1.0$ Adc, $I_B = 0.1$ Adc)		$V_{CE}(sat)$	— —	0.3 0.6	Vdc
Base-Emitter Saturation Voltage(1) ($I_C = 1.0$ Adc, $I_B = 0.1$ Adc)		$V_{BE}(sat)$	—	1.5	Vdc
Base-Emitter On Voltage(1) ($I_C = 250$ mAdc, $V_{CE} = 1.0$ Vdc)		$V_{BE}(on)$	—	1.0	Vdc
SMALL-SIGNAL CHARACTERISTICS					
Output Capacitance ($V_{CB} = 10$ Vdc, $I_C = 0$, $f = 0.1$ MHz)		C_{obo}	—	100	pF
Small Signal Current Gain ($I_C = 100$ mAdc, $V_{CE} = 10$ Vdc, $f = 1.0$ kHz)		h_{fe}	30	—	—
Current Gain — High Frequency ($V_{CE} = 10$ V, $I_C = 100$ mA, $f = 1$ MHz)		$|h_{fe}|$	1.0	—	—

(1) Pulse Test: Pulse Width ≤ 300 μs, Duty Cycle 2.0%.
*Indicates Data in addition to JEDEC Requirements.

APPENDIX 2   BJT Data Sheets   (Continued)

2N4237, 2N4238, 2N4239

FIGURE 1 — POWER-TEMPERATURE DERATING CURVE

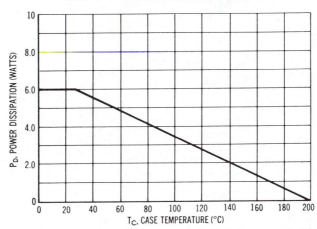

Safe Area Curves are indicated by Figure 5. All limits are applicable and must be observed.

SWITCHING CHARACTERISTICS

FIGURE 2 — SWITCHING TIME EQUIVALENT CIRCUIT

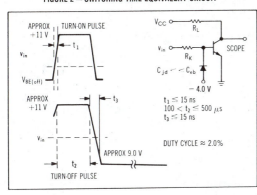

FIGURE 3 — TURN-ON TIME

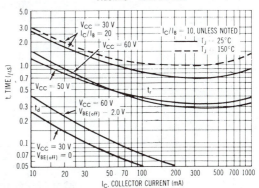

FIGURE 4 — THERMAL RESPONSE

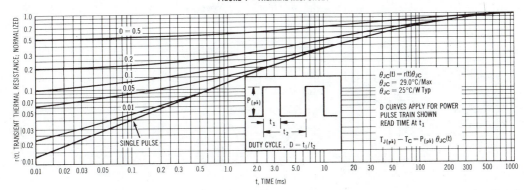

(continues)

APPENDIX 2   BJT Data Sheets   (Continued)

2N4237, 2N4238, 2N4239

FIGURE 5 — ACTIVE-REGION SAFE OPERATING AREAS

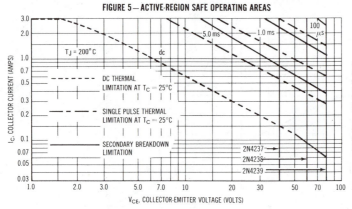

There are two limitations on the power handling ability of a transistor: junction temperature and secondary breakdown. Safe operating area curves indicate $I_C - V_{CE}$ limits of the transistor that must be observed for reliable operation; i.e., the transistor must not be subjected to greater dissipation than the curves indicate.

For this particular transistor family, the thermal curves are the limiting design values, except for a small portion of the dc curve. The pulse secondary breakdown curves are shown for information only.

FIGURE 6 — STORAGE TIME

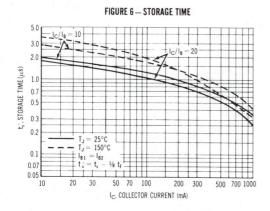

FIGURE 7 — FALL TIME

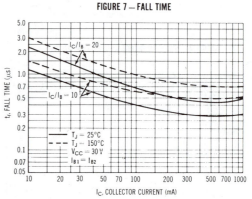

TYPICAL DC CHARACTERISTICS

FIGURE 8 — CURRENT GAIN

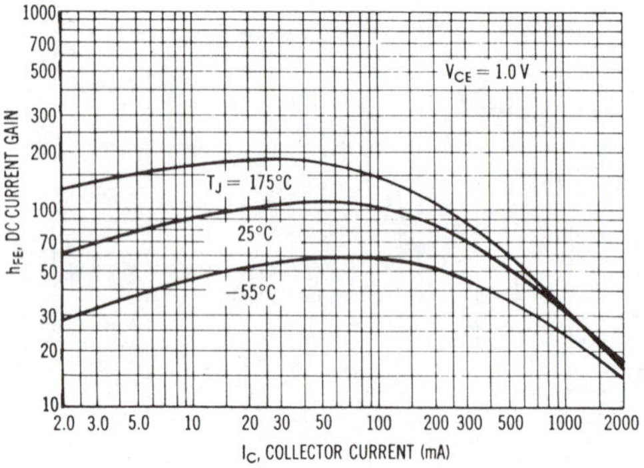

FIGURE 9 — COLLECTOR SATURATION REGION

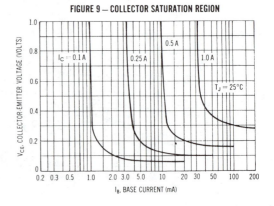

APPENDIX 2   BJT Data Sheets   (Continued)

2N4237, 2N4238, 2N4239

FIGURE 10 — EFFECTS OF BASE-EMITTER RESISTANCE

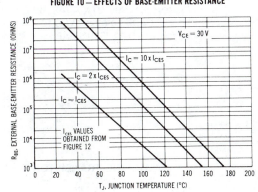

FIGURE 11 — "ON" VOLTAGE

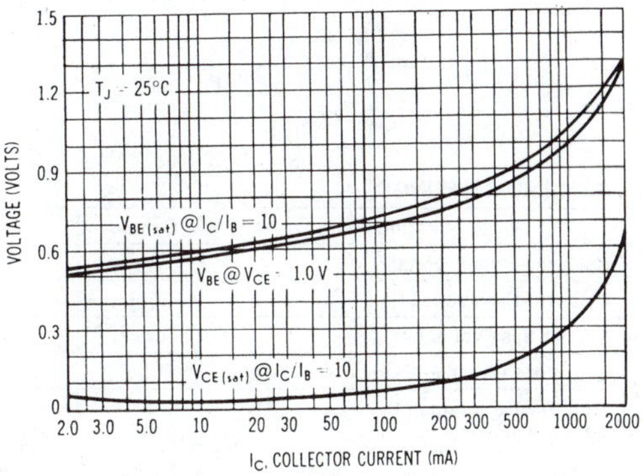

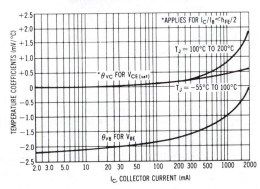

FIGURE 12 — COLLECTOR CUTOFF REGION

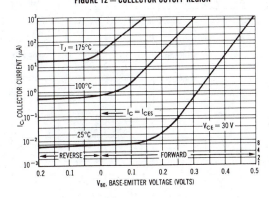

FIGURE 13 — TEMPERATURE COEFFICIENTS

(continues)

2N4260
2N4261

2N4261 JAN, JTX AVAILABLE
CASE 20-03, STYLE 10
TO-72 (TO-206AF)

3 Collector

2 Base

4 Case

1 Emitter

SWITCHING TRANSISTOR

PNP SILICON

MAXIMUM RATINGS

Rating	Symbol	Value	Unit
Collector-Emitter Voltage	V_{CEO}	15	Vdc
Collector-Base Voltage	V_{CBO}	15	Vdc
Emitter-Base Voltage	V_{EBO}	4.5	Vdc
Collector Current — Continuous	I_C	30	mAdc
Total Device Dissipation @ $T_A = 25°C$ Derate above 25°C	P_D	200 1.14	mW mW/°C
Operating and Storage Junction Temperature Range	T_J, T_{stg}	−65 to +200	°C

ELECTRICAL CHARACTERISTICS ($T_A = 25°C$ unless otherwise noted.)

Characteristic	Symbol	Min	Max	Unit
OFF CHARACTERISTICS				
Collector-Emitter Breakdown Voltage ($I_C = 10$ mAdc, $I_E = 0$)	$V_{(BR)CEO}$	15	—	Vdc
Collector-Base Breakdown Voltage ($I_C = 10$ μAdc, $I_E = 0$)	$V_{(BR)CBO}$	15	—	Vdc
Emitter-Base Breakdown Voltage ($I_E = 10$ μAdc, $I_C = 0$)	$V_{(BR)EBO}$	4.5	—	Vdc
Collector Cutoff Current ($V_{CE} = 10$ Vdc, $V_{BE(off)} = 2.0$ Vdc) ($V_{CE} = 10$ Vdc, $V_{BE(off)} = 2.0$ Vdc, $T_A = 150°C$) ($V_{CE} = 10$ Vdc, $V_{EB(on)} = 0.4$ Vdc)	I_{CEX}	— — —	0.005 5.0 0.05	μAdc
Base Cutoff Current ($V_{CE} = 10$ Vdc, $V_{BE(off)} = 2.0$ Vdc)	I_{BL}	—	0.005	μAdc
ON CHARACTERISTICS				
DC Current Gain ($I_C = 1.0$ mAdc, $V_{CE} = 1.0$ Vdc) ($I_C = 10$ mAdc, $V_{CE} = 1.0$ Vdc) ($I_C = 30$ mAdc, $V_{CE} = 2.0$ Vdc)	h_{FE}	25 30 20	— 150 —	—
Collector-Emitter Saturation Voltage ($I_C = 1.0$ mAdc, $I_B = 0.1$ mAdc) ($I_C = 10$ mAdc, $I_B = 1.0$ mAdc)	$V_{CE(sat)}$	— —	0.15 0.35	Vdc
Base-Emitter On Voltage ($I_C = 1.0$ mAdc, $V_{CE} = 1.0$ Vdc) ($I_C = 10$ mAdc, $V_{CE} = 1.0$ Vdc)	$V_{BE(on)}$	— —	0.8 1.0	Vdc
SMALL-SIGNAL CHARACTERISTICS				
Current-Gain — Bandwidth Product ($I_C = 5.0$ mAdc, $V_{CE} = 4.0$ Vdc, $f = 100$ MHz) 2N4260 2N4261 ($I_C = 10$ mAdc, $V_{CE} = 10$ Vdc, $f = 100$ MHz) 2N4260 2N4261	f_T	1200 1500 1600 2000	— — — —	MHz
Output Capacitance ($V_{CB} = 4.0$ Vdc, $I_E = 0$, $f = 100$ kHz)	C_{obo}	—	2.5	pF
Input Capacitance ($V_{BE} = 0.5$ Vdc, $I_C = 0$, $f = 100$ kHz)	C_{ibo}	—	2.5	pF
Current Gain — High Frequency ($I_C = 10$ mAdc, $V_{CE} = 10$ Vdc, $f = 100$ MHz) 2N4260 2N4261	$|h_{fe}|$	16 20	— —	—

APPENDIX 2   BJT Data Sheets   (Continued)

MHQ918, MHQ918H, MHQ918HX, MHQ918HXV

QUAD DUAL IN-LINE
NPN HERMETIC SILICON
HIGH FREQUENCY AMPLIFIER TRANSISTORS

. . . designed for low-level, high-gain amplifier applications.

● Low Noise Figure — NF = 4.0 dB (Typ) @ I_C = 1.0 mAdc

● High Current-Gain — Bandwidth Product —
   f_T = 850 MHz (Typ) @ I_C = 4.0 mAdc

● Transistors Similar to 2N918

● TO-116 Ceramic Package — Compact Size, Compatible with
   IC Automatic Insertion Equipment

● ''H'' Series for Hi-Rel Applications
   (See Tables 1 thru 3)

QUAD DUAL IN-LINE
NPN SILICON
HIGH FREQUENCY
AMPLIFIER TRANSISTORS

MAXIMUM RATINGS

Rating	Symbol	Value		Unit
Collector-Emitter Voltage	V_{CEO}	15		Vdc
Collector-Base Voltage	V_{CB}	30		Vdc
Emitter-Base Voltage	V_{EB}	3.0		Vdc
Collector Current — Continuous	I_C	50		mAdc
		Each Transistor	**Total Device**	
Power Dissipation @ T_A = 25°C Derate above 25°C	P_D	0.65 3.72	1.9 10.88	Watts mW/°C
Power Dissipation @ T_C = 25°C Derate above 25°C	P_D	1.3 7.43	4.6 26.3	Watts mW/°C
Operating and Storage Junction Temperature Range	T_J, T_{stg}	−65 to +200		°C

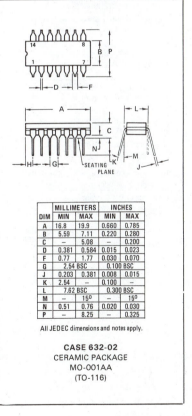

	MILLIMETERS		INCHES	
DIM	MIN	MAX	MIN	MAX
A	16.8	19.9	0.660	0.785
B	5.59	7.11	0.220	0.280
C	—	5.08	—	0.200
D	0.381	0.584	0.015	0.023
F	0.77	1.77	0.030	0.070
G	2.54 BSC		0.100 BSC	
J	0.203	0.381	0.008	0.015
K	2.54	—	0.100	—
L	7.62 BSC		0.300 BSC	
M	—	15⁰	—	15⁰
N	0.51	0.76	0.020	0.030
P	—	8.25	—	0.325

All JEDEC dimensions and notes apply.

CASE 632-02
CERAMIC PACKAGE
MO-001AA
(TO-116)

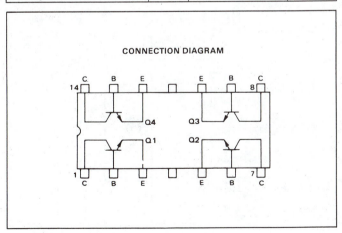

CONNECTION DIAGRAM

(continues)

MAXIMUM RATINGS

Rating	Symbol	2N2218 2N2219 2N2221 2N2222	2N2218A 2N2219A 2N2221A 2N2222A	2N5581 2N5582	Unit
Collector-Emitter Voltage	V_{CEO}	30	40	40	Vdc
Collector-Base Voltage	V_{CBO}	60	75	75	Vdc
Emitter-Base Voltage	V_{EBO}	5.0	6.0	6.0	Vdc
Collector Current — Continuous	I_C	800	800	800	mAdc
		2N2218,A 2N2219,A	2N2221,A 2N2222,A	2N5581 2N5582	
Total Device Dissipation @ T_A = 25°C Derate above 25°C	P_D	0.8 4.57	0.5 2.28	0.6 3.33	Watt mW/°C
Total Device Dissipation @ T_C = 25°C Derate above 25°C	P_D	3.0 17.1	1.2 6.85	2.0 11.43	Watts mW/°C
Operating and Storage Junction Temperature Range	T_J, T_{stg}	− 65 to + 200			°C

2N2218,A/2N2219,A
2N2221,A/2N2222,A
2N5581/82

JAN, JTX, JTXV AVAILABLE

2N2218,A
2N2219,A
CASE 79-02
TO-39 (TO-205AD)
STYLE 1

2N2221,A
2N2222,A
CASE 22-03
TO-18 (TO-206AA)
STYLE 1

2N5581
2N5582
CASE 26-03
TO-46 (TO-206AB)
STYLE 1

3 Collector

2 Base

1 Emitter

GENERAL PURPOSE TRANSISTOR

NPN SILICON

ELECTRICAL CHARACTERISTICS (T_A = 25°C unless otherwise noted.)

Characteristic	Symbol	Min	Max	Unit
OFF CHARACTERISTICS				
Collector-Emitter Breakdown Voltage (I_C = 10 mAdc, I_B = 0) Non-A Suffix	$V_{(BR)CEO}$	30	—	Vdc
A-Suffix, 2N5581, 2N5582		40	—	
Collector-Base Breakdown Voltage (I_C = 10 μAdc, I_E = 0) Non-A Suffix	$V_{(BR)CBO}$	60	—	Vdc
A-Suffix, 2N5581, 2N5582		75	—	
Emitter-Base Breakdown Voltage (I_E = 10 μAdc, I_C = 0) Non-A Suffix	$V_{(BR)EBO}$	5.0	—	Vdc
A-Suffix, 2N5581, 2N5582		6.0	—	
Collector Cutoff Current (V_{CE} = 60 Vdc, $V_{EB(off)}$ = 3.0 Vdc) A-Suffix, 2N5581, 2N5582	I_{CEX}		10	nAdc
Collector Cutoff Current (V_{CB} = 50 Vdc, I_E = 0) Non-A Suffix	I_{CBO}	—	0.01	μAdc
(V_{CB} = 60 Vdc, I_E = 0) A-Suffix, 2N5581, 2N5582		—	0.01	
(V_{CB} = 50 Vdc, I_E = 0, T_A = 150°C) Non-A Suffix		—	10	
(V_{CB} = 60 Vdc, I_E = 0, T_A = 150°C) A-Suffix, 2N5581, 2N5582		—	10	
Emitter Cutoff Current (V_{EB} = 3.0 Vdc, I_C = 0) A-Suffix, 2N5581, 2N5582	I_{EBO}	—	10	nAdc
Base Cutoff Current (V_{CE} = 60 Vdc, $V_{EB(off)}$ = 3.0 Vdc) A-Suffix	I_{BL}	—	20	nAdc
ON CHARACTERISTICS				
DC Current Gain (I_C = 0.1 mAdc, V_{CE} = 10 Vdc) 2N2218,A, 2N2221,A, 2N5581(1)	h_{FE}	20	—	—
2N2219,A, 2N2222,A, 2N5582(1)		35	—	
(I_C = 1.0 mAdc, V_{CE} = 10 Vdc) 2N2218,A, 2N2221,A, 2N5581		25	—	
2N2219,A, 2N2222,A, 2N5582		50	—	
(I_C = 10 mAdc, V_{CE} = 10 Vdc) 2N2218,A, 2N2221,A, 2N5581(1)		35	—	
2N2219,A, 2N2222,A, 2N5582(1)		75	—	
(I_C = 10 mAdc, V_{CE} = 10 Vdc, T_A = −55°C) 2N2218A, 2N2221A, 2N5581		15	—	
2N2219A, 2N2222A, 2N5582		35	—	
(I_C = 150 mAdc, V_{CE} = 10 Vdc)(1) 2N2218,A, 2N2221,A, 2N5581		40	120	
2N2219,A, 2N2222,A, 2N5582		100	300	

2N4856,A–2N4861,A

N-CHANNEL JUNCTION FIELD-EFFECT TRANSISTORS

Depletion Mode symmetrical Field-Effect transistors designed for low-power switching and chopper applications.

- Low Drain-Source "ON" Resistance —
  $r_{ds(on)}$ = 25 Ohms (Max) @ f = 1.0 kHz — 2N4856,A, 2N4859,A

- Low Drain Cutoff Current —
  $I_{D(off)}$ = 250 pAdc (Max) @ V_{DS} = 15 Vdc

N-CHANNEL JUNCTION FIELD-EFFECT TRANSISTORS

DS 5378 R1

*MAXIMUM RATINGS

Rating	Symbol	2N4856,A 2N4857,A 2N4858,A	2N4859,A 2N4860,A 2N4861,A	Unit
Drain-Gate Voltage	V_{DG}	+40	+30	Vdc
Drain-Source Voltage	V_{DS}	+40	+30	Vdc
Reverse Gate-Source Voltage	V_{GSR}	–40	–30	Vdc
Forward Gate Current	I_{GF}	50		mAdc
Total Device Dissipation @ T_A = 25°C Derate above 25°C	P_D	360 2.4		mW mW/°C
Storage Temperature Range	T_{stg}	–65 to +200		°C

*Indicates JEDEC Registered Data.

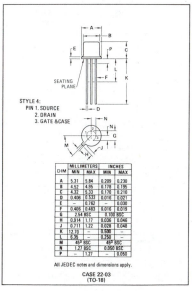

STYLE 4:
PIN 1. SOURCE
  2. DRAIN
  3. GATE &CASE

SEATING PLANE

DIM	MILLIMETERS MIN	MILLIMETERS MAX	INCHES MIN	INCHES MAX
A	5.31	5.84	0.209	0.230
B	4.52	4.95	0.178	0.195
C	4.32	5.33	0.170	0.210
D	0.406	0.533	0.016	0.021
E	–	0.762	–	0.030
F	0.406	0.483	0.016	0.019
G	2.54 BSC		0.100 BSC	
H	0.914	1.17	0.036	0.046
J	0.711	1.22	0.028	0.048
K	12.70	–	0.500	–
L	6.35	–	0.250	–
M	45° BSC		45° BSC	
N	1.27 BSC		0.050 BSC	
P	–	1.27	–	0.050

All JEDEC notes and dimensions apply.

CASE 22-03
(TO-18)

FIGURE 1 — SWITCHING TIMES TEST CIRCUIT

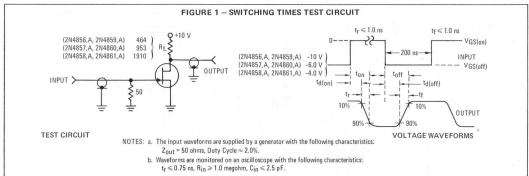

(2N4856,A, 2N4859,A)   464
(2N4857,A, 2N4860,A)   953
(2N4858,A, 2N4861,A)   1910

INPUT

TEST CIRCUIT

(2N4856,A, 2N4859,A) –10 V
(2N4857,A, 2N4860,A) –6.0 V
(2N4858,A, 2N4861,A) –4.0 V

VOLTAGE WAVEFORMS

NOTES: a. The input waveforms are supplied by a generator with the following characteristics:
  Z_{out} = 50 ohms, Duty Cycle ≈ 2.0%.
  b. Waveforms are monitored on an oscilloscope with the following characteristics:
  t_r ≤ 0.75 ns, R_{in} ≥ 1.0 megohm, C_{in} ≤ 2.5 pF.

(*continues*)

APPENDIX 3   FET Data Sheets   (Continued)

2N4856,A–2N4861,A

*ELECTRICAL CHARACTERISTICS (T_A = 25°C unless otherwise noted)

Characteristic		Symbol	Min	Max	Unit
OFF CHARACTERISTICS					
Gate-Source Breakdown Voltage		$V_{(BR)GSS}$			Vdc
(I_G = 1.0 μAdc, V_{DS} = 0)	2N4856,A,2N4857,A,2N4858,A		–40	—	
	2N4859,A,2N4860,A,2N4861,A		–30	—	
Gate-Source Cutoff Voltage		$V_{GS(off)}$			Vdc
(V_{DS} = 15 Vdc, I_D = 0.5 nAdc)	2N4856,A,2N4859,A		–4.0	–10	
	2N4857,A,2N4860,A		–2.0	–6.0	
	2N4858,A,2N4861,A		–0.8	–4.0	
Gate Reverse Current		I_{GSS}			
(V_{GS} = –20 Vdc, V_{DS} = 0)	2N4856,A,2N4857,A,2N4858,A		—	0.25	nAdc
(V_{GS} = –15 Vdc, V_{DS} = 0)	2N4859,A,2N4860,A,2N4861,A		—	0.25	
(V_{GS} = –20 Vdc, V_{DS} = 0, T_A = 150°C)	2N4856,A,2N4857,A,2N4858,A		—	0.5	μAdc
(V_{GS} = –15 Vdc, V_{DS} = 0, T_A = 150°C)	2N4859,A,2N4860,A,2N4861,A		—	0.5	
Drain Cutoff Current		$I_{D(off)}$			
(V_{DS} = 15 Vdc, V_{GS} = –10 Vdc)			—	0.25	nAdc
(V_{DS} = 15 Vdc, V_{GS} = –10 Vdc, T_A = 150°C)			—	0.5	μAdc
ON CHARACTERISTICS					
Zero-Gate Voltage Drain Current (1)	2N4856,A,2N4859,A	I_{DSS}	50	—	mAdc
(V_{DS} = 15 Vdc, V_{GS} = 0)	2N4857,A,2N4860,A		20	100	
	2N4858,A,2N4861,A		8.0	80	
Drain-Source "ON" Voltage		$V_{DS(on)}$			Vdc
(I_D = 20 mAdc, V_{GS} = 0)	2N4856,A,2N4859,A		—	0.75	
(I_D = 10 mAdc, V_{GS} = 0)	2N4857,A,2N4860,A		—	0.5	
(I_D = 5.0 mAdc, V_{GS} = 0)	2N4858,A,2N4861,A		—	0.5	
SMALL-SIGNAL CHARACTERISTICS					
Drain-Source "ON" Resistance		$r_{ds(on)}$			Ohms
(V_{GS} = 0, I_D = 0, f = 1.0 kHz)	2N4856,A,2N4859,A		—	25	
	2N4857,A,2N4860,A		—	40	
	2N4858,A,2N4861,A		—	60	
Input Capacitance	2N4856 thru 2N4861	C_{iss}	—	18	pF
(V_{DS} = 0, V_{GS} = –10 Vdc, f = 1.0 MHz)	2N4856 A thru 2N4861 A			10	
Reverse Transfer Capacitance		C_{rss}			pF
(V_{DS} = 0, V_{GS} = –10 Vdc, f = 1.0 MHz)	2N4856 thru 2N4861		—	8.0	
	2N4856 A,2N4859 A		—	4.0	
	2N4857 A,2N4858 A,2N4860 A,2N4861 A		—	3.5	

SWITCHING CHARACTERISTICS (See Figure 1) (2)

Turn-On Delay Time	Conditions for 2N4856,A, 2N4859,A:	2N4856, 2N4859	$t_{d(on)}$	—	6.0	ns
		2N4856A, 2N4859A		—	5.0	
	(V_{DD} = 10 Vdc, $I_{D(on)}$ = 20 mAdc,	2N4857, 2N4860		—	6.0	
	$V_{GS(on)}$ = 0, $V_{GS(off)}$ = –10 Vdc)	2N4857A, 2N4860A		—	6.0	
		2N4858, 2N4861		—	10	
		2N4858A, 2N4861A		—	8.0	
Rise Time	Conditions for 2N4857,A, 2N4860,A:	2N4856,A, 2N4859,A	t_r	—	3.0	ns
		2N4857,A, 2N4860,A		—	4.0	
	(V_{DD} = 10 Vdc, $I_{D(on)}$ = 10 mAdc,	2N4858, 2N4861		—	10	
	$V_{GS(on)}$ = 0, $V_{GS(off)}$ = –6.0 Vdc)	2N4858A, 2N4861A		—	8.0	
Turn-Off Time		2N4856, 2N4859	t_{off}	—	25	
	Conditions for 2N4858,A, 2N4861,A:	2N4856A, 2N4859A		—	20	ns
		2N4857, 2N4860		—	50	
	(V_{DD} = 10 Vdc, $I_{D(on)}$ = 5.0 mAdc,	2N4857A, 2N4860A		—	40	
	$V_{GS(on)}$ = 0, $V_{GS(off)}$ = –4.0 Vdc)	2N4858, 2N4861		—	100	
		2N4858A; 2N4861A		—	80	

*Indicates JEDEC Registered Data.

(1) Pulse Test: Pulse Width = 100 ms, Duty Cycle $\leq$ 10%.

(2) The $I_{D(on)}$ values are nominal; exact values vary slightly with transistor parameters.

APPENDIX 3   FET Data Sheets   (Continued)

3N128

CASE 20-03, STYLE 7
TO-72 (TO-206AF)

3 Source

Gate
2

4 Case
&
Substrate

1 Drain

MOSFET
AMPLIFIER

N-CHANNEL — DEPLETION

MAXIMUM RATINGS

Rating	Symbol	Value	Unit
Drain-Source Voltage	V_{DS}	+20	Vdc
Drain-Gate Voltage	V_{DG}	+20	Vdc
Gate-Source Voltage	V_{GS}	±10	Vdc
Drain Current	I_D	50	mAdc
Total Device Dissipation @ T_A = 25°C Derate above 25°C	P_D	330 2.2	mW mW/°C
Operating and Storage Junction Temperature Range	T_J, T_{stg}	−65 to +175	°C

ELECTRICAL CHARACTERISTICS (T_A = 25°C unless otherwise noted.)

Characteristic	Symbol	Min	Max	Unit
OFF CHARACTERISTICS				
Gate-Source Breakdown Voltage(1) (I_G = −10 μAdc, V_{DS} = 0)	$V_{(BR)DSS}$	−50	—	Vdc

Gate Reverse Current
  (V_{GS} = −8.0 Vdc, V_{DS} = 0)
  (V_{GS} = −8.0 Vdc, V_{DS} = 0, T_A = 125°C)
	I_{GSS}	—
		—
			0.05
			5.0
				nAdc

Gate Source Cutoff Voltage (V_{DS} = 15 Vdc, I_D = 50 μAdc)	$V_{GS(off)}$	−0.5	−8.0	Vdc
ON CHARACTERISTICS				
Zero-Gate-Voltage Drain Current(2) (V_{DS} = 15 Vdc, V_{GS} = 0)	I_{DSS}	5.0	25	mAdc
SMALL-SIGNAL CHARACTERISTICS				
Forward Transfer Admittance (V_{DS} = 15 Vdc, I_D = 5.0 mAdc, f = 1.0 kHz)	$|Y_{fs}|$	5000	12,000	μmhos
Input Admittance (V_{DS} = 15 Vdc, I_D = 5.0 mAdc, f = 200 MHz)	$Re(y_{is})$	—	800	μmhos
Output Conductance (V_{DS} = 15 Vdc, I_D = 5.0 mAdc, f = 200 MHz)	$Re(y_{os})$	—	500	μmhos
Forward Transconductance (V_{DS} = 15 Vdc, I_D = 5.0 mAdc, f = 200 MHz)	$Re(y_{fs})$	5000	—	μmhos
Input Capacitance (V_{DS} = 15 Vdc, I_D = 5.0 mAdc, f = 1.0 MHz)	C_{iss}	—	7.0	pF
Reverse Transfer Capacitance (V_{DS} = 15 Vdc, I_D = 5.0 mAdc, f = 1.0 MHz)	C_{rss}	0.05	0.35	pF
FUNCTIONAL CHARACTERISTICS				
Noise Figure (V_{DS} = 15 Vdc, I_D = 5.0 mAdc, f = 200 MHz)	NF	—	5.0	dB
Power Gain (V_{DS} = 15 Vdc, I_D = 5.0 mAdc, f = 200 MHz)	P_G	13.5	23	dB

(1) Caution Destructive Test, can damage gate oxide beyond operation.
(2) Pulse Test: Pulse Width = 300 μs, Duty Cycle = 2.0%.

(*continues*)

3N128

TYPICAL CHARACTERISTICS
(T_A = 25°C)

FIGURE 1 – DRAIN CHARACTERISTICS

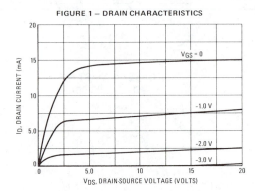

FIGURE 2 – TRANSFER CHARACTERISTICS

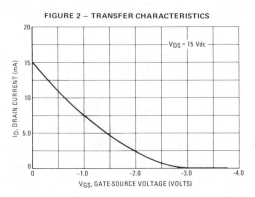

TYPICAL 1 kHz DRAIN CHARACTERISTICS
(T_A = 25°C, V_{DS} = 15 Vdc, f = 1.0 kHz)

FIGURE 3 – FORWARD TRANSADMITTANCE
versus GATE BIAS VOLTAGE

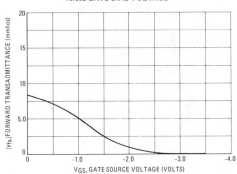

FIGURE 4 – FORWARD TRANSADMITTANCE
versus DRAIN CURRENT

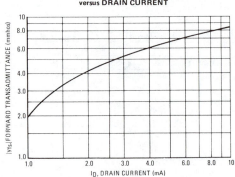

TYPICAL 200 MHz COMMON-SOURCE ADMITTANCE CHARACTERISTICS
(T_A = 25°C, V_{DS} = 15 Vdc, f = 200 MHz)

FIGURE 5 – INPUT ADMITTANCE (y_{is}) COMPONENTS

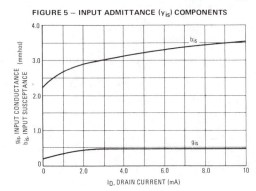

FIGURE 6 – FORWARD TRANSADMITTANCE (y_{fs}) COMPONENTS

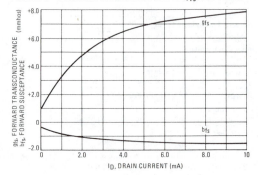

646

APPENDIX 3   FET Data Sheets   (Continued)

2N4351

**CASE 20-03, STYLE 2
TO-72 (TO-206AF)**

3 Drain

2
Gate

4
Case

1 Source

**MOSFET
SWITCHING**

N-CHANNEL — ENHANCEMENT

MAXIMUM RATINGS

Rating	Symbol	Value	Unit
Drain-Source Voltage	V_{DS}	25	Vdc
Drain-Gate Voltage	V_{DG}	30	Vdc
Gate-Source Voltage*	V_{GS}	30	Vdc
Drain Current	I_D	30	mAdc
Total Device Dissipation @ $T_A = 25°C$ Derate above 25°C	P_D	300 1.7	mW mW/°C
Total Device Dissipation @ $T_C = 25°C$ Derate above 25°C	P_D	800 4.56	mW mW/°C
Junction Temperature Range	T_J	175	°C
Storage Temperature Range	T_{stg}	−65 to +175	°C

*Transient potentials of ±75 Volt will not cause gate-oxide failure.

ELECTRICAL CHARACTERISTICS ($T_A = 25°C$ unless otherwise noted.)

Characteristic	Symbol	Min	Max	Unit
OFF CHARACTERISTICS				
Drain-Source Breakdown Voltage ($I_D = 10\ \mu A$, $V_{GS} = 0$)	$V_{(BR)DSX}$	25	—	Vdc
Zero-Gate-Voltage Drain Current ($V_{DS} = 10$ V, $V_{GS} = 0$) $T_A = 25°C$ $T_A = 150°C$	I_{DSS}	— —	10 10	nAdc μAdc
Gate Reverse Current ($V_{GS} = \pm 15$ Vdc, $V_{DS} = 0$)	I_{GSS}	—	±10	pAdc
ON CHARACTERISTICS				
Gate Threshold Voltage ($V_{DS} = 10$ V, $I_D = 10\ \mu A$)	$V_{GS(Th)}$	1.0	5	Vdc
Drain-Source On-Voltage ($I_D = 2.0$ mA, $V_{GS} = 10$ V)	$V_{DS(on)}$	—	1.0	V
On-State Drain Current ($V_{GS} = 10$ V, $V_{DS} = 10$ V)	$I_{D(on)}$	3.0	—	mAdc
SMALL-SIGNAL CHARACTERISTICS				
Forward Transfer Admittance ($V_{DS} = 10$ V, $I_D = 2.0$ mA, f = 1.0 kHz)	$|y_{fs}|$	1000	—	μmho
Input Capacitance ($V_{DS} = 10$ V, $V_{GS} = 0$, f = 140 kHz)	C_{iss}	—	5.0	pF
Reverse Transfer Capacitance ($V_{DS} = 0$, $V_{GS} = 0$, f = 140 kHz)	C_{rss}	—	1.3	pF
Drain-Substrate Capacitance ($V_{D(SUB)} = 10$ V, f = 140 kHz)	$C_{d(sub)}$	—	5.0	pF
Drain-Source Resistance ($V_{GS} = 10$ V, $I_D = 0$, f = 1.0 kHz)	$r_{ds(on)}$	—	300	ohms
SWITCHING CHARACTERISTICS				
Turn-On Delay (Fig. 5)	t_{d1}	—	45	ns
Rise Time (Fig. 6)	t_r	—	65	ns
Turn-Off Delay (Fig. 7)	t_{d2}	—	60	ns
Fall Time (Fig. 8)	t_f	—	100	ns

For the switching characteristics: $I_D = 2.0$ mAdc, $V_{DS} = 10$ Vdc, $V_{GS} = 10$ Vdc) (See Figure 9; Times Circuit Determined)

(continues)

APPENDIX 3   FET Data Sheets   (Continued)

2N4351

FIGURE 1 — FORWARD TRANSFER ADMITTANCE

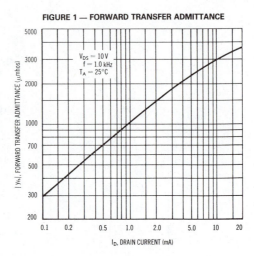

FIGURE 2 — TRANSFER CHARACTERISTICS

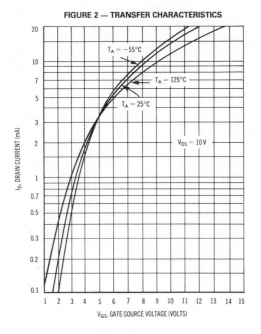

FIGURE 3 — DRAIN-SOURCE "ON" RESISTANCE

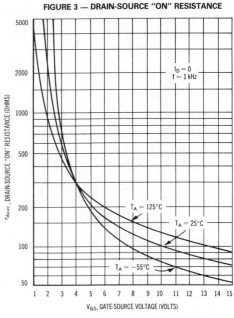

2N4351

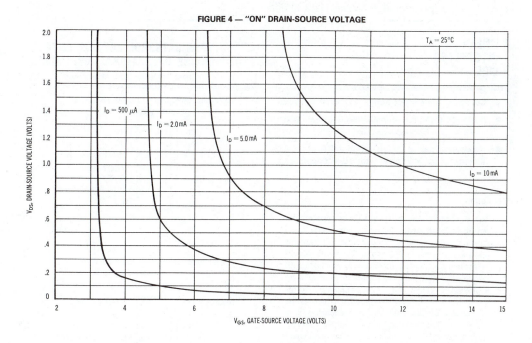

FIGURE 4 — "ON" DRAIN-SOURCE VOLTAGE

SWITCHING CHARACTERISTICS
($T_A = 25°C$)

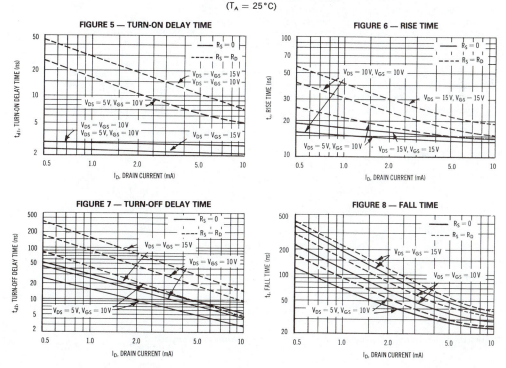

FIGURE 5 — TURN-ON DELAY TIME

FIGURE 6 — RISE TIME

FIGURE 7 — TURN-OFF DELAY TIME

FIGURE 8 — FALL TIME

APPENDIX 4   Operational Amplifier Data Sheets

**LINEAR
INTEGRATED
CIRCUITS**

**TYPES uA709AM, uA709M, uA709C
GENERAL-PURPOSE OPERATIONAL AMPLIFIERS**

D942, FEBRUARY 1971—REVISED AUGUST 1983

- Common-Mode Input Range . . . ±10 V Typical
- Designed to be Interchangeable with Fairchild μA709A, μA709, and μA709C
- Maximum Peak-to-Peak Output Voltage Swing . . . 28-V Typical with 15-V Supplies

**uA709AM, uA709M . . . J OR W PACKAGE
(TOP VIEW)**

```
        NC [ 1  U 14 ] NC
        NC [ 2    13 ] NC
FREQ COMP B [ 3    12 ] FREQ COMP A
       IN− [ 4    11 ] VCC+
       IN+ [ 5    10 ] OUT
      VCC− [ 6     9 ] OUT FREQ COMP
        NC [ 7     8 ] NC
```

description

These circuits are general-purpose operational amplifiers, each having high-impedance differential inputs and a low-impedance output. Component matching, inherent with silicon monolithic circuit-fabrication techniques, produces an amplifier with low-drift and low-offset characteristics. Provisions are incorporated within the circuit whereby external components may be used to compensate the amplifier for stable operation under various feedback or load conditions. These amplifiers are particularly useful for applications requiring transfer or generation of linear or nonlinear functions.

The uA709A circuit features improved offset characteristics, reduced input-current requirements, and lower power dissipation when compared to the uA709 circuit. In addition, maximum values of the average temperature coefficients of offset voltage and current are guaranteed.

The uA709AM and uA709M are characterized for operation over the full military temperature range of −55°C to 125°C. The uA709C is characterized for operation from 0°C to 70°C.

**uA709AM, uA709M . . . JG PACKAGE
uA709C . . . JG OR P PACKAGE
(TOP VIEW)**

```
FREQ COMP B [ 1  U 8 ] FREQ COMP A
        IN− [ 2    7 ] VCC+
        IN+ [ 3    6 ] OUT
       VCC− [ 4    5 ] OUT FREQ COMP
```

**uA709AM, uA709M . . . U FLAT PACKAGE
(TOP VIEW)**

```
        NC [●1   10 ] NC
FREQ COMP B [ 2    9 ] FREQ COMP A
        IN− [ 3    8 ] VCC+
        IN+ [ 4    7 ] OUT
       VCC− [ 5    6 ] OUT FREQ COMP
```

NC—No internal connection

symbol

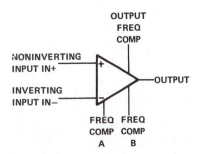

TYPES uA709AM, uA709M, uA709C
GENERAL-PURPOSE OPERATIONAL AMPLIFIERS

schematic

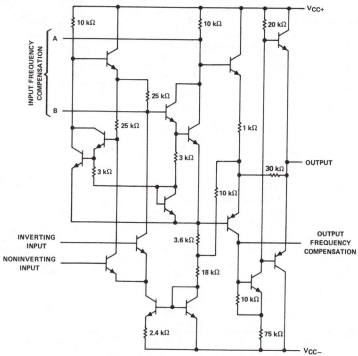

Component values shown are nominal.

absolute maximum ratings over operating free-air temperature range (unless otherwise noted)

	uA709AM uA709M	uA709C	UNIT
Supply voltage V_{CC+} (see Note 1)	18	18	V
Supply voltage V_{CC-} (see Note 1)	−18	−18	V
Differential input voltage (see Note 2)	±5	±5	V
Input voltage (either input, see Notes 1 and 3)	±10	±10	V
Duration of output short-circuit (see Note 4)	5	5	s
Continuous total dissipation at (or below) 70°C free-air temperature (see Note 5)	300	300	mW
Operating free-air temperature range	−55 to 125	0 to 70	°C
Storage temperature range	−65 to 150	−65 to 150	°C
Lead temperature 1,6 mm (1/16 inch) from case for 60 seconds J, JG, U, or W package	300	300	°C
Lead temperature 1,6 mm (1/16 inch) from case for 10 seconds P package		260	°C

NOTES: 1. All voltage values, unless otherwise noted, are with respect to the midpoint between V_{CC+} and V_{CC-}.
      2. Differential voltages are at the noninverting input terminal with respect to the inverting input terminal.
      3. The magnitude of the input voltage must never exceed the magnitude of the supply voltage or 10 volts, whichever is less.
      4. The output may be shorted to ground or either power supply.
      5. For operation of uA709AM and uA709M above 70°C free-air temperature, refer to the Dissipation Derating Curves, Section 2. In the J and JG packages, uA709AM and uA709M chips are alloy-mounted; uA709C chips are glass-mounted.

(continues)

APPENDIX 4   Operational Amplifier Data Sheets   (Continued)

TYPES uA709AM, uA709M, uA709C
GENERAL-PURPOSE OPERATIONAL AMPLIFIERS

electrical characteristics at specified free-air temperature, $V_{CC\pm} = \pm 9$ V to ± 15 V (unless otherwise noted)

PARAMETER		TEST CONDITIONS[†]		uA709AM MIN	uA709AM TYP[‡]	uA709AM MAX	uA709M MIN	uA709M TYP[‡]	uA709M MAX	UNIT
V_{IO}	Input offset voltage	$V_O = 0$, $R_S \leq 10$ kΩ	25°C		0.6	2		1	5	mV
			Full range			3			6	
α_{VIO}	Average temperature coefficient of input offset voltage	$V_O = 0$, $R_S = 50$ Ω	Full range		1.8	10		3		µV/°C
		$V_O = 0$, $R_S = 10$ kΩ	Full range		4.8	25		6		
I_{IO}	Input offset current	$V_O = 0$	25°C		10	50		50	200	nA
			−55°C		40	250		100	500	
			125°C		3.5	50		20	200	
α_{IIO}	Average temperature coefficient of input offset current	$V_O = 0$	−55°C to 25°C		0.45	2.8				nA/°C
			25°C to 125°C		0.08	0.5				
I_{IB}	Input bias current	$V_O = 0$	25°C		0.1	0.2		0.2	0.5	µA
			−55°C		0.3	0.6		0.5	1.5	
V_{ICR}	Common-mode input voltage range	$V_{CC\pm} = \pm 15$ V	25°C	±8	±10		±8	±10		V
			Full range	±8			±8			
V_{OPP}	Maximum peak-to-peak output voltage swing	$V_{CC\pm} = \pm 15$ V, $R_L \geq 10$ kΩ	25°C	24	28		24	28		V
			Full range	24			24			
		$V_{CC\pm} = \pm 15$ V, $R_L = 2$ kΩ	25°C	20	26		20	26		
		$V_{CC\pm} = \pm 15$ V, $R_L \geq 2$ kΩ	Full range	20			20			
A_{VD}	Large-signal differential voltage amplification	$V_{CC\pm} = \pm 15$ V, $R_L \geq 2$ kΩ, $V_O = \pm 10$ V	25°C		45			45		V/mV
			Full range	25		70	25		70	
r_i	Input resistance		25°C	350	750		150	400		kΩ
			−55°C	85	185		40	100		
r_o	Output resistance	$V_O = 0$ See Note 6	25°C		150			150		Ω
CMRR	Common-mode rejection ratio	$V_{IC} = V_{ICR}$ min	25°C	80	110		70	90		dB
			Full range	80			70			
k_{SVS}	Power supply sensitivity ($\Delta V_{IO}/\Delta V_{CC}$)	$V_{CC} = \pm 9$ V to ± 15 V	25°C		40	100		25	150	µV/V
			Full range			100			150	
I_{CC}	Supply current	$V_{CC\pm} = \pm 15$ V, No load, $V_O = 0$	25°C		2.5	3.6		2.6	5.5	mA
			−55°C		2.7	4.5				
			125°C		2.1	3				
P_D	Total power dissipation	$V_{CC\pm} = \pm 15$ V, No load, $V_O = 0$	25°C		75	108		78	165	mW
			−55°C		81	135				
			125°C		63	90				

[†] All characteristics are specified under open-loop with zero common-mode input voltage unless otherwise specified. Full range for uA709AM and uA709M is −55°C to 125°C.

[‡] All typical values are at $V_{CC\pm} = \pm 15$ V.

NOTE 6: This typical value applies only at frequencies above a few hundred hertz because of the effects of drift and thermal feedback.

APPENDIX 4   Operational Amplifier Data Sheets   (Continued)

TYPES uA709AM, uA709M, uA709C
GENERAL-PURPOSE OPERATIONAL AMPLIFIERS

electrical characteristics at specified free-air temperature (unless otherwise noted $V_{CC\pm} = \pm 15$ V)

	PARAMETER	TEST CONDITIONS[†]		uA709C			UNIT
				MIN	TYP	MAX	
V_{IO}	Input offset voltage	$V_{CC\pm} = \pm 9$ V to ± 15 V, $V_O = 0$	25 °C		2	7.5	mV
			Full range			10	
I_{IO}	Input offset current	$V_{CC\pm} = \pm 9$ V to ± 15 V, $V_O = 0$	25 °C		100	500	nA
			Full range			750	
I_{IB}	Input bias current	$V_{CC\pm} = \pm 9$ V to ± 15 V, $V_O = 0$	25 °C		0.3	1.5	μA
			Full range			2	
V_{ICR}	Common-mode input voltage range		25 °C	± 8	± 10		V
V_{OPP}	Maximum peak-to-peak output voltage swing	$R_L \geq 10$ kΩ	25 °C	24	28		V
			Full range	24			
		$R_L = 2$ kΩ	25 °C	20	26		
		$R_L \geq 2$ kΩ	Full range	20			
A_{VD}	Large-signal differential voltage amplification	$R_L \leq 2$ kΩ, $V_O = \pm 10$ V	25 °C	15	45		V/mV
			Full range	12			
r_i	Input resistance		25 °C	50	250		kΩ
			Full range	35			
r_o	Output resistance	$V_O = 0$, See Note 6	25 °C		150		Ω
CMRR	Common-mode rejection ratio	$V_{IC} = V_{ICR}$ min	25 °C	65	90		dB
k_{SVS}	Supply voltage sensitivity	$V_{CC} = \pm 9$ V to ± 15 V	25 °C		25	200	μV/V
P_D	Total power dissipation	$V_O = 0$ No load	25 °C		80	200	mW

[†]All characteristics are specified under open-loop operation with zero volts common-mode voltage unless otherwise specified. Full range for uA709C is 0 °C to 70 °C.

NOTE 6: This typical value applies only at frequencies above a few hundred hertz because of the effects of drift and thermal feedback.

operating characteristics $V_{CC\pm} = \pm 9$ V to ± 15 V, $T_A = 25$ °C

	PARAMETER	TEST CONDITIONS		uA709AM uA709M uA709C			UNIT
				MIN	TYP	MAX	
t_r	Rise time	$V_I = 20$ mV, $R_L = 2$ kΩ, See Figure 1	$C_L = 0$		0.3	1	μs
	Overshoot factor		$C_L = 100$ pF		6%	30%	

PARAMETER MEASUREMENT INFORMATION

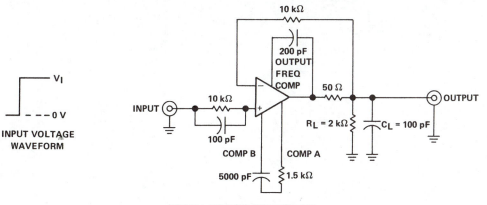

FIGURE 1—RISE TIME AND SLEW RATE

(*continues*)

APPENDIX 4   Operational Amplifier Data Sheets   (Continued)

**LINEAR
INTEGRATED
CIRCUITS**

**TYPES uA741M, uA741C
GENERAL-PURPOSE OPERATIONAL AMPLIFIER**

D920, NOVEMBER 1970—REVISED AUGUST 1983

- Short-Circuit Protection
- Offset-Voltage Null Capability
- Large Common-Mode and
  Differential Voltage Ranges
- No Frequency Compensation Required
- Low Power Consumption
- No Latch-up
- Designed to be Interchangeable with Fairchild
  μA741M, μA741C

description

The uA741 is a general-purpose operational amplifier featuring offset-voltage null capability.

The high common-mode input voltage range and the absence of latch-up make the amplifier ideal for voltage-follower applications. The device is short-circuit protected and the internal frequency compensation ensures stability without external components. A low potentiometer may be connected between the offset null inputs to null out the offset voltage as shown in Figure 2.

The uA741M is characterized for operation over the full military temperature range of −55 °C to 125 °C; the uA741C is characterized for operation from 0 °C to 70 °C.

symbol

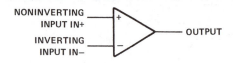

uA741M . . . J PACKAGE
(TOP VIEW)

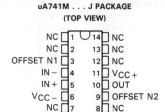

uA741M . . . JG PACKAGE
uA741C . . . D, P, OR JG PACKAGE
(TOP VIEW)

uA741M . . . U FLAT PACKAGE
(TOP VIEW)

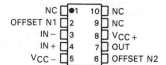

uA741M . . . FH, FK PACKAGE
(TOP VIEW)

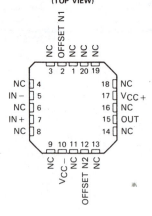

NC—No internal connection

TYPES uA741M, uA741C
GENERAL-PURPOSE OPERATIONAL AMPLIFIERS

schematic

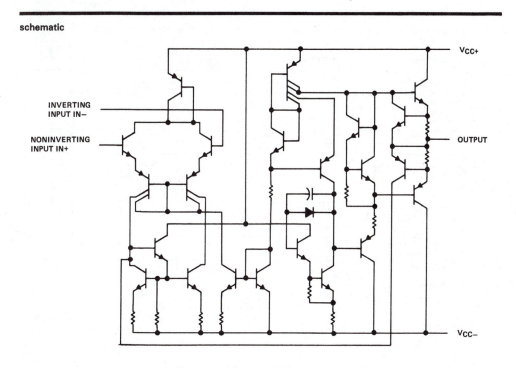

absolute maximum ratings over operating free-air temperature range (unless otherwise noted)

	uA741M	uA741C	UNIT
Supply voltage V_{CC+} (see Note 1)	22	18	V
Supply voltage V_{CC-} (see Note 1)	−22	−18	V
Differential input voltage (see Note 2)	±30	±30	V
Input voltage any input (see Notes 1 and 3)	±15	±15	V
Voltage between either offset null terminal (N1/N2) and V_{CC-}	±0.5	±0.5	V
Duration of output short-circuit (see Note 4)	unlimited	unlimited	
Continuous total power dissipation at (or below) 25°C free-air temperature (see Note 5)	500	500	mW
Operating free-air temperature range	−55 to 125	0 to 70	°C
Storage temperature range	−65 to 150	−65 to 150	°C
Lead temperature 1,6 mm (1/16 inch) from case for 60 seconds FH, FK, J, JG, or U package	300	300	°C
Lead temperature 1,6 mm (1/16 inch) from case for 10 seconds D, N or P package		260	°C

NOTES: 1. All voltage values, unless otherwise noted, are with respect to the midpoint between V_{CC+} and V_{CC-}.

2. Differential voltages are at the noninverting input terminal with respect to the inverting input terminal.

3. The magnitude of the input voltage must never exceed the magnitude of the supply voltage or 15 volts, whichever is less.

4. The output may be shorted to ground or either power supply. For the uA741M only, the unlimited duration of the short-circuit applies at (or below) 125°C case temperature or 75°C free-air temperature.

5. For operation above 25°C free-air temperature, refer to Dissipation Derating Curves, Section 2. In the J and JG packages, uA741M chips are alloy mounted; uA741C chips are glass mounted.

(continues)

APPENDIX 4   Operational Amplifier Data Sheets   (Continued)

TYPES uA741M, uA741C
GENERAL-PURPOSE OPERATIONAL AMPLIFIERS

electrical characteristics at specified free-air temperature, $V_{CC+} = 15$ V, $V_{CC-} = -15$ V

PARAMETER		TEST CONDITIONS[†]		uA741M			uA741C			UNIT
				MIN	TYP	MAX	MIN	TYP	MAX	
V_{IO}	Input offset voltage	$V_O = 0$	25°C		1	5		1	6	mV
			Full range			6			7.5	
$\Delta V_{IO(adj)}$	Offset voltage adjust range	$V_O = 0$	25°C		±15			±15		mV
I_{IO}	Input offset current	$V_O = 0$	25°C		20	200		20	200	nA
			Full range			500			300	
I_{IB}	Input bias current	$V_O = 0$	25°C		80	500		80	500	nA
			Full range			1500			800	
V_{ICR}	Common-mode input voltage range		25°C	±12	±13		±12	±13		V
			Full range	±12			±12			
V_{OM}	Maximum peak output voltage swing	$R_L = 10$ kΩ	25°C	±12	±14		±12	±14		V
		$R_L \geq 10$ kΩ	Full range	±12			±12			
		$R_L = 2$ kΩ	25°C	±10	±13		±10	±13		
		$R_L \geq 2$ kΩ	Full range	±10			±10			
A_{VD}	Large-signal differential voltage amplification	$R_L \geq 2$ kΩ	25°C	50	200		20	200		V/mV
		$V_O = \pm10$ V	Full range	25			15			
r_i	Input resistance		25°C	0.3	2		0.3	2		MΩ
r_o	Output resistance	$V_O = 0$, See Note 6	25°C		75			75		Ω
C_i	Input capacitance		25°C		1.4			1.4		pF
CMRR	Common-mode rejection ratio	$V_{IC} = V_{ICR}$ min	25°C	70	90		70	90		dB
			Full range	70			70			
k_{SVS}	Supply voltage sensitivity ($\Delta V_{IO}/\Delta V_{CC}$)	$V_{CC} = \pm9$ V to ±15 V	25°C		30	150		30	150	μV/V
			Full range			150			150	
I_{OS}	Short-circuit output current		25°C	±25	±40		±25	±40		mA
I_{CC}	Supply current	No load, $V_O = 0$	25°C		1.7	2.8		1.7	2.8	mA
			Full range			3.3			3.3	
P_D	Total power dissipation	No load, $V_O = 0$	25°C		50	85		50	85	mW
			Full range			100			100	

[†]All characteristics are measured under open-loop conditions with zero common-mode input voltage unless otherwise specified. Full range for uA741M is
−55°C to 125°C and for uA741C is 0°C to 70°C.

NOTE 6: This typical value applies only at frequencies above a few hundred hertz because of the effects of drift and thermal feedback.

APPENDIX 5   Voltage Comparator Data Sheets

**LINEAR
INTEGRATED
CIRCUITS**

**TYPES LM106, LM206, LM306
DIFFERENTIAL COMPARATORS WITH STROBES**

D1108, OCTOBER 1979—REVISED JULY 1983

- **Fast Response Times**
- **Improved Gain and Accuracy**
- **Fan-Out to 10 Series 54/74 TTL Loads**
- **Strobe Capability**
- **Short-Circuit and Surge Protection**
- **Designed to be Interchangeable with National Semiconductor LM106, LM206, and LM306**

**J OR N DUAL-IN-LINE
OR W FLAT PACKAGE
(TOP VIEW)**

NC	1	14	NC
GND	2	13	NC
IN +	3	12	NC
IN −	4	11	$V_{CC}+$
NC	5	10	NC
$V_{CC}-$	6	9	OUT
STROBE 1	7	8	STROBE 2

NC—No internal connection

description

The LM106, LM206, and LM306 are high-speed voltage comparators with differential inputs, a low-impedance high-sink-current (100 mA) output, and two strobe inputs. These devices detect low-level analog or digital signals and can drive digital logic or lamps and relays directly. Short-circuit protection and surge-current limiting is provided.

**D, JG OR P DUAL-IN-LINE PACKAGE
(TOP VIEW)**

GND	1	8	$V_{CC}+$
IN +	2	7	OUT
IN −	3	6	STROBE 2
$V_{CC}-$	4	5	STROBE 1

The circuit is similar to a TL810 with gated output. A low-level input at either strobe causes the output to remain high regardless of the differential input. When both strobe inputs are either open or at a high logic level, the output voltage is controlled by the differential input voltage. The circuit will operate with any negative supply voltage between −3 volts and −12 volts with little difference in performance.

The LM106 is characterized for operation over the full military temperature range of −55 °C to 125 °C, the LM206 is characterized for operation from −25 °C to 85 °C, and the LM306 from 0 °C to 70 °C.

functional block diagram

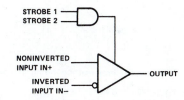

Reprinted by permission of copyright holder. © 1984 Texas Instruments Incorporated.

(continues)

APPENDIX 5   Voltage Comparator Data Sheets   (Continued)

TYPES LM106, LM206, LM306
DIFFERENTIAL COMPARATORS WITH STROBES

schematic

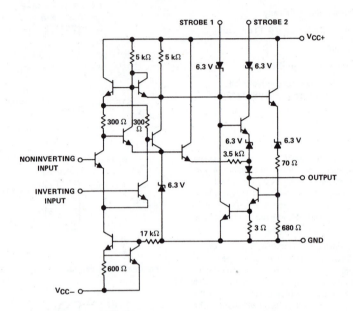

Resistor values are nominal in ohms.

absolute maximum ratings over operating free-air temperature range (unless otherwise noted)

Supply voltage V_{CC+} (see Note 1) . . . . 15 V
Supply voltage V_{CC-} (see Note 1). . . . . −15 V
Differential input voltage (see Note 2) . . . . ±5 V
Input voltage (either input, see Notes 1 and 3) . . . . ±7 V
Strobe voltage range (see Note 1). . . . . 0 V to V_{CC+}
Output voltage (see Note 1). . . . . 24 V
Voltage from output to V_{CC-} . . . . 30 V
Duration of output short-circuit (see Note 4) . . . . 10 s
Continuous total power dissipation at (or below) 25°C free-air temperature (see Note 5) . . . . 600 mW
Operating free-air temperature range: LM106 Circuits. . . . . −55°C to 125°C
                                                       LM206 Circuits. . . . . −25°C to 85°C
                                                       LM306 Circuits . . . . 0°C to 70°C
Storage temperature range . . . . −65°C to 150°C
Lead temperature 1,6 mm (1/16 inch) from case for 60 seconds: J, JG or W package . . . . 300°C
Lead temperature 1,6 mm (1/16 inch) from case for 10 seconds: D, N, or P package . . . . 260°C

NOTES: 1. All voltage values, except differential voltages and the voltage from the output to V_{CC-}, are with respect to the network ground terminal.
2. Differential voltages are at the noninverting input terminal with respect to the inverting input terminal.
3. The magnitude of the input voltage must never exceed the magnitude of the supply voltage or 7 volts, whichever is less.
4. The output may be shorted to ground or either power supply.
5. For operation above 25°C free-air temperature, refer to Dissipation Derating Curves, Section 2. In the J and JG packages, LM106 chips are alloy-mounted; LM206 and LM306 chips are glass-mounted.

APPENDIX 5 Voltage Comparator Data Sheets (Continued)

TYPES LM106, LM206, LM306
DIFFERENTIAL COMPARATORS WITH STROBES

electrical characteristics at specified free-air temperature, V_{CC+} = 12 V, V_{CC-} = −3 V to 12 V (unless otherwise noted)

PARAMETER		TEST CONDITIONS[†]		LM106, LM206			LM306			UNIT
				MIN	TYP	MAX	MIN	TYP	MAX	
V_{IO}	Input offset voltage	$R_S \leq 200\ \Omega$,	See Note 6	25 °C		0.5‡	2		1.6‡	5	mV
				Full range			3			6.5	
α_{VIO}	Average temperature coefficient of input offset voltage	$R_S = 50\ \Omega$,	See Note 6	Full range		3	10		5	20	μV/°C
I_{IO}	Input offset current		See Note 6	25 °C		0.7‡	3		1.8‡	5	μA
				MIN		2	7		1	7.5	
				MAX		0.4	3		0.5	5	
α_{IIO}	Average temperature coefficient of input offset current		See Note 6	MIN to 25 °C		15	75		24	100	nA/°C
				25 °C to MAX		5	25		15	50	
I_{IB}	Input bias current	V_O = 0.5 V to 5 V		MIN to 25 °C			45			40	μA
				25 °C to MAX		7‡	20		16‡	25	
$I_{IL(S)}$	Low-level strobe current	$V_{(strobe)}$ = 0.4 V		Full range		−1.7‡	−3.2		−1.7‡	−3.2	mA
$V_{IH(S)}$	High-level strobe voltage			Full range	2.2			2.2			V
$V_{IL(S)}$	Low-level strobe voltage			Full range			0.9			0.9	V
V_{ICR}	Common-mode input voltage range	V_{CC-} = −7 V to −12 V		Full range	±5			±5			V
V_{ID}	Differential input voltage range			Full range	±5			±5			V
A_{VD}	Large-signal differential voltage amplification	No load, V_O = 0.5 V to 5 V		25 °C		40‡			40‡		V/mV
V_{OH}	High-level output voltage	I_{OH} = −400 μA	V_{ID} = 5 mV	Full range	2.5		5.5				V
			V_{ID} = 8 mV	Full range				2.5		5.5	
V_{OL}	Low-level output voltage	I_{OL} = 100 mA	V_{ID} = −5 mV	25 °C		0.8‡	1.5				V
			V_{ID} = −7 mV	25 °C					0.8‡	2	
		I_{OL} = 50 mA	V_{ID} = −5 mV	Full range			1				
			V_{ID} = −8 mV	Full range						1	
		I_{OL} = 16 mA	V_{ID} = −5 mV	Full range			0.4				
			V_{ID} = −8 mV	Full range						0.4	
I_{OH}	High-level output current	V_{OH} = 8 V to 24 V	V_{ID} = 5 mV	MIN to 25 °C		0.02‡	1				μA
				25 °C to MAX			100				
			V_{ID} = 7 mV	MIN to 25 °C					0.02‡	2	
			V_{ID} = 8 mV	25 °C to MAX						100	
I_{CC+}	Supply current from V_{CC+}	V_{ID} = −5 mV, No load		Full range		6.6‡	10		6.6‡	10	mA
I_{CC-}	Supply current from V_{CC-}	No load		Full range		−1.9‡	−3.6		−1.9‡	−3.6	mA

[†]Unless otherwise noted, all characteristics are measured with both strobes open.

‡ These typical values are at V_{CC+} = 12 V, V_{CC-} = −6 V, T_A = 25 °C. Full range (MIN to MAX) for LM106 is −55 °C to 125 °C; for LM206 is −25 °C to 85 °C; and for LM306 is 0 °C to 70 °C.

NOTE 6: The offset voltages and offset currents given are the maximum values required to drive the output down to the low range (V_{OL}) or up to the high range (V_{OH}). Thus these parameters actually define an error band and take into account the worst-case effects of voltage gain and input impedance.

switching characteristics, V_{CC+} = 12 V, V_{CC-} = −6 V, T_A = 25 °C

PARAMETER	TEST CONDITIONS[†]	LM106, LM206			LM306			UNIT
		MIN	TYP	MAX	MIN	TYP	MAX	
Response time, low-to-high-level output	R_L = 390 Ω to 5 V, C_L = 15 pF, See Note 7		28	40		28	40	ns

NOTE 7: The response time specified is for a 100-mV input step with 5-mV overdrive and is the interval between the input step function and the instant when the output crosses 1.4 V.

(continues)

APPENDIX 5   Voltage Comparator Data Sheets   (Continued)

TYPES LM106, LM206, LM306
DIFFERENTIAL COMPARATORS WITH STROBES

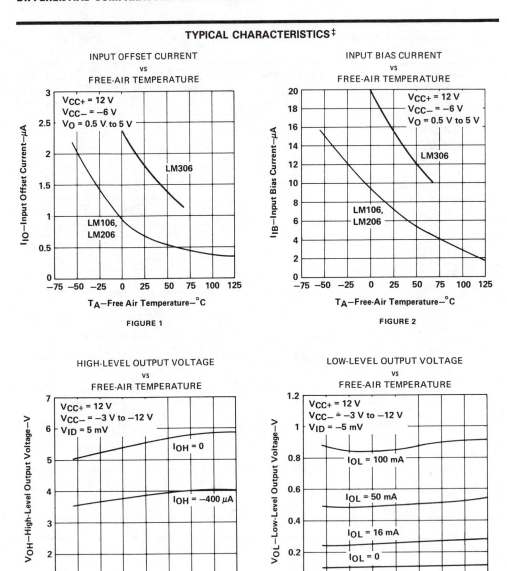

TYPICAL CHARACTERISTICS‡

FIGURE 1

FIGURE 2

FIGURE 3

FIGURE 4

‡Data for free-air temperature outside the range specified in the absolute maximum ratings for LM206 or LM306 is not applicable for those types.

APPENDIX 5   Voltage Comparator Data Sheets   (Continued)

TYPES LM106, LM206, LM306
DIFFERENTIAL COMPARATORS WITH STROBES

TYPICAL CHARACTERISTICS‡

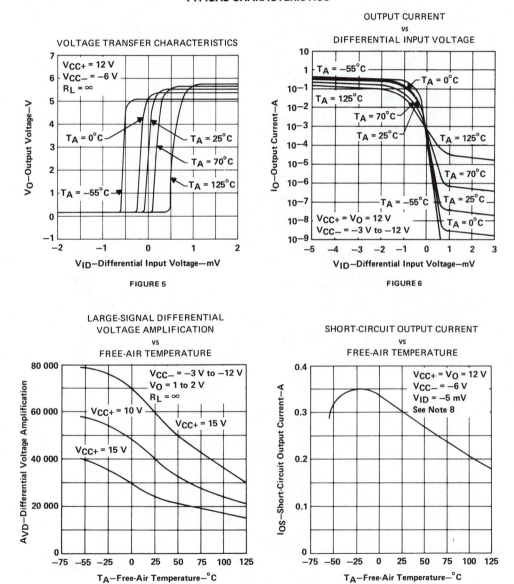

FIGURE 5

FIGURE 6

FIGURE 7

FIGURE 8

‡Data for free-air temperature outside the range specified in the absolute maximum ratings for LM206 or LM306 is not applicable for those types.
NOTE 8: This parameter was measured using a single 5-ms pulse.

(continues)

APPENDIX 5 Voltage Comparator Data Sheets (Continued)

TYPES LM106, LM206, LM306
DIFFERENTIAL COMPARATORS WITH STROBES

TYPICAL CHARACTERISTICS‡

OUTPUT RESPONSE FOR
VARIOUS INPUT OVERDRIVES

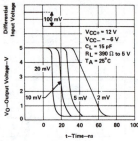

FIGURE 9

OUTPUT RESPONSE FOR
VARIOUS INPUT OVERDRIVES

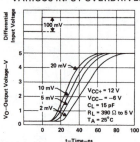

FIGURE 10

SUPPLY CURRENT FROM V_{CC+}
vs
SUPPLY VOLTAGE V_{CC+}

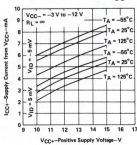

FIGURE 11

SUPPLY CURRENT FROM V_{CC-}
vs
SUPPLY VOLTAGE V_{CC-}

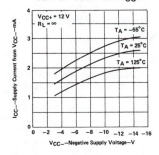

FIGURE 12

TOTAL POWER DISSIPATION
vs
FREE-AIR TEMPERATURE

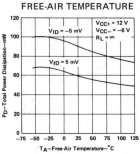

FIGURE 13

‡Data for free-air temperature outside the range specified in the absolute maximum ratings for LM206 or LM306 is not applicable for those types.

APPENDIX 6   Precision Timer 555 Data Sheets

**LINEAR
INTEGRATED
CIRCUITS**

**TYPES SE555, SE555C, SA555, NE555
PRECISION TIMERS**

D1669, SEPTEMBER 1973–REVISED OCTOBER 1983

- **Timing from Microseconds to Hours**
- **Astable or Monostable Operation**
- **Adjustable Duty Cycle**
- **TTL-Compatible Output Can Sink or Source up to 200 mA**
- **Functionally Interchangeable with the Signetics SE555, SE555C, SA555, NE555; Have Same Pinout**

description

These devices are monolithic timing circuits capable of producing accurate time delays or oscillation. In the time-delay or monostable mode of operation, the timed interval is controlled by a single external resistor and capacitor network. In the astable mode of operation, the frequency and duty cycle may be independently controlled with two external resistors and a single external capacitor.

The threshold and trigger levels are normally two-thirds and one-third, respectively, of V_{CC}. These levels can be altered by use of the control voltage terminal. When the trigger input falls below the trigger level, the flip-flop is set and the output goes high. If the trigger input is above the trigger level and the threshold input is above the threshold level, the flip-flop is reset and the output is low. The reset input can override all other inputs and can be used to initiate a new timing cycle. When the reset input goes low, the flip-flop is reset and the output goes low. Whenever the output is low, a low-impedance path is provided between the discharge terminal and ground.

The output circuit is capable of sinking or sourcing current up to 200 milliamperes. Operation is specified for supplies of 5 to 15 volts. With a 5-volt supply, output levels are compatible with TTL inputs.

The SE555 and SE555C are characterized for operation over the full military range of $-55\,°C$ to $125\,°C$. The SA555 is characterized for operation from $-40\,°C$ to $85\,°C$, and the NE555 is characterized for operation from $0\,°C$ to $70\,°C$.

NE555, SE555, SE555C . . . JG DUAL-IN-LINE PACKAGE
SA555, NE555 . . . D, JG, OR P DUAL-IN-LINE PACKAGE
(TOP VIEW)

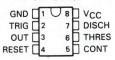

```
GND  [1   8] VCC
TRIG [2   7] DISCH
OUT  [3   6] THRES
RESET[4   5] CONT
```

SE555, SE555C . . . FH OR FK CHIP CARRIER PACKAGE
(TOP VIEW)

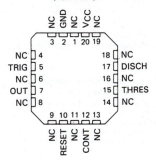

NC—No internal connection

functional block diagram

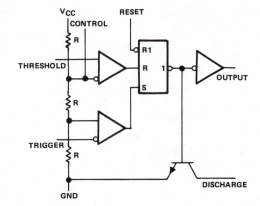

Reset can override Trigger, which can override Threshold.

(continues)

TYPES SE555, SE555C, SA555, NE555
PRECISION TIMERS

FUNCTION TABLE

RESET	TRIGGER VOLTAGE[†]	THRESHOLD VOLTAGE[†]	OUTPUT	DISCHARGE SWITCH
Low	Irrelevant	Irrelevant	Low	On
High	$< 1/3\ V_{DD}$	Irrelevant	High	Off
High	$> 1/3\ V_{DD}$	$> 2/3\ V_{DD}$	Low	On
High	$> 1/3\ V_{DD}$	$< 2/3\ V_{DD}$	As previously established	

[†]Voltage levels shown are nominal.

absolute maximum ratings over operating free-air temperature range (unless otherwise noted)

Supply voltage, V_{CC} (see Note 1) . . . . 18 V
Input voltage (control voltage, reset, threshold, trigger) . . . . V_{CC}
Output current . . . . ±225 mA
Continuous total dissipation at (or below) 25°C free-air temperature (see Note 2) . . . . 600 mW
Operating free-air temperature range: SE555, SE555C . . . . −55°C to 125°C
                                       SA555 . . . . −40°C to 85°C
                                       NE555 . . . . 0°C to 70°C
Storage temperature range . . . . −65°C to 150°C
Lead temperature 1,6 mm (1/16 inch) from case for 60 seconds: FH, FK, or JG package . . . 300°C
Lead temperature 1,6 mm (1/16 inch) from case for 10 seconds: D or P package . . . . 260°C

NOTES: 1. All voltage values are with respect to network ground terminal.
        2. For operation above 25°C free-air temperature, refer to Dissipation Derating Curves, Section 2. In the JG package, SE555 and SE555C chips are alloy mounted, SA555 and NE555 chips are glass mounted.

recommended operating conditions

	SE555		SE555C		SA555		NE555		UNIT
	MIN	MAX	MIN	MAX	MIN	MAX	MIN	MAX	
Supply voltage, V_{CC}	4.5	18	4.5	16	4.5	16	4.5	16	V
Input voltage (control voltage, reset, threshold, trigger)	V_{CC}		V_{CC}		V_{CC}		V_{CC}		V
Output current	±200		±200		±200		±200		mA
Operating free-air temperature, T_A	−55	125	−55	125	−40	85	0	70	°C

APPENDIX 6   Precision Timer 555 Data Sheets   (Continued)

TYPES SE555, SE555C, SA555, NE555
PRECISION TIMERS

electrical characteristics at 25 °C free-air temperature, V_{CC} = 5 V to 15 V (unless otherwise noted)

PARAMETER	TEST CONDITIONS		SE555			SE555C, SA555 NE555			UNIT
			MIN	TYP	MAX	MIN	TYP	MAX	
Threshold voltage level	V_{CC} = 15 V		9.4	10	10.6	8.8	10	11.2	V
	V_{CC} = 5 V		2.7	3.3	4	2.4	3.3	4.2	
Threshold current (see Note 3)				30	250		30	250	nA
Trigger voltage level	V_{CC} = 15 V		4.8	·5	5.2	4.5	5	5.6	V
	V_{CC} = 5 V		1.45	1.67	1.9	1.1	1.67	2.2	
Trigger current	Trigger at 0 V			0.5	0.9		0.5	2	μA
Reset voltage level			0.4	0.7	1	0.4	0.7	1	V
Reset current	Reset at V_{CC}			0.1	0.4		0.1	0.4	mA
	Reset at 0 V			−0.4	−1		−0.4	−1	
Discharge switch off-state current				20	100		20	100	nA
Control voltage (open circuit)	V_{CC} = 15 V		9.6	10	10.4	9	10	11	V
	V_{CC} = 5 V		2.9	3.3	3.8	2.6	3.3	4	
Low-level output voltage	V_{CC} = 15 V	I_{OL} = 10 mA		0.1	0.15		0.1	0.25	V
		I_{OL} = 50 mA		0.4	0.5		0.4	0.75	
		I_{OL} = 100 mA		2	2.25		2	3.2	
		I_{OL} = 200 mA		2.5			2.5		
	V_{CC} = 5 V	I_{OL} = 5 mA		0.05	0.15		0.05	0.25	
		I_{OL} = 8 mA		0.1	0.2		0.25	0.3	
High-level output voltage	V_{CC} = 15 V	I_{OH} = −100 mA	13	13.3		12.75	13.3		V
		I_{OH} = −200 mA		12.5			12.5		
	V_{CC} = 5 V	I_{OH} = −100 mA	3	3.3		2.75	3.3		
Supply current	Output low, No load	V_{CC} = 15 V		10	12		10	15	mA
		V_{CC} = 5 V		3	5		3	6	
	Output high, No load	V_{CC} = 15 V		9	10		9	13	
		V_{CC} = 5 V		2	4		2	5	

NOTE 3:   This parameter influences the maximum value of the timing resistors R_A and R_B in the circuit of Figure 13. For example, when V_{CC} = 5 V the maximum value is R = R_A + R_B ≈ 3.4 MΩ and for V_{CC} = 15 V the maximum value is 10 MΩ.

operating characteristics, V_{CC} = 5 V and 15 V

PARAMETER		TEST CONDITIONS[†]	SE555			SE555C, SA555 NE555			UNIT
			MIN	TYP	MAX	MIN	TYP	MAX	
Initial error of timing interval[‡]	Each timer, monostable[§]	T_A = 25 °C		0.5	1.5		1	3	%
	Each timer, astable[¶]			1.5			2.25		
Temperature coefficient of timing interval	Each timer, monostable[§]	T_A = MIN to MAX		30	100		50		ppm/°C
	Each timer, astable[¶]			90			150		
Supply voltage sensitivity of timing interval	Each timer, monostable[§]	T_A = 25 °C		0.05	0.2		0.1	0.5	%/V
	Each timer, astable[¶]			0.15			0.3		
Output pulse rise time		C_L = 15 pF, T_A = 25 °C		100	200		100	300	ns
Output pulse fall time				100	200		100	300	

[†]For conditions shown as MIN or MAX, use the appropriate value specified under recommended operating conditions.
[‡]Timing interval error is defined as the difference between the measured value and the nominal value computed by the formula: t_w = 1.1 $R_A C$.
[§]Values specified are for a device in a monostable circuit similar to Figure 10, with component values as follow: R_A = 2 kΩ to 100 kΩ, C = 0.1 μF.
[¶]Vallues specified are for a device in an astable circuit similar to Figure 1, with component values as follow: R_A = 1 kΩ to 100 kΩ, C = 0.1 μF.

(continues)

APPENDIX 6   Precision Timer 555 Data Sheets   (Continued)

TYPES SE555, SE555C, SA555, NE555
PRECISION TIMERS

TYPICAL CHARACTERISTICS†

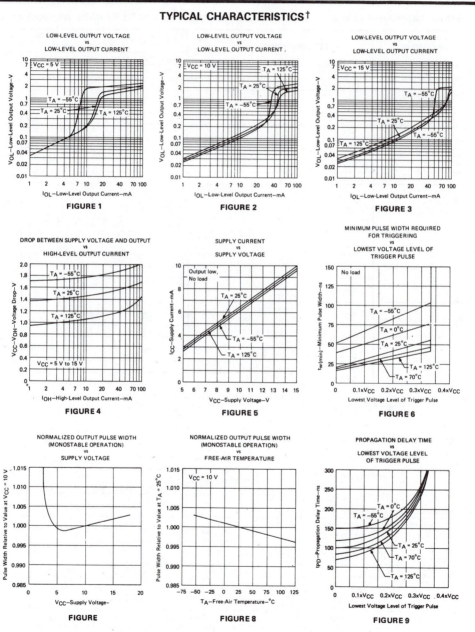

†Data for temperatures below 0°C and above 70°C are applicable for SE555 circuits only.

APPENDIX 7   TTL Logic Gate Data Sheets

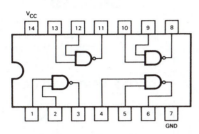

J Suffix — Case 632-07 (Ceramic)
N Suffix — Case 646-05 (Plastic)

SN54LS00
SN74LS00

QUAD 2-INPUT NAND GATE

LOW POWER SCHOTTKY

GUARANTEED OPERATING RANGES

SYMBOL	PARAMETER		MIN	TYP	MAX	UNIT
V_{CC}	Supply Voltage	54	4.5	5.0	5.5	V
		74	4.75	5.0	5.25	
T_A	Operating Ambient Temperature Range	54	−55	25	125	°C
		74	0	25	70	
I_{OH}	Output Current — High	54 , 74			−0.4	mA
I_{OL}	Output Current — Low	54			4.0	mA
		74			8.0	

DC CHARACTERISTICS OVER OPERATING TEMPERATURE RANGE (unless otherwise specified)

SYMBOL	PARAMETER		LIMITS			UNITS	TEST CONDITIONS
			MIN	TYP	MAX		
V_{IH}	Input HIGH Voltage		2.0			V	Guaranteed Input HIGH Voltage for All Inputs
V_{IL}	Input LOW Voltage	54			0.7	V	Guaranteed Input LOW Voltage for All Inputs
		74			0.8		
V_{IK}	Input Clamp Diode Voltage			−0.65	−1.5	V	V_{CC} = MIN, I_{IN} = −18 mA
V_{OH}	Output HIGH Voltage	54	2.5	3.5		V	V_{CC} = MIN, I_{OH} = MAX, V_{IN} = V_{IH} or V_{IL} per Truth Table
		74	2.7	3.5		V	
V_{OL}	Output LOW Voltage	54,74		0.25	0.4	V	I_{OL} = 4.0 mA $\quad V_{CC}$ = V_{CC} MIN,
		74		0.35	0.5	V	I_{OL} = 8.0 mA $\quad V_{IN}$ = V_{IL} or V_{IH} per Truth Table
I_{IH}	Input HIGH Current				20	μA	V_{CC} = MAX, V_{IN} = 2.7 V
					0.1	mA	V_{CC} = MAX, V_{IN} = 7.0 V
I_{IL}	Input LOW Current				−0.4	mA	V_{CC} = MAX, V_{IN} = 0.4 V
I_{OS}	Short Circuit Current		−20		−100	mA	V_{CC} = MAX
I_{CC}	Power Supply Current Total, Output HIGH Total, Output LOW				1.6	mA	V_{CC} = MAX
					4.4		

AC CHARACTERISTICS: T_A = 25°C

SYMBOL	PARAMETER	LIMITS			UNITS	TEST CONDITIONS
		MIN	TYP	MAX		
t_{PLH}	Turn Off Delay, Input to Output		9.0	15	ns	V_{CC} = 5.0 V
t_{PHL}	Turn On Delay, Input to Output		10	15	ns	C_L = 15 pF

Copyright of Motorola, Inc. Used by permission.

(continues)

APPENDIX 7   TTL Logic Gate Data Sheets   (Continued)

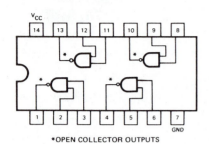

*OPEN COLLECTOR OUTPUTS

J Suffix — Case 632-07 (Ceramic)
N Suffix — Case 646-05 (Plastic)

SN54LS01
SN74LS01

QUAD 2-INPUT NAND GATE

LOW POWER SCHOTTKY

GUARANTEED OPERATING RANGES

SYMBOL	PARAMETER		MIN	TYP	MAX	UNIT
V_{CC}	Supply Voltage	54	4.5	5.0	5.5	V
		74	4.75	5.0	5.25	
T_A	Operating Ambient Temperature Range	54	−55	25	125	°C
		74	0	25	70	
V_{OH}	Output Voltage — High	54 , 74			5.5	V
I_{OL}	Output Current — Low	54			4.0	mA
		74			8.0	

DC CHARACTERISTICS OVER OPERATING TEMPERATURE RANGE (unless otherwise specified)

SYMBOL	PARAMETER		LIMITS			UNITS	TEST CONDITIONS	
			MIN	TYP	MAX			
V_{IH}	Input HIGH Voltage		2.0			V	Guaranteed Input HIGH Voltage for All Inputs	
V_{IL}	Input LOW Voltage	54			0.7	V	Guaranteed Input LOW Voltage for All Inputs	
		74			0.8			
V_{IK}	Input Clamp Diode Voltage			−0.65	−1.5	V	V_{CC} = MIN, I_{IN} = −18 mA	
I_{OH}	Output HIGH Current	54,74			100	μA	V_{CC} = MIN, V_{OH} = MAX	
V_{OL}	Output LOW Voltage	54,74		0.25	0.4	V	I_{OL} = 4.0 mA	V_{CC} = V_{CC} MIN, V_{IN} = V_{IL} or V_{IH} per Truth Table
		74		0.35	0.5	V	I_{OL} = 8.0 mA	
I_{IH}	Input HIGH Current				20	μA	V_{CC} = MAX, V_{IN} = 2.7 V	
					0.1	mA	V_{CC} = MAX, V_{IN} = 7.0 V	
I_{IL}	Input LOW Current				−0.4	mA	V_{CC} = MAX, V_{IN} = 0.4 V	
I_{CC}	Power Supply Current Total, Output HIGH Total, Output LOW				1.6	mA	V_{CC} = MAX	
					4.4			

AC CHARACTERISTICS: T_A = 25°C

SYMBOL	PARAMETER		LIMITS		UNITS	TEST CONDITIONS
		MIN	TYP	MAX		
t_{PLH}	Turn Off Delay, Input to Output		17	32	ns	V_{CC} = 5.0 V
t_{PHL}	Turn On Delay, Input to Output		15	28	ns	C_L = 15 pF, R_L = 2.0 kΩ

APPENDIX 7   TTL Logic Gate Data Sheets   (Continued)

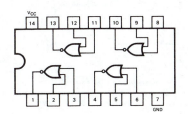

J Suffix — Case 632-07 (Ceramic)
N Suffix — Case 646-05 (Plastic)

SN54LS02
SN74LS02

QUAD 2-INPUT NOR GATE

LOW POWER SCHOTTKY

GUARANTEED OPERATING RANGES

SYMBOL	PARAMETER		MIN	TYP	MAX	UNIT
V_{CC}	Supply Voltage	54	4.5	5.0	5.5	V
		74	4.75	5.0	5.25	
T_A	Operating Ambient Temperature Range	54	−55	25	125	°C
		74	0	25	70	
I_{OH}	Output Current — High	54 , 74			−0.4	mA
I_{OL}	Output Current — Low	54			4.0	mA
		74			8.0	

DC CHARACTERISTICS OVER OPERATING TEMPERATURE RANGE (unless otherwise specified)

SYMBOL	PARAMETER		MIN	TYP	MAX	UNITS	TEST CONDITIONS	
V_{IH}	Input HIGH Voltage		2.0			V	Guaranteed Input HIGH Voltage for All Inputs	
V_{IL}	Input LOW Voltage	54			0.7	V	Guaranteed Input LOW Voltage for All Inputs	
		74			0.8			
V_{IK}	Input Clamp Diode Voltage			−0.65	−1.5	V	V_{CC} = MIN, I_{IN} = −18 mA	
V_{OH}	Output HIGH Voltage	54	2.5	3.5		V	V_{CC} = MIN, I_{OH} = MAX, V_{IN} = V_{IH} or V_{IL} per Truth Table	
		74	2.7	3.5		V		
V_{OL}	Output LOW Voltage	54,74		0.25	0.4	V	I_{OL} = 4.0 mA	V_{CC} = V_{CC} MIN, V_{IN} = V_{IL} or V_{IH} per Truth Table
		74		0.35	0.5	V	I_{OL} = 8.0 mA	
I_{IH}	Input HIGH Current				20	µA	V_{CC} = MAX, V_{IN} = 2.7 V	
					0.1	mA	V_{CC} = MAX, V_{IN} = 7.0 V	
I_{IL}	Input LOW Current				−0.4	mA	V_{CC} = MAX, V_{IN} = 0.4 V	
I_{OS}	Short Circuit Current		−20		−100	mA	V_{CC} = MAX	
I_{CC}	Power Supply Current Total, Output HIGH Total, Output LOW				3.2 5.4	mA	V_{CC} = MAX	

AC CHARACTERISTICS: T_A = 25°C

SYMBOL	PARAMETER	MIN	TYP	MAX	UNITS	TEST CONDITIONS
t_{PLH}	Turn Off Delay, Input to Output		10	15	ns	V_{CC} = 5.0 V
t_{PHL}	Turn On Delay, Input to Output		10	15	ns	C_L = 15 pF

(*continues*)

APPENDIX 7 TTL Logic Gate Data Sheets (Continued)

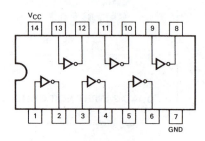

J Suffix — Case 632-07 (Ceramic)
N Suffix — Case 646-05 (Plastic)

SN54LS04
SN74LS04

HEX INVERTER

LOW POWER SCHOTTKY

GUARANTEED OPERATING RANGES

SYMBOL	PARAMETER			MIN	TYP	MAX	UNIT
V_CC	Supply Voltage		54	4.5	5.0	5.5	V
			74	4.75	5.0	5.25	
T_A	Operating Ambient Temperature Range		54	−55	25	125	°C
			74	0	25	70	
I_OH	Output Current — High		54, 74			−0.4	mA
I_OL	Output Current — Low		54			4.0	mA
			74			8.0	

DC CHARACTERISTICS OVER OPERATING TEMPERATURE RANGE (unless otherwise specified)

SYMBOL	PARAMETER		LIMITS			UNITS	TEST CONDITIONS	
			MIN	TYP	MAX			
V_IH	Input HIGH Voltage		2.0			V	Guaranteed Input HIGH Voltage for All Inputs	
V_IL	Input LOW Voltage	54			0.7	V	Guaranteed Input LOW Voltage for All Inputs	
		74			0.8			
V_IK	Input Clamp Diode Voltage			−0.65	−1.5	V	V_{CC} = MIN, I_{IN} = −18 mA	
V_OH	Output HIGH Voltage	54	2.5	3.5		V	V_{CC} = MIN, I_{OH} = MAX, V_{IN} = V_{IH} or V_{IL} per Truth Table	
		74	2.7	3.5		V		
V_OL	Output LOW Voltage	54,74		0.25	0.4	V	I_{OL} = 4.0 mA	V_{CC} = V_{CC} MIN, V_{IN} = V_{IL} or V_{IH} per Truth Table
		74		0.35	0.5	V	I_{OL} = 8.0 mA	
I_IH	Input HIGH Current				20	µA	V_{CC} = MAX, V_{IN} = 2.7 V	
					0.1	mA	V_{CC} = MAX, V_{IN} = 7.0 V	
I_IL	Input LOW Current				−0.4	mA	V_{CC} = MAX, V_{IN} = 0.4 V	
I_OS	Short Circuit Current		−20		−100	mA	V_{CC} = MAX	
I_CC	Power Supply Current Total, Output HIGH Total, Output LOW				2.4 6.6	mA	V_{CC} = MAX	

AC CHARACTERISTICS: T_A = 25°C

SYMBOL	PARAMETER	LIMITS			UNITS	TEST CONDITIONS
		MIN	TYP	MAX		
t_PLH	Turn Off Delay, Input to Output		9.0	15	ns	V_{CC} = 5.0 V
t_PHL	Turn On Delay, Input to Output		10	15	ns	C_L = 15 pF

APPENDIX 7   TTL Logic Gate Data Sheets   (Continued)

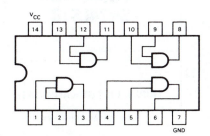

J Suffix — Case 632-07 (Ceramic)
N Suffix — Case 646-05 (Plastic)

**SN54LS08
SN74LS08**

QUAD 2-INPUT AND GATE

LOW POWER SCHOTTKY

GUARANTEED OPERATING RANGES

SYMBOL	PARAMETER		MIN	TYP	MAX	UNIT
V_{CC}	Supply Voltage	54	4.5	5.0	5.5	V
		74	4.75	5.0	5.25	
T_A	Operating Ambient Temperature Range	54	−55	25	125	°C
		74	0	25	70	
I_{OH}	Output Current — High	54 , 74			−0.4	mA
I_{OL}	Output Current — Low	54			4.0	mA
		74			8.0	

DC CHARACTERISTICS OVER OPERATING TEMPERATURE RANGE (unless otherwise specified)

SYMBOL	PARAMETER		LIMITS			UNITS	TEST CONDITIONS
			MIN	TYP	MAX		
V_{IH}	Input HIGH Voltage		2.0			V	Guaranteed Input HIGH Voltage for All Inputs
V_{IL}	Input LOW Voltage	54			0.7	V	Guaranteed Input LOW Voltage for All Inputs
		74			0.8		
V_{IK}	Input Clamp Diode Voltage			−0.65	−1.5	V	V_{CC} = MIN, I_{IN} = −18 mA
V_{OH}	Output HIGH Voltage	54	2.5	3.5		V	V_{CC} = MIN, I_{OH} = MAX, V_{IN} = V_{IH} or V_{IL} per Truth Table
		74	2.7	3.5		V	
V_{OL}	Output LOW Voltage	54,74		0.25	0.4	V	I_{OL} = 4.0 mA V_{CC} = V_{CC} MIN, V_{IN} = V_{IL} or V_{IH} per Truth Table
		74		0.35	0.5	V	I_{OL} = 8.0 mA
I_{IH}	Input HIGH Current				20	μA	V_{CC} = MAX, V_{IN} = 2.7 V
					0.1	mA	V_{CC} = MAX, V_{IN} = 7.0 V
I_{IL}	Input LOW Current				−0.4	mA	V_{CC} = MAX, V_{IN} = 0.4 V
I_{OS}	Short Circuit Current		−20		−100	mA	V_{CC} = MAX
I_{CC}	Power Supply Current Total, Output HIGH Total, Output LOW				4.8	mA	V_{CC} = MAX
					8.8		

AC CHARACTERISTICS: T_A = 25°C

SYMBOL	PARAMETER	LIMITS			UNITS	TEST CONDITIONS
		MIN	TYP	MAX		
t_{PLH}	Turn Off Delay, Input to Output		8.0	15	ns	V_{CC} = 5.0 V
t_{PHL}	Turn On Delay, Input to Output		10	20	ns	C_L = 15 pF

(continues)

APPENDIX 7   TTL Logic Gate Data Sheets   (Continued)

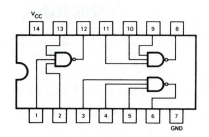

J Suffix — Case 632-07 (Ceramic)
N Suffix — Case 646-05 (Plastic)

**SN54LS10
SN74LS10**

TRIPLE 3-INPUT NAND GATE

LOW POWER SCHOTTKY

GUARANTEED OPERATING RANGES

SYMBOL	PARAMETER		MIN	TYP	MAX	UNIT
V_{CC}	Supply Voltage	54	4.5	5.0	5.5	V
		74	4.75	5.0	5.25	
T_A	Operating Ambient Temperature Range	54	−55	25	125	°C
		74	0	25	70	
I_{OH}	Output Current — High	54 , 74			−0.4	mA
I_{OL}	Output Current — Low	54			4.0	mA
		74			8.0	

DC CHARACTERISTICS OVER OPERATING TEMPERATURE RANGE (unless otherwise specified)

SYMBOL	PARAMETER		MIN	TYP	MAX	UNITS	TEST CONDITIONS
V_{IH}	Input HIGH Voltage		2.0			V	Guaranteed Input HIGH Voltage for All Inputs
V_{IL}	Input LOW Voltage	54			0.7	V	Guaranteed Input LOW Voltage for All Inputs
		74			0.8		
V_{IK}	Input Clamp Diode Voltage			−0.65	−1.5	V	V_{CC} = MIN, I_{IN} = −18 mA
V_{OH}	Output HIGH Voltage	54	2.5	3.5		V	V_{CC} = MIN, I_{OH} = MAX, V_{IN} = V_{IH} or V_{IL} per Truth Table
		74	2.7	3.5		V	
V_{OL}	Output LOW Voltage	54,74		0.25	0.4	V	I_{OL} = 4.0 mA $\quad V_{CC}$ = V_{CC} MIN, V_{IN} = V_{IL} or V_{IH} per Truth Table
		74		0.35	0.5	V	I_{OL} = 8.0 mA
I_{IH}	Input HIGH Current				20	μA	V_{CC} = MAX, V_{IN} = 2.7 V
					0.1	mA	V_{CC} = MAX, V_{IN} = 7.0 V
I_{IL}	Input LOW Current				−0.4	mA	V_{CC} = MAX, V_{IN} = 0.4 V
I_{OS}	Short Circuit Current		−20		−100	mA	V_{CC} = MAX
I_{CC}	Power Supply Current Total, Output HIGH Total, Output LOW				1.2 3.3	mA	V_{CC} = MAX

AC CHARACTERISTICS: T_A = 25°C

SYMBOL	PARAMETER		MIN	TYP	MAX	UNITS	TEST CONDITIONS
t_{PLH}	Turn Off Delay, Input to Output			9.0	15	ns	V_{CC} = 5.0 V
t_{PHL}	Turn On Delay, Input to Output			10	15	ns	C_L = 15 pF

APPENDIX 7   TTL Logic Gate Data Sheets   (Continued)

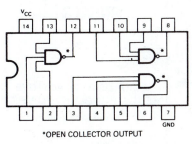

*OPEN COLLECTOR OUTPUT

J Suffix — Case 632-07 (Ceramic)
N Suffix — Case 646-05 (Plastic)

**SN54LS12
SN74LS12**

TRIPLE 3-INPUT NAND GATE

LOW POWER SCHOTTKY

GUARANTEED OPERATING RANGES

SYMBOL	PARAMETER		MIN	TYP	MAX	UNIT
V_{CC}	Supply Voltage	54	4.5	5.0	5.5	V
		74	4.75	5.0	5.25	
T_A	Operating Ambient Temperature Range	54	−55	25	125	°C
		74	0	25	70	
V_{OH}	Output Voltage — High	54 , 74			5.5	V
I_{OL}	Output Current — Low	54			4.0	mA
		74			8.0	

DC CHARACTERISTICS OVER OPERATING TEMPERATURE RANGE (unless otherwise specified)

SYMBOL	PARAMETER		LIMITS			UNITS	TEST CONDITIONS	
			MIN	TYP	MAX			
V_{IH}	Input HIGH Voltage		2.0			V	Guaranteed Input HIGH Voltage for All Inputs	
V_{IL}	Input LOW Voltage	54			0.7	V	Guaranteed Input LOW Voltage for All Inputs	
		74			0.8			
V_{IK}	Input Clamp Diode Voltage			−0.65	−1.5	V	V_{CC} = MIN, I_{IN} = −18 mA	
I_{OH}	Output HIGH Current	54,74			100	μA	V_{CC} = MIN, V_{OH} = MAX	
V_{OL}	Output LOW Voltage	54,74		0.25	0.4	V	I_{OL} = 4.0 mA	V_{CC} = V_{CC} MIN, V_{IN} = V_{IL} or V_{IH} per Truth Table
		74		0.35	0.5	V	I_{OL} = 8.0 mA	
I_{IH}	Input HIGH Current				20	μA	V_{CC} = MAX, V_{IN} = 2.7 V	
					·0.1	mA	V_{CC} = MAX, V_{IN} = 7.0 V	
I_{IL}	Input LOW Current				−0.4	mA	V_{CC} = MAX, V_{IN} = 0.4 V	
I_{CC}	Power Supply Current Total, Output HIGH Total, Output LOW				1.4 3.3	mA	V_{CC} = MAX	

AC CHARACTERISTICS: T_A = 25°C

SYMBOL	PARAMETER	LIMITS			UNITS	TEST CONDITIONS
		MIN	TYP	MAX		
t_{PLH}	Turn Off Delay, Input to Output		17	32	ns	V_{CC} = 5.0 V
t_{PHL}	Turn On Delay, Input to Output		15	28	ns	C_L = 15 pF, R_L = 2.0 kΩ

(continues)

APPENDIX 7   TTL Logic Gate Data Sheets   (Continued)

SN54LS74A
SN54LS74A

DESCRIPTION - The SN54LS/74LS74A dual edge-triggered flip-flop utilizes Schottky TTL circuitry to produce high speed D-type flip-flops. Each flip-flop has individual clear and set inputs, and also complementary Q and Q̄ outputs.

Information at input D is transferred to the Q output on the positive-going edge of the clock pulse. Clock triggering occurs at a voltage level of the clock pulse and is not directly related to the transition time of the positive-going pulse. When the clock input is at either the HIGH or the LOW level, the D input signal has no effect.

DUAL D-TYPE POSITIVE
EDGE-TRIGGERED FLIP-FLOP

LOW POWER SCHOTTKY

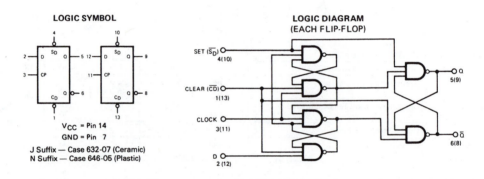

LOGIC SYMBOL

V_{CC} = Pin 14
GND = Pin 7

J Suffix — Case 632-07 (Ceramic)
N Suffix — Case 646-05 (Plastic)

LOGIC DIAGRAM
(EACH FLIP-FLOP)

SET ($\overline{S}_D$) 4(10)
CLEAR ($\overline{CD}$) 1(13)
CLOCK 3(11)
D 2(12)
Q 5(9)
Q̄ 6(8)

DC CHARACTERISTICS OVER OPERATING TEMPERATURE RANGE (unless otherwise specified)

SYMBOL	PARAMETER		LIMITS			UNITS	TEST CONDITIONS
			MIN	TYP	MAX		
V_{IH}	Input HIGH Voltage		2.0			V	Guaranteed Input HIGH Voltage for All Inputs
V_{IL}	Input LOW Voltage	54			0.7	V	Guaranteed Input LOW Voltage for All Inputs
		74			0.8		
V_{IK}	Input Clamp Diode Voltage			−0.65	−1.5	V	V_{CC} = MIN, I_{IN} = −18 mA
V_{OH}	Output HIGH Voltage	54	2.5	3.5		V	V_{CC} = MIN, I_{OH} = MAX, V_{IN} = V_{IH} or V_{IL} per Truth Table
		74	2.7	3.5		V	
V_{OL}	Output LOW Voltage	54,74		0.25	0.4	V	I_{OL} = 4.0 mA	V_{CC} = V_{CC} MIN, V_{IN} = V_{IL} or V_{IH} per Truth Table
		74		0.35	0.5	V	I_{OL} = 8.0 mA	
I_{IH}	Input High Current Data, Clock Set, Clear				20 40	μA	V_{CC} = MAX, V_{IN} = 2.7 V
	Data, Clock Set, Clear				0.1 0.2	mA	V_{CC} = MAX, V_{IN} = 7.0 V
I_{IL}	Input LOW Current Data, Clock Set, Clear				−0.4 −0.8	mA	V_{CC} = MAX, V_{IN} = 0.4 V
I_{OS}	Output Short Circuit Current		−20		−100	mA	V_{CC} = MAX
I_{CC}	Power Supply Current				8.0	mA	V_{CC} = MAX

APPENDIX 7   TTL Logic Gate Data Sheets   (Continued)

SN54LS/74LS74A

MODE SELECT — TRUTH TABLE

OPERATING MODE	INPUTS			OUTPUTS	
	$\overline{S}_D$	$\overline{C}_D$	D	Q	$\overline{Q}$
Set	L	H	X	H	L
Reset (Clear)	H	L	X	L	H
*Undetermined	L	L	X	H	H
Load "1" (Set)	H	H	h	H	L
Load "0" (Reset)	H	H	l	L	H

*Both outputs will be HIGH while both $\overline{S}_D$ AND $\overline{C}_D$ are LOW, but the output states are unpredictable if $\overline{S}_D$ and $\overline{C}_D$ go HIGH simultaneously. If the levels at the set and clear are near V_{IL} maximum then we cannot guarantee to meet the minimum level for V_{OH}.

H, h = HIGH Voltage Level
L, l = LOW Voltage Level
X = Don't Care
l, h (q) = Lower case letters indicate the state of the refer-
  enced input (or output) one set-up time prior to the
  LOW to HIGH clock transition.

GUARANTEED OPERATING RANGES

SYMBOL	PARAMETER		MIN	TYP	MAX	UNIT
V_{CC}	Supply Voltage	54	4.5	5.0	5.5	V
		74	4.75	5.0	5.25	
T_A	Operating Ambient Temperature Range	54	−55	25	125	°C
		74	0	25	70	
I_{OH}	Output Current — High	54 , 74			−0.4	mA
I_{OL}	Output Current — Low	54			4.0	mA
		74			8.0	

AC CHARACTERISTICS: T_A = 25°C, V_{CC} = 5.0 V

SYMBOL	PARAMETER	LIMITS			UNITS	TEST CONDITIONS	
		MIN	TYP	MAX			
f_{MAX}	Maximum Clock Frequency	25	33		MHz	Fig. 1	V_{CC} = 5.0 V, C_L = 15 pF
t_{PLH}	Clock, Clear, Set to Output		13	25	ns	Fig. 1	
t_{PHL}			25	40	ns		

AC SETUP REQUIREMENTS: T_A = 25°C, V_{CC} = 5.0 V

SYMBOL	PARAMETER	LIMITS			UNITS	TEST CONDITIONS	
		MIN	TYP	MAX			
$t_{W(H)}$	Clock	25			ns	Fig. 1	
$t_{W(L)}$	Clear, Set	25			ns	Fig. 2	
t_s	Data Setup Time — HIGH	20			ns	Fig. 1	V_{CC} = 5.0 V
	LOW	20			ns		
t_h	Hold Time	5.0			ns	Fig. 1	

(continues)

APPENDIX 7   TTL Logic Gate Data Sheets   (Continued)

SN54LS/74LS74A

AC WAVEFORMS

**Fig. 1 CLOCK TO OUTPUT DELAYS,
DATA SET-UP AND HOLD TIMES, CLOCK PULSE WIDTH**

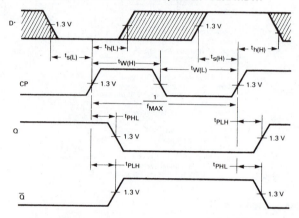

*The shaded areas indicate when the input is permitted to change for predicatable output performance.

**Fig. 2 SET AND CLEAR TO OUTPUT DELAYS,
SET AND CLEAR PULSE WIDTHS**

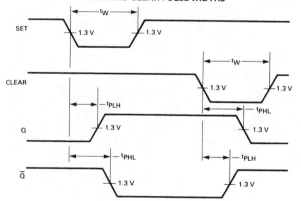

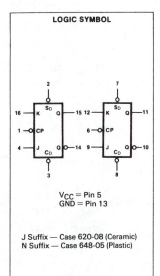

SN54LS76A
SN74LS76A

DUAL JK FLIP-FLOP
WITH SET AND CLEAR

LOW POWER SCHOTTKY

DESCRIPTION — The SN54LS/74LS76A offers individual J, K, Clock Pulse, Direct Set and Direct Clear inputs. These dual flip-flops are designed so that when the clock goes HIGH, the inputs are enabled and data will be accepted. The Logic Level of the J and K inputs will perform according to the Truth Table as long as minimum set-up times are observed. Input data is transferred to the outputs on the HIGH-to-LOW clock transitions.

MODE SELECT — TRUTH TABLE

OPERATING MODE	INPUTS				OUTPUTS	
	$\overline{S}_D$	$\overline{C}_D$	J	K	Q	$\overline{Q}$
Set	L	H	X	X	H	L
Reset (Clear)	H	L	X	X	L	H
*Undetermined	L	L	X	X	H	H
Toggle	H	H	h	h	$\overline{q}$	q
Load "0" (Reset)	H	H	l	h	L	H
Load "1" (Set)	H	H	h	l	H	L
Hold	H	H	l	l	q	$\overline{q}$

*Both outputs will be HIGH while both $\overline{S}_D$ and $\overline{C}_D$ are LOW, but the output states are unpredictable if $\overline{S}_D$ and $\overline{C}_D$ go HIGH simultaneously.

H,h = HIGH Voltage Level
L,l = LQW Voltage Level
X = Immaterial
l,h (q) = Lower case letters indicate the state of the referenced input (or output) one set-up time prior to the HIGH-to-LOW clock transition.

LOGIC SYMBOL

V_{CC} = Pin 5
GND = Pin 13

J Suffix — Case 620-08 (Ceramic)
N Suffix — Case 648-05 (Plastic)

LOGIC DIAGRAM

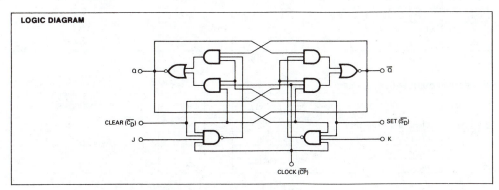

(*continues*)

APPENDIX 7   TTL Logic Gate Data Sheets   (Continued)

SN54LS/74LS76A

GUARANTEED OPERATING RANGES

SYMBOL	PARAMETER		MIN	TYP	MAX	UNIT
V_{CC}	Supply Voltage	54	4.5	5.0	5.5	V
		74	4.75	5.0	5.25	
T_A	Operating Ambient Temperature Range	54	−55	25	125	°C
		74	0	25	70	
I_{OH}	Output Current — High	54, 74			−0.4	mA
I_{OL}	Output Current — Low	54			4.0	mA
		74			8.0	

DC CHARACTERISTICS OVER OPERATING TEMPERATURE RANGE (unless otherwise specified)

SYMBOL	PARAMETER		LIMITS MIN	LIMITS TYP	LIMITS MAX	UNITS	TEST CONDITIONS
V_{IH}	Input HIGH Voltage		2.0			V	Guaranteed Input HIGH Voltage for All Inputs
V_{IL}	Input LOW Voltage	54			0.7	V	Guaranteed Input LOW Voltage for All Inputs
		74			0.8		
V_{IK}	Input Clamp Diode Voltage			−0.65	−1.5	V	V_{CC} = MIN, I_{IN} = −18 mA
V_{OH}	Output HIGH Voltage	54	2.5	3.5		V	V_{CC} = MIN, I_{OH} = MAX, V_{IN} = V_{IH} or V_{IL} per Truth Table
		74	2.7	3.5		V	
V_{OL}	Output LOW Voltage	54,74		0.25	0.4	V	I_{OL} = 4.0 mA, V_{CC} = V_{CC} MIN, V_{IN} = V_{IL} or V_{IH} per Truth Table
		74		0.35	0.5	V	I_{OL} = 8.0 mA
I_{IH}	Input HIGH Current	J, K / Clear / Clock			20 / 60 / 80	μA	V_{CC} = MAX, V_{IN} = 2.7 V
		J, K / Clear / Clock			0.1 / 0.3 / 0.4	mA	V_{CC} = MAX, V_{IN} = 7.0 V
I_{IL}	Input LOW Current	J, K / Clear, Clock			−0.4 / −0.8	mA	V_{CC} = MAX, V_{IN} = 0.4 V
I_{OS}	Short Circuit Current		−20		−100	mA	V_{CC} = MAX
I_{CC}	Power Supply Current				6.0	mA	V_{CC} = MAX

AC CHARACTERISTICS: T_A = 25°C, V_{CC} = 5.0 V

SYMBOL	PARAMETER	LIMITS MIN	LIMITS TYP	LIMITS MAX	UNITS	TEST CONDITIONS
f_{MAX}	Maximum Clock Frequency	30	45		MHz	V_{CC} = 5.0 V, C_L = 15 pF
t_{PLH}	Clock, Clear, Set to Output		15	20	ns	
t_{PHL}			15	20	ns	

AC SETUP REQUIREMENTS: T_A = 25°C, V_{CC} = 5.0 V

SYMBOL	PARAMETER	LIMITS MIN	LIMITS TYP	LIMITS MAX	UNITS	TEST CONDITIONS
t_W	Clock Pulse Width High	20			ns	V_{CC} = 5.0 V
t_W	Clear Set Pulse Width	25			ns	
t_S	Setup Time	20			ns	
t_h	Hold Time	0			ns	

APPENDIX 7   TTL Logic Gate Data Sheets   (Continued)

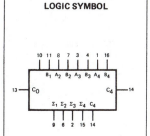

SN54LS83A
SN74LS83A

4-BIT BINARY FULL ADDER WITH FAST CARRY

LOW POWER SCHOTTKY

DESCRIPTION — The SN54LS/74LS83A is a high-speed 4-Bit Binary Full Adder with internal carry lookahead. It accepts two 4-bit binary words ($A_1 - A_4$, $B_1 - B_4$) and a Carry Input (C_0). It generates the binary Sum outputs $\Sigma_1 - \Sigma_4$ and the Carry Output (C_4) from the most significant bit. The LS83A operates with either active HIGH or active LOW operands (positive or negative logic). The SN54LS/74LS283 is recommended for new designs since it is identical in function with this device and features standard corner power pins.

PIN NAMES

		LOADING (Note a)	
		HIGH	LOW
$A_1 - A_4$	Operand A Inputs	1.0 U.L.	0.5 U.L.
$B_1 - B_4$	Operand B Inputs	1.0 U.L.	0.5 U.L.
C_0	Carry Input	0.5 U.L.	0.25 U.L.
$\Sigma_1 - \Sigma_4$	Sum Outputs (Note b)	10 U.L.	5(2.5) U.L.
C_4	Carry Output (Note b)	10 U.L.	5(2.5) U.L.

NOTES:
a. 1 TTL Unit Load (U.L.) = 40 µA HIGH/1.6 mA LOW.
b. The Output LOW drive factor is 2.5 U.L. for Military (54) and 5 U.L. for commercial (74) Temperature Ranges.

LOGIC SYMBOL

LOGIC DIAGRAM

V_{CC} = Pin 5
GND = Pin 12
○ = Pin Numbers

CONNECTION DIAGRAM DIP (TOP VIEW)

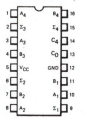

J Suffix — Case 620-08 (Ceramic)
N Suffix — Case 648-05 (Plastic)

NOTE:
The Flatpak version has the same pinouts (Connection Diagram) as the Dual In-Line Package.

(continues)

APPENDIX 7   TTL Logic Gate Data Sheets   (Continued)

SN54LS/74LS83A

FUNCTIONAL DESCRIPTION — The LS83A adds two 4-bit binary words (A plus B) plus the incoming carry. The binary sum appears on the sum outputs ($\Sigma_1 - \Sigma_4$) and outgoing carry (C_4) outputs.

$$C_0 + (A_1+B_1)+2(A_2+B_2)+4(A_3+B_3)+8(A_4+B_4) = \Sigma_1+2\Sigma_2+4\Sigma_3+8\Sigma_4+16C_4$$

Where: (+) = plus

Due to the symmetry of the binary add function the LS83A can be used with either all inputs and outputs active HIGH (positive logic) or with all inputs and outputs active LOW (negative logic). Note that with active HIGH Inputs, Carry Input can not be left open, but must be held LOW when no carry in is intended.

Example:

	C_0	A_1	A_2	A_3	A_4	B_1	B_2	B_3	B_4	Σ_1	Σ_2	Σ_3	Σ_4	C_4	
Logic Levels	L	L	H	L	H	H	L	L	H	H	H	L	L	H	
Active HIGH	0	0	1	0	1	1	0	0	1	1	1	0	0	1	(10+9 = 19)
Active LOW	1	1	0	1	0	0	1	1	0	0	0	1	1	0	(carry+5+6 = 12)

Interchanging inputs of equal weight does not affect the operation, thus C_0, A_1, B_1, can be arbitrarily assigned to pins 10, 11, 13, etc.

FUNCTIONAL TRUTH TABLE

$C(n-1)$	A_n	B_n	Σ_n	C_n
L	L	L	L	L
L	L	H	H	L
L	H	L	H	L
L	H	H	L	H
H	L	L	H	L
H	L	H	L	H
H	H	L	L	H
H	H	H	H	H

$C_1 - C_3$ are generated internally
C_0 — is an external input
C_4 — is an output generated internally

GUARANTEED OPERATING RANGES

SYMBOL	PARAMETER			MIN	TYP	MAX	UNIT
V_{CC}	Supply Voltage	54		4.5	5.0	5.5	V
		74		4.75	5.0	5.25	
T_A	Operating Ambient Temperature Range	54		−55	25	125	°C
		74		0	25	70	
I_{OH}	Output Current — High	54 , 74				−0.4	mA
I_{OL}	Output Current — Low	54				4.0	mA
		74				8.0	

APPENDIX 7 TTL Logic Gate Data Sheets (Continued)

SN54LS/74LS83A

DC CHARACTERISTICS OVER OPERATING TEMPERATURE RANGE (unless otherwise specified)

SYMBOL	PARAMETER		LIMITS			UNITS	TEST CONDITIONS	
			MIN	TYP	MAX			
V_{IH}	Input HIGH Voltage		2.0			V	Guaranteed Input HIGH Voltage for All Inputs	
V_{IL}	Input LOW Voltage	54			0.7	V	Guaranteed Input LOW Voltage for All Inputs	
		74			0.8			
V_{IK}	Input Clamp Diode Voltage			−0.65	−1.5	V	V_{CC} = MIN, I_{IN} =−18 mA	
V_{OH}	Output HIGH Voltage	54	2.5	3.5		V	V_{CC} = MIN, I_{OH} = MAX, V_{IN} = V_{IH} or V_{IL} per Truth Table	
		74	2.7	3.5		V		
V_{OL}	Output LOW Voltage	54,74		0.25	0.4	V	I_{OL} = 4.0 mA	V_{CC} = V_{CC} MIN, V_{IN} = V_{IL} or V_{IH} per Truth Table
		74		0.35	0.5	V	I_{OL} = 8.0 mA	
I_{IH}	Input HIGH Current C$_O$ A or B				20 40	μA	V_{CC} = MAX, V_{IN} = 2.7 V	
	C$_O$ A or B				0.1 0.2	mA	V_{CC} = MAX, V_{IN} = 7.0 V	
I_{IL}	Input LOW Current C$_O$ A or B				−0.4 −0.8	mA	V_{CC} = MAX, V_{IN} = 0.4 V	
I_{OS}	Output Short Circuit Current		−20		−100	mA	V_{CC} = MAX	
I_{CC}	Power Supply Current All Inputs Grounded All Inputs at 4.5 V, Except B All Inputs at 4.5 V				39 34 34	mA	V_{CC} = MAX	

AC CHARACTERISTICS: T_A = 25°C

SYMBOL	PARAMETER		LIMITS		UNITS	TEST CONDITIONS
		MIN	TYP	MAX		
t_{PLH} t_{PHL}	Propagation Delay, C$_O$ Input to any Σ Output		16 15	24 24	ns	
t_{PLH} t_{PHL}	Propagation Delay, Any A or B Input to Σ Outputs		15 15	24 24	ns	V_{CC} = 5.0 V C_L = 15 pF Figures 1 and 2
t_{PLH} t_{PHL}	Propagation Delay, C$_O$ Input to C$_4$ Output		11 15	17 22	ns	
t_{PLH} t_{PHL}	Propagation Delay, Any A or B Input to C$_4$ Output		11 12	17 17	ns	

AC WAVEFORMS

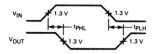

Fig. 1

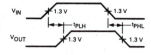

Fig. 2

B-SUFFIX SERIES CMOS GATES

The B Series logic gates are constructed with P and N channel enhancement mode devices in a single monolithic structure (Complementary MOS). Their primary use is where low power dissipation and/or high noise immunity is desired.

- Supply Voltage Range = 3.0 Vdc to 18 Vdc
- All Outputs Buffered
- Capable of Driving Two Low-power TTL Loads or One Low-power Schottky TTL Load Over the Rated Temperature Range.
- Double Diode Protection on All Inputs
- Pin-for-Pin Replacements for Corresponding CD4000 Series B Suffix Devices

L SUFFIX
CERAMIC PACKAGE
CASE 632

P SUFFIX
PLASTIC PACKAGE
CASE 646

ORDERING INFORMATION

A Series: −55°C to +125°C
MC14XXXBAL (Ceramic Package Only)

C Series: −40°C to +85°C
MC14XXXBCP (Plastic Package)
MC14XXXBCL (Ceramic Package)

MAXIMUM RATINGS* (Voltages Referenced to V_{SS})

Symbol	Parameter	Value	Unit
V_{DD}	DC Supply Voltage	− 0.5 to + 18.0	V
V_{in}, V_{out}	Input or Output Voltage (DC or Transient)	− 0.5 to V_{DD} + 0.5	V
I_{in}, I_{out}	Input or Output Current (DC or Transient), per Pin	± 10	mA
P_D	Power Dissipation, per Package†	500	mW
T_{stg}	Storage Temperature	− 65 to + 150	°C
T_L	Lead Temperature (8-Second Soldering)	260	°C

*Maximum Ratings are those values beyond which damage to the device may occur.
†Temperature Derating: Plastic "P" Package: − 12mW/°C from 65°C to 85°C
Ceramic "L" Package: − 12mW/°C from 100°C to 125°C

This device contains protection circuitry to guard against damage due to high static voltages or electric fields. However, precautions must be taken to avoid applications of any voltage higher than maximum rated voltages to this high-impedance circuit. For proper operation, V_{in} and V_{out} should be constrained to the range $V_{SS} \leq (V_{in}$ or $V_{out}) \leq V_{DD}$.
Unused inputs must always be tied to an appropriate logic voltage level (e.g., either V_{SS} or V_{DD}). Unused outputs must be left open.

MC14001B
Quad 2-Input NOR Gate

MC14002B
Dual 4-Input Nor Gate

MC14011B
Quad 2-Input NAND Gate

MC14012B
Dual 4-Input NAND Gate

MC14023B
Triple 3-Input NAND Gate

MC14025B
Triple 3-Input NOR Gate

MC14068B
8-Input NAND Gate

MC14071B
Quad 2-Input OR Gate

MC14072B
Dual 4-Input OR Gate

MC14073B
Triple 3-Input AND Gate

MC14075B
Triple 3-Input OR Gate

MC14078B
8-Input NOR Gate

MC14081B
Quad 2-Input AND Gate

MC14082B
Dual 4-Input AND Gate

CMOS SSI

(LOW-POWER COMPLEMENTARY MOS)

B-SERIES GATES

APPENDIX 8   CMOS Logic Gate Data Sheets   (Continued)

CMOS B-SERIES GATES

LOGIC DIAGRAMS

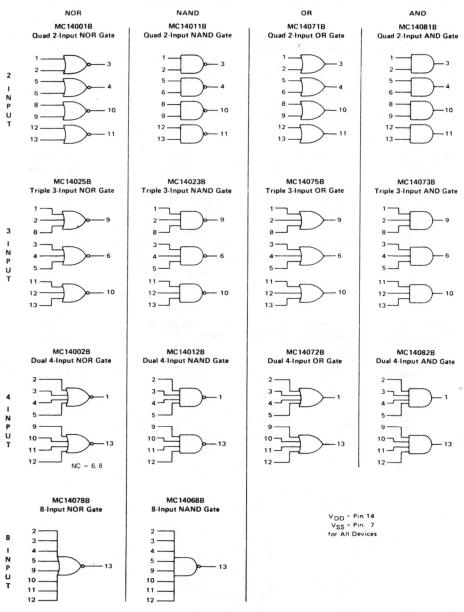

V_{DD} = Pin 14
V_{SS} = Pin 7
for All Devices

(continues)

CMOS B-SERIES GATES

PIN ASSIGNMENTS

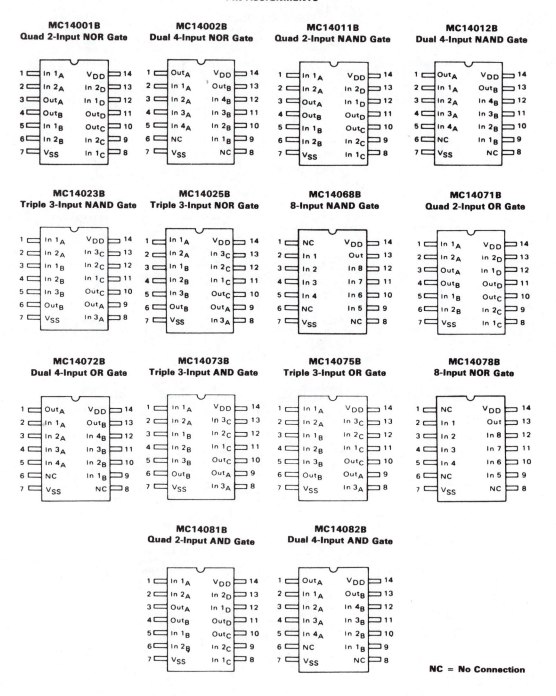

NC = No Connection

APPENDIX 8   CMOS Logic Gate Data Sheets   (Continued)

CMOS B-SERIES GATES

ELECTRICAL CHARACTERISTICS (Voltages Referenced to V_{SS})

Characteristic	Symbol	V_{DD} Vdc	T_{low}* Min	T_{low}* Max	25°C Min	25°C Typ #	25°C Max	T_{high}* Min	T_{high}* Max	Unit
Output Voltage "0" Level	V_{OL}	5.0	—	0.05	—	0	0.05	—	0.05	Vdc
$V_{in} = V_{DD}$ or 0		10	—	0.05	—	0	0.05	—	0.05	
		15	—	0.05	—	0	0.05	—	0.05	
"1" Level	V_{OH}	5.0	4.95	—	4.95	5.0	—	4.95	—	Vdc
$V_{in} = 0$ or V_{DD}		10	9.95	—	9.95	10	—	9.95	—	
		15	14.95	—	14.95	15	—	14.95	—	
Input Voltage "0" Level	V_{IL}									Vdc
(V_O = 4.5 or 0.5 Vdc)		5.0	—	1.5	—	2.25	1.5	—	1.5	
(V_O = 9.0 or 1.0 Vdc)		10	—	3.0	—	4.50	3.0	—	3.0	
(V_O = 13.5 or 1.5 Vdc)		15	—	4.0	—	6.75	4.0	—	4.0	
"1" Level	V_{IH}									Vdc
(V_O = 0.5 or 4.5 Vdc)		5.0	3.5	—	3.5	2.75	—	3.5	—	
(V_O = 1.0 or 9.0 Vdc)		10	7.0	—	7.0	5.50	—	7.0	—	
(V_O = 1.5 or 13.5 Vdc)		15	11.0	—	11.0	8.25	—	11.0	—	
Output Drive Current (AL Device)	I_{OH}									mAdc
(V_{OH} = 2.5 Vdc) Source		5.0	−3.0	—	−2.4	−4.2	—	−1.7	—	
(V_{OH} = 4.6 Vdc)		5.0	−0.64	—	−0.51	−0.88	—	−0.36	—	
(V_{OH} = 9.5 Vdc)		10	−1.6	—	−1.3	−2.25	—	−0.9	—	
(V_{OH} = 13.5 Vdc)		15	−4.2	—	−3.4	−8.8	—	−2.4	—	
(V_{OL} = 0.4 Vdc) Sink	I_{OL}	5.0	0.64	—	0.51	0.88	—	0.36	—	mAdc
(V_{OL} = 0.5 Vdc)		10	1.6	—	1.3	2.25	—	0.9	—	
(V_{OL} = 1.5 Vdc)		15	4.2	—	3.4	8.8	—	2.4	—	
Output Drive Current (CL/CP Device)	I_{OH}									mAdc
(V_{OH} = 2.5 Vdc) Source		5.0	−2.5	—	−2.1	−4.2	—	−1.7	—	
(V_{OH} = 4.6 Vdc)		5.0	−0.52	—	−0.44	−0.88	—	−0.36	—	
(V_{OH} = 9.5 Vdc)		10	−1.3	—	−1.1	−2.25	—	−0.9	—	
(V_{OH} = 13.5 Vdc)		15	−3.6	—	−3.0	−8.8	—	−2.4	—	
(V_{OL} = 0.4 Vdc) Sink	I_{OL}	5.0	0.52	—	0.44	0.88	—	0.36	—	mAdc
(V_{OL} = 0.5 Vdc)		10	1.3	—	1.1	2.25	—	0.9	—	
(V_{OL} = 1.5 Vdc)		15	3.6	—	3.0	8.8	—	2.4	—	
Input Current (AL Device)	I_{in}	15	—	±0.1	—	±0.00001	±0.1	—	±1.0	µAdc
Input Current (CL/CP Device)	I_{in}	15	—	±0.3	—	±0.00001	±0.3	—	±1.0	µAdc
Input Capacitance (V_{in} = 0)	C_{in}	—	—	—	—	5.0	7.5	—	—	pF
Quiescent Current (AL Device)	I_{DD}	5.0	—	0.25	—	0.0005	0.25	—	7.5	µAdc
(Per Package)		10	—	0.50	—	0.0010	0.50	—	15.0	
		15	—	1.00	—	0.0015	1.00	—	30.0	
Quiescent Current (CL/CP Device)	I_{DD}	5.0	—	1.0	—	0.0005	1.0	—	7.5	µAdc
(Per Package)		10	—	2.0	—	0.0010	2.0	—	15.0	
		15	—	4.0	—	0.0015	4.0	—	30.0	
Total Supply Current**† (Dynamic plus Quiescent, Per Gate, C_L = 50 pF)	I_T	5.0	$I_T = (0.3 \ \mu A/kHz) \ f + I_{DD}/N$							µAdc
		10	$I_T = (0.6 \ \mu A/kHz) \ f + I_{DD}/N$							
		15	$I_T = (0.9 \ \mu A/kHz) \ f + I_{DD}/N$							

*T_{low} = −55°C for AL Device, −40°C for CL/CP Device.
T_{high} = +125°C for AL Device, +85°C for CL/CP Device.

#Data labelled "Typ" is not to be used for design purposes but is intended as an indication of the IC's potential performance.

**The formulas given are for the typical characteristics only at 25°C.

†To calculate total supply current at loads other than 50 pF:

$$I_T(C_L) = I_T(50 \ pF) + (C_L - 50) \ Vfk$$

where: I_T is in µA (per package), C_L in pF, $V = (V_{DD} - V_{SS})$ in volts, f in kHz is input frequency, and k = 0.001 × the number of exercised gates per package.

(continues)

APPENDIX 8   CMOS Logic Gate Data Sheets   (Continued)

CMOS B-SERIES GATES

B-SERIES GATE SWITCHING TIMES

SWITCHING CHARACTERISTICS* ($C_L = 50$ pF, $T_A = 25°C$)

Characteristic	Symbol	V_{DD} Vdc	Min	Typ #	Max	Unit
Output Rise Time, All B-Series Gates	t_{TLH}					ns
$t_{TLH} = (1.35$ ns/pF$) C_L + 33$ ns		5.0	—	100	200	
$t_{TLH} = (0.60$ ns/pF$) C_L + 20$ ns		10	—	50	100	
$t_{TLH} = (0.40$ ns/pF$) C_L + 20$ ns		15	—	40	80	
Output Fall Time, All B-Series Gates	t_{THL}					ns
$t_{THL} = (1.35$ ns/pF$) C_L + 33$ ns		5.0	—	100	200	
$t_{THL} = (0.60$ ns/pF$) C_L + 20$ ns		10	—	50	100	
$t_{THL} = (0.40$ ns/pF$) C_L + 20$ ns		15	—	40	80	
Propagation Delay Time	t_{PLH}, t_{PHL}					ns
MC14001B, MC14011B only						
$t_{PLH}, t_{PHL} = (0.90$ ns/pF$) C_L + 80$ ns		5.0	—	125	250	
$t_{PLH}, t_{PHL} = (0.36$ ns/pF$) C_L + 32$ ns		10	—	50	100	
$t_{PLH}, t_{PHL} = (0.26$ ns/pF$) C_L + 27$ ns		15	—	40	80	
All Other 2, 3, and 4 Input Gates						
$t_{PLH}, t_{PHL} = (0.90$ ns/pF$) C_L + 115$ ns		5.0	—	160	300	
$t_{PLH}, t_{PHL} = (0.36$ ns/pF$) C_L + 47$ ns		10	—	65	130	
$t_{PLH}, t_{PHL} = (0.26$ ns/pF$) C_L + 37$ ns		15	—	50	100	
8-Input Gates (MC14068B, MC14078B)						
$t_{PLH}, t_{PHL} = (0.90$ ns/pF$) C_L + 155$ ns		5.0	—	200	350	
$t_{PLH}, t_{PHL} = (0.36$ ns/pF$) C_L + 62$ ns		10	—	80	150	
$t_{PLH}, t_{PHL} = (0.26$ ns/pF$) C_L + 47$ ns		15	—	60	110	

*The formulas given are for the typical characteristics only at 25°C.

#Data labelled "Typ" is not to be used for design purposes but is
intended as an indication of the IC's potential performance.

FIGURE 1 – SWITCHING TIME TEST CIRCUIT AND WAVEFORMS

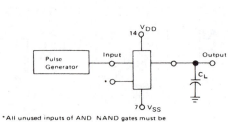

*All unused inputs of AND/NAND gates must be
connected to V_{DD}.
All unused inputs of OR/NOR gates must be
connected to V_{SS}.

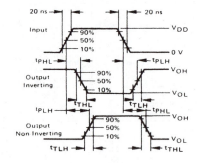

APPENDIX 8   CMOS Logic Gate Data Sheets   (Continued)

CMOS B-SERIES GATES

CIRCUIT SCHEMATIC
NOR, OR Gates

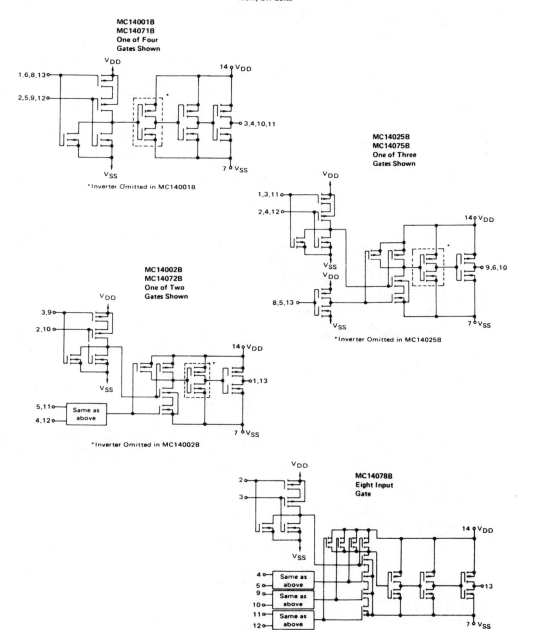

MC14001B
MC14071B
One of Four
Gates Shown

*Inverter Omitted in MC14001B

MC14025B
MC14075B
One of Three
Gates Shown

*Inverter Omitted in MC14025B

MC14002B
MC14072B
One of Two
Gates Shown

*Inverter Omitted in MC14002B

MC14078B
Eight Input
Gate

(continues)

APPENDIX 8 CMOS Logic Gate Data Sheets (Continued)

CMOS B-SERIES GATES

CIRCUIT SCHEMATICS
NAND, AND Gates

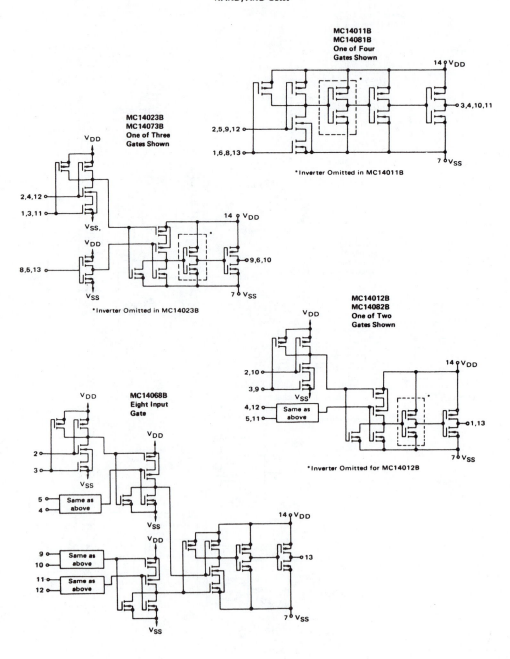

APPENDIX 8   CMOS Logic Gate Data Sheets   (Continued)

CMOS B-SERIES GATES

TYPICAL B-SERIES GATE CHARACTERISTICS

N-CHANNEL DRAIN CURRENT (SINK)	P-CHANNEL DRAIN CURRENT (SOURCE)

FIGURE 2 – V_{GS} = 5.0 Vdc   **FIGURE 3 – V_{GS} = –5.0 Vdc**

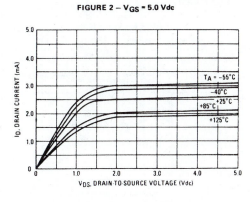
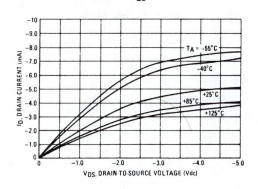

FIGURE 4 – V_{GS} = 10 Vdc   **FIGURE 5 – V_{GS} = –10 Vdc**

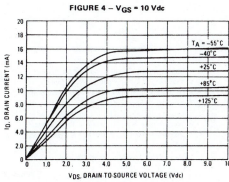
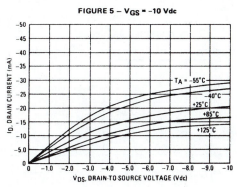

FIGURE 6 – V_{GS} = 15 Vdc   **FIGURE 7 – V_{GS} = –15 Vdc**

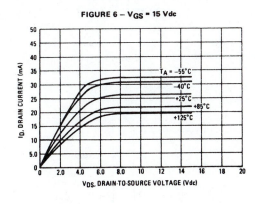
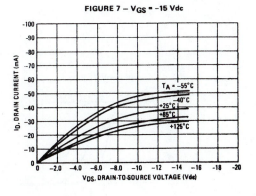

These typical curves are not guarantees, but are design aids.
Caution: The maximum rating for output current is 10 mA per pin.

(continues)

CMOS B-SERIES GATES

TYPICAL B-SERIES GATE CHARACTERISTICS (cont'd)

VOLTAGE TRANSFER CHARACTERISTICS

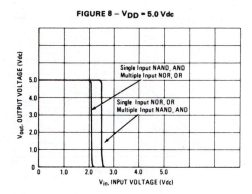

FIGURE 8 – V$_{DD}$ = 5.0 Vdc

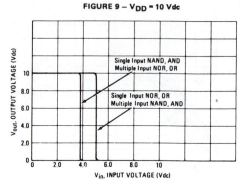

FIGURE 9 – V$_{DD}$ = 10 Vdc

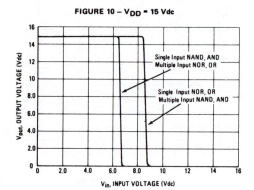

FIGURE 10 – V$_{DD}$ = 15 Vdc

DC NOISE MARGIN

The DC noise margin is defined as the input voltage range from an ideal "1" or "0" input level which does not produce output state change(s). The typical and guaranteed limit values of the input values V$_{IL}$ and V$_{IH}$ for the output(s) to be at a fixed voltage V$_O$ are given in the Electrical Characteristics table. V$_{IL}$ and V$_{IH}$ are presented graphically in Figure 11.

Guaranteed minimum noise margins for both the "1" and "0" levels =

    1.0 V with a 5.0 V supply
    2.0 V with a 10.0 V supply
    2.5 V with a 15.0 V supply

FIGURE 11 – DC NOISE IMMUNITY

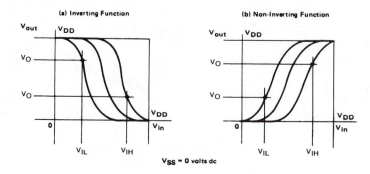

(a) Inverting Function   (b) Non-Inverting Function

V$_{SS}$ = 0 volts dc

MC14011B, MC14012B
See Page 6-5

MC14011UB, MC14012UB
See Page 6-14

MC14013B

CMOS SSI

(LOW-POWER COMPLEMENTARY MOS)

DUAL TYPE D FLIP-FLOP

DUAL TYPE D FLIP-FLOP

The MC14013B dual type D flip-flop is constructed with MOS P-channel and N-channel enhancement mode devices in a single monolithic structure. Each flip-flop has independent Data, (D), Direct Set, (S), Direct Reset, (R), and Clock (C) inputs and complementary outputs (Q and $\overline{Q}$). These devices may be used as shift register elements or as type T flip-flops for counter and toggle applications.

- Static Operation
- Diode Protection on All Inputs
- Supply Voltage Range = 3.0 Vdc to 18 Vdc
- Logic Edge-Clocked Flip-Flop Design
  Logic state is retained indefinitely with clock level either high or low; information is transferred to the output only on the positive-going edge of the clock pulse
- Capable of Driving Two Low-power TTL Loads or One Low-power Schottky TTL Load Over the Rated Temperature Range
- Pin-for-Pin Replacement for CD4013B

L SUFFIX
CERAMIC PACKAGE
CASE 632

P SUFFIX
PLASTIC PACKAGE
CASE 646

ORDERING INFORMATION

A Series: −55°C to +125°C
MC14XXXBAL (Ceramic Package Only)

C Series: −40°C to +85°C
MC14XXXBCP (Plastic Package)
MC14XXXBCL (Ceramic Package)

MAXIMUM RATINGS* (Voltages Referenced to V_{SS})

Symbol	Parameter	Value	Unit
V_{DD}	DC Supply Voltage	−0.5 to +18.0	V
V_{in}, V_{out}	Input or Output Voltage (DC or Transient)	−0.5 to V_{DD} +0.5	V
I_{in}, I_{out}	Input or Output Current (DC or Transient), per Pin	±10	mA
P_D	Power Dissipation, per Package†	500	mW
T_{stg}	Storage Temperature	−65 to +150	°C
T_L	Lead Temperature (8-Second Soldering)	260	°C

*Maximum Ratings are those values beyond which damage to the device may occur.
†Temperature Derating: Plastic "P" Package: −12mW/°C from 65°C to 85°C
  Ceramic "L" Package: −12mW/°C from 100°C to 125°C

BLOCK DIAGRAM

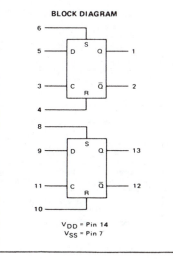

V_{DD} = Pin 14
V_{SS} = Pin 7

TRUTH TABLE

INPUTS				OUTPUTS		
CLOCK†	DATA	RESET	SET	Q	$\overline{Q}$	
⌐	0	0	0	0	1	
⌐	1	0	0	1	0	
⌐	X	0	0	Q	$\overline{Q}$	No Change
X	X	1	0	0	1	
X	X	0	1	1	0	
X	X	1	1	1	1	

X = Don't Care
† = Level Change

691

(continues)

APPENDIX 8 CMOS Logic Gate Data Sheets (Continued)

MC14013B

ELECTRICAL CHARACTERISTICS (Voltages Referenced to V_{SS})

Characteristic	Symbol	V_{DD} Vdc	T_{low}* Min	T_{low}* Max	25°C Min	25°C Typ #	25°C Max	T_{high}* Min	T_{high}* Max	Unit
Output Voltage "0" Level	V_{OL}	5.0	—	0.05	—	0	0.05	—	0.05	Vdc
$V_{in} = V_{DD}$ or 0		10	—	0.05	—	0	0.05	—	0.05	
		15	—	0.05	—	0	0.05	—	0.05	
"1" Level	V_{OH}	5.0	4.95	—	4.95	5.0	—	4.95	—	Vdc
$V_{in} = 0$ or V_{DD}		10	9.95	—	9.95	10	—	9.95	—	
		15	14.95	—	14.95	15	—	14.95	—	
Input Voltage "0" Level	V_{IL}									Vdc
($V_O = 4.5$ or 0.5 Vdc)		5.0	—	1.5	—	2.25	1.5	—	1.5	
($V_O = 9.0$ or 1.0 Vdc)		10	—	3.0	—	4.50	3.0	—	3.0	
($V_O = 13.5$ or 1.5 Vdc)		15	—	4.0	—	6.75	4.0	—	4.0	
"1" Level	V_{IH}									Vdc
($V_O = 0.5$ or 4.5 Vdc)		5.0	3.5	—	3.5	2.75	—	3.5	—	
($V_O = 1.0$ or 9.0 Vdc)		10	7.0	—	7.0	5.50	—	7.0	—	
($V_O = 1.5$ or 13.5 Vdc)		15	11.0	—	11.0	8.25	—	11.0	—	
Output Drive Current (AL Device)	I_{OH}									mAdc
($V_{OH} = 2.5$ Vdc) Source		5.0	−3.0	—	−2.4	−4.2	—	−1.7	—	
($V_{OH} = 4.6$ Vdc)		5.0	−0.64	—	−0.51	−0.88	—	−0.36	—	
($V_{OH} = 9.5$ Vdc)		10	−1.6	—	−1.3	−2.25	—	−0.9	—	
($V_{OH} = 13.5$ Vdc)		15	−4.2	—	−3.1	−8.8	—	−2.4	—	
($V_{OL} = 0.4$ Vdc) Sink	I_{OL}	5.0	0.64	—	0.51	0.88	—	0.36	—	mAdc
($V_{OL} = 0.5$ Vdc)		10	1.6	—	1.3	2.25	—	0.9	—	
($V_{OL} = 1.5$ Vdc)		15	4.2	—	3.4	8.8	—	2.4	—	
Output Drive Current (CL/CP Device)	I_{OH}									mAdc
($V_{OH} = 2.5$ Vdc) Source		5.0	−2.5	—	−2.1	−4.2	—	−1.7	—	
($V_{OH} = 4.6$ Vdc)		5.0	−0.52	—	−0.44	−0.88	—	−0.36	—	
($V_{OH} = 9.5$ Vdc)		10	−1.3	—	−1.1	−2.25	—	−0.9	—	
($V_{OH} = 13.5$ Vdc)		15	−3.6	—	−3.0	−8.8	—	−2.4	—	
($V_{OL} = 0.4$ Vdc) Sink	I_{OL}	5.0	0.52	—	0.44	0.88	—	0.36	—	mAdc
($V_{OL} = 0.5$ Vdc)		10	1.3	—	1.1	2.25	—	0.9	—	
($V_{OL} = 1.5$ Vdc)		15	3.6	—	3.0	8.8	—	2.4	—	
Input Current (AL Device)	I_{in}	15	—	±0.1	—	±0.00001	±0.1	—	±1.0	μAdc
Input Current (CL/CP Device)	I_{in}	15	—	±0.3	—	±0.00001	±0.3	—	±1.0	μAdc
Input Capacitance ($V_{in} = 0$)	C_{in}	—	—	—	—	5.0	7.5	—	—	pF
Quiescent Current (AL Device)	I_{DD}	5.0	—	1.0	—	0.002	1.0	—	30	μAdc
(Per Package)		10	—	2.0	—	0.004	2.0	—	60	
		15	—	4.0	—	0.006	4.0	—	120	
Quiescent Current (CL/CP Device)	I_{DD}	5.0	—	4.0	—	0.002	4.0	—	30	μAdc
(Per Package)		10	—	8.0	—	0.004	8.0	—	60	
		15	—	16	—	0.006	16	—	120	
Total Supply Current**† (Dynamic plus Quiescent, Per Package) ($C_L = 50$ pF on all outputs, all buffers switching)	I_T	5.0			$I_T = (0.75\,\mu A/kHz)\,f + I_{DD}$					μAdc
		10			$I_T = (1.5\,\,\mu A/kHz)\,f + I_{DD}$					
		15			$I_T = (2.3\,\,\mu A/kHz)\,f + I_{DD}$					

*T_{low} = −55°C for AL Device, −40°C for CL/CP Device.
T_{high} = +125°C for AL Device, +85°C for CL/CP Device.

#Data labelled "Typ" is not to be used for design purposes but is intended as an indication of the IC's potential performance.

**The formulas given are for the typical characteristics only at 25°C.

†To calculate total supply current at loads other than 50 pF:

$$I_T(C_L) = I_T(50\ pF) + (C_L - 50)\ Vfk$$

where: I_T is in μA (per package), C_L in pF, $V = (V_{DD} - V_{SS})$ in volts, f in kHz is input frequency, and k = 0.002.

This device contains protection circuitry to guard against damage due to high static voltages or electric fields. However, precautions must be taken to avoid applications of any voltage higher than maximum rated voltages to this high-impedance circuit. For proper operation, V_{in} and V_{out} should be constrained to the range $V_{SS} \leq (V_{in}$ or $V_{out}) \leq V_{DD}$.
Unused inputs must always be tied to an appropriate logic voltage level (e.g., either V_{SS} or V_{DD}). Unused outputs must be left open.

PIN ASSIGNMENT

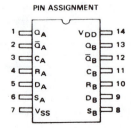

1	Q_A	V_{DD}	14
2	$\overline{Q}_A$	Q_B	13
3	C_A	$\overline{Q}_B$	12
4	R_A	C_B	11
5	D_A	R_B	10
6	S_A	D_B	9
7	V_{SS}	S_B	8

MC14013B

SWITCHING CHARACTERISTICS* (C_L = 50 pF, T_A = 25°C)

Characteristic	Symbol	V_{DD}	Min	Typ #	Max	Unit
Output Rise and Fall Time	t_{TLH}, t_{THL}			100	200	ns
t_{TLH}, t_{THL} = (1.5 ns/pF) C_L + 25 ns		5.0	—	100	200	
t_{TLH}, t_{THL} = (0.75 ns/pF) C_L + 12.5 ns		10	—	50	100	
t_{TLH}, t_{THL} = (0.55 ns/pF) C_L + 9.5 ns		15	—	40	80	
Propagation Delay Time	t_{PLH} t_{PHL}					ns
Clock to Q, $\overline{Q}$						
t_{PLH}, t_{PHL} = (1.7 ns/pF) C_L + 90 ns		5.0	—	175	350	
t_{PLH}, t_{PHL} = (0.66 ns/pF) C_L + 42 ns		10	—	75	150	
t_{PLH}, t_{PHL} = (0.5 ns/pF) C_L + 25 ns		15	—	50	100	
Set to Q, $\overline{Q}$						
t_{PLH}, t_{PHL} = (1.7 ns/pF) C_L + 90 ns		5.0	—	175	350	
t_{PLH}, t_{PHL} = (0.66 ns/pF) C_L + 42 ns		10	—	75	150	
t_{PLH}, t_{PHL} = (0.5 ns/pF) C_L + 25 ns		15	—	50	100	
Reset to Q, $\overline{Q}$						
t_{PLH}, t_{PHL} = (1.7 ns/pF) C_L + 265 ns		5.0	—	350	450	
t_{PLH}, t_{PHL} = (0.66 ns/pF) C_L + 67 ns		10	—	100	200	
t_{PLH}, t_{PHL} = (0.5 ns/pF) C_L + 50 ns		15	—	75	150	
Setup Times**	t_{su}	5.0	40	20	—	ns
		10	20	10	—	
		15	15	7.5	—	
Hold Times**	t_h	5.0	40	20	—	ns
		10	20	10	—	
		15	15	7.5	—	
Clock Pulse Width	t_{WL}, t_{WH}	5.0	250	125	—	ns
		10	100	50	—	
		15	70	35	—	
Clock Pulse Frequency	f_{cl}	5.0	—	4.0	2.0	MHz
		10	—	10	5.0	
		15	—	14	7.0	
Clock Pulse Rise and Fall Time	t_{TLH} t_{THL}	5.0	—	—	15	µs
		10	—	—	5.0	
		15	—	—	4.0	
Set and Reset Pulse Width	t_{WL}, t_{WH}	5.0	250	125	—	ns
		10	100	50	—	
		15	70	35	—	
Removal Times	t_{rem}					ns
Set		5	80	0	—	
		10	45	5	—	
		15	35	5	—	
Reset		5	50	−35	—	
		10	30	−10	—	
		15	25	−5	—	

*The formulas given are for the typical characteristics only at 25°C.

#Data labelled "Typ" is not to be used for design purposes but is intended as an indication of the IC's potential performance.

**Data must be valid for 250 ns with a 5 V supply, 100 ns with 10 V, and 70 ns with 15 V.

LOGIC DIAGRAM
(1/2 of Device Shown)

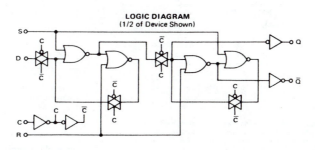

APPENDIX 9   ECL Logic Gate Data Sheets

MC10H100

MECL 10KH HIGH-SPEED EMITTER-COUPLED LOGIC

The MC10H100 is a MECL 10KH part which is a functional/pinout duplication of the standard MECL 10K family part, with 100% improvement in propagation delay, and no increases in power-supply current.

- Propagation Delay, 1.0 ns Typical
- 25 mW Typ/Gate (No Load)
- Improved Noise Margin 150 mV (Over Operating Voltage and Temperature Range)
- Voltage Compensated
- MECL 10K-Compatible

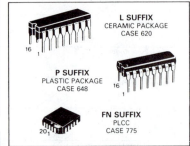

L SUFFIX
CERAMIC PACKAGE
CASE 620

P SUFFIX
PLASTIC PACKAGE
CASE 648

FN SUFFIX
PLCC
CASE 775

MAXIMUM RATINGS

Characteristic	Symbol	Rating	Unit
Power Supply (V_{CC} = 0)	V_{EE}	−8.0 to 0	Vdc
Input Voltage (V_{CC} = 0)	V_I	0 to V_{EE}	Vdc
Output Current — Continuous — Surge	I_{out}	50 100	mA
Operating Temperature Range	T_A	0 to +75	°C
Storage Temperature Range — Plastic — Ceramic	T_{stg}	−55 to +150 −55 to +165	°C

ELECTRICAL CHARACTERISTICS (V_{EE} = −5.2 V ±5%) (See Note)

Characteristic	Symbol	0° Min	0° Max	25° Min	25° Max	75° Min	75° Max	Unit
Power Supply Current	I_E	—	29	—	26	—	29	mA
Input Current High Pin 9 All Other Inputs	I_{inH}	— —	900 500	— —	560 310	— —	560 310	μA
Input Current Low	I_{inL}	0.5	—	0.5	—	0.3	—	μA
High Output Voltage	V_{OH}	−1.02	−0.84	−0.98	−0.81	−0.92	−0.735	Vdc
Low Output Voltage	V_{OL}	−1.95	−1.63	−1.95	−1.63	−1.95	−1.60	Vdc
High Input Voltage	V_{IH}	−1.17	−0.84	−1.13	−0.81	−1.07	−0.735	Vdc
Low Input Voltage	V_{IL}	−1.95	−1.48	−1.95	−1.48	−1.95	−1.45	Vdc

AC PARAMETERS

	Symbol	0° Min	0° Max	25° Min	25° Max	75° Min	75° Max	Unit
Propagation Delay Pin 9 Only Exclude Pin 9	t_{pd}	0.65 0.4	1.6 1.3	0.7 0.45	1.7 1.35	0.7 0.5	1.8 1.5	ns
Rise Time	t_r	0.5	2.0	0.5	2.1	0.5	2.2	ns
Fall Time	t_f	0.5	2.0	0.5	2.1	0.5	2.2	ns

NOTE:
Each MECL 10KH series circuit has been designed to meet the dc specifications shown in the test table, after thermal equilibrium has been established. The circuit is in a test socket or mounted on a printed circuit board and transverse air flow greater than 500 lfpm is maintained. Outputs are terminated through a 50-ohm resistor to −2.0 volts.

**Quad 2-Input
NOR Gate With Strobe**

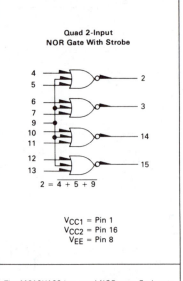

$2 = \overline{4 + 5 + 9}$

V_{CC1} = Pin 1
V_{CC2} = Pin 16
V_{EE} = Pin 8

The MC10H100 is a quad NOR gate. Each gate has 3 inputs, two of which are independent and one of which is tied common to all four gates.

APPENDIX 9 ECL Logic Gate Data Sheets (Continued)

MC10H104

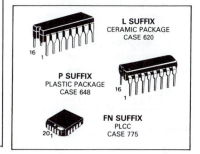

QUAD 2-INPUT AND GATE

The MC10H104 is a quad 2-input AND gate. One of the gates has both AND/NAND outputs available. This MECL 10KH part is a functional/pinout duplication of the standard MECL 10K family part, with 100% improvement in propagation delay, and no increase in power-supply current.

- Propagation Delay, 1.0 ns Typical
- Power Dissipation 25 mW/Gate (same as MECL 10K)
- Improved Noise Margin 150 mV (Over Operating Voltage and Temperature Range)
- Voltage Compensated
- MECL 10K-Compatible

L SUFFIX
CERAMIC PACKAGE
CASE 620

P SUFFIX
PLASTIC PACKAGE
CASE 648

FN SUFFIX
PLCC
CASE 775

MAXIMUM RATINGS

Characteristic	Symbol	Rating	Unit
Power Supply (V_{CC} = 0)	V_{EE}	–8.0 to 0	Vdc
Input Voltage (V_{CC} = 0)	V_I	0 to V_{EE}	Vdc
Output Current — Continuous — Surge	I_{out}	50 100	mA
Operating Temperature Range	T_A	0–75	°C
Storage Temperature Range — Plastic — Ceramic	T_{stg}	–55 to 150 –55 to 165	°C °C

ELECTRICAL CHARACTERISTICS (V_{EE} = – 5.2 V ±5%) (See Note)

Characteristic	Symbol	0° Min	0° Max	25° Min	25° Max	75° Min	75° Max	Unit
Power Supply Current	I_E	—	39	—	35	—	39	mA
Input Current High	I_{inH}	—	425	—	265	—	265	μA
Input Current Low	I_{inL}	0.5	—	0.5	—	0.3	—	μA
High Output Voltage	V_{OH}	– 1.02	– 0.84	– 0.98	– 0.81	– 0.92	– 0.735	Vdc
Low Output Voltage	V_{OL}	– 1.95	– 1.63	– 1.95	– 1.63	– 1.95	– 1.60	Vdc
High Input Voltage	V_{IH}	– 1.17	– 0.84	– 1.13	– 0.81	– 1.07	– 0.735	Vdc
Low Input Voltage	V_{IL}	– 1.95	– 1.48	– 1.95	– 1.48	– 1.95	– 1.45	Vdc

AC PARAMETERS

		0° Min	0° Max	25° Min	25° Max	75° Min	75° Max	
Propagation Delay	t_{pd}	0.4	1.6	0.45	1.75	0.45	1.9	ns
Rise Time	t_r	0.5	1.6	0.5	1.7	0.5	1.8	ns
Fall Time	t_f	0.5	1.6	0.5	1.7	0.5	1.8	ns

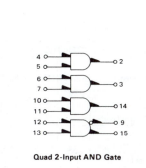

Quad 2-Input AND Gate

NOTE:
Each MECL 10KH series circuit has been designed to meet the dc specifications shown in the test table, after thermal equilibrium has been established. The circuit is in a test socket or mounted on a printed circuit board and transverse air flow greater then 500 linear fpm is maintained. Outputs are terminated through a 50-ohm resistor to –2.0 volts.

(continues)

APPENDIX 9   ECL Logic Gate Data Sheets   (Continued)

MC10H105

TRIPLE 2-3-2-INPUT OR/NOR GATE

The MC10H105 is a triple 2-3-2-input OR/NOR gate. This MECL 10KH part is a functional/pinout duplication of the standard MECL 10K family part, with 100% improvement in propagation delay, and no increases in power-supply current.

- Propagation Delay, 1.0 ns Typical
- Power Dissipation 25 mW/Gate (same as MECL 10K)
- Improved Noise Margin 150 mV (Over Operating Voltage and Temperature Range)
- Voltage Compensated
- MECL 10K-Compatible

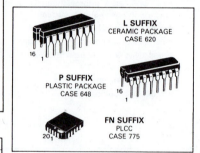

L SUFFIX
CERAMIC PACKAGE
CASE 620

P SUFFIX
PLASTIC PACKAGE
CASE 648

FN SUFFIX
PLCC
CASE 775

MAXIMUM RATINGS

Characteristic	Symbol	Rating	Unit
Power Supply (V_{CC} = 0)	V_{EE}	-8.0 to 0	Vdc
Input Voltage (V_{CC} = 0)	V_I	0 to V_{EE}	Vdc
Output Current — Continuous — Surge	I_{out}	50 100	mA
Operating Temperature Range	T_A	0-75	°C
Storage Temperature Range — Plastic — Ceramic	T_{stg}	-55 to 150 -55 to 165	°C °C

ELECTRICAL CHARACTERISTICS (V_{EE} = -5.2 V ±5%) (See Note)

Characteristic	Symbol	0° Min	0° Max	25° Min	25° Max	75° Min	75° Max	Unit
Power Supply Current	I_E	—	23	—	21	—	23	mA
Input Current High	I_{inH}	—	425	—	265	—	265	μA
Input Current Low	I_{inL}	0.5	—	0.5	—	0.3	—	μA
High Output Voltage	V_{OH}	-1.02	-0.84	-0.98	-0.81	-0.92	-0.735	Vdc
Low Output Voltage	V_{OL}	-1.95	-1.63	-1.95	-1.63	-1.95	-1.60	Vdc
High Input Voltage	V_{IH}	-1.17	-0.84	-1.13	-0.81	-1.07	-0.735	Vdc
Low Input Voltage	V_{IL}	-1.95	-1.48	-1.95	-1.48	-1.95	-1.45	Vdc

AC PARAMETERS

Propagation Delay	t_{pd}	0.4	1.2	0.4	1.2	0.4	1.3	ns
Rise Time	t_r	0.5	1.5	0.5	1.6	0.5	1.7	ns
Fall Time	t_f	0.5	1.5	0.5	1.6	0.5	1.7	ns

NOTE:
Each MECL 10KH series circuit has been designed to meet the dc specifications shown in the test table, after thermal equilibrium has been established. The circuit is in a test socket or mounted on a printed circuit board and transverse air flow greater then 500 linear fpm is maintained. Outputs are terminated through a 50-ohm resistor to -2.0 volts.

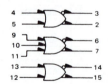

Triple 2-3-2 Input OR/NOR Gate

APPENDIX 9   ECL Logic Gate Data Sheets   (Continued)

TRIPLE 2-INPUT EXCLUSIVE "OR"/EXCLUSIVE "NOR"

The MC10H107 is a triple 2-input exclusive OR/NOR gate. This MECL 10KH part is a functional/pinout duplication of the standard MECL 10K family part, with 100% improvement in propagation delay, and no increase in power-supply current.

- Propagation Delay, 1.0 ns Typical
- Power Dissipation 35 mW/Gate Typical (same as MECL 10K)
- Improved Noise Margin 150 mV (Over Operating Voltage and Temperature Range)
- Voltage Compensated
- MECL 10K-Compatible

MC10H107

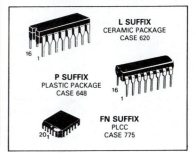

L SUFFIX
CERAMIC PACKAGE
CASE 620

P SUFFIX
PLASTIC PACKAGE
CASE 648

FN SUFFIX
PLCC
CASE 775

MAXIMUM RATINGS

Characteristic	Symbol	Rating	Unit
Power Supply (V_{CC} = 0)	V_{EE}	–8.0 to 0	Vdc
Input Voltage (V_{CC} = 0)	V_I	0 to V_{EE}	Vdc
Output Current — Continuous — Surge	I_{out}	50 100	mA
Operating Temperature Range	T_A	0–75	°C
Storage Temperature Range — Plastic — Ceramic	T_{stg}	–55 to 150 –55 to 165	°C °C

ELECTRICAL CHARACTERISTICS (V_{EE} = – 5.2 V ±5%) (See Note)

Characteristic	Symbol	0° Min	0° Max	25° Min	25° Max	75° Min	75° Max	Unit
Power Supply Current	I_E	—	31	—	28	—	31	mA
Input Current High	I_{inH}	—	425	—	265	—	265	µA
Input Current Low	I_{inL}	0.5	—	0.5	—	0.3	—	µA
High Output Voltage	V_{OH}	– 1.02	– 0.84	– 0.98	– 0.81	– 0.92	– 0.735	Vdc
Low Output Voltage	V_{OL}	– 1.95	– 1.63	– 1.95	– 1.63	– 1.95	– 1.60	Vdc
High Input Voltage	V_{IH}	– 1.17	– 0.84	– 1.13	– 0.81	– 1.07	– 0.735	Vdc
Low Input Voltage	V_{IL}	– 1.95	– 1.48	– 1.95	– 1.48	– 1.95	– 1.45	Vdc

AC PARAMETERS

Propagation Delay	t_{pd}	0.4	1.5	0.4	1.6	0.4	1.7	ns
Rise Time	t_r	0.5	1.5	0.5	1.6	0.5	1.7	ns
Fall Time	t_f	0.5	1.5	0.5	1.6	0.5	1.7	ns

NOTE:
Each MECL 10KH series circuit has been designed to meet the dc specifications shown in the test table, after thermal equilibrium has been established. The circuit is in a test socket or mounted on a printed circuit board and transverse air flow greater then 500 linear fpm is maintained. Outputs are terminated through a 50-ohm resistor to –2.0 volts.

**Triple 2-Input
Exclusive OR/NOR Gate**

(continues)

APPENDIX 9   ECL Logic Gate Data Sheets   (Continued)

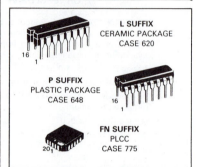

MC10131

MECL 10K SERIES

DUAL TYPE D MASTER-SLAVE FLIP-FLOP

DUAL TYPE D MASTER-SLAVE FLIP-FLOP

The MC10131 is a dual master-slave type D flip-flop. Asynchronous Set (S) and Reset (R) override Clock (C_C) and $\overline{\text{Clock Enable}}$ (C_E) inputs. Each flip-flop may be clocked separately by holding the common clock in the low state and using the enable inputs for the clocking function. If the common clock is to be used to clock the flip-flop, the $\overline{\text{Clock Enable}}$ inputs must be in the low state. In this case, the enable inputs perform the function of controlling the common clock.

The output states of the flip-flop change on the positive transition of the clock. A change in the information present at the data (D) input will not affect the output information at any other time due to master slave construction.

P_D = 235 mW typ/pkg (No Load)
f_{Tog} = 160 MHz typ
t_{pd} = 3.0 ns typ
t_r, t_f = 2.5 ns typ (20%–80%)

L SUFFIX
CERAMIC PACKAGE
CASE 620

P SUFFIX
PLASTIC PACKAGE
CASE 648

FN SUFFIX
PLCC
CASE 775

LOGIC DIAGRAM

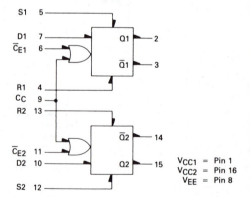

V_{CC1} = Pin 1
V_{CC2} = Pin 16
V_{EE} = Pin 8

PIN ASSIGNMENT

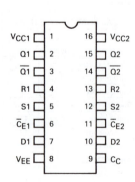

V_{CC1}	1	16	V_{CC2}
$Q1$	2	15	$Q2$
$\overline{Q1}$	3	14	$\overline{Q2}$
$R1$	4	13	$R2$
$S1$	5	12	$S2$
$\overline{C}_{E1}$	6	11	$\overline{C}_{E2}$
$D1$	7	10	$D2$
V_{EE}	8	9	C_C

CLOCKED TRUTH TABLE

C	D	Q_{n+1}
L	ϕ	Q_n
H	L	L
H	H	H

ϕ = Don't Care
$C = C_E + C_C$.
A clock H is a clock transition
from a low to a high state.

R-S TRUTH TABLE

R	S	Q_{n+1}
L	L	Q_n
L	H	H
H	L	L
H	H	N.D.

N.D. = Not Defined

MC10131
ELECTRICAL CHARACTERISTICS

Each MECL 10,000 series circuit has been designed to meet the dc specifications shown in the test table, after thermal equilibrium has been established. The circuit is in a test socket or mounted on a printed circuit board and transverse air flow greater than 500 linear fpm is maintained. Outputs are terminated through a 50-ohm resistor to −2.0 volts. Test procedures are shown for only one input, or for one set of input conditions. Other inputs tested in the same manner.

TEST VOLTAGE VALUES (Volts)

@ Test Temperature	V_{IH} max	V_{IL} min	V_{IHA} min	V_{ILA} max	V_{EE}	(V$_{CC}$) Gnd
−30°C	−0.890	−1.890	−1.205	−1.500	−5.2	1, 16
+25°C	−0.810	−1.850	−1.105	−1.475	−5.2	1, 16
+85°C	−0.700	−1.825	−1.035	−1.440	−5.2	1, 16

MC10131 Test Limits / Voltage Applied to Pins Listed Below

Characteristic	Symbol	Pin Under Test	−30°C Min	−30°C Max	+25°C Min	+25°C Typ	+25°C Max	+85°C Min	+85°C Max	Unit	V_{IH} max	V_{IL} min	V_{IHA} min	V_{ILA} max	V_{EE}	(V$_{CC}$) Gnd
Power Supply Drain Current	I_E	8	—	62	45	—	56	—	62	mAdc					8	1, 16
Input Current	I_{inH}	4	—	525	—	—	330	—	330	μAdc	4				8	1, 16
		5	—	525	—	—	330	—	330		5				8	1, 16
		6	—	350	—	—	220	—	220		6				8	1, 16
		7	—	390	—	—	245	—	245		7				8	1, 16
		9	—	425	—	—	265	—	265		9				8	1, 16
Input Leakage Current	I_{inL}	4,5,*	0.5	—	0.5	—	—	0.3	—	μAdc		*			8	1, 16
		6,7,9*	0.5	—	0.5	—	—	0.3	—	μAdc		*			8	1, 16
Logic "1" Output Voltage	V_{OH}	2	−1.060	−0.890	−0.960	—	−0.810	−0.890	−0.700	Vdc	5				8	1, 16
		2†	−1.060	−0.890	−0.960	—	−0.810	−0.890	−0.700	Vdc	7				8	1, 16
Logic "0" Output Voltage	V_{OL}	3	−1.890	−1.675	−1.850	—	−1.650	−1.825	−1.615	Vdc	5				8	1, 16
		3†	−1.890	−1.675	−1.850	—	−1.650	−1.825	−1.615	Vdc	7				8	1, 16
Logic "1" Threshold Voltage	V_{OHA}	2	−1.080	—	−0.980	—	—	−0.910	—	Vdc			5		8	1, 16
		2†	−1.080	—	−0.980	—	—	−0.910	—	Vdc			7	9	8	1, 16
Logic "0" Threshold Voltage	V_{OLA}	3	—	−1.655	—	—	−1.630	—	−1.595	Vdc			5		8	1, 16
		3†	—	−1.655	—	—	−1.630	—	−1.595	Vdc			7	9	8	1, 16
											+1.11 Vdc		Pulse In	Pulse Out	**−3.2 Vdc**	**+2.0 Vdc**
Switching Times — Clock Input Propagation Delay	t9+2−, t9+2+, t6+2−, t6+2+	2, 2, 2, 2	1.7	4.6	1.8	3.0	4.5	1.8	5.0	ns	7, 7		9, 9, 6, 6	2, 2, 2, 2	8	1, 16
Rise Time (20 to 80%)	t2+	2	1.0	—	1.1	2.5	4.5	1.1	4.9	ns	7		9	2		
Fall Time (20 to 80%)	t2−	2	1.0	—	1.1	2.5	4.5	1.1	4.9	ns	7		9	2		
Set Input Propagation Delay	t5+2+, t12+15+, t5+3−, t12+14−	2, 15, 3, 14	1.7	4.4	1.8	2.8	4.3	1.8	4.8	ns	—, 6, —, 9		5, 12, 5, 12	2, 15, 3, 14	8	1, 16
Reset Input Propagation Delay	t4+2−, t13+15−, t4+3−, t13+14−	2, 15, 3, 14	1.7	4.4	1.8	2.8	4.3	1.8	4.8	ns	—, 6, —, 9		4, 13, 4, 13	2, 15, 3, 14	8	1, 16
Setup Time	t_{setup}	7	2.5	—	2.5	—	—	2.5	—	ns			6, 7	2	8	1, 16
Hold Time	t_{hold}	7	1.5	—	1.5	—	—	1.5	—	ns			6, 7	2	8	1, 16
Toggle Frequency (Max)	f_{tog}	2	125	—	125	160	—	125	—	MHz			6	2	8	1, 16

* Individually test each input; apply V_{IL} min to pin under test.

† Output level to be measured after a clock pulse has been applied to the $\overline{C}_E$ input (pin 6) ⎍ V_{IH} max / V_{IL} min

(continues)

APPENDIX 9   ECL Logic Gate Data Sheets   (Continued)

MC10101

MECL 10K *SERIES*

QUAD OR/NOR GATE

QUAD OR/NOR GATE

The MC10101 is a quad 2-input OR/NOR gate with one input from each gate common to pin 12.

P_D = 25 mW typ/gate (No Load)
t_{pd} = 2.0 ns typ
t_r, t_f = 2.0 ns typ (20%–80%)

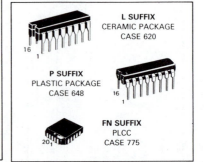

L SUFFIX
CERAMIC PACKAGE
CASE 620

P SUFFIX
PLASTIC PACKAGE
CASE 648

FN SUFFIX
PLCC
CASE 775

LOGIC DIAGRAM

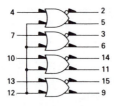

V_{CC1} = Pin 1
V_{CC2} = Pin 16
V_{EE} = Pin 8

PIN ASSIGNMENT

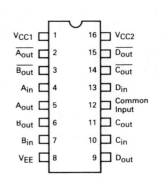

V_{CC1}	1	16	V_{CC2}
$\overline{A}_{out}$	2	15	$\overline{D}_{out}$
$\overline{B}_{out}$	3	14	$\overline{C}_{out}$
A_{in}	4	13	D_{in}
A_{out}	5	12	Common Input
B_{out}	6	11	C_{out}
B_{in}	7	10	C_{in}
V_{EE}	8	9	D_{out}

MC10101

ELECTRICAL CHARACTERISTICS

Each MECL 10,000 series circuit has been designed to meet the dc specifications shown in the test table, after thermal equilibrium has been established. The circuit is in a test socket or mounted on a printed circuit board and transverse air flow greater than 500 linear fpm is maintained. Outputs are terminated through a 50-ohm resistor to −2.0 volts. Test procedures are shown for only one gate. The other gates are tested in the same manner.

MC10101 Test Limits

Characteristic	Symbol	Pin Under Test	−30°C Min	−30°C Max	+25°C Min	+25°C Typ	+25°C Max	+85°C Min	+85°C Max	Unit
Power Supply Drain Current	I_E	8	—	29	—	20	26	—	29	mAdc
Input Current	I_{inH}	4	—	425	—	—	265	—	265	µAdc
		12	—	850	—	—	535	—	535	µAdc
	I_{inL}	4	0.5	—	0.5	—	—	0.3	—	µAdc
		12	0.5	—	0.5	—	—	0.3	—	µAdc
Logic "1" Output Voltage	V_{OH}	5	−1.060	−0.890	−0.960	—	−0.810	−0.890	−0.700	Vdc
		5	−1.060	−0.890	−0.960	—	−0.810	−0.890	−0.700	→
		2	−1.060	−0.890	−0.960	—	−0.810	−0.890	−0.700	
		2	−1.060	−0.890	−0.960	—	−0.810	−0.890	−0.700	
Logic "0" Output Voltage	V_{OL}	5	−1.890	−1.675	−1.850	—	−1.650	−1.825	−1.615	Vdc
		5	−1.890	−1.675	−1.850	—	−1.650	−1.825	−1.615	→
		2	−1.890	−1.675	−1.850	—	−1.650	−1.825	−1.615	
		2	−1.890	−1.675	−1.850	—	−1.650	−1.825	−1.615	
Logic "1" Threshold Voltage	V_{OHA}	5	−1.080	—	−0.980	—	—	−0.910	—	Vdc
		5	−1.080	—	−0.980	—	—	−0.910	—	→
		2	−1.080	—	−0.980	—	—	−0.910	—	
		2	−1.080	—	−0.980	—	—	−0.910	—	
Logic "0" Threshold Voltage	V_{OLA}	5	—	−1.655	—	—	−1.630	—	−1.595	Vdc
		5	—	−1.655	—	—	−1.630	—	−1.595	→
		2	—	−1.655	—	—	−1.630	—	−1.595	
		2	—	−1.655	—	—	−1.630	—	−1.595	
Switching Times (50-ohm load) Propagation Delay	t_{4+2-}	2	1.0	3.1	1.0	2.0	2.9	1.0	3.3	ns
	t_{4-2+}	2	→	→	→	→	→	→	→	
	t_{4+5+}	5	→	→	→	→	→	→	→	
	t_{4-5-}	5	→	→	→	→	→	→	→	
Rise Time (20 to 80%)	t_{2+}	2	1.1	3.6	1.1	—	3.3	1.1	3.7	→
	t_{5+}	5	→	→	→	—	→	→	→	
Fall Time (20 to 80%)	t_{2-}	2	1.1	3.6	1.1	—	3.3	1.1	3.7	→
	t_{5-}	5	→	→	→	—	→	→	→	

TEST VOLTAGE VALUES (Volts)

@ Test Temperature	V_{IH} max	V_{IL} min	V_{IHA} min	V_{ILA} max	V_{EE}	(Vcc) Gnd
−30°C	−0.890	−1.890	−1.205	−1.500	−5.2	
+25°C	−0.810	−1.850	−1.105	−1.475	−5.2	
+85°C	−0.700	−1.825	−1.035	−1.440	−5.2	

TEST VOLTAGE APPLIED TO PINS LISTED BELOW

Characteristic	V_{IH} max	V_{IL} min	V_{IHA} min	V_{ILA} max	V_{EE}	(Vcc) Gnd
I_E	—	—	—	—	8	1,16
I_{inH}	4	—	—	—	8	1,16
	12	—	—	—	8	1,16
I_{inL}	—	4	—	—	8	1,16
	—	12	—	—	8	1,16
V_{OH}	12	—	—	—	8	1,16
	4	—	—	—	→	→
	—	12	—	—		
	—	4	—	—		
V_{OL}	12	—	—	—	8	1,16
	4	—	—	—	→	→
	—	12	—	—		
	—	4	—	—		
V_{OHA}	—	—	12	—	8	1,16
	—	—	4	12	→	→
	—	—	—	4		
V_{OLA}	—	—	—	12	8	1,16
	—	—	12	4	→	→
	—	—	4	—		
Propagation Delay	—	—	Pulse In 4	Pulse Out	−3.2 V	+2.0 V
Rise Time / Fall Time	—	—	→	→	8	1,16

(continues)

MC1670   MASTER-SLAVE FLIP-FLOP

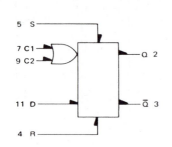

5 S

7 C1
9 C2
        Q 2

11 D
        $\bar{Q}$ 3

4 R

Master slave construction renders the MC1670 relatively insensitive to the shape of the clock waveform, since only the voltage levels at the clock inputs control the transfer of information from data input (D) to output.

When both clock inputs (C1 and C2) are in the low state, the data input affects only the "Master" portion of the flip-flop. The data present in the "Master" is transferred to the "Slave" when clock inputs (C1 "OR" C2) are taken from a low to a high level. In other words, the output state of the flip-flop changes on the positive transition of the clock pulse.

While either C1 "OR" C2 is in the high state, the "Master" (and data input) is disabled.

Asynchronous Set (S) and Reset (R) override Clock (C) and Data (D) inputs.

Power Dissipation = 220 mW typical (No Load)
f_{Tog} = 350 MHz typ

TRUTH TABLE

R	S	D	C	Q_{n+1}
L	H	ϕ	ϕ	H
H	L	ϕ	ϕ	L
H	H	ϕ	ϕ	N.D.
L	L	L	L	Q_n
L	L	L	⌐	L
L	L	L	H	Q_n
L	L	H	L	Q_n
L	L	H	⌐	H
L	L	H	H	Q_n

ϕ = Don't Care
ND = Not Defined
C = C1 + C2

V_{CC1} = Pin 1
V_{CC2} = Pin 16
V_{EE} = Pin 8

L SUFFIX
CERAMIC PACKAGE
CASE 620

Number at end of terminal denotes pin number for L package

Characteristic	Symbol	−30°C Min	−30°C Max	+25°C Min	+25°C Max	+85°C Min	+85°C Max	Unit
Power Supply Drain Current	I_E	—	—	—	48	—	—	mAdc
Input Current	I_{inH}							µAdc
Set, Reset		—	—	—	550	—	—	
Clock		—	—	—	250	—	—	
Data		—	—	—	270	—	—	
Switching Times								ns
Propagation Delay	t_{pd}	1.0	2.7	1.1	2.5	1.1	2.9	
Rise Time (10% to 90%)	t+	0.9	2.7	1.0	2.5	1.0	2.9	ns
Fall Time (10% to 90%)	t−	0.5	2.1	0.6	1.9	0.6	2.3	ns
Setup Time	$t_S"1"$	—	—	0.4	—	—	—	ns
	$t_S"0"$	—	—	0.5	—	—	—	
Hold Time	$t_H"1"$	—	—	0.3	—	—	—	ns
	$t_H"0"$	—	—	0.5	—	—	—	
Toggle Frequency	f_{Tog}	270	—	300	—	270	—	MHz

MC1670

FIGURE 1 – TOGGLE FREQUENCY WAVEFORMS

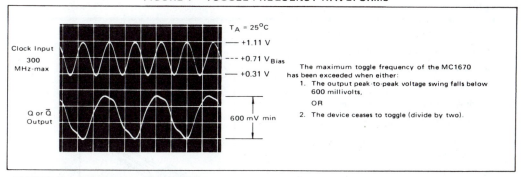

$T_A = 25^\circ C$

Clock Input
300
MHz-max

— +1.11 V
--- +0.71 V_{Bias}
— +0.31 V

Q or $\overline{Q}$
Output

600 mV min

The maximum toggle frequency of the MC1670 has been exceeded when either:

1. The output peak-to-peak voltage swing falls below 600 millivolts,

    OR

2. The device ceases to toggle (divide by two).

FIGURE 2 – MAXIMUM TOGGLE FREQUENCY (TYPICAL)

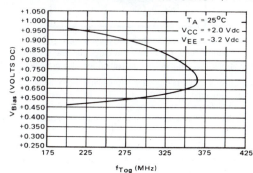

V_{Bias} (VOLTS DC)

f_{Tog} (MHz)

$T_A = 25^\circ C$
$V_{CC} = +2.0$ Vdc
$V_{EE} = -3.2$ Vdc

Figure 2 illustrates the variation in toggle frequency with the dc offset voltage (V_{Bias}) of the input clock signal.

Figures 4 and 5 illustrate minimum clock pulse width recommended for reliable operation of the MC1670.

FIGURE 3 – TYPICAL MAXIMUM TOGGLE FREQUENCY
versus TEMPERATURE

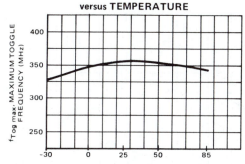

$f_{Tog\ max}$, MAXIMUM TOGGLE FREQUENCY (MHz)

T_A, AMBIENT TEMPERATURE ($^\circ$C)

Temperature	-30°C	+25°C	+85°C
V_{Bias}	+0.660 Vdc	+0.710 Vdc	+0.765 Vdc

Note: All power supply and logic levels are
shown shifted 2 volts positive.

APPENDIX 10   Voltage Regulator Data Sheets

**LINEAR
INTEGRATED
CIRCUITS**

**TYPES LM2930-5, LM2930-8
3-TERMINAL POSITIVE REGULATORS**

D2733, APRIL 1983

- **Input-Output Differential Less than 0.6 V**
- **Output Current of 150 mA**
- **Reverse Battery Protection**
- **Line Transient Protection**
- **40-Volt Load-Dump Protection**
- **Internal Short-Circuit Current Limiting**
- **Internal Thermal Overload Protection**
- **Mirror-Image Insertion Protection**
- **Direct Replacement for National LM2930 Series**

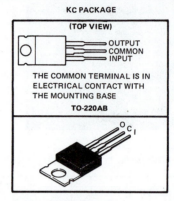

KC PACKAGE

(TOP VIEW)

→ OUTPUT
→ COMMON
→ INPUT

THE COMMON TERMINAL IS IN
ELECTRICAL CONTACT WITH
THE MOUNTING BASE

TO-220AB

description

The LM2930-5 and LM2930-8 are 3-terminal positive regulators that provide fixed 5-volt and 8-volt regulated outputs. Each features the ability to source 150 milliamperes of output current with an input-output differential of 0.6 volt or less. Familiar regulator features such as current limit and thermal overload protection are also provided.

The LM2930 series has low voltage dropout making it useful for certain battery applications. For example, the low voltage dropout feature allows a longer battery discharge before the output falls out of regulation; the battery supplying the regulator input voltage may discharge to 5.6 volts and still properly regulate the system and load voltage. Supporting this feature, the LM2930 series protects both itself and the regulated system from reverse battery installation or two-battery jumps.

Other protection features include line transient protection for load-dump of up to 40 volts. In this case the regulator shuts down to avoid damaging internal and external circuits. The LM2930 series regulator cannot be harmed by temporary mirror-image insertion.

schematic diagram

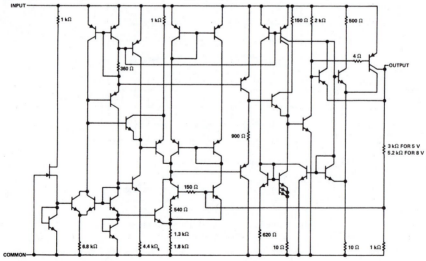

All component values are nominal.

Reprinted by permission of copyright holder. © 1984 Texas Instruments Incorporated.

APPENDIX 10   Voltage Regulator Data Sheets   (Continued)

TYPES LM2930-5, LM2930-8
3-TERMINAL POSITIVE REGULATORS

absolute maximum ratings over operating free-air temperature range (unless otherwise noted)

Continuous input voltage	26 V
Transient input voltage: t = 1 s	40 V
Continuous reverse input voltage	−6 V
Transient reverse input voltage: t = 100 ms	−12 V
Continuous total dissipation at 25°C free-air temperature (see Note 1)	2 W
Continuous total dissipation at (or below) 25°C case-temperature (see Note 1)	20 W
Operating free-air, case, or virtual junction temperature	−40°C to 150°C
Storage temperature range	−65°C to 150°C
Lead temperature 1,6 mm (1/16 inch) from case to 10 seconds	260°C

NOTE 1: For operation above 25°C free-air or case temperature, refer to Figures 1 and 2. To avoid exceeding the design maximum virtual junction temperature, these ratings should not be exceeded. Due to variation in individual device elecrical characteristics and thermal resistance, the bult-in thermal overload protection may be activated at power levels slightly above or below the rated dissipation.

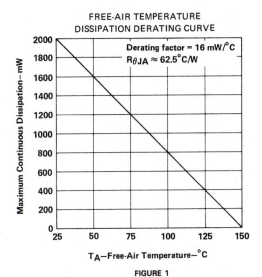

FREE-AIR TEMPERATURE
DISSIPATION DERATING CURVE

Derating factor = 16 mW/°C
$R_{\theta JA} \approx 62.5°C/W$

T_A—Free-Air Temperature—°C

FIGURE 1

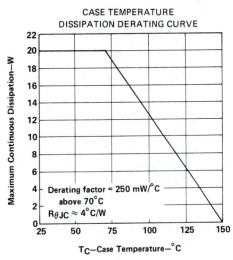

CASE TEMPERATURE
DISSIPATION DERATING CURVE

Derating factor = 250 mW/°C above 70°C
$R_{\theta JC} \approx 4°C/W$

T_C—Case Temperature—°C

FIGURE 2

recommended operating conditions

		MIN	MAX	UNIT
I_O	Output current		150	mA
T_J	Operating virtual junction temperature	−40	125	°C

(continues)

APPENDIX 10   Voltage Regulator Data Sheets   (Continued)

TYPES LM2930-5, LM2930-8
3-TERMINAL POSITIVE REGULATORS

LM2930-5 electrical characteristics at 25 °C virtual junction temperature, V_I = 14 V, I_O = 150 mA, (unless otherwise noted)

PARAMETER	TEST CONDITIONS[†]		MIN	TYP	MAX	UNIT
Output voltage	V_I = 6 V to 26 V, T_J = −40 °C to 125 °C	I_O = 5 mA to 150 mA,	4.5	5	5.5	V
Input regulation	I_O = 5 mA	V_I = 9 V to 16 V		7	25	mV
		V_I = 6 V to 26 V		30	80	
Ripple rejection	f = 120 Hz			56		dB
Output regulation	I_O = 5 mA to 150 mA			14	50	mV
Output voltage long-term drift[‡]	After 1000 h at T_J = 125 °C			20		mV
Dropout voltage	I_O = 150 mA			0.32	0.6	V
Output noise voltage	f = 10 Hz to 100 kHz			60		μV
Output voltage during line transients	V_I = −12 V to 40 V, R_L = 100 Ω		−0.3		5.5	V
Output impedance	I_O = 100 mA, I_O = 10 mA (rms), f = 100 Hz to 10 kHz			200		mΩ
Bias current	I_O = 10 mA			4	7	mA
	I_O = 150 mA			18	40	
Peak output current			150	300	700	mA

LM2930-8 electrical characteristics at 25 °C virtual junction temperature, V_I = 14 V, I_O = 150 mA, (unless otherwise noted)

PARAMETER	TEST CONDITIONS[†]		MIN	TYP	MAX	UNIT
Output voltage	V_I = 9.4 V to 26 V, T_J = −40 °C to 125 °C	I_O = 5 mA to 150 mA,	7.2	8	8.8	V
Input regulation	I_O = 5 mA	V_I = 9.4 V to 16 V		12	50	V
		V_I = 9.4 V to 26 V		50	100	
Ripple rejection	f = 120 Hz			52		dB
Output regulation	I_O = 5 mA to 150 mA			25	50	mV
Output voltage long-term drift[‡]	After 1000 h at T_J = 125 °C			30		mV
Dropout voltage	I_O = 150 mA			0.32	0.6	V
Output noise voltage	f = 10 Hz to 100 kHz			90		μV
Output voltage during line transients	V_I = −12 V to 40 V, R_L = 100 Ω		−0.3		8.8	V
Output impedance	I_O = 100 mA, I_O = 10 mA (rms), f = 100 Hz to 10 kHz			300		mΩ
Bias current	I_O = 10 mA			4	7	mA
	I_O = 150 mA			18	40	
Peak output current			150	300	700	mA

[†] Unless otherwise specified, all characteristics, except ripple rejection and noise voltage measurements, are measured using pulse techniques (t_w ≤ 10 ms, duty cycle ≤ 5%) with a capacitor of 0.1 μF across the input and a capacitor of 10 μF across the output. Output voltage changes due to changes in internal temperature must be taken into account separately.

[‡] Since long-term drift cannot be measured on the individual devices prior to shipment, this specification is not intended to be a guarantee or warranty. It is an engineering estimate of the average drift to be expected from lot to lot.

SELECTED ANSWERS

1.1	5 mV	2.35	$5.7 + 0.0029 \sin \omega t$ V	5.1	peak-to-peak = 14.82 mA
1.3	0.35 nm/s	2.36	$0.33 \sin \omega t$ V	5.5	peak-to-peak = 10.98 mA
1.4	0.145 A, 0.1 W	2.37	$0.225 + 0.009 v_s$ V	5.6	7.818 mA, 6.1 V, peak-to-
1.8	2 Ω	2.39	$V_L = 0.33 \ V_s + 1.2$ V		peak = 15.6 mA
1.9	a. 29.84 Ω, b. 214 Ω	2.40	$V_L = 0.83 \ V_s - 0.22$ V	5.8	18 kΩ, 83 kΩ
1.12	a. 20 A, b. 415.83 W,	2.43	$V_L = 0.182 \ V_s$	5.9	5 kΩ, 4.825 V
	c. 4.16 mJ	2.44	$V_L = 2.5 \sin \omega t$	5.11	peak-to-peak = 20 mA
1.14	a. 0, b. 87.5 J, c. 0	3.1	100, 0.99	5.13	3.45 V, 2.9 kΩ, 6.5 kΩ
1.16	1000 rad/s or 159 Hz	3.3	0.05 mA, 5 mA, 15 V	5.15	$h_{fe} = 120$, $h_{ie} = 250$ Ω
1.21	16.82 W	3.6	"chopped" sine waves	5.19	−400
1.22	172 kΩ	3.7	0.1 mA, 10 mA, 10 V	5.22	$R_S = 1$ kΩ, −32.8
1.23	22%	3.11	100, $i_c = 1.93$ mA;	5.25	−7.9
1.26	a. 25 V, b. 64 V		300, $i_c = 5$ mA	5.27	$R_C = 20$ kΩ, −43.3
1.27	1/11 A, − 6/11 A, 7/11 A	3.14	3 mA, −4 V	5.30	14.8
1.29	16.84 V	3.18	−2 V, 2.5 mA, 7.5 V	5.34	9.4
1.30	33.5 V	3.22	5 mA, 15 V	5.38	−9.8
1.31	3 V, 5.2 kΩ; 2.86 V,	3.24	−3 V, 7.2 mA, −12.8 V	5.40	−44.6
	3.43 kΩ	3.26	−0.6 V, 0.75 kΩ, 11.4 V	5.42	−8.26
1.33	0.31 A	3.28	−2.5 V, 5 V	5.46	1 MΩ, −200
1.35	1.25 mA, − 8.69 V	4.1	5 V, 10 mA, saturation	5.49	6.25%
1.36	3.08 V	4.2	saturation, 2.8 mA	6.1	2.7×10^4
1.37	− 3 V	4.3	4.23 mA, 11.5 V	6.2	0.76×10^4
1.39	1515 V, 1525 V	4.4	2 mA, 4 V	6.5	5170
1.42	0.107 mA, 15.5°	4.5	300, 1.96 mA, 4.32 V	6.8	114
1.43	2.2 V, −6.3°	4.7	100, 10 mA, 3.3 V	6.11	7.5×10^4
2.1	230, 260, 160, 560 μΩ	4.9	100, 5.4 mA, 4.6 V	6.13	3.8
2.2	SiO_2: $10^{16} - 10^{18}$ Ω	4.11	0.41 mA, 5.49 V	6.14	14.9
2.3	GaAs: $10^4 - 10^9$ Ω	4.15	$V_{CE} = 5.32$ V	6.15	22.2
	Ge: 10 kΩ to 10^3 MΩ	4.16	$R_C = R_B = 0.4$ kΩ	6.17	9.5, 0.11
2.5	$N = N_a$, p = n_i^2/N_a		$R_E = 0.1$ kΩ, $V_{BB} =$	6.18	9.5, 0.5×10^{-3}
2.9	0.216/Ω cm		1.8 V	6.20	1, 46.8
2.12	50 V, 0.7 V, 150 Ω	4.18	$R_C = 0.5$ kΩ, $R_E =$	6.24	−99
2.13	3.74 mA, 1.26 V		0.13 kΩ	6.25	−19.6
2.15	3.74 mA, 1.26 V		$R_B = 1.3$ kΩ, $V_{BB} =$	6.28	500, 0.05
2.19	5 mA		2.13 V	6.31	30, 3, 0.3 mA
2.21	−0.02 mA, −0.2 V	4.23	5 V, 10 V	6.35	0, 3333
2.22	1 mA, 5 V	4.26	1.9 V, 8.6 V, 11.4 mA	6.37	0.5, 500
2.29	$V_L = |V_s|$	4.30	2 kΩ, 6 kΩ	6.41	0.1, 500

7.1 10^3 MΩ, 1 Ω, 10
7.2   1 kΩ, 1 Ω, −9
7.3   −14 V
7.4 10^3 MΩ, 1.09 Ω, 10
7.8   10 V
7.11 1 kΩ, 5 kΩ
7.14 amplify, differentiate,
      integrate
7.16 10^7
7.17 10^3
7.19 10^4, 10^6
7.21 0.01 Ω, 0.5 kΩ
7.23 1 mA
7.32 $H = 0.05$
7.33 $H = A_d/2$, 2.5 V
7.35 $H = A_d/2$; $V_s = 15$ V;
      $v_o = -20$ V
7.37 2 s
7.40 $v_o = -0.025 \ln v_s$
8.7   49.5 db, 10^4 rad/s
8.8   39.5 db, 2.5 rad/s, 60 rad/s
8.9   34 db, 100 rad/s, 300 rad/s
8.11 35 db, 4040 and 2000 rad/s
8.14 31.2 db, 1000 and 1,550
      rad/s
8.16 26 db, 500 rad/s
8.17 26 db, 10^3 and 5×10^3
      rad/s
8.19 19.7 db, 10^3 and 6×10^3
      rad/s
8.21 21.86 db, 500 and 1850
      rad/s
8.22 40 db, 10^8 rad/s
8.25 39.39 db, 2.1×10^7 rad/s
8.27 21.9 db, 5.2×10^9 rad/s
8.29 31.2 db, 2.2×10^9 rad/s
8.31 250, 3.64×10^4, and 2.2
      $\times 10^9$ rad/s
8.32 54 db, 10^6 and 5×10^8
      rad/s
8.33 30 db, 1.34×10^7 rad/s
8.35 26 db, 500 and 10^6 rad/s
8.37 19.8 db, 9.1 and 5×10^5
      rad/s
9.1   0, 5 V; 5 V, 0.2 V
9.8   4.8 V, 5.346 V
9.9 $I_B = 0.5$ mA, $I_C =$
      4.8 mA
9.17 0.48 kΩ, 13 kΩ

9.18 30.05
9.19 0.5 V, 4.72 V
9.20 3
9.21 0.798 V
9.23 2.16 kΩ
9.25 1.18 V, 80
9.29 1.72 ms
9.31 0.44 mA, 9.8 mA, 0.2 V
10.2 3.72 mA, 2.56 V
10.5 4.3 and 4.8 mA, 0.2 V;
      $I = 3.44$ mA, 1.56 V
10.7 $v_L = 2.3$ V
10.10 NAND
10.11 NOR
10.13 0.395 and 9.8 mA, 0.2 V;
      0 mA, 10 V
10.15 17
10.17 47
10.23 0.0625 and 5 mA, 0.2 V
10.28 53
10.29 51
10.30 a. 1 mW b. 3 mW
11.1 0.4 V, 0.05 mA, 5.05
      mA, 0.45 mA; Active
11.3 0.5 mA, 5 mA, 0 V;
      2 mA, 5 mA, 0 V;
      saturation
11.5 47.7 mW
11.9 15 mW
11.11 v_{L2}: −3.15 V, 0 V
      v_{L1}: 0, −3.15 V
11.13 −0.75 V, −1.74 V
11.15 −0.76 V, −1.74 V
11.17 326
11.19 15
11.21 −2.6 V
11.23 20 mW
11.27 7.56
11.28 4.995 V
11.31 5 V, 0 V, 4 V and 1 V
11.33 0.275 mW
11.35 0.15 mA
11.39 22.58 mW
11.41 135.2 mW
12.1 2, 0.001
12.2 0.12 Ω
12.5 4.96 V
12.8 2250 °K
12.11 5.59 pF

12.13 8.854 pF
12.15 0.004 cm
12.17 24 mV, 125.1 mV
12.19 8.33×10^{16}/cm^3
12.21 5 μA
12.23 3.33 V, 1.4 mA
12.24 0.455 V, −0.21 μA
12.25 0.345 V, 2 mA
12.26 0.245 V
12.28 0.72 kΩ
12.29 3.3 mA
12.31 0.625
12.32 $-150 (1 + 0.2 V_m)V_C$
12.36 6.74 Ω
13.1 0.318 A
13.3 0.5 J, 0.0025 J
13.5 8 μWb, 14.3 mA
13.7 10 V, 0.5 A, 0.5 N, 5 W
13.9 159 turns
13.11 −0.03 sin ωt V,
      3×10^{-6} sin^2 ωt N-M
13.13 1.25 W
13.16 120 V
13.17 243 V
13.19 200 V, 220 V
13.20 70.1 V
13.23 4.95 A, 247.5 V
13.25 1247 rpm, 476 rpm
13.27 3600 rpm, 375 hp
13.29 977 W, 5023 W, 83.8%
13.31 84%
13.33 1200 rpm, 617 V, 10.8°
13.36 77.25 V, 41.5°
13.37 64.15 V
13.39 1800 rpm, 1728 rpm,
      82.5 N-M
14.1 0.006 Ω
14.2 60 kΩ
14.4 556 Ω, 50.5 Ω
14.6 10 Ω, 1.111 Ω
14.9 6.67 V, 5.714 V, 14.3%
14.11 11.1%
14.13 59 kΩ, 39 kΩ
14.15 36.7%, 6 mA
14.17 4.94 mA
14.20 1.5 V
14.21 30 V, 1.25 kΩ
14.22 5 V, 0.95 mA
14.24 5 kHz, −40

14.26   16 kHz

14.27   16 kHz

14.29   2 kΩ, 3.2 Ω

14.31   2.5 μF

14.33   28.7 μH

14.35   0.1 ms, 0.05 ms, 10 nF

14.37   2.85 V

14.41 -18 V, -1.75 V, -0.25 V

14.43 -28.1, -2.73 V, -0.39 V

INDEX